# NEUROMETHODS □ 9

## The Neuronal Microenvironment

# NEUROMETHODS

## Program Editors: Alan A. Boulton and Glen B. Baker

# NEUROMETHODS

**Program Editors: Alan A. Boulton and Glen B. Baker**

# The Neuronal Microenvironment

Edited by

## Alan A. Boulton,

*University of Saskatchewan, Saskatchewan, Canada*

## Glen B. Baker,

*University of Alberta, Alberta, Canada*

and

## Wolfgang Walz

*University of Saskatchewan, Saskatchewan, Canada*

## Humana Press • Clifton, New Jersey

**Library of Congress Cataloging in Publication Data**

The Neuronal microenvironment.

(Neuromethods ; 9.)
Includes bibliographies and index.
1. Brain chemistry—Research—Technique.
2. Electrolytes—Analysis.   3. Brain—Ventricles.
4. Cerebrospinal fluid.   5. Choroid plexus.   6. Water-
electrolyte balance. I. Boulton, A. A. (Alan A.)
II. Baker, Glen B., 1947–     .   III. Walz, Wolfgang.
IV. Series: Neuromethods ; 9. V. Series: Neuromethods.
Series I, Neurochemistry. [DNLM: 1. Body Fluids.
2. Brain—metabolism. 3. Neurochemistry—methods.
4. Water-Electrolyte Balance. W1 NE337G v.9 /
WL 300 N493817]
QP356.3.N478   1987        599'.0188        87-3668
ISBN 0-89603-115-2

# Foreword

Techniques in the neurosciences are evolving rapidly. There are currently very few volumes dedicated to the methodology employed by neuroscientists, and those that are available often seem either out of date or limited in scope. This series is about the methods most widely used by modern-day neuroscientists and is written by their colleagues who are practicing experts.

Volume 1 will be useful to all neuroscientists since it concerns those procedures used routinely across the widest range of sub-disciplines. Collecting these general techniques together in a single volume strikes us not only as a service, but will no doubt prove of exceptional utilitarian value as well. Volumes 2 and 3 describe all current procedures for the analyses of amines and their metabolites and of amino acids, respectively. These collections will clearly be of value to all neuroscientists working in or contemplating research in these fields. Similar reasons exist for Volume 4 on receptor binding techniques, since experimental details are provided for all types of ligand-receptor binding, including chapters on general principles, drug discovery and development, and a most useful appendix on computer programs for Scatchard, nonlinear, and competitive displacement analyses. Volume 5 provides procedures for the assessment of enzymes involved in biogenic amine synthesis and catabolism.

Volumes in the NEUROMETHODS series will be useful to neurochemists, -pharmacologists, -physiologists, -anatomists, psychopharmacologists, psychiatrists, neurologists, and chemists (organic, analytical, pharmaceutical, medicinal); in fact, everyone involved in the neurosciences, both basic and clinical.

# Preface to the Series

When the President of Humana Press first suggested that a series on methods in the neurosciences might be useful, one of us (AAB) was quite skeptical; only after discussions with GBB and some searching both of memory and library shelves did it seem that perhaps the publisher was right. Although some excellent methods books have recently appeared, notably in neuroanatomy, it is a fact that there is a dearth in this particular field, a fact attested to by the alacrity and enthusiasm with which most of the contributors to this series accepted our invitations and suggested additional topics and areas. After a somewhat hesitant start, essentially in the neurochemistry section, the series has grown and will encompass neurochemistry, neuropsychiatry, neurology, neuropathology, neurogenetics, neuroethology, molecular neurobiology, animal models of nervous disease, and no doubt many more "neuros." Although we have tried to include adequate methodological detail and in many cases detailed protocols, we have also tried to include wherever possible a short introductory review of the methods and/or related substances, comparisons with other methods, and the relationship of the substances being analyzed to neurological and psychiatric disorders. Recognizing our own limitations, we have invited a guest editor to join with us on most volumes in order to ensure complete coverage of the field and to add their specialized knowledge and competencies. We anticipate that this series will fill a gap; we can only hope that it will be filled appropriately and with the right amount of expertise with respect to each method, substance or group of substances, and area treated.

Alan A. Boulton
Glen B. Baker

# Preface

The concept of a neuronal microenvironment is a relatively
new one. It is based on the view that the extracellular space, which
constitutes about 20% of the brain volume, is a functional compartment in its own right. It is substantially involved in neurotransmission in such a way that as the electrolyte composition and volume
are changed they themselves influence neuronal excitability. Many
pathological processes as, for example, stroke, ischemia, anoxia,
and epilepsy, involve dramatic changes in these factors. The concept of a neuronal microenvironment acting as a communication
channel was first formulated clearly by C. Nicholson in 1979. Since
then new techniques and concepts have led to dramatic and new
developments in our understanding of the neuronal microenvironment. Of central interest is the electrolyte composition. These
electrolytes act as charge carriers and determine normal and abnormal neuronal excitability; they are the bulk osmoles that are shifted
from one compartment to the other, thus causing normal and
abnormal volume changes in different structures of the brain, and
they are a signal system tightly coupled to neuronal and glial
function. This view of the neuronal microenvironment cannot,
however, be completely understood without analysis of the brain's
macroenvironment, which is regulated at the blood–brain barrier.
The arrangement of this Neuromethods volume reflects this
approach.

The first two chapters are devoted to the techniques used in
assessing the pathways and fluid dynamics connected with the
blood–brain barrier and the cerebrospinal fluid. Ultrastructural
methods to examine normal and abnormal extracellular space dimensions are described in the next chapter, and this is followed by
an article describing modern physical methods, such as computer
tomography and nuclear magnetic resonance, that are used by
clinicians to estimate brain edema. Energy metabolism is closely
interrelated with ion and water movements, and this has been

reviewed in the next chapter, in which the measurement of metabolic activity associated with ion and water shifts has been described.

Three new methods introduced in the last 10 years have revolutionized our concept of the neuronal microenvironment, and their application continues to increase our knowledge. These are reviewed in the following two chapters. Ion-selective microelectrodes permit the measurement of small changes in ion activity in small compartments. In addition, voltammetric microsensors can now be applied to many substances, whereas patch-clamp techniques permit the investigation of ionic currents mediated by single modules in cellular membranes.

Brain slices are an increasingly popular biological preparation; they permit us to study the impact of a controlled microenvironment on processes within a normal cellular architecture without interference by the blood–brain barrier. They were first used for compartmental analyses and the study of cell swelling, and these aspects are reviewed in the following chapter, whereas an analysis of ionic movements and excitability properties by modern electrophysiological techniques within brain slices are dealt with in the following chapter.

Cell cultures are preparations that have become popular with neuroscientists in recent years. These consist of cells of a single or mixed origin, and exhibit an artificial, yet very simple, cellular architecture, which can be manipulated. The cell culture approach to the cellular microenvironment is reviewed in the following two chapters; the former concerns volume and ion transport measurements, and the latter describes electrophysiological techniques.

Up to this point, this volume has dealt mainly with the movements of bulk ions and water; some electrolytes, however, require special attention because they appear only in small concentrations in all compartments, yet these ions experience dramatic relative changes during activity and disease, and the shifts in their compartments possess many of the properties of a cofactor-like function-regulating cellular processes. These substances are calcium ions, magnesium ions, and hydrogen ions. The different methodological approaches necessary for their study are dealt with in the last three chapters of the volume.

As far as I am aware, this is the first monograph that is dedicated exclusively to a description of the methodological approach as used to study the new concept of the "neuronal microenvironment." It is our hope that this volume will serve as a useful guide for researchers who wish to start work in this promising, new, and rather fast-developing area and will offer to clinicians and others who need to analyze information about edemic states and abnormal electrolyte balances the information necessary for a critical evaluation of the changes found in some disease states.

**Wolfgang Walz**

# Contents



CEREBROVASCULAR WATER AND ION TRANSPORT
Joseph E. Melton



THE CHOROID PLEXUS–ARACHNOID MEMBRANE–
CEREBROSPINAL FLUID SYSTEM
Conrad E. Johanson

Contents *xv*



**FLUID COMPARTMENTS:** *Ultrastructural Methods to Identify Extracellular Spaces in the Central Nervous System*
Asao Hirano and Takeo Kato



**PHYSICAL AND BIOCHEMICAL METHODS FOR ANALYSIS OF FLUID COMPARTMENTS**
K. G. Go

## MEASUREMENT OF METABOLIC ACTIVITY ASSOCIATED WITH ION SHIFTS

Myron Rosenthal and Thomas J. Sick



## USE OF ION-SELECTIVE MICROELECTRODES AND VOLTAMMETRIC MICROSENSORS TO STUDY BRAIN CELL MICROENVIRONMENT

Charles Nicholson and Margaret E. Rice

PATCH CLAMP RECORDING METHODOLOGY
Jerry M. Farley

## ANALYSIS OF ION FLUXES AND FLUID COMPARTMENTATION IN BRAIN SLICES
### Wolfgang Walz

## ION TRANSPORT AND VOLUME MEASUREMENTS IN CELL CULTURES
H. K. Kimelberg and Wolfgang Walz

ELECTROPHYSIOLOGICAL METHODS APPLIED IN NERVOUS SYSTEM CULTURES
H. Kettenmann



ELECTROPHYSIOLOGICAL METHODS FOR STUDYING IONIC CURRENTS IN BRAIN SLICES AND CELL CULTURES
Brian A. MacVicar and Maeve O'Beirne

CALCIUM IONS
R. Pumain

## ACID–BASE BALANCE

Bo K. Siesjö, Fredrik Boris-Möller, and Eduardo Martins



## MAGNESIUM IONS

Jerry G. Chutkow

# Contributors

GLEN B. BAKER • *Neurochemical Research Unit, Department of Psychiatry, University of Alberta, Edmonton, Alberta, Canada*

FREDRIK BORIS-MÖLLER • *Laboratory for Experimental Brain Research, Lund University Hospital, Lund, Sweden*

ALAN A. BOULTON • *Neuropsychiatric Research Unit, University of Saskatchewan, Saskatoon, Saskatchewan, Canada*

JERRY G. CHUTKOW • *Department of Neurology, School of Medicine, State University of New York at Buffalo, Buffalo, New York*

JERRY M. FARLEY • *Department of Pharmacology and Toxicology, University of Mississippi Medical Center, Jackson, Mississippi*

K. G. GO • *Kliniek Voor Neuro-chrurgie, Academisch Ziekenhuis Groningen, Groningen, The Netherlands*

ASAO HIRANO • *Bluestone Laboratory, Division of Neuropathology, Department of Pathology, Montefiore Medical Center, Albert Einstein College of Medicine, Bronx, New York*

CONRAD E. JOHANSON • *Program in Neurosurgery, Brown University and Rhode Island Hospital, Providence, Rhode Island*

TAKEO KATO • *Bluestone Laboratory, Division of Neuropathology, Department of Pathology, Montefiore Medical Center, Albert Einstein College of Medicine, Bronx, New York*

H. KETTENMANN • *Department of Neurobiology, University of Heidelberg, Heidelberg, Federal Republic of Germany*

H. K. KIMELBERG • *Division of Neurosurgery, Albany Medical College, Albany, New York*

BRIAN A. MACVICAR • *Neuroscience Research Group, University of Calgary, Calgary, Alberta, Canada*

EDUARDO MARTINS • *Laboratory for Experimental Brain Research, Lund University Hospital, Lund, Sweden*

Joseph E. Melton • *Department of Medicine, Division of Pulmonary and Critical Care Medicine, UMDNJ-Robert Wood Johnson Medical School, New Brunswick, New Jersey*

Charles Nicholson • *Department of Physiology and Biophysics, New York University Medical Center, New York, New York*

Maeve O'Beirne • *Neuroscience Research Group, University of Calgary, Calgary, Alberta, Canada*

R. Pumain • *Unite de Recherches sur l'Epilepsie, Paris, France*

Margaret E. Rice • *Department of Physiology and Biophysics, New York University Medical Center, New York, New York*

Myron Rosenthal • *Department of Neurology, University of Miami School of Medicine, Miami, Florida*

Thomas J. Sick • *Department of Neurology, University of Miami School of Medicine, Miami, Florida*

Bo K. Siesjö • *Laboratory for Experimental Brain Research, Lund University Hospital, Lund, Sweden*

Wolfgang Walz • *Department of Physiology, College of Medicine, University of Saskatchewan, Saskatoon, Saskatchewan, Canada*

# Cerebrovascular Water and Ion Transport

Joseph E. Melton

## 1. Introduction

The capillaries of the central nervous system (CNS) are unique among the blood vessels of the body in having tight junctions between adjacent endothelial cells, no fenestrae, and a complete pericapillary investment (the glial foot processes) between parenchymal cells and capillary endothelium. The net result of these singular properties is to transform an ordinary capillary endothelium into a formidable restriction to the passage of both solute and solvent, a blood–brain barrier (BBB). The BBB serves to passively isolate the brain interstitial microenvironment from the plasma and, by means of carrier transport systems resident on the capillary endothelial wall, to actively regulate the passage of substrates and metabolites between the blood and the brain extracellular space. Taken together, these active and passive regulatory functions of the BBB serve to maintain an optimal environment for neuronal function and for transfer of information within the brain parenchyma (Nicholson, 1980; Cserr and Bundgaard, 1984).

The very properties of the capillary endothelium that make it such an effective barrier serve to thwart many of the studies that the inquisitive physiologist would like to perform. In particular, it would be of great interest to know how various classes of substrates and metabolites move across the BBB from plasma to brain interstitium and vice versa. The impermeability of the BBB to specific classes of solutes limits the utility of conventional tracer techniques in making such measurements. Additionally, the complex geometry of brain tissue and the presence of several potential intracerebral compartments for distribution [i.e., intracellular, extracellular, and bulk cerebrospinal fluid (CSF) spaces] serve to increase the difficulty of analyses that are, in other tissues, relatively straightforward.

This presentation will briefly review general concepts of BBB permeability and will critically assess the techniques currently in use to monitor movement of water and electrolytes across the cerebral vasculature. Readers desiring more general information about the BBB are advised to consult the monograph by Bradbury (1979) and the recent review by Fenstermacher and Rapoport (1984). Many of the concepts presented in this paper rely heavily on an earlier related review of methodology in BBB transport by Fenstermacher et al. (1981).

## 2. General Considerations in Measuring Blood-to-Brain Transfer of Solute or Water

A general protocol for measuring blood-to-brain transfer across the BBB involves administration of tracer substance into the vasculature, determination of tracer concentration in the blood at carefully timed intervals and, perhaps, measurement of tracer concentration in the brain following sacrifice of the test animal. The information collected in such experiments can be used to calculate an extraction fraction $(E)$ or an influx constant $(K_i)$. Both $E$ and $K_i$ are transfer constants rather than permeability constants, since they are determined partially by factors other than capillary permeability $(P)$ and capillary surface area $(S)$ (Fenstermacher, 1983).

The extraction fraction $(E)$ is defined as

$$E = B/A \tag{1}$$

where $B$ is the amount of tracer that moves unidirectionally from blood to brain during initial passage of a tracer bolus through the cerebral vasculature and $A$ is the total amount of tracer passing through the brain capillaries during this period. $E$ is a dimensionless fraction and, for purposes of the methods described in this review, is useful only when measured during a period when no backflux of tracer from brain to blood occurs. Values of $E$ ranging from 0.1 to 0.9 can be measured accurately (Fenstermacher et al., 1981).

If backflux has been negligible during measurement of $E$, this fraction can be related to $PS$, the permeability-surface area product for the capillary wall, by the following expression (Crone, 1963):

$$E = 1 - \exp(-PS/\dot{Q}) \tag{2}$$

where $\dot{Q}$ is the brain blood flow (volume/time/tissue mass). Since $E$ is dimensionless, $PS$ has the same units as $\dot{Q}$. Fenstermacher et al. (1981) have subsequently modified Eq. (2) by substituting a term $FV_c$ for $\dot{Q}$. $F$ is identical to $\dot{Q}$ as used by Crone (1963). $V_c$ is defined as the fractional volume of the blood that serves as an uptake source for the tracer of interest during the period when $E$ is measured. The use of $FV_c$ rather than $\dot{Q}$ in Eq. (2) permits allowance to be made for that portion of the tracer that is unavailable for tissue exchange during the measurement of $E$.

$K_i$, the unidirectional influx constant, differs from $E$ in that it is normally measured after multiple passes of the tracer through the brain vasculature. The tracer of interest is generally given intravenously, either by single injection or infusion, blood samples are taken, and after a time period ranging from minutes to hours, the amount of tracer in the brain is measured. Assuming that transfer has been unidirectional from blood to brain during the measurement period, $K_i$ is defined by the following expression:

$$K_i = \frac{B}{\int_0^t C_a(t)dt} \tag{3}$$

where $B$, as in Eq. (1), is the amount of solute that has passed only once from blood to brain during the experimental period, and $C_a(t)$ is the tracer activity of arterial plasma. The integral comprising the denominator of Eq. (3) represents the total amount of tracer available for movement from blood to brain parenchyma during the experiment and is directly analogous to the $A$ term of Eq. (1).

$K_i$, like $E$, may be related to $PS$.

$$K_i = FV_c[1 - \exp(-PS/FV_c)] \tag{4}$$

From Eqs. (2) and (4) it is obvious that the values of $E$ and $K_i$ depend both on the permeability-surface area product ($PS$) of the BBB and the effective brain blood flow ($FV_c$). Choice of a proper method for measurement of the BBB transfer of a particular compound requires some knowledge of the relative magnitude of the $PS/FV_c$ ratio. When $PS/FV_c$ is less than 0.2, the value of $K_i$ approaches that of $PS$ with an error of less than 10%. Alternatively, when $PS/FV_c$ is greater than 3, $K_i$ approaches $FV_c$ (Fenstermacher and Rapoport, 1984). This is another way of saying that when the $PS/FV_c$ ratio of a compound is high, transfer across the BBB is largely limited by brain blood flow. Conversely, when the $PS/FV_c$ ratio of a compound is low, transfer across the BBB is largely limited by the permeability-surface area product of the BBB. It is important to remember that changes in the $PS/FV_c$ ratio for a given compound may result from changes in either $PS$ or $FV_c$ and, in fact, these two quantities may be interrelated. A recent study by Phelps et al. (1981) suggests that $PS$ increases in a nonlinear fashion as $FV_c$ increases. This increase in $PS$ with increased brain blood flow presumably results from capillary recruitment in the brain at high flows with resultant increases in $S$, the surface area term of the permeability-surface area product.

In practice, selection of a technique to measure BBB transfer of a particular compound is largely based on an educated guess of the BBB permeability of the compound. The value of $PS$ for a given compound can be estimated knowing its chemical structure. Charged, lipophobic compounds move across the BBB very slowly (Bradbury, 1979) and, as a result, their transfer across the BBB is usually measured with multiple pass techniques that produce a value for $K_i$ after fitting the appropriate data into Eq. (3). The BBB permeability of an uncharged compound increases directly as a function of the lipid solubility of the compound (generally expressed as the octanol/water partition coefficient) and inversely as a function of molecular weight (Ohno et al., 1978; Bradbury, 1979; Rapoport et al., 1979; Levin, 1980). Knowledge of these two factors thus allows selection of an appropriate tracer method to be employed for transfer studies.

The compounds to be considered in detail in this review represent two extremes of the permeability range. The BBB is two to three orders of magnitude less permeable to water than the capillaries of systemic beds (Bradbury, 1979). Thus the BBB, unlike "typical" capillaries, has a water permeability roughly equivalent to that of a high-resistance epithelium. Compared to electrolytes, however, water moves freely across the BBB. The brain capillary $PS$ product for water in monkeys is 66 mL/h/g (Raichle et al., 1976). By contrast, the brain capillary $PS$ product for $Na^+$, measured in a variety of mammals, ranges from 0.1 to 0.2 mL/h/g (Bradbury, 1979). The other major plasma electrolytes are similarly impermeable. Obviously, different techniques have to be employed to measure transfer of water and electrolytes across the BBB.

## 3. Tracer Techniques for Quantifying Diffusive Water Movement Across the BBB

### 3.1. Indicator Diffusion

The indicator-diffusion method of measuring BBB permeability as developed by Crone (1963, 1965) provides a single-passage estimate of $E$ that is then normally used to calculate $PS$ or $P$ (Eq. 2). The basic methodology of the technique is very simple. Two tracers, the compound whose extraction is to be measured and a nonpermeating reference compound, are simultaneously injected into an artery that directly feeds the brain, usually a branch of the carotid. Brain venous outflow is subsequently collected from the cannulated sagittal sinus for a time period approximating a single arterial passage, and the tracer activity in the venous samples measured.

Calculation of $E$ is based on Eq. (1) and the Fick principle. For any given venous sample, the $A$ term of Eq. (1) is equal to $C_a$, the arterial activity of the test tracer, multiplied by $V_s$, the venous sample volume. The $B$ term of Eq. (1), the amount of test tracer extracted by the brain, can be calculated as the arterial–venous activity difference $[(C_a) - (C_v)]$ of the test tracer times the venous sample volume, $V_s$. Thus, Eq. (1) may be rewritten as

$$E = \frac{V_s[(C_a)_{\text{test}} - (C_v)_{\text{test}}]}{V_s(C_a)_{\text{test}}} = \frac{(C_a)_{\text{test}} - (C_v)_{\text{test}}}{(C_a)_{\text{test}}} \tag{5}$$

The arterial activity of the test tracer cannot be measured directly, but is estimated by $(C_v)_{\text{ref}}$, the concentration of the impermeable reference tracer in the venous sample. The reference tracer serves as an indicator of the degree of dilution occurring within the vascular system between injection and sampling sites. Thus, the $(C_v)_{\text{ref}}$ term is assumed to represent what $(C_a)_{\text{test}}$ would have been in venous blood if no exchange of test solute had taken place (Fenstermacher et al., 1981). Additionally, since the activities of test and reference compounds in the injection bolus usually differ, all tracer activities in venous samples are normalized to their respective activities in the injectate. Thus, the final form of Eq. (5) becomes

$$E = \frac{(C_v)_{\text{ref}}/(C_i)_{\text{ref}} - (C_v)_{\text{test}}/(C_i)_{\text{test}}}{(C_v)_{\text{ref}}/(C_i)_{\text{ref}}} \tag{6}$$

where $C_v$ and $C_i$ refer to concentrations in the venous samples and injectate, respectively.

The indicator diffusion technique has several advantages for measurement of BBB exchange. It is relatively easy to perform, it can be adapted for small animal models (e.g., Murray and Plioplys, 1972), and several serial measurements can be made in the same animal. Because there is no requirement for brain tissue sampling, this technique has been used for several studies of BBB permeability in humans (Lassen et al., 1971; Paulson et al., 1977; Hertz and Paulson, 1980).

The technical problems associated with the indicator diffusion technique are primarily concerned with the time over which venous sampling takes place and the choice of a proper reference tracer. The sampling time is important because the $E$ value of a given solute as defined in Eq. (1) is valid and measurable only during the period of unidirectional tracer movement from blood to brain. In practice, this problem can be dealt with by plotting percent extraction of the test compound, calculated for each venous sample, versus time. If the value of $E$ is selected from the early plateau phase of such a plot, backflux is minimized (Fenstermacher et al., 1981).

Choice of a proper reference tracer becomes a consideration since separation of test and tracer compounds during passage through the capillary bed can result in alterations of the test/reference ratio of the venous blood samples that are not associated with changes in extraction of the test compound and, hence, spurious values for $E$ (Eq. 6). The potential occurrence of this problem when using test and reference compounds of markedly different diffusability (the Taylor effect) was presented theoretically by Lassen and Crone (1970) and subsequently demonstrated experimentally by Yudilevich and De Rose (1971). Use of a relatively small reference compound (e.g., sodium, mannitol, or chloride) effectively corrects this problem.

Perhaps more important than the technical problems associated with the indicator dilution technique are its inherent methodological limitations. A tacit assumption of this technique is that the brain can be treated as an assemblage of homogeneous capillaries, i.e., that all capillaries have the same extraction fraction for a given solute. Hertz and Paulson (1980) have demonstrated that, in fact, transit time for a given solute may vary from capillary to capillary in the brain, the mean value of $E$ being biased toward capillaries with shorter transit times. Additionally, use of measured values of $E$ to calculate $PS$, a variable of great interest in describing the permeability properties of the BBB, requires some estimate of simultaneous brain blood flow, $\dot{Q}$ (Eq. 2). This value is usually estimated from separate studies or from the literature, and the resulting value of $PS$ may fail to represent the inherent variability in $\dot{Q}$ caused by the physiological state of a given animal. Even when brain blood flow is measured simultaneously with $E$ (e.g., Paulson et al., 1977), a technically difficult task, it is impossible to measure regional values of $PS$ in the brain since only a mean organ value of $E$ can be measured with the indicator diffusion technique.

Finally, the indicator dilution technique has a limitation that is specific for water and other compounds of similar permeability. Such tracers equilibrate with the tissue very quickly, resulting in the chance of appreciable backflux during the venous sampling period and subsequent underestimation of $E$ (Crone, 1965). For this reason, the upper limit of $E$ that can be measured accurately with the indicator dilution technique is 0.9 (Fenstermacher et al., 1981), a value very close to that expected for water. This may not be a limitation for all species. Bolwig and Lassen (1975) used the indicator dilution technique to measure brain extraction of water in the rat and report an $E$ value of 0.5 at normal brain blood flows.

## 3.2. Brain Uptake Index (BUI) Measurements

The brain uptake index (BUI) method of measuring BBB transport of labeled compounds was first described by Oldendorf (1970) and has subsequently been used to study brain capillary permeability of many compounds including water. Like the indicator diffusion technique, the BUI method measures unidirectional transfer of a tracer compound from blood to brain during a single passage through the capillary bed and relates this uptake to that of a reference tracer. Unlike the indicator diffusion method, however, the BUI normally uses a highly diffusable, flow-limited tracer as the reference compound and requires brain tissue sampling for determination of tracer influx into the brain.

As originally proposed by Oldendorf (1970), BUI measurements were performed by quickly injecting a bolus of labeled reference and test compounds into the carotid artery of rats, decapitating 15 s later, and removing the brain for subsequent counting. The reference compound, $^3$H-water in the original experiments, was considered to be totally extracted by the brain during a single pass through the cerebral vasculature and thus defined the amount of injectate distributing to any given portion of the brain. The brain uptake index (BUI) of the test compound could then be calculated with the following formula:

$$\text{BUI} = \frac{(C_b/C_i)_t \,(100)}{(C_b/C_i)_r} \tag{7}$$

where $C_b$ and $C_i$ refer to activity per sample of brain and injectate, respectively, of test (t) or reference (r) materials. If the reference compound is completely extracted during a single brain capillary passage and the time of decapitation is chosen so as to allow a sufficient interval for the bolus to pass completely through the brain, but not long enough to allow backflux of tracers, then the BUI, as defined in Eq. (7), is equivalent to $E$ in Eq. (1). If these conditions are not met, the BUI is undefined and useful only for comparisons of relative permeabilities of groups of compounds whose BUI values were measured in the same way and under the same physiological conditions.

Since its introduction, the BUI technique has undergone numerous changes in protocol designed to correct the technical problems that have become evident. Thus, for example, the time of decapitation, originally 15 s after injection, has been reduced to 5 s because several studies indicated that significant backflux of reference tracer from brain to blood occurred if longer times were used (Oldendorf and Braun, 1976; Furlow and Bass, 1976; Hardebo and Nilsson, 1979). Similarly, the volume of the bolus injection, originally 200 μL, has been subsequently reduced to 10 μL or less based on the observation that the larger injectate volume results in cerebral blood flow artifacts that can affect the measurement of the BUI (Hardebo and Nilsson, 1979; Clark et al., 1981). Reduction of the injectate volume, coupled with the use of smaller needles for injection, results in a more normal pattern of blood flow through the brain during the measurement of the BUI.

The most serious technical problem associated with the use of the BUI technique is the choice of a proper reference tracer. Uptake of $^3$H-water, the original reference tracer used by Oldendorf (1970), was subsequently demonstrated to be partially diffusion-limited in cerebral capillaries (Eichling et al., 1974; Raichle et al., 1974, 1976; Bolwig and Lassen, 1975). This problem becomes particularly apparent at high cerebral blood flows. Several other compounds whose uptake is flow-limited over a wide range of blood flows have subsequently been used as reference tracers (e.g., iodoantipyrine, butanol, ethanol, and nicotine), but virtually all appear to be partially diffusion-limited at high cerebral blood flows. If the reference tracer is not completely extracted by the brain during the first pass through the cerebral capillaries, BUI, as calculated in Eq. (7) is not equal to $E$. Even if BUI is shown to closely approximate $E$, $PS$ still cannot be calculated in the absence of some measure of cerebral blood flow (Eq. 2). A technique for simultaneous measurement of

BUI and cerebral blood flow has been developed (Pollay and Stevens, 1979), but is technically difficult and requires the use of an external radiation monitor for measurement of the blood flow tracer.

The BUI method, with appropriate reference tracers, has been used to measure brain uptake of water in several studies (Hardebo and Nilsson, 1979; Clark et al., 1981). Aside from the question of whether the BUI values measured in such experiments meet the criteria for being equated with $E$ (Eq. 1), such studies have the same disadvantages as the indicator diffusion studies. The measured value of $E$ cannot be used to calculate $PS$, the quantity of interest, in the absence of measurements of cerebral blood flow (Eq. 2), and such calculations are totally inappropriate in situations in which cerebral blood flow is changing. Additionally, as with the indicator diffusion technique, reliable estimates of BUI and $E$ can be expected only in the range of extractions from 0.1 to 0.9. The value of $E$ for water in cerebral capillaries may be at the higher end of this range. At the low end of this extraction range ($E < 0.1$), the use of this technique to measure brain uptake is particularly limited (Fenstermacher et al., 1981).

### 3.3. External Registration

The external registration technique for measuring BBB permeability was developed by Raichle and coworkers (Raichle et al., 1974, 1976; Eichling et al., 1974) and has subsequently been used to measure BBB permeability to water under a variety of circumstances (Raichle et al., 1975; Raichle and Grubb, 1978; Go et al., 1981). This technique overcomes a major disadvantage of both the indicator diffusion and BUI techniques by permitting simultaneous measurement of both $E$ and $\dot{Q}$, thus allowing direct determination of $PS$ from Eq. (2); however, the degree of technological support required to perform external registration measurements is available only in select settings.

The labels most commonly used in these experiments are $^{15}$O and $^{11}$C. Both are very short-lived isotopes that are prepared in a cyclotron immediately before use. Compounds prepared with these labels are injected into test animals, generally via the carotid artery, and the transit of the tracer through the brain is followed by means of an external NaI scintillation detector. The short half-life of the tracers permits several measurements of $E$ to be made in the same animal sequentially.

A semilogarithmic plot of activity vs time following injection of $H_2^{15}O$ is shown in Fig. 1. Immediately after injection, brain tracer activity reaches a peak value that is considered to represent total tracer activity in the detector field. Activity subsequently falls in a fashion that can be described as a biphasic clearance. The fast component of this clearance presumably represents movement of tracer still in the cerebral vascular compartment into the systemic circulation. The secondary slow component of clearance represents backflux of tracer from tissue to blood with subsequent removal. Interpreted in this manner, the clearance plot can be used to calculate $E$ from Eq. (1). $A$ is set equal to peak tracer activity and $B$ is determined by extrapolating the slow clearance component back to the time of peak tracer activity.

Figure 1 can also be used to calculate cerebral blood flow ($\dot{Q}$). The time–activity curve for tracer washout is first used to calculate the mean tracer transit time ($\bar{t}$).

$$\bar{t} = \frac{\int_0^t q(t)dt}{q_0} \tag{8}$$

where $q(t)$ is the tracer activity in the brain as a function of time and $q_o$ is the tracer activity of the injectate. From the central volume principle (Roberts et al., 1973)

$$\bar{t} = V/\dot{Q} \tag{9}$$

where $V$ is the volume of distribution of the tracer. Eq. (9) can be solved for $\dot{Q}$ and expressed in the following form:

$$\dot{Q} = \frac{(\lambda)\,(100)}{\bar{t}} \tag{10}$$

where $\lambda$ is the mean equilibrium blood–brain partition coefficient for the tracer in question. The only parameter not measured directly in Eq. (10) is $\lambda$. For $H_2^{15}O$, the most frequently used tracer in external registration measurements, $\lambda = 0.95$ to $0.96$ (Herscovitch and Raichle, 1985). Once $E$ and $\dot{Q}$ are determined, $PS$ can be calculated directly from Eq. (2).

The external registration technique has several attractive features. It provides quick, simultaneous measurements of $E$ and $\dot{Q}$ with only a single tracer. All of the aforementioned problems associated with reference tracers and vascular corrections are effectively eliminated. The major disadvantage in using the external

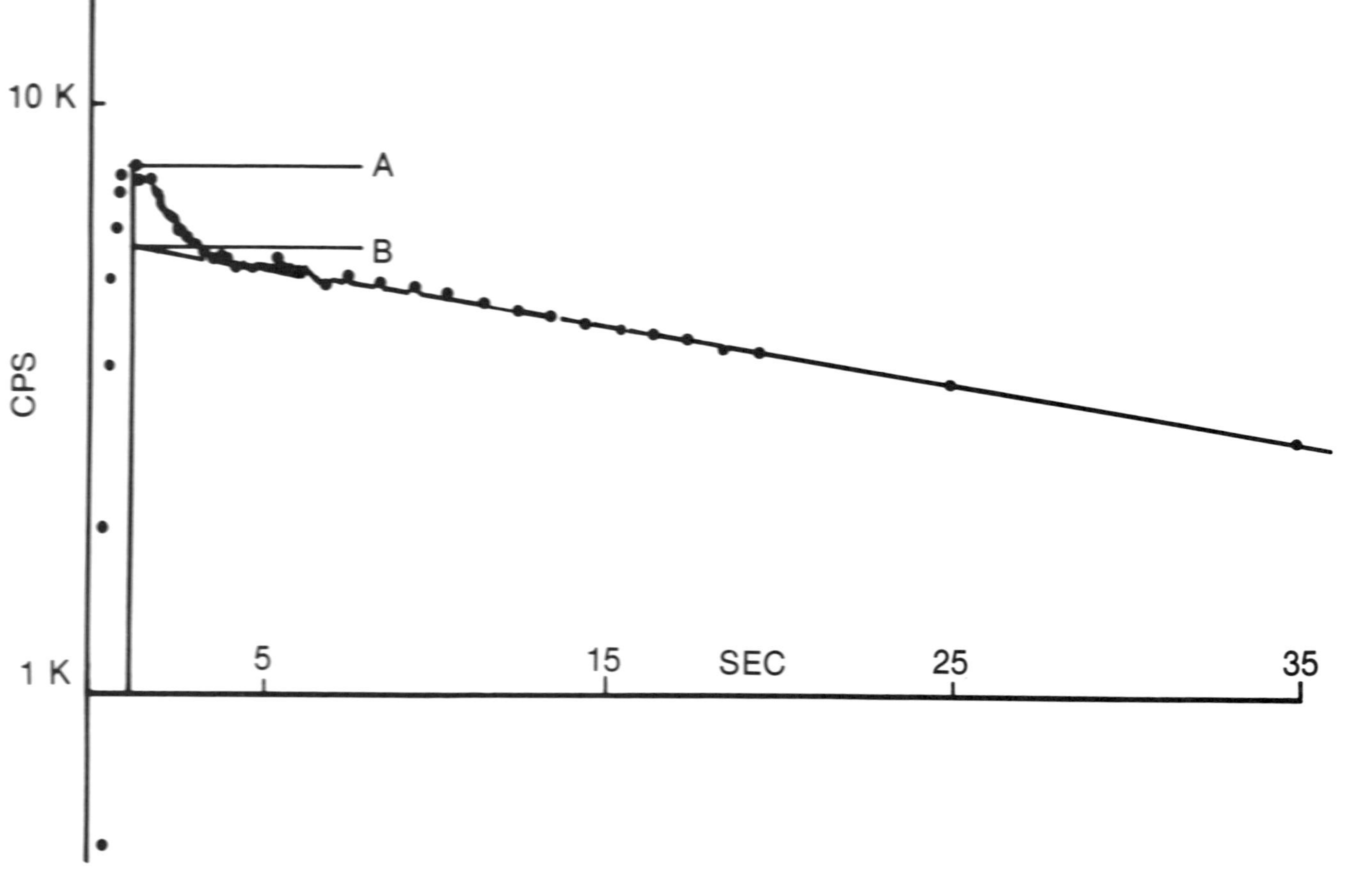


Fig. 1. The initial time course of $H_2^{15}O$ in brain following injection of a tracer bolus into the carotid artery. The fraction of $H_2^{15}O$ extracted by the brain is determined by the ratio $B/A$ (from Raichle et al., 1974).

registration technique to measure brain water permeability, aside from cost, is the fact that only a whole brain average value of $E$ can be determined. Thus, regional differences in $PS$ for water cannot be detected using this method.

### 3.4. IV Double-Label Technique

Gjedde et al. (1980) and Sage et al. (1981) have described and tested a technique for simultaneously measuring brain blood flow and brain extraction of a test compound with independent tracers. The brain blood flow marker (butanol) and a test compound are injected into a femoral vein. Blood is continuously withdrawn from an artery during the experimental period (10–25 s) and the experiment is terminated by decapitating the test animal. The tracer activity of the entire arterial sample is measured, thus providing a direct measure of the arterial time–activity integral over the time course of the experiment. Upon completion of the experiment the brain is removed and monitored for tracer activity, thus providing a measure of $B$, the amount of tracer that has moved from blood to brain during a single capillary passage. Butanol is assumed to be flow-limited over the cerebral blood flow range present during the experiment and thus 100% extracted. Under these conditions, $E$ is calculated using a variant of Eq. (3) (Sage et al., 1981):

$$E = \frac{B}{\dot{Q} \int_0^t C_a(t)dt} \tag{11}$$

where $C_a$ is the arterial concentration of the test tracer and $\dot{Q}$ is the cerebral blood flow as measured by the butanol uptake. This technique has subsequently been used to measure regional, rather than whole brain, glucose uptake kinetics (Betz and Iannotti, 1983).

The major problem associated with use of the double-label technique involves the potential limitation of butanol as a flow-limited blood flow marker. At high blood flows butanol may be less than completely extracted during its passage through the cerebral vasculature. Under these conditions $\dot{Q}$ must be corrected for the effect of incomplete brain extraction of butanol (Sage et al., 1981). On the positive side, the double-label technique is simple to perform and relatively cheap. It can be adapted to provide regional data, thus providing a rare opportunity to simultaneously study regional variations in blood flow and tracer extraction. The utility of this technique to study water permeability of the brain has not been evaluated.

## 4. Bulk Flow of Water Across the BBB

All of the techniques presented thus far have been used to study the diffusive movement of water across the BBB. Bulk flow of water across the BBB (i.e., water movement driven by osmotic and/or hydrostatic gradients) has been less frequently studied. Such studies can be performed with or without the use of labeled tracers and could provide fundamental information concerning processes affecting fluid movement in brain during both normal and pathological states.

The general expression for bulk flow of water across a membrane may be written as follows (House, 1974):

$$J_v = L_p (\Delta_p - \Delta\pi_i - \Sigma\sigma_s\Delta\pi_s) \tag{12}$$

where $J_v$ is the net volume flux of water (mL/cm$^2$/s), $L_p$ is the hydraulic conductivity of the membrane (mL/cm$^2$/s/atm), $\Delta_p$ is the hydrostatic pressure difference across the membrane (atm), $\Delta\pi_i$ and $\Delta\pi_s$ are the osmotic pressure gradients across the membrane caused by impermeant and permeant solutes, respectively (atm), and $\sigma_s$ is the reflection coefficient, a unitless number between 0 and 1.0 that relates the observed osmotic effectiveness of a given solute to its predicted osmotic effectiveness, assuming that it is totally impermeable across the membrane. The $\pi$ terms in Eq. (12) are calculated from the van't Hoff equation

$$\pi = RT\Delta C \tag{13}$$

where $R$ is the gas constant, $T$ is the absolute temperature, and $\Delta C$ is the concentration difference of the solute in question across the membrane.

Quantitative measurements of the hydraulic conductivity of the BBB have been made in the rabbit (Fenstermacher and Johnson, 1966) and in humans (Paulson et al., 1977). Both studies chose to manipulate $\Delta\pi$, measure $J_v$, and calculate $L_p$ (Eq. 12). Under normal circumstances the $\Delta p$ term of Eq. (12) is minimal and for experimental purposes it is difficult to manipulate. Fenstermacher and Johnson (1966) infused various solutes into the blood in such a way as to achieve a steady-state osmotic gradient over a relatively short time and measured the resultant volume changes in the brain by observing the meniscus of a saline-filled capillary in contact with the subarachnoid fluid of the brain. $L_p$ was determined by setting the $\sigma$ value of the largest solute used, raffinose, equal to 1, imposing an osmotic gradient with this solute, and measuring $J_v$. $L_p$ can then be calculated directly from Eq. (12). The $L_p$ value calculated for CNS capillaries in such experiments, $1.1 \times 10^{-6}$ mL/cm$^2$/min/mOsm, is much lower than that reported for equivalent systemic capillaries (Fenstermacher, 1985).

Paulson et al. (1977) measured the diffusion permeability and the filtration permeability of water across the BBB in humans. Diffusion permeability ($P_d$) was calculated by measuring the extraction of $^3$H$_2$O with the indicator diffusion technique, while simultaneously measuring brain blood flow. The resultant values, along with an estimate of brain capillary surface area, were used to solve Eq. (2) for $P_d$. Filtration permeability ($P_f$) was measured by imposing an osmotic gradient of radiographic contrast media or mannitol across the BBB and measuring fluid removal from the brain by monitoring dilution of impermeant radioactive tracers in cerebral venous blood samples. The $P_f/P_d$ ratio calculated from these experiments was 4.3, a value low enough to lead these authors to suggest that the hydraulic conductivity and the tracer water permeability of the BBB were identical, a situation to be expected if the BBB is aporous. A subsequent modeling study using the same data (Patlak and Paulson, 1981) suggested that deviations of the $P_f/P_d$ ratio from unity could be explained by assuming that an unstirred layer effect acts within the cerebral endothelial cell.

The above studies are simply examples of the sorts of information that can be derived from relatively simple experiments in which one of the variables determining bulk flow across the BBB is modified. The development of protocols for achieving square-wave changes in plasma solute concentration (Daniel et al., 1975; Patlak and Pettigrew, 1976) permits $\Delta\pi$ to be changed in a predictable and relatively instantaneous manner. A change in $\Delta\pi$ should always be measured in terms of plasma osmolality rather than plasma concentration of the infused solute. This restriction is necessary because the infusion of large impermeant solutes to raise plasma osmolality also results in extraction of water from tissue so that the change in steady-state plasma osmolality after infusion is less than $\Delta\pi$ calculated from the plasma concentration of the impermeant solute alone. Step changes in $\Delta\pi$, in concert with careful tissue water determinations, should permit regional estimations of brain capillary $L_p$, with the caveat that control tissue water content would have to be determined in separate animals. Such measurements would be of great interest given the current use of hyperosmotic solutions for reversible opening of the BBB and the apparent differences in fluid flow through gray and white matter (Rosenberg et al., 1980; Szentistvanyi et al., 1984).

One final point should be made concerning studies of water movement and distribution during osmotic perturbations. Unless such experiments are of very short duration, the possibility of volume regulatory ionic events cannot be precluded. Redistribution of brain electrolyte across the BBB occurs within 30 min following hyposmotic stress (Melton and Nattie, 1983) and, given the high reflection coefficient for electrolytes across the BBB (1.0 for NaCl) (Fenstermacher, 1985), will affect the osmotic gradient driving $J_v$ (Eq. 12). Such volume regulation appears to be intrinsic to the CNS in general (Fenstermacher, 1984) and cerebrovascular endothelium in particular (Kempski et al., 1985). Failure to recognize this possibility may lead to miscalculation of $L_p$.

## 5. Movement of Electrolytes Across the BBB

### 5.1. Blood-to-Brain Measurements

As previously emphasized, all physiologically important electrolytes, in marked contrast to water, cross the BBB at a very slow rate. Values for $K_i$ for these solutes, expressed as mL/h/g, range from 0.05 to 0.8 (Bradbury, 1979). This extremely low permeability of the BBB to electrolytes limits the variety of techniques that can be used to measure transfer of this class of solutes across the BBB. Single-pass tracer techniques are virtually useless in this permeability range since the amount of solute that enters the brain parenchyma over the time course of the experiment is negligible. Hence, all of the methods employed to study ion permeation of the BBB are multiple-pass techniques with subsequent sampling of brain tissue. Within this general constraint, techniques may vary as to whether they involve intravenous (iv) or intraarterial (ia) tracer injection, whether they maintain a constant plasma level of tracer or involve a single injection and falling plasma concentration of test solute and, finally, as to mode of data analysis.

The most commonly used techniques for measurement of ion transfer at the BBB employ iv introduction of tracer. Preparation of experimental animals in these studies begins with cannulation of a vein, usually the femoral, for injection/infusion and cannulation of an artery for blood sampling. For constant plasma level studies, iv infusion protocols must be determined (Daniel et al., 1975; Patlak and Pettigrew, 1976). For declining plasma level techniques, the tracer is given as a bolus. Arterial blood samples are drawn to establish plasma tracer concentration at fixed times after the initiation of infusion or injection. If the tracer is given as an iv bolus, arterial samples should be initially taken at frequent intervals, since the period immediately following injection may represent an appreciable portion of the activity–time integral that comprises the denominator of Eq. (3). At a fixed period after introduction of the tracer, the test animal is killed and the brain removed and assayed for activity. This may be done for the whole brain or for select brain regions.

The total activity of the tracer in brain tissue should be corrected for that portion confined to the brain vascular volume. This correction, expressed as a fraction of total brain activity, increases as the BBB permeability of the tracer decreases. Measurement of the intravascular correction is usually done by injecting a separate vascular volume marker (e.g., $^{125}I$ serum albumin) just prior to sacrifice, calculating the vascular space of the tissue, and determining the correction factor by multiplying the vascular volume times the activity of the tracer under study in the terminal plasma sample. This correction factor is then subtracted from the uncorrected brain tracer activity to yield the parenchymal tracer activity of brain, the $B$ term of Eq. (3). Alternatively, the brain vasculature can be flushed with saline immediately prior to sacrifice to remove all extraparenchymal tracer (Preston et al., 1983); however, backflux of test tracer from brain tissue to the flushing solution should be considered a potential problem in this procedure. For complete accuracy, the vascular space of each individual animal should be calculated.

As in all procedures in which the amount of test material available for exchange is determined from plasma activity, the potential for binding of tracer to plasma components should be realized and carefully evaluated. Such binding can appreciably reduce the amount of tracer actually available for exchange. This problem may not be significant in studies of monovalent electrolyte exchange across the BBB, but could considerably modify the results of studies employing, for example, divalent cations that show significant plasma binding.

Of the two plasma concentration time courses used in iv tracer experiments, constant plasma tracer activity or falling plasma tracer activity, the former is much less commonly used. Examples of this approach may be seen in the studies of Davson and Welch (1971), Levin and Patlak (1972), and Hawkins et al. (1982). In all of these studies, tracer is infused in such a way as to achieve a constant plasma activity and then groups of animals are sacraficed at periods ranging from minutes to hours after the beginning of the infusion. Tracer activity in the brain is subsequently measured by conventional counting or by autoradiography (Hawkins et al., 1982) and corrected for intravascular tracer. Data collected in such experiments may be analyzed in a variety of ways that will be described below.

The ease of iv bolus injection/falling plasma tracer concentration experiments has led to their widespread use in measuring ion permeation across the BBB. As in the constant plasma level experiments, large numbers of animals are used since separate groups must be sacrificed at each time point after isotope injection. This approach was used to study the kinetics of $^{24}$Na movement from blood to brain by Stulc (1967a,b). A variation on this approach that uses fewer animals has been published by Ohno et al. (1978). In this protocol, the so-called single time analysis (Fenstermacher et al., 1981), the test animal is killed at a fixed time after iv bolus injection of tracer. By application of initial rate analysis and with several crucial assumptions (discussed below), $K_i$ for the tracer in question can be calculated for each individual animal from this single time point. This method has been subsequently used to study the kinetics of BBB transport of several ions (Smith et al., 1981; Smith and Rapoport, 1984).

A final experimental protocol that should be mentioned is the *in situ* brain perfusion technique developed by Takasato et al. (1984). This technique for measuring cerebrovascular transport differs in several ways from procedures previously discussed. The experimental animal used in these studies is the rat. Tracer is introduced into the brain vasculature by retrograde perfusion of the external carotid artery of one side after ligation of the common carotid and branches of the internal and external carotid (Fig. 2). The net effect of this procedure is to reduce the nonhemispheric blood flow on the perfused side of the brain to less than 5% of the total perfusate flow. Thus, the composition of the fluid that the brain vasculature actually sees is effectively controlled by controlling the composition of the perfusate. Control of perfusate concentration allows this technique to be used for the study of transport saturation and competition experiments (e.g., Smith et al., 1984). The duration of these experiments ranges from 20 s to 5 min, less

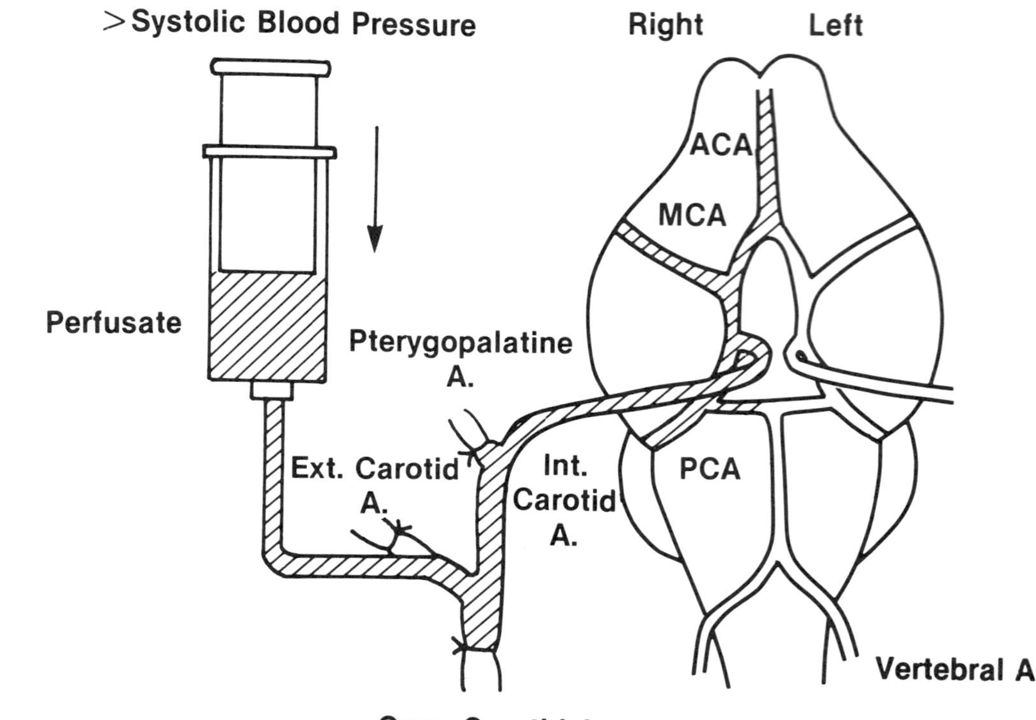


Fig. 2.   Diagram of the surgical preparation used for *in situ* perfusion of the right cerebral hemisphere of the rat. Abbreviation: A, artery; ACA, anterior cerebral artery; MCA, middle cerebral artery; PCA, posterior cerebral artery (from Takasato et al., 1984).

permeable solutes requiring longer perfusion times. After completion of the perfusion, the brain is removed and assayed for tracer activity. The resultant data, when analyzed by single-time analysis (see below), can be used to generate a *PS* value for the entire perfused hemisphere or, if more detailed dissection is performed, for specific brain regions. By varying the time of brain perfusion, permeability coefficients in the range of $10^{-8}$ to $10^{-4}$ cm/s may be measured (Takasato et al., 1984). The major drawback of the technique is that calculation of *PS* requires an estimate of the regional cerebral perfusate flow ($F_{pf}$), i.e., the perfusate flow to the region whose permeability is being measured. This is usually estimated from plasma clearance of a cerebral blood flow marker (e.g., $^{14}$C-iodoantipyrine) in a separate series of animals. Direct measurement of individual values of $F_{pf}$ would obviously be preferable, albeit technically difficult.

Although the method of data analysis in experiments of the sort mentioned thus far is left, to some degree, to the investigator; two general options are available: compartmental analysis or initial rate analysis. Both approaches may be used for the same data with the exception of single time point experiments, which can only be analyzed using initial rate analysis. Compartmental analysis yields a series of transfer constants that are model-specific and hence difficult to compare with results from experiments analyzed with different models. For estimates of $K_i$, initial rate analysis is preferred.

The essential assumption of initial rate analysis is that transport of the tracer in question is unidirectional from blood to brain over the time course of the experiment. When a constant plasma level of tracer is maintained, a linear increase in brain tracer concentration is sufficient proof that this requirement is met (Bradbury and Kleeman, 1967). Single-time analysis, a specialized form of initial rate analysis that has become increasingly popular, unfortunately does not permit a clear test of this assumption. In single-time analyses, accurate estimation of $K_i$ depends heavily on a careful selection of an appropriate experimental time. The time must be short enough that no backflux from brain to plasma takes place, but long enough that sufficient tracer for accurate counting gets into the brain parenchyma. As a general rule experiments should be sufficiently long for the amount of tracer accumulated in the brain to be at least twice the amount of tracer in the vascular compartment in order to accurately estimate $K_i$ (Blasberg et al., 1983b). The brain vascular space should be calculated in each individual test animal so that total brain activity can be accurately corrected for tracer in the vascular space. When the requirement for negligible backflux is met and tracer activity in the brain parenchyma is accurately measured, $K_i$ can be calculated directly from Eq. (3) (Ohno et al., 1978).

Another more generalized form of initial rate analysis, the so-called multiple-time/graphical analysis, has recently been developed by Patlak et al. (1983). In this form of analysis, data from either infusion or injection experiments of varying duration are graphically analyzed by plotting the ratio of brain tracer activity to the final plasma tracer activity against the ratio of the arterial activity-time integral to final plasma tracer activity for each experiment (Blasberg et al., 1983a) (Fig. 3). As long as the resultant plot is linear, movement of tracer is essentially unidirectional from blood to brain. The linear portion of the plot has a slope equal to $K_i$ and an intercept equal to the vascular fraction of the tissue. This form of analysis requires a large number of time points, but has the marked advantage of eliminating the need for a vascular marker and a correction for intravascular tracer (Patlak et al., 1983).

## 5.2. Brain-to-Blood Measurements

Brain-to-blood transfer (efflux) of solutes is not as commonly measured as influx. This is particularly true of substances with BBB permeabilities as low as those of electrolytes. Efflux can be estimated by compartmental analysis of brain uptake data (e.g., Levin and Patlak, 1972), but, as with all modeling approaches, the values determined are largely a function of the parameters of the model. Direct measurement of brain solute efflux has been approached in three basic ways: (1) brain washout; (2) ventriculo-cisternal (VC) perfusion; and (3) brain parenchymal microinjection.

### 5.2.1. Brain Washout

The brain washout technique, a variant of the BUI method of measuring BBB permeability, was used by Bradbury et al. (1975) to measure brain efflux of radiolabeled solutes following intracarotid injection. The general premise of this technique is that if the amount of tracer that has entered the brain during passage of the tracer bolus through the cerebral circulation can be estimated,

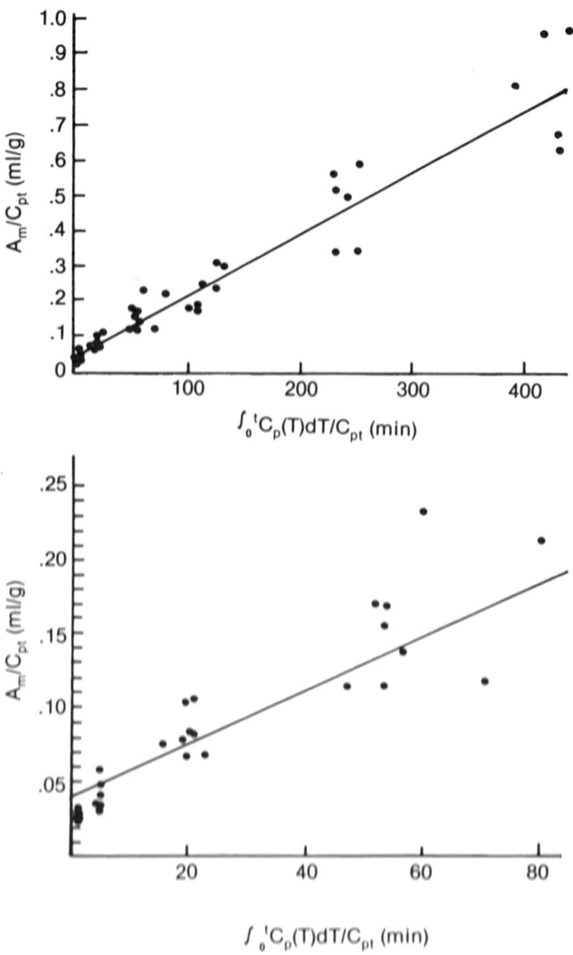

Fig. 3.   Graphical analysis of 49 experiments in the rat over a time course of 1 to 240 min. A plot of thalamic tissue $^{14}$C-aminoisobutyric acid (AIB) radioactivity/final plasma radioactivity (ordinate) versus plasma arterial radioactivity integral/final plasma radioactivity (abscissa) is shown (top). An expanded-scale plot of the early time points is also shown (bottom). The best linear fit of the data is shown as a solid line in both plots. The slope of the line is equivalent to the unidirectional influx coefficient ($K_i$) for AIB. The ordinate intercept is equivalent to the vascular fraction of the tissue, (from Blasberg et al., 1983a).

tissue sampling at subsequent times should demonstrate a progressive reduction of brain tracer activity that can be related to a rate constant for efflux. As in the BUI, the ratio of the activity of test tracer in brain to that in injectate is normalized by use of a permeable reference tracer (Eq. 7). Aside from the previously mentioned problems involved in selecting a proper reference material, this technique is totally inappropriate for measuring efflux of electrolytes because the BBB permeability of these solutes is so low that significant brain loading of tracer from the intracarotid bolus cannot be achieved.

### 5.2.2. Ventriculocisternal Perfusion

The technique of using VC perfusion to measure brain-to-blood transfer constants was developed by Patlak and Fenstermacher (1975). The brain ventricles are perfused with a physiological solution containing the radiolabeled test substance and a reference extracellular marker. Animals are sacraficed at various times after the beginning of infusion, the brains are removed, and the tissue sampled serially from the perfused surface outward. The samples are assayed for tracer activity and plots of tissue activity versus perpendicular distance from the perfused surface constructed. This protocol requires relatively large animals since accurate measurement of the tracer concentration profile requires a thick strip of brain tissue. If the perfusion has been sufficiently long to establish a tissue steady state and the test substance is sufficiently permeable across the BBB, a brain-to-blood transfer constant can be calculated. The crucial assumption in this calculation is that at steady state all of those influx processes that result in an increase in brain tracer activity are matched by efflux across the BBB into the cerebral vasculature. During this steady state, the following relationship holds if the movement of the test substance through the extracellular volume is primarily diffusional (Fenstermacher et al., 1981):

$$C_x/C_0 = S \exp(-\sqrt{k/D_t}\, x) \qquad (14)$$

where $C_0$ is the activity of the test material in the ventricular perfusate at the ependymal surface; $C_x$ is the activity of the test material in the brain at a point perpendicular to, and $x$ distance from, the ependymal surface; $S$ is the apparent tissue distribution space, a function of the ependymal permeability of the tracer and its tissue distribution space; $k$ is the brain-to-blood transfer time constant (1/s); and $D_t$ is the effective diffusion coefficient of the tracer in brain tissue ($cm^2/s$). From Eq. (14) it can be seen that a semilogarithmic plot of $C_x/C_0$ vs $x$ will yield a plot whose slope is $\sqrt{k/D_t}$. If $D_t$ is known or can be estimated, $k$ can be calculated.

Aside from the formidable technical problems associated with the VC perfusion experiments, use of this technique to calculate brain-to-blood transfer constants is fraught with analytical problems. These difficulties are discussed in detail in Fenstermacher et al. (1981) and will not be dealt with in this presentation. Suffice it to say that Fenstermacher et al. (1981) note that the time required to establish a steady state during VC perfusion experiments is roughly 4–5 times longer than the transport half-time ($t_{1/2}$) of the test tracer. Since most estimates of $t_{1/2}$ of electrolytes (e.g., $Na^+$) in brain range from 1–2 h, this constraint effectively limits the use of this technique to measure efflux of electrolyte from brain to blood.

### 5.2.3. Brain Parenchymal Microinjection

This technique for measuring solute efflux from brain to blood has been used to measure efflux of amino acids from brain (Lajtha and Toth, 1962) and, more recently, has been employed extensively by Cserr and coworkers to measure efflux of large molecular weight tracers and $Na^+$ from brain interstitium (Cserr et al., 1981; Szentistvanyi et al., 1984). As used by Cserr et al. (1981), this technique involves injection of tracer directly into brain tissue via a stereotaxically implanted cannula. The cannula is implanted 1 wk prior to injection to permit time for BBB healing, and tracers are injected in very small volumes (0.5 µL) and very slowly (over 8–30 min) to avoid production of intracranial pressure gradients. Groups of animals are killed at various times after injection and the activity of tracer in brain tissue measured.

Analysis of the data collected in such experiments is based on the following relationship:

$$N_b = N_i e^{-kt} \tag{15}$$

where $N_b$ is the tracer activity measured in the brain at time $t$; $N_i$ is the total activity of tracer in the brain at $t = 0$; and $k$ is the rate constant for total efflux of tracer from brain tissue. Use of Eq. (15) assumes that the tracer is neither bound nor metabolized and that efflux from brain occurs passively and with first-order kinetics. Under these conditions a semilogarithmic plot of the percent of injected tracer remaining in the brain versus time will yield a straight line with a slope equal to $k$. If clearance of tracer from brain to CSF can be measured or estimated, $k$ can be corrected to yield $k_{i,}$

the rate constant for tracer efflux across the BBB from brain to blood (Cserr et al., 1981). Values of $k_p$ for $^{22}$Na measured using this technique closely match independently measured values for the blood-to-brain influx coefficient of this tracer, a situation to be expected if brain $Na^+$ is in steady state.

The brain microinjection technique permits efflux to be measured from specified local brain regions and has the additional advantage of being relatively simple and straightforward. On the negative side, use of this protocol requires a large number of animals for determination of $k$. Additionally, calculation of $k_p$ from $k$ requires estimates of bulk CSF volume and CSF turnover that are, at best, imperfect. Even with these limitations, this method of measuring brain-to-blood efflux is probably the most appropriate technique for compounds of very limited BBB permeability.

## 6. In Vitro Methods

The last decade has seen the development of in vitro methods that promise new insights into the basic transport functions of the BBB. Research employing isolated brain microvessels, recently reviewed by Joo (1985), has clearly demonstrated that the brain capillary endothelium has transport properties more commonly associated with epithelium, specifically carrier systems capable of transporting ions and metabolic substrates (Betz, 1983; Dick et al., 1984). Eisenberg et al. (1980), based on findings of $Na^+$-$K^+$ ATPase on isolated brain microvessels, have suggested that the BBB plays an active role in formation and ionic homeostasis of brain extracellular fluid.

Isolated brain capillaries are particularly useful for studying and characterizing the biochemical properties of the BBB and its associated enzyme systems; however, the use of this model for studies of BBB solute and water flux is currently problematical. Difficulties relate first to the relatively impure preparations that result from available brain microvessel isolation techniques and, second, to the damaged state of the microvessels following isolation and purification. Current separation techniques result in preparations of both arterioles and capillaries, and contamination of preparations with glia and smooth muscle is not uncommon. Even relatively pure capillary fragments have open ends that presumably leave both luminal and abluminal plasma membranes accessible to tracer, complicating analysis of transport studies (Bradbury,

1985). The ability to seal and perfuse microvessel fragments, in the way that renal tubules are perfused, could provide more specific information concerning the transport properties of the brain capillary endothelium.

The ability to isolate cerebral microvessels has also provided a starting point for another in vitro approach to BBB permeability studies, the cultured brain capillary endothelial cell (Bowman et al., 1983). Pure cultures of bovine brain capillary endothelial cells have been established, tested for purity by verification of the presence of endothelium-specific enzyme systems, and characterized microscopically. The cultured cells aggregate to form structures resembling tight junctions and respond to selected physiological stimuli in a manner similar to that of the BBB in vivo (Bowman et al., 1983). Bradbury (1985) has suggested that monolayers of these cells, in effect a cultured BBB, could be placed in an Ussing chamber and subjected to standard epithelial transport techniques. This method would allow the investigator to control the fluid environment on both sides of an intact and functioning BBB and has the potential to yield significant information concerning the transport properties of this membrane system.


## 7. Summary

The major goal of this review has been to describe and critically assess techniques used to obtain quantitative estimates of BBB permeability to water and ions. Such techniques measure either the unidirectional extraction or the unidirectional transfer constant of the compound in question across the BBB. The effective permeability of the BBB, generally expressed as the permeability-surface area product ($PS$), can be calculated from either of these quantities. The emphasis throughout the review has been on in vivo techniques. The major in vitro methods, isolated brain capillaries and capillary endothelial cell culture, although offering much promise for the future, have not been used to date to measure transport of material across the entire endothelial cell and thus across the BBB.

One final point should be made. When considering the effective permeability of the BBB to a given substance, it is important to remember that $PS$ is the product of both a true permeability term ($P$), which appears to be largely a function of the physical structure of the BBB, and a capillary surface area term ($S$), which has considerable potential for physiological modification. *PS* is not a fixed value and, in situations in which *PS* has been demonstrated to change in response to an experimental manipulation, changes in both permeability and surface area must be considered as the potential source of the variability.

## Acknowledgments

The author would like to thank Dr. Joseph Fenstermacher for reading and critically commenting on various portions of this review and Marcella Spioch for her excellent assistance in production of this manuscript. Preparation of this review was supported by National Heart, Lung and Blood Institute Grant HL 07467.

## References

Betz A. L. (1983) Sodium transport in capillaries isolated from rat brain. *J. Neurochem.* **41,** 1150–1157.

Betz A. L. and Iannotti F. (1983) Simultaneous determination of regional cerebral blood flow and blood-brain glucose transport kinetics in the gerbil. *J. Cereb. Blood Flow Metab.* **3,** 193–199.

Blasberg R. G., Fenstermacher J. D., and Patlak C. S. (1983a) Transport of α-aminoisobutyric acid across brain capillary and cellular membranes. *J. Cereb. Blood Flow Metab.* **3,** 8–32.

Blasberg R. G., Patlak C. S., and Fenstermacher J. D. (1983b) Selection of experimental conditions for the accurate determination of blood-brain transfer constants from single-time experiments: A theoretical analysis. *J. Cereb. Blood Flow Metab.* **3,** 215–225.

Bolwig T. G. and Lassen N. A. (1975) The diffusion permeability to water of the rat blood–brain barrier. *Acta Physiol. Scand.* **93,** 415–422.

Bowman P. D., Ennis S. R., Rarey K. E., Betz A. L., and Goldstein G. W. (1983) Brain microvessel endothelial cells in tissue culture: A model for study of blood–brain barrier permeability. *Ann. Neurol.* **14,** 396–402.

Bradbury M. W. B. (1985) Critique: The blood–brain barrier *in vitro.* *Neurochem. Int.* **7,** 27–28.

Bradbury M. W. B. (1979) *The Concept of a Blood Brain Barrier.* John Wiley, Chichester.

Bradbury M. W. B. and Kleeman C. R. (1967) Stability of the potassium content of cerebrospinal fluid and brain. *Am. J. Physiol.* **213,** 519–528.

Bradbury M. W. B., Patlak C. S., and Oldendorf W. H. (1975) Analysis of brain uptake and loss of radiotracers after intracarotid injection. *Am. J. Physiol.* **229**, 1110–1115.

Clark H. B., Hartman B. K., Raichle M. E., Preskorn S. H., and Larson K. B. (1981) Measurement of cerebral vascular extraction fractions in the rat using intracarotid injection techniques. *Brain Res.* **208**, 311–323.

Crone C. (1963) Permeability of capillaries in various organs as determined by use of the indicator diffusion method. *Acta Physiol. Scand.* **58**, 292–305.

Crone C. (1965) The permeability of brain capillaries to non-electrolytes. *Acta Physiol. Scand.* **104**, 407–417.

Cserr H. F. and Bundgaard M. (1984) Blood–brain interfaces in vertebrates: A comparative approach. *Am. J. Physiol.* **246**, R277–R288.

Cserr H. F., Cooper N., Suri K., and Patlak C. S. (1981) Efflux of radiolabeled polyethylene glycols and albumin from rat brain. *Am. J. Physiol.* **240**, F319–F328.

Daniel P. M., Donaldson J., and Pratt O. E. (1975) A method for injecting substances into the circulation to react rapidly and to maintain a steady level. With examples of its application in the study of carbohydrate and amino acid metabolism. *Med. Biol. Eng.* **13**, 214–227

Davson H. and Welch K. (1971) The permeation of several materials into the fluids of the rabbit's brain. *J. Physiol.* **218**, 337–351.

Dick A. P., Harik S. I., Klip A., and Walker D. M. (1984) Identification and characterization of the glucose transporter of the blood–brain barrier by cytochalasin B binding and immunological reactivity. *Proc. Natl. Acad. Sci. USA* **81**, 7233–7237.

Eichling J. O., Raichle M. E., Grubb R. L., and Ter-Pogossian M. M. (1974) Evidence of the limitations of water as a freely diffusable tracer in the brain of the rhesus monkey. *Circ. Res.* **35**, 358–364.

Eisenberg H. M., Suddith R. L., and Crawford J. S. (1980) Transport of Sodium and Potassium Across the Blood–Brain Barrier, in *The Cerebral Microvasculature* (Eisenberg H. M. and Suddith R. L., eds.) Plenum, New York.

Fenstermacher J. D. (1985) Flow of Water and Solutes Across the Blood–Brain Barrier, in *Trauma of the Central Nervous System* (Dacey R. G., ed.) Raven, New York.

Fenstermacher J. D. (1984) Volume Regulation of the Central Nervous System, in *Edema* (Staub N. C. and Taylor A. E., eds.) Raven, New York.

Fenstermacher J. D. (1983) Drug Transfer Across the Blood–Brain Barrier, in *Topics in Pharmaceutical Sciences 1983* (Breimer D. and Speiser P., eds.), Elsevier, Amsterdam.

Fenstermacher J. D. and Johnson J. A. (1966) Filtration and reflection coefficients of the rabbit blood–brain barrier. *Am. J. Physiol.* **211**, 341–346.

Fenstermacher J. D. and Rapoport S. I. (1984) The Blood–Brain Barrier, in *Handbook of Physiology* section 2. *The Cardiovascular System* vol. IV (Renkin E. M. and Michel C. C., eds.) American Physiological Society, Bethesda, Maryland.

Fenstermacher J. D., Blasberg R. G., and Patlak, C. S. (1981) Methods for quantifying the transport of drugs across brain barrier systems. *Pharmacol. Ther.* **14,** 217–248.

Furlow T. W. and Bass N. H. (1976) Cerebral hemodynamics in the rat assessed by a non-diffusible indicator-dilution technique. *Brain Res.* **110,** 366–370.

Gjedde A. J., Hansen A. J., and Siemkowicz E. (1980) Rapid simultaneous determination of regional blood flow and blood-brain glucose transfer in brain of rat. *Acta Physiol. Scand.* **108,** 321–330.

Go K. G., Lammertsma A., Paans A., Vaalburg W., and Woldring M. (1981) Extraction of water labeled with oxygen 15 during single capillary transit. Influence of blood pressure, osmolarity and blood–brain barrier damage. *Arch. Neurol.* **38,** 581–584.

Hardebo J. E. and Nilsson B. (1979) Estimation of cerebral extraction of circulating compounds by the brain uptake index method: Influence of circulation time, volume injection, and cerebral blood flow. *Acta Physiol. Scand.* **107,** 153–159.

Hawkins R. A., Mans A. M., and Biebuyck J. F. (1982) Amino acid supply to individual cerebral structures in awake and anesthetized rats. *Am. J. Physiol.* **242,** E1–E11.

Herscovitch P. and Raichle M. E. (1985) What is the correct value for the brain-blood partition coefficient for water? *J. Cereb. Blood Flow Metab.* **5,** 65–69.

Hertz M. M. and Paulson O. B. (1980) Heterogeneity of cerebral capillary flow and its consequences for estimation of blood–brain barrier permeability. *J. Clin. Invest.* **65,** 1145–1151.

House C. R. (1974) *Water Transport in Cells and Tissues* pp. 36–76. Williams and Wilkins, Baltimore, Maryland.

Joo F. (1985) The blood–brain barrier *in vitro:* Ten years of research on microvessels isolated from the brain. *Neurochem. Int.* **7,** 1–25.

Kempski O., Spatz M., Valet G., and Baethmann A. (1985) Cell volume regulation of cerebrovascular endothelium *in vitro. J. Cell. Physiol.* **123,** 51–54.

Lajtha A. and Toth J. (1962) The efflux of intracerebrally administered amino acids from the brain. *J. Neurochem.* **9,** 199–212.

Lassen N. A. and Crone C. (1970) The Extraction Fraction of a Capillary Bed to Hydrophilic Molecules; Theoretical Considerations Regarding the Single Injection Technique With a Discussion of the Role of Diffusion Between Laminar Streams (Taylor's Effect), in *Capillary Permeability* (Crone C. and Lassen N. A., eds.) Alfred Benzon Symposium II, Academic, New York.

Lassen N. A., Trap-Jensen J., Alexander S. C., Olesen J., and Paulson O. B. (1971) Blood–brain barrier studies in man using the double-indicator method. *Am. J. Physiol.* **220,** 1627–1631.

Levin V. A. (1980) Relationship of octanol/water partition coefficient and molecular weight to rat brain capillary permeability. *J. Med. Chem.* **23,** 682–684.

Levin V. A. and Patlak C. S. (1972) A compartmental analysis of $^{24}$Na in rat cerebrum, sciatic nerve, and cerebrospinal fluid. *J. Physiol.* **224,** 559–581.

Melton J. E. and Nattie E. E. (1983) Brain and CSF water and ions during dilutional and isosmotic hyponatremia in the rat. *Am. J. Physiol.* **244,** R724–R732.

Murray J. E. and Plioplys A. (1972) An indicator dilution technique for study of blood-to-brain solute passage in the rat. *J. Appl. Physiol.* **33,** 681–683.

Nicholson C. (1980) Dynamics of the brain cell microenvironment. *Neurosci. Res. Prog. Bull.* **18,** 177–322.

Ohno K., Pettigrew K. D., and Rapoport S. I. (1978) Lower limits of cerebrovascular permeability to electrolytes in the conscious rat. *Am. J. Physiol.* **235,** H299–H307.

Oldendorf W. H. (1970) Measurement of brain uptake of radiolabeled substances using a tritiated water internal standard. *Brain Res.* **24,** 372–376.

Oldendorf W. H. and Braun L. D. (1976) [$^{3}$H]-tryptamine and [$^{3}$H]-water as diffusible internal standards for measuring brain extraction of radiolabeled substances following carotid injection. *Brain Res.* **113,** 219–224.

Patlak C. S. and Fenstermacher J. D. (1975) Measurements of dog blood-brain transfer constants by ventriculocisternal perfusion. *Am. J. Physiol.* **229,** 877–884.

Patlak C. S. and Paulson O. B. (1981) The role of unstirred layers for water exchange across the blood–brain barrier. *Microvasc. Res.* **21,** 117–127.

Patlak C. S. and Pettigrew K. D. (1976) A method to obtain infusion schedules for prescribed blood concentration time courses. *J. Appl. Physiol.* **40,** 458–463.

Patlak C. S., Blasberg R. G., and Fenstermacher J. D. (1983) Graphical evaluation of blood-to-brain transfer constants from multiple-time uptake data. *J. Cereb. Blood Flow Metabol.* **3,** 1–7.

Paulson O. B., Hertz M. M., Bolwig T. G., and Lassen N. A. (1977) Filtration and diffusion of water across the blood–brain barrier in man. *Microvasc. Res.* **13,** 113–124.

Phelps M. S., Huang E. J., Selin C., and Kuhl D. E. (1981) Cerebral extraction of N-13 ammonia: Its dependence on cerebral blood flow and capillary permeability-surface area product. *Stroke* **12,** 607–619.

Pollay M. and Stevens A. (1979) Simultaneous measurement of regional blood flow and glucose extraction in rat brain. *Neurochem. Res.* **4**, 109–123.

Preston E., Allen M. and Haas N. (1983) A modified method for measurement of radiotracer permeation across the rat blood–brain barrier: The problem of correcting brain uptake for intravascular tracer. *J. Neurosci. Meth.* **9**, 45–55.

Raichle M. E. and Grubb R. L. (1978) Regulation of brain water permeability by centrally released vasopressin. *Brain Res.* **143**, 191–194.

Raichle M. E., Eichling J. O., and Grubb R. L. (1974) Brain permeability of water. *Arch. Neurol.* **30**, 319–321.

Raichle M. E., Eichling J. O., Straatman M. G., Welch M. J., Larsen K. B., and Ter-Pogossian M. M. (1976) Blood–brain barrier permeability of $^{11}$C-labeled alcohols and $^{15}$O-labeled water. *Am. J. Physiol.* **230**, 543–552.

Raichle M. E., Hartman B. K., Eichling J. O., and Sharpe L. G. (1975) Central noradrenergic regulation of cerebral blood flow and vascular permeability. *Proc. Natl. Acad. Sci. USA* **72**, 3726–3730.

Rapoport S. I., Ohno K., and Pettigrew K. D. (1979) Drug entry into brain. *Brain Res.* **172**, 354–359.

Roberts G. W., Larsen K. B., and Spaeth E. E. (1973) Interpretation of mean transit measurements for multiphase systems. *J. Theoret. Biol.* **39**, 447–475.

Rosenberg G. A., Kyner W. T., and Estrada E. (1980) Bulk flow of brain interstitial fluid under normal and hyperosmolar conditions. *Am. J. Physiol.* **238**, F42–F49.

Sage J. I., Van Uitert R. L., and Duffy T. E. (1981) Simultaneous measurement of cerebral blood flow and unidirectional movement of substances across the blood–brain barrier: Theory, method and application to leucine. *J. Neurochem.* **36**, 1731–1738.

Smith Q. R. and Rapoport S. I. (1984) Carrier-mediated transport of chloride across the blood–brain barrier. *J. Neurochem.* **42**, 754–763.

Smith Q. R., Johanson C. E., and Woodbury D. M. (1981) Uptake of $^{36}$Cl and $^{22}$Na by the brain-cerebrospinal fluid system: Comparison of the permeability of the blood-brain and blood-cerebrospinal fluid barriers. *J. Neurochem.* **37**, 117–124.

Smith Q. R., Takasato Y., and Rapoport S. I. (1984) Kinetic analysis of L-leucine transport across the blood–brain barrier. *Brain Res.* **311**, 167–170.

Stulc J. (1967a) The entry of $^{24}$Na from blood into the brain of mice during 30 minutes after intravenous isotope injection. *Life Sci.* **6**, 85–95.

Stulc J. (1967b) The permeability of mouse cerebral capillaries to sodium. *Life Sci.* **6**, 1837–1846.

Szentistvanyi I., Patlak C. S., Ellis R. A., and Cserr H. F. (1984) Drainage of interstitial fluid from different regions of rat brain. *Am. J. Physiol.* **246,** F835–F844.

Takasato Y., Rapoport S. I., and Smith Q. R. (1984) An *in situ* brain perfusion technique to study cerebrovascular transport in the rat. *Am. J. Physiol.* **247,** H484–H493.

Yudilevich D. L. and De Rose N. (1971) Blood-brain transfer of glucose and other molecules measured by rapid indicator diffusion. *Am. J. Physiol.* **220,** 841–846.

# The Choroid Plexus–Arachnoid Membrane–Cerebrospinal Fluid System

Conrad E. Johanson

## 1. Historical Background

The choroid plexus traditionally has been considered the major, but not sole, component of the blood–CSF barrier (BCFB). Most analyses of the BCFB have been directed to the choroid plexuses, rather than to the arachnoid, because of the predominant function of the former in CSF secretion and homeostasis. The physiological literature on the choroid plexus (CP) is much more extensive than on the arachnoid membrane; consequently this review emphasizes methodologies employed to evaluate choroidal function.

Mammalian studies of the epithelial components of the BCFB have been much less extensive than those of other epithelia (kidney, small intestine, and so on), and have even lagged behind those of the blood–brain barrier. The size, shape, and location of choroid plexuses have presumably all been factors discouraging research efforts (Johanson, 1987b).

The diminutive size of CP tissues, i.e., about 0.5% of brain weight, has thwarted some technical approaches in the smaller murine species. For example, unless milligram-weight specimens are pooled from many rats, it has not been feasible to utilize techniques requiring considerable amounts of tissue (e.g., drug concentration-binding studies, membrane separation techniques, and so on). The relatively small size of the choroid epithelial cells (8- to 10-$\mu$m cubes in many mammals) has not favored microelectrode impalements. Moreover, the shape of mammalian CP is not conducive to mounting this tissue in an Ussing-type chamber; thus, the lack of information on short-circuit currents and solute fluxes between serosal and mucosal compartments, in vitro. Finally, the relative inaccessibility of the plexuses, buried deep in the interior of the brain, has necessitated laborious surgical procedures for *in situ* isolation of the choroidal tissues.

In spite of the above-mentioned hindrances, considerable methodological progress has been made over the past decade.

Substantial evidence has been gradually accumulating to support that the BCFB is the main transport interface for moving many water-soluble drugs, ions, and micronutrients into the CNS (Spector and Eells, 1984; Johanson, 1987b). The purpose of this review is to describe and evaluate the advantages and disadvantages of numerous CP and arachnoid preparations. The physiological concepts gained from each preparation are briefly discussed.

## 2. Mammalian vs Nonmammalian Models

Although the scope of this review is mainly about mammalian preparations, it is important to point out that other vertebrate species not in the mammalian class have been useful models of BCFB function (Cserr et al., 1980). The in vitro bullfrog CP, used extensively by Wright and colleagues, has yielded prolific information about the transport of $Na^+$, $K^+$, $Cl^-$, $HCO_3^-$, sugars, and amino acids across the choroidal membrane mounted in an Ussing-type chamber (Wright, 1978). Measurements of electrophysiological parameters for amphibian CP have made possible the delineation of electrochemical gradients for ions across the external limiting membranes of this epithelium. Since the frog CP normally operates at a temperature lower than that of poikilothermic mammals, this factor has to be considered for interspecies comparisons of active transport and CSF formation. Also, carbonic anhydrase, the enzyme that catalyzes the hydration of $CO_2$ to $HCO_3^-$, is present at substantially lower levels in amphibians than in mammals; this may partly explain the lower secretion rate (per unit weight of CP) by the former. Frogs do not have a lateral ventricle CP (LVCP), i.e., the region most widely studied in mammals.

The dogfish shark is conspicuous for its relatively large-size CP, i.e., about 100 mg. Such an ample source of CP tissue would be useful for in vitro techniques requiring sizeable amounts for membrane separations. Dogfish CP function may significantly differ from that of mammals, however. For example, carrier transport of urea across the BCFB, to maintain osmotic balance in the CNS, has been demonstrated in the dogfish (Cserr et al., 1968), but not in the rat (Parandoosh and Johanson, 1982).

The hagfish is a primitive vertebrate that lacks a CP (Cserr et al., 1980). Thus transport studies of the hagfish barrier system, i.e., cerebral capillaries, may furnish clues, deductively, about the role of the CP in the overall economy of brain extracellular fluid homeostasis in the higher vertebrates. In general, phylogenetic information sheds insight on the nature of CP function in poorly developed organisms.

Most of the information summarized below is for adult mammals commonly used in laboratory investigations, i.e., rats, rabbits, cats, dogs, and sheep. More recently, plexus tissues from cattle have provided generous amounts of material needed for ultracentrifugation techniques. The anatomic similarity among choroid plexuses of vertebrates points to a common denominator of basic function.

## 3. Locations and Structure of the Blood–CSF Barrier (BCFB)

The BCFB is composed mainly of the CP villous structures in the ventricular system and the arachnoid membrane as the roof over the subarachnoid space. Because the size, structure, and location of the choroid plexus tissues have greatly influenced technical approaches, it is pertinent to consider salient histological and gross anatomical features of this secretory and reabsorptive epithelium (*see* Fig. 1; Tennyson and Pappas, 1968). Arachnoidal structure is described by Spector and Johanson (1987).

Tufts of choroid plexus tissues are suspended in the CSF of each of the four ventricles of the brain. There are regional differences in the size and configuration of CP tissues in the various ventricular cavities. In mammals, the third ventricle CP (3VCP) comprises about 10% of the total weight of all the plexuses; the 4VCP and each LVCP make up about 40 and 25%, respectively, of the total mass (Fig. 2). The gross shape of the LVCP is a narrow band of tissue, whereas that of the 4VCP is more like a sheet.

When comparing the cellular composition of the barrier systems, it is important to consider that the parenchymal cells of the CP consist of a *single* layer of epithelium, whereas the arachnoid membrane is composed of *multiple* layers of cells (Spector and Johanson, 1987). Tight junctions (*zonulae occludentes*) are present between the constituent cells of both barriers. Thus, barrier tightness, together with transport mechanisms, assure CSF of specialized and stable composition (Cserr, 1971).

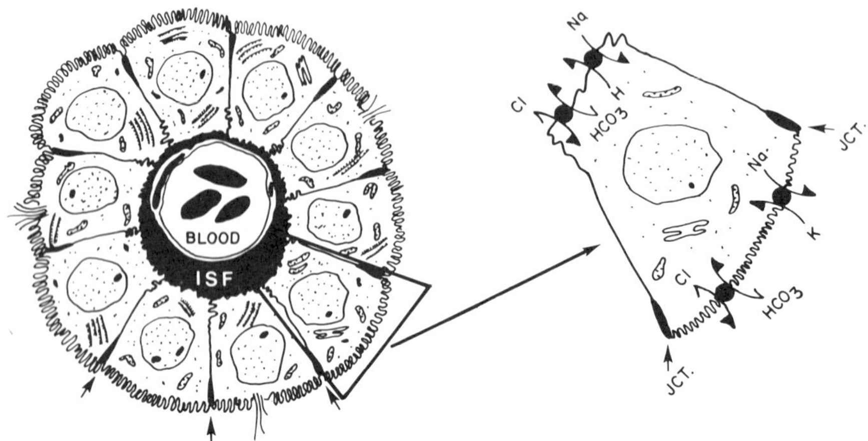


Fig. 1.   Cross-sectional schematic of the structure and some ion transport systems in the choroid plexus of the rat lateral ventricle. The interstitial fluid (ISF) compartment is the darkened area between the blood and the circumferentially lying epithelial cells. The basolateral (ISF-facing) membrane, together with the tight junctions (arrows), constitute the blood–CSF barrier. Drawn approximately to scale. Normally, in vivo, there is net secretion of $Na^+$, $Cl^-$, and $HCO_3^-$, from plasma to CSF. Evidence for the various antiport systems has been furnished by Wright's laboratory for the frog and by Johanson and colleagues for the rat. Arms of the anion exchangers can run in either direction, depending upon the relative concentrations of $Cl^-$ and $HCO_3^-$ in cell and CSF (figure reproduced, with permission, from Johanson et al., 1985).

## 4. Research on the Choroid Plexus vs the Blood–Brain Barrier

The BCFB works in concert with the blood–brain barrier (BBB) to assure homeostasis of the composition and volume of the extracellular fluid bathing the cells of the brain. A comparison of the structure and function of these two barriers, i.e., CP vs. cerebral capillaries, reveals similarities as well as interesting differences. One important example of similarity is the predominance of ATPase activity on the non-blood- (i.e., CSF-) facing membrane of the barrier, i.e., the apical membrane of CP (Smith and Johanson, 1980b) and the abluminal membrane of cerebral endothelium (Eisenberg et al., 1980). Thus, the predominant ATPase membrane in both choroid plexus and the BBB is involved with clearing $K^+$

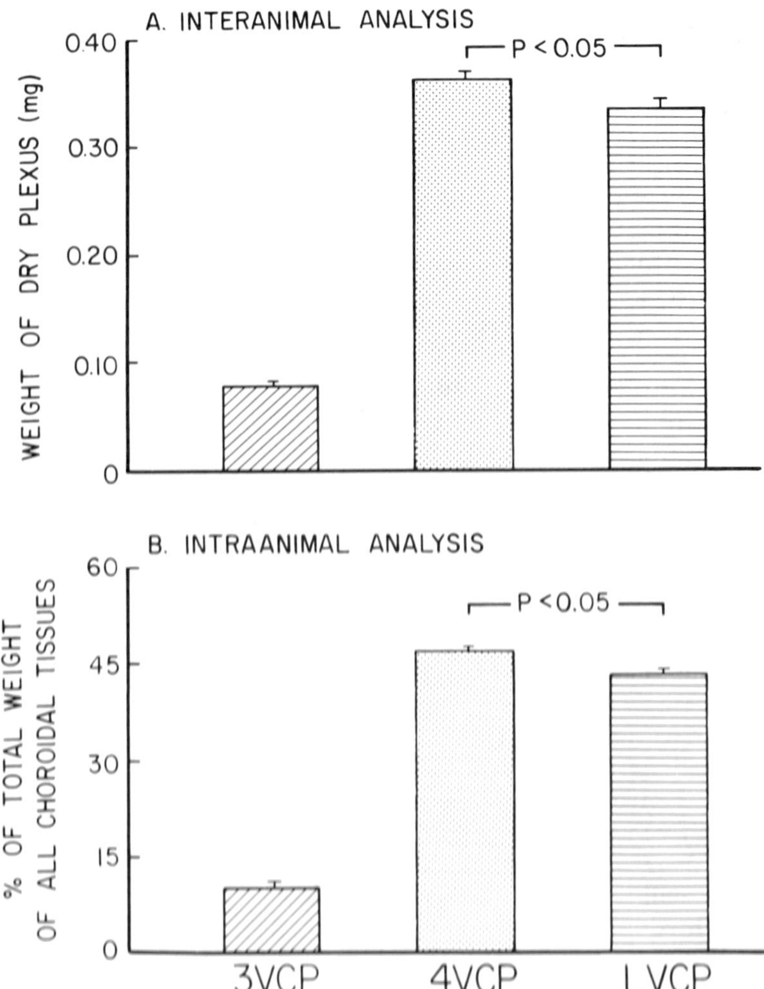



Fig. 2.   Intraventricular distribution of choroidal tissues. (A) Mean ± SEM weight (mg) of adult rat choroid plexus tissues dried to constant weight ($n = 32$ for each region). 3VCP and 4VCP refer to choroid plexus tissues in third and fourth ventricles, respectively. CP tissues from right and left lateral ventricles of each animal were pooled for the LVCP analysis. 4VCP weight is significantly greater than LVCP or 3VCP, $p < 0.05$ by Student's $t$-test. (B) The CP weight in each region is expressed as % of total dry weight of all CP tissues in the same animal. The two LVCP tissues in each rat were pooled. Intraanimal analysis was performed on each of 32 rats, for which mean ± SEM is presented. Statistical analysis and abbreviations are the same as in (A) (figure reproduced, with permission, from Harbut and Johanson, 1986).

and inorganic anions (such as I$^-$) from the CNS extracellular fluid to blood (Bradbury, 1979). This suggests that the CP epithelium could serve as a model for some BBB transport systems, especially those involving astroglial-endothelial translocating mechanisms. Indeed, Bradbury (1975) has clearly described the common embryological origin of choroidal and astrocytic cells. One technical advantage in cellular analysis of the barriers is that the CP (composed mainly of epithelium) is readily excised as a discrete parenchyma (Fig. 1), in contrast to the complexity of isolating endothelial and astrocytic elements from homogenized brain tissue.

On the other hand, the structural uniqueness of the BCFB imposes limitations on the methodologies applicable to CP *per se.* The anatomical interposition of the large compartment of interstitial fluid between the leaky choroidal capillaries and the basolateral–membrane barrier of the epithelium precludes the use of conventional single-pass bolus and arterial integral techniques popularized by Oldendorf, Crone, Rapoport, Fenstermacher, and others for the BBB (Rapoport, 1976; Bradbury, 1979).

There are marked differences in the functions of the two barriers. First and foremost, the choroid epithelia, unlike the cerebral endothelia, are equipped for large-capacity secretion of ions and water into the CNS (Johanson, 1987b). Additionally, there are specialized carrier transport systems in CP for facilitating the movement of vitamins, nucleotide precursors, and other micronutrients from blood to CSF. Several of these carriers in CP, e.g., ascorbic acid, appear to be absent or negligible in the cerebral capillary wall (Spector and Lorenzo, 1974). Such selectivity in the BCFB affords pharmacological opportunities for manipulating fluxes of many water-soluble solutes across the mammalian CP (Johanson, 1987b).


## 5. Polarity of the Choroidal Membrane

There is both ultrastructural and microfunctional sidedness to the epithelial membrane of the choroid plexus. The respective poles of the cell have their distinctive features. Thus, the basolateral (blood-facing) and apical (CSF-facing) membranes have their peculiar array of receptors, enzyme activities, ion channels, and transport systems (Fig. 1). Such polarization allows for net secretion of solutes and water from blood to CSF simultaneously with net reabsorption of other substances in the reverse direction. This asymmetric two-way traffic of solutes across the BCFB is finely coordinated so that there is resultant homeostasis of both choroid cellular fluid and the elaborated CSF. Clearly one of the biggest challenges is to delineate the complex, coordinated interaction of various transport systems in the external limiting membranes of the epithelium.

A model for inorganic ion transport systems is presented in Fig. 1 for the adult rat CP. Much of the information for this scheme was obtained by a battery of in vitro and in vivo techniques, many of which are described in sections 6–10 of this chapter. Models for the transport of several organic compounds (Spector and Eells, 1984) by mammalian CP (e.g., nucleosides) have been presented by Spector (1982).

To characterize basolateral and apical transport mechanisms, one approach has been to combine compartmentation techniques (section 6) with kinetic analyses (sections 7 and 8). Promising subcellular methods, e.g., vesicular membrane preparations, allow for the precise characterization of membrane transport systems (section 10).

## 6. Compartmental Analysis of Choroid Plexus Function

Physiological analyses of either the intact (in vivo) or isolated (*in situ* or in vitro) CP must take into account the three major compartments of this tissue: vascular, interstitial (extracellular), and parenchymal (epithelial).

### 6.1. Vascular Compartment

The great vascularity of the CP indicates that brisk blood flow is essential to supporting its diverse secretory and reabsorptive functions. It is often necessary to assess either choroid plexus blood flow or the residual blood left in the extirpated tissue.

#### 6.1.1. Choroid Plexus Blood Flow (CPBF)

Blood flow to the CP is among the highest values quantified for numerous regions in the CNS. Mammalian plexuses are typically perfused at rates of 3–6 mL/min/g. Welch (1963) calculated that CPBF in anesthetized, adult rabbits is about 3 mL/min/g; this approach involves a laborious micropuncture technique and is therefore limited to larger laboratory mammals. The invasive surgery (craniotomy and vessel exposure) may have affected the flow being measured. Alternatively, the microsphere approach has furnished values of 3–6 mL/min/g for CPBF. Some of the initial attempts with microspheres resulted in considerable variation in the flow measurements (Hayward and Vogh, 1979). More recent microsphere investigations in sheep (Page et al., 1980), cats (Nakamura and Hochwald, 1983), and piglets (Stonestreet et al., 1983) have yielded flow rates with acceptable levels of variability (standard errors < 10% of respective means). It is important to have at least 400 dpm per CP specimen. Third, the popular iodoantipyrine (IAP) method has been applied to rat and rabbit CP, resulting in determinations of choroidal perfusion rates of 1–1.5 mL/min/g (Pollay et al., 1979; Tyson et al., 1982). Such lower values can likely be attributed to loss of IAP, from choroid cell to CSF, during the short interval of time that the plexuses are being exposed and removed postexperimentally from the animal. Thus, there is the obvious need for a flow indicator that remains trapped inside the epithelium, so that the perfusion rate will not be underestimated. The identification of such a tracer would facilitate flow measurements, especially in the smaller laboratory animals.

### 6.1.2. Residual Blood Volume

In connection with in vivo studies of the CP parenchymal content of ions, tracers, or drugs, it is essential either to wash out the residual blood or to quantify it in the sampled tissues. This procedure is necessary because as much as 20% of the wet weight of CP tissues may be from blood content (Johanson et al., 1974b). Perfusion displacement of blood by saline is straightforward, but care must be taken not to leach out test substances (especially highly diffusible, low mw substances) from the thin choroidal membrane. Percent washouts of 96–98% can be obtained with both lateral and fourth ventricle choroid plexuses (Murphy and Johanson, 1985). In other experiments, it may be desirable to quantify experimental or developmental changes in blood content, which are readily determined by assaying for hemoglobin or appropriate radioisotopic tracers. Either residual plasma and/or residual erythrocytes (RBCs) can be easily determined from the steady-state volume of distribution ($V_d$) in CP of radioiodinated serum albumin or $^{51}$Cr-tagged RBCs (Johanson et al., 1976). The mean hematocrit of the CP vasculature was determined by a dual-label technique utilizing these indicators (Heisey, 1968).

The $^{131}$I-RISHA $V_d$ in adult rat CP is about 10% (unpublished data). For compartmentation analysis, the $^{51}$Cr-RBC $V_d$ together with the radioinulin space are sufficient, since the tracer inulin estimates the residual plasma volume as well as the size of the interstitial space. When comparing $^{51}$Cr-RBC data, it should be noted whether or not the animals were killed by exsanguination. In exsanguinated rats (Sprague-Dawley, 200–400 g), the $^{51}$Cr-RBC $V_d$ is approximately 6–9% of the fresh CP weight (Johanson et al., 1974b; 1976). Plexuses from older animals (350–500 g) generally have a greater proportion of blood than young adults.

In in vitro studies of the CP, the issue of the residual RBC contribution to tissue (epithelial) transport has been discussed (Smith and Johanson, 1985). The calculated maximal error is about 10%. Following 30 min of incubation in artificial CSF, the CP appears blanched because of loss of red cells from the tissue vessels to the medium; under such experimental conditions, contributions from the red cells to tissue transport parameters seem to be negligible and thus do not need correction.

## 6.2. Extracellular Fluid Volume

The extracellular fluid volume (ECFV) in the CP is 15–20% of the tissue wet weight (Johanson et al., 1974b; 1976). There is generally small variation among species, stage of development, and experimental treatments. The greatest alteration in ECFV occurs when the CP is removed from the ventricular system and immersed in synthetic CSF medium. Therefore, the in vivo and in vitro measurements are discussed separately.

### 6.2.1. In Vivo Measurements

To obtain reliable measurements of ECFV in vivo, it is essential to maintain a stable level of radioactive tracer, commonly a saccharide, in the plasma. Radiolabeled inulin, raffinose, sucrose, or mannitol have all been used to indicate ECFV (Smith et al., 1981a). Steady-state volume of distribution (%), i.e., 100 times dpm/g CP divided by dpm/g plasma $H_2O$, is obtained when 30–60 min are allowed for tracer penetration into the interstitium. There is a tendency of the lower molecular weight compounds to distribute into a slightly higher volume of tissue fluid. When ECFV is used to calculate the choroidal cell concentration of a solute at high concentration in the ECF (e.g., [Na]), it becomes critically important to minimize the variation in the ECFV measurement. One way to accomplish this is to inject sufficient amount of radioactivity (e.g., 50 µCi of $^3$H per 200 g body weight) to obtain dpm in the diminutive CP specimens at least 10 times the background level.

The normal ECFV of about 15% can be "dissected" into its constituent parts, i.e., the residual plasma volume (about 10% of CP weight) and the true interstitial fluid volume [(about 5% of the tissue weight (Murphy and Johanson, 1985)]. Substantial alterations in the amount of blood in the CP, occurring either experimentally or developmentally, can effect significant changes in ECFV. Thus the CP is unlike tissues such as skeletal muscle (an order of magnitude less vascular), in which alterations in ECFV reflect primarily the changes in interstitial rather than residual plasma volume.

### 6.2.2. In Vitro Quantifications

When the CP is removed from the ventricular system and suspended in artificial CSF, there is at least a doubling of the ECFV (Smith and Johanson, 1985). Inulin and mannitol spaces ($V_d$) of 40–45% are typically found after 30 min of incubation. Such rapid swelling is similar to that observed in many other tissues, including brain slices. The time course of the swelling has been described for the lateral ventricle CP of the rat (Fig. 3). Lowering the temperature of the CSF medium serves to minimize the expansion of the extracellular compartment relative to the cellular space (Table 2 in Smith and Johanson, 1985). The acid–base balance of the CSF medium needs to be carefully controlled, since shifting of fluid between cells and ECF occurs when the pH and bicarbonate composition of CSF are altered (Table 1 in Johanson et al., 1985). Following removal of CP from the test tube, the amount of CSF adhering to the tissue can introduce error in the measurement of ECFV. It is important to adopt a uniform procedure for wiping adherent fluid from the in vitro CP. An effective technique in the author's laboratory has been to drag the forcep-held tissue along a uniform length (say 5 cm) of glass slide. This consistent approach helps to minimize the variation in both ECFV and tissue water content measurements.

Inulin and mannitol have been used as ECFV indicators in vitro. The primary prerequisite is that the test marker does not enter into the parenchymal cells; thus cellular viability must be maintained throughout the in vitro experiment. Sixty minutes of incubation of rat CP does not interfere with the choroid cells' ability to exclude inulin; this was demonstrated by the comparable [3]H-inulin $V_d$ values obtained when the tracer was presented either 5 or 60 min prior to the end of the 1-h incubations in synthetic CSF (Smith and Johanson, 1985).

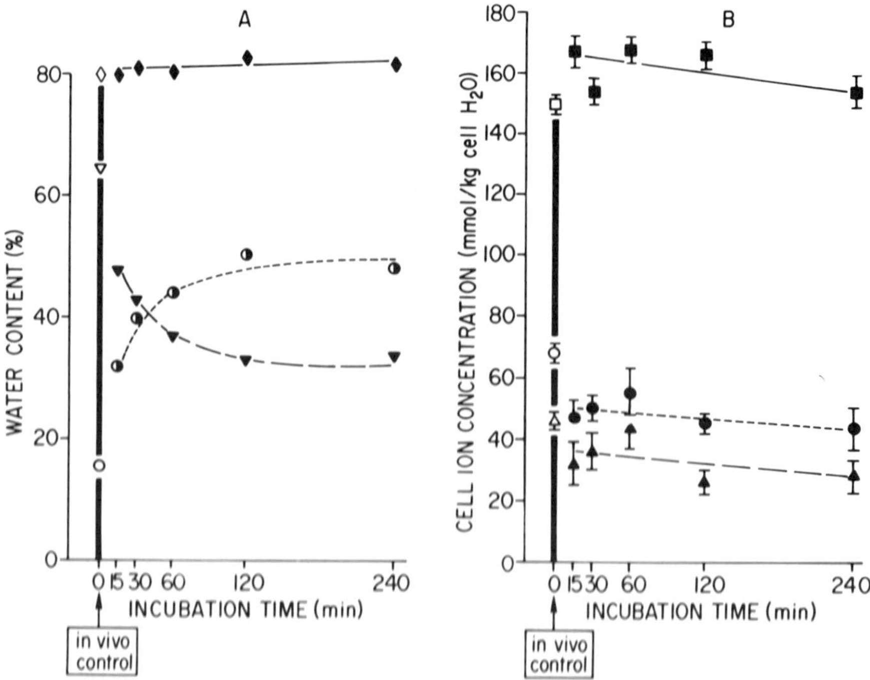



Fig. 3. Time course of choroid plexus fluid compartment size and cell ion concentrations during in vitro incubation of lateral ventricle tissue from adult rats. (A) Time course of tissue (diamonds), and extracellular (half-filled circles) and intracellular water (triangles) contents. (B) Time course of cell concentrations of $K^+$ (squares), $Cl^-$ (circles), and $Na^+$ (triangles). Each point represents mean ± SEM for six to nine animals (reproduced, with permission, from Smith and Johanson, 1985).

Preincubation of CP in CSF is advantageous in certain protocols. In a study of $^{36}Cl$ penetration kinetics, it was pertinent to employ an ECFV marker that rapidly equilibrated between CSF and tissue. Mannitol was selected as an ECFV marker because of its relatively small mol. wt. (i.e., 180), and thus its rapid diffusivity. Even so, mannitol requires about 10 min to attain steady-state distribution in the in vitro plexus (Deng and Johanson, 1984).

## 6.3. Parenchymal (Epithelial) Compartment

With respect to solute distribution from the blood, the rate-limiting barrier for transport into the choroidal cells is the basolateral membrane of the epithelium, not the capillary wall, as in the brain tissue (Parandoosh and Johanson, 1982). Thus, in vivo analyses of CP content of tracers, ions, drugs, and so on should include a deductive correction for the amount of test substance in the ECF. This procedure enhances the signal-to-noise ratio; in other words, the primary uptake by parenchymal cells can then be distinguished from the total tissue uptake (cellular plus extracellular). Similarly, steady-state compositional analyses should involve measurement of interstitial and vascular components so that the resultant cellular concentration can be obtained by subtraction from the tissue total. The general equation is: cell = tissue − ECF − RBC. The expanded equation, with its assumptions and limitations, has been described elsewhere (Johanson and Woodbury, 1977). It applies to both in vivo and in vitro systems. Practically, though, it is not necessary to correct the negligible vascular (RBC) component in the in vitro tissues (*see* section 6.1.2.).

### 6.3.1. Water

To quantify the amount of water in the cellular compartment, the first step is to determine the tissue water content of the CP. The following is an overview of water measurements and distribution analyses, determined in several thousand specimens of CP by the author and colleagues over the past decade. The small size of the rat CP (mg or less) necessitates an electrobalance weighing technique to furnish accurate values for the fresh and dry weights (Johanson et al., 1976). Typically, a value close to 80% is found for the water content [i.e., $100 \times$ (wet wt. − dry wt.)/(wet wt.)] of CP removed from intact rats. The *cellular* water of 60% (of tissue wet wt.) is found by subtracting 15 (ECFV) and 5% (RBC $H_2O$) from the 80% water in the entire tissue (Johanson et al., 1974b). Redistribution of water occurs when the extirpated tissue is incubated in artificial CSF, even though the total water content remains stable. Thus, the in vitro cellular water is only about 35%, i.e., 80% total less the mean ECFV of approximately 45% (*see* Fig. 3).

### 6.3.2. Ions

The concentration of $K^+$, $Na^+$, and $Cl^-$ in the CP of intact rats (i.e., in vivo experiments) have been extensively determined by compartmentation analysis. Tissue $K^+$ and $Na^+$ are readily measured by flame photometry, and the resulting values are close to 90 and 50 mEq/kg wet CP, respectively (Johanson et al., 1974b). [$Cl^-$] is quantified by amperometric titration, and it is usually between 60 and 65 mEq/kg (Smith et al., 1981b; Murphy, 1984). There are regional differences in ion concentration among the lateral, third, and fourth ventricle plexuses; so these tissues should not be pooled for electrolyte analyses (Harbut and Johanson, 1986). When compartmentalizing these ions in the calculations, it is pertinent to multiply the extracellular (plasma) concentration by the appropriate Donnan factor to arrive at the true interstitial concentration. Correction for the residual erythrocyte content of $K^+$ is especially important, since about 10% of the $K^+$ in CP is in red cells trapped in the tissue at the time of sampling (Johanson et al., 1974b). Lateral ventricle CP has been the most extensively analyzed, and it has yielded values (mmol/kg cell $H_2O$) of about 150–160 for $[K^+]_i$, 30–40 for $[Na^+]_i$, and 60–70 for $[Cl^-]_i$ in the epithelium.

In vitro analyses have furnished comparable values for levels of cellular ions. In spite of apparent redistribution of water between cellular and extracellular compartments (Fig. 3A), the concentrations of the major ions in the isolated plexus remain close to the in vivo values, for up to 4 h of incubation (Fig. 3B). Such ionic homeostasis is an indication of the viability of the in vitro preparation.

### 6.3.3. Organic Electrolytes

One of the earliest in vivo studies of organic solute distribution in the CP was with antipyrine and barbital, compounds with comparable molecular weights, but with diverse lipid solubilities (Johanson and Woodbury, 1977). More recent studies with DMO (dimethadione) can exemplify the compartmentalization of a drug in the CP (Johanson, 1984). DMO, a derivative of trimethadione, is an organic acid used to indicate cell pH. The steady-state distribution of ionized and nonionized DMO in the vascular, interstitial, and parenchymal (epithelial) compartments of adult rat CP is depicted in Fig. 4. Total DMO activity is counted in plasma and in CP. The fraction of ionized and nonionized DMO in the plasma is then determined from arterial pH and the $pK_a'$ 6.1 of DMO. An important assumption is that, at steady state, the concentration of nonionized DMO is the same in extracellular and intracellular water. With the DMO technique, pH values close to 7.0 are obtained for the CP from intact animals (Johanson, 1978). The Henderson-Hasselbalch equation is used to calculate the choroid cell $[HCO_3]$ of about 9–10 mM.

The DMO method has also been used in the in vitro CP (Johanson et al., 1985). Mean control values of 6.95 and 8 mM for

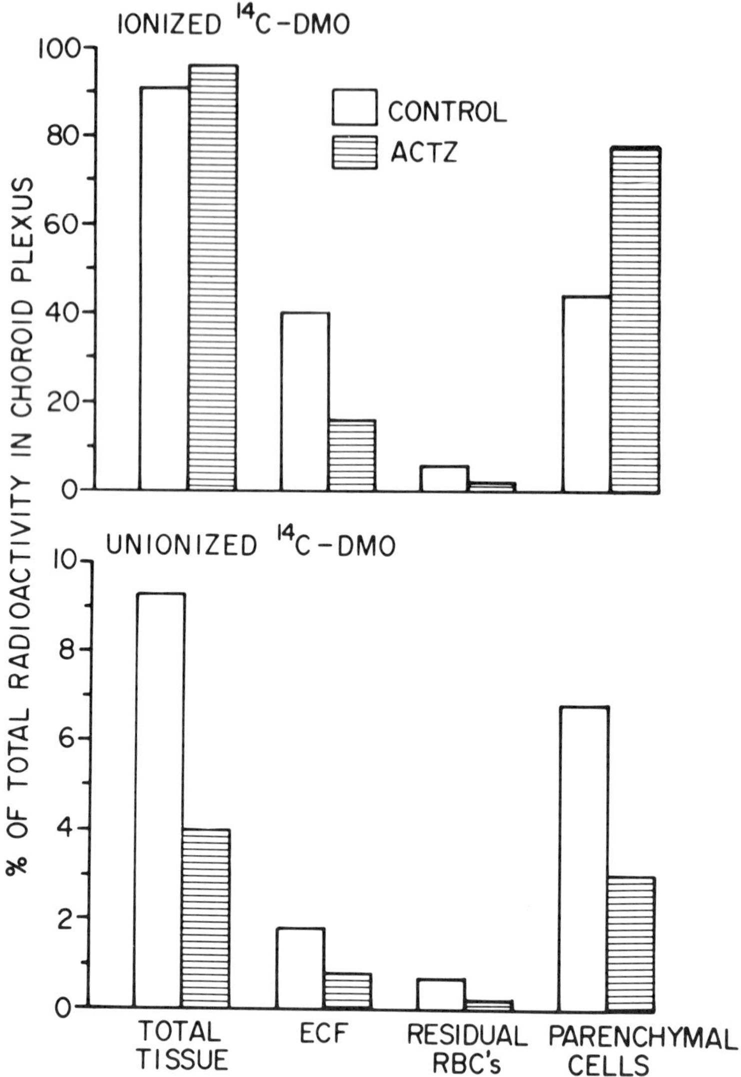


Fig. 4.   Distribution of ionized and unionized [¹⁴C]DMO (dimethadione) among tissue compartments in lateral ventricle CP of adult nephrectomized rats treated for 1 h with acetazolamide (20 mg/kg) or vehicle control (unfilled bars). Data for one treated animal and its matched control are representative of others in the respective groups. Ionized radioactive DMO in parenchymal cells determined as difference between total tissue amount and that in extracellular fluid (ECF) and residual erythrocyte (RBC) compartments. With this information, cell pH can be calculated from the Henderson-Hasselbalch equation (reproduced, with permission, from Johanson, 1984).

pH and [HCO$_3$] are slightly lower than the corresponding in vivo determinations. These small differences may be accounted for by anesthesia and pCO$_2$ levels. The in vitro measurements that enable CP pH determinations are: artificial CSF pH (measured directly by microelectrode in the medium), the $^{14}$C-DMO volume of distribution (with the CSF as the reference ECF), and the ECFV (radioinulin or -mannitol $V_d$). The equation for in vitro pH calculation has been given by Schousboe (1972).

## 7. In Vivo Choroid Plexus Models (Noninvasive)

Investigations of the CP and CSF vary considerably in the degree to which the system is physically perturbed by surgical preparation or otherwise mechanically altered in order to obtain experimental data. All in vivo models are noninvasive in the sense that the CP is surgically neither *isolated* in, nor *removed* from, the cranial cavity. Since the CPs are not mechanically disturbed, they continue to function with uninterrupted blood flow and neural tone, and are bathed with endogenous CSF. Another significant advantage of the noninvasive in vivo approach is that the normal ventricular volume and hydrostatic pressure relationships exist in the animal. These factors are of primary importance in establishing accurate control information for transport and fluid secretion.

A disadvantage with the in vivo experimentation is the complex set of variables relating to anesthesia, diurnal rhythms, and neurohumoral modulation. In designing in vivo protocols, a large number of experiments (or measurements) may be necessary to control for the numerous variables. Some of these factors discussed below, e.g., anesthesia, are not necessarily of concern in all the in vivo models; and they may also need consideration for some other models (e.g., anesthesia for the *in situ*, but not the in vitro, preparations).

### 7.1. Anesthetics

Many anesthetics have depressant effects on tissue metabolism and blood flow (e.g., barbiturates). Some agents stimulate CP metabolism [e.g., enflurane (Myers and Shapiro, 1978)], however, or enhance blood flow to the CNS (e.g., ketamine). In our in vivo experiments with CP, we have used either ether, ketamine, or pentobarbital to anesthetize the animal for tissue and CSF sampling. These agents have different effects on respiration, and thus on arterial $pCO_2/pH$; consequently, baseline levels of CP [K$^+$] and [Na$^+$] can vary with the degree of arterial acidosis (*see* Table 4 in Smith and Johanson, 1980a). One way to evaluate the direct dependence (or not) of an experimental alteration upon the anesthetic is to attempt to reproduce the finding in further experimentation with another anesthetic from a different class. For example, acetazolamide substantially elevates CP cell pH in vivo whether ether or ketamine anesthesia is used (Johanson, 1984; Murphy, 1984). Moreover, systemic acidosis leads to marked retention of K$^+$ in CP whether pentobarbital, ether, or ketamine is utilized (Smith and Johanson, 1980a; Pershing and Johanson, 1982; Murphy, 1984; Harbut and Johanson, 1986). Thus, anesthetics need to be evaluated in conjunction with baseline as well as primary drug effects. Neither ketamine nor pentobarbital seem to affect CSF formation rate in rats (Mann and Mann, 1983), probably because of the balancing of the numerous factors affecting CSF secretion. In cats the formation of CSF waxed and waned, however, respectively, with ketamine and pentobarbital (Feldman et al., 1980).

## 7.2. Diurnal Rhythms

The circadian nature of the concentration of various hormones in the CSF is becoming firmly established. For example, CSF [melatonin] can vary 10-fold over the 24-h period (Hedlund et al., 1977). There are receptors and transport mechanisms in CP for hormones like prolactin (Walsh et al., 1984) and thyroxine (Spector and Levy, 1975). Insulin and corticosteroids can modulate Na$^+$ transport and Na$^+$-K$^+$-ATPase activity in the CP (Murphy, 1984; Rychter and Stastny, 1980). Vasopressin and melatonin substantially alter the morphology of the choroidal epithelium (Kozlowski et al., 1978; Decker and Quay, 1982). Thus, the CP must significantly affect, and be affected by, hormones in the CSF. Since blood and CSF levels of many hormones fluctuate daily, and because many of these hormonal agents can regulate CP function, it is important to carry out a given series of experiments at the same periodicity. Such regularity should reduce experimental variability.

## 7.3. Sex Differences

There is a paucity of information for physiological and biochemical parameters for CP in males vs. females. Sexual differences in the CSF levels of certain hormones, e.g., prolactin, could form the basis for differential responsivity (Alaranta et al., 1983); in clinical studies it was noted that women secrete more prolactin into CSF than men. Prolactin as well as other hormones have been implicated in CSF electrolyte homeostasis. Overall, there is enough anecdotal evidence in the literature to justify the concern that sexual differences in baseline hormonal profiles may, in some investigations, complicate the array of experimental variables. More systematic studies are needed to evaluate male/female differences in hormonal and prostaglandin modulation of CP function (*see* Fig. 5 for a male/female study of prostaglandin E2 effects on ion transport in the CP–CSF system in rats).

## 7.4. Neurohumoral Modulation

The marked effects of the autonomic nervous system and neurotransmitters in plasma and CSF on CP function have been reviewed by Edvinsson et al. (1983). Sympathetic tone exerts a prominent influence on CP blood flow, epithelial enzyme activity, ion transport, and CSF secretion. Altered levels of circulating catecholamines can substantially affect CP function; therefore, surgical and other experimental procedures that modify norepinephrine (noradrenaline) concentrations (Pershing and Johanson, 1982) need to be carefully regarded in experimental protocols. The normal trophic or homeostatic effects of the sympathetic nerves to CP can be deductively evaluated by performing sympathectomy (Johanson and Harbut, 1984). Mammalian CPs receive sympathetic fibers mainly from the superior cervical ganglia.

## 7.5. Baseline Physiological Analyses of the Choroid Plexus–CSF–Brain System

Kinetic analyses of the penetration of radioisotopic nonelectrolytes and ions, from plasma into various regions of the CNS, have been instructive for comparing distributional phenomena at the three major transport interfaces: blood–CSF (CP), brain–CSF (ependyma), and blood–brain (cerebral capillaries). Lipid-soluble agents like antipyrine display flow-limited distribution, so $^{14}$C-antipyrine attains steady-state distribution in CP and cerebral cortex by 3–6 min, five times faster than in muscle (*see* Table IA in Johanson and Woodbury, 1977). Antipyrine diffuses rapidly across all barriers, but it still needs 10–15 min to distribute entirely into the large-cavity CSF. Examples of water-soluble agents, the distribution of which is not primarily limited by blood flow, are given below for nonelectrolytes and ions.

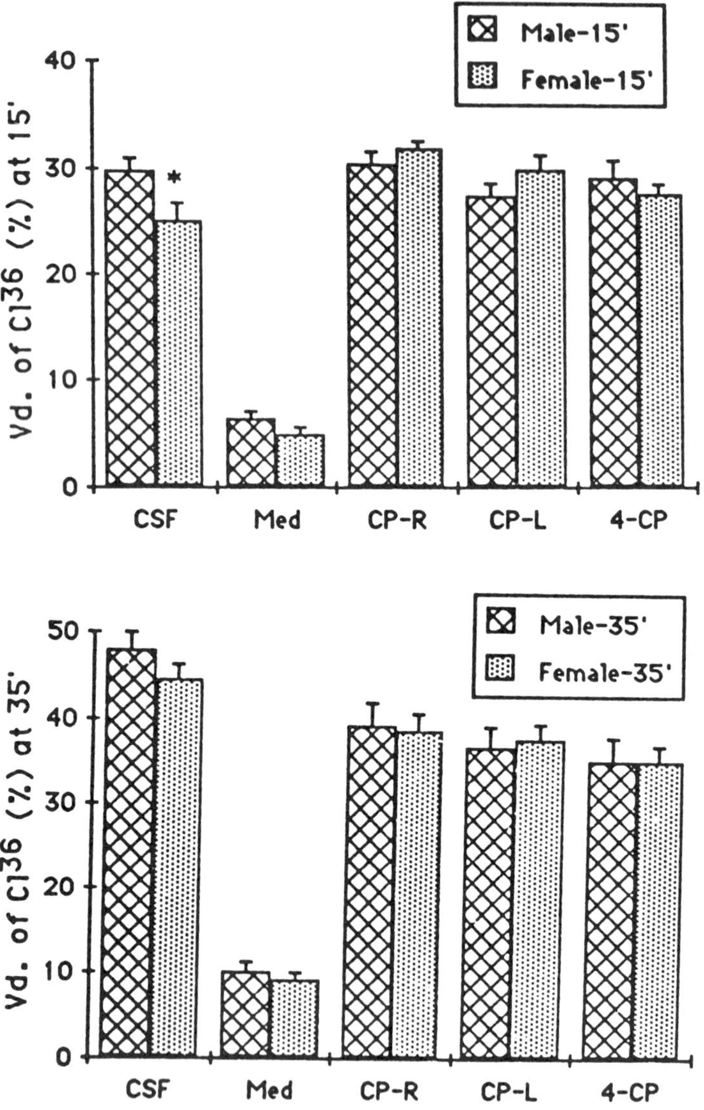



Fig. 5. Comparison of $^{36}Cl^-$ transport in male and female rats. $^{36}Cl$ was injected intravenously and maintained at steady-state level. The rats were killed after 15 min (top) or 35 min (bottom), then samples were rapidly taken and analyzed. Each bar is the mean ± SEM for five animals. Statistical comparisons were done with Student's $t$-test. The volume of distribution ($V_d$) of $^{36}Cl^-$ in female CSF is lower than that of males at 15 min. All other differences, male vs. female, are generally not significantly different. Abbreviations: Med, medulla; CSF, cisternal cerebrospinal fluid; CP, choroid plexus from right (R), left (L), or fourth (4) ventricle (V) (unpublished data by Q. Deng and C. Johanson).

### 7.5.1. Nonelectrolytes

Polysaccharides with mol. wt. ranging from 5500 (inulin) to (340) sucrose penetrate cells negligibly, and so their movement across the blood–CSF and BBB must be mainly paracellular. Even mannitol, a hexahydric alcohol (mol. wt. = 180), does not penetrate the CP epithelium (Johanson, 1980). Urea (mol. wt. = 60) gains access to cell water; consequently it is a useful agent to identify sites of molecular sieiving. The time course of uptake of $^{14}$C-urea by the rat CNS is depicted in Fig. 6. Two compartmentation analyses have furnished evidence that the rate-limiting barrier to hydrophilic solutes diffusing through the CP is the basolateral membrane (Johanson and Woodbury, 1978; Parandoosh and Johanson, 1982). The steady-state volume of distribution of urea in the CP remained remarkably constant between 8 and 16 h postnephrectomy (Fig. 6), a reflection of nonchanging CP permeability

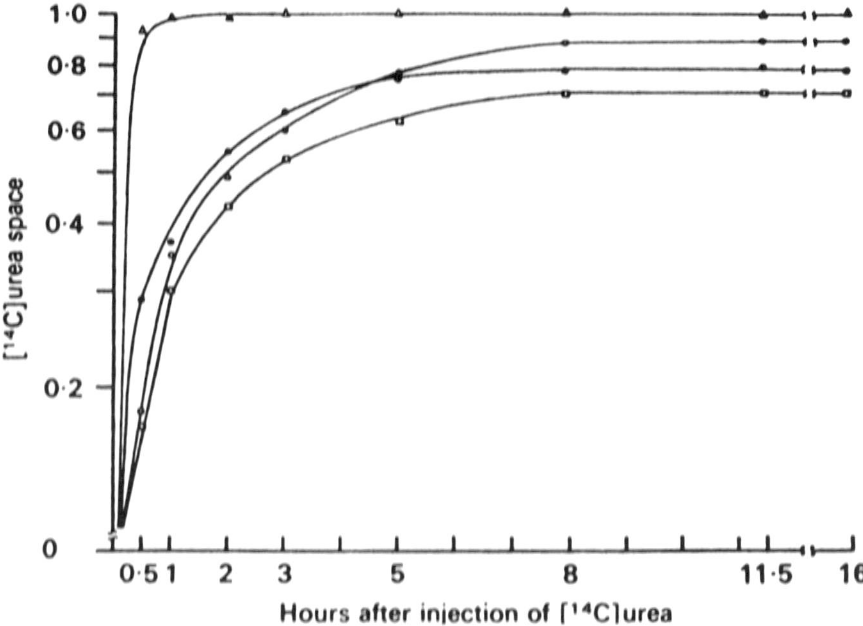


Fig. 6. Time course of uptake of $^{14}$C-urea by tissues and CSF in the adult rat. $^{14}$C-urea space = dpm/g tissue (or CSF) divided by dpm/g plasma water. Each plot on graph for skeletal muscle (triangle), cerebral cortex (unfilled circle), and CP (filled circle) is a mean value for five or six rats; for CSF (square), each plotted mean value represents four or five animals. Standard errors are generally <5% of the respective means (reproduced, with permission, from Johanson and Woodbury, 1978).

as well as stability of $^{14}$C-urea activity in the plasma (this was confirmed by analyzing the $^{14}$C activity directly in the plasma samples). It is of interest to point out the similarity in uptake curves, CP vs. brain, for urea and other solutes. Again, this is consistent with the lipoid nature of both barriers (CP epithelium and BBB endothelium).

### 7.5.2. Ions

The kinetics of permeation of radioactive $^{22}$Na$^+$ and $^{36}$Cl$^-$ into the CP–CSF system has been extensively characterized (Smith et al., 1982). Since Na$^+$ and Cl$^-$ are the main ions involved in CSF secretion, it is important to understand their movements across both the basolateral and apical membranes. When administered ip or iv, these ions require about 10 min to distribute completely into the interstitial fluid of the CP. Thereafter, the relatively slow increase in $V_d$ of $^{22}$Na$^+$ and $^{36}$Cl$^-$ in CP epithelium represents the net accumulation resulting from basolateral uptake concurrently with apical extrusion. Consequently, it requires several hours for the attainment of in vivo steady-state distribution [in contrast to 5 min in vitro, where there is no CSF sink (Parandoosh and Johanson, 1979)].

Steady-state levels of $^{22}$Na$^+$ and $^{36}$Cl$^-$ in rat CP epithelium, determined by compartmentation analyses, agree well with cellular concentrations measured chemically for the respective stable ions. Such concurrence in values, "hot" vs. "cold," indicates that there is negligible binding of Na$^+$ and Cl$^-$ to intracellular macromolecules. Microelectrode-determined activities for Na$^+$ and Cl$^-$ have apparently not been done for mammalian CP. At any rate, many CNS papers dealing with ion-sensitive measurements state values in concentrations, probably because activity coefficients are difficult to determine in mixed solutions (Hansen, 1985).

It is useful to relate choroid cell concentration of ions (and other solutes) to the corresponding concentration in CSF. One application is in the estimation of apical transmembrane gradients. Ideally, the best sample of CSF for this purpose would be nascent CSF collected directly from the secreting membrane. Lacking the newly formed fluid, though, many investigators have sampled CSF from the cisterna magna and considered it as representative of CP-elaborated CSF. Miner and Reed (1972) compared the ionic composition of cisternal fluid with that of CSF collected directly from the CP membrane. Generally, the concentration differences between the two fluids were 5–10% or less, and in some cases not statistically significant. Thus, it appears that, as a first approximation, CSF sampled from the cisterna magna is a reliable estimate of CP fluid, especially when the fourth ventricle CP (4VCP) is compared to CSF in the adjacent cistern (Murphy, 1984).

Permeability surface-area products (PAs) have been calculated for $^{22}Na^+$ and $^{36}Cl^-$ uptake by the in vivo CP, but their values likely underestimate the true uptake because of the CSF sink action on the tracer (loss from choroid cell to CSF) during the approach to steady-state distribution. On the other hand, the PA values for the fast-component of CSF uptake of $^{22}Na^+$ and $^{36}Cl^-$ (and nonelectrolytes) appear to reflect accurately the movement of these tracers across the CPs, rather than the BBB. There are developmental, pharmacological, and geometrical evidences that the CSF fast component of uptake represents transchoroidal flux (Johanson, 1987b). It is fortunate that the CSF fast-component PA reliably estimates (albeit indirectly) CP membrane permeability, because direct in vivo measurements of the CP barrier are not feasible with Oldendorf-type techniques.

### 7.6. Pharmacological Modification

The cerebral endothelia and the CP epithelia together form a cellular "line of defense" for the *internal milieu* of the CNS. The luminal and basolateral membranes of the BBB and CP, respectively, are not only relatively impervious to diffusing water-soluble solutes, but also contain specialized carrier transport systems for regulating entry of inorganic and organic substrates into the barrier cells. For example, $Na^+$-$H^+$ and $Cl^-$-$HCO_3^-$ antiport systems on the basolateral membrane of CP have been described (Murphy, 1984; Deng, 1986). The discussion below summarizes the types of experiments providing evidence for this conclusion.

#### 7.6.1. Na$^+$-H$^+$ Exchange

Alteration of $^{22}Na^+$ uptake by drugs that inhibit (amiloride) or stimulate (insulin) $Na^+$-$H^+$ antiport in other tissues has furnished evidence for $Na^+$-$H^+$ exchange in the CP. In acute experiments (1 h) with ketamine-anesthetized rats, Murphy (1984) found that amiloride markedly reduced CP cell (and CSF) uptake of $^{22}Na^+$ from plasma. CP stable $[Na^+]$ was also lowered. These in vivo findings confirmed comparable in vitro analyses (Murphy and Johanson, 1983). The significance of the in vivo experiment lies in the ability to determine changes in the composition of the parenchymal cells composing the blood–CSF barrier, secondary to alterations in the activity of a $Na^+$-translocating mechanism. The comparable experiment with the BBB in vivo appears to be not technically feasible.

### 7.6.2. $Cl^-$-$HCO_3^-$ Exchange

Disulfonic stilbenes, DIDS and SITS, are commonly used as specific inhibitors of $Cl^-$-$HCO_3^-$ exchange in many cells. Following DIDS administration ip (25 mg/kg), there is a substantial reduction in the penetration of $^{36}Cl^-$ from plasma into CP cells (lateral and fourth ventricle) and into CSF (Deng, 1986). Such findings are consistent with in vitro kinetic (Deng and Johanson, 1984) and steady-state (Smith and Johanson, 1985) studies of $Cl^-$ accumulation by CP. CP $[Cl^-]$ is also a function of extracellular $[HCO_3^-]$ (Johanson et al., 1985). The ability to demonstrate stilbene-sensitive $Cl^-$ uptake in vivo as well as in vitro makes a stronger case for the relevance of the $Cl^-$-$HCO_3^-$ antiport mechanism in the CSF secretory process.

## 7.7. Biochemical Inhibition Studies

$Na^+$-$K^+$-ATPase and carbonic anhydrase are key transport enzymes in mammalian CP. Maximal inhibition of each enzyme can curtail CSF production by about 50%. $Na^+$-$K^+$-ATPase has a predominantly apical location, whereas carbonic anhydrase is at both poles of the cell and in the cytoplasm (Masuzawa, 1981). Translocation of $Na^+$, $K^+$, $H^+$, $HCO_3^-$, and $Cl^-$ ions are affected at both membranes upon enzymatic inhibition of the $Na^+$-$K^+$-ATPase (ATP hydrolysis) and/or carbonic anhydrase (hydration of $CO_2$). The disruption of the ion transport is reflected in altered cell levels of cations and anions. The analysis of the induced changes in cellular ions (and those in CSF) have provided insight on location and coordination of the basolateral and apical transporters. In vivo studies with ouabain and acetazolamide (Smith and Johanson, 1980a,b) indicate that the CSF secretion rate is not a sensitive function of CP cellular $[K^+]/[Na^+]$.

### 7.7.1. $Na^+$-$K^+$-ATPase

The activity of the $Na^+$-$K^+$ pump, mediated by $Na^+$-$K^+$-ATPase, keeps CP cell $[Na^+]$ and $[K^+]$ at approximately 30–40 m$M$ and 145–155 m$M$, respectively. These values are comparable to those of other secretory epithelia (*see* review of literature by Johanson et al., 1974b). CP tissues are removed as quickly as possible after killing the rats (1–2 min), in order to obtain control values close to those existing in the living animal. Uncorrected factors, such as binding of $Na^+$ and $K^+$ to connective tissue (if it exists) or the presence of other cell types (e.g., a glial cell is sometimes observed in electron micrographs), would introduce a small error into the cell ion calculations. In our work, however, the main emphasis has not been on absolute values as much as concentration differences between control and treated animals.

The cardiac glycoside ouabain is a potent inhibitor of $Na^+$-$K^+$-ATPase when applied to the apical surface of the epithelium in vivo. Intraperitoneally injected ouabain (10 mg/kg) does not inhibit the CP $Na^+$-$K^+$ pump because the drug cannot penetrate the blood–CSF barrier sufficiently to affect the ATPase apically. In fact, the *rise* in CP $[K^+]/[Na^+]$ after ip ouabain is likely the result of increased $K^+$ transport out of the CSF to compensate for the dissipative effects associated with the acute hyperkalemia. Thus, these in vivo experiments functionally demonstrate the predominance of the $Na^+$-$K^+$-ATPase on the CSF-facing membrane (Smith and Johanson, 1980b). For the same reason, CP cell pH, which is sensitive to the transmembrane $Na^+$ gradient, is altered by ouabain in vitro (Johanson et al., 1985), but not when systemically administered (Johanson, 1979).

### 7.7.2. Carbonic Anhydrase

Acetazolamide is the classical inhibitor of carbonic anhydrase, which helps to buffer cellular $CO_2$ by converting $OH^-$ to $HCO_3^-$. The generated $HCO_3^-$ is then available for cellular extrusion, as is $H^+$ formed by protolysis of water when the $OH^-$ is consumed. In vivo studies of CP pH by the DMO method have provided insight on the mechanism by which acetazolamide inhibits CSF secretion (Johanson, 1984). The cellular alkalinization induced by acetazolamide limits the availability of $H^+$ for exchange with $Na^+$, resulting in less uptake of $Na^+$ by the cell for subsequent secretion into CSF. Under in vitro control conditions, it has not been possible to alkalinize CP pH by acetazolamide (Johanson et al., 1985). This points out the necessity to use the in vivo *secreting* CP, with its functional polarity, for demonstrating certain drug effects.

Cell $[HCO_3^-]$ can be calculated from DMO-derived pH and the estimated $pCO_2$ for the tissue. Choroid cell $pCO_2$ has been assumed to be one half the difference between arterial and CSF $pCO_2$ (the latter determined from the measurement of CSF total $CO_2$). It seems reasonable that CSF $pCO_2$ is greater than CP $pCO_2$, since the former is also affected by $CO_2$ production from neurons and glia. The CP, because of its small size and brisk blood flow, would eliminate $CO_2$ more rapidly than large-cavity CSF (Murphy, 1984). Errors in the $[HCO_3^-]$ calculation have been discussed elsewhere (Johanson, 1984). Electrode measurements of $pCO_2$ have been done on brain tissue (Brezezinski et al., 1967); and they need to be performed on the in vivo CP.

## 7.8. Pathophysiological Effects

The CNS has remarkable ability to protect itself against large fluctuations in plasma $K^+$ and $H^+$ ions. The epithelial cells of the CP display significant changes in ion composition when this tissue is challenged with ionic distortions in the plasma. Such alterations in the choroidal epithelium are reflections of homeostatic transport activity to assure stability of CSF composition.

### 7.8.1. Hypokalemia and Hyperkalemia

Two models of potassium deficiency have been tested in regard to the CP involvement in CSF $[K^+]$ homeostasis (Johanson et al., 1974b). Either a diet deficient in $K^+$ (for 4 wk) or daily injections of desoxycorticosterone (for 2 wk) lowers the plasma $[K^+]$ in rats from 4 to 2.5 mmol/L. With both hypokalemic regimens, CP tissue $[K^+]$ was depleted by 10–15%, whereas CSF $[K^+]$ decreased by only 5%. Experimental changes in CP cell $[K^+]$ are compared with those in skeletal muscle (Fig. 7). Thus, this model demonstrated that choroid cell $[K^+]$ as well as muscle $[K^+]$, unlike CSF $[K^+]$, is not independent of plasma $[K^+]$. Both acute and long-term hyperkalemia have also been examined in rats. One hour after ip KCl (4.5 mmol/kg), the choroid cell $[K^+]$ was elevated by 15 mmol/L (Smith and Johanson, 1980b). Various durations of hyperkalemia (induced by bilateral nephrectomy) from 1 to 24 h also lead to increases in choroid cell $[K^+]$ from 30 to 40 m$M$ (Johanson et al., 1974b; Hise and Johanson, 1979). Such ligation of the renal pedicles is one model for kidney failure. The resulting uremia brings about marked changes in transport and permeability phenomena in the cells composing the BCFB.

### 7.8.2. Acidosis and Alkalosis

The common procedure for inducing metabolic acid–base distortions is to administer (ip or iv) solutions of HCl, $NH_4Cl$, NaCl (control) or $NaHCO_3$ (Pershing and Johanson, 1982; Harbut, 1982; Murphy, 1984; Johanson et al., 1987). By 15 min the distortion is fully developed, and it is sustained for several hours (Harbut and Johanson, 1986). By dose–response analysis, we determined that salt doses of 4–5 mmol/kg brought about peak effects in $Na^+/K^+$

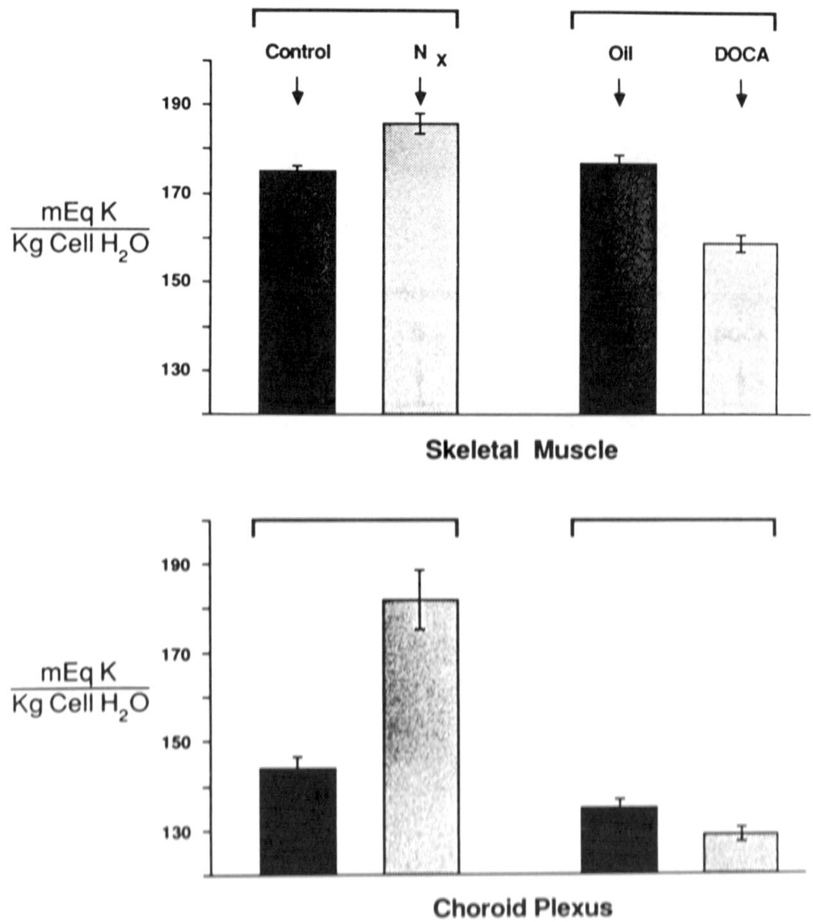


Fig. 7.   Concentration of $K^+$ in the cells of CP vs. skeletal muscle in adult rats subjected to various treatments to alter plasma $[K^+]$. Each bar is the mean ±SEM for at least 10 animals. $N_x$, 24-h bilateral nephrectomy (plasma $[K^+]$ = 9 mmol/L). DOCA, desoxycorticosterone acetate daily injections for 2 wk to lower plasma $[K^+]$ to 2.6 mmol/L, from control of 4.0 mmol/L. Oil, sesame oil vehicle for DOCA injections (the plotted data are reproduced, with permission, from tables published by Johanson et al., 1974b).

redistribution in the CP without morbid effects to the animals. In response to systemic acidosis, there is marked retention of $K^+$ concurrently with depletion of $Na^+$ (secondary to decreased uptake of $Na^+$ by the cell from the plasma). A time-course analysis (up to 8 h) of this phenomenon in CP (lateral vs. third ventricle) has recently been described (Harbut and Johanson, 1986). This acidosis model has been helpful in correlating plexus $Na^+$ and $K^+$ transport phenomena with plasma acid–base balance and CSF secretion.

## 8. In Vivo Choroid Plexus Models (Partially Invasive)

In this category will be discussed models that are noninvasive because the CP is not surgically isolated in or removed from the ventricular system, but that are somewhat invasive in the sense that the ambient fluid (ventricular CSF) is either barbotaged (intrathecal injection) or continuously displaced (ventriculocisternal perfusion). Thus, endogenous CSF is mixed to a variable degree with artificial CSF, and there is modification of the normal CSF pressure/volume relationships. The CP itself is not surgically acted upon or encapsulated as in the *in situ* preparations (*see* section 9), however. One variation or another of the injections and perfusions discussed below have been utilized by scores of laboratories. In this section, however, they are discussed in the limited context of inclusion of CP tissues in postexperimental analyses.

### 8.1. Intrathecal Injections

To ascertain the effects of water-soluble drugs on apical membrane function in the CP, it is expedient to administer some agents directly into the CSF in order to circumvent the BCFB. In addition, placement of many drugs into the CSF minimizes problems often encountered with systemic injections, i.e., renal elimination, hepatic degradation, plasma protein binding, and so on. CSF renewal gradually dilutes the concentration of drug. Thus, when planning for the length of the experiment, it should be kept in mind that after several hours most of the drug will probably be absorbed into the blood, via bulk flow of CSF across arachnoid villi into the dural venous sinuses. If the drug is toxic to regulatory centers in the medulla, as ouabain is, it may be necessary to maintain the animal's respiration and blood pressure.

A common procedure is to inject drug into one lateral ventricle. We have used the small animal stereotaxic instrument (David Kopf Instrument, Tujuanja, CA) and the atlas coordinates by Pellegrino and Cushman (1967) for guiding the cannula (27-gage needle) through the burr hole into the rat brain (Deng and Johanson, 1985). A 100-μL Hamilton syringe is seated in a portable infusion pump (e.g., Model 1100, Harvard Apparatus) to deliver artificial CSF at 5 μL/min. The normal rate of CSF production by adult rats is 3–4 μL/min. The slow infusion of a total of 25–50 μL of injectate seems to have negligible untoward effects on heart rate, MABP, respiration, and intracranial pressure. Barbotage should be considered if the rate of infusion or total volume of injectate greatly exceeds the above parameters. Otherwise, fluid retention in the CNS could lead to undesired pathophysiological variables, i.e., increased intracranial pressure, cerebral edema, and so on.

Once intraventricularly injected, where does the drug go and in what concentration? Trypan blue has been used to trace the path of the injected CSF. Thirty to 45 min after injection into the left lateral ventricle, there is blue staining of brain tissue underlying both lateral ventricles, the third ventricle, and to a lesser extent, the fourth ventricle. Both lateral and third plexuses stain extensively, the 4VCP less so. To estimate the concentration of injected drug at the end of the experiment, it is straightforward to put a radioactive tracer (like tritiated sucrose or inulin) in the injectate. Immediately before sacrifice, CSF is sampled from the cisterna magna and then analyzed for radioactivity. It is not uncommon to have the drug in the injected CSF eventually diluted after 30–60 min by one to two orders of magnitude. Moreover, if a particular agent is actively transported by the CP, its concentration in the choroidal tissue may be higher than that calculated by the CSF dilution analysis.

Deng (1986) has used the intrathecal administration of drugs to characterize the nature of $Cl^-$ transport by rat CP. Thus, $PGE_2$ and the disulfonic stilbene DIDS enhance and inhibit, respectively, the uptake of $^{36}Cl^-$ by CP and CSF from plasma. Because DIDS does not readily penetrate the BCFB, it has been presented intrathecally in some experiments, intraperitoneally in others, to furnish kinetic evidence for the existence of $Cl^-$-$HCO_3^-$ exchange at both poles of the choroidal epithelium.

## 8.2. Ventriculocisternal Perfusions

In 1958, Manuilov described the technique of ventriculocisternal perfusion (VCP) in dogs, thus allowing fluids to be perfused from the ventricular cavities downstream to the cisterna magna. Pappenheimer and colleagues later refined the methodology in unanesthetized goats (1961), and they extensively characterized pressure–volume relationships under various conditions of perfusion. Steady-state perfusion is assessed from the constancy of the composition and volume (per unit time) of perfusate collected cisternally. The strength of the VCP technique lies in the ability of the investigator to infuse an artificial CSF, the composition of which is readily manipulable, over the CP and ependymal linings. The calculation of transport constants for tracers removed from perfusion fluid is of limited value because it is difficult to distinguish the choroidal and ependymal fluxes.

VCP can be used to advantage, however, in some studies specifically designed to analyze regional and overall CP function. Pappenheimer et al. (1961) initially showed that organic acids are removed from CSF by carrier transport in the fourth ventricle region, most likely by CP. CSF formation rates calculated by indicator dilution in perfusion fluid can be correlated with weight of CP (Johanson et al., 1974a) or with ionic composition of the CP (Smith and Johanson, 1980b). Salicylate (1 m$M$ in ventricularly perfused rabbits) reduced the clearance of $^{14}$C-penicillin from CSF to blood by nearly 60%; in this regard, the involvement of the CP in this transport process is indicated by corresponding in vitro studies showing that 1 m$M$ salicylate decreased the 5-min tissue to medium ratio for $^{14}$C-penicillin by almost 60% (Spector and Lorenzo, 1974). Ventricular perfusions have also been used to expose the CP to hypertonic fluids (e.g., CSF plus 100–150 mOsm NaCl) to quantify the enhanced permeativity of tracers across the BCFB (Johanson et al., 1974a) as has similarly been done with the BBB (Thompson, 1970). After VCP with hypertonic sucrose or urea (0.5$M$), the CP permeability to horseradish peroxidase and the structure of the tight junctions were both modified. There was fragmentation of the ridges and expansion of the intercellular spaces (Bouchaud and Bouvier, 1978).

McComb et al. (1977) have utilized VCP in conjunction with venous sinus sampling at different sites to separate reabsorptive fluxes across the blood–CSF and BBB. Cannulation of the torcular in the rabbit allows sampling of tracers (cleared from VCP fluid) absorbed mainly by the choroid plexuses. On the other hand, a cannula in the superior saggital sinus picks up tracer that has been reabsorbed primarily through the capillaries in cerebral cortex. The variation in activity of tracer in the blood was much less in machine-withdrawn than in hand-drawn samples. Their aspiration rate of blood at 0.1 mL/min from the torcular probably did not alter normal patterns of blood flow because this volume loss is only about 1/70 of the total. This interesting approach is only semiquantitative because some degree (unknown) of subependymal drainage (including deep cerebral capillaries) must mix with blood draining the plexuses.

Most VCP systems have the inflow cannula in one or both lateral ventricles. A modified technique for selective perfusion of the fourth ventricle in conscious dogs has been developed, however, by Freye and Gupta (1979). Continuous perfusion through the fourth ventricle leads to a steady-state concentration gradient between the ventricular space and adjacent 4VCP. Such perfusion, together with postexperimental analysis of 4VCP, appears to be a promising in vivo model of the blood–CSF barrier.

Generally, the VCP technique permits direct contact between the in vivo choroidal epithelium and continually renewed solutions of defined composition. Overall, the VCP is convenient for steadily administering CSF-like solutions into the interior of the brain. By sequential perfusion, animals can often serve as their own controls. Practically, though, the interval for viable experiments is probably limited to just a few hours because both sloughing of cells and diminishing production of CSF are complications of extended perfusions.

## 9. *In Situ* Choroid Plexus Models

Several preparations have been developed allowing the CP to remain situated in the ventricular cavities of the intact (Rougemont et al., 1960; Welch, 1962; Miner and Reed, 1972) or extricated (Pollay et al., 1972) brain. Thus, the *in situ* preparations have the advantages accruing from the CP being left attached to the brain. Functional data for the BCFB are then obtained by analysis of CP nascent fluid or arteriovenous differences (*see* Table 1).

### *9.1. Collection of Exudate Under Oil*

An attempt was made by Bowsher (1958) to collect fluid directly from the exposed CP by means of lens paper strips. In this early experiment the tissue was probably injured since the exudate $[K^+]$ was high. Soon thereafter, Rougemont et al. (1960) devised an important modification of this basic approach, by bathing the exposed plexus of cats in Pantopaque oil. The choroidal tissue, lighter than the oil, tends to rise in it, thereby facilitating collection

Table 1
*In Situ* Preparations of Lateral Ventricle Choroid Plexus

| | Investigators | | | |
|---|---|---|---|---|
| | Rougemont et al., 1960 | Welch, 1963 | Miner and Reed, 1972 | Pollay et al., 1972 |
| Anesthetic | Pentobarbital | Urethane | Pentobarbital | None[a] |
| Ambient environment of CSF | Pantopaque oil | CSF | Mineral oil | CSF |
| Arterial inflow | Natural | Natural | Natural | Perfused |
| Venous outflow analyzed? | No | Yes | No | Yes |
| Nascent CSF collected | Yes | No | Yes | No |
| CSF formation ($\mu$L/min/mg) | n.m.[b] | 0.37 | 0.4 | 0.13 |
| Acetazolamide inhibition, of CSF | n.m. | Approximately 100% | Approximately 100%[c] | Approximately 100% |
| CSF solute determinations | $[Na^+]$, $[Cl^-]$, $[K^+]$ (measured) | n.m. | $[Na^+]$, $[Cl^-]$, $[K^+]$, $[Ca^{2+}]$, $[Mg^{2+}]$, [protein] (measured) | $Na^+$ and $Cl^-$ fluxes (calculated) |

[a]Brain (containing CP) was removed from decapitated animal.
[b]n.m. = not measured
[c]Melby et al., 1982.

of fluid from the upper surface of the epithelium. A micromanipulator is used to lower the sampling pipet (fine glass) under the oil in order to collect approximately 5 $\mu$L of fluid in 10 min. The relatively low values of fluid $[K^+]$, i.e., close to concentrations in endogenous CSF, are indicative of negligible damage to the tissue. Miner and Reed (1972) later reproduced the values of $[K^+]$, $[Na^+]$, and $[Cl^-]$ for nascent fluid.

Rougemont's collection technique was systematically characterized (1960), and thus it served as a useful starting point for other variations of *in situ* methodology. The technique gained credibility because it was demonstrated that the nascent CSF contained less [$K^+$], but more [$Cl^-$] and [$Na^+$] than that expected in an ultrafiltrate of serum. Such a distribution of ions was consistent with the orthodox view (gained largely from ionic data obtained from cisterna magna and ventricular fluid samples) that CSF was a finely regulated secretion. The elegant experiments by Ames and colleagues that soon followed used this preparation to determine the effects of enzyme inhibitors and perturbations in plasma chemistry on the composition of nascent fluid (Ames et al., 1964, 1965a, b). Welch (1963) discussed the likelihood (but not absolute proof) that the nascent fluid was derived completely from choroidal rather than extrachoroidal sources.

## 9.2. *Choroid Plexus Fluid Sampled in Chambers*

The isolation of a part of the CP in a container or chamber would virtually assure that the fluid exudate has its origin from the blood supply to the choroidal epithelium. A technique for surgical encapsulation of the canine CP was briefly described by Howarth and Jowett (1962). The CP was folded within a square of thin Teflon to which a tube was fixed. The open edges were sutured, leaving room for entrance and exit of major vessels serving the plexus. After a month, the capsule became embedded in fibrous tissue. One or 2 mL was collected every 24 h into a small polythene bag draining the tube. Unfortunately, Howarth and Jowett did not present figures for CSF ion composition or flow rate per unit of wet tissue.

At about the same time, Clark (1962) also described the attempt to isolate a portion of the plexus in a closed chamber constructed of polyethylene. The procedure involved extensive craniotomy in anesthetized dogs, with much of the lateral surface of the hemisphere removed. Sealing compound was applied to the outer side of the capsule. For sealing compounds, sterile vaseline was used for acute isolation; and methacrylate for chronic isolation. Successful isolation was accomplished in about 60% of the experiments. Failures were caused mainly by tearing damage to the fragile tissue or contamination of the entire CP by the sealing compound. Surprisingly, inability to obtain adequate sealing was the least of the technical problems.

A decade later Miner and Reed (1972) and Husted and Reed (1976) refined the chamber isolation for acute experiments in cats, and they extensively characterized the properties of the secreted fluid. Their isolation chamber is fabricated from plexiglass and consists of a concave base forming the bottom of the chamber and a top resting on the base and constituting the wall (Fig. 8). The base ($3 \times 6 \times 3$ mm) has a handle 2.5 cm long to allow for clamping to a micromanipulator. Semisolid white petrolatum is placed completely around the chamber to stabilize it in position. Approximately 2.5 mg of CP are isolated in the chamber. The chamber wall is made water-tight to its base by Bio-Bond (Takeda Chemical Industries, Osaka, Japan). Preparation viability is assessed by adequate blood flow, lack of visible tissue damage, and minimal protein content of the chamber fluid. Reed and colleagues have identified the two crucial steps in the whole procedure as placing the tissue onto the chamber base and then sealing the chamber.

Chamber fluid (CF) is then sampled under oil to minimize evaporative loss (Miner and Reed, 1972). The small size of the samples (40–50 μL) necessitates gravimetric determination to improve accuracy. Chamber fluid formation rate was 0.4 μL/min/mg, comparable to CSF formation rates obtained in vivo by the ventriculocisternal perfusion method. The finding that the chamber fluid has a chemical composition similar to the nascent fluid collected freely from CP in the contralateral ventricle has been offered as evidence that the former fluid is unmodified by contact with adjacent neural tissues. CF protein content undergoes an average two-fold increase (from 0.05 to 0.11%) throughout the experiment, even as plasma protein is decreasing (from 8 to 6.4%). The range of values of CF [protein] is 1.7–30% of that in serum. Both $[Ca^{2+}]$ and $[Mg^{2+}]$ in CF show correlation with [protein] in a manner suggesting that plasma is the main source of all three constituents. Even the maximum amount of plasma contaminating the CF, however, does not appear to seriously alter the ionic levels.

The *in situ* chamber sampling method developed by the Utah group is potentially a powerful tool for delineating CP transport phenomena. The ability to manipulate the composition and volume of the chamber fluid (Husted and Reed, 1976, 1977) is a decided advantage in transport modeling. Practically, though, there has been considerable limitation in the preparation's ability to consistently secrete fluid at a normal rate. Future experiments need to assess the effects of surgery and isolation on the blood flow and neural tone in the *in situ* CP. To evaluate permeability of the

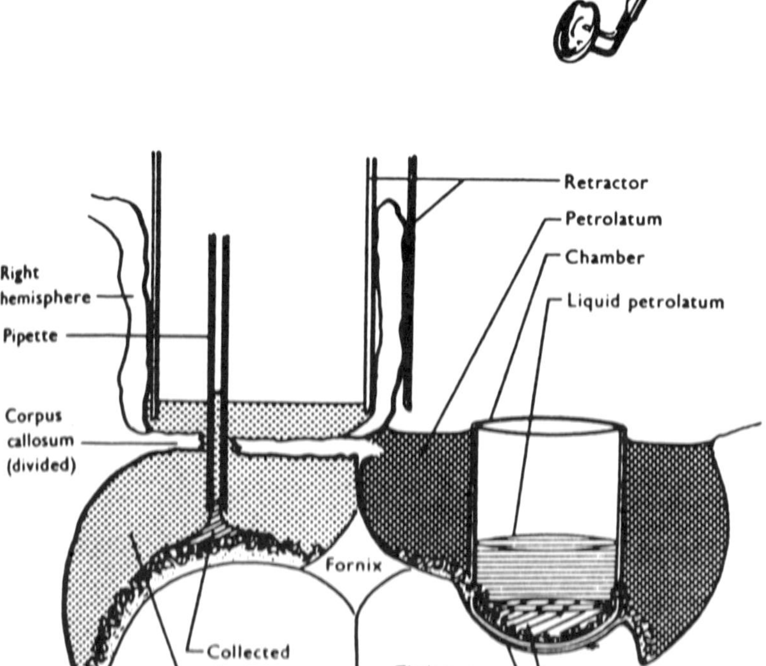



Fig. 8.    Frontal section of cat brain through the thalamus and body of the lateral ventricles, illustrating two techniques for sampling choroid plexus fluid. On left, the exuding nascent fluid (covered by oil) is sampled by micropipet from the surface of the CP. On right, CSF is sampled from a segment of CP isolated in a chamber, the construction of which is depicted in the upper right hand corner (reproduced, with permission, from Miner and Reed, 1972).

exposed membrane, it seems worthwhile to ascertain if there is molecular sieving of small nonelectrolytes (like urea) in the chamber fluid. Hydrostatic pressure values for the fluid column in the chamber, and their effect on CP function, need to be systematically analyzed.

### 9.3. Secretion Estimated by Loss of Volume from Plasma Compartment During Transit of Blood

In the steady state, the loss of volume from the choroidal plasma during transit of blood through the tissue should be equal to the volume of CSF secreted. Welch (1963) used this conservation principle as the theoretical basis for his *in situ* preparation of the rabbit lateral ventricle CP. The novel measurement in his physiological studies was the linear velocity of blood in the main vein draining the CP, quantitated by analysis of motion pictures made during the injection of spherules of octanol. Blood flow was determined as cross-sectional area of the vein times the velocity of the octanol spheruli. Lack of precise methods led to the assumptions that (1) the main choroidal vein has a circular cross section and (2) the inserted pipet tip in the vein negligibly alters flow. With additional measurements of hematocrit (H) in arterial (a) and venous (v) blood, he calculated the CSF formation rate, $Q_{csf}$, as equal to venous blood flow, Qv, times [(Hv/Ha) − 1].

Welch and colleagues used their elegant *in situ* preparation to analyze rates of CSF secretion (volume flow) and solute transport across the CP. By varying the osmotic pressure of fluid bathing the plexus, they could calculate coefficients of osmotic flow. Net transport of solutes (ions, nonelectrolytes, and so on) between plasma and artificial CSF bathing solution was described in terms of a nondiffusional flow and a diffusional component, the latter proportional to the concentration gradient across the epithelium. Welch's pioneering efforts were significant for their quantitation of transepithelial flow rates. Although limited by errors in the measurements of the tissue blood flow and in the choroidal artery hematocrit, the transport data generated by this *in situ* method nevertheless served to spur the development of other techniques for more effectively controlling arterial inflow rate and composition.

### 9.4. Fluid Production Calculated by Arteriovenous Difference During Extracorporeal Perfusion

Cannulation of main arteries and veins allows for fine control and measurements of volume and composition of fluids traversing the CP. The reliability of the coefficients for net transchoroidal transport is directly related to the accuracy of the inflow and outflow parameters. One of the earliest attempts to perfuse the CP was by Csaky and Rigor (1967), who utilized CP from a horse's lateral ventricle. Glucose transport was studied by comparing concentrations of sugar in the perfusate and in the bath in which the plexus was suspended during the arteriovenous perfusion.

In 1972, Pollay et al. described an extracorporeal perfusion system as an effective way to quantify secretory and reabsorptive phenomena in sheep CP. Perfusion was initiated within 10 min of the animal's death. Rate of inflow from the reservoir into the tissue was controlled by a roller pump. The time required to fill 100-$\mu$L volumetric pipets was used to compute blood flow into and out of the perfused CP. CSF secretion (volume flow) was determined as a function of hematocrit values and blood flow (similar to Welch's equations above). With data for net $Na^+$, $Cl^-$, and water fluxes, they calculated that the virtual concentration of $Na^+$ and $Cl^-$ in nascent CSF should be 159 and 130 mEq/L.

The main technical feature of the Pollay approach is the extensive surgical isolation of the choroidal vasculature. Why not totally extricate the CP and its blood supply from the brain? Because it is difficult to separate the proximal anterior choroidal branches from the brain and to avoid consequent irreversible spasm of the anterior choroidal artery (ACA). Alternatively, the CP can be left *in situ* while the ACA and vein of Galen are cannulated; moreover, with this approach, the exposed ventricular system serves as a natural conduit for percolation of artificial CSF. Figure 9 shows the placement of cannulae and ligatures necessary for isolation of the CP inflow and outflow.

The *in situ* CP remains viable for up to 3 h of perfusion, having a $Q_{O_2} = 3.2$. In some preparations there was a 5% error (or less) in the loss of inflowing blood because of the lack of feasibility of coagulating small arterial branches deep in the choroidal fissure. In a few experiments longer than 3 h, there was thrombosis of arterioles with resultant rise in perfusion pressure. This problem can be minimized by doubly filtering the perfusate through glass wool and by using blood diluted with low-molecular-weight dextran. One important criterion for viability of CP function is its ability to secrete CSF at a rate obtaining in vivo. CSF formation rate per weight of CP is generally in the range of 0.3–0.5 $\mu$L/min/mg for most mammals. The mean value of 0.13 obtained for extracorporeally perfused sheep CP seems to be half or less of the conventional figure for in vivo preparations, but is not dissimilar to that reported for monkeys (Milhorat et al., 1971; Milhorat, 1969).

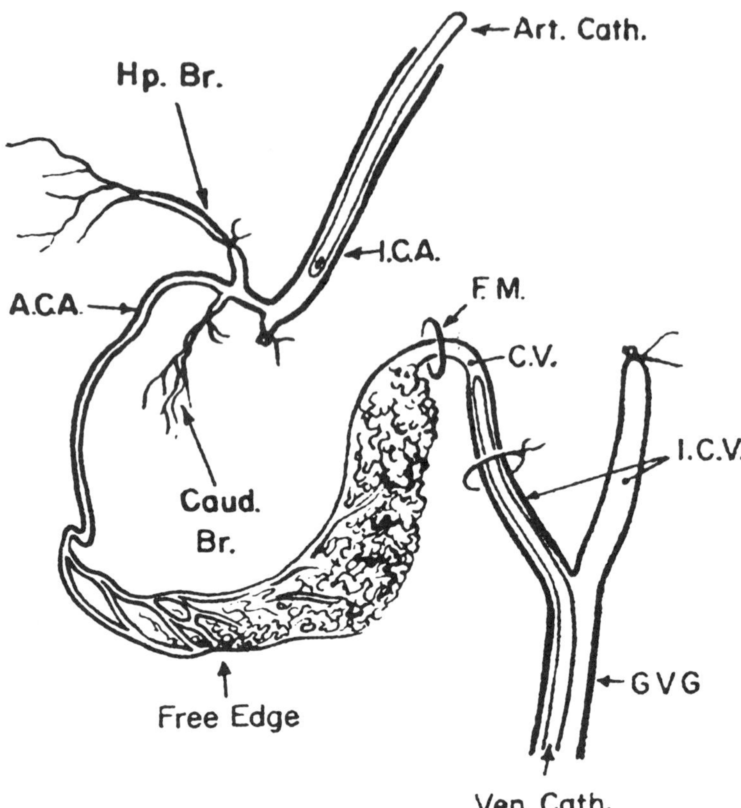


Fig. 9. Vascular surgical preparation for extracorporeal perfusion of sheep choroid plexus. Diagram of vascular system and placement of arterial (art. cath.) and venous catheters (ven. cath.). Abbreviations: ICA, internal carotid artery; Hp Br and Caud Br, hippocampal and caudate branches of anterior choroidal artery (ACA); FM, foramen of Munro; CV, choroidal vein; ICV, internal cerebral vein; GVG, great vein of Galen (reproduced, with permission, from Pollay et al., 1972).

Pollay and colleagues have concluded that the *in situ* CP has a significant volume flow, but that it is equivocal as to whether it represents normal flow rate.

In spite of its considerable promise, the extracorporeal perfusion technique has not been extensively employed. Presumably, the laborious setup has discouraged its wide application, although other investigators have pursued its usage (Blount et al., 1973).

## 10. In Vitro Choroid Plexus Models

Most studies of CP have been carried out with in vitro preparations, which allow for intensive analysis of a single barrier system (i.e., the BCFB uncomplicated by the interactive effects of the BBB). The in vitro systems encompass diverse methodologies, ranging from tissue cultures to membrane fractionations. The most common isolation approach, however, has been to excise the choroidal tissues for incubation in artificial CSF at 37°C. The composition of the CSF medium is close to endogenous CSF (m$M$): NaCl (130), $NaHCO_3$(18), $KH_2PO_4$(3.0), $CaSO_4$(1.4), $MgSO_4$(0.8), and glucose (12.1). Organic solutes like urea, citrate, and amino acids are sometimes added in small quantities. Total osmolality is 290 mosm. CSF pH and $pCO_2$ are approximately 7.35 and 30 torr, respectively.

The volume ratio of CP tissue to CSF medium should be at least 1:1000, i.e., 1 mg CP to 1 mL CSF. Therefore, CP loss or uptake of solute or tracer would not significantly alter CSF composition throughout the experimental incubation. This problem becomes relevant when analyzing transmembrane distribution ratios of ions like $K^+$ by use of agents such as ouabain, which extensively deplete the CP of its $K^+$ content (Smith and Johanson, 1980b). The unstirred layer effect also needs to be carefully considered (Grady and Blaumanis, 1983).

The potential difference (PD) of the CP is approximately –50 mV (see footnote 1 in Smith and Johanson, 1985). Definitive, systematic PD measurements of mammalian CP are lacking because of the difficulty in holding microelectrodes inside the small epithelial cells. Attempts to measure PD of lateral ventricle tissue have been complicated by movement of the in vitro CP in CSF bubbled vigorously with $O_2$ to maintain the high rate of metabolism (Q. Smith, personal communication). Nevertheless, PD values between –50 and –55 mV can be calculated by the Goldman equation for both in vitro and in vivo rat CP preparations, when certain assumptions are made about cytosolic ion binding and permeability ratios ($P_{Na^+ (or Cl^-)}/P_k^+ < 0.1$). The distribution of voltage-sensitive compounds is discussed in section 10.1.2.

Our laboratory has had more success incubating CP tissues from the lateral rather than the fourth ventricle. 4VCP from adult Sprague-Dawley rats tends to disintegrate somewhat in vitro. On the other hand, lateral ventricle choroid plexus (LVCP) does not physically disintegrate throughout the incubations; this conclusion was drawn from observations about the time course of change (negligible) in the weight of incubated plexuses (Deng and Johanson, 1984). Another advantage of LVCP is that tissue from one lateral ventricle can serve as a control for the CP removed from the contralateral ventricle; or one LVCP can be used for water and electrolyte analyses, and the other for radioisotope distribution. In vitro preparations permit variation of ionic environment and exposure to pharmacologic agents not feasible in the in vivo brain (Johanson et al., 1985). Thus, the in vitro plexus is more amenable to dose–response analysis, without interference from drug metabolites or other pharmacokinetic variables in vivo. Moreover, the in vitro preparation is not complicated by neural tone, blood flow, or circulating hormones.

The main criterion for viability is that the in vitro choroidal epithelium content of $K^+$, $Na^+$, $Cl^-$, pH, and membrane potential remain stable over the initial hour (or more) of incubation. Also, the metabolic rate (Quay, 1966) and ultrastructural integrity (Yuen and Agnew, 1978) of mammalian CP are maintained during in vitro incubation.

## 10.1. Steady-State Incubations

### 10.1.1. Inorganic Ions

Inorganic ion transport by mammalian CP has been investigated predominantly in vivo. Wright (1978) has extensively used in vitro amphibian CP (fourth ventricle) to examine epithelial ion transport. Recently, a mammalian (rat) in vitro system has been characterized for LVCP (Smith and Johanson, 1985). The time course of in vitro LVCP tissue, extracellular and intracellular contents, is shown in Fig. 3. Serving as in vivo controls were nonincubated tissues removed from intact animals. Four hours of incubation does not significantly alter the total tissue water content. The marked rise in extracellular fluid volume ($^3$H-inulin $V_d$) was accompanied by a decrease in the fractional cellular volume (calculated as the difference between total and extracellular water). Cell $Cl^-$, $Na^+$, and $K^+$ reached stable levels by 15 min; thus, a 30-min incubation assures steady-state intracellular concentrations.

Extracellular ion substitutions are carried out in vitro readily in contrast to the difficulty (or lack of feasibility) of substantially altering plasma (or CSF) ion levels. The relationship between choroid cell $[Cl^-]$ and $[HCO_3^-]$ to the respective concentrations of these anions in CSF medium (37°C) is presented in Fig. 10 (bottom).

There is a linear relationship between cellular and CSF ions over a six-fold range of extracellular concentrations (20–120 m$M$ for $Cl^-$, and 1–6 m$M$ for $HCO_3^-$). The slopes of the least-squares regressions are 0.3 to 0.4. Such anion ratios in vitro are comparable to steady-state transmembrane distributions obtained in vivo. The nature of the anion distribution indicates the existence of $Cl^-$-$HCO_3^-$ exchange in CP (Johanson et al., 1985).

The transmembrane distribution of cations has also been examined in the in vitro substitution experiments (Fig. 10). Baseline analysis revealed that $[Na^+]i$ vs $[Na^+]_{csf}$ is rectilinear, whereas $[K^+]i$ vs $[K^+]_{csf}$ is curvilinear (Fig. 10, top). The inhibitor of $Na^+$-$K^+$-ATPase, ouabain, was used to determine dose–response effects on $Na^+$ and $K^+$ transmembrane distribution (Fig. 11, right panel). The time course of the ouabain effect is followed to steady state (Fig. 11, left); the relatively high concentration of $10^{-3}$ $M$ is necessary because $Na^+$-$K^+$-ATPase in CP and other tissues of the rat is resistant to inhibition by the cardiac glycosides. These types of experiments have been useful in characterizing $Na^+$-$K^+$ (Smith and Johanson, 1985) and $Na^+$-$Ca^{2+}$ antiport (Burton, 1982) systems in the choroidal epithelium.

### 10.1.2. *Organic Electrolytes and Drugs*

A popular approach, introduced several decades ago, has been to incubate CP tissues in synthetic CSF containing one or more radioisotopes. Tracer uptake is then expressed as the tissue-to-medium ratio, i.e., T/M. Thus, T/M (%) = 100 times dpm/g CP divided by dpm/g CSF medium. Steady-state accumulation for many organic acids and bases is attained by 30 min. The CP, like the proximal tubule, has active transport systems that mediate the uptake of organic anions and cations. Thus, many drugs are accumulated above electrochemical equilibrium by the choroidal epithelium. In evaluating T/M ratios, care must be taken to correct for possible binding of organic electrolytes to tissue macro-molecules. T/M ratios of greater than unity are consistent with, but not proof for, *net* translocation of test substance across the CP membrane. Examples can be cited for organic compounds that are extensively concentrated by CP, but do not undergo *net* secretion or reabsorption across cells separating CSF from blood.

In spite of its limitations, the T/M approach has yielded useful information about the kinetic characteristics of the respective organic acid and base systems in CP and the ability of these trans-

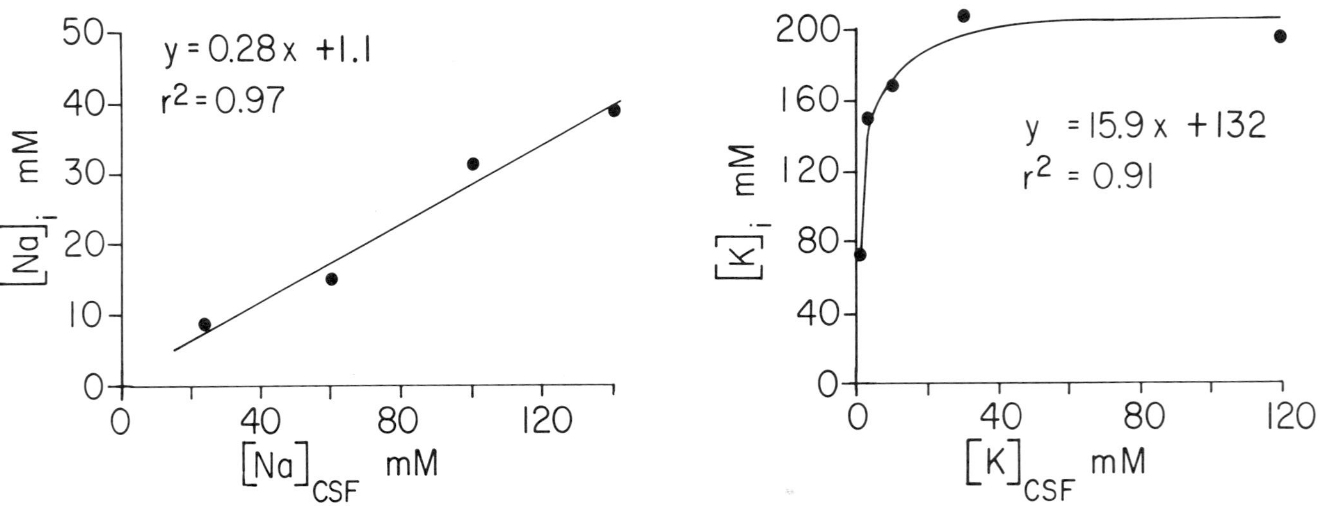

y = 0.28 x + 1.1
r2 = 0.97
[Na]i mM
[Na]CSF mM
y = 15.9 x + 132
r2 = 0.91
[K]i mM
[K]CSF mM

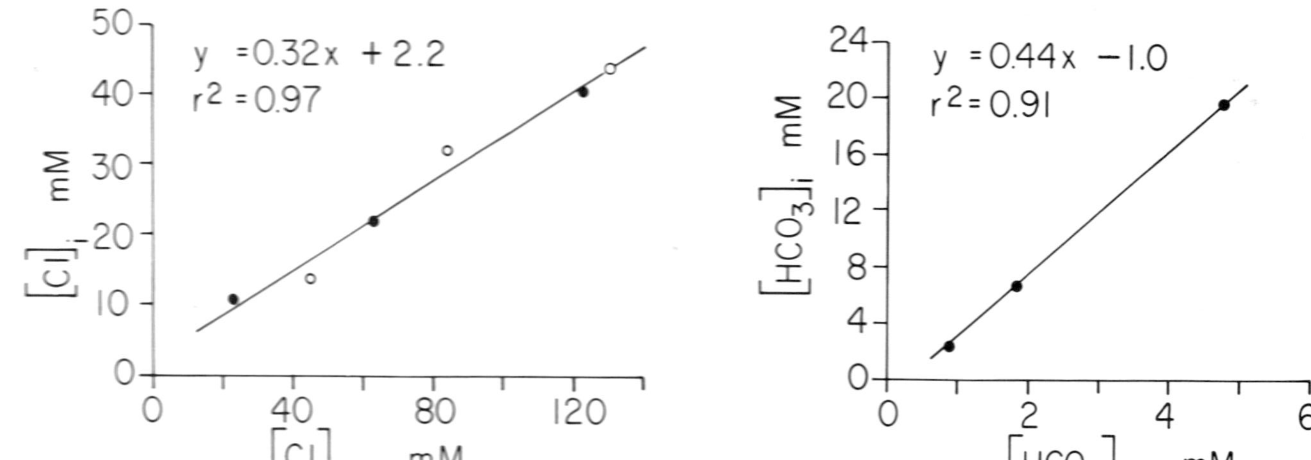



Fig. 10.   In vitro steady-state cellular levels of ions in adult rat CP, as a function of extracellular ionic concentrations (37°C). Uptake curves were constructed from rectilinear regression (least squares) for $Na^+$, $Cl^-$, and $HCO_3^-$, and from curvilinear regression ($y = b \ln x + a$) for $K^+$. $Na^+$, $Cl^-$, and $HCO_3^-$ data, respectively; obtained from Burton (1982), Smith and Johanson (1985), and Johanson et al. (1985). $K^+$ data were pooled from Burton (1982) and from Smith and Johanson (1985). Modified figures are reproduced, with permission, from the *Am. J. Physiol.*

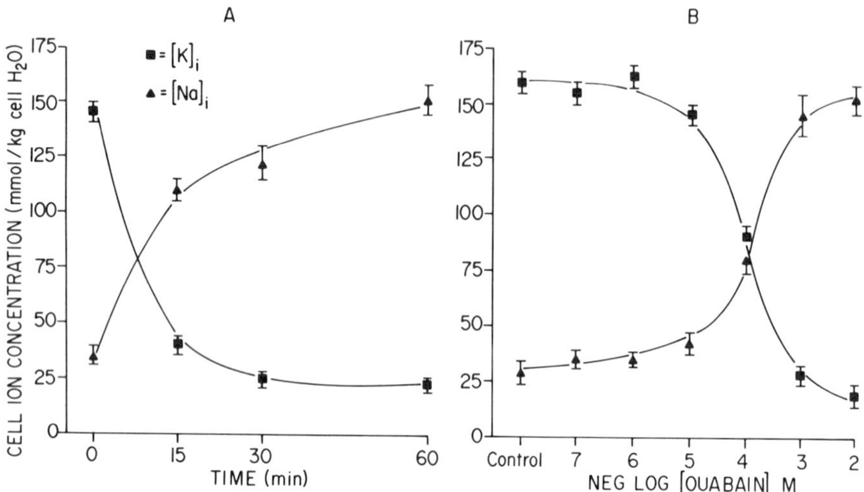

Fig. 11.    Effect of ouabain on in vitro cell concentrations of $K^+$ and $Na^+$ in adult rat lateral ventricle CP. (A) Time course of CP cell ion concentrations during incubation with 1 m$M$ ouabain. (B) Effect of 30 min of incubation with varying concentrations of ouabain on CP $[K^+]_i$ and $[Na^+]_i$. Each point is mean ± SEM for five animals. $I_{50}$ is $10^{-4}M$ (reproduced, with permission, from Smith and Johanson, 1985).

port systems to clear toxic organic compounds from the CSF. Particularly noteworthy is the unusually high capacity of the in vitro CP to concentrate inorganic iodide; e.g., T/M ratios as great as 30 have been obtained. In this case, the high T/M ratio found in isolated tissue reliably reflects the ability of the in vivo CP to maintain low levels of iodide in CSF. Steady-state CSF-to-plasma ratios of 0.02 or less have been described (Woodbury et al., 1974).

When dealing with drugs extensively accumulated by the CP epithelium, i.e., those with T/M ratios of >5, it is not so critical to correct for the extracellular distribution. For compounds with a T/M ratio of unity or less, however, it is essential to subtract the T/M extracellular component (inulin $V_d$) of about 0.4–0.5.

The great ability of the CP to actively accumulate organic electrolytes can cause problems when the epithelial PD is estimated from the steady-state distribution of a voltage-sensitive compound like methyltriphenylphosphonium Br (MTPP). Under control conditions, the steady-state T/M ratio of $^{14}$C-MTPP in LVCP is close to 60. The calculated *cellular* $V_d$ is 149 mL/g $H_2O$; thus membrane potential $(E_m) = -61.5 \log 149 = -134$ mV, clearly too high. Choline (10 m$M$) does not act as a competitive inhibitor of tracer MTPP. $^{14}$C-MTPP $V_d$ is dependent upon concentration of stable MTPP, suggesting carrier transport. Interestingly, the MTPP-calculated PD is significantly lowered by depolarization because of elevated CSF [K$^+$]. In general, however, we have found that the cell-to-CSF distribution ratio for MTPP is 5- to 10-fold too high to be explained by passive distribution (Q. Smith and C. Johanson, unpublished data). Cellular active transport rather than binding is the more likely explanation for the inability of MTPP to indicate PD reliably.

## 10.2. Uptake Kinetics

### 10.2.1. Michaelis-Menten Analyses

Radioisotope uptake curves can be resolved into components of half-times and rate constants. Alternatively, another popular approach is to describe the uptake by CP in terms of $V_{max}$ and $K_m$, if the solute is transported by a carrier system in the membrane. Carrier transport can be either by facilitated diffusion or by active transport; both types of transport exist in the CP (Spector and Johanson, 1987).

Typically, a radiolabeled organic compound is incubated with CP isolated in artificial CSF. Tissue uptake, expressed as T/M ratio, is determined at time intervals sufficient for construction of the complete uptake curve. T/M is then plotted vs. incubation time or vs. concentration (Fig. 12). Competitive and metabolic inhibitors are used to ascertain stereospecificity, saturability, and dependence of transport upon cellular energy. To assure meaningfulness of the calculated $K_m$ and $V_{max}$ values, it is important to assess the tissue radioactivity for purity and possible binding of the tracer test substance to tissue components.

It is advantageous to relate the $K_m$ and $V_{max}$ data obtained in vitro to corresponding values found in vivo. Comparability of the kinetic constants determined by the methodologies for isolated and intact choroid plexuses results in a cogent argument for the functional significance in the living organism. For a given solute, it is often instructive to compare $K_m$ (and $V_{max}$) values for the CP with those obtained for cerebral capillaries. This furnishes perspective on the relative roles of the two major barrier systems in moving the solute in question into the CNS (Spector and Johanson, 1987).

### 10.2.2. Inorganic Ion Influx ($K_{in}$)

The kinetics of penetration of several inorganic ions has been analyzed in the in vitro LVCP. The incubation time required for

                                                                 *Johanson*

V (μmoles / min / ml)
0.8
0.6
0.4
0.2
0
2.0
1.6
1.2
0.8
0.4
CONCENTRATION (mM)
(3)
(5)
(6)
(5)
(6)
(6)
(6)
(5)
(5)
(7)
(6)
(5)
(6)
0.0 mM Salicylate
1.0 mM Salicylate
$V_{net}$
$V_S$
$K_D = 0.33$ min$^{-1}$
12
10
8
6
4
2
0
T / M
$K_I$ (Salicylate) = 0.143 mM
$K_T$ = 0.043 mM
(0.0 mM Salicylate)
$K_T$ = 0.143 mM
(1.0 mM Salicylate)
$V_{max}$ = 75.212
$V_S$ / S (min$^{-1}$)
0.4  0.8  1.2  1.6  2.0
20  40  60  80  100
$V_S$ ( $\frac{μmoles}{ml}$ /min)

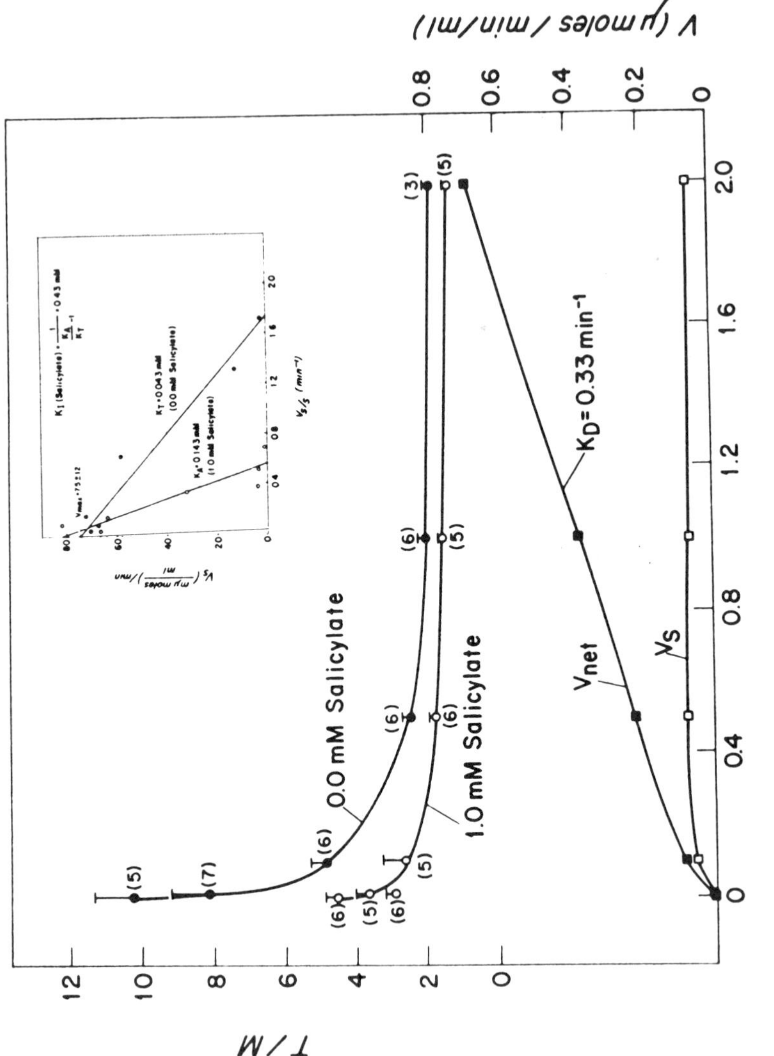

Fig. 12.   In vitro uptake of $^{14}$C-penicillin by rabbit CP as a function of medium [penicillin]. (Upper 2 curves) Incubations were 5 min, with and without 1 m$M$ salicylate. Tissue uptake is expressed as T/M ratios and presented as means ±SEM. Number of CPs used for each concentration is given in parentheses. (Lower curves) The net transport ($V_{net}$) and saturable transport ($V_{sat}$) of penicillin into cellular $H_2O$ as a function of medium [penicillin]. Salicylate was not added to the medium. The slope ($K_d$) of $V_{net}$ was determined by least squares regression, with values of $V_{net}$ at 0.5, 1, and 2 m$M$ medium concentrations. The inset shows the Hofstee transformations of values for the saturable component ($V_s$) of penicillin uptake, with and without 1 m$M$ salicylate. The derived $K_t$ for salicylate (KI) is given (reproduced, with permission, from Spector and Lorenzo, 1974).

attainment of steady-state distribution at 37°C varies from about 15 min (for $^{86}Rb^+$ and $^{45}Ca^{2+}$) to approximately 5 min for $^{22}Na^+$ and $^{36}Cl^-$. The uptake of these cations and anion occurs by antiport mechanisms, i.e., $Na^+$-$K^+$, $Na^+$-$Ca^{2+}$, $Na^+$-$H^+$, and $Cl^-$-$HCO_3^-$ exchangers. Specific inhibitory agents are available for each transport mechanism.

10.2.2.1. $^{86}Rb^+$     A useful isotopic analog for $K^+$ in transport studies is rubidium. For tracer studies, $^{86}Rb^+$ ($t_{1/2}$ = 18 d) is technically more advantageous than "hot" $^{42}K^+$ ($t_{1/2}$ = 12 h). In amphibian CP the affinities of $^{42}K^+$ and $^{86}Rb^+$ for transport are both 1.23 relative to 1 for $Na^+$ (*see* review by Bradbury, 1979). The rate of permeation of $^{86}Rb^+$ into CP epithelium is depicted in Fig. 13. The

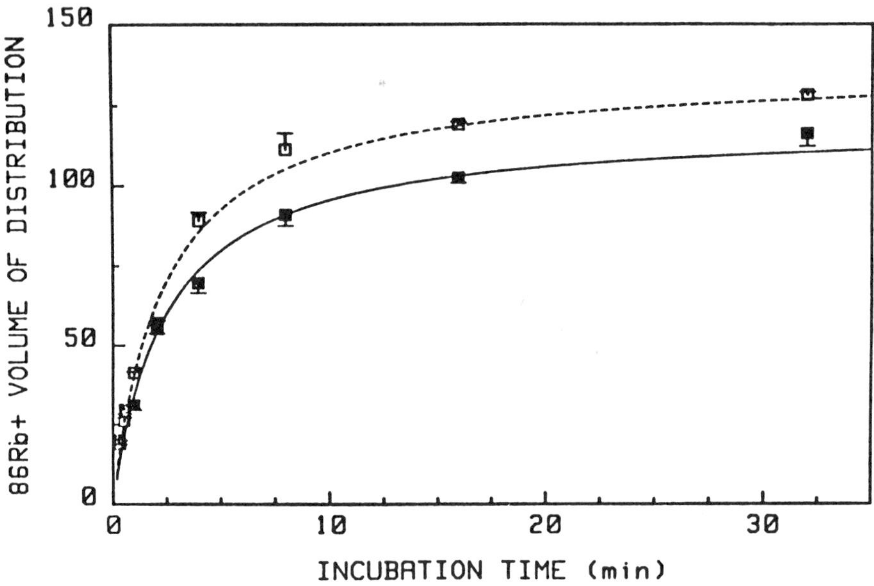


Fig. 13.    Effect of cerebrospinal fluid pH on the uptake of $^{86}Rb^+$ by the in vitro CP of adult rats. Lateral ventricle tissues were incubated in simulated CSF containing $^{86}Rb^+$ for 0.25, 0.5, 1, 2, 4, 8, 16, or 32 min after being preincubated for 20 min in a similar solution not containing $^{86}Rb^+$. All solutions were equilibrated with humidified 95% $O_2$ and 5% $CO_2$ at 37°C. Two CSF conditions were investigated: control pH (7.4) and acidosis (7.0). All values depicted represent the mean ± SEM of two to three LVCP samples. Filled squares and solid line correspond to control data; unfilled squares and dashed curve represent acidosis data (R. Harbut and C. Johanson, unpublished data; for further description, *see* Harbut, 1982).

uphill movement of $^{86}Rb^+$ is mediated by the $K^+$ arm of the $Na^+$-$K^+$ pump. The experiment shown in Fig. 13 sought to find whether or not moderate acidosis, extracellularly, affects $^{86}Rb^+$ uptake (Harbut, 1982). The induction of in vitro acidosis is more straightforward than in the intact animal, in which: (1) CSF metabolic acidosis is difficult to induce because of homeostatic BCFB mechanisms and (2) compensatory changes in blood $pCO_2$ in response to systemic metabolic acidosis could introduce complicating variables to the experiment. The early part of the $^{86}Rb^+$ uptake slope ($K_{in}$ or influx rate) was not significantly altered by reduction in CSF pH and $[HCO_3^-]$.

10.2.2.2. $^{45}CA^{2+}$     $^{45}Ca^{2+}$ has the suitable half life of 164 d and is relatively inexpensive. The rate of uptake of "hot" $Ca^{2+}$ by LVCP is similar to that of $^{86}Rb^+$. The early part of the uptake curve represents mainly influx (back diffusion should be minimal); thus, Burton (1982) analyzed the slope (linear regression by the least-squares method) obtained from the plot of $^{45}Ca^{2+}$ volume of distribution (i.e., dpm $^{45}Ca^{2+}$/mg CP divided by dpm $^{45}Ca^{2+}$/μL CSF) vs. uptake times of 2, 4, and 6 min ($n = 2$ for each time point). Control "slopes" were compared to slopes obtained with data from experiments in which certain drugs were added to the CSF to test effects on $^{45}Ca^{2+}$ uptake. For example, amiloride (which inhibits $Na^+$-$Ca^{2+}$ exchange in some systems) caused a reduction of 45% in the $K_{in}$ slope for $^{45}Ca^{2+}$ (Burton, 1982).

10.2.2.3. $^{22}NA^+$ AND $^{36}CL^-$     Radioactive $^{22}Na^+$ has a much longer half life (2.6 yr) than $^{24}Na^+$ (15 h); thus, the experimental protocol will dictate which isotope is more convenient. $^{22}Na^+$ penetrates LVCP rapidly, presumably by basolateral $Na^+$-$H^+$ exchange because the in vitro uptake is reduced 50% by amiloride (V. Murphy and C. Johanson, unpublished data). $^{36}Cl^-$ uptake from artificial CSF is similarly rapid, with steady-state distribution of 55% attained in 5–10 min. $^{36}Cl^-$ permeation into the choroidal epithelium occurs by $Cl^-$-$HCO_3^-$ exchange; the disulfonic stilbenes DIDS and SITS decrease the uptake and steady-state concentration (Deng and Johanson, 1984). Both $Na^+$ and $Cl^-$ distributions in CP have been investigated at 27 and 37°C. The hypothermia considerably slows down uptake of tracers and allows the calculation of $Q_{10}$ values.

## 10.3. Tracer Efflux Phenomena

Cations and anions are extruded at both poles of the choroidal epithelium. To distinguish between antiport and symport extrusion mechanisms, it is necessary to utilize sensitive efflux techniques. Although it is not always possible to distinguish an extrusion mechanism existing at one or both poles of the cell, it has been feasible to characterize active ($Na^+$-$K^+$ pump) and facilitated ($Cl^-$-$HCO_3^-$) efflux mechanisms (Saito and Wright, 1982; Johanson and Smith, 1984). The general procedure is to load the CP with tracer, then quantitate serially the released activity in the artificial CSF samples.

### 10.3.1. Superfusion of CSF Across Secured CP

Structurally, the posterior CP of the bullfrog lends itself to being stretched flat, hooked, and held to the lower end of the superfusion apparatus, i.e., a 10-cm long vertical glass tube of 5 mm inside diameter (Saito and Wright, 1982). Superfusate is pumped into the upper end of the glass tube at 4 mL/min and then collected at the lower end, directly into scintillation vials. Samples can be collected at intervals ranging from 15 to 120 s. The tracer-free superfusate picks up activity as it courses over the tissue previously loaded to the steady-state level.

An example can be given for $^{22}Na^+$ release from the amphibian CP (Saito and Wright, 1982). Since efflux can occur actively as well as passively, it is expedient to have agents capable of stimulating or inhibiting the carrier-mediated transport. $Na^+$ extrusion from the cell occurs largely via the $Na^+$-$K^+$ pump, the activity of which is enhanced by extracellular $[K^+]$ elevation and reduced by ouabain inhibition of $Na^+$-$K^+$-ATPase. Kinetics of $^{22}Na^+$ efflux from the CP, with and without $Na^+$ pump activity, was investigated by compartmental analysis. The $^{22}Na^+$ washout can be fitted by computer analysis to the equation: $Y = a_j \exp(-k_j t)$, where $Y$ is the $^{22}Na^+$ in CP at time $t$; $a_j$ is the $^{22}Na^+$ in the $j$th compartment at time 0; and $k_j$ is the rate constant of $^{22}Na^+$ washout from the $j_{th}$ compartment.

With this technique, Saito and Wright (1982) determined rate constants $(k)$ for $Na^+$ efflux from frog CP, when various cations substituted for extracellular $K^+$. The rank order for substitution by various monovalent cations was: $Rb^+$, $K^+ > NH_4^+ > Cs^+ > Li^+$. Active $Na^+$ transport ($\Delta k \times Na^c$) increased sigmoidally with extracellular $[K^+]$ and intracellular $[Na^+]$.

### 10.3.2. Sequential Transfer of CP into CSF-Filled Vials

In vitro efflux of anions has been analyzed in cultured astrocytes (Kimelberg, 1981), and has provided substantial kinetic data about ion exchange. Recently, an efflux preparation has been devised for mammalian CP, and it is particularly useful because it allows for analysis of the entire time course of release of tracers from a single specimen of CP (Johanson and Smith, 1984). Lateral ventricle CP is loaded with tracer by incubating the tissue in artificial CSF (37°C) containing radioisotope (e.g., 10 $\mu$Ci$^{36}$Cl$^-$). After the LVCP is loaded with tracer to a steady-state level (approximately 30 min), it is removed from the medium containing the tracer and briefly rinsed (1 s) in medium free of radioactivity, to remove CSF adhering to the external surface of the tissue. With fine opthalmological forceps, the "hot" plexus is then transferred and suspended in a small beaker containing either CSF of normal (control) or altered composition (drug, pH, anion concentration, and so on). Subsequently, at certain time intervals, the plexus is transferred and suspended, in turn, into a series of baths each with tracer-free CSF of the same volume (1 mL) and composition.

For each control and experimental series, the tissue is transferred and suspended at 0, 5, 10, 15, 20, 30, 40, 50, 60, 70, 80, and 90 s. CP from one lateral ventricle serves as a convenient control for the contralateral tissue. For each time interval, the amount of dpm appearing in the CSF bath is, of course, equal to the amount lost from the tissue. Efflux curves are generated from a time course plot of dpm in CSF bath vs. time interval. Control efflux curves for $^{36}$Cl$^-$ resolve consistently into two components, with $t_{1/2}$ values of 3 s (extracellular) and 12 s (cellular), respectively. Efflux of $^{14}$C-sucrose, an extracellular marker, resolves into a single component ($t_{1/2}$ = 3 s). Both acidic bath pH (6.7) and SITS (a disulfonic inhibitor of Cl$^-$-HCO$_3^-$ exchange) increase by 50% the slow component $t_{1/2}$ (i.e., cellular) for $^{36}$Cl$^-$ from about 12 to 18 s (Johanson and Smith, 1984).

These results demonstrate sufficient sensitivity of the model to detect drug effects on ion exchange systems that extrude ions from the choroidal epithelium. Extrusion of ions and drugs from the plexus is difficult to study in vivo. Thus, this in vitro preparation is probably the most useful for analyzing ion effluxes via antiport or by symport mechanisms, e.g., Na$^+$-HCO$_3^-$ (Johanson, 1984).

## 10.4. Ussing Chamber: Bidirectional Transport

Many technical advantages accrue by being able to mount an epithelial tissue as a sheet between halves of an Ussing chamber. This popular in vitro approach has provided a wealth of information about ion fluxes, potential difference (PD), conductances, and steady-state levels of solutes in various epithelial cells. The amphibian CP is structurally suited for chamber mounting. Wright

(1978) has extensively used the bullfrog fourth ventricle plexus for transport analyses. He has summarized several advantages of the amphibian preparation: stability of active transport rates for up to 8 h, the capability of measuring unidirectional fluxes of radioisotopic solutes and water, and the ability to quantify transepithelial short-circuit currents and intracellular potentials. Other aspects of the amphibian CP preparation need further evaluation. For example, how do the glial-type cells between the capillaries and epithelial cells affect the transepithelial solute movements? And what are the similarities and differences between the respective CP functions of poikilothermic amphibians vs. homeothermic mammals?

Mammalian plexuses from either the lateral or fourth ventricle do not have the structural configurations readily conducive for the Ussing-type chamber preparation. Thus, there is a paucity of measurements of unidirectional fluxes of solutes (apical to baso-lateral or vice versa) for in vitro mammalian choroidal tissues. There have been a few attempts, however, to mount mammalian CP between chamber halves in order to obtain electrical measure-ments. In fact, most of the bioelectric information procured for CP epithelium has been analyzed with Ussing preparations.

One of the earliest attempts to mount a sheet of CP between halves of an Ussing-type chamber was carried out by Patlak et al. (1966). They mounted the fourth ventricle CP of the elasmobranch *Squalus acanthias* (dogfish) in a chamber with an orifice of 10 mm$^2$; their measured PD of 2–4 mV was always positive, i.e., ventricular to extradural fluid, and stable for at least 3 h. Subsequently, Ussing-type preparations were used in rabbits and cats (Schone and Loeschcke, 1969; Welch et al., 1972). Small (1–3 mV) or negli-gible potentials have been measured across the choroidal mem-brane of mammals as well as amphibians. With identical Ringer's solution on both sides of the frog CP, the electrical resistance is about 200 ohm cm$^2$ (Wright, 1972). Cellular microelectrode im-palements show that the choroidal epithelium is about 45–60 mV negative to extracellular fluid, whether measured in the chamber or *in situ* (Welch and Sadler, 1965). Utilization of the Goldman equation along with data for transmembrane distribution of Na$^+$, K$^+$, and Cl$^-$ yield a calculated PD of –50 mV for adult rat CP incubated to steady state in artificial CSF (Smith and Johanson, 1985).

## 10.5. Choroid Cell Separation Techniques

Choroidal transport processes, and the regulation thereof, are affected by tissue blood flow as well as by epithelial carrier mechanisms. Neurohumoral modulation of secretory phenomena in the CP can be achieved by alterations in either receptor activity in the walls of the blood vessels or, in some cases, by innervation and receptors associated with the parenchymal epithelium (Nathanson, 1983). Thus, when a drug inhibits or facilitates a particular function of the CP, it is often difficult to ascribe the effect locally to one or the other compartment. Cellular function in vivo and in vitro has been determined indirectly by compartmentation analysis. The advantages and limitations of the compartmentalization approach are discussed in section 6. A more direct method is to separate the choroidal cells from the vascular/interstitial elements by chemical or mechanical teasing techniques. Cellular isolation is obviously limited to in vitro experiments, and it cannot distinguish basolateral and apical membrane phenomena. Some of the separation procedures show hormone sensitivity in the choroid vascular fraction, suggesting that such activity is caused by some undissociated secretory cells. Still, there are many expedient applications for isolated, purified secretory cells and partly deepithelialized vascular networks. Analysis of ligand binding and enzyme activities in cell preparations have considerable benefits over corresponding studies of entire tissue homogenates. Nathanson (1979) has systematically developed three methods for separating epithelial from vascular elements in several mammalian species. These methods are described in Fig. 14.

## 10.6. Tissue Membrane Fragments: Receptor Binding

The binding of ligands to receptors in CP is a problem integral to understanding modulatory phenomena at the BCFB. Yet, there is little information for $B_{max}$ and $K_d$ parameters in this secretory and regulatory tissue. The conventional radioligand binding techniques require considerable mass of tissue for the assays. Presumably, the small amount of choroidal tissue available per animal, especially in the commonly used smaller rodent models, has discouraged detailed pharmacological analyses of CP. Pooling of CP specimens from lateral, third, and fourth ventricles increases sample size, but this approach is not ideal since there is heterogeneity of structure and function in different regions. The fourth CP tissue contains a higher proportion of epithelial elements than does the lateral, a factor that has to be considered when interpreting data derived from homogenate analyses (Nathanson, 1980). Moreover, the limited availability of tissue is exacerbated in ontogenetic studies, because the CP mass in perinatal animals is only a fraction of that in adults (Johanson et al., 1976). Adult rats vs cats have about 4 vs 80 mg of total CP tissue per animal (Harbut and Johanson, 1986; Nathanson, 1980).

Specimen size is not the only consideration when selecting the species for investigation. Receptor sensitivity varies considerably among species. For example, in response to beta-adrenergic stimulation of adenyl cyclase, the CP of the cat is quite sensitive, whereas choroidal tissues in the dog, rabbit, and calf are less so; rat CP has the least sensitivity (Nathanson, 1980). Given the species variation in receptor function, it is probably desirable to survey receptor properties in a variety of species in order to place mammalian binding data in better perspective, especially since the information may relate to the human condition.

The great vascularity of the CP necessitates displacement of blood cells by cold, buffered saline. Transcardial perfusion in animals anesthetized with either ether or barbiturate gave comparable biochemical results, suggesting that the anesthetic did not complicate binding phenomena in a study of adrenergic receptors by Nathanson (1980). To reduce further the number of cell types in the homogenate, it is desirable to remove the larger choroidal vessels from the tissue specimens.

Recently, Pazos et al. (1984) have characterized the binding of serotonergic ligands to the porcine CP. For each saturation or competition experiment, they combined plexuses from four to six hogs. Homogenates (tissue plus $0.32M$ sucrose) were centrifuged for 15 min at $70,000g$. After resuspending the pellet and incubating at 37°C for 15 min, the resultant pellet was harvested for resuspension in storage buffer (50 m$M$ Tris-HCl, pH 7.7, 4 m$M$ $CaCl_2$, 0.1% ascorbic acid).

The protocol for their binding assays, described below, serves to illustrate typical step-by-step procedures. Incubations are carried out in triplicate:

1. Add 750 μL of tissue suspension to tubes containing 50 μL of radioligands plus 200 μL of the different drugs (agonists, inhibitors, and so on). The final tissue concentration for CP is 20 mg/mL.
2. Incubate tubes at 37°C for 30 min.
3. Filter rapidly through Whatman filters under reduced pressure.
4. Wash filters quickly with ice-cold buffer twice, then transfer them to counting vials containing scintillation fluid.

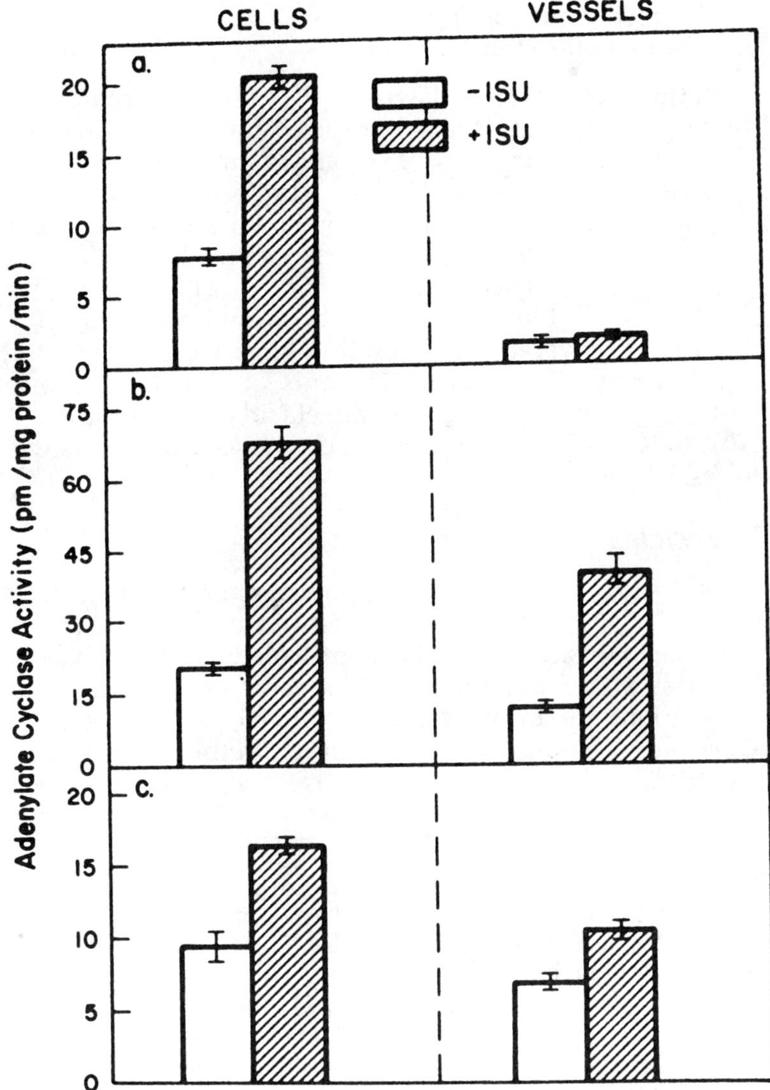



Fig. 14. Three methods to localize basal and isoproterenol (ISO)-stimulated adenylate cyclase activity in *epithelial cell* and *vascular* fractions obtained from intact choroid plexus. The results described below gave similar results for calf CP. (a) This method gave the most complete dissociation of cells from vessels. Cat CP was gently fragmented by hand in homogenizer in cold 25 m*M* Tris maleate buffer, pH 7.4, containing 140 m*M* NaCl and 0.5% BSA, and then centrifuged on a discontinuous su-

5.  Extract radioactivity from filters by incubation and subsequent cooling.
6.  Count radioactivity on filters.

With this methodology, Pazos et al. (1984) have characterized the kinetic and pharmacological properties of the binding of tritiated 5-hydroxytryptamine (5-HT, serotonin), ketanserin, and mesulergine to membranes of CP and brain tissue (for comparison). Binding parameters for saturation kinetics and drug competition were calculated by the SCTFIT program (DeLean et al., 1980). Ligands supposedly selective for $5\text{-HT}_{1A}$, $5\text{-HT}_{1B}$, or $5\text{-HT}_2$ subtypes did not show high affinity for choroidal binding sites; thus, these newly discovered 5-HT binding sites were designated as $5\text{-HT}_{1C}$. The physiological function and specific sites associated with $5\text{-HT}_{1C}$ receptor activity await elucidation. Serotoninergic innervation to the plexus terminates on the vasculature as well as the epithelial cells.

## *10.7. Vesicles*

Over a dozen years ago, plasma membrane vesicles were introduced by Hopfer as a new method to elucidate epithelial transport mechanisms. The vesicles are isolated from the basolateral as well as apical membranes. Transport of ions and organic molecules can be monitored by electrophysiological and radioisotopic methods. The very small size of the vesicles, i.e., $< 5$

---

crose gradient. Vessel fraction ($1.7–2.25M$ interface) was washed and collected against a 100-$\mu$m Nitex screen. The remaining nonvessel fractions were diluted with medium and collected at $50,000g$. (b) To obtain cells, cat CP was incubated in 10 m$M$ Hepes buffer, pH 7.4, containing 0.05% hyaluronidase, 137 m$M$ NaCl, 2.7 m$M$ KCl, 0.7 m$M$ $Na_2HPO_4$, 5.6 m$M$ glucose, 1% BSA, and 0.012% collagenase (three changes, 15 min each, at 37°C), after which 10% horse serum was added. The cells in suspension were washed, filtered through Nitex, pelleted, and homogenized. The remaining partially de-epithelialized capillary network was washed and homogenized. (c) This method caused the least disruption of the capillary network, but left the most epithelium on the vessels. Cells were dissociated from capillaries by incubating rabbit CP in 10 m$M$ Hepes (pH 7.4) containing 137 m$M$ NaCl, 2.7 m$M$ KCl, 0.7 m$M$ $Na_2HPO_4$, 5.6 m$M$ glucose, 1% BSA, and 2.25 m$M$ EDTA (three changes, 20 min each, 4°C), with intermittent trituration (reproduced, with permission, from Nathanson, 1979) (copyright 1979 by the AAAS).

$\times 10^3$ Å, has necessitated the development of special nonclassical methods. Although most of our knowledge about vesicles has been advanced by gastrointestinal and renal physiologists, recently there have been reports about the transport properties of vesicles isolated from CPs, mainly from the laboratories of Wright and of Kinne.

## 10.7.1. Radioisotope Studies

The uptake of L-amino acids into ovine and bovine CP vesicles has been analyzed by Ross and Wright (1984). The source of the CP tissue for vesicular preparation can be either the local slaughterhouse (e.g., for cow brains) or a commercial house, like Pel Freez Biologicals (Rogers, AR), which has ample supplies of rabbit plexuses. Typically, the following basic preparative procedures are involved in isolating and purifying the membranes. First, the plexuses are freed of blood and debris by washing with ice-cold normal saline. Centrifugation ($300g$ for 5 min) removes capillaries from the whole tissue homogenates, the latter made by conventional procedures (Wright, 1980). To the supernatant is added 10 m$M$ CaCl$_2$, and the mixture is then stirred at 4°C for 15 min. Subsequently, a series of steps accomplishes vesicle isolation: (1) pellet the Ca$^{2+}$-aggregated subcellular organelles at low force ($2500g$ for 10 min), (2) from the resulting supernatant, harvest the membranes by centrifuging at 50,000$g$ for 30 min, (3) with a homogenizer, resuspend the membrane in 300 m$M$ mannitol/pH 7.4/10 m$M$ HEPES, and (4) then, pellet at 50,000$g$ for 30 min and resuspend in the same buffer, also containing 4.5 mg protein/mL.

The membranes are durable and can be stored for up to a month in liquid nitrogen. With the procedures described above, Ross and Wright (1985) obtained vesicles with a 4% protein yield and a three-fold enrichment of alkaline phosphatase. Na$^+$-K$^+$-ATPase is a reliable marker for evaluating purity of the apical membrane of mammalian CP. The specific activity of Na$^+$-K$^+$-ATPase was enriched three-fold for the rabbit and nine- to 14-fold for cow membranes. Langenbeck and Kinne (1980) reported a four- to five-fold enriched fraction of Na$^+$-K$^+$-ATPase in preparations of hog CP.

The uptake of radioisotope tracers into membrane vesicles can be done at room temperature, by techniques using rapid mixing and filtration (Stevens et al., 1982). In general terms, the method involves the following:

1. Mixing rapidly the membranes with buffer containing the tracer.
2. Incubating the mixture, for a few seconds up to 2 h.
3. Quenching the reaction mixture with ice-cold buffer.
4. Filtering and washing the "hot" membranes.
5. Counting the radioactivity trapped in membranes on filters.

It is convenient to express uptake as mol per mg protein. With this approach, Ross and Wright (1984) have demonstrated $Na^+$-dependent, concentrative uptake of histidine, proline, and MeAIB by vesicles of CP. Saturation and competition phenomena have been quantitated for the amino acids. Proline and histidine probably share a common transport pathway in CP, and this carrier system may contribute to the regulation of amino acids in the CSF.

### 10.7.2. Electrophysiological Analyses

The diminutive size of the vesicle precludes the use of classical electrophysiological recording techniques. Yet it is important to have measurements of transvesicular potential difference because membrane potential is a significant driving force in the transport of ions, metabolic intermediates, amino acids, and even in the translocation of neutral molecules by electrogenic processes (Wright, 1984). Thus, indirect optical techniques to record membrane potentials in vesicles have been developed, mainly by the application of voltage-sensitive dyes.

Extrinsic optical probes of membrane potential have taken the form of rapidly and slowly responding dyes with response times of milliseconds and a few seconds, respectively. Cyanine dyes are popular because of their large signals, i.e., about 0.5% change in fluorescense (F) per millivolt. Diffusion potentials, generated with known ion gradients (e.g., $\Delta K^+$) in the presence of ionophores (e.g., valinomycin), are used to calculate optical signals. Vesicles are preloaded with solution containing known $[K^+]$; then the $F$ value is recorded when these same vesicles are added to polystyrene cuvets containing a wide range of $[K^+]$. The inclusion of the appropriate ionophore ensures the full development of $K^+$ equilibrium potentials. Dye signals should not be allowed to vary with solution pH or ionic strength, and so buffers and impermeant ions need to be employed. It is desirable to have a linear relationship between $F$ and $E_K$. $E_K = E_m = (RT/F) \ln K_o/K_i$, where $E_m$ is the membrane potential, i.e., that of the intravesicular space relative to the outside solution.

In addition to $K^+$ diffusion potentials, $H^+$ as well as $Na^+$ diffusion potentials have been used to calibrate the dyes. Commercially available cyanine dyes have been the most popular, including [diS-$C_3$-(5)], [diS-$C_2$-(5)], and [diO-$C_2$-(5)]. In CP brush borders, Wright (1984) found that the calibration of the [diS-$C_3$-(5)] signals with diffusion potentials was independent not only of the ion employed, but also of the ionophore used to generate the potential.

Optically derived membrane potentials have been used to (1) determine the ion permeability of brush border membranes and (2) delineate the kinetics and specificity of sugar- and organic acid-$Na^+$ cotransport systems. The permeability of brush border membranes from the cow CP has been determined by a standard $Ca^{2+}$ differential centrifugation procedure. The details of the protocol, including the insertion of concentration and potential data into the constant-field equation (to solve for ion permeabilities), have been outlined by Wright (1984). With respect to the kinetic description of the organic solute-$Na^+$ cotransport systems, the information obtained for the CP vesicles is substantially less than that procured for renal and intestinal membranes. It is quite likely, however, that the choroidal vesicle cotransporters are also electrogenic and associated with augmented conductance of $Na^+$ in the membrane.

For complete electrophysiological description, it is highly desirable to have information about membrane currents. Wright has proposed the feasibility of extending the patch-clamp pipet technique to epithelial vesicles, in order to record the kinetics of $Na^+$-cotransporters and ion channels. Recently, Brown et al. (1986) have used the patch clamp technique to study the nature of conductance in the CP of *Necturus* and of the bullfrog. Highly selective $K^+$ channels were identified in whole tissue preparations from both species. They concluded that the CP apical membranes contain $Ca^{2+}$-activated, voltage-dependent, large $K^+$-channels, perhaps accounting for the great conductance of the brush border.

## 11. Arachnoid Membrane Function and Preparations

The thin arachnoid membrane on the surface of the brain and spinal cord directly covers the CSF in the subarachnoid space. The role of the arachnoid membrane in the CSF system has not been nearly so extensively studied as that of the choroid plexus. Wood's two recent volumes on the *Neurobiology of Cerebrospinal Fluid* contain 107 chapters on various facets of CSF function; although there is a single chapter on the arachnoid villi, there are no presentations on the arachnoid membrane. Thus, our knowledge of arachnoid membrane function is substantially less than that of the CP.

Over a large region of the cerebral cortex, there is anatomical similarity between the arachnoid and pia maters. About three to four layers of distinctive leptomeningeal cells make up the arachnoid. Although the pia closely hugs the contours of the cortex, the arachnoid remains as a layer in contact with the dura. In some regions the arachnoid is continuous with the pia. Arachnoid cells are large and contain numerous mitochondria in a relatively great volume of cytoplasm (reviewed by Bradbury, 1979). They are joined together by gap junctions and desmosomes in the innermost layer and by tight junctions in the middle and outer layers. Dermietzel (1975) has used the freeze-etch method to depict the extensive zonulae occludentes. As is the case with the CP epithelium, the arachnoid cells have phagocytotic potential.

Three basic methodological approaches have been attempted to assess arachnoidal function: (1) in vivo injections of probes into the fluids surrounding the membrane, (2) excision of the tissue for transport studies, and (3) tissue cultures of the pia-arachnoid membrane.

## 11.1. Intravascular and Intrathecal Injections

For over a half century, it has been known that intravascularly administered tracers like Evans-blue albumin and fluorescein readily stain the dura that overlies the arachnoid membrane, but do not gain access to the CSF (reviewed by Rapoport, 1976). Microperoxidase, when introduced into the subarachnoid space, does not reach the dural tissue because of the imperviousness of the middle layer of the arachnoid–dura complex (Nabeshima et al, 1975). Thus, there is little doubt that the tightness of the arachnoid membrane is consistent with a barrier function that contributes to homeostasis of CSF composition.

There is strong clinical interest in the effects on the meninges caused by intrathecal agents and solutions that traverse the subarachnoid space. For example, V'allfors and colleagues (1983a,b) have investigated the effects of local anesthetics and radiological contrast media on the leptomeninges. Methiodal sodium extensively destroyed leptomeningeal cells. In another study of cats, they examined by scanning electron microscopy the integrity of the mesothelium of the arachnoid surface following exposure to air and neurosurgical irrigation fluids. Exposure to saline or air caused perforation and detatchment of mesothelial cells from underlying connective tissue. Such in vivo analyses are important in view of possible deleterious consequences to CSF homeostasis, brought about by iatrogenic disruptions of arachnoidal membrane function.

## 11.2. In Vitro Arachnoid Preparations

The arachnoid membrane of the bullfrog is a sheet about 25 $\mu$m thick, consisting of about a dozen layers of flattened epithelial-like cells. Thus, this amphibian membrane lends itself to mounting in Ussing-type chambers (Perez-Gomez et al., 1976). The electrical resistance of the arachnoid preparation is approximately 2000 ohm cm$^2$, comparable to the "tight" epithelia (like urinary bladder) rather than to "leaky" epithelia (like gall bladder). With such high resistance electrically, it is not surprising that the steady-state transmural PD can be as high as 45 mV, CSF positive. On the assumption of correctly estimated surface areas over the chamber orifice, it appears that the absolute permeability coefficients (calculated from membrane conductance and relative permeabilities) for Na$^+$, K$^+$, and Cl$^-$ are about 10-fold lower in arachnoid than in CP (Perez-Gomez et al., 1976).

Organic solute transport by the arachnoid membrane has not been extensively analyzed, but there have been descriptions of amino acid transport. One of the initial studies, by Wright et al. (1971), demonstrated that the disappearance of $^{14}$C-leucine from fluid in cups placed over the intact cerebral cortical surface was more rapid when the arachnoid-pial membrane was intact than when stripped. Shortly thereafter, more direct evidence for transport by mammalian arachnoid was obtained by O'Tuama et al. (1973), who found that the dura–arachnoid of dogs accumulated amino acids when incubated in vitro. Subsequently, Wright (1974) observed that the unidirectional flux of glycine across the chamber-isolated frog arachnoid membrane was greater from CSF to serosa than in the reverse direction.

## 11.3. Tissue Culture

The pia-arachnoid of the newborn rat brain can be cultured either on collagen-coated coverslips in the Maximow assembly or on glass coverslips for fluorescence histochemistry. Spatz et al.

(1975) have described the organotypic pia–arachnoid cultures. The composition of the cultured membrane consists of at least two distinct cellular elements, i.e., the epithelial (arachnoidal) and the endothelial.

Hervonen et al. (1981) have investigated barrier mechanisms in relation to the transport or uptake of markers like horseradish peroxidase (HRP) or L-3,4-dihydroxyphenylalanine (L-Dopa) by the pia–arachnoid. After 2–4 wk of culturing, the tissues were washed and preincubated for 10 min in HEPES-buffered Locke's solution (pH 7.4). Subsequent incubation for 10 min at 36 or 20°C was carried out in the presence of various amines and other probes. They found that the pia–arachnoid, but not the endothelium, was able to markedly pinocytose HRP label; this suggested that the arachnoid cells normally degrade proteinaceous material absorbed from the CSF in the subarachnoid space. On the other hand, Hervonen and colleagues (1981) noted that the endothelium selectively took up L-Dopa. The culture technique allows "physiological dissection" of the barrier components as the respective properties of the cells are being characterized in the living state.

## 12. Future Research of the Choroid Plexus–Arachnoid–CSF System

From vesicles to in vivo preparations, the various methodologies are furnishing new insights about CP function. In general, membrane vesicles have proven to be a powerful tool for elucidating antiport function (e.g., $Na^+$-$H^+$ exchange) in epithelia. Yet it is always important to relate results from fragmented membranes to significance in physiological systems in the intact organism. Recently, Murphy and Johanson furnished in vivo evidence for the existence of $Na^+$-$H^+$ exchange in the basolateral membrane of the CP (*see* Fig. 15); their approach was comparable to that in vesicular studies, i.e., analyzing $^{22}Na^+$ uptake in response to altered transmembrane gradients for $H^+$ and $Na^+$. $Na^+$-$H^+$ exchange is a topic of wide interest to molecular biologists. New analytical opportunities will steadily appear as molecular biological techniques are more frequently applied to BCFB studies. However, to gain the most benefit from these advancing state-of-the-art techniques, it is essential to continue to enhance our knowledge of the basic physiology of the CP.

Each method described in this review has inherent strengths and weaknesses. To maximize the advantages, it has been the

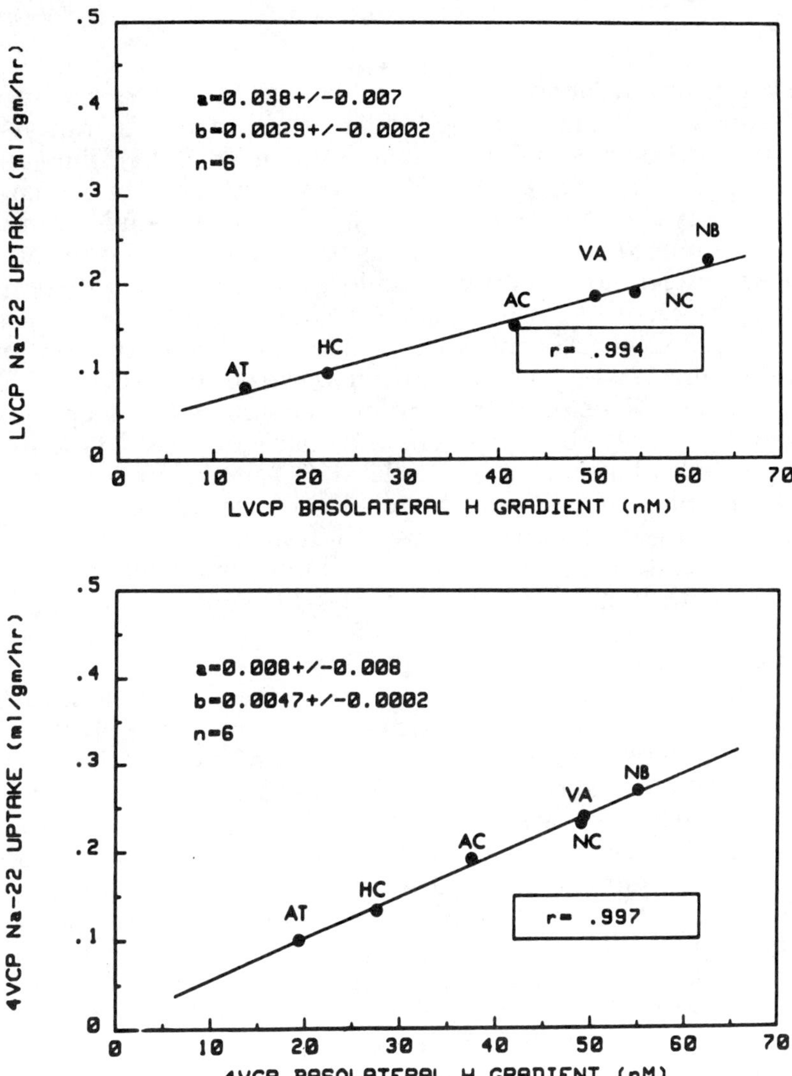



Fig. 15.    In vivo rat CP (lateral and fourth ventricle) uptake of $^{22}$Na$^+$ from blood, as a function of the H$^+$ ion gradient across the basolateral membrane. Cell pH determined by the DMO method. 1-*h* treatments: AT, acetazolamide, 20 mg/kg; HC, hydrochloric acid; AC, ammonium chloride; NC, sodium chloride control; NB, sodium bicarbonate; VA, vehicle for acetazolamide. All salts were given ip, 4 mmol/kg. Slopes determined by linear regression. Each circle represents the mean slope of $^{22}$Na$^+$ uptake curve, determined from several animals for each treatment. a, *y*-intercept; b, slope (from V. Murphy and C. Johanson, unpublished data; for details of the method *see* Murphy, 1984).

strategy of our laboratory to use a battery of in vitro and in vivo methods to characterize the location and function of various ion transport systems in CP (Deng and Johanson, 1985). This can be exemplified for the $Cl^--HCO_3^-$ exchange system, for which evidence has been marshalled from in vivo kinetic studies, in vitro incubations to monitor uptake as well as efflux, the intrathecal injection technique, and ventriculocisternal perfusions (*see* references by Deng, Johanson, and by Smith). There are reasons to think that $Cl^--HCO_3^-$ antiporters exist at both poles of the choroid epithelium. It would be highly desirable to be able to determine $Cl^-$ activity by microelectrode recording in mammalian CP epithelium (10-$\mu$m cubical cells) as Wright and colleagues have done for the larger cells of amphibians. The $Cl^--HCO_3^-$ exchanger may participate in CSF secretion, but additional evidence is needed to support this hypothesis (Johanson, 1984). $Cl^-$ transport in the CP is not as well understood as $Na^+$ transport. Chloride movement paracellularly and through channels can not be ruled out. Another problem to be resolved is the question of the existence of $Na^+-Cl^-$ cotransport in the mammalian CP epithelium.

New approaches are needed to analyze CP function. Choroidal tissue was initially cultured nearly 40 yr ago. Recently there has been renewed interest in tissue cultures of plexus (Crook et al., 1981). Agnew and colleagues (1984) have worked up a serum-free culture system for evaluating solute exchange in the CP. Progress has been made in defining media and supplements to optimize growth and survival of the cultured tissue. In vivo polarity (basilar end of cells toward luminal cavity and apical end facing the medium) has also been maintained. Nathanson's laboratory has succeeded in culturing isolated choroid epithelial cells from the calf. The acutely isolated cells have 70–95% viability, and have been kept in sound condition for culture periods of up to 4 d (Gabuzda et al., 1983). The tissue culture approach will likely circumvent some of the technical problems (related to CP inaccessibility) that have previously hampered biochemical analyses, e.g., adenylate cyclase measurements.

An interesting avenue for future pursuit is the pharmacological manipulation of transport capabilities of the CP. The clinical implications of this problem have been thoroughly treated (Johanson, 1987b). There are also some intriguing possibilities for basic experiments. For example, acetazolamide causes a marked cellular alkalinization of CP cell pH in vivo (Johanson, 1984). If the substantial elevation in CP pH could be sustained, this model might be used to advantage in other paradigms. For example, what are the effects of alkaline pH in the choroid epithelium on the microtubular system, and on the role of $pH_i$ generally as a third messenger? In this regard, acetazolamide induces generalized atrophic changes in the secretory epithelium of the plexus, including denuding of the microvilli (Azzam et al., 1978). Choudhury et al. (1979) have considered that the acetazolamide-linked atrophy model would be valuable in evaluating the controversy of choroidal vs. extrachoroidal secretion; apparently, the ventricular ependyma compensate for the atrophic CP. At any rate, there is a variety of drugs that selectively ablate or hypertrophy the ultrastructure of the choroidal cells (Wenk et al., 1979; Santolaya and Echandia, 1968; Levine and Sowinski, 1973).

The ontogenetic approach should prove to be a powerful tool in understanding the role of CP transport in the overall economy of brain cell metabolism (Johanson, 1987a). CSF secretion by CP is incompletely developed in infant rats, and so this model has been fruitful in sorting out age-related changes in physiological factors such as blood flow, neural tone, receptor density, and enzyme activities. Methods are being sought to bridge our phenotypic analyses with genotypic phenomena to answer the question: What is the time course and nature of genetic control of active transport systems in the CP? Thus the long-term goal of our laboratory is to relate cellular physiology to molecular biological processes, in order to appreciate the function and therapeutic manipulation of this important transport interface.

Finally, what is the future of research on the arachnoid membrane? Wright has advocated a significant role for the arachnoid membrane in CSF dynamics and homeostasis (at least in amphibians), citing the ample blood supply to the arachnoid as well as its transport capabilities (Perez-Gomez et al., 1976). The situation in mammals is less clear because of limited information. Spector and Johanson (1987) have de-emphasized, however, a major active role of the arachnoid in CSF function; in part, their conclusion was drawn from the apparent lack of facilitated transport of many micronutrients in the in vitro arachnoid membrane. Certainly, more physiological information is needed. At the ultrastructural and molecular level, there have been recent arachnoidal studies of intermediate filaments (Michaels and Tornheim, 1984), cytokeratin (Frank et al., 1983), and substance P (Liu-Chen et al., 1983). It remains to be seen how these recent findings affect the barrier role of the arachnoid membrane, and how they may relate to corresponding phenomena in the CP.

## References


Agnew W. F., Alvarez R. B., Yuen T. G. H., Abramson S. B., and Kirk D. (1984) A serum-free culture system for studying solute exchanges in the choroid plexus. *In Vitro* **20**, 712–722.

Alaranta H., Hurme M., Lahtela K., and Hyyppa M. T. (1983) Prolactin and cortisol in cerebrospinal fluid: Sex-related associations with clinical and psychological characteristics of patients with low back pain. *Psychoneuroendocrinology* **8**, 333–341.

Ames A., III, Higashi K., and Nesbett F. B. (1965a) Relation of potassium concentration in choroid plexus fluid to that in plasma. *J. Physiol.* **181**, 506–515.

Ames A., III, Higashi K., and Nesbett F. B. (1965b) Effects of $pCO_2$, acetazolamide and ouabain on volume and composition of choroid-plexus fluid. *J. Physiol.* **181**, 516–524.

Ames A., III, Sakanoue M., and Endo S. (1964) Na, K, Ca, Mg and Cl concentrations in choroid plexus fluid and cisternal fluid compared with plasma ultrafiltrate. *J. Neurophysiol.* **27**, 672–681.

Azzam N. A., Choudhury S. R., and Donohue J. M. (1978) Changes in the surface fine structure of choroid plexus epithelium following chronic acetazolamide treatment. *J. Anat.* **127**, 333–342.

Blount R., Foreman P., Harding M., and Segal M. (1973) The perfusion of the isolated choroid plexus of the sheep. *J. Physiol.* **232**, 12–13P.

Bouchaud C. and Bouvier D. (1978) Fine structure of tight junctions between rat choroidal cells after osmotic opening induced by urea and sucrose. *Tiss. Cell* **10**, 331–342.

Bowsher D. (1958) A Possible Mechanism of Hydrocephus: The Osmotic Regulation of Cerebrospinal Fluid Volume, in *Ciba Foundation Symposium on Cerebrospinal Fluid* (Wolstenholme G. and O'Connor C., eds.) Little, Brown, Boston.

Bradbury M. W. B. (1975) Ontogeny of Mammalian Brain-Barrier Systems, in *Fluid Environment of the Brain* (Cserr H. F., Fenstermacher J. D., and Fencl V., eds.) Academic, New York.

Bradbury M. (1979) Energy-Dependent Transport at the Barriers, in *The Concept of a Blood-Brain Barrier* John Wiley, New York.

Brown P. D., Loo D. D. F., Sachs G., and Wright E. M. (1986) Calcium-activated K channels in amphibian choroid plexus. *Fed. Proc.* **45**, 740.

Brzezinski J., Kjallquist A., and Siesjo B. K. (1967) Mean carbon dioxide tension in the brain after carbonic anhydrase inhibition. *J. Physiol.* **188**, 13–23.

Burton S. (1982) Carrier-mediated transport of calcium into the *in vitro* choroid plexus: Inhibitory effects of sodium, potassium and pharmacological agents. Ph.D. Thesis, University of Utah, Salt Lake City, Utah.

Choudhury S. R., Azzam N. A., and Donohue J. M. (1979) Changes in the surface fine structure of rat third ventricular ependyma following chronic acetazolamide treatment. *J. Anat.* **129,** 51–62.

Clark K. (1962) Isolation of the choroid plexus *in vivo. J. Neurosurg.* **19,** 1004–1006.

Crook R. B., Kasagami H., and Prusiner S. B. (1981) Culture and characterization of epithelial cells from bovine choroid plexus. *J. Neurochem.* **37,** 845–854.

Cserr H. (1971) Physiology of the choroid plexus. *Physiol. Rev.* **51,** 273–311.

Cserr H. F., Bundgaard M., Ashby J. K., and Murray M. (1980) On the anatomic relation of choroid plexus to brain: A comparative study. *Am. J. Physiol.* **238,** R76–R81.

Cserr H., Fenstermacher J. D., and Rall D. P. (1968) Permeabilities of the Choroid Plexus and Blood–Brain Barrier to Urea, in *Excerpta Medica International Congress* series No. 195 *Urea and the Kidney* Elsevier, New York.

Czaky T. Z. and Rigor B. M. (1967) The Choroid Plexus as a Glucose Barrier, in *Progress in Brain Research. Brain Barrier Systems* vol. 29 (Lajtha A. and Ford D. H., eds.) Elsevier, Amsterdam.

Decker J. F. and Quay W. B. (1982) Stimulatory effects of melatonin on ependymal epithelium of choroid plexuses in golden hamsters. *J. Neural Transm.* **55,** 53–67.

DeLean A., Stadel J., and Lefkowitz R. J. (1980) A ternary complex model explains the agonist-specific binding properties of the adenylate cyclase-coupled $\beta$-adrenergic receptor. *J. Biol. Chem.* **255,** 7108.

Deng Q. S. (1986) Drug modification of chloride transport in the choroid plexus-cerebrospinal fluid system of the rat. Ph.D. Thesis, University of Utah, Salt Lake City, Utah.

Deng Q. S. and Johanson C. E. (1984) Effects of different temperatures, pH and pharmacological agents on chloride transport in rat choroid plexus. *Fed. Proc.* **43,** 1088.

Deng Q. S. and Johanson C. E. (1985) Stilbene and autonomic agents alter Cl penetration into the *in vivo* choroid plexus-CSF system. *Fed. Proc.* **44,** 1746.

Dermietzel R. (1975) Junctions in the central nervous system of the cat. V. The junctional complex of the pia-arachnoid membrane. *Cell Tiss. Res.* **164,** 309–329.

Edvinsson L., Lindvall M., Owman C., and West K. A. (1983) Autonomic Nervous Control of Cerebrospinal Fluid Production and Intracranial Pressure, in *Neurobiology of Cerebrospinal Fluid* vol. 2 (Wood J. H., ed.) Plenum, New York.

Eisenberg H. M., Suddith R. L., and Crawford J. S. (1980) Transport of Sodium and Potassium Across the Blood–Brain Barrier, in *The Cere-*

*bral Microvasculature-Investigation of the Blood–Brain Barrier* (Eisenberg H. and Suddith R., eds.) Plenum, New York.

Feldman A. M., Epstein M. H., and Brusilow S. W. (1980) Role of Cyclic AMP in Cerebrospinal Fluid Production, in *Neurobiology of Cerebrospinal Fluid* vol. 1 (Wood J. H., ed.) Plenum, New York.

Frank E. H., Burge B. W., Liwnicz B. H., Lotspeich L. J., and White J. C. (1983) Cytokeratin provides a specific marker for human arachnoid cells. *Exp. Cell Res.* **146,** 371–376.

Freye E. and Gupta B. N. (1979) A modified technique for the selective perfusion of the fourth cerebral ventricle in conscious dogs. *J. Pharmacol. Meth.* **2,** 305–314.

Gabuzda D. H., Hunnicut E. J., Owen C. J., and Nathanson J. A. (1983) Choroid plexus epithelial cells in culture: Biochemical and pharmacological characteristics. *Soc. Neurosci. Abstr.* **9,** 118.

Grady P. A. and Blaumanis O. R. (1983) Structural evidence for unstirred layers in the choroid plexus epithelium. *Soc. Neurosci. Abstr.* **9,** 885.

Hansen A. J. (1985) Effect of anoxia on ion distribution in the brain. *Physiol. Rev.* **65,** 101–149.

Harbut R. E. (1982) Investigation of the primary stimulus and mechanism of the ammonium chloride-induced increase in the content of potassium in choroid plexus epithelial cells. Ph.D. Thesis, University of Utah, Salt Lake City, Utah.

Harbut R. E. and Johanson C. E. (1986) Third ventricle choroid plexus function and its response to acute perturbations in plasma chemistry. *Brain Res.* **374,** 137–146.

Hayward J. R. and Vogh B. P. (1979) Some measurements of autonomic nervous system influence on production of cerebrospinal fuid in the cat. *J. Pharmacol. Exp. Ther.* **208,** 341–346.

Hedlund L., Lischko M. M., Rollag M. D., and Niswender G. D. (1977) Melatonin daily cycle in plasma and cerebrospinal fluid of calves. *Science* **195,** 686–687.

Heisey S. R. (1968) Brain and choroid plexus blood volumes in vertebrates. *Comp. Biochem. Physiol.* **26,** 489–498.

Hervonen H., Spatz M., Bembry J., and Murray M. R. (1981) Studies related to the blood–brain barrier to monoamines and protein in pia-arachnoid cultures. *Brain Res.* **210,** 449–454.

Hise M. A. and Johanson C. E. (1979) The sink action of the cerebrospinal fluid in uremia. *Eur. Neurol.* **18,** 328–337.

Howarth F. and Jowett A. (1962) A technique for surgical encapsulation of a canine choroid plexus. *J. Physiol.* **162,** 20P.

Husted R. F. and Reed D. J. (1976) Regulation of cerebrospinal fluid potassium by the cat choroid plexus. *J. Physiol.* **259,** 213–221.

Husted R. F. and Reed D. J. (1977) Regulation of cerebrospinal fluid bicarbonate by the cat choroid plexus. *J. Physiol.* **267,** 411–428.

Johanson C. E. (1978) Choroid epithelial cell pH. *Life Sci.* **23**, 861–868.

Johanson C. E. (1979) Effect of enzyme inhibitors on epithelial cell pH in choroid plexus and salivary gland. *Pharmacologist* **21**, 242.

Johanson C. E. (1980) Permeability and vascularity of the developing brain: Cerebellum vs. cerebral cortex. *Brain Res.* **190**, 3–16.

Johanson C. E. (1984) Differential effects of acetazolamide, benzolamide and systemic acidosis on hydrogen and bicarbonate gradients across the apical and basolateral membranes of the choroid plexus. *J. Pharmacol. Exp. Ther.* **231**, 502–511.

Johanson C. E. (1987a) Ontogeny and Phylogeny of the Blood–Brain Barrier, in *The Clinical Impact of the Blood–Brain Barrier and Its Manipulation* (Neuwelt E., ed.) Plenum, New York, in press.

Johanson C. E. (1987b) Potential for Pharmacological Manipulation of the Blood–Cerebrospinal Fluid Barrier, in *The Clinical Impact of the Blood–Brain Barrier and Its Manipulation* (Neuwelt E., ed.) Plenum, New York, in press.

Johanson C. E. and Harbut R. E. (1984) Ionic homeostasis of the choroid plexus–csf system in ganglionectomized or adrenalectomized rats stressed with acidosis. *Soc. Neurosci. Abstr.* **10**, 1162.

Johanson C. E. and Smith Q. R. (1984) Efflux of Cl-36 from choroid plexus by chloride-bicarbonate exchange. *Physiologist* **27**, 272.

Johanson C. E. and Woodbury D. M. (1977) Penetration of C-14 barbital and C-14 antipyrine into the choroid plexus and cerebrospinal fluid of the rat. *Exp. Brain Res.* **30**, 65–74.

Johanson C. E. and Woodbury D. M. (1978) Uptake of C-14 urea by the *in vivo* choroid plexus–cerebrospinal fluid–brain system: Identification of sites of molecular sieving. *J. Physiol.* **275**, 167–176.

Johanson C. E., Parandoosh Z., and Smith Q. R. (1985) Chloride-bicarbonate exchange in the choroid plexus: Analysis by the DMO method for cell pH. *Am. J. Physiol.* **249**, F470–F477.

Johanson C. E., Reed D. J., and Woodbury D. M. (1976) Developmental studies of the compartmentalization of water and electrolytes in the choroid plexus of neonatal rat brain. *Brain Res.* **116**, 35–48.

Johanson C. E., Allen J., and Withrow C. D. (1987) Regulation of brain and CSF pH in developing mammalian central nervous system. *Dev. Brain Res.*, submitted.

Johanson C. E., Foltz F. M., and Thompson A. M. (1974a) The clearance of urea and sucrose from isotonic and hypertonic fluids perfused through the ventriculo-cisternal system. *Exp. Brain Res.* **20**, 18–31.

Johanson C. E., Reed D. J., and Woodbury D. M. (1974b) Active transport of sodium and potassium by the choroid plexus of the rat. *J. Physiol.* **241**, 359–372.

Kimelberg H. K. (1981) Active and exchange transport of chloride in astroglial cells in culture. *Biochim. Biophys. Acta* **646**, 179–184.

Kozlowski G. P., Brownfield M. S., and Hostetter G. (1978) Neurosecretory Supply to Extrahypothalamic Structures: Choroid Plexus, Circumventricular Organs, and Limbic System, in *Neurosecretion and Neuroendocrine Activity. Proceedings of VII International Symposium* Leningrad (Bargmann W. and Giessen O., eds.) Springer-Verlag, Berlin.

Langenbeck U. and Kinne R. (1980) Enrichment and preliminary characterization of a plasma membrane fraction from hog choroid plexus. *Hoppe Seyler's Z. Physiol. Chem.* **361,** 1311.

Levine S. and Sowinski R. (1973) Choroid plexitis produced in rats by cyclophosphamide. *J. Neuropathol. Exp. Neurol.* **32,** 365–370.

Liu-Chen L. Y., Han D. H., and Moskowitz M. A. (1983) Pia arachnoid contains substance P originating from trigeminal neurons. *Neuroscience* **9,** 803–808.

Mann J. D. and Mann E. S. (1983) Differential Effects of Pentobarbital, Ketamine Hydrochloride, Enflurane and Halothane on Cerebrospinal Fluid Dynamics, in *Neurobiology of Cerebrospinal Fluid* vol. 2 (Wood J. W., ed.) Plenum, New York.

Manuilov I. A. (1958) Technique for perfusion of brain ventricles in dogs under chronic experimental conditions. *Sechenov Physiol J. USSR* (Eng. transl.) **44,** 458–462.

Masuzawa T., Shimabukuro H., Sato F., and Saito T. (1981) Ultrastructural localization of carbonic anhydrase activity in the rat choroid plexus epithelial cell. *Histochemistry* **73,** 201–209.

McComb J. G., Davson H., and Hollingsworth J. R. (1977) Attempted separation of blood–brain and blood–cerebrospinal fluid barriers in the rabbit. *Exp. Eye Res.* **25** (suppl.), 333–343.

Melby J. M., Miner L. C., and Reed D. J. (1982) Effect of acetazolamide and furosemide on the production and composition of cerebrospinal fluid from the cat choroid plexus. *Can. J. Physiol. Pharmacol.* **60,** 405–409.

Michaels J. E. and Tornheim P. A. (1984) Arachnoid matter of the bullfrog, *Rana catesbeiana.* A potential model for the study of intermediate filaments. *Cell Tiss Res.* **236,** 693–697.

Milhorat T. H. (1969) Choroid plexus and cerebrospinal fluid production. *Science* **166,** 1514–1516.

Milhorat T. H., Hammock M. K., Fenstermacher J. D., Rall D. P., and Levin V. A. (1971) Cerebrospinal fluid production by the choroid plexus and brain. *Science* **173,** 330–332.

Miner L. C. and Reed D. J. (1972) Composition of fluid obtained from choroid plexus tissue isolated in a chamber *in situ. J. Physiol.* **227,** 127–139.

Murphy V. A. (1984) Sodium-hydrogen exchange in the rat choroid plexus. Ph.D. Thesis, University of Utah, Salt Lake City, Utah.

Murphy V. A. and Johanson C. E. (1983) Amiloride and insulin alter choroid plexus (CP) sodium. *Pharmacologist* **25,** 251.

Murphy V. A. and Johanson C. E. (1985) Adrenergic-induced enhancement of brain barrier system permeability to small non-electrolytes: Choroid plexus vs. cerebral capillaries. *J. Cereb. Blood Flow Metab.* **5,** 401–412.

Myers R. R. and Shapiro H. M. (1978) Paradoxical effect of enflurane on choroid plexus metabolism: Clinical implications. *Proc. Ann. Meet. Am. Soc. Anesthesiol.* 489–490.

Nabeshima S., Reese T. S., Landis D. M. D., and Brightman M. W. (1975) Junctions in the meninges and marginal glia. *J. Comp. Neurol.* **164,** 127–170.

Nakamura S. and Hochwald G. M. (1983) Effects of arterial $pCO_2$ and cerebrospinal fluid volume flow rate changes on choroid plexus and cerebral blood flow in normal and experimental hydrocephalic cats. *J. Cereb. Blood Flow Metab.* **3,** 369–375.

Nathanson J. A. (1979) β-Adrenergic-sensitive adenyl cyclase in secretory cells of choroid plexus. *Science* **204,** 843–44.

Nathanson J. A. (1980) β-Adrenergic-sensitive adenylate cyclase in choroid plexus: Properties and cellular localization. *Mol. Pharmacol.* **18,** 199–209.

Nathanson J. A. (1983) Adrenergic-Receptor Mechanisms in Mammalian Choroid Plexus, in *Neurobiology of Cerebrospinal Fluid* vol. 2 (Wood J. H., ed.) Plenum, New York.

O'Tuama L. A., Remler M. P., and Nichols H. N. (1973) Accumulation of Radiolabelled Neutral Amino Acids by Canine Dura-Arachnoid *Soc. Neurosci. Abstr., 3rd Ann. Meet.* 374.

Page R. B., Funsch D. J., Brennan R. W., and Hernandez M. J. (1980) Choroid plexus blood flow in the sheep. *Brain Res.* **197,** 532–537.

Pappenheimer J. R., Heisey S. R., and Jordan E. F. (1961) Active transport of Diodrast and phenolsulfonphthalein from cerebrospinal fluid to blood. *Am. J. Physiol.* **200,** 1–10.

Parandoosh Z. and Johanson C. E. (1979) Effect of vasopressin on the penetration of C-14 urea into brain compartments protected by barrier systems. *Soc. Neurosci. Abstr.* **5,** 308.

Parandoosh Z. and Johanson C. E. (1982) Ontogeny of the blood–brain barrier permeability to, and cerebrospinal fluid sink action on, C-14 urea. *Am. J. Physiol.* **243,** R400–R407.

Patlak C. S., Adamson R. H., Oppelt W. W., and Rall D. P. (1966) Potential difference of the ventricular fluid *in vivo* and *in vitro* in the dogfish. *Life Sci.* **5,** 2011–2015.

Pazos A., Hoyer D., and Palacios J. M. (1984) The binding of serotonergic ligands to the porcine choroid plexus: Characterization of a new type of serotonin recognition site. *Eur. J. Pharmacol.* **106,** 539–546.

Pellegrino L. J. and Cushman A. J. (1967) in *A Stereotaxic Atlas of the Rat Brain* Meredith, New York.

Perez-Gomez J., Bindslev N., Orkand P. M., and Wright E. M. (1976) Electrical properties and structure of the frog arachnoid membrane. *J. Neurobiol.* **7,** 259–270.

Pershing L. K. and Johanson C. E. (1982) Acidosis-induced enhanced activity of the Na-K exchange pump in the *in vivo* choroid plexus: An ontogenetic analysis of possible role in cerebrospinal fluid pH homeostasis. *J. Neurochem.* **38,** 322–332.

Pollay M., Stevens A., Estrada E., and Kaplan R. (1972) Extracorporeal perfusion of choroid plexus. *J. Appl. Physiol.* **32,** 612–617.

Pollay M., Stevens F. A., and Welch J. (1979) Choroid plexus blood flow in rat and rabbit. *Acta Neurol. Scand.* suppl. **72 60,** 596–597.

Quay W. B. (1966) Regional differences in metabolism and composition of choroid plexuses. *Brain Res.* **2,** 378–389.

Rapoport S. I. (1976) *The Blood–Brain Barrier in Physiology and Medicine* Raven, New York.

Ross H. J. and Wright E. M. (1984) Neutral amino acid transport by plasma membrane vesicles of the rabbit choroid plexus, *Brain Res.* **295,** 155–160.

Rougemont J., Ames A., Nesbett F. B., and Hofmann H. F. (1960) Fluid formed by choroid plexus. A technique for its collection and a comparison of its electrolyte composition with serum and cisternal fluids. *J. Neurophysiol.* **23,** 485–495.

Rychter Z. and Stastny F. (1980) Stimulatory and Inhibitory Effect of Hydrocortisone on the Morphogenesis of the Choroid Plexus in Chick Embryo, in *Ontogenesis of the Brain* (Trojan S. and Stastny F., eds.) University of Karlova, Praha, Czechoslovakia.

Saito Y. and Wright E. M. (1982) Kinetics of the sodium pump in the frog choroid plexus. *J. Physiol.* **328,** 229–243.

Santolaya R. C. and Echandia E. L. R. (1968) Induced changes in choroid plexus cells fine structure. *Acta Physiol. Latino-Americana* **18,** 194–198.

Schone H. and Loeschcke H. H. (1969) Bestandspotentiale am plexus choroideus des 4. ventrikels von katze und kaninchen *in vitro. Pflugers Arch.* **306,** 195–209.

Schousboe A. (1972) Development of potassium effects on ion concentrations and indicator spaces in rat brain-cortex slices during postnatal ontogenesis. *Exp. Brain Res.* **15,** 521–531.

Smith Q. R. and Johanson C. E. (1980a) Effect of carbonic anhydrase inhibitors and acidosis on choroid plexus epithelial cell sodium and potassium. *J. Pharmacol. Exp. Ther.* **215,** 673–680.

Smith Q. R. and Johanson C. E. (1980b) Effect of ouabain and potassium on ion concentrations in the choroidal epithelium. *Am. J. Physiol.* **238,** F399–F406.

Smith Q. R. and Johanson C. E. (1985) Active transport of chloride by lateral ventricle choroid plexus of the rat. *Am. J. Physiol.* **249**, F470–F477.

Smith Q. R., Pershing L. K., and Johanson C. E. (1981a) A comparative analysis of extracellular fluid volume of several tissues as determined by six different markers. *Life Sci.* **29**, 449–456.

Smith Q. R., Woodbury D. M., and Johanson C. E. (1981b) Uptake of Cl-36 and Na-22 by the choroid plexus-cerebrospinal fluid system: Evidence for active chloride transport by the choroidal epithelium. *J. Neurochem.* **37**, 107–116.

Smith Q. R., Woodbury D. M., and Johanson C. E. (1982) Kinetic analysis of Cl-36, Na-22 and H-3 mannitol uptake into the *in vivo* choroid plexus–cerebrospinal fluid system: Ontogeny of the blood–brain and blood–CSF barriers. *Dev. Brain Res.* **3**, 181–198.

Spatz M., Renkawek K., Murray M. R., and Klatzo I. (1975) Uptake of radiolabeled glucose analogues by organotypic pia arachnoid cultures. *Brain Res.* **100**, 710–715.

Spector R. (1982) Nucleoside transport in choroid plexus: Mechanism and specificity. *Arch. Biochem Biophys.* **216**, 693–703.

Spector R. and Eells J. (1984) Deoxynucleoside and vitamin transport into the central nervous system. *Fed. Proc.* **43**, 196–200.

Spector R. and Johanson C. E. (1987) Choroid plexus: Structure, development and function. *Sci. Am.*, in press.

Spector R. and Levy P. (1975) Thyroxine transport by the choroid plexus *in vitro*. *Brain Res.* **98**, 400–404.

Spector R. and Lorenzo A. V. (1974) Specificity of ascorbic acid transport system of the central nervous system. *Am. J. Physiol.* **226**, 1468–1473.

Stevens B. R., Ross H. J., and Wright E. M. (1982) Multiple transport pathways for neutral amino acids in rabbit jejunal brush border vesicles. *J. Mem. Biol.* **66**, 213–225.

Stonestreet B. S., Nowicki P. T., Hansen N. B., Petit R., and Oh W. (1983) Effect of aminophylline on brain blood flow in the newborn piglet. *Dev. Pharmacol. Ther.* **6**, 248–258.

Tennyson V. M. and Pappas G. D. (1968) The fine structure of the choroid plexus: Adult and developmental stages. *Prog. Brain Res.* **29**, 63–85.

Thompson A. M. (1970) Hyperosmotic Effects on Brain Uptake of Nonelectrolytes, in *Capillary Permeability* (Crone C. and Lassen N. A., eds.) Munksgaard, Copenhagen.

Tyson G., Kelly P., McCulloch J., and Teasdale G. (1982) Autoradiographic assessment of choroid plexus blood flow and glucose utilization in the unanesthetized rat. *J. Neurosurg.* **57**, 543–547.

V'allfors B., Hansson H. A., and Belghmaidi M. (1983a) Mesothelial cell integrity of the subdural and arachnoid surfaces of the cat brain after exposure to neurosurgical irrigation fluids and air: A scanning electron microscopic study. *Neurosurgery* **12**, 35–39.

V'allfors B., Hansson H. A., Belghmaidi M., and Persson L. I. (1983b) Effect of radiologic contrast media and local anaesthetics on the blood–brain barrier and on the leptomeninges. *Acta Neurol. Scand.* **68,** 164–170.

Walsh R. J., Posner B. I., and Patel B. (1984) Binding and uptake of [$^{125}$]iodoprolactin by epithelial cells of the rat choroid plexus: An *in vivo* autoradiographic analysis. *Endocrinology* **114,** 1496–1505.

Welch K. (1962) Active transport of iodide by choroid plexus of the rabbit *in vitro. Am. J. Physiol.* **202,** 757–760.

Welch K. (1963) Secretion of cerebrospinal fluid by choroid plexus of the rabbit. *Am. J. Physiol.* **205,** 617–624.

Welch K. and Sadler K. (1965) Electrical potentials of the choroid plexus of the rabbit. *J. Neurosurg.* **22,** 344–351.

Welch K., Araki H., and Arkins T. (1972) Electrical potentials of the *lamina epithelialis choroidea* of the fourth ventricle of the cat *in vitro:* Relationship to the CSF blood potential. *Dev. Med. Child Neurol.* **14,** (suppl. 27), 146–151.

Wenk E. J., Levine S., and Hoenig E. M. (1979) Fine structure of contrasting choroid plexus lesions caused by tertiary amines or cyclophosphamide. *J. Neuropathol. Exp. Neurol.* **38,** 1–9.

Woodbury D. M., Johanson C. E., and Brondsted H. (1974) Maturation of the Blood–Brain and Blood–CSF Barriers and Transport Systems, in *Narcotics and the Hypothalamus* (Zimmermann E. and George R., eds.) Raven, New York.

Wright E. M. (1972) Mechanisms of ion transport across the choroid plexus. *J. Physiol.* **226,** 545–571.

Wright E. M. (1974) Active transport of glycine across the frog arachnoid membrane. *Brain Res.* **76,** 354–358.

Wright E. M. (1978) Transport processes in the formation of the cerebrospinal fluid. *Rev. Physiol. Biochem. Pharmacol.* **83,** 1–34.

Wright E. M. (1984) Electrophysiology of plasma membrane vesicles. *Am. J. Physiol.* **246,** F363–F372.

Wright P. M., Nogueira G. J., and Levin E. (1971) Role of the pia mater in the transfer of substances in and out of the cerebrospinal fluid. *Exp. Brain Res.* **13,** 294–305.

Wright S. H., Kippen I., Klinenberg J. R., and Wright E. M. (1980) Specificity of the transport system for tricarboxylic acid cycle intermediates in renal brush borders. *J. Membrane Biol.* **57,** 73–82.

Yagaloff K. A. and Hartig P. R. (1985) $^{125}$I-Lysergic acid diethylamide binds to a novel serotonergic site on rat choroid plexus epithelial cells *J. Neurosci.,* **5**(12), 3178–3183.

Yuen T. G. H. and Agnew W. F. (1978) Ultrastructural alterations during choroid plexus incubations. *Exp. Neurol.* **60,** 96–115.

# Fluid Compartments

## Ultrastructural Methods to Identify Extracellular Spaces in the Central Nervous System

Asao Hirano and Takeo Kato

## 1. Introduction

Before the advent of the electron microscope, the extracellular space of the central nervous system (CNS) was considered very wide. Since that time, however, fine structural studies have revealed that the CNS is crowded with numerous cell processes and cell bodies and there is virtually no identifiable extracellular space. Therefore, tracer methods are usually required to visualize the extracellular space as a positive image.

In choosing an appropriate tracer, several considerations must be kept in mind. Perhaps the most important is the size of the tracer molecule, since that will determine, to a great extent, the lower limits of the spaces through which the tracer can permeate. In addition, one must consider the physiological effects of the tracer in order to avoid introducing artifacts in the tissue under study. Other considerations include the ease of visualization of the tracer and the compatibility of the methods used for visualization with good preservation. Finally, it is often useful if the same tracer can be used in both the light and electron microscopes. Unfortunately, most tracers do not meet this criterion.

In this chapter we shall describe a number of tracers that have been used in the CNS and will discuss them from the points of view of the considerations mentioned above. For the sake of clarity we have divided them into naturally electron-opaque tracers, tracers in which enzymatic properties are exploited, radioactive tracers, fluorescent tracers, and those techniques in which the tracer material evolved directly out of the experimental methods used.

## 2. Electron-Opaque Tracers

Tracers within this group are directly visible as electron-dense granular material. Unfortunately, however, these tracers cannot be used for light microscopy unless a special procedure is performed, such as binding a fluorescent marker to electron-dense dextrans (Olsson et al., 1975).

Popular tracers in this group include lanthanum, ferritin, thorotrast, and dextran.

### 2.1. Lanthanum

Ionic or colloidal lanthanum has been used frequently for the study of the extracellular space and the junctional complexes between adjacent endothelial cells or choroid plexus epithelium in the CNS. Colloidal lanthanum (lanthanum hydroxide) is formed from an ionic solution of a lanthanum salt, such as lanthanum nitrate, by making its pH slightly basic (Karnovsky and Revel, 1966; Brightman and Reese, 1969; Milhorat et al., 1975). At neutral pH, the lanthanum salts remain ionic. Lanthanum has been demonstrated to be toxic to living tissue (Cuddihy and Boecker, 1970), and it may introduce artifacts.

A fairly typical procedure is one used by Bouldin and Krigman (1975). Anesthetized adult rats were transcardially perfused, first with 15 mL of 1% $NaNO_2$ in physiologic saline, and then with 250–300 mL of $La^{3+}$ solution containing 20 m$M$ La $(NO_3)_3$, 80 m$M$ NaCl, 3.5 mM KCl, 1.0 mM $CaCl_2$, 1.0 mM $MgCl_2$, and 1.0 m$M$ glucose at a pH of 7.3–7.4 for 15 min. The $La^{3+}$ solution was followed by an additional 15 min perfusion with 300–350 mL of a balanced salt solution containing 43 m$M$ $Na_2SO_4$, 16 mM $NaHCO_3$, 10 m$M$ sodium acetate, 3.5 mM KCl, 1.0 mM $CaCl_2$, 1.0 mM $MgCl_2$, 1.0 m$M$ glucose, 1.6 mM $Na_2HPO_4$, 0.4 mM $NaH_2PO_4$, and 33 m$M$ sucrose (pH 7.3–7.4). Glutaraldehyde sufficient to make either a 1 or 2.5% solution was added to this salt solution, and the animal was subjected to a third 15-min perfusion. Thirty minutes after finishing the third perfusion, the brain tissue was removed and put in 2.5% glutaraldehyde in the balanced salt solution for 4–6 h. Then the tissue was washed out overnight in the balanced salt solution and postfixed in 1% $OsO_4$ in the same solution. Routine processes for electron microscopy were followed. Although $La(NO_3)_3$ and $LaCl_3$ are soluble in water at pH 7.3–7.4, $La_2(SO_4)_3$ is insoluble and precipitated as an electron-dense substance, easily visualized in the electron microscope.

By the use of these methods, the zonulae occludentes of the capillaries and choroid epithelia in the brain have come to be considered to constitute the anatomic basis of the blood–brain and blood–cerebrospinal fluid barriers (Brightman and Reese, 1969). Machen et al. (1972) have demonstrated that in the rabbit gall bladder and intestine there are two main groups of zonulae occludentes; one group has low resistance to ion permeation and is penetrated by $La^{3+}$ and the other group has high resistance and is not permeated by $La^{3+}$.

According to Castel et al. (1974) and Bouldin and Krigman (1975), $La^{3+}$ injected into the blood stream cannot penetrate the zonulae occludentes of the endothelia of the cerebral capillaries, but can penetrate those of the epithelium of the choroid plexus. Results obtained from freeze-fracture studies indicate that an impermeable zonula occludentes consists of a meshwork of many anastomosing and branching continuous strands, whereas a permeable one is composed of a much smaller number of discontinuous strands (Claude and Goodenough, 1973; Connell and Mercer, 1974; Bohr and Mollgard, 1974).

A variation of these colloidal or ionic methods is the use of solid lanthanum nitrate. When implanted into the CNS, it dissolves slowly and spreads through the extra-cellular spaces, which then appear as electron-dense regions (Fig. 1). Small pellets of lanthanum nitrate are implanted into the forebrains of anesthetized, young adult albino rats (about 250 g) through a burr hole in the skull. The animals were sacrificed by perfusion through the heart with 5% glutaraldehyde in 1/15$M$ phosphate buffer, pH 7.4, within 5 min of implantation. Brain tissue at various distances from the implant was removed and postfixed in Dalton's chrome-osmium, pH 7.4. Small blocks of tissue were dehydrated in ethanol and embedded in Epon 812 after two changes in propylene oxide. Thin sections were prepared and observed, unstained, in the electron microscope.

This method has been particularly successful in visualizing a special extracellular space that is lined by the transverse bands, the lateral loops of the myelin sheath, and the axon (Fig. 2). The transverse bands in the paranodes of the myelinated fibers constitute a unique cell junction between two different types of cells: the axon and the oligodendrocyte. Early studies of thin sections through the paranodes suggested that the transverse bands were parallel to the lateral loops and described a helical path around the axon (Hirano and Dembitzer, 1967). Later studies using the ex-

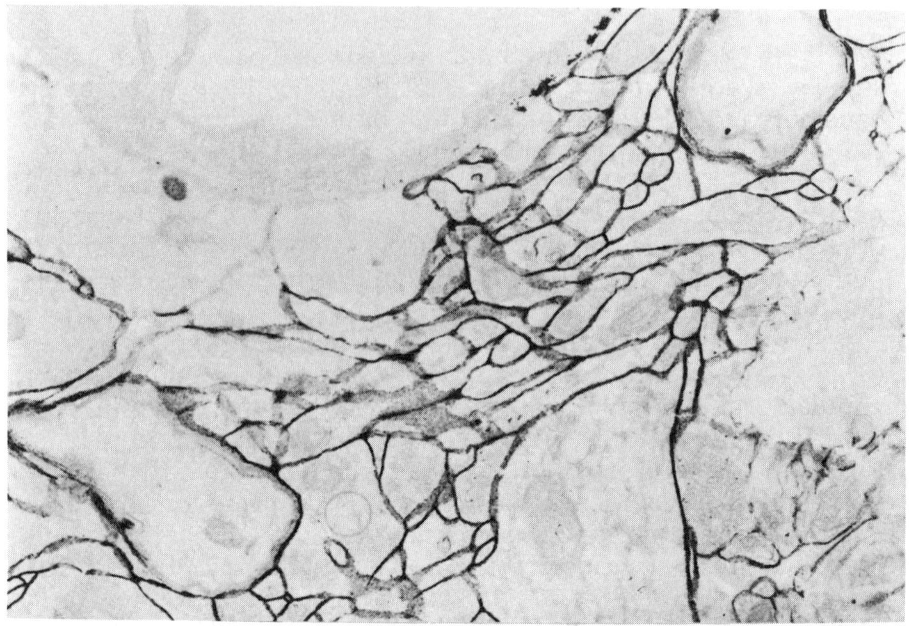

Fig. 1.   Lanthanum demonstrated as an electron-dense substance that occupies the extracellular space around neuronal and glial processes (× 18,900).

tracellular tracer lanthanum were interpreted as confirming this view, and it was suggested that the spaces between the transverse bands, as well as the spaces between the lateral loops, constitute a means of access between the parenchymatous extracellular space and the periaxonal space (Hirano and Dembitzer, 1969). Subsequently, freeze-fracture studies of turtle (Schnapp et al., 1976) and frog (Rosenbluth, 1976) CNS suggested that the transverse bands were arranged oblique rather than parallel to the lateral loops, and therefore, the spaces between the transverse bands communicated directly with the spaces between the lateral loops. In 1982, Hirano and Dembitzer reconciled the apparent discrepancy by demonstrating the coexistence of transverse bands that were both parallel and oblique to the lateral loops, even within a single paranode. Similar results were also demonstrated in a recent freeze-fracture study (Rosenbluth, 1985).

## 2.2. Ferritin

Ferritin is an electron-dense particle that has a molecular weight of about 400,000. In the electron microscope these appear as

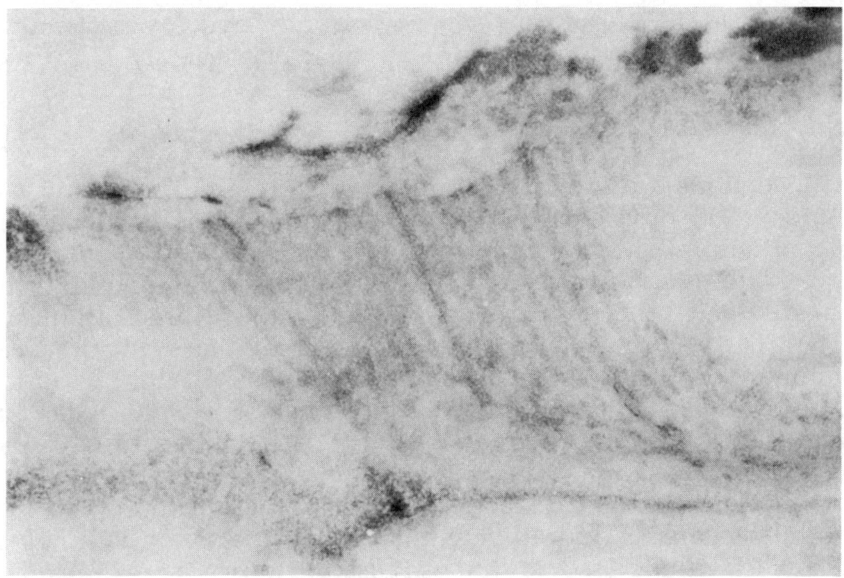

Fig. 2. Deposits of lanthanum between the transverse bands at the paranodal area (×91,500) (reproduced by permission from Hirano and Dembitzer, 1982).

discrete individual particles about 10 nm in diameter. Ferritin has been employed as a large molecule probe for studies of vascular permeability (Farrant, 1954; Farquhar et al., 1961; Farquhar and Palade, 1961; Clementi and Palade, 1969). Intravenously injected ferritin does not pass beyond the endothelial lining of the cerebral capillaries, but does traverse the endothelium of the cerebral arterioles, mainly in the form of pinocytotic vesicles (Westergaard and Brightman, 1973). A cationized ferritin has been also used to detect the regional difference of electric charge on the surface of the vascular endothelium (Simionescu et al., 1982).

## 2.3. Thorotrast

Thorotrast (Testagar and Company, Inc.) is not a single substance, but contains 24–26% thorium dioxide by volume, 25% aqueous dextrin, and 0.15% methyl parasept as a preservative (Lampert and Carpenter, 1965). Electron microscopically, thorotrast is visualized as irregular, very electron-dense particles mixed with larger, much less dense material. The size of the electron-dense particles ranges from 7 to 15 nm (Lampert and Carpenter, 1965). According to Rowley (1963), dextrin, a constituent of thorotrast, can induce the release of histamine and 5-hydroxytryptamine from mast cells and may change vascular permeability.

## *2.4. Dextran*

Dextrans are stable, electron-dense polysaccharide particles that are more physiological than metal tracers such as ferritin, gold, or lanthanum. They have been widely used as electron microscopic tracers for studies on vascular permeability (Simionescu and Palade, 1971; Simionescu et al., 1972; Caulfield and Farquhar, 1974). The electron density of dextrans in the tissue is enhanced by staining with lead acetate after aldehyde-$OsO_4$ fixation. In addition, various sized particles of dextran (molecular weight: 10,000 to over 500,000) can be obtained by fractionation procedures. Thus various fractions can be used to provide clues as to the relative sizes of narrow extracellular spaces. Dextrans labeled with fluorescein isothiocyanate (FITC) can be employed as tracers for both light- and electron-microscopic studies (Olsson et al., 1975).

## 3. Enzymatic Tracers

This group of tracers consists of proteins with enzymatic activity. Visualization of the tracers in the tissue preparation depends on a histochemical reaction. The tracer substance itself cannot be visualized, but the reaction product is visible. Within this group we include horseradish peroxidase, microperoxidase, cytochrome C, myoglobin, and acetylcholinesterase.

### *3.1 Peroxidatic Tracers*

Except for acetylcholinesterase, all of the enzymes mentioned above display peroxidatic activity and depend on the same fundamental reaction for their visualization. In the presence of low concentrations of hydrogen peroxide and a chromogen (Chr-$H_2$), the enzymatic activity promotes the polymerization of the chromogen to a visible colored reaction product as follows (Seligman et al., 1968):

$$Enzyme + H_2O_2 \rightleftharpoons Enzyme \cdot H_2O_2$$

$$Enzyme \cdot H_2O_2 + Chr\text{-}H_2 \rightleftharpoons Enzyme + (Chr)_n + 2H_2O$$

Most often the chromogen used is 3,3'-diaminobenzidine tetrahydrochloride (DAB), which results in the formation of osmium black after treatment with $OsO_4$. Other chromogens used have included benzidine dihydrochloride (BDHC) and tetramethyl benzidine (TMB). Of these, TMB is the most sensitive and can result in visualization of amounts of horseradish peroxidase that are not detectable by using either DAB or BDHC. Furthermore, TMB does not seem to show the carcinogenic activity displayed by DAB and is, therefore, safer to handle (Mesulam, 1978). Ultrastructural demonstration of the TMB reaction product has recently been shown by a number of laboratories (Sakumoto et al., 1980; Schonitzer and Hollander, 1981; Sturmer et al., 1981; Carson and Mesulam, 1982; Turner and Marfurt, 1983).

In using any of the peroxidatic enzymes as tracers, great attention must be paid to proper controls. Most tissues show endogenous peroxidatic activity. Red cells, as well as mitochondria and peroxisomes, normal constituents of cells, will all show reaction products when subjected to the histochemical methods used to visualize the peroxidatic tracers. The close comparison of controls that were not subjected to exogenous peroxidatic enzymes must be made with the tracer-treated tissues.

### 3.1.1. Horseradish Peroxidase (HRP)

Graham and Karnovsky (1966) first reported a method for identifying exogenous HRP ultrastructurally in the epithelium of the renal tubules. Since then, the method and its modifications have been applied to studies in various organs and tissues, including the nervous system.

HRP has a molecular weight of about 40,000 and an apparent diameter of about 5 nm. A typical procedure using HRP as a tracer of the extracellular space in the CNS is one that was used by Hirano et al. (1970) in a study of experimental allergic encephalomyelitis (EAE) in rats.

A quantity of 25 mg of HRP (Sigma, Type II) in 1 mL of saline was injected into the femoral veins of both EAE and control rats. At various intervals after administration of the tracer (usually 5 to 15 min), lumbosacral spinal cords from both animals were fixed by immersion in 5% glutaraldehyde in 1/15$M$ phosphate buffer at pH 7.4 for 2 to 3 h. After fixation and overnight rinsing in buffer, 10- to 15-$\mu$m thick frozen sections were cut on a freezing microtome. The frozen sections were incubated for peroxidatic activity for 20 min at room temperature in a saturated solution of DAB in $0.05M$ Tris-HCl buffer (pH 7.6) containing 0.01% $H_2O_2$ (2–3 mg DAB and 0.1 mL of 1% $H_2O_2$ in 10 mL Tris-HCl buffer). The sections were then washed in three changes of distilled water and postfixed in 1% osmic acid in Millonig's buffer for 1 h, and embedded for electron microscopy. After suitable areas were selected from thick sections, thin sections were cut and viewed either with or without lead staining.

One great advantage of HRP is its ability to be used at both the light and electron microscopic levels. In the light microscope, the HRP reaction product is visible as a brown material; it appears quite dense in the electron microscope. It is, however, quite sensitive to fixation. Therefore, both the choice of fixatives and the duration of fixation are of great importance in order to obtain optimal results. A mixture of glutaraldehyde and paraformaldehyde, such as 1.25% glutaraldehyde and 1.0% paraformaldehyde, is still popular as a fixative (Mesulam et al., 1980). According to Jones and Leavitt (1974), a concentration of paraformaldehyde exceeding 0.4% compromised the demonstration of HRP. They recommended 1.25% glutaraldehyde and 0.4% paraformaldehyde. Kim and Strick (1976) and Malmgren and Olsson (1977, 1978) showed that even low concentrations of paraformaldehyde considerably reduced the number of detectable HRP-labeled structures, and recommended fixation without paraformaldehyde (1.5–2.5% glutaraldehyde only).

Rosene and Mesulam (1978) examined the relationship between fixation time and the preservation of enzymatic activity of HRP in nervous tissue and showed that it was greatly reduced by prolonged fixation of several hours and was completely abolished after 12 h. Accordingly, they limited fixation time to 30 min and immediately thereafter washed out excess fixative from the tissue by perfusing with a cold sucrose-buffer solution.

The antigenicity of HRP as a protein molecule is apparently much more stable than the enzymatic activity of HRP after fixation. For example, the enzymatic activity of HRP incorporated into CNS tissue is completely lost after formalin-fixation and paraffin-embedding procedures, whereas its antigenicity is well preserved (Vacca et al., 1975). These authors therefore chose a novel method using immunoperoxidase techniques after fixation in order to visualize the exogenous HRP.

The use of cobalt-glucose oxidase instead of hydrogen peroxide has been recommended to facilitate the electron microscopic demonstration of HRP (Itoh et al., 1979).

One caution concerning HRP is its apparent pharmacological effect on living tissue. According to Cotran and Karnovsky (1967), HRP induces the release of vasoactive amines from mast cells in some mammals and thereby increases vascular permeability.

HRP has been used in the CNS as both an intracellular and extracellular tracer. HRP is taken up by neuronal perikarya and by axon terminals and is transported along the axons anterogradely or retrogradely. HRP methods have therefore been used frequently in studies demonstrating the neuronal connections between distant structures (Kristensson and Olsson, 1971; Kato, 1983; Kato et al., 1985 a,b). Intravenously administered HRP has also been employed for tracing the pathway of hematogenous fluid from the lumen of the vessels to the extracellular spaces of the perivascular area and the brain parenchyma in various pathologic conditions (Brightman et al., 1970; Hirano et al., 1969, 1970; Shivers et al., 1984).

In most areas of the CNS in normal mammals, intravascularly injected HRP does not pass beyond the endothelial lining of the cerebral vessels. Penetration of HRP is stopped at the zonulae occludentes between adjacent endothelial cells (Reese and Karnovsky, 1967). In pathologic conditions, however, that produce vasogenic brain edema HRP, along with the edema fluid, can traverse the endothelial lining from the vascular lumen to the perivascular space and spread through the extracellular space of the brain parenchyma (Fig. 3).

In order to demonstrate the time sequence of movement and the distribution of HRP across the endothelium, a small amount of the tracer was administered intravenously, and it was followed by perfusion fixation at various time intervals (Clementi and Palade, 1969).

### 3.1.2. Microperoxidase

Microperoxidase, a heme-peptide, is substantially smaller than HRP. It is only 1900 in molecular weight and about 2 nm in diameter. The peroxidatic activity of microperoxidase, however, is only approximately 1/4 that of HRP on a weight basis.

Intravenously injected microperoxidase behaves similarly to HRP in that it is blocked by the tight junctions between adjacent endothelial cells in the normal CNS and does not enter the extracellular space of the parenchyma. However, unlike HRP, intraventricularly injected microperoxidase penetrates into the periaxonal spaces as well as the usual extracellular spaces. This higher

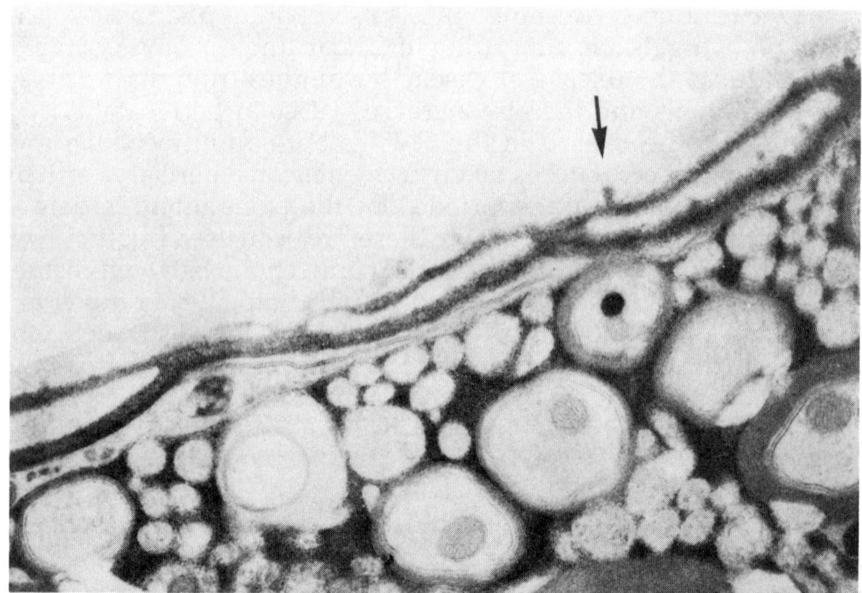

Fig. 3.   Penetration of horseradish peroxidase is evident in the extracellular space as an electron-dense tracer. The arrow indicates pinocytosis in the endothelium of the edematous brain ($\times 28,560$) (reproduced by permission from Hirano, 1969).

permeability may be explained by the small molecular size of microperoxidase (Feder et al., 1969; Feder, 1970, 1971).

### 3.1.3. Myoglobin

Myoglobin is readily available as a highly purified protein tracer that has peroxidatic activity. The size of myoglobin is about 4 nm in diameter and is therefore intermediate between HRP (about 5 nm in diameter) and microperoxidase (about 2 nm in diameter). Another advantage over HRP is the lack of any apparent pharmacological effect on vascular permeability (Cotran and Karnovsky, 1967). The peroxidatic activity of myoglobin can be maintained over a broad pH range (pH 6.0–8.0), with pH 6.8 being optimal for the reaction (Anderson, 1972). However, the peroxidatic activity is considerably weaker than that of HRP, and it requires higher concentrations of $H_2O_2$ and DAB in the reaction medium and a longer reaction time.

### 3.1.4. Cytochrome C

Cytochrome C has certain advantages over HRP. It is a nontoxic physiological substance and a smaller protein (about 13,000 in molecular weight and about 3 nm in diameter) (Milhorat et al., 1975). Cytochrome C has been visualized electron microscopically by the procedure of Karnovsky and Rice (1969). However, the peroxidatic activity is much lower than that of HRP (Karnovsky and Rice, 1969) and, furthermore, fine structural preservation may be impaired during histochemical incubation for cytochrome C, which requires a pH below 4.0 (Milhorat et al., 1973, 1975).

## 3.2. Acetylcholinesterase

Kreutzberg and Kaiya (1974) used exogenous acetylcholinesterase as a tracer for electron microscopic visualization of the extracellular pathways in the brain. They used the histochemical method of Lewis and Shute (1969). The acetylcholinesterase is prepared from the electric organ of the electric eel. It is a glycoprotein-containing glucosamine and galactosamine and has a molecular weight of 250,000 (Leuzinger and Baker, 1967). Before injecting the acetylcholinesterase, endogenous activity of acetylcholinesterase must be inhibited by the irreversible inhibitor diisopropylfluorophosphate.

# 4. Radioactive Tracers

Visualization of radioactive tracers in the tissue depends on autoradiographic methods. Tritiated amino acids such as $^3$H-leucine, $^3$H-proline, or $^3$H-lysine, have been used for studies on axonal flow and neuronal connections (Droz, 1975; Droz et al., 1975; Graybiel, 1975). Injected radio-active amino acids are incorporated into proteins synthesized in the nerve cell body and transported along the axon to the terminals. $^3$H-Thymidine, which is incorporated into DNA during replication, has been employed for determining the time of origin of neurons in the developing CNS (Jacobson, 1978).

With respect to the vascular permeability of the brain, $^{131}$I- or $^{125}$I-labeled albumin or $\gamma$-globulin have been used as radioactive tracers. Intravenously injected $^{131}$I-albumin was extravasated in pathologic areas of the brain (Rozdilsky and Olszewski, 1957;

Vulpe et al., 1960). Cutler et al. (1967) used [125]I, instead of [131]I, because of its higher autoradiographic resolution.


## 5. Fluorescent Tracers

Tracers in this group have the ability to fluoresce under UV light. They are therefore limited to the optical microscope. Fluorescein isothiocyanate (FITC), Evans blue, rhodamine B, trypaflavine, and so on are included in this group of tracers.

By choosing different color fluorochromes, two or more tracers can be employed in a single animal at the same time. For example, albumin labeled with Evans blue or rhodamine B can be identified by the red fluorescence, and $\gamma$-globulin labeled with FITC is visualized as green fluorescence under UV light. When both of these are simultaneously injected intravascularly in an animal with a damaged blood–brain barrier, albumin appears earlier in the perivascular space and spreads more extensively in the brain parenchyma than does $\gamma$-globulin (Steinwall and Klatzo, 1965).


## 6. Miscellaneous

In some cases pathologically induced extracellular spaces of the CNS can be visualized without recourse to specific tracers.

Brain edema occurs most commonly in the form of hematogenous or vasogenic brain edema. This type of edema is caused by traumatic brain injuries, cerebrovascular lesions, brain tumors, inflammations, and other pathologic conditions that produce disruption of the blood–brain barrier. The edema fluid is derived from blood plasma, and has a similar texture and electron density (Hirano et al., 1965). The edema fluid traverses the endothelial cells of the capillary, spreads to the perivascular space, and further infiltrates into the brain parenchyma between the surrounding astrocytic perivascular foot processes.

There are a large number of myelinated and unmyelinated axons in the white matter. These axons are arranged in parallel to form fiber bundles. There is no junctional apparatus between individual fibers. When the hematogenous edema fluid reaches the extracellular space of the white matter, it spreads between the axons (Fig. 4). In this condition, the individual myelinated axons are widely separated from each other, but retain their integrity remarkably well (Fig. 5). The hematogenous edema fluid is usually confined to the extracellular space between the myelinated axons and other cell processes and does not infiltrate into the myelin sheath, except for special conditions (Hirano, 1981; Hirano and Llena, 1983).

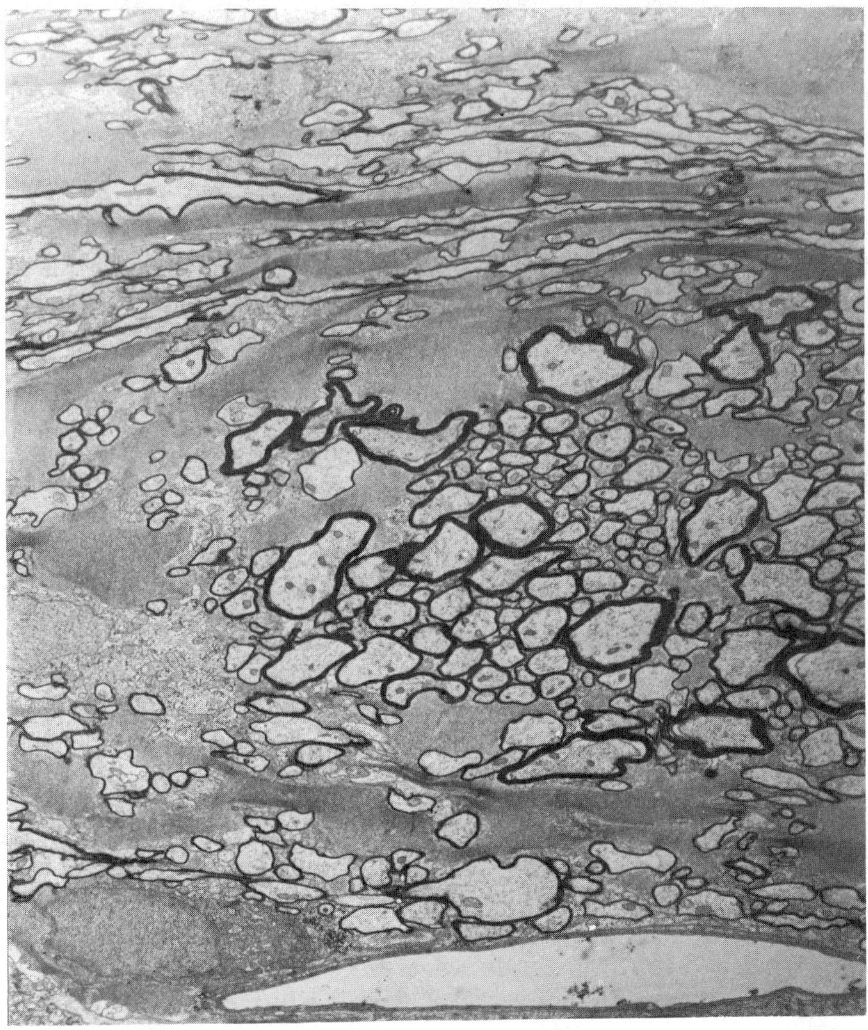

Fig. 4. Edematous cerebral white matter. Individual nerve fibers are separated from one another by electron-dense edema fluid (×3,720) (reproduced by permission from Hirano et al., 1964a).

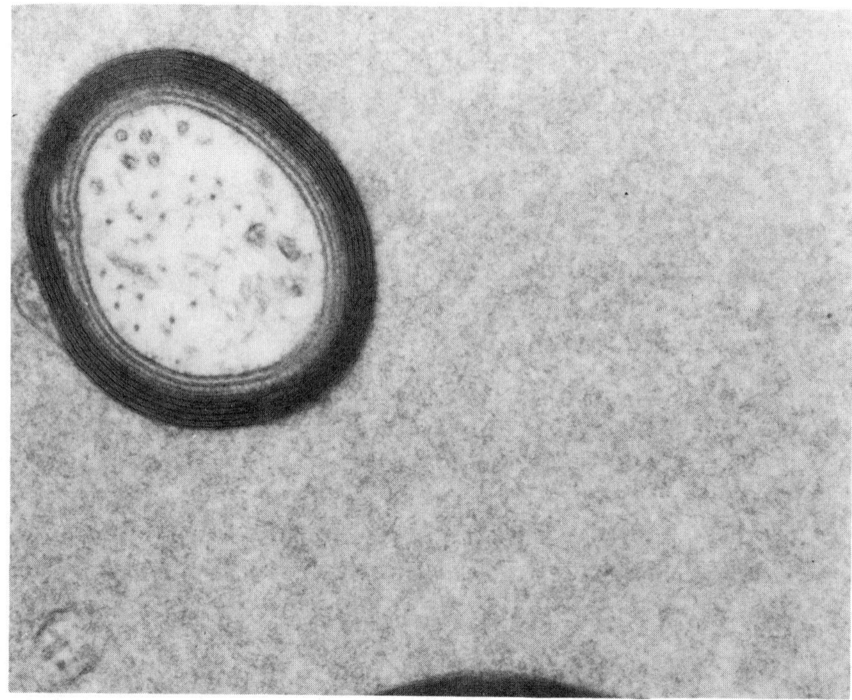

Fig. 5.   A myelinated nerve fiber surrounded by edema fluid. The integrity of the fiber is well preserved ($\times$72,210) (reproduced by permission from Hirano et al., 1967).

In contrast to the white matter, the gray matter has numerous synaptic connections between the neuronal processes and soma. In addition, the dendritic trees each occupy their own individual territories and the processes spread in all directions. The astrocytes have sheet-like processes connected by various junctions that surround neuronal elements. All of these serve to maintain contact between adjacent cells and therefore limit the expansion of the extracellular space by edema fluid.

In a rather unique experimental setting, cryptococcal polysaccharide has been directly visualized in the extracellular spaces of the rat CNS. Cryptococcal polysaccharide implanted into the animal brain spreads through the extracellular space. It is difficult to identify this substance in unstained specimens, but after staining with uranyl acetate or lead hydroxide, it becomes dramatically electron dense with a reticular appearance and is easily identifiable in the electron microscope (Fig. 6). Interestingly, the electron

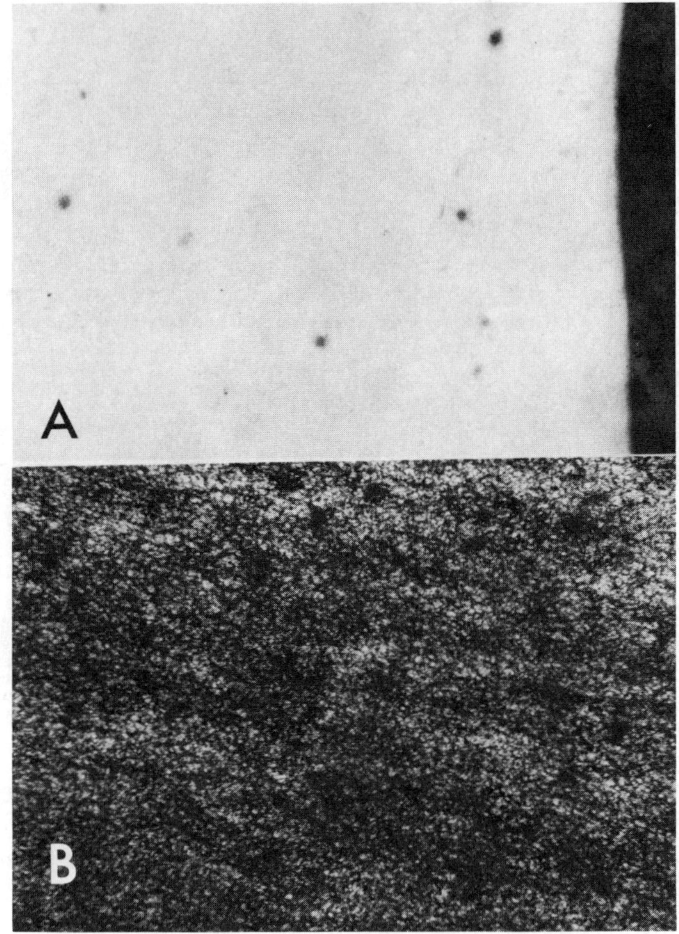

Fig. 6.   Site of cryptococcal polysaccharide implantation. Part of a red cell (right) and minute electron-dense particles are seen in unstained sections (A). Cryptococcal polysaccharide becomes distinctly visible after staining with lead hydroxide (B) (A, ×27,500; B, ×22,000) (reproduced by permission from Hirano et al., 1964b).


microscopic appearance of the cryptococcal polysaccharide is different from that of hematogenous plasma. The electron density of the accompanying extravasated blood plasma is higher than that of the cryptococcal polysaccharide in unstained sections, but this relative difference of electron density is completely reversed when the sections are stained with uranyl acetate or lead hydroxide (Hirano et al., 1964 a,b) (Fig. 7).

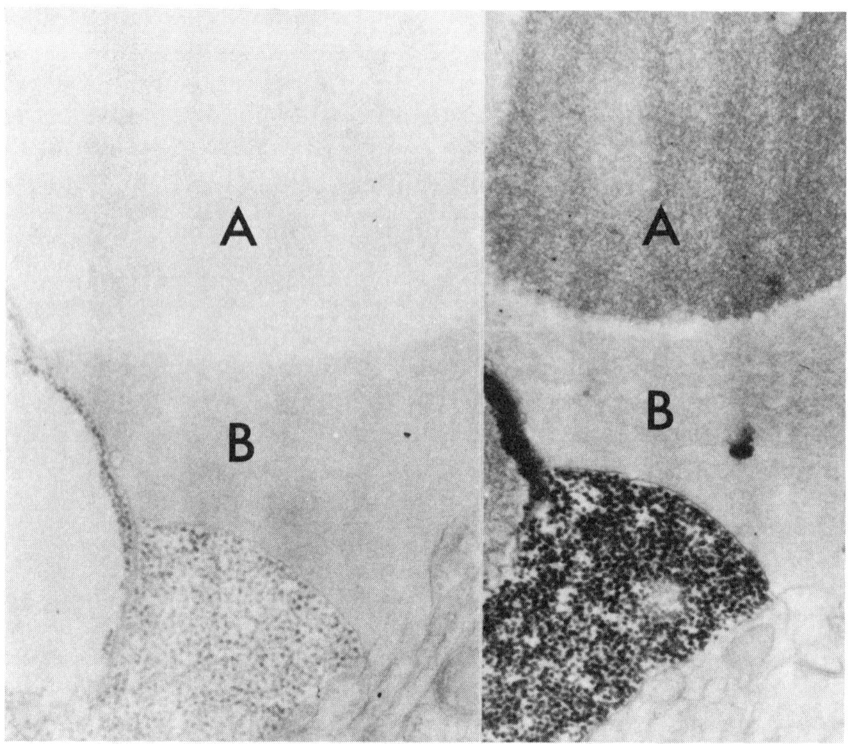

Fig. 7. Two types of fluid (A) Cryptococcal polysaccharide and (B) plasma, with different structural appearances and staining reactions in the cerebral tissue adjacent to an implant. The relative difference of electron density in an unstained section (left) is completely reversed by staining with uranyl acetate (right). Many of the glycogen granules within the plasma membrane are also stained more clearly (×22,900) (reproduced by permission from Hirano et al., 1964b).

## 7. Concluding Remarks

The use of tracers for the visualization of the extracellular spaces of the central nervous system has been a fruitful enterprise. By these means many of the details of capillary permeability and the route of hematogenous fluid within the parenchyma has been elucidated in normal, experimental, and pathological conditions. Further work in this field can be aided by drawing on the experience of previous investigators who have shown the nature and usefulness of a variety of tracer materials.

## Acknowledgment

The authors wish to thank Dr. Herbert M. Dembitzer for his helpful suggestions during the preparation of this manuscript.


## References


Anderson W. A. (1972) The use of exogenous myoglobin as an ultrastructural tracer. *J. Histochem. Cytochem.* **20**, 672–684.

Bohr V. and Mollgard K. (1974) Tight junctions in human fetal choroid plexus visualized by freeze-etching. *Brain Res.* **81**, 314–318.

Bouldin T. W. and Krigman M. R. (1975) Differential permeability of cerebral capillary and choroid plexus to lanthanum ion. *Brain Res.* **99**, 444–448.

Brightman M. W. and Reese T. S. (1969) Junctions between intimately opposed cell membranes in the vertebrate brain. *J. Cell Biol.* **40**, 648–677.

Brightman M. W., Klatzo I., Olsson Y., and Reese T. S. (1970) The blood–brain barrier to proteins under normal and pathological conditions. *J. Neurol. Sci.* **10**, 215–239.

Carson K. A. and Mesulam M.-M. (1982) Electron microscopic demonstration of neural connections using horseradish peroxidase: A comparison of the tetramethyl-benzidine procedure with seven other histochemical methods. *J. Histochem. Cytochem.* **30**, 425–435.

Castel M., Sahar A., and Erlij D. (1974) The movement of lanthanum across diffusion barriers in the choroid plexus of the cat. *Brain Res.* **67**, 178–184.

Caulfield J. P. and Farquhar M. G. (1974) The permeability of glomerular capillaries to graded dextrans. Identification of the basement membrane as the primary filtration barrier. *J. Cell Biol.* **63**, 883–903.

Claude P. and Goodenough D. A. (1973) Fracture faces of zonulae occludentes from 'tight' and 'leaky' epithelia. *J. Cell Biol.* **58**, 390–400.

Clementi F. and Palade G. E. (1969) Intestinal capillaries. I. Permeability to peroxidase and ferritin. *J. Cell Biol.* **41**, 33–58.

Connell C. J. and Mercer K. L. (1974) Freeze-fracture appearance of the capillary endothelium in the cerebral cortex of mouse brain. *Am. J. Anat.* **140**, 595–599.

Cotran R. S. and Karnovsky M. J. (1967) Vascular leakage induced by horseradish peroxidase in the rat. *Proc. Soc. Exp. Biol. Med.* **126**, 557–561.

Cuddihy R. G. and Boecker B. B. (1970) Kinetics of lanthanum retention and tissue distribution in the beagle dog following administration of

[140] LaCl$_3$ by inhalation, gavage and injection. *Health Phys.* **19**, 419–426.

Cutler R. W. P., Lorenzo A. V., and Barlow C. F. (1967) Brain vascular permeability to I$^{125}$ gamma globulin and leukocytes in allergic encephalomyelitis. *J. Neuropathol. Exp. Neurol.* **26**, 558–571.

Droz B. (1975) Autoradiography as a Tool for Visualizing Neurons and Neuronal Processes, in *The Use of Axonal Transport for Studies of Neuronal Connectivity* (Cowan W. M. and Cuenod M., eds.) Elsevier, Amsterdam.

Droz B., Rambourg A., and Koenig H. L. (1975) The smooth endoplasmic reticulum, structure and role in the renewal of axonal membrane and synaptic vesicles by fast axonal transport. *Brain Res.* **93**, 1–14.

Farquhar M. G. and Palade G. E. (1961) Glomerular permeability. II. Ferritin transfer across the glomerular capillary wall in nephrotic rats. *J. Exp. Med.* **114**, 699–716.

Farquhar M. G., Wissig S. L., and Palade G. E. (1961) Glomerular permeability. I. Ferritin transfer across the normal glomerular capillary wall. *J. Exp. Med.* **113**, 47–66.

Farrant J. L. (1954) An electron microscopic study of ferritin. *Biochem. Biophys. Acta* **13**, 569–576

Feder N. (1970) A heme-peptide as an ultrastructural tracer. *J. Histochem. Cytochem.* **18**, 911–913.

Feder N. (1971) Microperoxidase. An ultrastructural tracer of low molecular weight. *J. Cell Biol.* **51**, 339–343.

Feder N., Reese T. S., and Brightman M. W. (1969) Microperoxidase, a new tracer of low molecular weight. A study of the interstitial compartments of the mouse brain. *J. Cell Biol.* **43**, 35A–36A.

Graham R. C. and Karnovsky M. J. (1966) The early stages of absorption of injected horseradish peroxidase in the proximal tubules of mouse kidney: Ultrastructural cytochemistry by a new technique. *J. Histochem. Cytochem.* **14**, 291–302.

Graybiel A. M. (1975) Wallerian Degeneration and Anterograde Tracer Methods, in *The Use of Axonal Transport for Studies of Neuronal Connectivity* (Cowan W. M. and Cuenod, M., ed.) Elsevier, Amsterdam.

Hirano A. (1969) The Fine Structure of Brain in Edema, in *The Structure and Function of Nervous Tissue* vol. 2 (Bourne, G. H., ed.) Academic, New York.

Hirano A. (1981) *A Guide to Neuropathology*. Igaku-Shoin, Tokyo.

Hirano A. and Dembitzer H. M. (1967) A structural analysis of the myelin sheath in the central nervous system. *J. Cell Biol.* **34**, 555–567

Hirano A. and Dembitzer H. M. (1969) The transverse bands as a means of access to the periaxonal space of the central myelinated nerve fibers. *J. Ultrast Res.* **28**, 141–149

Hirano A. and Dembitzer, H. M. (1982) Further studies on the transverse bands. *J. Neurocytol.* **11,** 861–866

Hirano A., Dembitzer, H. M., Becker N. H., Levine S., and Zimmerman H. M. (1970) Fine structural alterations of the blood–brain barrier in experimental allergic encephalomyelitis. *J. Neuropathol. Exp. Neurol.* **29,** 432–440.

Hirano A. and Llena J. F. (1983) Morphological Aspects of Brain Edema, in *Advances in Cellular Neurobiology* vol. 4 (Federoff S. and Hertz L., eds.) Academic, New York.

Hirano A., Becker N. H., and Zimmerman H. M. (1969) Pathological alterations in the cerebral endothelial cell barrier to peroxidase. *Arch. Neurol.* **20,** 300–308.

Hirano A., Zimmerman H. M., and Levine S. (1964a) The fine structure of cerebral fluid accumulation. III. Extracellular spread of cryptococcal polysaccharides in the acute state. *Am. J. Pathol.* **45,** 1–19.

Hirano A., Zimmerman H. M., and Levine S. (1964b) The fine structure of cerebral fluid accumulation. IV. On the nature and origin of extracellular fluids following cryptococcal polysaccharide implantation. *Am. J. Pathol.* **45,** 195–207.

Hirano A., Zimmerman H. M., and Levine S. (1965) The fine structure of cerebral fluid accumulation. IX. Edema following silver nitrate implantation. *Am. J. Pathol.* **47,** 537–548.

Hirano A., Zimmerman H. M., and Levine S. (1967) Fine Structure of Cerebral Fluid Accumulation, in *Brain Edema* (Klatzo I. and Seitelberger F., eds.) Springer-Verlag, New York.

Itoh K., Konishi A., Nomura S., Mizuno N., Nakamura Y., and Sugimoto T. (1979) Application of coupled oxidation reaction to electron microscopic demonstration of horseradish peroxidase: Cobalt-glucose oxidase method. *Brain Res.* **175,** 341–346.

Jacobson M. (1978) *Developmental Neurobiology* 2nd Ed., Plenum, New York.

Jones E. G. and Leavitt R. Y. (1974) Retrograde axonal transport and the demonstration of non-specific projections to the cerebral cortex and striatum from thalamic intralaminer nuclei in the rat, cat and monkey. *J. Comp. Neurol.* **154,** 349–378.

Karnovsky M. J. and Revel J. P. (1966) Hexagonal pattern in tight junctions as revealed by neutral lanthanum suspensions. *J. Cell Biol.* **31,** 56A–57A.

Karnovsky M. J. and Rice O. F. (1969) Exogenous cytochrome C as an ultrastructural tracer. *J. Histochem. Cytochem.* **17,** 751–753.

Kato T. (1983) Transient retinal fibers to the inferior colliculs in the newborn albino rat. *Neurosci. Lett.* **37,** 7–9.

Kato T., Hirano A., Honda K., Katagiri T., and Sasaki H. (1985a) Transient retino-inferior collicular fibers in neonatal rats: Their persistence after removal of one eye at birth. *J. Neuropath. Appl. Neurobiol.* **11,** 265–272.

Kato T., Hirano A., Katagiri T., and Sasaki H. (1985b) Transient uncrossed corticospinal fibers in the newborn rat. *J. Neuropath. Appl. Neurobiol.* **11,** 171–178.

Kim C. C. and Strick P. L. (1976) Critical factors involved in the demonstration of horseradish peroxidase retrograde transport. *Brain Res.* **103,** 356–361.

Kreutzberg G. W. and Kaiya H. (1974) Exogenous acetylcholinesterase as tracer for extracellular pathways in the brain. *Histochemistry* **42,** 233–237.

Kristensson K. and Olsson Y. (1971) Retrograde axonal transport of protein. *Brain Res.* **29,** 313–365.

Lampert P. and Carpenter S. (1965) Electron microscopic studies on the vascular permeability and the mechanism of demyelination in experimental allergic encephalomyelitis. *J. Neuropath. Exp. Neurol.* **24,** 11–24.

Leuzinger W. and Baker A. L. (1967) Acetylcholinesterase. I. Large-scale purification, homogeneity, and amino acid analysis. *Proc. Natl. Acad. Sci. USA* **57,** 446–451

Lewis P. R. and Shute C. C. (1969) An electron-microscopic study of cholinesterase in the rat adrenal medulla. *J. Microsc.* **89,** 181–193.

Machen T. E., Erlig D., and Wooding F. B. P. (1972) Permeable junctional complexes. The movement of lanthanum across rabbit gallbladder and intestine. *J. Cell Biol.* **54,** 302–312.

Malmgren L. T. and Olsson Y. (1977) A sensitive histochemical method for light- and electron-microscopic demonstration of horseradish peroxidase. *J. Hostochem. Cytochem.* **25,** 1280–1283.

Malmgren L. T., and Olsson Y. (1978) A sensitive method for histochemical demonstration of horseradish peroxidase in neurons following retrograde axonal transport. *Brain Res.* **148,** 279–294.

Mesulam M.-M. (1978) Tetramethyl benzidine for horseradish peroxidase neurohistochemistry: A non-carcinogenic blue reaction-product with superior sensitivity for visualizing neural afferents and efferents. *J. Histochem. Cytochem.* **26,** 106–117.

Mesulam M.-M., Hegarty E., Barbar H., Carson K. A., Gower E. C., Knapp A. G., Moss M. B., and Mufson E. J. (1980) Additional factors influencing sensitivity in the tetramethyl benzidine method for horseradish peroxidase neurohistochemistry. *J. Histochem. Cytochem.* **28,** 1255–1259.

Milhorat T. H., Davis D. A., and Hammock M. K. (1975) Experimental intracerebral movement of electron microscopic tracers of various molecular sizes. *J. Neurosurg.* **42,** 315–329.

Milhorat T. H., Davis D. A., and Lloyd B. J., Jr. (1973) Two morphologically distinct blood–brain barriers preventing entry of cytochrome C into cerebrospinal fluid. *Science* **180,** 76–78.

Olsson Y., Svensjo E., Arfors K.-E., and Hultstrom D. (1975) Fluorescein labelled dextrans as tracers for vascular permeability studies in the nervous system. *Acta Neuropathol.* (Berl.) **33,** 45–50.

Reese T. S. and Karnovsky M. J. (1967) Fine structural localization of a blood–brain barrier to exogenous peroxidase. *J. Cell Biol.* **34,** 207–217.

Rosenbluth J. (1976) Intramembranous particle distribution at the node of Ranvier and adjacent axolemma in myelinated axons of the frog brain. *J. Neurocytol.* **5,** 731–745.

Rosenbluth J. (1985) Normal and Abnormal Axolemmal Structure in Freeze-Fractured Myelinated Fibers, in *The Pathology of the Myelinated Axon* (Adachi M., Hirano A., and Aronson S. M., eds.) Igaku-Shoin, New York.

Rosene D. R. and Mesulam M.-M. (1978) Fixation variables in horseradish peroxidase neurohistochemistry. I. The effects of fixation time and perfusion procedures upon enzyme activity. *J. Histochem. Cytochem.* **26,** 28–39.

Rowley D. A. (1963) Mast cell damage and vascular injury in the rat: An electron microscopic study of a reaction produced by Thorotrast. *Br. J. Exp. Pathol.* **44,** 284–290.

Rozdilsky B. and Olszewsky J. (1957) Permeability of cerebral blood vessels studied by radioactive iodinated bovine albumin. *Neurology* **7,** 270–279.

Sakumoto T., Nagai T., Kimura H., and Maeda T. (1980) Electron microscopic visualization of tetramethylbenzidine reaction product on horseradish peroxidase neurohistochemistry. *Cell. Mol. Biol.* **26,** 211–216.

Schnapp B., Peracchia C., and Mugnaini E. (1976). The paranodal axoglial junction in the central nervous system studied with thin sections and freeze-fracture. *Neuroscience* **1,** 181–190.

Schonitzer K. and Hollander H. (1981) Anterograde tracing of horseradish peroxidase (HRP) with the electron microscope using the tetramethylbenzidine reaction. *J. Neurosci. Meth.* **4,** 373–383.

Seligman A. M., Karnovsky M. J., Wasserkrug H. L., and Hanker J. S. (1968) Nondroplet ultrastructural demonstration of cytochrome oxidase activity with a polymerizing osmiophilic reagent, diaminobenzidine (DAB). *J. Cell Biol.* **38,** 1–14.

Shivers R. R., Edmonds C. L., and DelMaestro R. F. (1984) Microvascular permeability in induced astrocytomas and peritumor neuropil of rat brain. *Acta Neuropathol.* (Ber.) **64,** 192–202.

Simionescu N. and Palade G. E. (1971) Dextrans and glycogens are particulate tracers for studying capillary permeability. *J. Cell Biol.* **50,** 616–624.

Simionescu N., Simionescu M., and Palade G. E. (1972) Permeability of intestinal capillaries. *J. Cell Biol.* **53,** 365–392.

Simionescu M., Simionescu N., and Palade G. E. (1982) Biochemically Differentiated Microdomains of the Cell Surface of Capillary Endothelium, in *Endothelium* (Fishman A. P., ed.) New York Academy of Sciences, New York.

Steinwall O. and Klatzo I. (1965) Selective vulnerability of the blood–brain barrier in chemically induced lesions. *J. Neuropathol. Exp. Neurol.* **25,** 542–559.

Sturmer C., Bielenberg K., and Spatz B. (1981) Electron-microscopical identification of 3,3',5,5'-tetra-methylbenzidine reacted horseradish peroxidase after retrograde axoplasmic transport. *Neurosci. Lett.* **23,** 1–6.

Turner D. F. and Marfurt C. F. (1983) Electron microscopic demonstration of horseradish peroxidase-tetramethyl-benzidine reaction product as a method for identifying sensory nerve fibers in the rat tooth pulp. *Neurosci. Lett.* **41,** 213–217.

Vacca L. L., Rosario S. L., Zimmerman E. A., Tomashefsky P., Ng P.-Y., and Hsu K. C. (1975) Application of immunoperoxidase techniques to localize horseradish peroxidase-tracer in the central nervous system. *J. Histochem. Cytochem.* **23,** 208–215.

Vulpe M., Hawkins A., and Rozdilsky B. (1960) Permeability of cerebral blood vessels in experimental allergic encephalomyelitis studied by radioactive iodinated bovine albumin. *Neurology* **10,** 171–177.

Westergaard E. and Brightman M. W. (1973) Transport of proteins across normal cerebral arterioles. *J. Comp. Neurol.* **152,** 17–44.

# Physical and Biochemical Methods for Analysis of Fluid Compartments

## K. G. Go

## 1. Introduction

The fluid environment of the central nervous system comprises the following compartments:

Tissue fluid, consisting of (1) the extracellular compartment and (2) the intracellular compartment
Cerebrospinal fluid
Vascular compartment

The extracellular fluid, cerebrospinal fluid (CSF), and plasma from the vascular compartment are characterized by a high content of sodium as the prevailing cation, whereas the intracellular compartment has a high potassium content. At the interfaces between the compartments, exchange processes take place, with those between the vascular compartment on one hand and brain tissue or cerebrospinal fluid on the other being governed by the blood–brain and blood–CSF barriers, respectively; and those between the extra- and intracellular compartments being governed by the Donnan equilibrium and the various pumps and transporting enzymes at the cell membrane.

Changes of the brain tissue electrolytes, such as the elevations of potassium concentration in the extracellular fluid following massive neuronal activity or neuronal damage, may be dissipated by exchange with the large volume of cerebrospinal fluid, while exchanges within the tissue between extra- and intracellular (glial) compartments also take place.

The normal water content of brain tissue differs among various brain structures, with the largest difference being between gray and white matter. There are also consistent minor differences among the various gray matter and white matter structures; they provide the basis for the detailed structure visible on proton nuclear magnetic resonance images.

In pathological conditions, the fluid content of the brain tends to increase. The clinical significance of brain edema as an increase of brain fluid content lies in the inherent swelling of the brain tissue, which, in view of its rigid bony envelopes, sooner or later leads to elevation of intracranial pressure, compromising tissue blood supply, and eventually resulting in lethal compression of vital brain stem structures. Depending on the underlying pathogenetic mechanism, the fluid accumulation may be largely located in a particular compartment. In vasogenic edema, in which breakdown of the blood–brain barrier results in extravasation of plasma constituents, the fluid collects and spreads in the extracellular space of the brain tissue. In cytotoxic types of brain edema, in which the function of the cellular membrane pumps is disturbed, it is the intracellular compartment that expands. In the osmotic type of brain edema caused by hyposmolarity of plasma relative to brain, fluid is osmotically shifted to the brain; the elevation of brain water content, however, is short-lasting, since the fluid drains into the cerebrospinal fluid spaces, significantly increasing cerebrospinal fluid formation rate (Go, 1981).

Brain water content may be assessed by a number of methods, some of which are destructive, such as the gravimetrical and tissue flotation method. These are therefore only applicable to experimental animals or tissue biopsies from human patients. Others, such as computerized tomography and proton nuclear magnetic resonance (NMR) imaging, are noninvasive and therefore allow for monitoring of changes of brain water in the clinical setting as well as in experimental animals.

Knowledge of the volume of the extracellular space is required for studies of the distribution of substances in brain tissue: This may pertain to the tissue ions and other physiological substances and also to drugs with pharmacological action. The measurement of intracellular pH using 5,5-dimethyloxazolidine also requires the size of the extracellular space to be known for the calculations (Roos, 1965). Before it was possible to isolate vasogenic edema fluid for direct analysis, its composition had to be inferred from data on whole brain tissue and on the size of the cerebral extracellular space.

The changes of extracellular space, such as expansion in vasogenic edema or reduction of its size in hypoxia, ischemia, and other varieties of cytotoxic edema, may be monitored by the measurement of electrical impedance of the tissue or by the application of indicators with an extracellular distribution. The latter technique may be performed in vitro using brain slices or in vivo by the administration of the marker into the blood or into the

CSF by ventriculocisternal perfusion. Ventriculocisternal perfusion has also allowed for the study of the dynamics of CSF circulation and of exchange processes at the brain CSF interface. Exchange processes between the extra- and intracellular tissue compartments may be studied in vitro using brain slices, and in vivo they may be inferred from changes in the composition of isolated vasogenic edema fluid. The intracellular volume is usually derived from extracellular space volume, as its complement in making up total tissue volume. Alternatively, it has been directly measured by flow cytometry of dissociated cells in studies on the cellular response to osmotic stress.

## 2. Fresh and Dry Tissue Weight Differences

The classical method of determining tissue water content by the difference between fresh and dry weight may be considered the standard by which the other methods of water content determination have been evaluated. It is based on the assumption that the weight difference between fresh and dry tissue constitutes only the tissue water that has evaporated during the process of drying.

For the determination of fresh tissue weight, the nervous tissue has to be dissected out of its bony encapsulation. To minimize weight loss by evaporation, it is recommended that the dissection be carried out in a chamber containing a humid atmosphere. Also, the weighing procedure should be conducted as quickly as possible. A preferable alternative is weighing the samples within preweighed plastic tubes that can be occluded by a cap. Care should be taken not to include fragments of bone or blood clots in the samples to be weighed. When larger pieces of tissue are available from which the samples can be taken, these should be extracted from deeper portions, where evaporation has not yet occurred. Others prefer to carry out the dissection procedure in a Petri dish on ice to minimize evaporation.

Drying the tissue is usually accomplished by placing the samples in an oven. It is recommended not to use too high a temperature (160°C), or loss of material by decomposition may occur. Enlarging the evaporative surface by mincing the tissue promotes drying (Fig. 1). The loss of weight by drying proceeds in two phases: an initial fast phase in which the weight of an entire rat brain hemisphere is reduced to around 21%, followed by a second, almost stationary phase, in which a very slow yet progressive

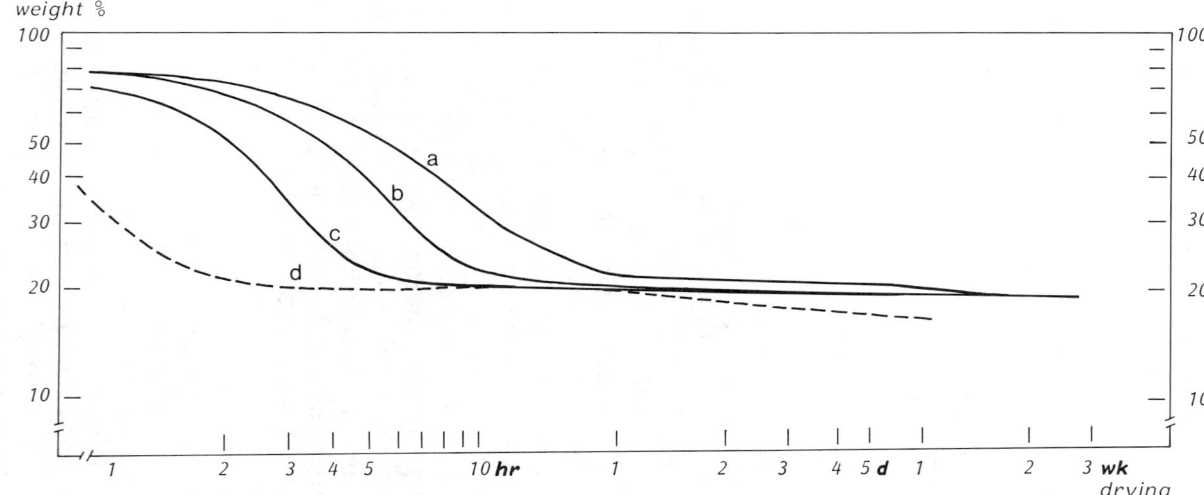


Fig. 1. The percentage loss of weight of rat cerebral hemispheres during drying (a) at 90°C of whole hemispheres, the curve plateaus at 21% and later at 19% (periods of slower weight loss), (b) at 90°C of minced hemispheres; enlargement of evaporative surface promotes drying, (c) at 120°C of whole hemispheres, drying proceeds faster down to a plateau at 19%, and (d) at 160°C of whole hemispheres; there is progressive weight loss to 17%.

weight loss occurs until a dry weight percentage of approximately 19% is attained.

In the literature, a drying time is reported, varying from overnight to 4 d. Adjuvant measures to promote the drying process comprise previous maceration of the tissue in acetone (Clasen et al., 1965) or drying in the presence of a dehydrating agent, such as phosphorus pentoxide (Kamman et al., 1984). Freeze-drying in a vacuum has also been performed (Shigeno et al., 1982). The weighing of the dried samples should be conducted immediately following the desiccation, since the dehydrated samples tend to regain weight after standing; this may amount to 0.8% in 3 d or 3% in 7 d.

The tissue water content is usually expressed as a percentage of fresh tissue weight:

$$\% \text{ Water content } (\%WC) = \frac{\text{fresh tissue weight} - \text{dry tissue weight}}{\text{fresh tissue weight}}$$

$$\% \text{ Dry weight } (\%DW) = 100\% - \%WC$$

The % WC indicates changes of tissue water content, since these may accompany various cerebral lesions. With other biochemical and morphological alterations, it is utilized to evaluate tissue damage and the efficacy of therapeutic measures in exerting an influence. The % WC does not indicate the absolute amount of water per unit mass of tissue, as may be required for correlations with other parameters. For that purpose it is preferable to express water content as:

$$\text{g water}/100 \text{ g dry tissue} = \% \text{ WC} \times 100 \text{ g}/\% \text{ DW}$$

This also enables the computation of changes of tissue volume in brain edema, since changes of volume of the intracranial contents may give rise to changes of intracranial pressure.

The amount of water, expressed per unit of dry tissue weight, also allows for correlation with tissue electrolyte contents, which similarly expressed are invariant to "dilution" effects as a result of the tissue swelling.

In other instances, in which comparison of tissues having different control values is required (such as between adult and immature individuals) for the evaluation of edema, the tissue swelling percentage (SP) calculated according to Elliott and Jasper (1949) may be utilized.

$$\text{SP} = \frac{(\% \text{ DW normal tissue} - \% \text{ DW edematous tissue})}{(\% \text{ DW edematous tissue} - \% \text{ DW edema fluid})}$$

For % DW of edema fluid it is recommended to take a value of 0 when the edema fluid is water (as in osmotic brain edema), 8 when the fluid is a plasma ultrafiltrate, and 11 when it consists of plasma (Pappius, 1974). For vasogenic edema fluid isolated from a freezing lesion in cats, we determined 6.57 ± 1.35% DW.

Because of the risk of evaporative water loss, especially in small samples with their comparatively large surface area, the determination of water content by fresh-dry weight difference is less suitable for small samples. Pieces of cerebral cortex weighing 10 mg lost 2.5%/min of their water during the first 2 min and 1.5%/min over subsequent 18 min, in an atmosphere with a humidity of 55% at 21°C (Nelson et al., 1971). For the cerebral hemispheres of Mongolian gerbils, weight loss caused by desiccation was less than 0.3% when the weighing procedure was performed within 30 s (Ito et al., 1979). The application of the gravimetric method to entire cerebral hemispheres of smaller animals allows for the assessment of the total amount of edema that is formed in the hemisphere as a consequence of the experimental procedure, e.g., the occlusion of a blood vessel. Focal increases of water content, however, may elude determination on whole hemispheres, since the changes may be "diluted" by the remainder of the brain having normal water content. Whenever an improved spatial resolution for detecting small foci of edema is required, the method based on specific gravity should be resorted to.

## 3. Determination of Specific Gravity

Following the method of Lowry and Hunter (1945) for the determination of protein concentration in plasma, the technique of tissue flotation for the determination of tissue specific gravity, as devised by Nelson et al. (1971), has proved to be a sensitive method that can be applied to small tissue samples. The sample is placed into a liquid gradient column consisting of a mixture of the liquids bromobenzene and kerosene, both of which are immiscible with water and therefore do not interfere with the water content of the sample. In the column the lighter kerosene is largely situated on top, whereas its concentration decreases and the heavier bromobenzene increasingly prevails downward. The specific gravity of the sample is determined by reading the equilibration depth to which it has sunk after 2 min. The relationship between depth and specific gravity (sp. gr.) has previously been calibrated by drops of standard salt solutions of various specific gravity introduced into the gradient column. As used for mouse brain, the gradient column was prepared from two stock solutions, the denser of which (sp. gr. 1.0650) consisted of 48.0 mL of bromobenzene (sp. gr. 1.49716) and 77.0 mL of kerosene (sp. gr. 0.78734), and the lighter (sp. gr. 1.0350) of 42.5 mL of bromobenzene and 82.5 mL of kerosene. The denser stock solution is placed in a 250 mL graduated cylinder and the lighter layered on top of it. The solutions are mixed by making short strokes with the coiled end of a wire, starting at the junction of the two solutions and subsequently proceeding above and below the previous limits of mixing until the entire column is continuously layered. The procedure has been applied to cat brain by Tornheim et al. (1976) for the study of contusion and to gerbil brain by Fujimoto et al. (1976) for the assessment of ischemic edema. A gradient column with a larger range was used by Ferszt et al. (1980) for the study of the dynamics of tissue flotation itself. Their denser stock solution (sp. gr. 1.070) was a mixture of 282.593 g of bromobenzene and 240.779 g of kerosene Roth-fraction; the lighter stock solution (sp. gr. 1.000) was a mixture of 174.829 g of bromobenzene and 234.445 g of kerosene. A gradient column with a modified liquid composition was that of Torack et al. (1976b) for formalin-fixed human brain samples. Their stock solutions consisted of bromobenzene (sp. gr. 1.5219) and *n*-decane (sp. gr. 0.7300) in the volume ratios of 66:87 for the denser and 35:65 for the lighter stock solution; mixing was performed by a spiral of wire through the entire length of the column. Generally the following equations apply for the calculation of the volume ratios of bromobenzene (BB) and kerosene (K) to prepare stock solutions with a desired weight:

$$\text{vol } \% \text{ BB} = \frac{\text{desired sp. gr.} - \text{sp. gr. of K}}{\text{sp. gr. of BB} - \text{sp. gr. of K}} \times 100$$

$$\text{vol } \% \text{ K} = 100 - \text{vol } \% \text{ BB}$$

For cat brain and human brain biopsies, Marmarou et al. (1978a) and Takagi et al. (1981) introduced an automated technique to prepare the gradient column. Their denser stock solution (sp. gr. 1.0650) was made up from 391.2 mL of bromobenzene and 608 mL of kerosene and their lighter stock solution (sp. gr. 0.9750) from 264.4 mL of BB and 735.6 mL of K. An empty 100-mL graduated cylinder was filled through two tubes of equal length (60 cm) from a 100-mL flask that contained the denser stock solution and was placed 40 cm above the graduated cylinder. Into this flask the lighter stock mixture emptied under constant stirring through a tube of equal length and diameter from a top flask, which was placed 43 cm higher. By simultaneous release of the clamps, which occluded all three outflow tubes, a linear gradient column built up in the graduated cylinder, with the bottom layer having the specific gravity of the denser stock solution (1.065) and the uppermost layer a specific gravity being the mean (1.020) of those of both solutions. It is essential, and therefore should be checked before building the gradient column, that the outflow from the bottom flask into the cylinder amounts to twice that from the top into the bottom flask.

A similar automated arrangement was described by Shigeno et al. (1982), with the difference that both containers of stock solutions were placed on the same level above the container in which the gradient column was prepared. As the lighter stock solution was forced out of its container by a cone-shaped floating weight (ensuring a linear increase of outflowing volume) into the container with the heavier stock solution through a single tube, the latter container emptied into the bottom container through double tubing mounted on a platform that floated on the surface of the gradient column.

The standard salt solutions for calibration of the gradients comprise ascending concentrations of $K_2SO_4$ or $Na_2SO_4$ dissolved in distilled water. The specific gravities of these standard solutions may be assessed using a pyknometer.

To assess brain water content and its changes, the specific gravity values of the tissue *per se* can be taken. It is also possible to calculate tissue water content from tissue specific gravity values by means of the fundamental relationship between tissue specific gravity, tissue water content, and the specific gravity of tissue solids, derived by Nelson et al. (1971):

$$gH_2O/g_t = 1 - [(sg_t - 1)/(1 - 1/sg_s)\, sg_t]$$

in which $sg_t$ is the tissue specific gravity in the individual case to be measured and $sg_s$ is the (mean) specific gravity of tissue solids in general, as it is computed from:

$$sg_s = 1/[1 - (sg_t - 1)g_t/(g_s \cdot sg_t)]$$

This requires the determination of tissue weight $g_t$ and tissue solids $g_s$ in a number of samples. The $sg_s$ value was found to be 1.30–1.32 for mouse cerebral cortex, 1.28–1.33 for cerebellum, 1.22–1.24 for pons.

The relationship between tissue water and tissue specific weight as expressed in the equation of Nelson et al. (1971) can be plotted as a straight line (Marmarou et al., 1982). At a water content of zero, the tissue specific gravity equals that of tissue solids, and with increasing water content, tissue specific gravity decreases linearly (Fig. 2). The theoretical relationship only holds when the tissue components involved are tissue solids and tissue water, as in osmotic brain edema following water intoxication, as studied by Nelson et al. (1971). The relationship will not be valid when the edema fluid contains protein or when there are additional constituents, such as blood occurring in hemorrhages or in dilated blood vessels, in which instance specific weight tends to increase. Indeed, cat brain, in which the vessels had previously been per-

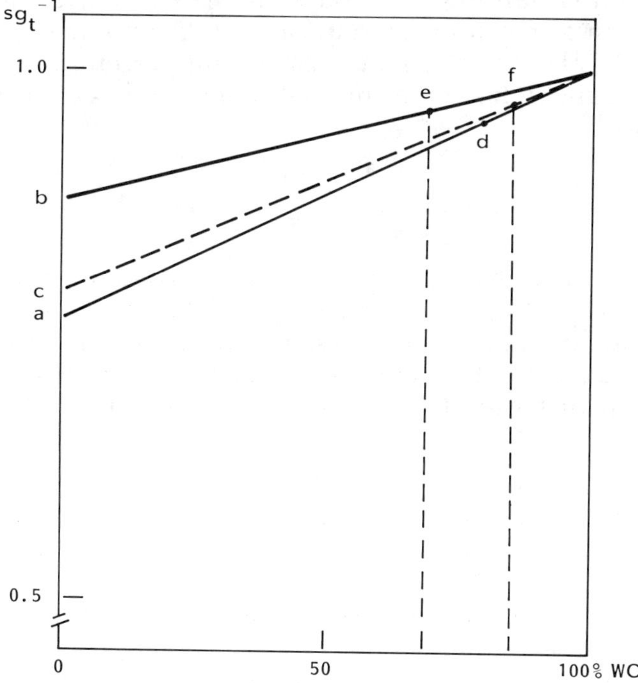

Fig. 2. Relation between tissue water content and tissue specific gravity (a) $1/sg_s$ of mouse cerebral cortex (0.7634); (b) $1/sg_s$ of cat cerebral white matter (0.8276); (c) $1/sg_s$ of human cerebral cortex (0.7838); (d) $1/sg_t$ of normal mouse cerebral cortex (0.9534); (e) $1/sg_t$ of normal cat white matter (0.9601) corresponding to water content of 68.7%; (f) $1/sg_t$ of normal human cerebral cortex (0.9656) corresponding to water content of 84.7%.

fused with saline showed lower specific gravity values than unperfused cat brain with blood in its vessels (Shigeno et al., 1982).

When the edema fluid contains protein, the specific gravity of the tissue solids changes, and a correction has to be applied to the relation, which then develops into:

$$gH_2O/g_t = \frac{(H_{nb} - H_{sr})(sg_{nb} - sg_{sr}) + sg_t(H_{sr}sg_{sr} - H_{nb}sg_{nb})}{(sg_{sr} - sg_{nb})sg_t}$$

in which $H_{nb}$ is the normal tissue water content (in g water/g tissue), $H_{sr}$ is the normal serum water content (0.9379 g water/g serum), $sg_{nb}$ is the normal specific gravity of fresh brain tissue (1.0416 for cat white matter), and $sg_{sr}$ is the normal specific gravity of cat serum (1.0248) (Marmarou et al., 1982). (For edema fluid derived from a freezing lesion in cats, we determined a specific weight of $1.0213 \pm 0.0030$ and a dry weight percentage of $6.57 \pm 1.35\%$.) By substitution of the normal values for the cat the relation reduced to:

$$\% \text{ g water}/g_t = \frac{1597.3}{sg_t} - 1464.9$$

This amounts to a steeper slope of the water content versus tissue specific gravity plot (Marmarou et al., 1982).

Swelling of brain can be caused both by an increase of water content (edema) and by an increase of blood volume (hyperemia). Whereas an increase of water content results in a reduction of

Table 1

Specific Gravity of Normal Brain Tissue in Various Species

| | Human | Cat | Mouse | Mongolian gerbil | Rat |
|---|---|---|---|---|---|
| Cerebral cortex | 1.0356[a] | 1.0408[b] | 1.0489[c] | 1.0482[d] | 1.0463[e] |
| White matter | 1.038[3a] | 1.0416[b] | | | |

[a]Takagi et al., 1981.
[b]Marmarou et al., 1978a.
[c]Nelson et al., 1971.
[d]Abe et al., 1980.
[e]Shigeno et al., 1982.

tissue specific gravity, increase of blood volume results in elevation of tissue specific gravity, as demonstrated by Ferszt et al. (1978) for the swelling of the brain following ultraviolet irradiation and by Tornheim and Mc Laurin (1984) following mechanical trauma.

Determination of tissue specific gravity by the flotation method is a sensitive technique that can be applied to samples as small as 5 mg. The risk of losing water by evaporation is limited to the dissection phase, although this may be further restricted by immersing sections of the brain in a Petri dish containing kerosene and proceeding with the dissection under kerosene (Tornheim et al., 1976).

Generally, a readout of the equilibration depth in the gradient column is made after 2 min, when the descent of the tissue sample has become nearly stationary (Nelson et al. 1971). An investigation into the dynamics of flotation by Ferszt et al. (1980) has confirmed the existence of an initial rapid phase and a second slower phase of descent, with the transition between the phases marking the point of equilibrium. Whereas large blocks of tissue (100–500 mg) showed a clear difference between a fast initial phase and an almost negligible second phase, samples of around 30 mg settled after 30 s to enter into a slow second phase; samples of around 30 mg settled after 30 s to enter into a slow second phase; very small samples of 1–5 mg showed a much retarded initial phase and a a faster second phase, which was ascribed to increased friction caused by their larger surface area. In practice, this amounted to a readout of higher specific gravity values for the smaller samples. Also, previous immersion in kerosene for 30 min seemed to reduce the specific gravity of the samples, which was ascribed to dissolving of the lipids by the kerosene.

Decrease of tissue specific gravity in brain pathology and brain edema was recognized by earlier authors, as reviewed by Schroder (1973). However, they generally failed to establish the fundamental relationship between tissue specific gravity and water content.

Fixation with 20% neutral buffered formalin was reported not to affect specific gravity of human brain tissue. Normal fresh cortical gray matter had a mean specific gravity of 1.0506 ± 0.0014 versus 1.0485 ± 0.0032 for normal fixed cortical gray matter; normal fresh white matter had a mean specific gravity of 1.0493 ± 0.0006, compared to 1.0487 ± 0.0038 for normal fixed white matter (Torack et al., 1976b).

## 4. Computer Tomography Images

Computer tomography (CT) is a noninvasive diagnostic technique that has found wide clinical application for the diagnosis of a multitude of lesions. Hypodense areas that often surround focal brain lesions can usually be ascribed to brain edema, and the increase of water content, as well as the spatial extension of the edema, can be assessed from the changes on CT images.

In CT imaging, sections of the organ are depicted as a matrix of picture elements (pixels), each pixel corresponding to a volume element (voxel) of tissue in the actual section of the organ. To this end, the organ is scanned by an array of parallel beams (or fanning beams in new generation of CT scanners) of X-rays in the plane of the section. The scanning is performed from various directions (i.e., with various angles, $\phi$), until a semicircle has been covered. During their passage through the tissue there is attenuation of the intensity of the X-rays, caused by deflection (Compton scattering) and capture (photoelectric absorption) of the X-ray photons by electrons in the tissue. As a result, the ray intensity $(I)$ at the end of the beam, as it is recorded by radiation detectors, is lower than the intensity of the incident ray $(I_0)$ by $I = I_0 e^{-p}$; with $p = -\ln(I/I_0)$. The symbol $p$ is called the ray projection or ray-sum; it represents the total attenuation after passage through a length of tissue $s$, and may be regarded as being composed of the local attenuation coefficients $\mu_{(x,y)}$ of volume elements of tissue along the ray path (at angle $\phi$) and at distance $r$ from the origin (Fig. 3).

$$p_{(r,\phi)} = \int_{r,\phi} \mu_{(x,y)} ds$$

Reconstruction of the image from the projections, obtained at

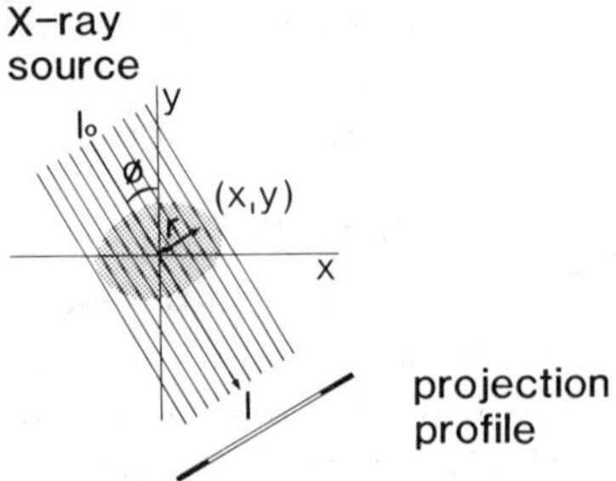


Fig. 3. Schematic presentation of projection profile with respect to object and X-ray path.

many angles, actually comprises the inversion of this equation, yielding the values of $\mu$ for the picture elements from the measured values of $p$ (Brooks and DiChiro, 1975).

To achieve this goal, various solutions have been devised. In one of the early procedures, which may be called simple back projection (Fig. 4), the projection profiles (i.e., the set of projections obtained by beams of the same angle) from various scanning directions are back-projected, i.e., projected in the opposite direction in the plane of imaging, which causes the pattern of attenuation to arise in the matrix of pixels at the center of the plane. A serious shortcoming of the simple back-projection algorithm is blurring of the reconstructed image, which is related to the finite width of the scanning beam and the finite number of scanning directions (Cho, 1974).

Another reconstruction algorithm, as applied by Hounsfield, involves improvement of the reconstructed image in a sequence of iterations. At each iteration, projections are made of the reconstructed image and compared with profiles actually measured; the differences are used in back projection to correct the reconstructed image. The technique may provide accurate reconstructions, although at the expense of much computing; therefore it is not widely applied currently.

At the present time, the algorithm of filtered back projection is usually employed. It is a Fourier-based technique, such as the two-dimensional Fourier reconstruction method. In the Fourier-based techniques the attenuation pattern is transformed from the spatial domain (in which attenuation is a function of spatial coordinates) to the frequency or Fourier domain, in which the attenuation function is expressed as a superposition of sinusoidal spatial

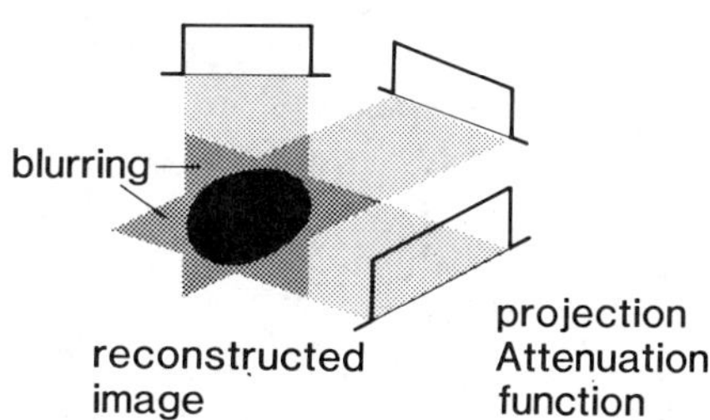


Fig. 4. Schematic drawing of the simple back-projection procedure (in which, for simplification, only three scan directions are presented). Shown is the overlap of neighboring profiles, which results in overall blurring of the central reconstructed image.

waves propagating in various directions across the Fourier plane with various (spatial) frequencies and amplitudes denoted by Fourier coefficients.

In the two-dimensional Fourier reconstruction technique the projections are first (one-dimensionally) Fourier transformed; then the subsequent synthesis in the two-dimensional Fourier plane is performed by integration of these transforms of the projections over the semicircle. According to the projection theorem of two-dimensional Fourier reconstruction, the Fourier coefficients obtained from a projection at an angle $\phi$ can be assigned to the spatial waves propagating at the same angle in the Fourier plane. Since these Fourier coefficients in the Fourier plane conform to the polar coordinate system of the projections, from which they originate, they have to be interpolated to obtain a rectangular array of Fourier coefficients to fit the rectangular matrix of pixels, in which the attenuation function, obtained from the inverse transform, is to be presented in the spatial domain (Fig. 5). Because of the two transformation and interpolation procedures involved and the fine sampling required, the technique tends to be time-consuming unless simplifying or economizing procedures, such as the fast Fourier transform algorithm, are applied (Brooks and DiChiro, 1976; Scudder, 1978; Cho, 1974).

The filtered back-projection algorithm involves filtering the projection profiles after their Fourier transform. Since the blurring of the image resulting from simple back projection can be interpreted as a loss of higher spatial frequency information, the filtering can be understood as an enhancement of the higher spatial frequencies. It gives an adequate image when the filtered profile is back-transformed to the spatial domain and then back-projected to form the image (Fig. 6) (Ter-Pogossian, 1977).

An alternative procedure, involving convolution of the projection profiles, remains in the spatial domain and does not require the transformation back and forth to the Fourier domain. It results in adequate reconstruction, since the convolution procedure is analogous to filtering in enhancing higher spatial frequencies (Zonneveld, 1983; Cho, 1974).

To express attenuation coefficients on CT scans, Hounsfield and Ambrose (1973) introduced a scale on which water as a reference was assigned the value 0 and air was assigned –500. Other substances with a lower attenuation coefficient than water were also denoted by a negative value, whereas substances with a higher $\mu$ than water were assigned positive attenuation numbers. The

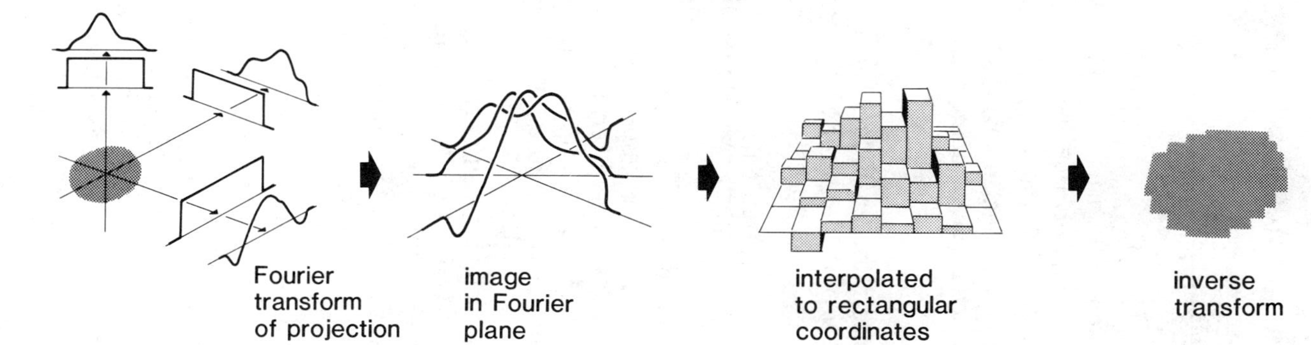



Fig. 5. Schematic presentation of the procedure of two-dimensional Fourier image reconstruction, showing the subsequent steps.

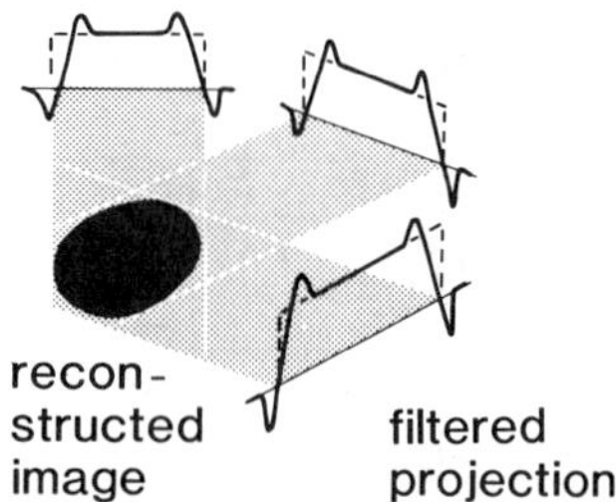



Fig. 6. Simplified drawing (with only three scan directions) of the filtered back-projection technique. The overlap of neighboring profiles is corrected by the filtering, which involves an attenuation of the edges of the profiles.

scale applies to the first generation EMI Mark I scanner, operated at 120 kV. Since water has a $\mu$ of 0.19/cm for this beam, the $\mu$ of any substance for the same beam can be transferred to the EMI scale by:

$$\text{EMI number} = \left( \frac{\mu \text{ substance}}{\mu \text{ water}} - 1 \right) \times 500$$

In dealing with substances with a high atomic number (Z), consequently exhibiting a prevalence of photoelectric absorption, it may be necessary to separate the attenuation coefficient into Compton ($\mu_c$) and photoelectric ($\mu_p$) components, and to compare them to those of water, being 0.173/cm and 0.0012/cm, respectively (Brooks et al., 1980).

On the EMI scale, cerebrospinal fluid has a density of about +8 EMI units; white matter, +15 EMI units; gray matter, +18 EMI-units; and bone, +200–+500 EMI units (Oldendorf, 1978). Later generation scanners express the attenuation numbers on a new Hounsfield scale, which has a range of –1000 Hounsfield units (HU) for air to +1000 HU for solid bone, with water being the reference at 0; thus the value for any substance on the new 1000 scale in Hounsfield units amounts to twice the value in EMI units.

On the new 1000 scale, the following Hounsfield numbers have been reported: CSF, +15 HU; white matter, +32 HU; gray matter, +39 HU; whole blood, +56 HU; erythrocytes, +98 HU; plasma, +33 HU. The higher attenuation number of gray matter of about 5 HU over white matter could be ascribed to the 8% lower content of carbon (Z = 6) and 8% higher content of oxygen (Z = 8) as bound constituents of gray matter, resulting in a higher photoelectric absorption (although equal Compton attenuation) of gray matter (Brooks et al., 1980). The true Hounsfield number for white matter was determined to be 32 ± 2 HU and was independent of X-ray energy (kilovoltage). The decrease in CT number of edematous white matter could be related to the water content; moreover, the CT number of edematous white matter appeared to be energy-dependent (Fullerton and Blanco, 1981).

By correlating CT readings of edematous brain areas with water content as assessed in tissue biopsies during operation, Lanksch et al. (1976) observed that a 10% increase of water content corresponded to a decrease of 3 EMI units. Using the newer generation EMI-CT1010 scanner, a 12.3% increase of water content correlated with a decrease of 9.5 HU on the 1000 scale (Lanksch et al., 1981). As to the correlation procedure, it may be remarked that, unless stereotaxic techniques are employed, the transfer of spatial coordinates from CT images to topography during operation tends to lack precision. It is also recommended to take CT readings from brain areas that are free from visible artifacts or CSF spaces, and preferably close to the center, to reduce bone-related errors of reconstruction. Sections with CSF spaces above or below should also be avoided (Zatz et al., 1982).

The investigation of whole brains obtained at autopsy allows for a better topographical correlation of CT findings with other regional tissue parameters such as water content, specific gravity, and histological aspects determined in samples from the same locations selected for the CT readings. A disadvantage would be the fixation procedure, which is required to solidify the brain and make it manageable and might cause fluid shifts. Apparently this did not result in significant changes in water content, specific gravity, or CT attenuation. Thus a CT number of 12.6 ± 2.9 EMI units was measured for perifocal edema in white matter of formalin-fixed brains, corresponding to a water content of 79.70 ± 4.77% and a tissue specific gravity of 1.0444 ± 0.005 versus a control attenuation number of 18.2 ± 1.5 EMI units, a control water content of 70.53 ± 1.4%, and a control specific gravity value of 1.0475 ± 0.0026 (Torack et al., 1976a).

The correlation between tissue water content and CT number is not straight-forward because of the role of other tissue constituents, such as lipids with their lower attenuation coefficients and blood with its higher attenuation resulting from its iron content. It is conceivable that in a lesion, in which the normal proportions of water, lipids, blood, and other tissue solids are disturbed, the attenuation may turn out differently than water content indicates. Out of 14 acute brain infarctions, 7 were not visualized on CT because of their isodense appearance, presumably because of the cumulative result of low-density edema and high-density blood. In subacute infarctions, on the other hand, a very low attenuation may be the result of a combination of increased water and an accumulation of free fatty acids (Torack et al., 1976a). Furthermore, a decrease of total tissue lipid content reported for perifocal edema may be expected to reduce the density-lowering effect of edema (Lanksch et al., 1976). Also, the protein content of the edema fluid may be significant in determining attenuation number. Edemas with higher attenuation values (16 EMI units) were histochemically characterized by a stronger periodic acid schiff staining (denoting the presence of globulins) in the histological sections compared to edemas with a CT number of 11 EMI units (Clasen et al., 1976). A plasma protein concentration of 2.5 g/100 mL in edematous tissue was estimated to contribute 5.9 HU to the attenuation number of edematous tissue (Clasen et al., 1981).

In dealing with CT numbers, it should be remembered that they do not represent physical units, but rather reflect the attenuation of a beam of X-rays in a given scanner system. The beam is composed of X-rays of various wave lengths, of which, during passage through substances of a high atomic number (such as bone or contrast media), the softer rays (i.e., those possessing longer wave lengths) are particularly attenuated by photoelectric absorption. This results in the residual prevalence of the X-rays of shorter wave length, so-called beam hardening. Although the effect is corrected for during reconstruction to prevent inhomogeneities in the image, changes of attenuation numbers may still occur. For example, CSF in the spinal canal may show Hounsfield numbers up to 30 HU because of the presence of the surrounding mass of vertebral bone. Within the cranial cavity, shifts of almost 15 HU may occur between maximum and minimum skull thickness, as a result of beam hardening (Payne and Latchaw, 1978). The Hounsfield numbers also depend on factors related to the system, such as the applied reconstruction algorithm, detector efficiency, and signal-to-noise ratio. Therefore, it is recommended that caution be exercised in comparing the results of different scanners, even among scanners of the same manufacturers and model (Levi et al., 1982). Uncertainties in representation comprise variations (noise) between pixels of a given scan in representing the same material, variations from scan to scan, as well as artifacts (McCullough, 1977). According to an evaluation of 23 CT scanners, the noise was generally between 0.2 and 1% (Speller et al., 1981). Variations from scan to scan by a given machine may be considered part of quality assurance and may be determined by the state of maintenance. In the calibration of a scanner, for example, it should be taken into account that, because of its different density, water at room temperature and water at body temperature differ by 5–6 HU (Bydder and Kreel, 1979). Variations from scan to scan may also be caused by a different positioning within the scan aperture (Levi et al., 1982; Ter-Pogossian, 1977).

Other factors that may influence the readings are movement artifacts and other artifacts arising from erroneous reconstruction caused by changing parameters such as changes of detector sensitivity. Incorrect attenuation numbers may also be the result of the so-called partial volume effect, which signifies that a density reading in a pixel may conceivably represent the average of divergent attenuation coefficients when different tissues occur together in the volume element depicted by the pixel (Zonneveld and Vijverberg, 1984).

In conclusion, CT imaging may be considered a valuable noninvasive means of assessing brain edema, especially in the clinical setting, with proper regard of the factors, both from the tissue and the scanning system, that may influence attenuation numbers.

## 5. Proton Nuclear Magnetic Resonance

The advent of proton NMR imaging has provided another noninvasive technique for both the diagnosis of various brain lesions and the assessment of brain water content, since the tissue NMR parameters are determined by the volume and the state of tissue water.

Placed in a steady magnetic field $B_0$, protons are considered to perform a precession movement, with a precession frequency or so-called Larmor frequency $v = \gamma B_o$ in which $\gamma$ is the gyromagnetic constant, being 42.6 MHz/Tesla for protons. Subjecting the protons to an additional rotating magnetic field perpendicular to the direction of the steady field may give rise to the phenomenon of resonance when the rotation frequency of the field equals the Larmor frequency. For $B_0$ ranging from 0.1 to 2 Tesla, this resonance frequency therefore is in the radio frequency range of 4–85 MHz. In the quantum description of the phenomenon, protons with their spin of $I = 1/2$ may be assumed to adopt $2I + 1 = 2$ energy levels when they are placed in a magnetic field. During resonance, the radio frequency field induces transitions between the two energy levels; the energy difference being $\Delta E = h\nu$, in which $h$ is Planck's constant and $\nu$ is the Larmor frequency. The lower energy level corresponds to a parallel magnetic moment, i.e., with an orientation parallel to the field $B_0$, whereas the higher energy level represents the antiparallel orientation. It can be estimated that among a population (Np + Na) of proton spins, Np of which have the parallel orientation and Na the antiparallel moment, according to Np/Na = exp ($\Delta E/kT$), in which $k$ is Boltzmann's constant and $T$ the absolute temperature, only seven proton spins out of a sample Np + Na = 2 million proton spins have the parallel in excess of those having the antiparallel moment (which still amounts to $\approx 3 \times 10^{16}$ proton spins constituting the signal in a gram of water).

During resonance, this net magnetic moment carries out a precession movement around the perpendicular field, with its orientation increasingly tilting to 90° or even 180° with respect to the original alignment; the exciting radio frequency pulses inducing this effect are termed 90 and 180° pulses (Fig. 7), respectively.

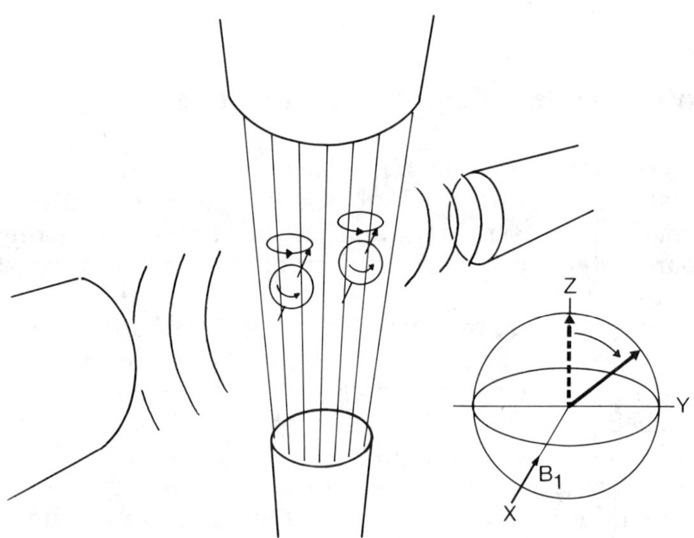


Fig. 7. The magnetization $M$ is usually considered in a coordinate system $(x, y, z)$ rotating with the Larmor frequency. The increasing tilt of $M$ at resonance can be understood as a rotation of $M$ around the direction of the radio frequency field $B_1$.

The precession of the magnetic moment around the steady field $B_0$ will induce an electrical potential in a coil, allowing for detection of the magnetization. At the end of the pulse the magnetization *(M)* of the spins returns to its original alignment; in this process of relaxation the amplitude of the induced electrical potential decays following an exponential function, and the signal is thus called free induction decay (FID). The relaxation process can be described in terms of a return of the longitudinal vector of $M$ (i.e., the vector $M_z$ along the direction of $B_0$), e.g. from an inverse orientation after a 180° pulse to the original orientation; and an independent decay of a transverse component of $M$ (i.e., in the $xy$-plane perpendicular to $B_0$), because of dephasing of the transverse spin vectors that all had been aligned in phase by the excitation pulse. The relaxation of the longitudinal component is characterized by a time constant $T_1$, also called the spin-lattice relaxation time since this component of relaxation is assumed to be caused by interaction with neighboring molecules (the lattice) $M_z(t) = M_z(0)\{1 - \exp(-t/T_1)\}$. The transverse relaxation time $T_2$ or spin–spin relaxation time characterizes the dephasing of the spins caused by spin–spin interactions $M_{x,y}(t) = M_{x,y}(0) \cdot \exp(-t/T_2)$. In these empirical equations, according to Bloch, *M(t)* denotes the magnetization at time $t$ and $M(0)$ at the beginning of relaxation (Darrow, 1953; Fullerton, 1982; Go et al., 1983).

Previous investigations have demonstrated that in biological tissues the proton spins display shorter relaxation times than in pure water, where both $T_1$ and $T_2$ are about 2.5 s. For rat cerebral cortex a $T_1$ of around 400 ms and a $T_2$ of about 80 ms was found in a NMR spectroscopic study in vitro at 14 MHz (Go and Edzes, 1975). The NMR behavior of water protons in tissue may be explained by a model in which the protons of water molecules in the vicinity of macromolecules tend to be restricted in their molecular motions and also interact with the macromolecules, to which the proton spins readily transfer their energy, quickening their relaxation to the equilibrium state (Go and Edzes, 1975; Raaphorst et al., 1975; Cottam et al., 1972). Tumors of the brain (Bakay et al., 1975; Parrish et al., 1974), as well as those of other organs (Damadian, 1971; Cottam et al., 1972), exhibited longer relaxation times than the corresponding normal tissues. In brain edema of various origins, a linear correlation was found between the relaxation times and tissue water content (Go and Edzes, 1975). The increase of relaxation times in brain tumors as well as other organ tumors appeared to be related to their higher water content (Bovee et al., 1974) and possibly to some other factors as well (Raaphorst and Kruuv, 1981).

In edematous tissue, the larger amounts of water constitute a liquid environment approaching that of pure water, in which the proton spins exhibit longer relaxation times. Since most focal lesions of the brain are associated with brain edema, and various cerebral structures have different water contents, these differences in water content may be reflected as differences in signal intensity in an NMR image. By modulating $T_1$ and $T_2$ effects into the signal, moreover, enhancement of the contrasts may be achieved. Thus in NMR imaging the signal must encode for both relaxation parameters of the tissue to provide the differences in signal intensity as well as for spatial coordinates to allow reconstruction into the image.

Spatial encoding is based upon the relationship $v = \gamma\, B_0$ between field strength and resonance frequency. By the application of field gradients in which field strength at a given location along the gradient is determined by the spatial coordinates, the resonance frequency corresponding to that particular field strength thus encodes for the spatial coordinates. This principle of spatial encoding was introduced by Lauterbur (1973) as an entirely novel method of image formation in which interaction of the object with a radiation field is restricted by a second field. It is therefore not prone to the shortcoming that the spatial resolution is limited by wave length, as is the case in conventional ways of imaging, in which the object interacts with one matter or radiation field, characterized by a wave length that should be smaller than the smallest feature to be discerned.

A number of techniques have been developed to reconstruct the image, comprising sequential point, sequential line, sequential plane, and simultaneous techniques. In the sequential point techniques, individual volume elements of the organ are sequentially observed and an image is built point by point. They tend to be time-consuming and comprise such methods as the Fonar technique developed by Damadian, in which a small volume element is selected by the modification of the steady as well as the radio frequency pulses. The sensitivity point technique according to Hinshaw (1976) uses three orthogonal independently oscillating field gradients that leave a small volume element where the field does not fluctuate with time. The sequential line techniques collect data simultaneously from one-dimensional sets of volume elements by a Fourier spectroscopy procedure. For example, the sensitive line is again selected by time-dependent magnetic field gradients in a way that the signal from all but the sensitive line is time-dependent; averaging of the response isolates the signal from only the voxels in the line (Holland et al., 1980). The sequential plane methods acquire data simultaneously from an entire plane of volume elements and are currently widely employed. A projection reconstruction technique, similar to the CT filtered back-projection algorithm, was proposed by Lauterbur (1973) for NMR imaging. An alternative approach is the Fourier zeugmatography procedure (Kumar et al., 1975) in which the plane is scanned by a stepwise phase-encoding gradient along the $Y$ axis and a fixed measuring gradient along the $X$ axis. Since the collected NMR data are already in the frequency domain, this two-dimensional Fourier technique is therefore a more direct reconstruction procedure than the filtered back-projection algorithm. Finally, simultaneous techniques collect data simultaneously from selectable planes or from the entire three-dimensional array of voxels. In the multiplanar imaging procedure (Mansfield, 1976), information from two or three dimensions is interlaced in a single frequency domain. In another currently applied volume acquisition procedure, the entire volume is excited and the signal is observed in the presence of a gradient; the Fourier transform of the signal constitutes the projection of the entire volume onto the axis of the gradient, which contains information on all three spatial dimensions; the entire volume is covered by switching the gradient stepwise around a hemisphere.

The modulation of $T_1$ and $T_2$ effects into the signal is effected by the application of pulse sequences. Currently widely employed for imaging are the spin echo and inversion recovery pulse sequences.

In the spin echo pulse sequence, the magnetization vector $M_z$ is rotated to the transverse $(X\text{-}Y)$ plane by a 90° pulse, where it is detectable by the coils, which are also in the transverse plane. Here the magnetization is subject to transverse relaxation as characterized by $T_2$, but also to additional dephasing caused by inhomogeneities of the magnetic field. Therefore, instead of using the FID as the imaging signal, the spin echo technique is applied, in which the dephasing caused by field inhomogeneities is cancelled by a so-called echo-forming 180° pulse about the X axis. This pulse, which is applied at time $T_E$ after the 90° pulse, reverses the order of slower and faster rotating spins, thus resulting in a rephasing and in the echo as a signal (Fig. 8). Other nomenclatures prefer to define $T_E$ as the time between the 90° pulse and the echo.

To encode $T_1$ influences, which govern the decay of longitudinal magnetization $M_z$ (Fig. 9), the inversion recovery pulse sequence is utilized. It starts with a 180° pulse that inverts $M_z$ from

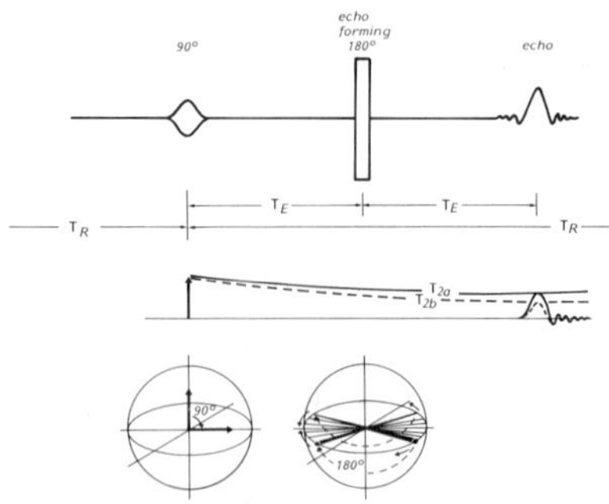

Fig. 8. The spin echo pulse sequence.

the $+Z$ to the $-Z$ axis. After a delay time $T_D$ (also called inversion time $T_I$, usually of the order of 200–500 ms) a 90° pulse is given to rotate $M_z$ to the transverse plane, followed by the echo-forming 180° pulse after time $T_E$ (of the order of 25–50 ms), as in the spin echo pulse sequence. During time $T_D$ relaxation of magnetization takes place; for rapidly relaxing spins (i.e., with short $T_1$), $M_z$ will have recovered to the $+Z$ axis, whereas for spins with longer $T_1$, $M_z$ may have remained on the $-Z$ axis. The 90° pulse rotates the vector in a clockwise direction, bringing a positive $M_z$ to the $+Y$ axis and a negative $M_z$ to the $-Y$ axis. In the transverse plane the $M$ vector then decays by dephasing. The longer $T_2$ of a given structure is, the less the signal (whether positive or negative) has decayed at the time $(2T_E)$ of the echo. On the image positive signals will be displayed as light and negative signals as the darkest structures.

From one gradient switch to the next, or also during the same gradient step or projection, when more data are required to improve signal-to-noise ratio, the pulse sequence is repeated, as characterized by a repetition time $T_R$ (of the order of 1 s). The repetition time allows the magnetization to recover in most structures before the next pulse sequence begins, except in those structures, such as CSF, with a very long $T_1$, in which the magnetization vector will not have recovered to its full length. As a result of this so-called saturation effect, the signal originating from these structures is very low and displayed as dark, both in the spin echo and inversion recovery images.

The signal intensity $I$ as a function of proton spin density $\rho$, and relaxation times $T_1$ and $T_2$, may be expressed in spin echo images as

$$I = \rho[1 - 2e^{-(T_R-T_E)/T_1} + e^{-T_R/T_1}] \cdot e^{-2T_E/T_2}$$

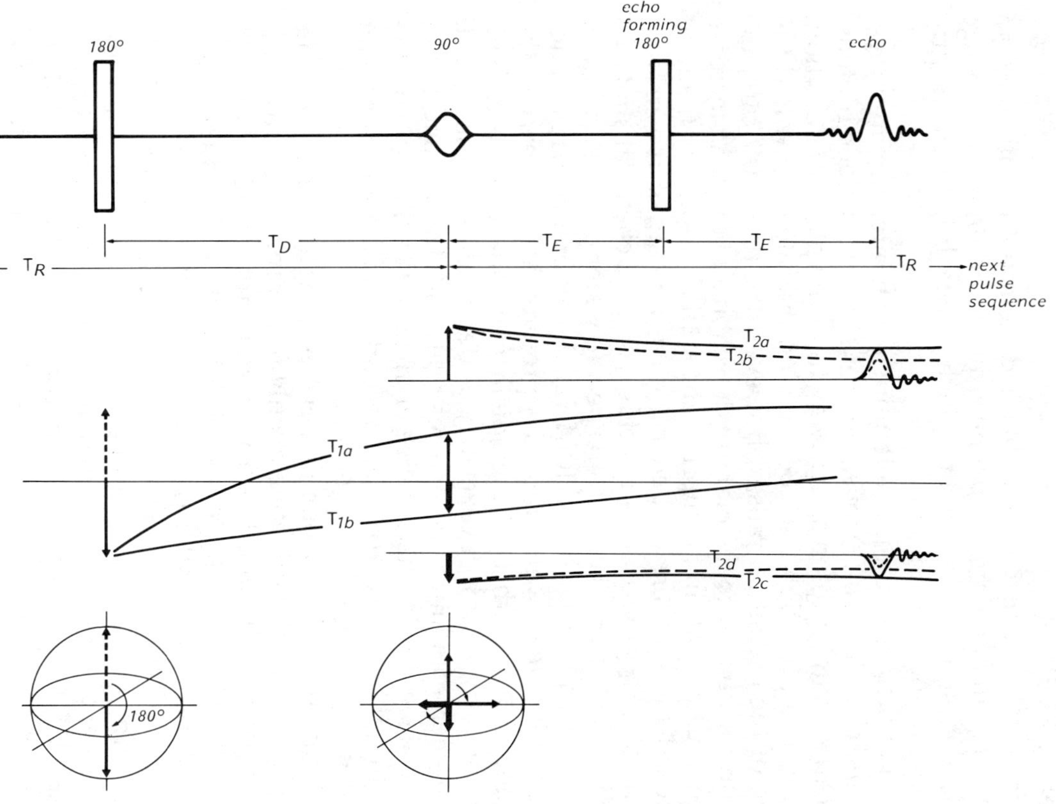


Fig. 9. The inversion recovery pulse sequence.

whereas on inversion recovery images

$$I = \rho[1 - e^{-T_D/T_1}\{2 - 2e^{-(T_R-T_E)/T_1}\} + e^{-T_R/T_1}] \cdot e^{-2T_E/T_2}$$

(Go et al. 1983).

From reconstructed data obtained from both sequences, an average $T_1$ can be computed per pixel and represented as an image that is denoted a $T_1$ map, on which image brightness is proportional to the $T_1$ value.

Other factors that may determine signal intensity are movement and chemical shift. Movement, for example of blood protons, may carry excited proton spins away from the plane of detection, thus reducing the signal. Depending on the rate of flow and the timing of the excitation pulses, however, the flow may carry excited proton spins into the plane of detection and cause an increase of signal. Chemical shift comprises the shift of resonance frequency caused by a different chemical environment of the proton. Protons of $CH_2$ groups in fatty acids display a chemical shift of 3 ppm (i.e., 3 Hz per MHz) with respect to water protons. At lower resonance frequencies (below 10 MHz), the difference in resonance frequency between water and $CH_2$ protons does not exceed the frequency bandwidth of the gradient over one pixel, and both water and fatty acids in a voxel may be depicted in the same pixel. With higher resonance frequencies ($\sim 100$ MHz), chemical shift may cause significant disturbance of images. On the basis of chemical shift, it has become possible recently to make separate water and fatty acid images.

Although cerebral white matter has a high lipid content, its NMR signal does not originate mainly from protons of fatty acids (contrary to fatty tissue), but from water protons, since replacement of the water by deuterium significantly reduces the signal from cerebral white matter (Kamman et al., 1984).

As to the permissible delay time after removal from the body for in vitro spectroscopic studies of nervous tissue biopsies, measurement of $T_1$ and $T_2$ at various intervals after removal and storage at 4°C demonstrated no consistent changes of these parameters for rat tissues after up to 24 h (Kamman et al., 1985). Fixation of brain tissue, by both immersion and perfusion, shortened $T_1$ and prolonged $T_2$. $T_1$ was considerably reduced, to 132 ms for rat cerebral cortex and to 195 ms for cerebral white matter after 5 d of fixation (as compared to values of 500 ms and 320 ms, respectively, for the native tissues) (Kamman et al., 1985).

There is also a significant dependence of the relaxation times upon the resonance frequency utilized. Rabbit cerebral gray matter displayed a $T_1$ of 332 ms at 2.5 MHz and 644 ms at 24 MHz (Ling et al., 1980). Rat cerebral white matter showed a $T_1$ of 1078 ms at 100 MHz (Naruse et al., 1982), whereas at 10.6 MHz a value of 364 ms was measured for $T_1$ (Kamman et al., 1985). The reported $T_2$ values for rat cerebral white matter were 75.6 ms at 100 MHz and 72 ms at 10.6 MHz.

Correlation of brain tissue relaxation times with tissue water content (Go and Edzes, 1975; Bakay et al., 1975) has been repeatedly observed (Naruse et al., 1982; Go et al., 1984; Bartkowski et al., 1984), although it appears that $T_1$ referred to the volume of water in the tissue, whereas $T_2$ reflected the distribution and composition of the edema fluid (Naruse et al., 1982).

As yet the relationship between relaxation times and brain water content has only been empirically established. A theoretical relationship was proposed by Fullerton et al., (1982) on the basis of a model developed by Berendsen (1975). According to their fast proton exchange model, the water in protein solutions, cell suspensions, and biological tissues may be divided into three fractions. A fraction of free water ($f_w$) comprises the bulk of unbound water, a fraction of hydrated water ($f_h$) occurs loosely bound as a mantle around macromolecules, and a fraction of crystalline water ($f_c$) is specifically and tightly bound to macromolecular elements. The relaxation rates of the tissue ($R_1 = 1/T_1$, $R_2 = 1/T_2$) may be considered weighted averages of the relaxation rates of the component fractions. Thus the longitudinal relaxation rate $R_1$ of the system $R_1 = f_w\,R_{1w} + f_h\,R_{1h} + f_c\,R_{1c}$; but because the crystalline fraction has a large $T_1$ (i.e., a small longitudinal relaxation rate $R_{1c}$), the role of the crystalline fraction is negligible in determining $T_1$ of the whole system. $R_{1c} = 1/T_{1c} \sim 0$, $R_1 = 1/T_1 = f_w R_{1w} + f_h\,R_{1h}$; in which $R_{1w} = 1/T_{1w}$ with $T_{1w}$ being about 2.7 s, the relaxation time of free water, and $R_{1h} = 1/T_{1h}$; the relaxation time of hydration water $T_{1h} = 64$ ms, as derived from data obtained from dried gelatin, in which the remaining water is assumed to consist of hydration water. With respect to transverse relaxation of the system, however, the crystalline water fraction constitutes a significant contribution. The transverse relaxation rate of the system $R_2 = 1/T_2 = f_w\,R_{2w} + f_h\,R_{2h} + f_c\,R_{2c}$ with $T_2$ of the free water fraction $T_{2w} = 1/R_{2w} \sim 100$ ms, $T_2$ of the hydration fraction $T_2h = 1/R_{2h} = 10$ ms and $T_2$ of the crystalline fraction $T_{2c} = 1/R_{2c} = 10$ μs.

The hydration water fraction was estimated to be 0.084 of total water for gray matter and 0.125 for white matter, as measured on the basis of the linear dependence of $T_{1h}$ on resonance frequency, whereas $T_{1w}$ of free water showed no dependence on the resonance frequency (Fullerton et al., 1984).

## 6. Measurement of Total Amount of Brain Edema and Brain Volume Increments

In lesions that are confined to one cerebral hemisphere, as in ischemia caused by occlusion of vessels supplying the hemisphere, it may be relevant to measure the total amount of edema developing in the affected hemisphere for the evaluation of the role of physiological parameters or of the efficacy of therapeutic measures. The total amount of hemispheric edema can be estimated from the weight difference of the two hemispheres, provided that equal boundaries of the hemispheres are defined and that the volume increment of the brain is only caused by water accumulation in the brain tissue and not by other space occupying masses, such as hematomas, hemorrhages, or tumors.

Alternatively, the amount of edema may be stated in terms of local elevations of water content and extent of the edematous area in various sections of the brain. Elevation of tissue water content at several locations in sections of the brain can be measured by fresh-dry weight difference, tissue specific gravity determination, or the reading of attenuation values on CT images. The extent of the edematous area can be measured planimetrically, as it is visualized on computer tomograms as hypodense areas or on NMR tomograms as areas of altered signal intensity indicating a raised water content. The effect of dexamethasone therapy on peritumoral edema was assessed in the course of treatment by comparing the volume of edema, as obtained by the multiplication of the thickness of CT slices with the sum of the surfaces of edematous areas on CT scans (Hatam et al., 1982). Also, programs have been devised for the computation of the volume of a particular structure within the brain, such as the ventricular system, on the basis of CT scans (Walser and Ackerman, 1977; Reid, 1983).

In vasogenic edema with inherent breakdown of the blood–brain barrier, the extent of the edematous area can be marked by systemically administered tracers, which are barred from normal brain areas by the intact blood–brain barrier, but accompany the fluid exudate accumulating in the tissue as edema. The tracers that have been utilized comprise vital stains allowing for macroscopic visualization of the edematous areas, as well as markers containing a radioactive tag. From the area of sodium fluorescein extravasation on sections of cat brain, the influence of arterial blood pressure was assessed (Klatzo et al., 1967). The edema from a freezing lesion in adult and immature cat brain was compared by planimetric measurement of Evans blue extravasation on brain sections (Go et al., 1973). Rovit and Hagan (1968) estimated the effect of dexamethasone on hemispheric edema by assessing the exudation of radioiodinated serum albumin in serial sections of cat brain.

By its space-occupying effect within the cranial vault, brain edema may give rise to intracranial pressure elevation. In the clinical setting, recording of intracranial pressure has been practiced to monitor the development of edema. However, elevation of intracranial pressure is an indirect criterion for the amount of edema. The accumulation of fluid in brain edema may not be the only cause of pressure rise; space-occupying masses of another nature as well, such as blood in hematomas or cerebrospinal fluid in the case of hydrocephalus, may also be causes.

The rise of intracranial pressure as a result of brain volume increments is characterized by the pressure–volume relationship of the craniospinal system. A pressure–volume relationship may be distinguished that pertains to acute changes, such as those brought about by the injection of a bolus of fluid. The pressure elevation $dP$ caused by such an instantaneous volume increment $dV$ is described by $dV/dP = C$, in which $C$ is called the compliance. The compliance appears to depend on the pressure level by $C = 1/kP$ (Marmarou et al., 1978b); $k$ is the constant denoting the steepness of the relation, also called the elasticity slope $\beta$ by Sklar and Elashvili (1977).

More gradual volume increments, such as those effected by fluid infusions into the CSF space or by the inflation of intracranial balloons, allow for accomodation of the volume increment by displacement and absorption of cerebrospinal fluid. Initial volume increments hardly induce elevation of intracranial pressure, because of buffering by spaces containing displaceable amounts of CSF. When these spaces are large, as in immature or old individuals, the initial stage may be protracted. Following the initial flat part of the pressure curve, there is a steeper stretch at larger volume additions, and eventually there may be a third, again flatter, stretch, when vasomotor auto-regulation begins to fail, followed by an ultimate steep end at extremely high pressures (Lofgren et al., 1973; Avezaat et al., 1979).

In most conditions the pressure–volume relationship may be considered a composite of both acute and subacute curves. Because of the complexity of the pressure–volume relationship, the derivation of volume increments from intracranial pressure responses generally tends to be subject to significant inaccuracies.

## 7. Tracers with an Extracellular Distribution

The assessment of the size of cerebral extracellular space is based upon a thorough permeation of the tissue with a solution of an indicator of extracellular space of known concentration. The indicator concentration increases in the tissue until the extracellular space is entirely saturated with the marker and the concentration reaches a plateau. In this steady-state situation the extracellular space (ECS) in percent of tissue volume may be calculated from:

$$\text{ECS}(\%) = \frac{\text{indicator concentrations in tissue}}{\text{indicator concentrations in medium}} \times 100$$

A number of substances has been applied as extracellular space markers. These include the ions thiocyanate, sulfate, thiosulfate, and iodide and the saccharides sucrose and inulin. Since these substances do not always behave as ideal indicators of extracellular space, but may penetrate to some extent into the intracellular compartment, the space is usually designated by the name of the indicator employed.

The application of the indicator may be accomplished in vitro using brain slices or in vivo by its administration into the blood circulation or by its administration into both the blood and CSF.

In vitro application of the marker involves incubation of brain slices in a medium containing the tracer in known concentration. The use of brain slices has the advantage that the blood–brain and blood–CSF barriers are circumvented. Thus in one of the early investigations with sucrose as an indicator, an extracellular space of 15.5% was measured in rabbit brain slices (Davson and Spaziani, 1959). Problems that may arise are mainly associated with the development of tissue swelling. Tissue swelling may attain significant degrees (70% in 2 h) with careless handling and preparation of the tissue or with anaerobic incubation (Katzman and Pappius, 1973). The swelling seems to mainly involve the intracellular compartment (or noninulin space). But even during optimal and aerobic conditions of incubation some swelling may still occur, related to tissue damage at the cut surfaces, especially when this proportion is significant with respect to total slice volume (Franck, 1970).

In vivo administration of the indicator by way of *blood* circulation is an adequate method when dealing with other tissues, such as skeletal muscle. In these cases the plasma concentration stands for the indicator concentration in the medium, as required in the equation for the calculation of indicator space. Application of the method to the central nervous system meets with the difficulty that capillary passage of the marker is impeded by the blood–brain barrier, whereas loss of indicator from nervous tissue may additionally occur by diffusion into CSF (the so-called sink action). As a consequence, a steady-state between plasma and nervous tissue is not usually reached. Thus a sucrose space of 2.66% was measured in rabbit brain (Oldendorf and Davson, 1967) and an iodide space of 2% (Ahmed and Van Harreveld, 1969) after intravenous administration of the markers to the animals. Since these values are of the same order of magnitude as the cerebral blood volume, it is conceivable that for some substances such as inulin no extravasation occurs. Extravasation of other substances, such as sucrose, was indicated by the decrease of sucrose space when the sink action of the CSF was enhanced by ventriculocisternal perfusion with fluid that was free of tracer (Oldendorf and Davson, 1967).

In patients with various types of brain tumors, sucrose space was assessed after intravenous administration of sucrose, since there were indications that the blood–brain barrier in the tumors was defective for sucrose. The measured sucrose space varied from 30 to 60% in glioblastomas and metastatic tumors and from 15 to 80% in astrocytomas. Many factors have been considered responsible for this scatter of values, such as diversity of tumor type within the same group, variability of blood volume, and intracellular uptake of the tracer in necrotic areas, but above all, probably, an inadequate equilibration of indicator concentration between plasma and tumor (Bakay, 1970).

For ions such as iodide and thiocyanate, which are removed from the CSF by active transport at the choroid plexus, an increase of indicator space, up to 12.4% or even to 38% for thiocyanate (Streicher et al., 1964; Pappius, 1968) and to 11.76% for iodide (Reed et al., 1965) could be achieved by raising the plasma levels and saturating the transport mechanism. The large value for thiocyanate is probably the result of intracellular uptake of the tracer.

In vivo administration of the indicator by way of both the *blood* circulation and the *CSF* in approximately equal concentrations counteracts the sink action of the cerebrospinal fluid and enables a steady state to be reached in brains of small size, such as that of the rat. Compared to the use of brain slices, the method has the advantage that the tissue is not deprived of its blood supply. With $^{14}$C-inulin as an indicator, an indicator space of 14% was measured in rat brain (Woodward et al., 1967), whereas a thiosulfate space of 12.9% was measured with $^{35}$S-thiosulfate as the marker (Baethmann et al., 1969). In brains of larger size, such as those of the dog and cat, only the structures in the immediate vicinity of the ventricles and cisterns are adequately permeated by the tracer, as has been demonstrated by autoradiography of brain sections made after the conclusion of experiments. When slabs radiating from the ventricle to the periphery are cut from these brain sections, the indicator concentration can be assessed in portions, into which these slabs are again divided. With increasing distance from the ventricle, the portions appeared to show a decreasing indicator space. The correct marker space has been derived by extrapolating to zero distance on the assumption that at this distance an equilibration of tissue extracellular fluid and CSF providing the tracer has taken place. Thus, in these periventricular structures a sulfate space of 12% was assessed in the cat (Cutler et al., 1986b) and a thiosulfate space of 12.66% was assessed in the dog (Reulen et al., 1970).

When it is assumed that the profile of marker concentration in the periventricular tissue is only determined by diffusion and not by convection (i.e., bulk flow) of fluid in the extracellular space, and moreover, no losses of indicator occurs from extracellular space either by uptake into cellular elements or by passage into the blood, then the following relationship would apply for marker concentration $C$ as a function of time $t$ and distance $x$ to the ventricle:

$$\frac{\partial C}{\partial t} = D\,\frac{\partial^2 C}{\partial x^2}$$

in which $D$ is the diffusion constant of the marker (Patlak and Fenstermacher, 1975).

This can be solved to yield a relationship between measurable entities such as the marker concentration at distance $x$ from the ventricle $C_x$, marker concentration in the cerebrospinal fluid $C_0$, and time $t$, of the type of a complementary error function:

$$\frac{C_x}{C_0} = \mathrm{erfc}\,\frac{x}{2\sqrt{Dt}}$$

It allows for the calculation of indicator space without the requirement that a steady state be attained. Thus a sucrose space of 17.5% was assessed for the caudate nucleus of the dog (Patlak and Fenstermacher, 1975).

When the assessment of extracellular space in cerebral cortex was required, administration of the marker was effected by subarachnoid-cisternal perfusion. By this method an enlarged thiosulfate space of 15% was measured in edematous freezing lesions of the cortex in dogs, as compared to a space of 10.2% in control cortex (Fenske et al., 1973).

## 8. Measurement of Tissue Electrical Impedance

Low-frequency (i.e., of a few kHz) alternating current passing through brain tissue may be considered to be conducted mainly by ions in the extracellular fluid, since the cell membranes enveloping the cellular contents impede the participation of intracellular ions. Therefore the specific conductance of the tissue, or its inverse, the specific impedance, can be used to determine the size of tissue extracellular space (Van Harreveld, 1966; Aladjova, 1964).

The impedance $(Z)$ of a tissue against alternating electric current is regarded as a vector composed of two independent vectors, a real component being the resistance $(R)$, and an imaginary component that is the reactance or capacitive component $(1/2\pi fC)$ (Fig. 10), $Z = R + j/2\pi fC$ in which $f$ = the frequency, $j = \sqrt{-1}$, and $C$ = the capacity. The angle between $Z$ and $R$ is called the phase angle $\phi$.

In many impedance measurements usually only the resistive component is measured to represent the total impedance. This is warranted by the small value of the phase angle (about 4°, and never exceeding 20°), at which this has been determined (Ranck, 1983; Van Harreveld et al., 1963), confirming the predominance of the resistance component. When both components are measured at various frequencies, it appears that the resistance component decreases with increasing frequency, but the capacitive reactance becomes almost negligible at high frequencies (50–100 kHz) (Shalit and Mahler, 1966; Aladjova, 1964), implying that at these frequencies the cell membranes lose their property of constituting a high-

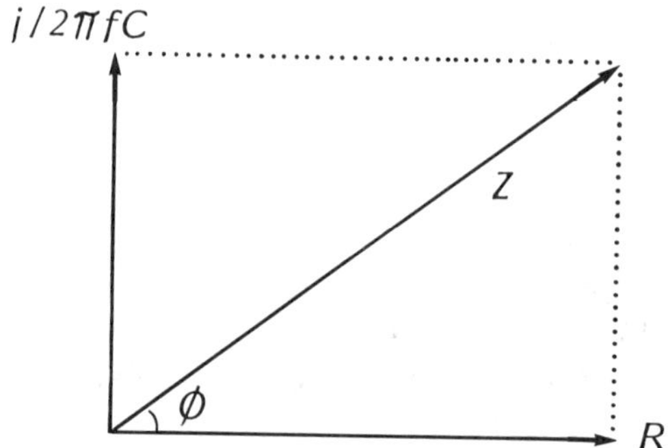



Fig. 10. Schematic drawing of the real and imaginary components of electrical impedance.

resistance barrier between intra- and extracellular compartments, and conductance of the current occurs through both the extra- and intracellular fluids. The use of frequencies below 1 kHz, however, is not recommended, because at these low frequencies polarization phenomena caused by changes of ionic concentrations at the many interfaces of the tissue become significant, resulting in an additional polarization impedance with resistance and capacitive components and in an increase of the noise level. The significance of polarization impedance effects at low frequencies became evident in a study comparing resistance and reactance of brain tissue and salt solutions, as measured at various frequencies with the same electrodes (Shalit and Mahler, 1966).

Brain tissue with its polymorphic cellular population and fiber systems in fact constitutes a complex medium in terms of its electrical properties. Simplified approximations have been proposed that consider the tissue a suspension of cells in a fluid. For such model systems, mathematical relations have been derived between the electrical resistance and various parameters of the suspension (Fricke, 1924, 1929; Cole, 1929), usually of the type as first derived by Maxwell (1873) for a homogeneous suspension of spheres in a medium of different resistance;

$$\frac{1 - r_1/r}{2 + r_1/r} = \rho \frac{1 - r_1/r_2}{2 + r_1/r_2}$$

in which $r$ is the specific resistance of the suspension, $r_1$ is that of the medium, $r_2$ is that of the spheres, and $\rho$ is the volume concentration of the spheres. The relation devised by Cole (1929) proved to be valid for a suspension of sea urchin eggs in sea water. Using a modification of the Maxwell equation, Van Harreveld and Schade (1960) computed a value of 27–33% for the size of the extracellular space on the basis of specific impedance data from rabbit cerebral cortex. Similarly, Fenstermachter et al. (1970) calculated an extracellular space of 15% for cat cerebral cortex.

Impedance measurements are usually conducted by means of bipolar or tetrapolar electrode arrangements. In the bipolar technique, the same electrodes introduce the current and measure the impedance. In the tetrapolar technique (Freygang and Landau, 1955; Ranck, 1963), one pair of electrodes introduces a constant current, and a second pair measures the potential difference across the tissue; this technique has the advantage that the measuring electrodes draw no current and are therefore less sensitive to contact impedance changes in the area and to polarization effects, as well (Shalit and Mahler, 1966). The measuring circuit may consist of a series ohmmeter or a bridge circuit arrangement. The series ohmmeter circuit can be expected to be more sensitive to fluctuations of the measuring current than is the bridge circuit (Gessert et al., 1970), although the modern ohmmeters possessing a high input impedance constitute no appreciable load on the system. The adjustment procedures, which rendered the use of bridge circuits cumbersome and less suitable to record dynamic changes, may be automated at the present time, employing modern electronic circuitry. The measuring current is usually in the 0.1–0.5 $\mu$A range, to avoid neural excitation.

Measurement of impedance by means of bipolar or tetrapolar arrangements of electrodes, all placed on the nervous tissue itself, yielded values of 218 ohm-cm for rabbit cerebral cortex and 872 ohm-cm for white matter at 1 kHz (Van Harreveld et al., 1963); at 5 kHz an impedance of 256 ohm-cm was measured for rabbit cerebral cortex by Ranck (1963). The specific impedance of white matter in the spinal cord was higher (800–1200 ohm-cm) when measured at right angles to the nerve fibers than when measured longitudinally (90–200 ohm-cm) (Ranck and BeMent, 1965). This was ascribed to the architecture of white matter, in which large extracellular spaces among nonmyelinated nerve fibers, together with the axoplasm of the fibers, form a longitudinal pathway of low resistance for the current, and a close packing of myelinated fibers and oligodendrocytes surrounding the tracts of nonmyelinated fibers offers high resistance against transverse electric currents (Van Harreveld, 1966).

In the clinical setting, measurements of electrical impedance in brain tissue are usually performed with a needle electrode inserted into the brain and a reference electrode connected to the body to minimize damage to the nervous tissue. A shortcoming is the participation of extracerebral tissues in determining the impedance value, which is probably responsible for the large interindividual variability of measured values. Moreover, systemic impedance is greatly influenced by the size of systemic extracellular space and total body water, which reflect the state of hydration of the body (Ducrot et al., 1970; Hoffer et al., 1970). The variations of values measured in different brain structures, however, tend to be quite consistent within the same individual; therefore comparison of individuals is possible after normalization.

Using a probe electrode, Organ and Kwan (1970) measured an impedance of 500–1000 $\Omega$ for white matter structures, 300–400 $\Omega$ for gray matter, and 66 $\Omega$ for cerebrospinal fluid spaces (at 1 kHz) in cats. During stereotaxic neurosurgery on patients, an impedance of 1180 $\Omega$ was measured in gray matter and 1440 $\Omega$ in white matter, versus 650 $\Omega$ in cerebrospinal fluid (Organ et al., 1968). Impedance measurement has proved very useful in stereotaxic neurosurgery to identify various nervous structures passed by the electrode (Robinson, 1962; Tachibana, 1971; Laitinen and Johansson, 1967). As early as 1923, Grant applied impedance measurement to localize brain tumors.

Tumors tend to have a lower specific impedance than normal tissue because of the presence of extracellular edema within and around the tumor (Tachibana, 1970; Organ et al., 1968; Shimabukuro, 1969). Apart from tumors, contusional lesions, hemorrhages, and infarctions associated with edema showed decreased impedance values (Go et al., 1972). In cats with freezing lesions, a mean impedance of 840 $\Omega$ was measured in the edematous cortical lesion, as compared to 1395 $\Omega$ in control cortex and 950 $\Omega$ in the edematous white matter versus 1565 $\Omega$ in normal white matter (Van der Veen et al., 1973). By means of a probe electrode provided with measuring rings at various depths, the extension of the edematous area from a freezing lesion in the cortex could be monitored in cats (Gazendam et al., 1979c).

Relevant to the dispute on the small size of cerebral extracellular space as assessed from conventional electron micrographs was the observation that during histological fixation procedures there was an elevation of electrical impedance of the tissue, indicating a diminution of tissue extracellular space (Van Harreveld, 1966).

In conclusion, the technique of electrical impedance measurement does not yield absolute values of cerebral extracellular space because of the complexity of nervous tissue architecture as an electrical medium. However, it is an excellent means of monitoring changes and recording topographic differences, and is well suited for application in a clinical setting.

## 9. Isolation and Analysis of Cerebral Edema Fluid

Because of the extensive diffusional exchange of small molecular substances across the ventricular ependyma between CSF and cerebral extracellular space, the composition of CSF is usually assumed to reflect the composition of CEF.

Direct procurement of cerebral extracellular fluid meets with the vulnerability of the blood–brain barrier. Following the intravenous administration of Evans blue or other indicators of blood–brain barrier disruption, it can be seen that the fluid obtained contains the indicator concomitant with plasma proteins, which implies that damage to the barrier has occurred by the insertion of the device by means of which the fluid was obtained. Also, after chronic implantation of capsular devices, the fluid that is collected within the capsule could not be considered normal brain extracellular fluid, since it resembled edema fluid in its composition (Adachi et al., 1974).

Edema fluid can be obtained from cat brain following a severe freezing injury to the cerebral cortex. The injury comprises freezing at a temperature of –30 or –40°C for 5 min, by means of a freezing probe measuring 13 mm in diameter applied onto the exposed dura. Edema fluid is collected by the insertion of 25-gage needles, measuring 4 mm and longer. The shorter needles reach into the cortex, and the longer penetrate into the underlying white matter. The needles have been mounted on an acrylic platform and have been filled with moistened nylon threads to serve as wicks that should prevent clogging of the needles by tissue debris. Fifteen centimeter lengths of polyethylene tubing, in which the edema fluid may collect, are connected to the needles (Fig. 11). The polyethylene tubes are filled with a volatile (boiling point 40°C) gasoline or kerosene fraction serving as a fluid seal to prevent evaporation of the edema fluid. As early as 30 min after the freezing injury, edema fluid (characterized by its blue staining, when Evans blue as a marker has previously been injected intravenously), may start to fill the polyethylene tubes. The tubes may be disconnected from the needles at regular intervals or when they are entirely filled with edema fluid. The tubes may be stored after plugging both their ends with pieces of stainless steel wire. Maximum rates of edema fluid production usually occur at $1\frac{1}{2}$–3 h after the freezing injury, and the collection of fluid may continue for over 5 h after freezing. The quantity of fluid accumulating is variable; it may amount to 40 µL in 30 min per tube. Since the driving force of the fluid extrusion seems to be the raised tissue pressure in the edematous areas, maintenance of arterial blood pressure is important to keep up edema fluid production. Edema fluid with blood admixture may be centrifuged and the blood sediment discarded (Patberg et al., 1977; Gazendam et al., 1979a).

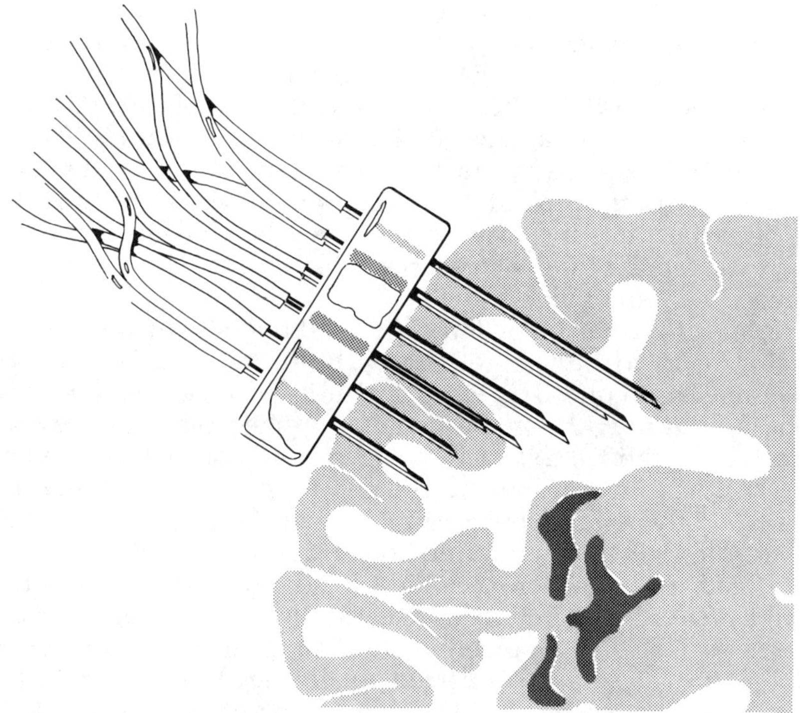

Fig. 11. Schematic drawing of a frontal section of cat brain, showing the location of needle tips with respect to the various brain structures, from which the needles derive their contents of edema fluid.

The composition of the edema fluid is characterized by its protein content, as compared to cerebrospinal fluid. In addition, its $Na^+$ content (142.54 mEq/L) more closely resembles that of plasma (144.58 mEq/L) than that of cat cerebrospinal fluid (159.20 mEq/L), indicating its vasogenic origin. Its $K^+$ content is elevated (7.05 mEq/L), as compared to plasma (4.24 mEq/L) or CSF (3.34 mEq/L), obviously because of the freezing damage to cells in the freezing lesion. Other constituents indicating cellular origin and presumably permeability disturbance of cellular membranes from freezing damage are the contents of enzymes such as lactate dehydrogenase and creatinephosphokinase (BB isoenzyme). The protein content of edema fluid amounts to $3.98 \pm 1.09$ g/100 mL.

Isolation of edema fluid from a freezing injury has been applied in the evaluation of the formation of edema mediators such as kinins, free fatty acids, and glutamate in brain injury (Unterberg et al., 1984; Maier-Hauff et al., 1984). The finding of an elevated glutamine content in the edema fluid indicates that the mechanisms for the clearance of extracellular glutamate by conversion into glutamine are operative along the pathway of edema fluid migration into the tissue extracellular spaces, without, however, attaining a normalization of the glutamate content (Go, 1986).

From the changes of edema fluid composition during hypoxia and following inhibition of the cellular membrane $Na^+,K^+$-ATPase by ouabain, shifts of extracellular fluid constituents to the intracellular compartment could be inferred (Go et al., 1979; Gazendam et al., 1979b).

Isolation of edema fluid has also been performed by centrifugation (2.500 rpm for 30 min) of edematous white matter from dog brain (Clasen et al., 1967). Its composition, however, indicated that the procedure had resulted in significant admixture by cellular contents.

## 10. Ventriculocisternal Perfusion

Ventriculocisternal perfusion allows for the replacement and control of a substantial portion of the fluid environment of the brain by an artificial or mock CSF of known composition, thus enabling deduction of the dynamics of CSF production and absorption as well as the exchange of various substances on the basis of changes in composition of the perfusate brought about by its passage through the ventricular system.

The cerebrospinal fluid is secreted within the ventricular system by the choroid plexuses in all the ventricles and probably also originates from brain extracellular fluid by inflow through the ventricular walls. Its absorption into the blood circulation takes place at the arachnoid villi and granulations of the convexity of the brain by a mechanism that shows a linear dependence of absorption rate upon fluid pressure (e.g., in the human, 7.6 $\mu$L/min/mm $H_2O$), as soon as an opening pressure (in the human, 68 mm $H_2O$) is exceeded (Lorenzo et al., 1974).

The measurement of CSF formation rate is essentially based upon dilution of a tracer that does not permeate the surrounding brain parenchyma. Although employed by a number of investigators (Heisey et al., 1962; Cserr, 1965; Bradbury and Davson, 1965), inulin does not qualify as a tracer impermeant to the ependyma, since ventriculocisternal perfusion can be used to introduce inulin into the brain for the determination of inulin space. Consequently, measured values of cerebrospinal fluid formation rate tend to be higher when inulin is used as a tracer (Martins et al., 1975). Later investigators preferred[131]I- or [125]I-radioiodinated serum albumin (RISA) or blue dextran, which can be colorimetrically assessed (Cutler et al., 1968b; Hochwald et al., 1974; Anderson and Heisey, 1975; Martins et al., 1975). In employing RISA as a tracer, it is important to add some nonradioactive albumin to the artifical CSF, or absorption of the radioactive albumin to the walls of the perfusion system may take place, resulting in significant errors (Hochwald and Sahar, 1971).

As an artificial CSF, the fluid composition according to Merlis (1940) has been used for cat and rabbit (Graziani et al., 1965; Cutler et al., 1968b; Ahmed and Van Harreveld, 1969). It contains (in mM/L): $Na^+$, 141; $K^+$, 3.3; $Ca^{2+}$, 1.25; $Mg^{2+}$, 1.2; $Cl^-$, 152; $HPO_4^{2-}$, 0.48; $HCO_3^-$, 21; glucose, 3.4; and urea, 2.2; pH was adjusted to 7.4 with $CO_2$. It may be necessary to prepare a stock solution without $Ca^{2+}$ and $Mg^{2+}$ (since these may precipitate upon autoclaving) and to add these constituents before the perfusion (Pappenheimer et al., 1962). Other investigators prefer Elliott's B solution for human and cat (Rubin et al., 1966; Lorenzo et al., 1974; Weiss and Wertman, 1978).

For the technique of ventriculocisternal perfusion, it is recommended to start with the insertion of the ventricular cannula, usually employing a stereotaxic device to guide and fix the cannula. The cannula is introduced into the lateral ventricle through a hole drilled into the skull, according to coordinates as specified for the animal species. The cannula is connected to thick-walled noncompressible polyvinyl tubing entirely filled with perfusion fluid and leading to the syringe of the perfusion pump. There should be an outlet on a joint connecting to a pressure transducer to check the perfusion pressure. Proper placement of the tip of the cannula within the ventricular cavity is indicated by the presence of pulsations on the pressure recording and the occurrence of a pressure peak upon compressing the animal's chest. Absence of pulsations may be caused by improper placement of the cannula, obstruction of the cannula by debris, or low ventricular fluid pressure following loss of CSF. In the latter case the ventricular fluid pressure may be restored by starting the perfusion at a low rate. Improper placement of the cannula or obstruction of the system should be suspected when the perfusion pressure rises despite a low perfusion rate or when leakage develops. Subsequently the cisternal cannula can be inserted, either through a skin incision, or after exposure of the atlantooccipital membrane. To this end, the back of the neck should be extended to stretch the membrane, although it is also essential to keep the CSF system filled during puncture of the membrane by proceeding with the perfusion at a low rate; otherwise the arachnoid mater may detach from the dura and plug the cannula's orifice. The effluent perfusate from the cisternal cannula may be collected in small test tubes provided with an occluding cap to prevent evaporation and allowing for the determination of outflow volume (gravimetrically) and tracer concentration at intervals.

Since CSF formation rate is dependent upon osmolarity and pH of the CSF, both with respect to plasma, it is paramount to control these parameters. Plasma hypoosmolarity notably causes an increased formation of CSF (DiMattio et al., 1975). Since the formation and pH homeostasis of the CSF are interrelated (possibly because the carbonic anhydrase in the choroid plexus subserves both functions), alkalosis has been reported to inhibit CSF formation rate (Oppelt et al., 1963). It is also important to control the outflow pressure at the cisternal end of the perfusion system, since cerebrospinal fluid absorption rate (and possibly also formation rate) (Eisenberg et al., 1974; Go et al., 1980) are dependent upon CSF pressure. To minimize absorption, the outflow pressure is usually kept slightly below zero. An exceedingly low outflow pressure, however, may cause collapse of the arachnoid mater against the medulla and jeopardize the perfusion. The perfusion rate applied in various animal species, is usually chosen to be about 10 times the cerebrospinal fluid formation rate (*see* Table 2).

Table 2
Ventriculocisternal Perfusion Rate With Respect to CSF Formation Rate

| Animal species | Perfusion rate, mL/min | CSF formation rate, mL/min | References |
|---|---|---|---|
| Rat | 0.02 | 0.0022 | Cserr, 1965 |
| Dog | 0.3 | 0.035 | Cserr, 1965 |
| Dog | 0.23–0.33 | 0.027 | Bering and Sato, 1963 |
| Goat | 1.5–2.0 | 0.154 | Pappenheimer et al., 1962 |
| Cat | 0.077 | 0.025 | Hochwald et al., 1974 |
| Cat | 0.1 | 0.022 | Go et al., 1980 |
| Rabbit | 0.068 | 0.0124 | Bradbury and Davson, 1965 |
| Rabbit | 0.077 | 0.0081 | Hochwald and Sahar, 1971 |
| Calf | 2.12 | 0.275 | Calhoun et al., 1967 |
| Human | 1.45–1.9 | 0.35 | Cutler et al., 1968a |
| Human | 2.5 | 0.313 | Lorenzo et al., 1974 |
| Macaca mulatta | 0.12–0.15 | 0.0148 | Levin et al., 1971 |
| Macaca mulatta | 0.18–0.24 | 0.031 | Martins et al., 1975 |

At steady-state conditions, CSF formation rate $V_f$ and cerebrospinal fluid absorption rate $V_a$ may be calculated according to the equations derived by Heisey et al. (1962).

$$V_f = \frac{V_i(C_i - C_o)}{C_o} \; ; \; V_a = V_i + V_f - V_o$$

in which $V_i$ = the inflow rate of perfusion fluid; $V_o$ = the outflow rate; $C_i$ = the tracer concentration of the inflowing fluid; $C_o$ = the tracer concentration of the outflowing fluid.
Alternatively, $V_f$ has been calculated by

$$V_f = \frac{V_i(C_i - C_o)}{\bar{C}}$$

in which $\bar{C}$ = the arithmetic mean or the exponential mean of $C_i$ and $C_o$, the latter being $C_o + 0.37\ (C_i - C_o)$ (Eisenberg et al. 1974; Hochwald and Sahar, 1971; Bering and Sato, 1963), when outflow is upstream of absorption sites (Welch, 1975).

It is conceivable that for clinical use the procedure, with its inherent risks of infection, will meet with objections. Therefore efforts have been undertaken to minimize the invasive character of the procedure or to make use of a situation in which the procedure is required for therapeutic purposes. Measurement of CSF formation and absorption rates was conducted by interventricular or lumboventricular perfusion in the course of treatment of children with tumors of the central nervous system (Rubin et al., 1966). In other clinical arrangements, assessment of CSF formation and absorption rates was not based upon tracer dilution, but upon the responses of CSF pressure, monitored during the withdrawal or addition of fluid.

As early as 1934 Masserman assessed CSF formation rate in humans by withdrawing a volume $\Delta V$ of 10–60 mL of CSF by lumbar puncture and measuring the time $T_r$ required for the return of CSF pressure to its initial value in the recumbent patient. The CSF formation rate was calculated from $V_f = \Delta V/T_r$ to be 0.319 mL/min, a value that is well in accordance with later values determined on the basis of tracer dilution.

Utilizing a similar procedure, Katzman and Hussey (1970) calculated a CSF formation rate of 0.323 mL/min in 11 humans. They also assessed the CSF absorption rate in these patients, using a constant infusion test in which sterile saline was infused into the CSF space through lumbar puncture at a rate of 0.76 mL/min, while the pressure was recorded at intervals of 2–10 min. As a result of the infusion, the pressure was reported to rise until by 40–60 min a steady pressure level was reached. Knowing the formation rate $V_f$ and the infusion rate $V_i$, the CSF absorption rate $V_a$ (at the steady-state pressure) could be calculated as $V_a = V_i + V_f$. These procedures, which are based upon pressure responses to induced alterations of CSF volume, may only be considered valid on the assumption that acute changes of pressure have no influence upon both cerebrospinal fluid absorption and formation (Katzman and Hussey, 1970).

In patients with hydrocephalus, Borgesen and Gjerris (1982)

determined CSF absorption rate $V_a$ by lumboventricular perfusion, combined with pressure monitoring.

$$V_a = V_f + V_i - V_o$$

in which $V_i$ was the lumbar inflow rate of 1.5–4.5 mL/min, $V_o$ was the ventricular outflow rate, gravimetrically determined, and $V_f$ was the CSF formation rate, assumed to be 0.40 mL/min. By conducting the procedure at various pressure levels, the conductance to outflow

$$C_{out} = \frac{dV_a}{dP} \text{ mL/min/mm Hg}$$

could be computed from the changes of CSF absorption rate $dV_a$ and the pressure differences $dP$. Methodologically, the technique hinges on the assumption of a uniform $V_f$ of 0.4 mL/min in all patients. From actual measurements it is known that CSF formation rate amounts to $0.35 \pm 0.02$ mL/min in normal humans (Cutler et al., 1968a), and $0.30 \pm 0.02$ mL/min in hydrocephalic children, with $V_f$ declining at a rate of 0.003 mL/min/mm of pressure (Lorenzo et al., 1970).

A procedure for the non-steady-state measurement of CSF formation rate in dogs employed a recirculatory spinal perfusion tracer clearance technique (Sklar et al., 1980). The recirculation system comprised an inflow into the lumbar subarachnoid space and an outflow from the cisterna magna, driven by a peristaltic pump at a rate of 4–5 mL/min. Fluorescein-tagged albumin was used as a tracer; its concentration was continuously measured in the perfusion system with a fluorometric flow cell. At resting pressure the CSF formation rate

$$V_f = \frac{-dC/dt}{C} V_t$$

in which $C$ represents the concentration of fluorescent tracer at any particular instant, $dC/dt$ is calculated as though the decay were linear, and $V_t$ represents the total volume of the recirculatory perfusion system estimated from tracer dilution after the initial introduction of a bolus of tracer. Since tracer concentration decreased because of CSF formation, $V_t$ was calculated after extrapolation of the concentration curve back to zero time. At resting pressure, the CSF absorption rate $V_a = V_f$. Using the pressure–volume relationship, the arrangement also enabled the estimation of CSF absorption and formation rates at different pressure levels. A syringe infusion pump connected in series with the perfusion system allowed for an adjustment of the infusion rate $dV_1/dt$ ($V_1$ representing the volume of the external part of the perfusion system) without changing tracer concentration, as the syringe was filled with recirculating fluid. Changing the infusion rate resulted in a change of intracranial volume, which was assessed from the pressure volume relationship $\Delta V_2 = \frac{1}{\beta} \ln P/Po$, in which $P$ and $Po$ represent the measured pressure and the resting pressure, respectively, and $\beta$ represents the slope constant [defined by $dP/dV = \beta P$, with an approximate value of 1.0 in dogs (Sklar and Elashvili, 1977)]. The CSF formation rate could then be calculated from

$$V_f = \frac{dC/dt}{C}\left(V_1 + V_{20} + \frac{1}{\beta}\ln P/Po\right)$$

in which $V_{20}$ represents the internal perfused volume at resting condition, calculated by $V_2 = V_t - V_1$. The CSF absorption rate $V_a$ was computed from $V_a = V_f - dV_1/dt$.

The introduction of the elasticity parameters of the pressure–volume relationship, required to obtain the contribution of intracranial volume (which furthermore does not partake into the recirculation system), probably also introduces uncertainties in the calculation of $V_f$.

Ventriculocisternal perfusion has been utilized for the study of exchange processes of various substances between ventricular CSF and the surrounding brain. The studies may pertain to radioactive isotopes of substances not normally occurring in the body, such as $^{35}SO_4$; they may comprise radioactive tracers of physiological substances such as $^{42}K$ or they concern the nonradioactive substances themselves such as $HCO_3^-$, $Cl^-$, or urea. A third compartment always participating in the relationships is the blood plasma. Analogous to exchange at the blood–brain barrier, transfer coefficients for efflux $k_o$ and influx $k_i$ may be derived. Generally, the material balance may be stated: total influx = total efflux.

$$\text{total influx} = V_i C_i + V_f C_f + \text{Infl}_{pl}$$

in which $V_i$ is the inflow rate of perfusion fluid; $C_i$ is the concentration of substance in the perfusion fluid; (hence $V_i C_i$ is the amount of substance entering with the perfusion fluid); $V_f$ is the CSF formation rate; $C_f$ is the concentration in newly formed CSF, when dealing with substances that are its normal constituents—its amount secreted with the CSF is therefore $V_f C_f$. Infl$_{pl}$ is the influx from plasma, other than by CSF secretion; it is usually expressed as a function of plasma concentration $C_{pl}$; $Infl_{pl} = k_i\, C_{pl}$.

$$\text{total efflux} = V_o C_o + V_a \bar{C}_v + Effl_{br}$$

in which $V_o$ is the rate of outflow of perfusate and $C_o$ is the substance concentration in the outflowing perfusate. Therefore $V_o C_o$ is the amount of substance leaving with the perfusate, and $V_a \bar{C}_v$ the amount of substance accompanying the cerebrospinal fluid being absorbed, since $V_a$ is the CSF absorption rate and $\bar{C}_v$ is the mean ventricular concentration of the substance; $Effl_{br}$ is the efflux of substance into surrounding brain as a function of mean ventricular concentration, $Effl_{br} = k_o\, \bar{C}_v$.

The parameters of cerebrospinal fluid kinetics $V_f$ and $V_a$ may be simultaneously assessed using an impermeant reference tracer, with inflow concentration $C'_i$ and outflow concentration $C'_o$.

$$V_f = \frac{V_i(C'_i - C'_o)}{C'_o} \; ; \; V_a = V_i + V_f - V_o$$

For the investigation of the transport of $^{35}SO_4$ between ventricular cerebrospinal fluid and brain, the following equations were derived (Cutler et al., 1968b).

Total influx $= V_i C_i$; since the substance was only administered in the perfusion fluid, total efflux $= V_o C_o + k_o\, \bar{C}_v$. Therefore

$$k_o = \frac{V_i C_i - V_o C_o}{\bar{C}_v},$$

determined to be 0.023 mL/min for adult cat brain. In a group of animals the tracer was only administered intravenously instead of in the perfusion fluid. Total influx $= k_i C_{pl}$; total efflux $= V_o C_o + k_o \bar{C}_v$. Therefore

$$k_i = \frac{V_o C_o + k_o\, \bar{C}_v}{C_{pl}},$$

which appeared to be 0.03 mL/min for the adult cat brain.

To study the exchange of $^{42}K$ between ventricular CSF and brain, $^{42}K$ was given in the perfusion fluid. Thus total influx $= V_i C_i$ and total efflux $= V_o C_o + V_a C_o + k_o \bar{C}_v$, resulting in the following balance: $V_i C_i = V_o C_o + V_a C_o + k_o \bar{C}_v$, for which $k_o = 0.096$ mL/min could be determined (Katzman and Pappius, 1973).

Similarly, the following equations were developed by Pappenheimer et al., (1965) for the exchange of $HCO_3^-$ and $Cl^-$ between tissue in which the substances normally occurred and ventricular CSF, where they were introduced by ventriculocisternal perfusion.

$$\text{total influx} = V_iC_i + V_fC_f + k_i\,C_{pl}$$

$$\text{total efflux} = V_oC_o + V_aC_o + k_o\,\bar{C}_v$$

Furthermore, a net transependymal flux rate $\dot{n}$ was defined as $\dot{n} = k_o\bar{C}_v - k_iC_{pl} = V_iC_i - V_oC_o + V_fC_f - V_aC_o$. For $HCO_3^-$ it was found to be 0.19 mL/min and for $Cl^-$, 0.14 mL/min.

Applying these equations to the exchange of $^{42}K$ between CSF and brain, Cserr (1965) found $k_i = 0.154$ mL/min and $k_o = 0.214$ mL/min in the dog.


## 11. Measurement of Cellular Volume

Changes of cellular volume in nervous tissue may occur in the scope of physiological responses to changes of the ionic environment, such as the swelling of astrocytes in response to an elevation of extracellular $K^+$ concentration; cell volume changes may also occur as a component of a pathological swelling of the tissue in brain edema.

Changes of cellular volume are generally considered in conjunction with complementary alterations of extracellular space. In brain slices, cell swelling may be inferred from a decrease of inulin space (Pappius, 1968). Similarly, in brain tissue *in situ*, expansion of the intracellular compartment has been deduced from the shrinking of the extracellular space, since this may be estimated by recording electrical impedance or studying tissue ultrastructure (with proper reserve).

Studies of cell volume proper, however, with adequate temporal resolution to follow sequences of events, has only been conducted using dissociated C6 glioma cells and F115 neuroblastoma cells from culture by passing them through a Metricell flow cytometric system (Chaussy et al., 1981). The principle of determination of the volume distribution of a suspension of cells in electrolyte solutions according to Coulter comprises the recording of electrical potential changes between electrodes in each of two chambers, since an electrical current passing with a flow of the suspension through an orifice between the two chambers is impeded by the passage of cellular elements through the orifice. The recorded potential change is related to the volume of the passing particle according to

$$V = \frac{q^2 f_k}{\rho\, i\, F}\, u$$

in which $V$ is the volume of the particle, $q$ is the cross section of the orifice, $\rho$ is the specific resistance of the medium, $i$ is the current flowing through the orifice, $f_k$ is the geometrical correction factor of the aperture, $F$ is a factor considering the shape and conductivity of the particle, and $u$ is the voltage pulse caused by passage of the particle through the orifice. The volume distribution curves obtained with the original Coulter arrangement, however, tended to be skewed to larger particle volumes, although their actual size distribution was Gaussian. The error in sizing appeared to be caused by distortion of the electrical field at the orifice because of changes of cell shape during passage through the orifice as well as the geometry of the fluid flow through the orifice. Corrections, achieving an accurate correlation of the voltage pulse with particle size, comprised hydrodynamic focusing of the particles by ensheathment of the flow of particles through the orifice by a mantle of fluid devoid of particles and, furthermore, by electrical calibration of the signal, as effectuated in the Metricell system (Kachel, 1976).

Thus the volume of C6 glioma cells was determined to be 980 $\mu m^3$, increasing to 1469 $\mu m^3$ under hypoosmolar conditions (Chaussy et al., 1981). The normalization of hypoosmolar cell volume could be shown to be associated with a stimulation of respiration and to be unaffected by inhibition of $Na^+K^+$ATP-ase by ouabain (Kempski et al., 1983).

## 12. Conclusions

Because the fluid environment of the brain with its possible derangements not only involves tissue electrolyte balance and exchange processes between the various tissue fluid compartments, but also may imply gross changes of brain volume and CSF volume, the study of brain fluid compartments possesses considerable clinical relevance, as manifested by the techniques of computerized X-ray tomography, proton nuclear magnetic resonance imaging, and electrical impedance measurement. Although these techniques may lack the precision in estimating water content, so common to the in vitro, but destructive procedures of gravimetry or specific weight determination, they have found extensive clinical application on account of their diagnostic merits. Also, in the study of CSF circulation, in which production is a biochemical process of active transport, but in which absorption is largely biomechanically determined by hydrostatic pressure, modifications bearing upon the craniospinal pressure/volume relationship have been applied to the classical procedure of ventriculocisternal perfusion, to suit the clinical situation. On the other end of the range of methods used to analyze brain fluid compartments, typically, is the application of flow cytometry to suspensions of cultured glial and neuronal cells, in the light of the recent emergence of interest in the study of cell volume regulatory processes. It is probably from these basic cell biological functions that a deeper understanding of the mechanisms of fluid secretion will be attained in the near future.

## Acknowledgments

I am greatly indebted to Dr. H. L. Journée, Msc, MD (Dept. of Neurosurgery, University of Groningen), Prof. H. J. C. Berendsen (Dept. of Physical Chemistry, University of Groningen) and Mr. G. P. Vijverberg, MSc (Philips Medical Systems Division, Eindhoven) for their critical comments and valuable suggestions.

## References

Abe K., Abe T., Klatzo I., and Spatz M. (1980) Effect of endogenous central nervous system depressants in ischemic cerebral edema of gerbils. *Adv. Neurol.* **28,** 429–443.

Adachi C., Mihara H., and Matsuo O. (1974) Analysis of fluid in capsules implanted into dog brain. *Jpn. J. Physiol.* **24,** 59–71.

Ahmed N. and Van Harreveld A. (1969) The iodide space in rabbit brain. *J. Physiol.* **204,** 31–50.

Aladjova N. A. (1964) Slow electrical processes in the brain. *Prog. Brain Res.* **7,** 156–206.

Anderson D. K. and Heisey S. R. (1975) Creatinine, potassium, and calcium flux from chicken cerebrospinal fluid. *Am. J. Physiol.* **228,** 415–419.

Avezaat C. J. J., Van Eijndhoven J. H. M., and Wyper D. J. (1979) Cerebrospinal fluid pulse pressure and intracranial volume–pressure relationships. *J. Neurol. Neurosurg. Psychiat.* **42,** 687–700.

Baethmann A., Steude V., Horsch S., and Brendel W. (1969) Einige Ergebnisse zur Bestimmung des Extrazellulärraumes (EZR) in ZNS von Ratten. *Pflugers Arch.* **307**, R113–114.

Bakay L. (1970) The extracellular space in brain tumours. II. The sucrose space. *Brain* **93**, 659–708.

Bakay L., Kurland R. J., Parrish R. G., Lee J. C., Peng R. J., and Bartkowski H. M. (1975) Nuclear magnetic resonance studies in normal and edematous brain tissue. *Exp. Brain. Res.* **23**, 241–248.

Bartkowski H. M., Bederson J., Nishimura M., Moon K., and Pitts L. H. (1984) Nuclear magnetic resonance imaging and spectroscopy in experimental brain edema. *Magn. Res. Med.* **1**, 98–99.

Berendsen H. J. C. (1975) Specific Interactions of Water With Biopolymers, in *Water, A Comprehensive Treatise* vol. 5 *Water in Disperse Systems* (Franks F, ed.) Plenum, New York.

Bering E. A. and Sato O. (1963) Hydrocephalus: Changes in formation and absorption of cerebrospinal fluid within the cerebral ventricles. *J. Neurosurg.* **20**, 1050–1063.

Borgesen S. E. and Gjerris F. (1982) The predictive value of conductance to outflow of CSF in normal pressure hydrocephalus. *Brain* **105**, 65–86.

Bovee W., Huisman P., and Smidt J. (1974) Tumor detection and nuclear magnetic resonance. *J. Natl. Cancer Inst.* **52**, 595–598.

Bradbury M. V. and Davson H. (1965) The transport of potassium between blood, cerebrospinal fluid and brain. *J. Physiol.* **181**, 151–174.

Brooks R. A. and DiChiro G. (1975) Theory of image reconstruction in computed tomography. *Radiology* **117**, 561–572.

Brooks R. A. and DiChiro G. (1976) Principles of computer-assisted tomography (CAT) in radiographic and radioisotopic imaging. *Phys. Med. Biol.* **21**, 689–732.

Brooks R. A., DiChiro G., and Keller M. R. (1980) Explanation of cerebral white-grey contrast in computed tomography. *J. Comput. Assist. Tomogr.* **4**, 489–491.

Bydder G. M. and Kreel L. (1979) The temperature dependence of computed tomography attenuation values. *J. Comput. Assist. Tomogr.* **3**, 506–510.

Calhoun M. C., Hurt H. D., Eaton H. D., Rousseau J. E., and Hall R. C. (1967) Rates of formation and absorption of cerebrospinal fluid in Holstein male calves. *Bull. Storrs. Agricult. Exp. Station.* **401**, 3–22.

Chaussy L., Baethmann A., and Lubitz W. (1981) Electrical Sizing of Nerve and Glia Cells in the Study of Cell Volume Regulation, in *Cerebral Microcirculation and Metabolism* (Cervos-Navarro J. and Fritschka E., eds.) Raven, New York.

Cho Z. H. (1974) General views on 3-D image reconstruction and computerized transverse axial tomography. *IEEE Trans Nucl. Sci.* **NS-21**, 44–54.

Clasen R. A., Cooke P. M., Pandolfi S., Carnecki G., and Bryar G. (1965) Hypertonic urea in experimental cerebral edema. *Arch. Neurol.* **12,** 424–434.

Clasen R. A., Sky-Peck H. H., Pandolfi S. Laing I., and Hass G. M. (1967) The Chemistry of Isolated Edema Fluid in Experimental Cerebral Injury. in *Brain Edema* (Klatzo I., and Seitelberger F., eds.) Springer, New York.

Clasen R. A., Huckman M. S., Pandolfi S., Laing I., and Jacobs J. (1976) Computed Tomography of Vasogenic Cerebral Edema, in *Dynamics of Brain Edema* (Pappius H. M. and Feindel W., eds.) Springer, New York.

Clasen R. A., Huckman M. S., Von Roenn K. A., Pandolfi S., Laing I., and Lobick J. J. (1981) A correlative study of computed tomography and histology in human and experimental vasogenic cerebral edema. *J. Comput. Assist. Tomogr.* **5,** 313–327.

Cole K. S. (1929) Electric impedance of suspensions of Arbacia eggs. *J. Gen. Physiol.* **12,** 37–54.

Cottam G. L., Vasek A., and Lusted D. (1972) Water proton relaxation rates in various tissues. *Res. Commun. Chem. Path. Pharmacol.* **4,** 495–502.

Cserr H. F. (1965) Potassium exchange between cerebrospinal fluid, plasma and brain. *Am. J. Physiol.* **209,** 1219–1226.

Cutler R. W. P., Page L., Galicich J., and Watters G. V. (1968a) Formation and absorption of cerebrospinal fluid in man. *Brain* **91,** 707–719.

Cutler R. W. P., Robinson R. J., and Lorenzo A. V. (1968b) Cerebrospinal fluid transport of sulfate in the cat. *Am. J. Physiol.* **214,** 448–454.

Damadian R. (1971) Tumor detection by nuclear magnetic resonance. *Science* **171,** 1151–1153.

Darrow K. K. (1953) Magnetic resonance. I. Nuclear magnetic resonance. *Bell. Syst. Techn. J.* **32,** 74–99.

Davson H. and Spaziani E. (1959) The blood–brain barrier and the extracellular space of brain. *J. Physiol.* **149,** 135–143.

DiMattio J, Hochwald G. M., Malhan C., and Wald A. (1975) Effects of changes in serum osmolarity on bulk flow of fluid into cerebral ventricles and on brain water content. *Pflugers Arch.* **359,** 253–264.

Ducrot H., Thomasset A., Joly R., Jungers P., Eyraud C., and Lenoir J. (1970) Détermination du volume des liquides extracellulaires chez l'homme par la mesure de l'impédance corporelle totale. *Presse Med.* **78,** 2269–2272.

Eisenberg H. M., Mc Lennan J. E., and Welch K. (1974) Ventricular perfusion in cats with kaolin-induced hydrocephalus. *J. Neurosurg.* **41,** 20–28.

Elliott K. A. C. and Jasper H. (1949) Measurement of experimentally induced brain swelling and shrinkage. *Am. J. Physiol.* **157,** 122–129.

Fenske A., Samii M., Reulen H. J., and Hey O. (1973) Extracellular space and electrolyte distribution in cortex and white matter of dog brain in cold-induced oedema. *Acta Neurochir.* **28,** 81–94.

Fenstermacher J. D., Li C. L., and Levin V. A. (1970) Extracellular space of the cerebral cortex of normothermic and hypothermic cats. *Exp. Neurol.* **27,** 101–114.

Ferszt R., Neu S., Cervos-Navarro J., and Sperner J. (1978) The spreading of focal brain edema induced by ultraviolet irradiation. *Acta Neuropathol.* **42,** 223–229.

Ferszt R., Hahm H., and Cervos-Navarro J. (1980) Measurement of the specific gravity of the brain as a tool in brain edema research. *Adv. Neurol.* **28,** 15–26.

Franck G. (1970) Echanges cationiques au niveau des neurones et des cellules gliales du cerveau. *Arch. Int. Physiol. Biochim.* **78,** 613–866.

Fricke H. (1924) A mathematical treatment of the electrical conductivity of colloids and cell suspensions. *J. Gen. Physiol.* **6,** 375–384.

Fricke H. (1929) The electric conductivity of disperse systems. *J. Gen. Physiol.* **6,** 741–746.

Freygang W. H. and Landau W. M. (1955) Some relations between resistivity and electrical activity in the cerebral cortex of the cat. *J. Cell. Comp. Physiol.* **45,** 377–391.

Fujimoto T., Walker J. T., Spatz M., and Klatzo I. (1976) Pathophysiologic Aspects of Ischemic Edema, in *Dynamics of Brain Edema* (Pappius H. M. and Feindel W., eds.) Springer, New York.

Fullerton G. D. (1982) Basic concepts for nuclear magnetic resonance imaging. *Magnet. Reson. Imag.* **1,** 39–55.

Fullerton G. D. and Blanco E. (1981) Fundamentals of computerized tomography (CT) tissue characterization of the brain. *Proc. SPIE* **273,** 256–266.

Fullerton G. D., Potter J. L., and Dornbluth N. C. (1982) NMR relaxation of protons in tissues and other macromolecular water solutions. *Magnet. Reson. Imag.* **1,** 209–228.

Fullerton G. D., Cameron I. L., and Ord V. A. (1984) Frequency dependence of magnetic resonance spin-lattice relaxation of protons in biological material. *Radiology* **151,** 135–138.

Gazendam J., Go K. G., and Van Zanten A. K. (1979a) Composition of isolated edema fluid in cold-induced edema. *J. Neurosurg.* **51,** 70–77.

Gazendam J., Go K. G., and Van Zanten A. K. (1979b) The effect of intracerebral ouabain administration on the composition of edema fluid isolated from cats with cold-induced brain edema. *Brain Res.* **175,** 279–290.

Gazendam J., Go K. G., Van der Meer J., and Zuiderveen F. (1979c) Changes of electrical impedance in edematous cat brain during hypoxia and after intracerebral ouabain injection. *Exp. Neurol.* **55,** 78–87.

Gessert W. L., Nijboer J., Reid K., and Liedtke R. (1970) Bio-impedance instrumentation. *Ann. NY Acad. Sci.* **170,** 520–531.

Go K. G. (1986) Disturbances of Extracellular Homeostasis After a Primary Insult as a Mechanism in Secondary Brain Damage, in *Mechanisms of Secondary Brain Damage* (Baethmann A., Go K. G., and Unterberg A., eds.) Plenum, New York.

Go K. G. (1981) The Classification of Brain Edema, in *Brain Edema* (de Vlieger M., de Lange S. A., and Beks J. W. F., eds.) Wiley, New York.

Go K. G. and Edzes H. T. (1975) Water in brain edema. Observations by the pulsed nuclear magnetic resonance technique. *Arch. Neurol.* **32,** 462–465.

Go K. G., Van der Veen P. H., Ebels E. J., and Van Woudenberg F. (1972) Study of electrical impedance of oedematous cerebral tissue during operations. *Acta Neurochir.* **27,** 113–124.

Go K. G., Ebels E. J., Van Woudenberg F., and Geerlings T. (1973) The development of oedema in the immature brain. A comparison of cold-induced oedema in young and adult cat brain. *Psychiat. Neurol. Neurochir.* **76,** 427–437.

Go K. G., Gazendam J., and Van Zanten A. K. (1979) Influences of hypoxia on the composition of isolated edema fluid in cold-induced brain edema. *J. Neurosurg.* **51,** 78–84.

Go K. G., Hochwald G. M., Koster-Otte L., Van Zanten A. K., and Gandhi M. (1980) The effect of cold-induced brain edema on cerebrospinal fluid formation rate. *J. Neurosurg.* **53,** 652–655.

Go K. G., Van Dijk P., Luiten A. L., Brouwer-Van Herwijnen A. A., Van der Leeuw Y. C. L., Kamman R. L., Vencken L. M., Wilmink J., and Berendsen H. J. C. (1983) Interpretation of nuclear magnetic resonance tomograms of the brain. *J. Neurosurg.* **59,** 574–584.

Go K. G., Van Dijk P., Luiten A. L., and Teelken A. W. (1984) Proton Spin Tomography in Brain Edema, in *Recent Progress in the Study and Therapy of Brain Edema* (Go K. G. and Baethmann A., eds.) Plenum, New York.

Grant F. C. (1923) Localization of brain tumors by determination of the electrical resistance of the growth. *J. Am. Med. Assoc.* **81,** 2169–2171.

Graziani L., Escriva A., and Katzman R. (1965) Exchange of calcium between blood, brain and cerebrospinal fluid. *Am. J. Physiol.* **208,** 1058–1064.

Hatam A., Yu Z. Y., Bergstrom M., Berggren B. M., and Greitz T. (1982) Effect of dexamethasone treatment on peritumoral brain edema:

Evaluation by computed tomography. *J. Comput. Assist. Tomogr.* **6,** 586–592.

Heisey S. R., Held D., and Pappenheimer J. R. (1962) Bulk flow and diffusion in the cerebrospinal fluid system of the goat. *Am. J. Physiol.* **203,** 775–781.

Hinshaw W. S. (1976) Image formation by nuclear magnetic resonance: The sensitive point method. *J. Appl. Phys.* **47,** 3709–3721.

Hochwald G. M. and Sahar A. (1971) Effect of spinal fluid pressure on cerebrospinal fluid formation. *Exp. Neurol.* **32,** 30–40.

Hochwald G. M., Wald A., DiMattio J., and Malhan C. (1974) The effects of serum osmolarity on cerebrospinal fluid volume flow. *Life Sci.* **15,** 1309–1316.

Hoffer E. C., Meador C. K., and Simpson D. C. (1970) A relationship between whole body impedance and total body water volume. *Ann. NY Acad. Sci.* **170,** 452–461.

Holland G. N., Moore W. S., and Hawkes R. C. (1980) Nuclear magnetic resonance tomography of the brain. *J. Comput. Assist. Tomogr.* **4,** 1–3.

Hounsfield G. N. and Ambrose J. (1973) Computerized transverse axial scanning (tomography). I. Description of system II clinical application. *Br. J. Radiol.* **46,** 1016–1047.

Ito U., Ohno K., Nakamura R., Suganuma F., and Inaba Y. (1979) Brain edema during ischemia and after restoration of blood flow. *Stroke* **10,** 542–547.

Kachel V. (1976) Basic principles of electrical sizing of cells and particles and their realization in the new instrument "Metricell". *J. Histochem. Cytochem.* **24,** 211–230.

Kamman R. L., Go K. G., Muskiet F. A. J., Stomp G. P., Van Dijk P., and Berendsen H. J. C. (1984) Proton spin relaxation studies of fatty tissue and cerebral white matter. *Magnet. Reson. Imag.* **2,** 211–220.

Kamman R. L., Go K. G., Stomp G., Hulstaert C., and Berendsen H. J. C. (1985) Changes of relaxation times T1 and T2 in rat tissues after biopsy and fixation. *Magnet. Reson. Imag.* **3,** 245–250.

Katzman R. and Hussey F. (1970) A simple constant-infusion manometric test for measurement of CSF absorption. I. Rationale and method. *Neurology* **20,** 534–544.

Katzman R. and Pappius H. M. (1973) *Brain Electrolytes and Fluid Metabolism* Williams & Wilkins, Baltimore.

Kempski O., Chaussy L., Gross U., Zimmer M., and Baethmann A. (1983) Volume regulation and metabolism of suspended $C_6$ glioma cells: An in vitro model to study cytotoxic brain edema. *Brain Res.* **279,** 217–228.

Klatzo I., Wisniewski H., Steinwall O., and Streicher E. (1967) Dynamics of Cold Injury Edema. in *Brain Edema* (Klatzo I. and Seitelberger F., eds.) Springer, New York.

Kumar A., Welti D., and Ernst R. R. (1975) NMR fourier zeugmatography. *J. Magnet. Reson.* **18,** 69–83.

Laitinen L. V. and Johansson G. G. (1967) Locating human cerebral structures by the impedance method. *Confin. Neurol.* **29,** 197–201.

Lanksch W., Oettinger W., Baethmann A., and Kazner E. (1976) CT Findings in Brain Edema Compared With Direct Chemical Analysis of Tissue Samples, in *Dynamics of Brain Edema* (Pappius H. M. and Feindel W., eds.) Springer, New York.

Lanksch W., Baethmann A., and Kazner E. (1981) Computed Tomography of Brain Edema, in *Brain Edema* (De Vlieger M., De Lange S. A., and Beks J. W. F., eds.) Wiley, New York.

Lauterbur P. C. (1973) Image formation by induced local interactions. Examples employing nuclear magnetic resonance. *Nature* **242,** 190–191.

Levi C., Gray J. E., Mc Cullough E. C., and Hattery R. R. (1982) The unreliability of CT numbers as absolute values. *Am. J. Roentgenol.* **139,** 443–447.

Levin V. A., Milhorat T. H., Fenstermacher J. D., Hammock M. K., and Rall D. P. (1971) Physiological studies on the development of obstructive hydrocephalus in the monkey. *Neurology* **21,** 238–246.

Ling C. R., Foster M. A., and Hutchison J. M. S. (1980) Comparison of NMR water proton $T_1$ relaxation times of rabbit tissues at 24 MHz and 25 MHz. *Phys. Med. Biol.* **25,** 748–751.

Lofgren J., Von Essen C., and Zwetnow N. N. (1973) The pressure–volume curve of the cerebrospinal fluid space in dogs. *Acta Neurol. Scand.* **49,** 557–574.

Lorenzo A. V., Page L., and Watters G. V. (1970) Relationship between cerebrospinal fluid formation, absorption and pressure in human hydrocephalus. *Brain* **93,** 679–692.

Lorenzo A. V., Bresnan M. J., and Barlow C. F. (1974) Cerebrospinal fluid absorption deficit in normal pressure hydrocephalus. *Arch. Neurol.* **30,** 387–393.

Lowry O. H. and Hunter T. H. (1945) The determination of serum protein concentration with a gradient tube. *J. Biol. Chem.* **159,** 465–474.

Maier-Hauff K., Lange M., Schürer L., Guggenbichler C., Vogt W., Jacob K., and Baethmann A (1984) Glutamate and Free Fatty Acid Concentrations in Extracellular Vasogenic Edema Fluid, in *Recent Progress in the Study and Therapy of Brain Edema* (Go K. G. and Baethman A., eds.) Plenum, New York.

Mansfield P. (1976) Proton spin imaging by nuclear magnetic resonance. *Contemp. Physics* **6,** 553–576.

Marmarou A., Poll W., Shulman K., and Bhagavan H. (1978a) A simple gravimetric technique for measurement of cerebral edema. *J. Neurosurg.* **49,** 530–537.

Marmarou A., Shulman K., and Rosende R. M. (1978b) A nonlinear analysis of the cerebrospinal fluid system and intracranial pressure dynamics. *J. Neurosurg.* **48**, 332–344.

Marmarou A., Tanaka K., and Shulman K. (1982) An improved gravimetric measure of cerebral edema. *J. Neurosurg.* **56**, 246–253.

Martins A., Ramirez A., and Doyle T. F. (1975) Comparison of radioiodinated serum albumin and blue dextran as indicators to measure rate of formation of cerebrospinal fluid. *Exp. Neurol.* **47**, 249–256.

Masserman J. H. (1934) Cerebrospinal hydrodynamics. IV. Clinical experimental studies. *Arch. Neurol. Psychiat.* **32**, 523–553.

Maxwell J. C. (1873) *Treatise on Electricity and Magnetism* Clarendon, Oxford.

McCullough E. C. (1977) Factors affecting the use of quantitative information from a CT-scanner. *Radiology* **124**, 99–107.

Merlis J. K. (1940) The effect of changes in the calcium content of the cerebrospinal fluid in spinal reflex activity in the dog. *Am. J. Physiol.* **131**, 67–72.

Naruse S., Horikawa Y., Tanaka C., Hirakawa K., Nishikawa H., and Yoshizaki K. (1982) Proton nuclear magnetic resonance studies on brain edema. *J. Neurosurg.* **56**, 747–752.

Nelson S. R., Mantz M. L., and Maxwell J. A. (1971) Use of specific gravity in the measurement of cerebral edema. *J. Appl. Physiol.* **30**, 268–271.

Oldendorf W. H. (1978) The quest for an image of brain: A brief historical and technical review of brain imaging techniques. *Neurology* **28**, 517–533.

Oldendorf W. H. and Davson H. (1967) Brain extracellular space and the sink action of cerebrospinal fluid. *Arch. Neurol.* **17**, 196–205.

Oppelt W. W., Maren T. H., Owens E. S., and Rall D. P. (1963) Effects of acid–base alterations on cerebrospinal fluid production. *Proc. Soc. Exp. Biol. Med.* **114**, 86–89.

Organ L. W. and Kwan H. C. (1970) Electrical impedance variation along a tract of brain tissue. *Ann. NY Acad. Sci.* **170**, 491–508.

Organ L. W., Tasker R. R., and Moody N. F. (1968) Brain tumor localization using an electrical impedance technique. *J. Neurosurg.* **28**, 35–44.

Pappenheimer J. R., Heisey S. R., Jordan E. F., and Downer J. de C. (1962) Perfusion of the cerebral ventricular system in unanesthesized goats. *Am. J. Physiol.* **203**, 763–774.

Pappenheimer J. R., Fencl V., Heisey S. R., and Held D. (1965) Role of cerebral fluids in control of respiration as studied in unanesthesized goats. *Am. J. Physiol.* **208**, 436–450.

Pappius H. M. (1968) Spaces in brain tissue in vitro and in vivo. *Prog. Brain Res.* **29**, 455–464.

Pappius H. M. (1974) Fundamental Aspects of Brain Edema, in *Handbook of Clinical Neurology* vol. 16 (Vinken P. J. and Bruyn B. W., eds.) North Holland, Amsterdam.

Parrish R. G., Kurland R. J., Janese W. W., and Bakay L. (1974) Proton relaxation rates of water in brain and brain tumors. *Science* **183**, 438–439.

Patberg W. R., Go K. G., and Teelken A. W. (1977) Isolation of edema fluid in cold induced cerebral edema for the study of colloid osmotic pressure, lactate dehydrogenase activity and electrolytes. *Exp. Neurol.* **54**, 141–147.

Patlak C. S. and Fenstermacher J. D. (1975) Measurements of dog blood–brain transfer constants by ventriculocisternal perfusion. *Am. J. Physiol.* **229**, 877–884.

Payne J. T. and Latchaw R. (1978) Variation and non uniform CT number response for intracranial contents as a function of skull thickness and head size. *J. Comput. Assist. Tomogr.* **2**, 509.

Raaphorst G. P. and Kruuv J. (1981) Nuclear magnetic resonance spin-lattice times of normal and transformed cultured mammalian cells and of normal and neoplastic animal tissues. *Physiol. Chem. Phys.* **13**, 251–258.

Raaphorst G. P., Kruuv J., and Pintar H. M. (1975) Nuclear magnetic resonance study of mammalian cell water. *Biophys. J.* **15**, 391–402.

Ranck J. W. (1963) Specific impedance of rabbit cerebral cortex. *Exp. Neurol.* **7**, 144–152.

Ranck J. B. and BeMent S. L. (1965) The specific impedance of the dorsal columns of cat: An anisotropic medium. *Exp. Neurol.* **11**, 451–463.

Reed D. J., Woodbury D. M., Jacobs L., and Squires R. (1965) Factors affecting distribution of iodide in brain and cerebrospinal fluid. *Am. J. Physiol.* **209**, 757–764.

Reid M. H. (1983) Organ and lesion volume measurements with computed tomography. *J. Comput. Assist. Tomogr.* **7**, 268–273.

Reulen H. J., Hase U., Fenske A., Samii M., and Schurmann K. (1970) Extrazelluläraum und Ionenverteilung in grauen und weissen Substanz des Hundehirns. *Acta Neurochir.* **22**, 305–325.

Robinson B. W. (1962) Localization of intracerebral electrodes. *Exp. Neurol.* **6**, 201–223.

Roos A. (1965) Intracellular pH and intracellular buffering power of the cat brain. *Am. J. Physiol.* **209**, 1233–1246.

Rovit R. L. and Hagan R. (1968) Steroids and cerebral edema. The effects of glucocorticoids on abnormal capillary permeability following cerebral injury in cats. *J. Neuropath. Exp. Neurol.* **27**, 277–299.

Rubin R. C., Henderson E. S., Ommaya A. K., Walker M. D., and Rall D. P. (1966) The production of cerebrospinal fluid in man and its modification by acetazolamide. *J. Neurosurg.* **25**, 430–436.

Schroder R. (1973) Vergleichende Untersuchung zum spezifischen Gewicht des peritumoralen menschlichen Hirngewebes. *Acta Neurochir.* **28**, 341–352.

Scudder H. J. (1978) Introduction to computer aided tomography. *Proc. IEEE* **66**, 628–637.

Shalit M. N. and Mahler Y. (1966) Brain impedance measurement by the use of small bipolar needle electrodes. *J. Appl. Physiol.* **21**, 1237–1242.

Shigeno T., Brock M., Shigeno S., Fritschka E., and Cervos-Navarro J. (1982) The determination of brain water content: Microgravimetry versus drying-weighing method. *J. Neurosurg.* **57**, 99–107.

Shimabukuro H. (1969) Electrical impedance method for localizing brain structures. *Arch. Japn. Chir.* **38**, 612–625.

Sklar F. H. and Elashvili I. (1977) The pressure–volume function of brain elasticity. Physiologic considerations and clinical applications. *J. Neurosurg.* **47**, 670–679.

Sklar F. H., Reisch J., Elashvili I., Smith T., and Long D. M. (1980) Effects of pressure on cerebrospinal fluid formation: Nonsteady-state measurements in dogs. *Am. J. Physiol.* **239**, R277–R284.

Speller R. D., White D. R., Showalter C. F., Rothenberg L. N., Pentlow K. S., Morgan T. J., and Shope T. B. (1981) An evaluation of CT systems from ten manufacturers. *Br. J. Radiol.* **54**, 1053–1061.

Streicher E., Ferris P. J., Prokop J. D., and Klatzo I. (1964) Brain volume and thiocyanate space in local cold injury. *Arch. Neurol.* **11**, 444–448.

Tachibana S. (1970) Impedography in tumor localization. *Tr. Am. Neurol. Ass.* **95**, 317–319.

Tachibana S. (1971) Impedance study of brain tissue changes after penetrating injury. *Exp. Neurol.* **32**, 206–217.

Takagi H., Shapiro K., Marmarou A., and Wisoff H. (1981) Microgravimetric analysis of human brain tissue. Correlation with computerized tomography scanning. *J. Neurosurg.* **54**, 797–801.

Ter-Pogossian M. M. (1977) Computerized cranial tomography: Equipment and physics. *Sem. Roentgen.* **12**, 13–25.

Torack R. M., Alcala H., and Gado M. (1976a) Water, Specific Gravity and Histology as Determinants of Diagnostic Computerized Cranial Tomography (CT), in *Dynamics of Brain Edema* (Pappius H. M. and Feindel W., eds.) Springer, New York.

Torack R. M., Alcala H., Gado M., and Burton R. (1976b) Correlative assay of computerized cranial tomography (CTT), water content and specific gravity in normal and pathological post mortem brain. *J. Neuropath. Exp. Neurol.* **35**, 385–392.

Tornheim P. A. and Mc Laurin R. L. (1984) Effects of Mechanical Impact to the Skull on Tissue Density of the Cerebral Cortex, in *Recent Progress in the Study and Therapy of Brain Edema* (Go K. G. and Baethmann A., eds.) Plenum, New York.

Tornheim P. A., Mc Laurin R. L., and Thorpe J. F. (1976) The edema of cerebral contusion. *Surg. Neurol.* **5**, 171–175.

Unterberg A., Maier-Hauff K, Wahl M., Lange M., and Baethmann A. (1984) Cerebral Uptake and Consumption of Plasma Kininogens in Vasogenic Brain Edema: Recent Findings of Kinin Mechanisms, in *Recent Progress in the Study and Therapy of Brain Edema* (Go K. G. and Baethmann A., eds.) Plenum, New York.

Van der Veen P. H., Go K. G., Zuiderveen F., Buiter D., and Van der Meer J. (1973) Electrical impedance of cat brain with cold-induced edema. *Exp. Neurol.* **40**, 675–682.

Van Harreveld A. (1966) *Brain Tissue Electrolytes.* Butterworths, London.

Van Harreveld A. and Schade J. P. (1960) On the Distribution and Movements of Water and Electrolytes in the Cerebral Cortex, in *Structure and Function of the Cerebral Cortex* (Tower D. B. and Schade J. P., eds.) Elsevier, Amsterdam.

Van Harreveld A., Murphy T., and Nobel K. W. (1963) Specific impedance of rabbit's cortical tissue. *Am. J. Physiol.* **205**, 203–207.

Walser R. L. and Ackerman L. V. (1977) Determination of volume from computerized tomograms. Finding the volume of fluid filled brain cavities. *J. Comput. Assist. Tomogr.* **1**, 117–130.

Weiss M. H. and Wertman N. (1978) Modulation of CSF production by alterations in cerebral perfusion pressure. *Arch. Neurol.* **35**, 527–529.

Welch K. (1975) The principles of physiology of the cerebrospinal fluid in relation to hydrocephalus including normal pressure hydrocephalus. *Adv. Neurol.* **13**, 247–332.

Woodward D., Reed D. J., and Woodbury D. M. (1967) Extracellular space of rat cerebral cortex. *Am. J. Physiol.* **212**, 367–370.

Zatz L. M., Jernigan T. L., and Ahumada A. J. (1982) White matter changes in cerebral computed tomography related to aging. *J. Comput. Assist. Tomogr.* **6**, 19–23.

Zonneveld F. W. (1983) *Computed Tomography* Philips Medical Systems Publishing, Eindhoven.

Zonneveld F. W. and Vijverberg G. P. (1984) The relationship between slice thickness and image quality in CT. *Medicamundi* **29**, 104–117.

# Measurement of Metabolic Activity Associated with Ion Shifts

## Myron Rosenthal and Thomas J. Sick

## 1. Introduction

Brain "function" comprises the activities that transfer and integrate information within and among brain cells. Such information consists of changes in electrical potentials that exist across cell membranes. Electrical potentials are characteristic of both neurons and glia because of the asymmetrical distribution of ions, particularly $Na^+$ and $K^+$, across the membranes of these cells, between the intracellular and extracellular milieu. Under normal circumstances, the intracellular $K^+$ activity is more than 30 times greater than that in the extracellular space, whereas the extracellular $Na^+$ activity is 10 or more times greater than that within these cells (e.g., Katz, 1966; Katzman and Pappius, 1973). These transmembrane ion gradients produce a situation equivalent to a battery between cells and their external environment, the inside of brain cells being negative with respect to their outside. Whenever ion gradients change, such as occurs in response to ionic or neurotransmitter-mediated changes in membrane conductances (permeability), there are changes in transmembrane electrical potentials. These voltage shifts are the "information" of the nervous system.

Maintenance of transmembrane ion gradients requires one or more $Na^+$-$K^+$ active transport pumps that require metabolic energy. Membrane conductances of $Na^+$ and $K^+$ are never zero. There is always some inward leakage of $Na^+$ and outward diffusion of $K^+$, which means that pumping continues even when the brain is in a quiescent state. Ion transport is also dependent upon the intensity of electrophysiological activity of the tissue. During action potentials and synaptic transmission, conductances of $Na^+$ and $K^+$ are transiently increased, and there is increased diffusion of $Na^+$ and $K^+$, the result of which is a net gain of $Na^+$ and a net loss of $K^+$ by neurons and glia. These changes must be actively reversed by the ion pumps in order to reestablish "resting" membrane potentials.

The tight couple between metabolism and pump activity has been demonstrated by the loss of ion transport during metabolic inhibition in vitro (Hodgkin and Keynes, 1955; Shanes and Berman, 1955) and in vivo (Dixon, 1949; Morris, 1974; Kirshner et al., 1975; Sick and Kreisman, 1981). This link was also demonstrated by studies indicating that the largest fraction of brain energy is expended for ion transport. Precise measurement of the fraction of energy required to maintain ion gradients is complicated, since the brain constantly consumes energy for "residual" and "activation" functions (Astrup, 1982). Calculations of "residual" metabolism have required either membrane stabilization or direct inhibition of ion transport. For example, inhibition of $Na^+,K^+$-ATPase with ouabain produced a 40% decrease in oxygen consumption of brain slices (Whittam, 1961). A similar 40–50% decline in oxygen and glucose consumption was recorded in canine brain when ion leakage was decreased by lidocaine or transport inhibited by ouabain (Astrup et al., 1981). Even larger decreases in glucose consumption were reported during pentobarbital-induced EEG suppression in rat brain (Crane et al., 1978).

Studies of "activated" metabolism have required direct measurement of stimulus-provoked increments in metabolic and ion transport activity. Such investigations have consistently recorded linear relationships among increments in $K^+_o$ and signals of oxidative metabolic activity, demonstrating that the predominant fraction of increased energy, consumed in response to brain activation, is expended for restoration of ion gradients (Lewis and Schuette, 1975, 1976; Lothman et al., 1975; Somjen et al., 1976; Rosenthal et al., 1979). Thus, it is likely that expenditure of energy for ion transport is well above 50% of the brain's total.

The couple between ion transport and metabolism provides a basis for understanding brain metabolic physiology. Neuronal activation results in displacement of ions across cell membranes, resulting in increased ion pumping and increased energy utilization. This, in turn, leads to increased glucose and oxygen consumption, and, under normal conditions, blood flow is increased to deliver these substrates in adequate amounts to match their consumption. When energy supply does not match demand, concentrations of high-energy intermediates are decreased, ion gradients are lost, and brain function is compromised. Unlike other tissues, brain is extremely intolerant of metabolic insufficiency. Brain cell damage occurs rapidly when energy supply is inadequate. Yet unknown is whether such damage occurs because ion gradients are lost or because of the onset of other pathological processes. The concept is well established, however, that electrophysiology, metabolism, and viability of brain are intimately related.

In early studies, brain ion transport was inferred from evaluation of electrophysiological signals such as EEG or extracellular voltages. More recently, development of ion-selective microelectrodes (Walker, 1971) has enhanced studies of brain physiology by providing means to directly and noninvasively record ion shifts. Evaluation of ion shifts without reference to metabolism, however, opens only a limited window on understanding the relationships between ion transport and metabolism and the effects of pharmacological or pathological processes. Whether one studies in vitro preparations such as homogenates, cell cultures, or brain slices, or whether the intact brain is being examined, investigators must constantly be concerned with oxygenation, substrate supply, and energy status.

To relate ion transport and metabolism, metabolic measuring techniques are required that complement and enhance the recording of ion activities. It is toward describing such metabolic methods that this review is aimed.

## *1.1. General Concepts of Brain Energy Metabolism*

A review of some general concepts of brain metabolism may be helpful as a prelude to consideration of metabolic measuring techniques, although excellent textbooks and reviews of this subject are available (*see*, for example, Siesjo, 1978; Passonneau et al., 1980, McCandless, 1985). Brain energy production (or conservation) can be described as a sequential series of steps comprising: (1) the cytoplasmic reactions of anerobic "glycolysis"; (2) the intramitochondrial oxidative reactions of the tricarboxylic (or Krebs) cycle; and (3) the transfer, within the mitochondrial respiratory chain, of hydrogen ions and electrons (reducing equivalents) to molecular oxygen, with the coupled conservation of energy through oxidative phosphorylation. In general, this pattern of glycolytic and oxidative activity in brain is similar to that of other tissues such as liver or muscle, with the exception that alternate pathway activity via the pyruvate-decarboxylation paths and the pentose phosphate shunt has little significance. Relationships among these reactions are depicted in Fig. 1.

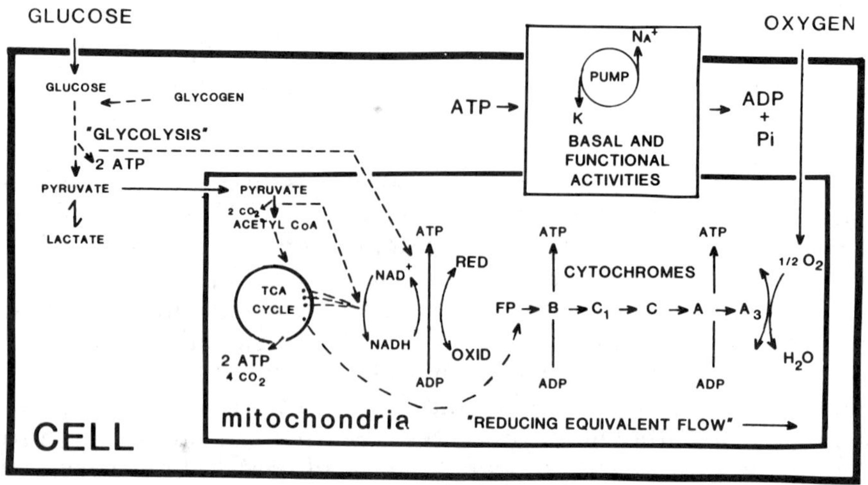



**SUMMARY**

GLUCOSE ⟶ 2 ATP VIA GLYCOLYSIS

30 ATP VIA NAD

4 ATP VIA FP

2 ATP FROM TCA CYCLE

_______________________

38 ATP TOTAL

Fig. 1. Relationships among the sequence of energy-conserving reactions comprising glycolysis, the TCA cycle, and oxidative phosphorylation and a summary of net ATP production. Dashed lines indicate the passage of reducing equivalents from intermediates to the electron carriers of the mitochondrial respiratory chain.


## 1.1.1. Glycolysis

In this series of reactions, often called the Embden-Meyerhof pathway, six-carbon glucose molecules are converted to two three-carbon molecules (pyruvate). These reactions occur in the cytoplasm and result in the net production of two ATP molecules and two molecules of reduced nicotinamide adenine dinucleotide (NADH). No oxygen is required for these reactions, but anaerobic glycolysis is a very inefficient mode of energy conservation. Under aerobic conditions, electrons from cytoplasmic NADH are carried to the intramitochondrial respiratory chain by the alpha-glycerophosphate shuttle system for further synthesis of ATP. When oxygen is unavailable, however, NADH must still be reoxidized to $NAD^+$ so that glycolytic production of ATP can continue.

Such NADH reoxidation occurs in a reaction that produces lactate from pyruvate, although lactate may have deleterious effects. It is estimated that brain glycolysis is increased 5–7-fold during anoxia, but glycolytic ATP production still falls far short of supplying mammalian brain's energy requirements.

### 1.1.2. Tricarboxylic Acid Cycle

Under aerobic conditions, pyruvate is converted into a two-carbon molecule (acetyl CoA) and $CO_2$. The tricarboxylic acid cycle comprises a sequence of intramitochondrial reactions by which acetyl CoA is converted into $CO_2$, electrons, and hydrogen ions ($H^+$). These reducing equivalents are accepted by the initial coenzyme of the respiratory chain (the intramitochondrial NAD), which couples the TCA cycle reactions with the flavoproteins, the cytochromes, and molecular oxygen. Fatty acid oxidation and amino acid catabolism are also linked to the respiratory chain via acetyl CoA or directly through the TCA cycle.

### 1.1.3. Respiratory Chain

Although all mammalian tissues are not at equilibrium, but at steady states that require metabolic energy for their maintenance, this is especially the situation for brain. In fact, mammalian brain is so far from equilibrium that anaerobic glycolysis is insufficient to account for energy demand. Brain relies upon free energy derived from the oxidative metabolic reactions that occur in the respiratory chain of mitochondria for its functional capability and survival. These respiratory chain reactions comprise the final steps in the oxidative production of ATP and involve transfer of reducing equivalents up the ascending redox potentials of the intramitochondrial electron carriers. Electron transfer is coupled to the phosphorylation of ADP and inorganic phosphate (Pi) to ATP through the free energy liberated by the reduction of oxygen to water. As shown, three sites are generally considered to be phosphorylation points between NAD and oxygen.

Tight coupling of electron transfer to oxidative phosphorylation provides a control mechanism for the entire metabolic system. When ADP levels are low, electron transport and oxygen consumption are greatly diminished. When ADP is added to the bathing medium of in vitro preparations such as mitochondria, cells, or tissues, or when intracellular ADP is increased, such as by increased tissue activity in vivo, a complex pattern of events occurs that results in changes in mitochondrial ultrastructure (Hackenbrock, 1968), increased electron transport, and oxygen consumption (Chance and Williams, 1956), and rephosphorylation of ADP to ATP.

Mitochondrial oxidation of glucose produces almost 20 times as much energy as glycolysis. For such oxidation to occur, each respiratory chain carrier must function continuously as one of a series of reduction-oxidation (redox) couples. Thus, NAD, flavoproteins, and cytochromes each exist in either an oxidized or reduced form dependent upon whether it is donating or accepting reducing equivalents. The chain acts as the "final common pathway" for the production of ATP from substrates.

## 1.2. Ion Shifts in Brain

For investigative purposes, ion shifts in brain can be grouped into four categories: (1) those provoked by "physiological" activation of brain tissues. These relatively small ion shifts (1–5 m$M$ increments in $K^+_o$, for example) can be evoked by direct application of stimulus pulses to CNS tissues or by pathway stimulation; (2) the larger ion shifts associated with epileptic seizures. Seizures are characterized by an apparent ceiling level of $K^+_o$ at around 10–12 m$M$ (Hotson et al., 1973; Moody et al., 1974; Fisher et al., 1976; Heinemann and Lux, 1977); (3) the massive ion movements that occur during spreading depression provoked either by intense electrical stimulation, mechanical deformation of tissue, or application of KCl (Leao, 1951; Grafstein, 1956; Marshall, 1959; Vyskocil et al., 1972). $K^+_o$ levels in excess of 70 m$M$ have been recorded in some instances; and (4) ion shifts provoked by metabolic inhibition such as hypoxia, ischemia, or hypoglycemia. These ion shifts are related to the intensity of the metabolic insult, but can be as large as those of spreading depression.

For reference purposes, representative ion shifts are shown in Figs. 2 and 3. In the example depicted in Fig. 2A (from Heinemann et al., 1977), tactile stimulation of a cat forepaw (Top left panel) elevated cerebral $K^+_o$ by less than 1 m$M$ and decreased $Ca^{2+}_o$. More synchronous activation of cortical neurons, produced by electrical stimulation of the thalamus (Fig. 2A, right panel), resulted in approximately a 2 m$M$ increase in $K^+_o$ and a 0.2 m$M$ decrement of $Ca^{2+}_o$. Cerebral epileptic activity produced a very intense neuronal discharge, a 4–5 m$M$ increase in $K^+_o$, and a 0.3 m$M$ decrease in $Ca^{2+}_o$ (Fig. 2B). Ion shifts accompanying cortical spreading depression are shown in Fig. 2C (compiled from Kraig and Nicholson, 1978 and Mutch and Hansen, 1984). During the peak of the wave,

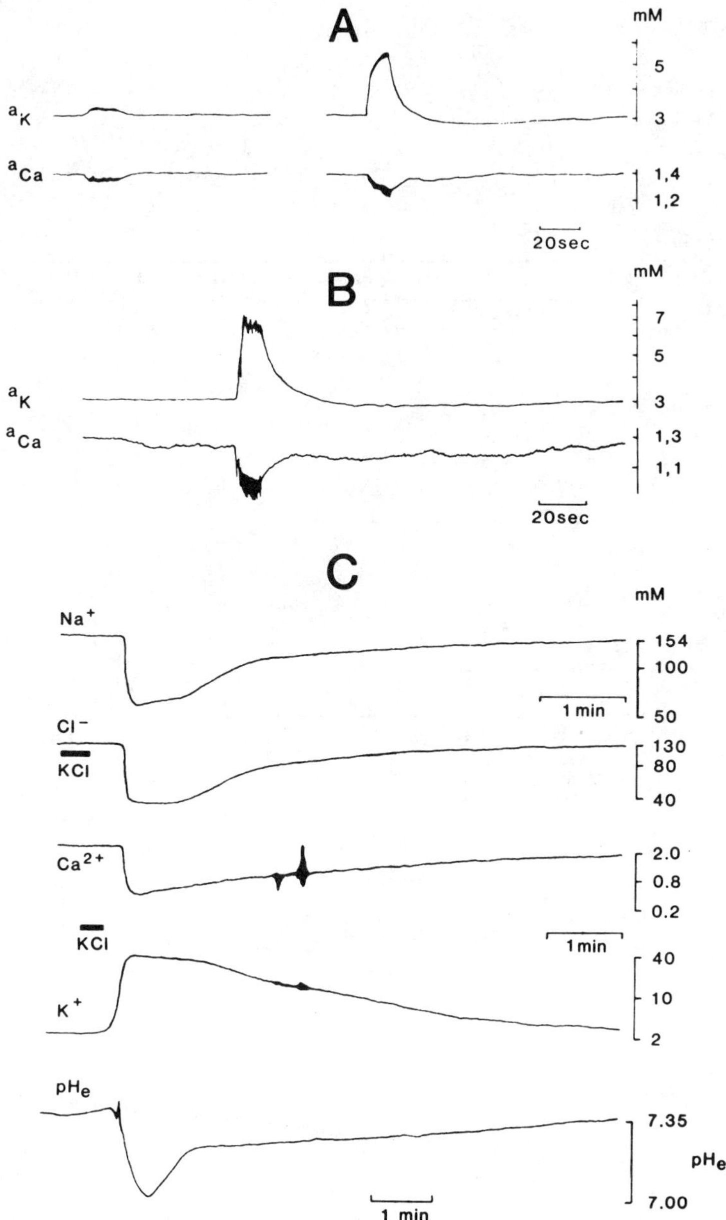



Fig. 2. Representative ion shifts in brain. (A) Left traces are ion shifts provoked by "physiological" stimulation of cat forepaw; right traces are responses to electrical stimulation of the thalamus (from Heinemann et al., 1977). (B) Ion shifts during seizure activity (from Heinemann et al., 1977). (C) Ion shifts accompanying cortical spreading depression [from Kraig and Nicholson, 1978 (with permission from Pergamon Press); Mutch and Hansen, 1984].

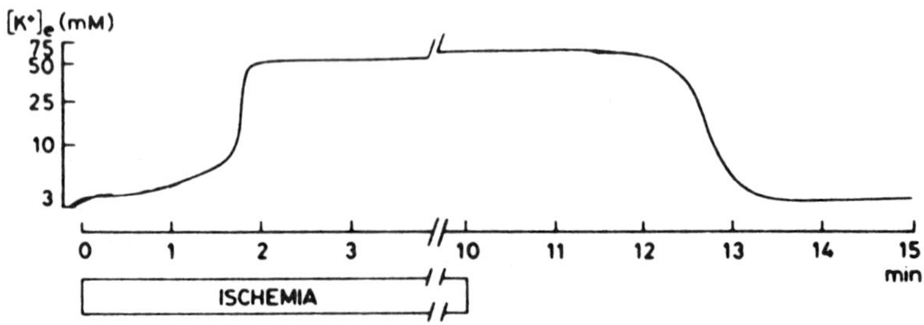


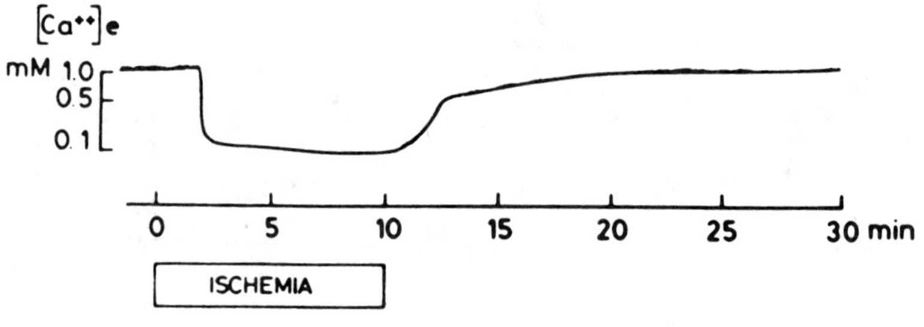


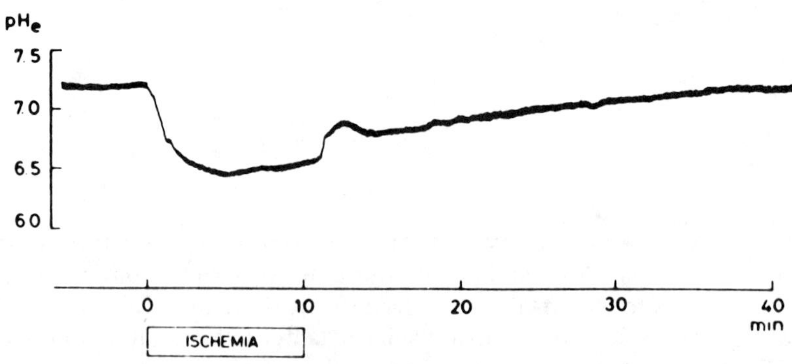


Fig. 3.   Ion shifts measured in cerebral cortex during brain ischemia (compiled from Siemkowicz and Hansen, 1981).

$Na^+_o$ and $Cl^-_o$ decreased approximately 100 m$M$, $Ca^{2+}_o$ decreased nearly 2.0 m$M$ $K^+_o$ increased approximately 50 m$M$, and extracellular pH decreased approximately 0.4 units.

When oxidative metabolism is compromised, ion shifts occur that are similar to, but often more intense than, those observed during neuronal activation. Examples of shifts in $K^+_o$, $Ca^{2+}_o$, and pH, which occur during brain ischemia, are shown in Fig. 3 (modified from Siemkowicz and Hansen, 1981). The duration of ion changes depends upon the duration of metabolic inhibition.

## 2. Metabolic Measuring Techniques

In sections below, techniques will be compared for their capability to measure metabolic activities that occur in association with ion shifts such as those described above. This will offer a means to identify the strengths and limitations of each metabolic method for studies of the coupling of metabolism and ion transport.

### 2.1. A-V Difference Methods

An early but very useful approach to brain metabolism was based upon the measurement of arteriovenous (A-V) differences to calculate, by Fick principles, the consumption of substrates or oxygen. This approach has been especially useful for the study of large changes in glucose and oxygen consumption, such as the increases in use of these substrates that occur during generalized seizures (Plum et al., 1968; Borgstrom et al., 1976). Measurement of arteriovenous differences is complicated by technical difficulties, however, and the fact that such differences are often very small and easily lost in system "noise." Also, spatial resolution is limited to the brain region in which arterial and venous blood supply is totally isolated. In most studies, A-V difference measurements are made from the whole brain, thereby making it most difficult to gain insight into metabolic activity associated with local ion shifts that result from physiological brain activation. For example, ion shifts associated with seizures are only slightly larger than those accompanying physiological activation. The ease with which A-V differences of oxygen and glucose are detected during seizures likely reflects a larger volume of neuronal activation that overcomes spatial limitations of the method. Also, early investigations indicated no change in brain oxygen consumption during periods of increased mentation (Sokoloff et al., 1955) or between periods of sleep and wakefulness (Mangold et al., 1955). Such small perturbations in ion homeostasis that occur during normal mental processes likely encompass a small proportion of total brain activity and are not detectable by A-V difference methods. Indeed, Lewis and Schuette (1976) did record changes in relative oxygen consumption provoked by direct cortical stimulation by measuring sagittal sinus blood flow and hemoglobin saturation. Although stimulation was applied to a wider area of cerebral cortex than the local (1–2 mm) areas sufficient for optical metabolic techniques (see below), these studies demonstrated that increases in relative oxygen consumption occurred in proportion to increments in $K^+_o$ and to the amplitude of shifts toward oxidation of intramitochondrial NADH.

A-V difference techniques are also limited in studies of metabolic rates necessary to prevent ionic and electrophysiological dysfunction during global metabolic insults. Under these circumstances, thresholds for ionic and electrical failures may vary among regions with differing metabolic rates. Global measurements may not reflect such regional differences, and ion homeostasis may be locally lost with little change in brain oxygen or glucose consumption.

Temporal resolution of A-V difference techniques is limited by the requirement to measure brain blood flow. Radiolabeled inert gas methods can require as much as 25–30 min (Lassen and Klee, 1965). Clearance techniques for hydrogen or dye indicators are much faster, but are still limited in their capability to follow physiological activities or even the rapid ionic and metabolic changes produced by acute insults such as ischemia. Also, arteriovenous difference methods are valid only for substances not synthesized in brain such as oxygen and glucose. Other metabolites of interest for investigations of ion homeostasis, such as ATP and PCr, are not amenable to measurement by these approaches.

## 2.2. Assay and Autoradiographic Methods

Many of the basic concepts of brain metabolism have emerged from assay sampling of the concentrations of intermediary metabolites present in global or local regions of tissue. Most assay protocols are based upon standard enzymatic fluorometric methods (Lowry and Passonneau, 1972). Advances in technology have made possible the increasing sensitivity of analysis procedures, however, so that local metabolite concentrations can be determined in increasingly small tissue samples (Chi et al., 1978).

Assay methods have been particularly useful in evaluation of the effects of metabolic inhibition and the consequences of seizures and spreading depression. During seizures, for example, specific metabolite changes have depended upon models, but the pattern is usually as expected from a tissue that is consuming energy at a rate faster than energy is being conserved. Although studies have not been specific to ion shifts during seizures, typical ictal bursts produced decreases in PCr and ATP and increases in ADP, AMP, and lactate. Relationships of metabolite concentration shifts to electrophysiology have been assumed (Minard and Davis, 1962; Folbergrova, 1974; Duffy et al., 1975; Chapman et al., 1977). Other investigators have attempted to account for the ion shifts of spreading depression by stopping metabolism at times determined by $K^+_o$ or by the progression of the wave of extracellular depolarization. Such studies have shown that spreading depression is associated with decreases in glucose and high-energy phosphates and increased tissue lactate (Krivanek, 1961; Quistorff et al., 1979; Csiba et al., 1985).

During metabolic insults, relationships between brain ion shifts and metabolism have also been described from evaluations based upon assay sampling. Decreases in blood flow, oxygen, or glucose beyond "critical" levels resulted in decreased PCr and then in decreased ATP with concomitant and expected effects on other representative metabolites (Siesjo, 1978). Nevertheless, these studies illustrate a limitation of invasive sampling procedures since it is difficult to directly relate biochemical data to brain activity, especially when this activity is rapidly changing. Assay techniques can measure only a single point in time. Conclusions on the time course of events and the relationships between metabolism and ion transport must rely upon statistical analyses of data derived from many animals.

Since assay-sampling the concentrations of brain metabolites is perhaps the simplest method conceptually for metabolic studies, it might at least be expected that an account of assay methods would be straightforward and results would not be controversial. Yet many publications in this field have been criticized for sampling artifacts caused by control variations or by difficulties in stopping metabolism and preventing postmortem changes. Physiological variables known to influence metabolite levels include body temperature, arterial blood pressure, blood glucose concentration, blood pH, and the arterial tensions of oxygen and carbon dioxide. Anesthesia provides an additional variable since anesthetics have large, and often differing, influences on brain metabolite data. Investigators in recent years have taken great care to maintain these variables in animals within standard (and presumably "normal") ranges.

Since metabolite values change rapidly when brain circulation or oxygen supply is impaired (e.g., Nilsson et al., 1975), there has been much effort exerted to improve methods for stopping metabolism as a first step in the assay of metabolite concentrations. As described by Siesjo (1978), early studies were based upon freezing after decapitation, immersion of animals in a coolant, rapid suction of tissue into a coolant, or tissue excision with cooled spoons. However, reports that circulation of deeper structures continues as the brain freezes in layers during surface application of coolants (Kerr, 1935; Richter and Dawson, 1948) led to the application of surface freezing techniques for assay studies (Ponten et al., 1973). Surface cooling has become the predominant method presently used in anesthetized animals, although "freeze clamping" (Quistorff, 1975) and microwave irradiation (Medina et al., 1975; Hampson et al., 1982) are also used, and, together with the "freeze-blowing" technique of Veech et al. (1973), have the advantage of being applicable to unanesthetized preparations. Microwave irradiation, and certainly the "freeze-blowing" procedure, have the disadvantage that structural integrity is lost, thereby precluding any possibility of regional analysis. For this and other questions related to the efficiency of "trapping" metabolic activity, there remains no consensus on the optimal approach to the problem of stopping metabolic processes for assays of metabolite levels.

Interpretation of assay data is complicated by the fact that the concentration of any metabolite reflects the difference between its supply and use. For example, ATP levels were restored after ischemia at a time when oxygen consumption was decreased. NAD and cytochrome $a,a_3$ were hyperoxidized and ECoG was suppressed during this time (Harrison et al., 1984). This likely indicates that ATP concentration is restored because ATP use is also decreased. Another disadvantage to this approach is that assay procedures for metabolites have very limited potential for purposes of neurological diagnosis or prognosis.

Application of radioactive glucose analogs, such as 2-deoxy-$d$-glucose (2-DG), has provided a means of estimating brain glucose phosphorylation and, thereby, methods to calculate the rate of local glucose consumption (Sokoloff et al., 1977). This approach offers regional assessments with high resolution. In addition, the tight couple between metabolism, ion transport, and electrophysiology has allowed the mapping of brain "function" on the basis of this metabolic activity (Sokoloff, 1981).

Autoradiographic procedures are based upon the fact that radioactive glucose analogs, injected into the blood stream, are taken up by brain and phosphorylated as glucose. Unlike glucose, the phosphorylated analog is not a suitable substrate for further reactions along the glycolytic pathway. Rather, the phosphorylated analog is trapped in the tissue and its accumulation is considered to be proportional to the rate of cell glucose phosphorylation. By using autoradiography after a suitable period of time (generally 45 min), the size and location of the radioisotope pool can be calculated through knowledge of the relative concentrations of glucose and its analog in the precursor pools and certain kinetic constants.

As is common to most nonassay metabolic methods, glucose autoradiography requires assumptions that complicate calculation of absolute values (Hawkins and Miller, 1978; Gjedde, 1982; Partridge et al., 1982). Nevertheless, this procedure has proven useful for qualitative studies. When spreading depression was induced by continuous KCl application to the cerebral surface in rats, for example, glucose utilization increased in the neocortex, but decreased in certain subcortical structures (Shinohara et al., 1979). During seizures, glucose consumption increased in affected areas (Kennedy et al., 1975; Caveness et al., 1980).

Although glucose autoradiography offers excellent spatial resolution, it is limited in temporal resolution by the requirement that a metabolic "steady state" be maintained for a prolonged period between injection of the radioactive glucose analog and decapitation. This period, which has been 45 min in most published studies, is necessary to allow the free, unreacted analog to be cleared from the tissue. This delay limits use of glucose autoradiography for the study of ionic events that occur over much shorter periods of time and precludes its use for analysis of transient activity such as evoked potentials. Recent studies using [14]C-glucose with autoradiography have been performed within 5 min periods in small animals, however (Hawkins et al., 1979; Lu et al., 1983). Interpretation of glucose autoradiography is also complicated by the fact that increases in glucose consumption could be indicative of conflicting events such as neuronal activation or hypoxia. Interpretation is best made in combination with brain physiological measurements and often with data derived from assays of metabolites. Nevertheless, autoradiographic principles have been a foundation for positron emission tomography, which has greatly enhanced the study of brain metabolic activities in animals and humans.

## *2.3. Positron Emission Tomography*

Positron emission tomography (PET) has proven useful for investigations that link electrophysiology, ion shifts, and brain metabolism, especially in large animals and humans. For PET analysis, chemical compounds with desired biological activity are labeled with a short-lived radioactive isotope and are introduced into the blood either by injection or inhalation. As the radioactive tracer decays, it emits a positron (positive electron) that almost immediately combines with an electron, causing annihilation of the positron and electron and the creation of two gamma photons that travel in opposite directions. Pairs of external photon-counting tubes are aligned in a circular array to detect the photons and determine the position of the annihilation event. The spatial distribution of the radioactivity is reconstructed by computer. Mathematical models allow for computation of tissue utilization of the tracer compounds (Raichle, 1979). This technique can provide a noninvasive, local image of many biochemical processes and even drug distributions.

Two tracers, $^{18}$F-deoxyglucose (FDG) and $^{15}O_2$, have commonly been employed for measurements of glucose and oxygen utilization, respectively. An advantage of PET is that quantitative estimates of glucose and oxygen utilization (rather than simply tissue content) are assumed to be made and there is improved spatial resolution compared to measurements of A-V differences. Although direct measurements of changes in ion activities have not yet been obtained in conjunction with PET, inferences may be made from studies in animal models of physiological activation, seizures, and metabolic insult. Interpretation of PET data is complicated, however, by the temporal constraints of isotope half-life, the necessity for long scan times, and questions of whether deoxyglucose phosphorylation in cells reflects only the glucose metabolic rate or other metabolic activities as well (Fox, 1984). During the 1–2 h period required to complete a PET scan with FDG, for example, the brain may undergo multiple cycles with ictal bursts followed by periods of postictal depression and periods of relatively normal electrical activity. Thus PET measurements will reflect a weighted average of ictal and interictal periods. This may explain findings that glucose use is decreased during interictal periods, but that glucose utilization may be increased or decreased during ictal episodes (Engel et al., 1985). This temporal limitation may be overcome to some extent in studies of electroconvulsive shock in which the seizure is under experimental control or in studies of $O_2$ consumption in which the PET scan time is significantly lessened.

PET measurements of glucose and oxygen utilization have been valuable in studies of large metabolic changes, such as those of human stroke, especially in those situations in which ischemia-induced changes are relatively invariant or are slowly evolving. As expected, glucose and oxygen utilization are depressed during ischemic insults at times when characteristic ion shifts are occurring as recorded in animal studies (Powers and Raichle, 1985; Frackowiak, 1985). Loss of ion homeostasis is accompanied by brain edema and likely contributes to tissue infarction and cell death. Although the critical level of glucose or oxygen utilization necessary to prevent loss of ion homeostasis or to promote recovery is unknown, PET methodology seems capable of addressing this important question.

An advantage of PET is the apparent sensitivity of this technique to metabolic changes that are associated with ion shifts smaller than those that accompany metabolic inhibition or seizures. In humans, prolonged activation of sensory pathways was found by PET to increase glucose utilization in brain structures appropriate for the particular sensory modality (Phelps et al., 1981; Reivich et al., 1985). Since sensory activation in experimental animals results in small ion shifts, including increased $K^+_o$, it is reasonable to assume that similar ion shifts accompany sensory stimulation in humans and that these ion changes are associated with the increases in glucose utilization. Recent studies of $O_2$ utilization by PET have failed to detect increased $O_2$ consumption with sensory stimulation, however. This result is surprising since $O_2$ consumption was shown to increase with neuronal activation that provoked only small increments in $K^+_o$ (Lewis and Schuette, 1976). Further PET studies will likely clarify this situation.

Already discussed is the limitation in temporal resolution placed upon PET by the half-life of the radionuclide. For example, the half-life of $^{18}$FDG is approximately 2 h. Therefore prolonged steady states with constant neuronal activity are required. This restricts study of ion shifts that occur with evoked potentials or even spreading depression. Also, quantitation of glucose and oxygen utilization is dependent upon adherence to the assumptions required for the mathematical models that may be violated under pathological conditions. Another disadvantage of PET technology is a pragmatic one that has also limited the application of magnetic resonance spectroscopy (MRS). Each of these techniques requires very expensive detecting and computational equipment. For PET, an immediately available cyclotron is also required for the production of short-lived radionuclides, and a trained staff must manage this complex equipment.

## 2.4. Magnetic Resonance Spectroscopy (MRS, NMR)

Magnetic resonance spectroscopy is another evolving technique that holds great potential, both experimentally and clinically, for measuring (and imaging) metabolic activity in biological tissues. Certain atomic nuclei (for example, $^1$H, $^{13}$C, and $^{31}$P) have magnetic dipole moments that will align in a strong magnetic field and precess (spin) along the axis of the dipole at characteristic frequencies. The presence of these nuclei can be detected by applying radiofrequency pulses to tissues that produce an oscillating magnetic field. When the pulses are presented at the resonant frequency of a particular compound, the dipole will realign itself at right angles to the fixed magnetic field. Upon termination of the radiofrequency pulse, the dipole will decay back into alignment along the lines of magnetic flux. This decay induces a measurable current in the magnetic field. Detection of any compound is dependent upon the strength of its dipole moment, the strength of the magnetic field, the tissue concentration of that molecule, and interference by compounds with similar resonant frequencies. Typical instruments consist of a superconducting magnet to polarize the magnetic nuclei in a tissue, a radiofrequency transmitter to supply excitation radiation, a radiofrequency receiver to detect the "resonance" or frequency shifts, and a computer system for experimental control, data acquisition, and calculations.

Although in vivo phosphorous-MRS is less advanced than PET for clinical applications, the availability of large-bore superconducting magnets has made possible MRS measurements of creatine phosphate, phosphodiesters, and inorganic orthophosphate and ATP, ADP, AMP, and other important metabolic intermediates in several organs, including the brains of animal models and humans (Radda and Seeley, 1979; Thulborn et al., 1982; Cady et al., 1983; Chance et al., 1983; Prichard et al., 1983; Hilberman et al., 1984). Intracellular pH may also be estimated from changes in inorganic phosphate (Petroff et al., 1985). In addition, recent investigations have shown that changes in other organic molecules can be measured by either $^{13}$C or $^{1}$H MRS, although special techniques are required (Prichard and Shulman, 1986).

An advantage of $^{31}$P MRS is that high-energy phosphate concentrations can be measured as a function of time. This measurement is noninvasive and can be accomplished repeatedly from the same organ. As with PET, direct studies of ion shifts in brain have not yet been reported, although MRS has also been advantageously used under conditions of known ion shifts produced by neuronal activation and metabolic insults. For example, MRS techniques demonstrated a decline in PCr, ATP, and pH during ischemia in gerbil brain (Thulborn et al., 1982), and similar insults were correlated with increases in brain lactate (Prichard and Shulman, 1986). These changes are likely related to elevation of $K^{+}_{o}$ and $H^{+}_{o}$ and decreased $Ca^{2+}_{o}$. Similar decreases in PCr, ATP, and pH occurred during seizures (Prichard et al., 1983). The more massive ion shifts associated with cortical spreading depression have not been studied with MRS, and the question of sensitivity of MRS to physiological activation has not yet been resolved.

The usefulness of MRS in studies of ion homeostasis cannot be overemphasized. Such questions as the threshold of metabolic insult for loss of ion homeostasis during hypoxia, or whether ion shifts of seizures or spreading depression result in brain energy failure, are amenable to study by this procedure. There are certain disadvantages of MRS that must be considered, however. One of these is its limited temporal resolution. The detectability of MRS signals depends upon concentrations, sample size, and field strength and uniformity of the magnetic field. The sampling volume may be controlled to some extent by varying the size of the radiofrequency coil. Signal detection may also be improved by using magnets with uniform and high field strength, but these add considerably to the cost of this expensive technique. Temporal resolution must be sacrificed, however, for measurements of compounds in low concentrations from small samples, although it may be improved if the experimental manipulation can be repeated so that spectra may be averaged. Another disadvantage of $^{31}$P MRS is that concentrations rather than turnover rates of high-energy phosphates are measured, although magnetization transfer has been used to follow the rates of some enzyme-catalyzed reactions in brain (Shoubridge et al., 1982; Balaban et al., 1983). Also, current procedures are limited to changes rather than absolute concentrations.

## 2.5. Optical Techniques

Of the available procedures for monitoring metabolic activities, optical techniques have been the most applied to date for the direct correlation of metabolism with ion shifts. These techniques take advantage of the optical properties of electron transport chain reactants to record reduction/oxidation (redox) transitions as these occur in the mitochondria of cells in intact brain and other tissues. The redox status of the mitochondrial reactants provides a sensitive index of the functional condition of the oxidative metabolic processes because this status is determined by the availability of reducing equivalents from glycolysis, oxygen availability and consumption, and the rate of ATP utilization (see below and Chance and Williams, 1956). Thus, measurement of mitochondrial redox shifts offers a direct window to many of the intracellular activities that comprise the intracellular energy conservation system.

### 2.5.1. Optical Properties of the Mitochondrial Electron Carriers

There are two fundamental optical properties of the mitochondrial respiratory chain members that are at the basis of optical technologies. One is that the electron carriers of the respiratory chain absorb light at specific wavelengths. In addition, light absorption by certain electron carriers (the pyridine nucleotides and flavoproteins) results in their fluorescence. The wavelengths of absorption and fluorescence are truly distinguishing features allowing for identification of each individual reactant (Keilin, 1966). The second is that the oxidized and reduced forms of the electron carriers show absorption and, in some cases, fluorescence differences. These properties are illustrated in Fig. 4, which contains an absorption spectrum derived from slices of cerebral cortex (from Jobsis, 1979). This spectrum represents differences in absorbance between oxygenated and anoxic slices plotted versus wavelength. Such a spectrum of absorption differences is typical of those that occur between oxygenated and anoxic preparations of mitochondria or tissues in vitro or organs in vivo.

As shown in Fig. 4, transitions from an oxygenated to a disoxygenated state result in changes in absorption that can be attrib-

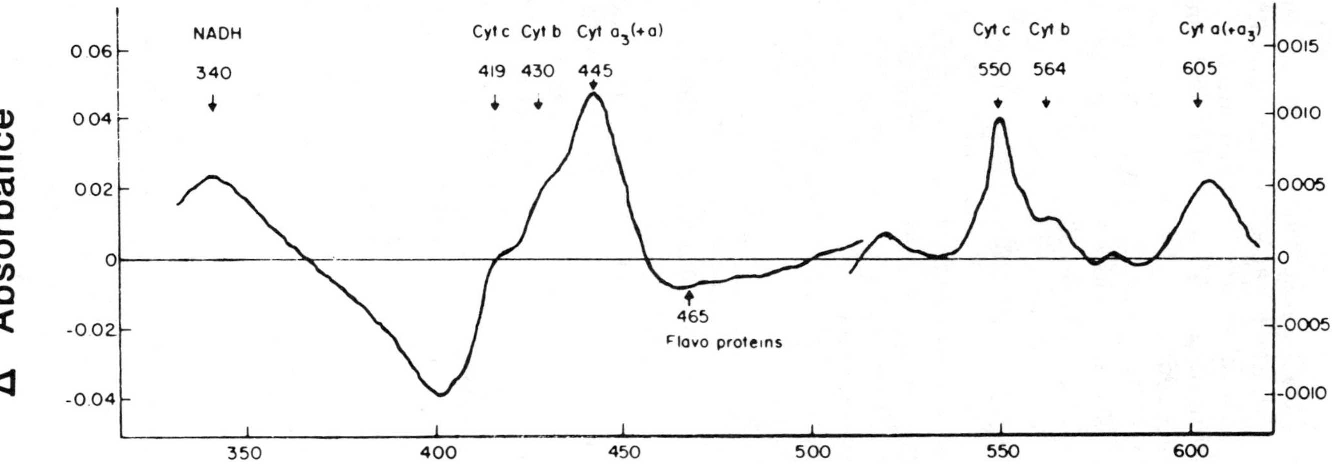


Fig. 4.  Absorption spectrum from cerebral slices. This spectrum represents the difference in absorbance between oxygenated and anoxic slices plotted as a function of wavelength. Increases in absorbance are plotted in the upward direction (from Jobsis, 1979).

uted to the reduction of specific respiratory chain reactants. For example, there is an absorption peak at 605 nm, indicating that this is a wavelength of maximal difference between the reduced and oxidized forms of cytochrome $a$ (with a small contribution from cytochrome $a_3$). Therefore, reduced cytochrome $a,a_3$ absorbs more light at 605 nm than does the oxidized form of this cytochrome. Absorption maxima at 564 and 550 nm indicate that reduced cytochrome $b$ and $c$, respectively, also absorb more light at these characteristic wavelengths than do the oxidized forms of these cytochromes. The negative peak (absorption minimum) at around 465 nm is attributable to the flavoproteins, indicating that these absorb more intensely when oxidized. The absorption maximum at 445 nm demonstrates that cytochrome $a_3$ (with a small contribution from cytochrome $a$) absorbs more intensely at this wavelength when reduced than when oxidized. Additional absorption maxima occur at approximately 430 nm (cytochrome $b$), 419 nm (cytochrome $c$), and 340 nm (NADH). Although not shown, there is also an absorption minimum around 840 nm, indicating that cytochrome $a,a_3$ absorbs more intensely in this infrared region of the spectrum when oxidized than when reduced. Relative heights and locations of these absorption peaks may vary somewhat among preparations for two reasons: (1) light-scattering properties and penetration varies in a wavelength-dependent manner and (2) there may be more than one species of certain cytochromes present in tissues with slight differences in optical properties. This is especially the situation for cytochrome $b$, the absorption peak of which, in many circumstances, is masked either by absorption of cytochrome $c$ or by other $b$-related cytochromes. These complications do not preclude the monitoring of cytochrome $a,a_3$ redox activity in vivo in coordination with electrophysiological and ionic events (c.f. Fig. 5). Although redox changes that occur during tissue "functional" activity are small compared to the oxygenated-anoxic shifts illustrated, these "physiological" changes are readily measurable.

The pyridine nucleotides (NAD) and flavoproteins also have fluorescence properties that have contributed to optical monitoring. When illuminated with light in the wavelength region from around 310–375 nm, the reduced form of NAD (NADH) will emit photons in the blue region of the spectrum, around 450 nm. The oxidized form of NAD (NAD$^+$) has minimal fluorescence efficiency at these wavelengths. In contrast, flavoproteins fluoresce more when oxidized than when reduced. If illuminated with light

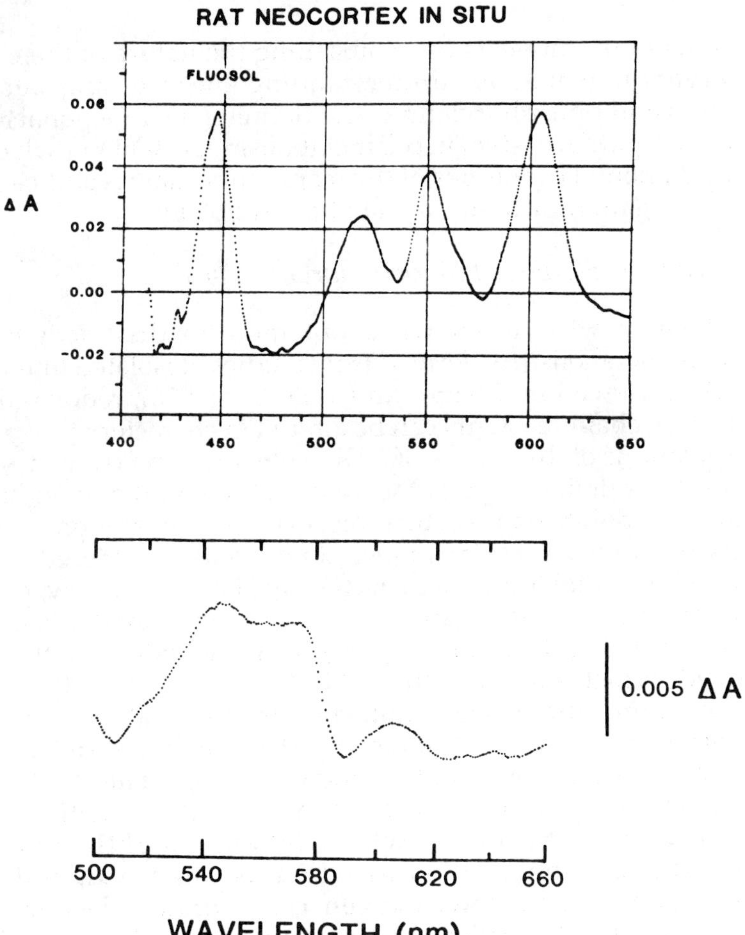



Fig. 5.   Reflection difference spectra from rat brain recorded with a rapid-scanning spectrophotometer. The upper spectrum is the difference between oxygenated and anoxic hemoglobin-free (Fluosol-perfused) rat brain. The lower spectrum is the difference between hemoglobin-perfused control brain and that stimulated by direct application of stimulus pulses to the cortical surface.

around 420 nm, oxidized flavoproteins fluoresce in the green portion of the spectrum, around 540 nm. Reduced flavoproteins contribute negligibly to this fluorescence.

The optical properties of the mitochondrial electron carriers have been the basis for nondestructive techniques applicable to in vitro and in vivo preparations. Although there has been much emphasis in recent years on establishing the nature of the energy conservation reactions, understanding the coupling between oxidation and phosphorylation, and defining the functional role of the mitochondrial system in cell metabolism, present knowledge of the fundamental properties of the respiratory chain is sufficient for the investigation of tissue metabolic physiology.

### 2.5.2. Optical Studies of Mitochondria In Vitro

Much of what is known about mitochondrial activity and oxidative metabolism is derived from studies of isolated mitochondria. As reviewed by Chance and Williams (1956), redox ratios of respiratory chain reactants can be altered in characteristic ways by manipulations of the supply of ADP, substrate, and oxygen. These investigators defined five mitochondrial "states" that have served as reference points for investigations in many preparations, including tissues in vivo. For example, state 4 was considered to be a "resting" condition under which ADP availability was low, oxygen consumption was slow, and a maximal redox gradient existed between the highly reduced pyridine nucleotide and the fully oxidized cytochrome $a$. When ADP was added to the medium containing the isolated mitochondria, a "state-3" condition was provoked that was characterized by an increased respiratory rate (oxygen consumption increased). In state 3, all coenzymes and cytochromes became more oxidized with the exception of cytochrome $a$, which became slightly reduced. When ADP phosphorylation to ATP was completed, and ADP levels again became low, oxygen consumption declined and mitochondrial redox ratios shifted to those found in state 4. All respiratory chain reactants became fully reduced during anoxia (state 5) and fully oxidized when substrate was unavailable (state 2).

Thus, for the cytochromes, flavoproteins, and pyridine nucleotide, specific changes in redox ratios accompany changes in the rate of oxygen utilization and oxygen and substrate availability. As extrapolated to tissues, when ATP is expended to provide energy for cell activity, ADP must be rephosphorylated. Consequently, there is an increased rate of oxygen utilization mediated by increased ADP [or by the decreased phosphate potential (ATP/ADP + Pi)], which initiates transport of reducing equivalents from substrate to oxygen. The system is well regulated and efficiently consumes oxygen in proportion to the need for energy production. Oxygen use, electron transfer, and ADP rephosphorylation are obligatorily coupled to one another, and it is this coupling that provides a basis for use of redox status to signal metabolic activity.

### 2.5.3. Development of Optical Methods

Development of optical methods for the study of respiratory chain activity has occurred in phases encompassing: (1) early procedures based upon microscope-based spectroscopic instruments; (2) spectrophotometer-based systems using electronic photodetectors for absorption measurement; (3) development of fluorometric and spectrophotometric systems capable of monitoring redox activity in intact, blood-perfused tissues; and (4) recent applications of diode array detection devices in "rapid-scanning" spectrophotometers.

2.5.3.1. SPECTROSCOPIC TECHNIQUES.      Descriptions of the light absorption properties of cells and tissue were provided by Mac-Munn (1884) and later by Keilin (1925, 1966). These were accomplished by simple microspectroscopic instruments by which tissues, cells, or even small insects were transilluminated with light at all visible wavelengths ("white" light). This light was collected by a microscope that was positioned at the surface of the tissue opposite the light source. The light that was passed through the tissue (light not absorbed by the tissue) was dispersed into a spectrum by a prism placed in the microscope's objective. With such instruments, MacMunn and Keilin noticed that light absorption did not occur equally at all wavelengths. Rather, they observed absorption bands that always appeared in the same spectral positions and were remarkably constant throughout a wide variety of biological preparations (*see* Fig. 6). MacMunn became convinced that these absorption bands were derived from intracellular pigments rather than from hemoglobin. This idea remained controversial until Keilin identified the absorbing pigments as intracellular "cytochromes" and determined that they compose an electron transport chain that plays an essential role in cell respiration. Keilin, in fact, reported perhaps the first experiment that related changes in light absorption of the respiratory pigments and functional activity when he observed the appearance of absorption bands in thorax muscles of moths as they moved their wings (Keilin, 1925). Although this was the first measurement of the time course of an intracellular metabolic event, the appearance and disappearance of absorption bands could only be visualized through the microspectroscope, which limited quantification. Also, these absorption changes could be stored permanently only on photographic film.

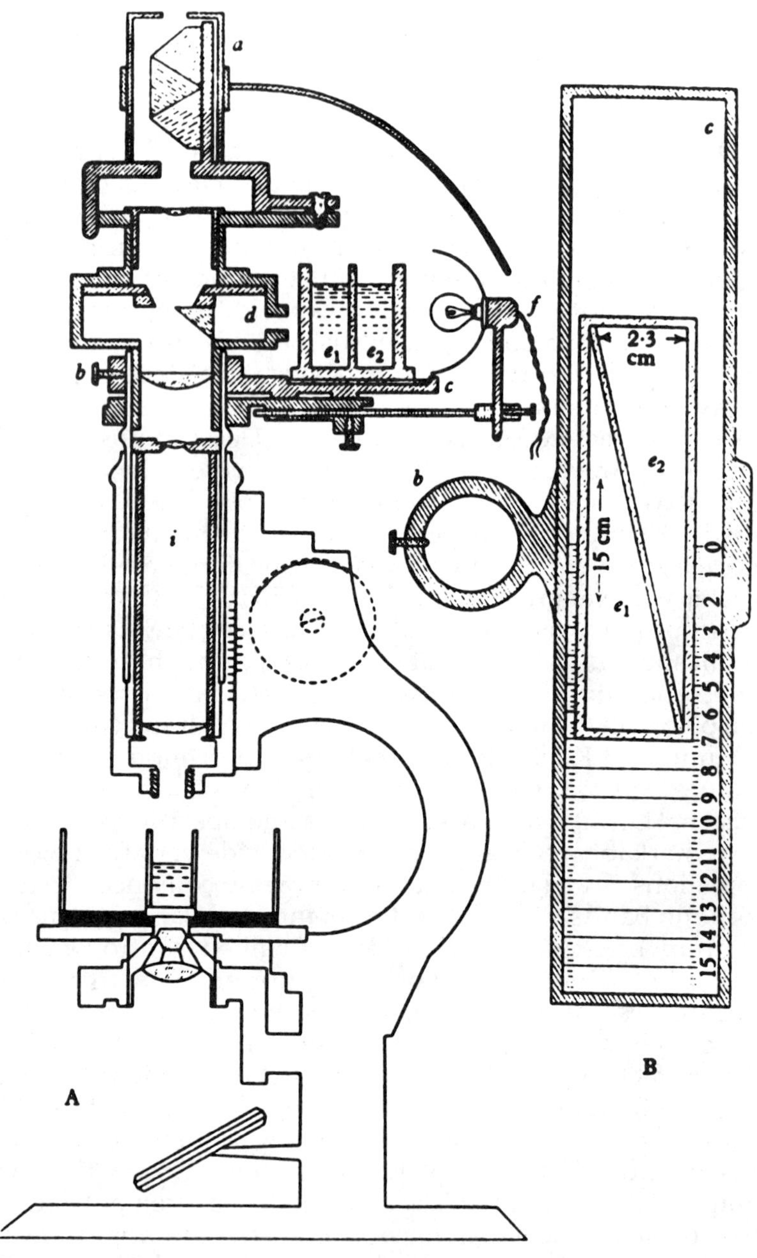

2.5.3.2. SPECTROPHOTOMETRIC TECHNIQUES.    Major advances in optical methods occurred as a result of developments in electronics and light detectors. These instruments standardized spectral measurement and facilitated the quantification of absorbance in solutions or tissues. Principles of these instruments are derived from the laws of Lambert and Beer, which relate the ability of a solution to absorb monochromatic light and the concentration of the absorbing molecules in this solution. As commonly stated, the Beer-Lambert law defines optical density or absorbance as:

$$A = \log(Io/I) = kcl$$

where $A$ is absorbance or the logarithm to the base 10 of the ratio $Io/I$, $Io$ is the intensity of light transmitted through the medium, and $I$ is the intensity of incident light. The law states that absorbance equals the product of the concentration of absorbing molecules $(c)$, the molar extinction coefficient $(k)$, and the path length through the medium $(l)$. The molar extinction coefficient varies with the wavelength of the light and is a characteristic of the particular molecular species in the absorbing solution. It must be emphasized, however, that this relationship is based upon absorption in a clear, nonabsorbing, homogeneously dispensed medium.

Noteworthy of early spectrophotometric instruments were the wavelength scanning systems of Baumberger (1939) and Arvanitaki and Chalazonitis (1947). These instruments relied upon a single light source to provide monochromatic light for scanning a spectral region of interest and electronic detection of transmitted light. Each provided a means to record absolute absorption spectra, but neither had the capability of storing or subtracting to provide difference spectra.

Many advances in optical instrumentation were pioneered in the 1950s by Britton Chance and collaborators at the University of Pennsylvania (for review, *see* Chance, 1957). Spectrophotometers designed by these investigators had broader spectral range to include infrared and ultraviolet absorption as well as that of visible light. Light detectors were increased in sensitivity, which improved the resolution of observable cellular pigments. Signal-to-noise ratios were greatly enhanced allowing greater wavelength resolution and the ability to quantitate absorption changes. Also, light detectors had greatly increased response times, thereby improving the temporal resolution of changes in mitochondrial redox activities.

Two types of spectrophotometers were developed during this period. One of these was a differential wavelength-scanning, "split-beam" spectrophotometer for the measurement of spectral changes of respiratory chain pigments (Chance, 1951a; Yang and Legallais, 1954). The other was a "dual wavelength" or "double-beam" spectrophotometer designed to record rates of absorption changes (Chance, 1951b).

2.5.3.2.1. Split-beam spectrophotometry.    Split-beam instruments have the advantageous capability of being able to identify the optically active molecules in a preparation and to determine their concentrations. Most of these types of instruments, whether built in laboratories or commercially constructed, have unique features requiring lenses, mirrors, and other optical devices that are described in specific publications. The diagram in Fig. 7, and subsequent depictions of apparatus, is meant to illustrate general features of design and application. In usual approaches of this instrumentation, light from an excitation source, such as a tungsten, xenon, or mercury arc lamp, is diffracted in a monochromator. This monochromator is most often under motor control to sweep through a spectrum of wavelengths. The output of this monochromator is a single light beam of sequentially varying wavelengths. This light beam is split ("chopped"), usually by a vibrating mirror, so that it is presented alternately along two parallel paths. Within each path is a preparation under investigation. One of these preparations is the "sample"; the other serves as a "reference" for the measurement of absorption differences between light in the two pathways. Light transmitted through the sample and reference preparations is measured by a photomultiplier tube. The necessary electronic circuits consist of amplifiers, a log converter, and demultiplexing circuits to identify whether the light being measured at any moment in time is derived from the "sample" or "reference" pathway. These signals are subtracted, and the output from this instrument is usually displayed logarithmically on a chart recorder as a "difference spectrum." In the more sophisticated instruments, there is synchronization between the "wavelength-drive" mechanisms changing excitation wavelengths by advancing the monochromator and the mechanisms advancing the paper in the chart recorder. This has provided the most exact means of recording changes in optical density (absorbance) as a function of wavelength.

An advantage of this type of instrument is that difference spectra between the "sample" and "reference" preparations can be generated as rapidly as the spectrum is scanned by advancing the monochromator. Thus, nonlabile absorption, common to both

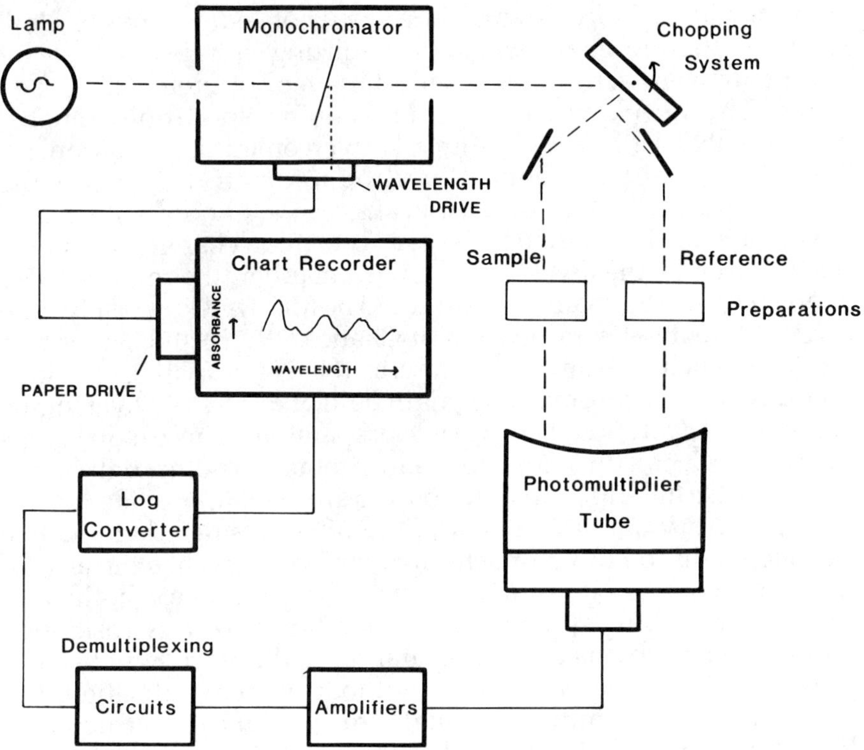



Fig. 7. General features of a split-beam spectrophotometer for monitoring spectral changes in light absorption.

specimens, is subtracted. In a typical experiment, mitochondria may be suspended in appropriate media and housed in cuvets within each light path. One preparation may then be dis-oxygenated, whereas the other is kept in an oxygenated environment. A difference spectrum can then be recorded by subtraction of the signals from each light path. The versatility of this type of instrumentation was shown by Jobsis et al. (1977), who used fiber bundles to deliver light from the chopping system alternately to the brain surfaces of two cats. Light reflected from each brain was collected with fiber bundles that were combined into a single bundle that presented the light to a photomultiplier tube.

A disadvantage of this type of instrument, however, is that acquisition of spectra commonly requires several minutes, which limits the ability to follow rapid changes of redox activity. Until recently, therefore, investigations utilizing split-beam type instruments were carried out after attainment of metabolic steady states. The development of rapid-scanning spectrophotometric instruments (see below) has greatly overcome this limitation.

2.5.3.2.2. Dual-wavelength Spectrophotometry.   Since the spectral locations of absorption peaks are well known, it became advantageous to continously record absorption changes at a single peak chosen by the investigator. In this manner, the redox activity of one respiratory chain reactant could be followed kinetically. The original "double-beam" system of Chance (1951b) still serves as a model for such instruments as described in Fig. 8. With this instrument, two monochromatic beams of light at different wavelengths are presented alternately to a single preparation. These light beams may be produced through two monochromators or through filters. In some instruments, separate sources are used to provide light to each monochromator or filters. More often, a single light source supplied light to both monochromators, since this has the advantage that changes in source intensity will influence both light beams. A chopping system, often a vibrating mirror, is positioned within the light paths to the preparation so that each light beam is presented for an equal duration without overlap with the other beam. Light transmitted through the preparation is detected by a photomultiplier tube and electronically processed for display on a chart recorder or oscilloscope. Output of such systems can be displayed as changes in absorbance as a function of time.

The recording of changes in absorption peak amplitudes requires subtraction of absorption at two wavelengths. One of these wavelengths is usually selected to be at the absorption maximum of the pigment being investigated. The other is chosen as a nearby reference wavelength to compensate for changes in light scattering that have an equal influence upon absorption at the two wavelengths. Differences in optical absorption between the sample and reference wavelengths are proportional to the redox activity of a respiratory pigment. Using such an instrument, Chance and Williams (1956) followed redox changes of the various respiratory chain components as a function of time and provided detailed descriptions of respiratory chain activity.

Instruments derived from these "split-beam" and "dual-beam" concepts provided means to sensitively probe metabolic activity of mitochondria in vitro, intact cells in suspension, tissue

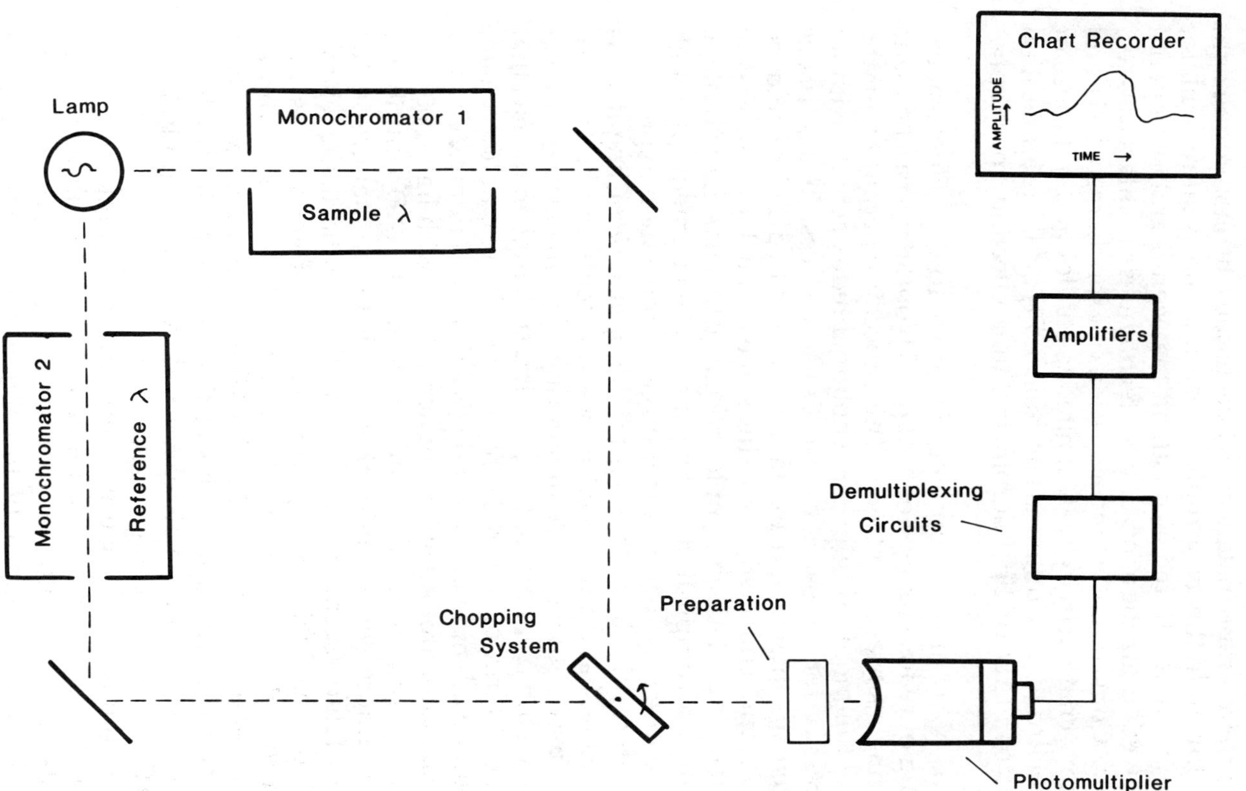


Fig. 8.    Basic diagram of a dual wavelength spectrophotometer for recording absorption changes at specific wavelengths as a function of time.

slices, and, in certain instances, whole tissues, by monitoring light passed through the absorbing medium. The requirement for transillumination offered a constraint that significantly hampered and slowed utilization of optical techniques in intact tissues, however. Consequently, investigations of ion homeostasis and respiratory chain activity were limited to studies of thin tissues or slices in vitro (e.g., Cummins and Bull, 1971) until the application to brain of NAD or Fp fluorometry and reflection spectrophotometry.

2.5.3.2.3. NAD and Flavoprotein Fluorometry.      Fluorometric recording of redox changes of NAD or flavoproteins can be made from a single surface of a tissue. This procedure permitted for the first time the monitoring of mitochondrial redox activity without the necessity for exposure of two tissue surfaces. Thus, tissues could remain intact, and excitation and emitted light could be delivered and collected from the same small location. Tissue fluorometry is a relatively simple procedure that has the advantage that it can be accomplished with inexpensive equipment. Although this procedure may require some surgical preparation to provide a direct path for the light to a tissue, the optical recording itself is "noninvasive," and fluorescence measurements can be made simultaneously with electrode monitoring of brain electrophysiology and ion shifts. For NAD, the technique is based on the fact that the reduced form of the initial member of the respiratory chain, nicotinamide adenine dinucleotide (NADH), will fluoresce with a broad maximum around 450 nm when excited with light between 310 and 370 nm. Since the oxidized form of NAD contributes negligible fluorescence under this illumination, changes in the intensity of fluorescence emitted from a tissue indicate the relative state of the $NADH/NAD^+$ ratio. NAD fluorometry was first adapted for intact tissues by Chance and Jobsis (1959), who recorded increased ratios of oxidized NAD with contractile activity in skeletal muscles. This approach was first used in brain by Chance et al. (1962), who recorded increased NAD reduction during anoxia and NAD oxidation during seizures.

Early fluorometers were composed of a single excitation source and one detector (Chance et al., 1962). In subsequent instruments, a photomultiplier tube was added to monitor the intensity of the excitation light to compensate for instabilities in source brightness (Jobsis et al., 1966). Major components of a typical fluorometer are depicted in Fig. 9 (Jobsis et al., 1971). A mercury arc lamp was originally used as light source because it has a high output level at 366 nm. Early applications used 1000-W

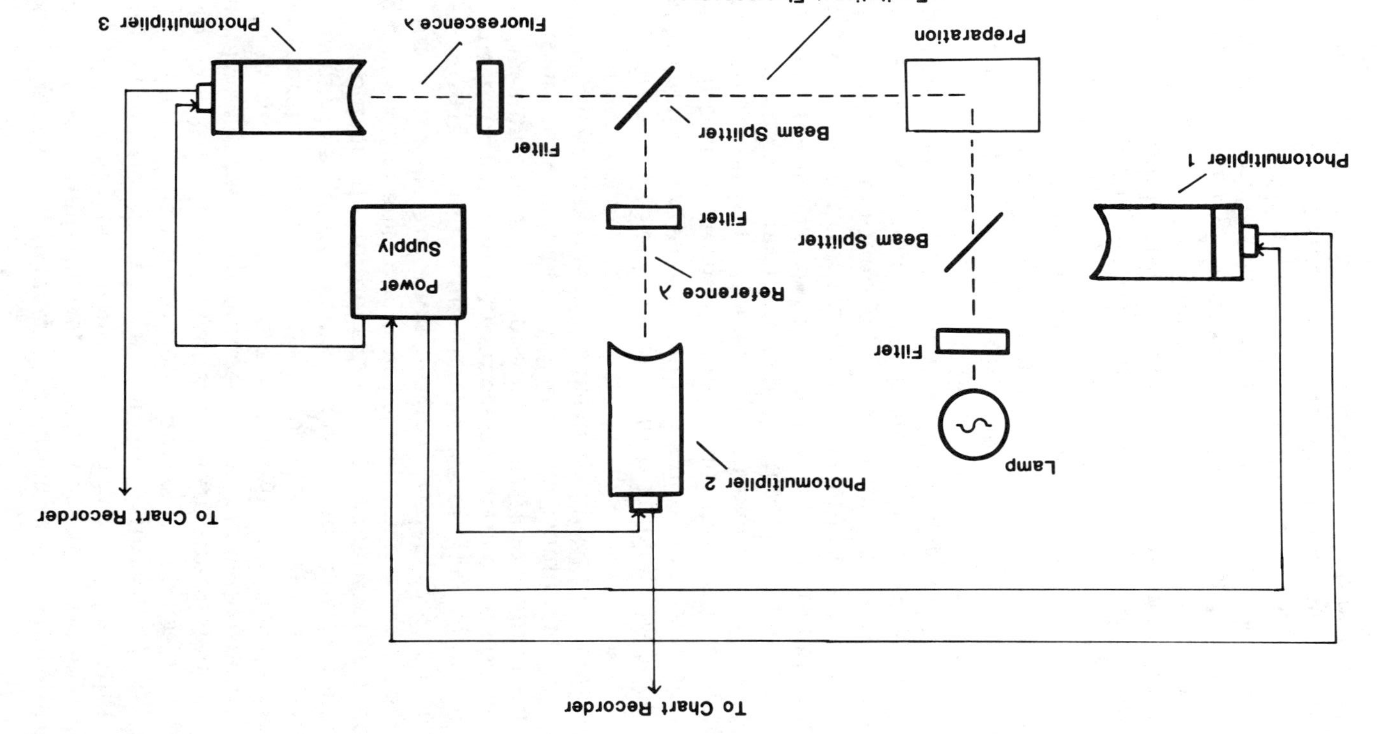


Fig. 9.   Major components of a typical fluorometer designed for the measurement of NAD or Fp fluorescence in tissues.

water-cooled lamps to provide the excitation energy, but lower wattage, air-cooled lamps or laser sources are now often used (Renault et al., 1982). Filters may be placed in the light path between the lamp and the preparation to assure that light at wavelengths greater than those necessary for excitation are not directed to the tissue. In instruments in which lamp intensity is monitored, a beam splitter is positioned in the light path to deflect a fraction of the excitation light to the photomultiplier tube (labeled as 1 in the figure). The output of this photomultiplier is used, in a feedback circuit, to regulate the high voltage output of the power supply that controls the amplification of photomultiplier tubes detecting fluorescence and reflectance to compensate for changes in source intensity. The excitation light may be delivered to tissue through microscope epi-illumination optics or fiber optic light guides. Reflected excitation light and fluorescence are collected either with fiber optics or microscope objectives, and the intensities of these signals are detected by individual photomultiplier tubes. Filters in the light path between the tissue and the PMT detectors assure that there is no overlap in these signals.

Among the more recent advances in technology have been those that led to the increased versatility required to allow a single instrument to measure fluorescence or tissue reflectance, or serve as a transmission spectrophotometer (Chance et al., 1975). This was accomplished by housing filters in rotating wheels to control light to tissues. Light was still detected with a photomultiplier tube. Another was the incorporation of a television camera into a fluorometric system to produce a portable instrument convenient for use under surgical conditions with the capability of recording fluorescence changes with sufficient spatial resolution to distinguish local events (Schuette et al., 1974). In this instrument, an ultraviolet light source with large focusing optics provided light to a circular tissue area, 5 cm in diameter. Output light was monitored with an image-intensified video camera with a filter beam splitter attached to the front of the camera lens to separate reflected from fluorescent light. The display consisted of two discrete images, one of reflectance, the other of fluorescence. A two-channel video densitometer allowed for spatially related quantification of light intensity changes. Another advance was the operation of fluorometers with light guides in direct contact with the brain, for delivery of excitation light and collection of fluorescence. The use of small fiber bundles in this manner has made possible experiments with conscious, mobile animals (Mayevsky, 1984).

Complications of NAD fluorometry include the origin of the labile tissue fluorescence and the potential artifact that changes in blood volume and hemoglobin oxygenation may impart to the signal. The importance of defining the labile fluorescence is derived from the fact that there are cytoplasmic as well as intramitochondrial NAD pools whose redox status may change independently. Also, there may be a second NADH fraction in mitochondria and NADPH has optical characteristics that are similar to those of NADH.

Much investigation has centered around the relative contributions of mitochondrial and cytoplasmic NADH to the labile fluorescence signal. Studies favoring the intramitochondrial origin of this signal include those showing that NADH fluorescence (quantum efficiency) is greatly enhanced when this coenzyme is bound to any of several enzymes (Boyer and Theorell, 1956; Velick, 1961). In cells, mitochondrial NADH exists in a bound form. The NADH that is not in mitochondria exists, for the most part, free in solution in cytoplasm. Of the NAD that is bound in cytoplasm, the majority is bound to glyceraldehyde phosphate dehydrogenase, which decreases, rather than increases, NADH fluorescence efficiency (Velick, 1961). More directly, Avi-Dor et al. (1962) and Estabrook (1962) showed that the fluorescence of mitochondrial NADH was greatly enhanced (approximately $10\times$) over that of an equal amount of NADH dissolved in vitro. Chance and Theorell (1959) found that the majority of fluorescence was emitted in the head end of a sperm cell where the mitochondria are in greatest abundance, whereas almost no signal was obtained from the remaining cytoplasm. Jobsis and Duffield (1967) reported, using iodoacetate in substate depleted and anoxic states in intact muscle, that mitochondrial NADH exhibited a fluorescence intensity at least 10 times higher than cytoplasmic NADH, and this was later followed by similar experiments, calculations, and conclusions in heart (Chapman, 1972; Nuutinen, 1984) and brain (Jobsis et al., 1971; O'Connor, 1977).

Another possible source of labile fluorescence is NADPH. Avi-Dor et al. (1962) showed, however, that mitochondrial NADH has a $3\text{--}4\times$ larger fluorescence efficiency than NADPH. Also, in rat liver, NADPH levels remained constant, although fluorescence varied with NADH, and in rat heart the correlation between NADH and fluorescence held with no correlation to NADPH during mitochondrial inhibition by amytal (Chance et al., 1965). In brain, fluorescence changes during seizures and anoxia were correlated with NADH changes, as assayed biochemically, and not with changes in NADPH (Jobsis et al., 1971). Also, reflection spectrophotometry of cytochrome $a,a_3$ demonstrates that this cytochrome responds in a similar manner, and with a similar time course, to neuronal activation in brain as does the NADH fluorescence signal (see below).

A second NADH fraction (the so-called NADH-2 of Chance and Hollunger, 1961) appears unlikely to contribute significantly to fluorescence signals. Jobsis and Duffield (1967) determined that addition of alpha-glycerophosphate to toad skeletal muscle mitochondria partially reduced this fraction, but that this contributed only 1–2% of the total state 4 signal. The function of this NAD fraction remains unknown, but it has been suggested that it functions in reductive biosynthetic reactions.

There has been much investigation centered upon the most accurate method for correcting the emitted fluorescence signal for tissue movement and other changes that alter light scattering properties and for disturbances created by changes in hemoglobin volume and oxygenation. Jobsis and Stainsby (1968) were the first investigators to use the intensity of reflected excitation light to compensate for these potential artifacts in the fluorescence signal, presuming that the hemodynamic influences upon light at 366 and 450 nm were approximately equal. A procedure for weighting the reflectance signal to compensate for hemodynamic effects on tissue fluorescence was suggested by Jobsis et al. (1971) and evaluated by Harbig et al. (1976). These latter investigators induced transient hemodilution by rapidly flushing an oxygenated saline solution into carotid or lingual arteries. They concluded that a linear relationship exists between NADH fluorescence and 366 nm reflectance. Studies by Kovach et al. (1977) suggested that proper correction of tissue fluorescence requires subtraction of the product of reflected excitation light times a factor (*"k"*), which was also determined by transient hemodilution. Use of 366 nm reflectance as reference in this manner, however, requires the assumption that the NAD redox state remains unchanged by hemodilution. Since the NADH/NAD ratio depends upon tissue oxygenation, this assumption carries with it the condition that tissue oxygen tension is also unchanged by transient hemodilution. Polarographic measurements of oxygen tension during hemodilution with oxygenation saline or perfluoronated hydrocarbon solutions indicate, however, that $tPO_2$ may not remain constant under these circumstances (Sick and Rosenthal, unpublished).

The use of correction other than that of the excitation source was first suggested by Kobayashi et al. (1971), who measured reflectance at a wavelength indifferent to the metabolic processes (720 nm) for compensation. Others have favored reflectance at isobestic wavelengths, at which the extinction coefficients (absorption) of oxygenated and deoxygenated hemoglobin are identical (Renault et al., 1984). Measurement of the fluorescence of a systemically injected, metabolically inert fluorochrome, rhodamine B, was also proposed (Kramer and Pearlstein, 1979). None of these procedures, however, has been shown by direct measurement of tissue oxygenation and fluorescence to be the best among all possible indices of tissue pyridine nucleotide redox status. One means to approach this problem is to examine changes in fluorescence that occur when hemoglobin is not present. Figure 10 shows fluorescence at 450 nm corrected for changes in 366 nm reflectance weighted by bolus injections of an oxygen-carrying fluorochrome (Fluosol-DA). In the left portion of this figure, the fluorometric measurements were made from the cerebral cortex of a pentobarbital-anesthetized rat with normal hematocrit. As expected, the substitution of nitrogen for oxygen in the respired gas mixture increased tissue fluorescence. This change was interpreted as a shift toward reduction of NAD. When Fluosol-DA was substituted for hemoglobin (hematocrit less than 1%), a similar fluorescence increase occurred during hypoxia. This fluorescence increase was not different in rate and only slightly different in amplitude from that recorded with hemoglobin perfusion. This difference in amplitude may indicate a small effect of reflectance changes not compensated for by the subtraction of 366 nm light. More likely, this difference may be caused by the fact that nitrogen was given terminally, thereby producing the maximal reduction of NAD to NADH. The similarity of the two traces derived with and without hemoglobin demonstrates that compensation by reflectance at 366 nm, and likely compensation by the other wavelengths proposed, serves equally well for the qualitative and, for all but the most extreme circumstances, the quantitative interpretation of NADH fluorescence.

The issue of specular reflectance must be considered both for fluorescence and for spectrophotometric measurements in tissues. Any tissue presents a mirror-like surface for light reflectance. Collection of the specularly reflected light can cause undue artifacts for metabolic signals since this light has not penetrated tissue and has not been available for absorption by molecules of metabolic inter-

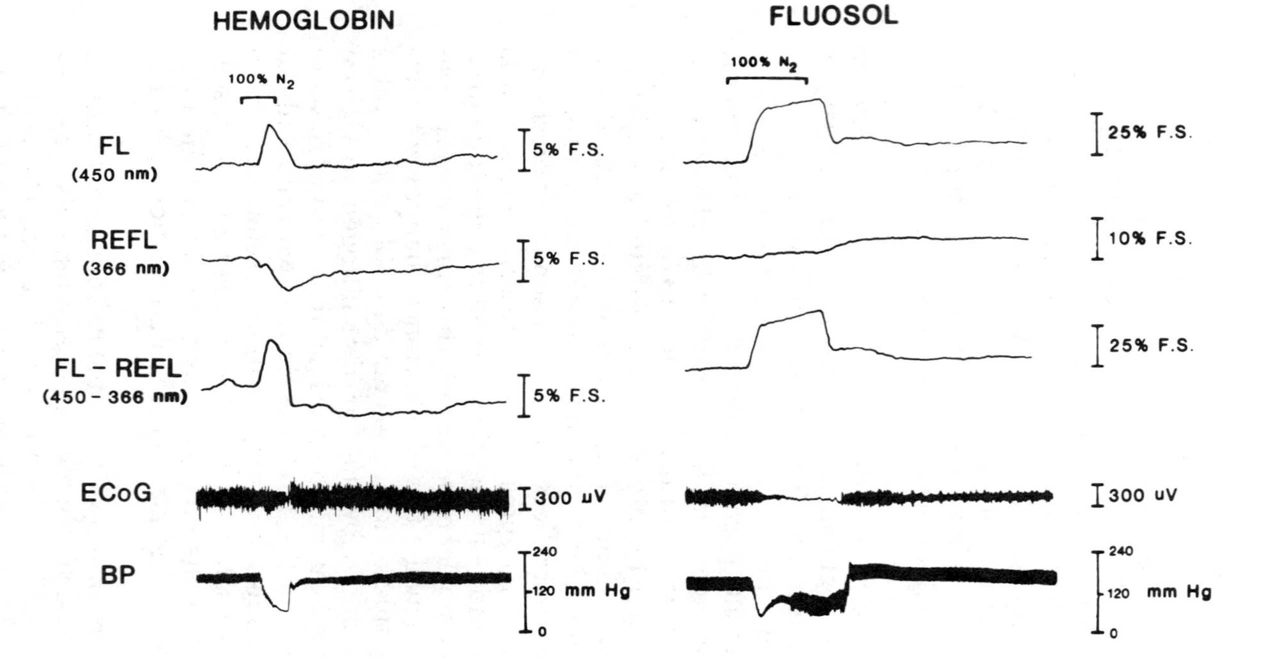


Fig. 10.    Fluorescence at 450 nm corrected for changes in 366 nm reflectance during hypoxia. Left traces were recorded in rat brain normally perfused with blood. Right traces were recorded after blood substitution by oxygenated, perfluorinated hydrocarbon solution (Fluosol). Top Traces on both sides are uncorrected fluorescence at 450 nm. Below these are traces representing reflectance at 366 nm, the corrected fluorescence, electrocorticographic (ECoG) activity, and systemic blood pressure. NADH reduction is shown as an upward deflection in the corrected fluorescence trace.

est. Such light does not contain information on tissue metabolism, yet it may be markedly influenced by changes in light scattering at the tissue surface. Rosenthal and Jobsis (1971) found, however, that reflectance changes during direct cortical stimulation were decreased with the incidence angle. A similar finding was also demonstrated by Renault et al. (1984). By presenting excitation light to the brain surface at an approximate 45° angle during direct cortical stimulation, reflectance changes were extensively decreased or absent. Specular reflectance is of great concern in fluorometry since light penetration is a function of wavelength (higher wavelengths penetrate much deeper) and also because tissue fluorescence offers a very weak signal for optical monitoring.

Although most investigations have concentrated upon the recording of NADH fluorescence, redox activity of the flavoproteins can also be monitored fluorometrically. Excitation light in the 420-nm region provokes fluorescence from oxidized flavoproteins that is recordable at approximately 540 nm (Chance and Shoener, 1966; Chance et al., 1978). Instrumentation is similar to that used for NADH fluorometry (Chance et al., 1971).

2.5.3.2.4. Reflection Spectrophotometry. The study of tissue physiology and redox activity of cytochromes required development of spectrophotometric procedures applicable to intact tissues. Early progress in this area was reported by Lubbers and collaborators (Lubbers and Wodick, 1969; Wodick and Lubbers, 1974), who recorded spectra from perfused guinea pig brain by a reflection technique. This was followed by the dual and triple wavelength instruments of Jobsis et al. (1977), which have proven adaptable for the measurement of redox activities of cytochrome $a,a_3$ in brain and other tissues, such as heart and kidney (Snow et al., 1981; Balaban and Sylvia, 1981). Initial spectrophotometric measurements from brain were based upon the reflection of incident light presented from two monochromators, but a less expensive system using interference filters and a motor-driven chopper blade has been described (Duckrow et al., 1982). Instruments for reflection spectrophotometry are similar in concept to those used in studies by transillumination techniques (c.f., Fig. 8). In most systems, tissues are illuminated alternately at sample and reference wavelengths. Timing circuits, triggered by detectors in the path between the light source and tissue, control the demodulation of the PMT output to continuous signals representing the intensity of light returning from the tissue at each wavelength. Other methods of monitoring cytochrome redox activity in brain have involved scanning spectroscopy of brain surfaces during periods in which these brains were freeze trapped by immersion in liquid nitrogen (Bashford et al., 1982). This has the advantage of providing a spectral read-out of data, but continuous monitoring is sacrificed. Reflection instruments have been designed for monitoring from brain in the infrared spectral region (Jobsis, 1977; Brazy et al., 1985) and for simultaneous measurement of reflectance and fluorescence changes (Hassinen and Jamsa, 1982).

Measurement of absorption by transillumination has the advantage that quantitative information on the concentrations of the light-absorbing molecules in a medium can be derived as long as Beer-Lambert principles apply. Although these principles describe the relationship between the log of the ratio of light entering and emerging from a preparation, they apply most specifically to a clear, nonabsorbing, homogeneously dispersed medium. Measurements in tissues by either transillumination or reflectance approaches are complicated by several factors including: (1) the complex composition of the tissue and therefore the likelihood that there is unequal distribution of the absorbing molecules; (2) the tissue is not clear so that much light is lost by scattering; (3) there may be many different molecules absorbing in the region of interest; and (4) the path length is difficult to measure precisely, and this path length may vary because of physiological effects or edema.

Since Beer-Lambert principles of light absorption are not directly applicable in the reflectance mode, spectrophotomers measuring tissue reflectance are calibrated to allow for relative monitoring of absorption shifts. The high stability of most systems, however, allow quantitative comparisons to be made very exactly within each animal. Many studies have been performed that demonstrate that redox activities monitored under "control" conditions can be used as a reference to define changes in these activities as monitored after physiological or pharmacological conditions are varied (e.g., Duckrow et al., 1981; LaManna et al., 1981; Novack et al., 1982).

Amplitudes of reflectance changes from brain have been described in many ways, including percentages of full-scale light levels and changes in optical density. In order to offer some standardization of reflectance data, Sylvia and Rosenthal (1978) defined the "total labile signal" as that difference in reflectance produced by changing the fraction of inspired oxygen from 0 to 100%. Subsequent reflectance shifts were expressed as a percentage of this signal. Brizzee and Kreisman (1984) pointed out, however, that although inspiration of nitrogen will presumably produce a fully reduced state of mitochondria as one extreme of the "labile signal," hyperoxygenation of brain beyond that produced by inspiration of 100% oxygen will produce greater mitochondrial oxidation (Hempel et al., 1977). Therefore, inspiration of 100% $O_2$ produces an oxidation of the mitochondrial reactants that reaches an "experimental" maximum that, when compared with the value produced by nitrogen inspiration, offers an "experimental labile signal" to which subsequent changes in reflectance can be compared.

Light reflectance from brain and other tissues is dependent not only upon absorption by cytochromes, but also upon light-scattering and hemodynamic changes. Examination of reflection spectra from brain indicates that spectral interference from heme-containing pigments severely complicates the measurement of reflectance due to cytochromes $b$ and $c$ in the 540–580 nm region. This is shown in spectra depicted in Fig. 5. The upper spectrum is the difference between oxygenated and deoxygenated, hemoglobin-free (Fluosol-perfused) rat brain. The lower spectrum is the difference between hemoglobin-perfused control brain and that stimulated by direct application of stimulus pulses to the cortical surface. A typical peak is evident at 605 nm attributable to a redox shift (toward oxidation) of cytochrome $a,a_3$. Changes in the 540–580 nm region are similar, however, to those evident in spectra of hemoglobin and do not provide a good indication of redox transitions of cytochromes in that spectral region.

As evident in Fig. 5, the choice of "reference" wavelengths is especially important to reflection spectrophotometry since absorption at the "sample" wavelengths is altered by factors other than cytochromes. Many studies have used as reference 590 nm, since light at this wavelength undergoes an equal optical density change as that at 605 nm when light scattering or blood volume is altered and when there are shifts in hemoglobin saturation (Jobsis et al., 1977). For lack of an existing term, such a reference wavelength was referred to as "equibestic." Studies such as those showing that an equal amplitude of cytochrome $a,a_3$ reduction during anoxia (with increased blood volume) and ischemia (blood volume decreased), (Rosenthal et al., 1976b; Harrison et al., 1985), reaction spectra (Rosenthal et al., 1979), and spectral changes in response to neuronal activation (LaManna et al., submitted) were used to validate this procedure (c.f., Rosenthal and LaManna, 1981). Recent studies, however, in bloodless rats (fluorocarbon-perfused) (Piantadosi and Jobsis-Vandervliet, 1984) and in hemoglobin-perfused rats with different reference wavelength algorithms (Karimen and Burkhart, 1985) suggest that, although data are quantitatively similar under physiological conditions, there may be some quantitative differences among the various compensating procedures under extreme conditions of hyperoxygenation and deoxygenation. What is fortuitously evident from all of these studies, however, is that quantitative differences are slight even under extreme conditions.

Of the various cytochrome absorption peaks, the one at 603–605 nm proved to be the most easily measureable in the presence of hemoglobin. One advantage of performing investigations in bloodless animals was suggested by Piantadosi and Jobsis-Vandervliet (1984). Without hemoglobin, the monitoring of redox activity of cytochromes other than $a,a_3$ may be routinely possible.

2.5.3.2.5. Rapid-Scanning Spectrophotometry.   As discussed previously, two modes of spectrophotometry have generally been employed for mitochondrial studies: (1) a "split-beam" system in which wavelengths of light are presented sequentially and the absorption of all cytochromes is recorded at relatively long intervals and (2) a "multi-wavelength" system in which monochromatic light at an absorption peak and at reference wavelengths is continuously presented so that redox shifts of a single reactant can be followed kinetically. Instruments combining each of these functions have been described (e.g., Schwab and Sies, 1978), but there continued to be a need for "rapid-scanning" instruments that could acquire spectra with speeds sufficient to record the kinetics of redox shifts of many reactants simultaneously. A double-beam photometer version of such an instrument was produced by Lubbers and Niesel (1959). Technological advances in light sources, detectors, and laboratory mini- and microcomputers have, more recently, allowed for the design and construction of sensitive "rapid-scanning" spectrophotometers capable of recording absorption changes in transmission or reflectance modes. In an early version of such an instrument, "white" light was presented to tissue, and transmitted or reflected light was split into a spectrum by a spectrograph, the spectral output of which was detected by vidicon camera (Mandel et al., 1977). A later version of this technology was based on detection by a linear photodiode array (LaManna et al., 1985). As illustrated in Fig. 11, light transmitted or diffusely reflected from a sample is collected by

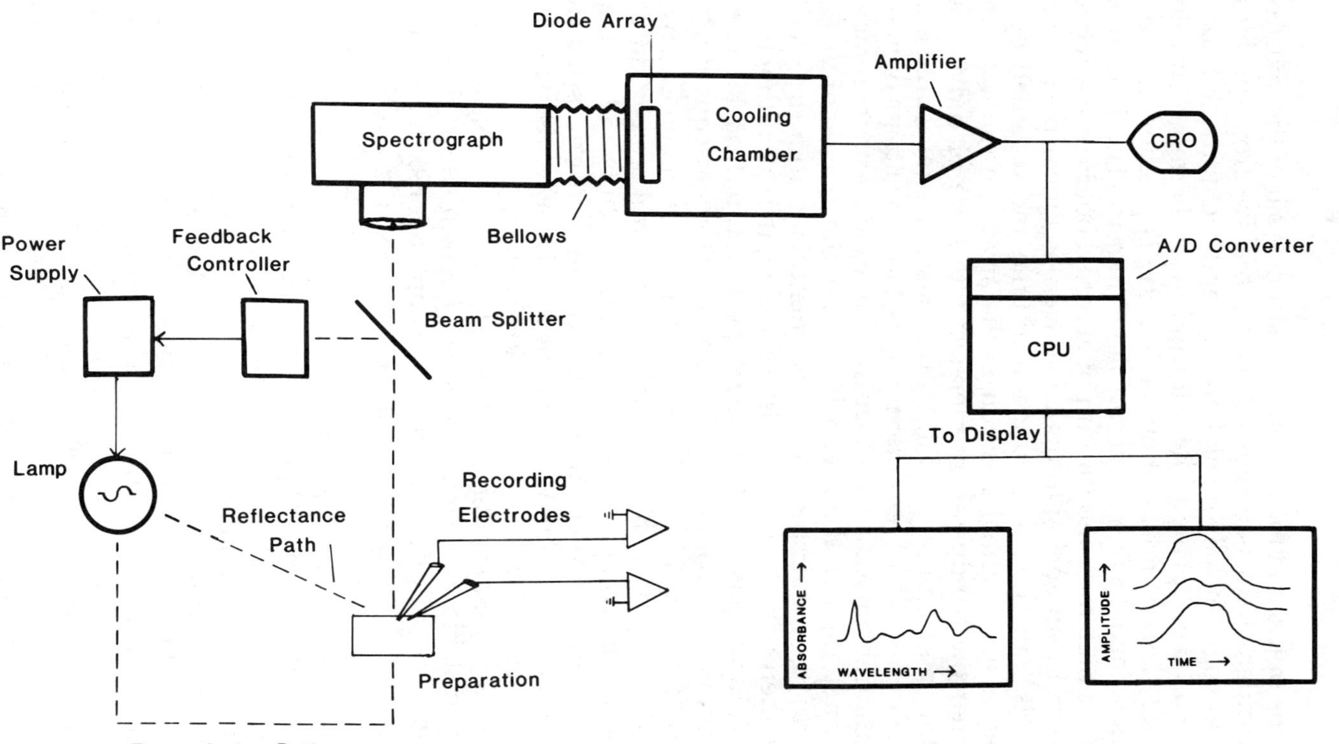


Fig. 11.   A rapid-scanning spectrophotometer designed for monitoring absorption or reflectance from tissues in vitro or in vivo.

microscope optics and focused on the entrance slit of the spectrograph. Incident light is dispersed within the spectrograph by a concave, holographically etched grating, and the diffracted spectrum is focused in a plane at the exit window. A thermoelectrically cooled, 512-element linear photodiode array is positioned in the exit window for detection of spectral light intensity. The detector is scanned at a high rate, and the output of each diode is converted to digital form by an analog-to-digital converter. Linear position in the array is translated as a function of time and recorded by a computer as a sequential position in the array. To calibrate the system, narrow bandpass, interference filters are placed in the path to the spectrograph, and light transmission is measured and stored. Computer routines determine the diode positions of maximum light transmission and calculate the wavelength increment (in nanometers) per diode position and the wavelength (in nanometers) of the first diode. Subsequent data from the diode array can then be presented against a wavelength scale rather than a sequential array position. Individual spectral scans from biological preparations are acquired by the computer for subsequent processing (Pikarsky et al., 1985). The computer functions in two modes to display data. By subtraction of spectral scans, "difference" spectra are compared for optical density shifts and evaluation of absolute concentrations. In the second, specific wavelengths are monitored to allow for the display of redox shifts of many molecular species as a function of time.

Diode array scanning spectrophotometers are now commercially available from several sources. Unfortunately, these instruments are usually designed for cuvet applications, and versatility for use in tissues in vitro or in vivo is presently limited. The ability to provide both spectral and kinetic data, however, and to monitor light absorption by endogenous pigments as well as dye indicators of such important parameters as pH, membrane voltage, or intracellular $Ca^{2+}$, should offer stimulus to increasing the versatility of these instruments.

2.5.3.2.6. Limitations.    There are two limitations of optical techniques that must be considered. One limitation is spatial resolution, since most instruments are designed to record at a single location in a tissue. There have been three approaches to overcome this limitation. One is the use of two- and three-dimensional representations of relative brain $NADH/NAD^+$ ratios as acquired by photographing tissues under 366 nm light illumination (Ji et al., 1977; Quistorff and Chance, 1977; Welsh et al., 1977). As with metabolite assays, temporal resolution is lost. A second method requires collecting NAD fluorescence with a videocamera (Schuette et al., 1974). Another is the "flying spot" fluorometer of Chance et al. (1978) based upon histogram analysis of the fluorescence intensities at multiple loci. Data acquisition and analysis for these latter methods are more complex than those of other procedures, and these have not been commonly employed for the study of metabolism and ion shifts.

A second limitation is derived from the fact that light penetration is wavelength-dependent, thereby requiring skull removal for studies in the ultraviolet and visible regions of the spectrum. This has severely hindered studies of mitochondrial redox activity in human brain, although this has been accomplished in a manner that suggests the usefulness of optical techniques as diagnostic as well as research tools. In the earliest application of this approach, Jobsis et al. (1972) compared redox shifts of NAD, by fluorometry, in brain areas judged to be within epileptic foci and surrounding regions in patients undergoing extirpation of such foci for intractable epilepsy. NAD oxidation, provoked by electrocortical stimulation, was smaller in epileptic foci and had a cyclic pattern. Mitochondrial redox activity was used to assess metabolic integrity of exposed human brain during vascular surgery (Austin et al., 1977; Fein, 1982) and in epileptic foci (Ramsay et al., 1984). To overcome the necessity for brain exposure, Jobsis (1977) has taken advantage of the greater penetration of infrared light to monitor the oxygenation status of hemoglobin and redox shifts of cytochrome oxidase at 820–840 nm through human skull and brain tissue. With diode laser excitation sources, infrared spectrophotometry confirmed that hypoxia is accompanied by hemoglobin deoxygenation and a shift of cytochrome $a,a_3$ to a more reduced state (Brazy et al., 1985). Since optical monitoring is less expensive and more easily accomplished than PET and NMR detection, these early data suggest that optical measurements may provide a rapid, inexpensive, and easily accomplished approach to the assessment of tissue oxygen sufficiency in clinical situations.

### 2.5.4. Mitochondrial Redox Activity and Ion Shifts

Many investigators have taken advantage of the temporal resolution of optical techniques to relate mitochondrial redox activity to brain electrophysiology and ion shifts. The following section is meant only to illustrate the potential for future studies in this area and is not included as a review of accomplishments in this field.

Simultaneous recording of the large effects upon ion homeostasis and mitochondrial activity of metabolic insults has been undertaken by many laboratories. For example, hypoxia is accompanied by reduction of respiratory chain carriers and increased $K^+_o$ (Fig. 12). Optical and electrode approaches are sufficiently sensitive to record smaller changes in metabolism and ion activity. Figure 13 shows that neuronal activation by afferent or direct stimulation is accompanied by a metabolic reaction that can be recorded as an oxidation of NAD or cytochrome $a,a_3$ in brain. Such oxidative shifts occurred in rat, cat, or human in acute or chronic preparations of cats and were recordable from exposed tissue surfaces or through cranial windows (Rosenthal and Jobsis, 1971; Jobsis et al., 1972; Jobsis et al., 1977; Ramsay et al., 1984). Such stimulation is also accompanied by increases in $K^+_o$, which reach peak amplitude slightly before peak oxidation is attained. Recovery of $K^+_o$ to baseline is also completed before re-reduction of the mitochondrial chain components to their baseline redox status. Measurements in hippocampus, cerebral cortex, and spinal cord demonstrate a consistent correlation between amplitude of $K^+_o$ increments and the mitochondrial oxidative responses, leading to the conclusion that excess $K^+_o$ represents a load that calls for proportionate expenditure of oxidative energy (Lewis and Schuette, 1975; Lothman et al., 1975; Rosenthal et al., 1979). This relationship also led to the conclusion that the major fraction of oxidative energy consumed by central gray matter after neuronal activation is expended in the restoration of transmembrane ion gradients (Somjen et al., 1976). Rates, amplitudes, and direction of activity-provoked redox shifts have proven to be sensitive indices of pharmcological effects (LaManna et al., 1977a,b) and pathophysiological changes in brain caused by hypoxia (LaManna et al., 1984), ischemia (Duckrow et al., 1981; Mayevsky and Zarchin, 1981), and recurrent seizures (Kreisman et al., 1981). In fact, metabolic responses to stimulation were more sensitive than was steady-state ("resting") redox status in signaling pathophysiology associated with aging (Sylvia et al., 1983) and cerebral norepinephrine depletion (LaManna et al., 1981).

Oxidative responses of NAD accompanied spreading depression (SD) induced electrically or by KCl superfusion (Rosenthal and Somjen, 1973; Mayevsky et al., 1974; Mayevsky and Chance, 1975). These responses led to rejection of the idea that a shortage of

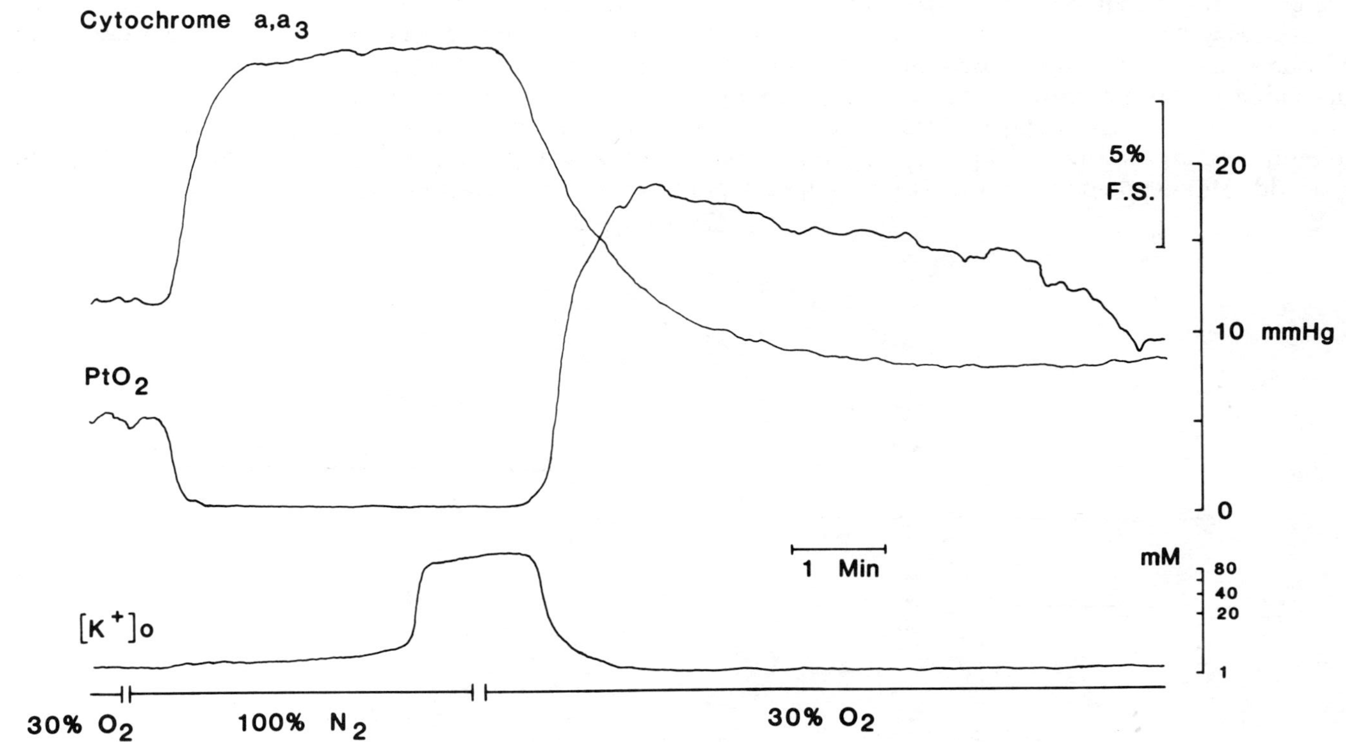


Fig. 12. Effects of hypoxia on $K^+_o$, tissue oxygen tension ($PtO_2$), and the redox state of cytochrome $a,a_3$ in rat cerebral cortex (from Sick et al., 1982).

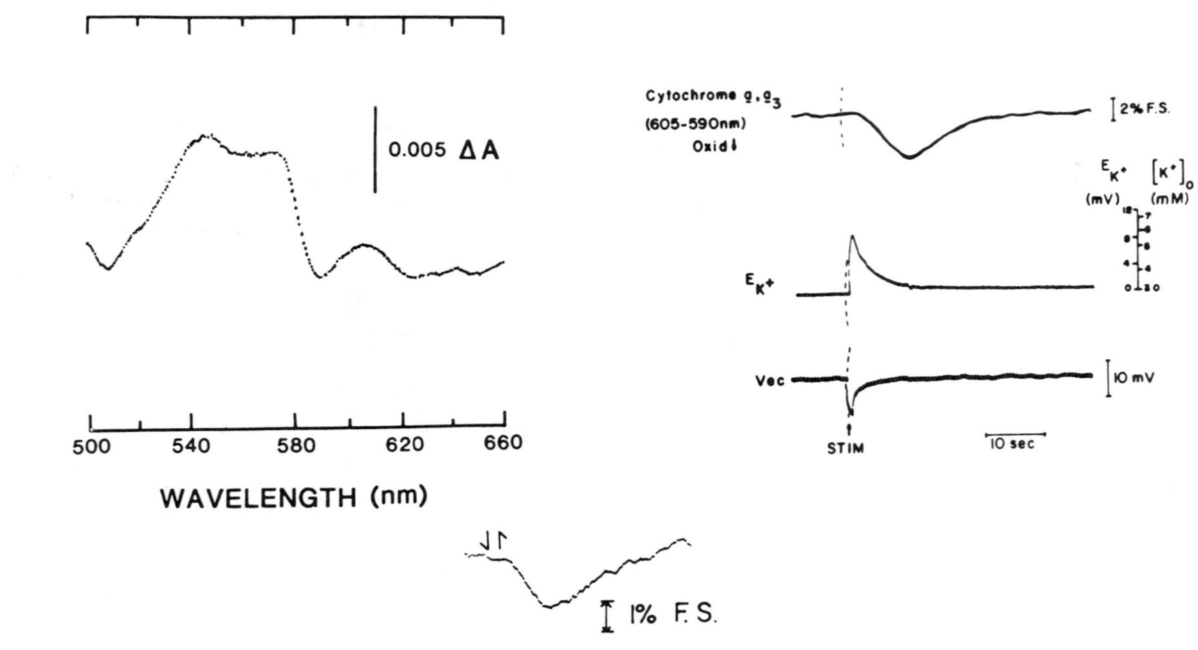



Fig. 13.   Effects of direct cortical stimulation on $K^+_o$ and mitochondrial electron transport carriers. (Upper left trace) Spectrum representing reflectance differences between "resting" brain and the brain 5 s after a 2-s train of stimulus pulses presented to the cortical surface. Unlike spectra presented previously, this spectrum is plotted so that the upward deflection at 605 nm represents an oxidation of cytochrome $a,a_3$. (Lower left trace) Changes in fluorescence of NAD produced by direct cortical stimulation in human neocortex. Decreased fluorescence is plotted as a downward deflection as a function of time, indicating oxidation of NADH. (Right traces) Responses of cytochrome $a,a_3$, $K^+_o$, and extracellular voltage to direct cortical stimulation. Decreased absorption at 605 nm (relative to 590 nm) is plotted as a downward deflection indicating an oxidative response of this cytochrome.

oxidizable substrate and, hence, of oxidative energy turnover played a role in causing SD. Responses of $K^+_o$ and mitochondrial redox reactants have been recorded during seizures in many laboratories. The precise correlation, however, between the amplitude of mitochondrial oxidation and the increment in $K^+_o$ broke down during both seizures and spreading depression (Lothman et al., 1975). Under these conditions, redox changes were greater than expected from extrapolation of the relationship found by mild stimulation. It may be that in these conditions, normal regulatory mechanisms of metabolic activity fail, or that extraordinary demands other than ion transport are placed on tissue respiration by these activities.

## 3. Summary

An understanding of tissue function requires an understanding of the couples that exist between ion shifts and metabolism. This is especially true in brain, in which maintenance of ion homeostasis consumes such a large fraction of the total tissue energy and in which small insults to metabolism have large ionic and functional consequences. The study of ion shifts and metabolism requires techniques that can monitor each without precluding measurement of the other. This requirement has led to the development of several approaches for laboratory investigation of brain oxidative metabolism that have already increased understanding of this important subject. Although the study of ion shifts in brain is one that likely encompases processes that are carried out in both neurons and glia, at this time no technique is capable of distinguishing between metabolic activity that occurs in neurons and in glia. Advances in technology have contributed significantly, however, to the study of brain metabolic physiology by providing means to correlate local metabolism with locally recorded changes in ion activity, and certain of these procedures have the advantage of being applicable to human patients to aid in clinical diagnosis. In an age considered to be "the era of the neurosciences," advances in the technologies of PET, magnetic resonance spectroscopy and optical monitoring will continue to occur that will allow these approaches to become more quantitative and thereby still more useful for the study of metabolic activities and ion shifts.

## Acknowledgments

Data presented as original in this review is derived from research programs supported by NIH grants HL 38657 and NS14325. The contributions to the development of this manuscript by Mr. Solomon Pikarsky are gratefully acknowledged.

## References

Arvanitake A. and Chalazonitis N. (1947) De la nature des. III. Recherces spectro kymographiques sur les cytochromes de neurons in vivo. *Arch Int. Physiol.* **54,** 441–457.

Astrup J. (1982) Energy-requiring cell functions in the ischemic brain. *J. Neurosurg.* **56,** 482–497.

Astrup J., Moller Sorenson P., and Rahbek Sorensen H. (1981) Oxygen and glucose consumption related to $Na^+$-$K^+$ transport in the canine brain. *Stroke* **12,** 726–730.

Austin G., Haugen G., and LaManna J. C. (1977) Cortical Oxidative Metabolism Following Microanastomosis for Brain Ischemia, in *Oxygen and Physiological Function* (Jobsis F. F., ed.) Professional Information Library, Dallas, Texas.

Avi-Dor J. J., Olson M., Doherty M. D., and Kaplan N. O. (1962) Fluorescence of pyridine nucleotides in mitochondria. *J. Biol. Chem.* **237,** 2377–2383.

Balaban R. S. and Sylvia A. L. (1981) Spectrophotometric monitoring of $O_2$ delivery to the exposed rat kidney. *Am. J. Physiol.* F257–F262.

Balaban R. S., Kantor H. L., and Ferretti J. A. (1983) In vivo flux between phosphocreatine and adenosine triphosphate determined by two-dimensional phosphorous NMR. *J. Biol. Chem.* **258,** 12787–12789.

Bashford C. L., Barlow C. H., Chance B., Haselgrove J., and Sorge J. (1982) Optical measurements of oxygen delivery and consumption in gerbil cerebral cortex. *Am. J. Physiol.* C265–271.

Baumberger J. P. (1939) The relation between the "oxidation-reduction potential" and the oxygen consumption rate of yeast cell suspensions. *Cold Spring Harbor Symp. Quant. Biol.* **7,** 195–215.

Borgstrom L., Chapman A. G., and Siesjo B. K. (1976) Glucose consumption in the cerebral cortex of rat during bicuculline-induced status epilepticus. *J. Neurochem.* **27,** 971–973.

Boyer P. D. and Theorell H. (1956) The change in reduced diphosphopyridine nucleotide (DPNH) fluorescecen upon combination with liver alcohol dehydrogenase (ADH). *Acta Chem. Scand.* **10,** 447–450.

Brazy J. E., Lewis D. V., Mitnick M. H., and Jobsis-Vandervliet, F. F. (1985) Noninvasive monitoring of cerebral oxygenation in preterm infants: Preliminary observations. *Pediatrics* **75,** 217–225.

Brizzee B. L. and Kreisman N. R. (1984) Quantification of cerebral oxygenation by in situ measurements of reduction/oxidation (redox) changes in cytochrome oxidase: Approaches and limitations. *Soc. Neurosci. Abst.* **10,** 1004.

Cady E. B., Dawson M. J., Hope P. L., Tofts P. S., Costello A. M., Delpy D. J., Reynolds E. O. R., and Wilkie D. R. (1983) Non-invasive investigation of cerebral metabolism in newborn infants by phosphorus nuclear magnetic resonance spectroscopy. *Lancet* **2,** 1059–1062.

Caveness W. F., Kato M., Malamut B., Hosokawa S., Wakisaka S., and O'Neill R. R. (1980) Propagation of focal motor seizures in the pubescent monkey. *Ann. Neurol.* **7,** 213–221.

Chance B. (1951a) Rapid and sensitive spectrophotometry. I. Accelerated and stopped flow methods for the measurement of the reaction kinetics and spectra of unstable compounds in the visible region of the spectrum. *Rev. Sci. Instrum.* **22,** 619–627.

Chance B. (1951b) Rapid and sensitive spectrophotometry. III. A double beam apparatus. *Rev. Sci. Instrum.* **22,** 634–638.

Chance B. (1957) Techniques for the Assay of the Respiratory Enzymes, in *Methods in Enzymology* vol. 4, (Colowick S. P. and Kaplan N. O., eds.) Academic, New York.

Chance B. and Hollunger G. (1961) The interaction of energy and electron transfer reaction in mitochondria. *J. Biol. Chem.* **236,** 1534–1543.

Chance B. and Jobsis F. F. (1959) Changes in fluorescence in a frog sartorius muscle following a twitch. *Nature* **184,** 195–196.

Chance B. and Schoener B. (1966) Fluorometric Studies of Flavin Component of the Respiratory Chain, in *Flavins and Flavoproteins* (Slater E. C., ed.) Elsevier, Amsterdam.

Chance B. and Theorell B. (1959) Localization and kinetics of reduced pyridine nucleotide in living cells by microfluorometry. *J. Biol. Chem.* **234,** 3044–3050.

Chance B. and Williams G. R. (1956) The respiratory chain and oxidative phosphorylation. *Adv. Enzymol.* **17,** 65–134.

Chance B., Barlow C., Nakase Y., Takeda H., Mayevsky A., Fischetti R., Graham N., and Sorge J. (1978) Heterogeneity of oxygen delivery in normoxic and hypoxic states: A fluorometer study. *Am. J. Physiol.* H809–H820.

Chance B., Cohen P., Jobsis F. F., and Schoener B. (1962) Intracellular oxidation-reduction states in vivo. *Science* **137,** 499–508.

Chance B., Graham N., and Mayer D. (1971) A time sharing fluorometer for the readout of intracellular oxidation-reduction states of NADH and flavo-protein. *Rev. Sci. Instrum.* **42,** 951–957.

Chance B., Legallais V., Sorge J., and Graham N. (1975) A versatile time-sharing multichannel spectrophotometer, reflectometer and fluorometer. *Anal. Biochem.* **66,** 498–514.

Chance B., Schoener B., Krejci K., Russmann W., Weseman W., Schnitger H., and Bucher T. (1965) Kinetics of fluorescence and metabolite changes in rat liver during a cycle of ischemia. *Biochem. Z.* **341,** 325–333.

Chance B., Younkin D., Eleff S., Warnell R., and Delivoria-Pappadoppolous M. (1983) [31]P-NMR of cortical oxidative metabolism in neonates. *Pediatric Res.* **17,** 397A.

Chapman A. G., Meldrum B. S., and Siesjo B. K. (1977) Cerebral metabolic changes during prolonged epileptic seizures in rats. *J. Neurochem.* **28,** 1025–1035.

Chapman J. B. (1972) Fluorometric studies of oxidative metabolism in isolated papillary muscle of the rabbit. *J. Gen. Physiol.* **59,** 135–154.

Chi M. M., Lowry C. V., and Lowry O. H. (1978) An improved enzymatic cycle for nicotinamide-adenine dinucleotide phosphate. *Anal. Biochem.* **89,** 119–129.

Crane P. D., Braun L. B., Cornford E., Cremer J. E., Glass M. J., and Oldendorf W. H. (1978) Dose dependent reduction of glucose utilization by pentobarbital in rat brain. *Stroke* **9,** 12–18.

Csiba L., Paschen W., and Mies G. (1985) Regional changes in tissue pH and glucose content during cortical spreading depression in rat brain. *Brain Res.* **336,** 167–170.

Cummins J. T. and Bull R. J. (1971) Spectrophotometric measurements of metabolic responses in isolated rat brain cortex. *Biochim. Biophys. Acta* **253,** 29–38.

Dixon K. C. (1949) Anaerobic leakage of potassium from brain. *Biochem. J.* **44,** 187–190.

Duckrow R. B., LaManna J. C., and Rosenthal, M. (1982) Sensitive and inexpensive dual wavelength reflection spectrophotometry using interference filters. *Anal. Biochem.* **125,** 13–23.

Duckrow R. B., LaManna J. C., and Rosenthal M. (1981) Disparate recovery of resting and stimulated oxidative metabolism following transient ischemia. *Stroke* **12,** 677–686.

Duffy T. E., Howse D. C., and Plum F. (1975) Cerebral energy metabolism during experimental status epilepticus. *J. Neurochem.* **24,** 925–934.

Engel J., Ackermann R. F., Kuhl D. E., and Phelps M. E. (1985) Brain Imaging of Glucose Utilization in Convulsive Disorders, in *Brain Imaging and Brain Function* (Sokoloff L., ed.) Raven, New York.

Estabrook R. W. (1962) Fluorometric measurements of reduced pyridine nucleotide in cellular and subcellular particles. *Anal. Biochem.* **4**, 231–245.

Fein J. M. (1982) Cortical nicotinamide adenine dinucleotide (NADH) kinetics in patients undergoing extracranial-intracranial bypass. *Neurosurgy* **10**, 428–436.

Fisher R. S., Pedley T. A., Moody W. J., and Prince D. A. (1976) The role of extracellular potassium in hippocampal epilepsy. *Arch. Neurol.* **33**, 76–83.

Folbergrova J. (1974) Energy metabolism of mouse cerebral cortex during homocysteine convulsions. *Brain Res.* **81**, 443–454.

Fox J. S. (1984) PET scan controversy aired. *Science* **224**, 145–146.

Frackowiak R. S. J. (1985) Pathophysiology of Human Cerebral Ischemia: Studies with Positron Emission Tomography and $^{15}$Oxygen, in *Brain Imaging and Brain Function* (Sokoloff L., ed.) Raven, New York.

Gjedde A. (1982) Calculation of cerebral glucose phosphorylation from brain uptake of glucose analogs in vivo: A re-examination. *Brain Res. Rev.* **4**, 237–274.

Grafstein B. (1956) Mechanism of spreading cortical depression. *J. Neurophysiol.* **19**, 154–171.

Hackenbrock C. R. (1968) Ultrastructural bases for metabolically linked mechanical activity in mitochondria. *J. Cell Biol.* **37**, 345–352.

Hampson R. K., Medina M. A., and Olson M. S. (1982) The use of high-energy microwave irradiation to inactivate mitochondrial metabolism. *Ann. Biochem.* **123**, 49–54.

Harbig K., Chance B., Kovach A. G., and Reivich M. (1976) In vivo measurement of pyridine nucleotide fluorescence from cat brain cortex. *J. Appl. Physiol.* **41**, 480–488.

Harrison M. B., Busto R., Ginsberg M., Rosenthal M., and Sick T. J. (1984) Correlation of mitochondrial oxidative activity, metabolites and ECoG after cerebral ischemia in rat. *Soc. Neurosci. Abst.* **10**, 1004.

Harrison M. B., Sick T. J., and Rosenthal M. (1985) Mitochondria redox responses to cerebral ischemia produced by the four-vessel occlusion model in the rat. *Neurolog. Res.* **7**, 142–148.

Hassinen I. and Jamsa T. (1982) A reflectance spectrophotometer-surface fluorometer suitable for monitoring changes in hemoprotein spectra and fluorescence of flavins and nicotinamide nucleotides in intact tissues. *Anal. Biochem.* **120**, 365–372.

Hawkins R. A. and Miller A. L. (1978) Loss of radioactive 2-deoxy-D-glucose-6-phosphate from brain of conscious rats: Implications for quantitative autoradiographic determination of regional glucose utilization. *Neuroscience* **3**, 251–258.

Hawkins R. A., Hass W. K., and Ransohoff J. (1979) Measurement of regional brain glucose utilization in vivo using 2-$^{14}$C glucose. *Stroke* **10**, 690–703.

Heinemann U. and Lux H. D. (1977) Ceiling of stimulus induced rises in extracellular potassium concentration in cerebral cortex of cat. *Brain Res.* **120,** 231–249.

Heinemann U., Lux H. D., and Gutnick M. J. (1977) Extracellular free calcium and potassium during paroxysmal activity in the cerebral cortex of the cat. *Exp. Brain Res.* **27,** 236–243.

Hempel F. G., Jobsis F. F., LaManna J. C., Rosenthal M., and Saltzman H. A. (1977) Oxidation of cerebral cytochrome a,a$_3$ by oxygen plus carbon dioxide at hyperbaric pressures. *J. Appl. Physiol.* **43,** 872–877.

Hilberman M., Subramanian J. H., Haselgrove J., Cone J. B., Egan J. W., Gyulai L., and Chance B. (1984) In vivo time-resolved brain phosphorus nuclear magnetic resonance. *Cereb. Blood Flow Metab.* **4,** 334–342.

Hodgkin A. L. and Keynes R. D. (1955) Active transport of cations in giant axons from Sepia and Loligo. *J. Physiol.* (Lond.) **128,** 28–60.

Hotson J. R., Sypert, G. W., and Ward A. A. (1973) Extracellular potassium concentration changes during propagated seizures. *Exp. Neurol.* **38,** 20–26.

Ji S. C., Chance B., Stuart B. H., and Nathan R. (1977) Two-dimensional analysis of the redox state of the rat cerebral cortex in vivo by NADH fluorescence photography. *Brain Res.* **119,** 357–373.

Jobsis F. F. (1977) Non-invasive, infrared monitoring of cerebral and myocardial oxygen sufficiency and circulatory parameters. *Science* **198,** 1264–1267.

Jobsis F. F. (1979) Oxidative Metabolic Effects of Cerebral Hypoxia, in *Advances in Neurology* vol. 23 (Fahn S., Davis J., and Rowland L., eds.) Raven, New York.

Jobsis F. F. and Duffield J. C. (1967) Oxidative and glycolytic recovery metabolism in muscle. *J. Gen. Physiol.* **50,** 1009–1047.

Jobsis F. F. and Stainsby W. N. (1968) Oxidation of NAD during contractions of circulated mammalian skeletal muscle. *Resp. Physiol.* **4,** 292–300.

Jobsis F. F., Keizer J. H., LaManna J. C., and Rosenthal M. (1977) In vivo reflectance spectrophotometry of cytochrome a,a$_3$ in the intact cerebral cortex of the cat. *J. Appl. Physiol.* **43,** 858–872.

Jobsis F. F., O'Connor M. J., Rosenthal M., and Van Buren J. M. (1972) Fluorometric Monitoring of Metabolic Activity in the Intact Cerebral Cortex, in *Neurophysiology Studied in Man* (Somjen G., ed.) Excerpta Medica, Amsterdam.

Jobsis F. F., O'Connor M. J., Vitale A., and Vreman H. (1971) Intracellular redox changes in functioning cerebral cortex. I. Metabolic effects of epileptiform activity. *J. Neurophysiol.* **34,** 735–749.

Jobsis F. F., Legallais V., and O'Connor M. J. (1966) A regulated differential fluorometer for the assay of oxidative metabolism in intact tissues. *IEEE Trans. Bio-Med. Electron.* **13,** 93–99, 1966.

Kariman K. and Burkhart D. S. (1985) Non-invasive in vivo spectrophotometric monitoring of brain cytochrome $a,a_3$ revisited. *Brain Res.* **360,** 203–213.

Katz B. (1966) *Nerve, Muscle and Synapse* McGraw-Hill, New York.

Katzman R. and Pappius H. M. (1973) *Brain Electrolytes and Fluid Metabolism* Williams and Wilkins, Baltimore.

Keilin D. (1925) On cytochrome, a respiratory pigment common to animals, yeast and higher plants. *Proc. Roy. Soc. London* **B98,** 312–339.

Keilin D. (1966) *The History of Cell Respiration and Cytochrome* Cambridge University Press, Cambridge.

Kennedy C., Des Rosiers M., Jehle J. W., Reivich M., Sharp F., and Sokoloff L. (1975) Mapping of functional neural pathways by autoradiographic survey of local metabolic rate with ($^{14}$C)deoxyglucose. *Science* **187,** 850–853.

Kerr S. E. (1935) Studies on the phosphorus compounds of brain. I. Phosphocreatine. *J. Biol. Chem.* **110,** 625–635.

Kirshner J. S., Blank W. F., and Myers R. E. (1975) Brain extracellular potassium activity during hypoxia in the cat. *Neurology* **25,** 1001–1005.

Kobayashi S., Katsuyuki N., Kaede K., and Ogata E. (1971) Optical consequences of blood substitution on tissue oxidation-reduction state microfluorometry. *J. Appl. Physiol.* **31,** 93–96.

Kovach A. G. B., Dora E., Eke A., and Gyulai L. (1977) Effects of Microcirculation on Microfluorometric Measurements, in *Oxygen and Physiological Function* (Jobsis F. F., ed.) Professional Information Library, Dallas, Texas.

Kraig R. P. and Nicholson C. (1978) Extracellular ionic variations during spreading depression. *Neuroscience* **3,** 1045–1059.

Kramer R. S. and Pearlstein R. D. (1979) Cerebral cortical microfluorometry at isobestic wavelengths for correction of vascular artefact. *Science* **205,** 693–696.

Kreisman N. R., LaManna J. C., Sick T. J., and Rosenthal M. (1981) Oxidative metabolic responses with recurrent seizures in rat cerebral cortex: Role of systemic factors. *Brain Res.* **218,** 175–188.

Krivanek J. (1961) Some metabolic changes accompanying Leao's spreading cortical depression in the rat. *J. Neurochem.* **6,** 183–189.

LaManna J. C., Cordingley G., and Rosenthal M. (1977a) Phenobarbital effects on extracellular potassium activity and respiratory chain metabolism in intact cerebral cortex of cats. *J. Pharmacol. Exp. Ther.* **200,** 560–569.

LaManna J. C., Harik, S. I., Light A. I., and Rosenthal M. (1981) Norepinephrine depletion alters cerebral oxidative metabolism in the "active" state. *Brain Res.* **204**, 87–101.

LaManna J. C., Light A. I., Peretsman S. P., and Rosenthal M. (1984) Oxygen insufficiency during hypoxic hypoxia in rat brain cortex. *Brain Res.* **293**, 313–318.

LaManna J. C., Lothman E., Rosenthal M., Somjen G. G., and Younts B. W. (1977b) Phenytoin, electric, ionic and metabolic responses in cortex and spinal cord. *Epilepsia* **18**, 317–329.

LaManna J. C., Pikarsky S. M., Sick T. J., and Rosenthal M. (1985) A rapid-scanning spectrophotometer designed for biological tissues in vitro or in vivo. *Anal. Biochem.* **144**, 483–493.

LaManna J. C., Sick T. J., Pikarsky S. M., and Rosenthal M. (submitted) Detection of an oxidizable fraction of cytochrome oxidase in intact rat brain.

Lassen N. A. and Klee A. (1965) Cerebral blood flow determined by saturation and desaturation with Krypton[85]: An evaluation of the validity of the inert gas method of Kety and Schmidt. *Circ. Res.* **16**, 26–32.

Leao A. A. A. P. (1951) The slow voltage variation of cortical spreading depression of activity. *Electroenceph. Clin. Neurophysiol.* **3**, 315–321.

Lewis D. V. and Schuette W. H. (1975) NADH fluorescence and $K^+_o$ changes during hippocampal stimulation. *J. Neurophysiol.* **38**, 405–417.

Lewis D. V. and Schuette W. H. (1976) NADH fluorescence, $K^+_o$ and oxygen consumption in cat cerebral cortex during direct cortical stimulation. *Brain Res.* **110**, 523–535.

Lothman E., LaManna J. C., Cordingley G., Rosenthal M., and Somjen G. G. (1975) Responses of electrical potential, potassium levels and oxidative metabolic activity of the cerebral neocortex of cats. *Brain Res.* **88**, 15–36.

Lowry O. H. and Passonneau J. V. (1972) *A Flexible System of Enzymatic Analysis*. Academic, New York.

Lu D. M., Davis D. W., Mans A. M., and Hawkins R. A. (1983) Regional cerebral glucose utilization measured with $^{14}C$ glucose in brief experiments. *Am. J. Physiol.* **245**, C428–C438.

Lubbers D. W. and Niesel W. (1959) Der kurzzeit-spektralanalysator. Ein schnellarbeitendes spektralphotometer zur laufenden messung von absorptions-bzw. extinktionspektren. *Pfluger's Arch.* **268**, 286–292.

Lubbers D. W. and Wodick R. (1969) The examination of multicomponent systems in biological materials by means of a rapid scanning photometer. *Appl. Opt.* **8**, 1055–1062.

MacMunn C. A. (1884) On myohaematin, an intrinsic muscle-pigment of vertebrates and invertebrates, on histohaematin and on the spectrum of the supra-renal bodies. *J. Physiol.* **5**, 24–26.

Mandel L. J., Riddle T. G., and LaManna J. C. (1977) A Rapid Scanning Spectrophotometer and Fluorometer for In Vivo Monitoring of Steady-State and Kinetic Optical Properties of Respiratory Enzymes, in *Oxygen and Physiological Function* (Jobsis F. F., ed.) Professional Information Library, Dallas, Texas.

Mangold R., Sokoloff L., Therman P. O., Conner E. H., Kleinerman J., and Kety S. S. (1955) The effects of sleep and lack of sleep on the cerebral circulation and metabolism of normal young men. *J. Clin. Invest.* **34,** 1092–1100.

Marshall W. H. (1959) Spreading cortical depression of Leao. *Physiol. Rev.* **39,** 239–279.

Mayevsky A. (1984) Brain NADH redox state monitored in vivo by fiber optic surface fluorometry. *Brain Res. Rev.* **7,** 49–68.

Mayevsky A. and Chance B. (1975) Metabolic responses of the awake cerebral cortex to anoxia, hypoxia, spreading depression and epileptiform activity. *Brain Res.* **98,** 149–165.

Mayevsky A. and Zarchin N. (1981) The effects of unilateral carotid occlusion on the responses to decapitation in the gerbil brain. *Brain Res.* **206,** 115–160.

Mayevsky A., Zeuthen T., and Chance B. (1974) Measurements of extracellular potassium, ECoG and pyridine nucleotide levels during cortical spreading depression in rats. *Brain Res.* **76,** 347–349.

McCandless D. W., Ed. (1985) *Cerebral Energy Metabolism and Metabolic Encephalopathy* Plenum, New York.

Medina M. A., Jones D. J., Stavinoha W. B., and Ross D. H. (1975) The levels of labile intermediary metabolites in mouse brain following rapid tissue fixation with microwave irradiation. *J. Neurochem.* **24,** 223–227.

Minard F. N. and Davis R. V. (1962) The effect of electroshock on the acid-soluble phosphates of rat brain. *J. Biol. Chem.* **237,** 1283–1289.

Moody W. J., Futamachi K. J., and Prince D. A. (1974) Extracellular potassium activity during epileptogenesis. *Exp. Neurol.* **42,** 248–262.

Morris M. E. (1974) Hypoxia and extracellular potassium activity in the guinea-pig cortex. *Can. J. Physiol. Pharmacol.* **52,** 872–882.

Mutch W. A. C. and Hansen A. J. (1984) Extracellular pH changes during spreading depression and cerebral ischemia: Mechanisms of brain pH regulation. *J. Cereb. Blood Flow Metab.* **4,** 17–27.

Nilsson B., Norberg K., Nordstrom C.-H., and Siesjo B. K. (1975) Rate of energy utilization in the cerebral cortex of rats. *Acta Physiol. Scand.* **93,** 569–571.

Novack R., LaManna J. C., and Rosenthal M. (1982) Ethanol and acetaldehyde alter brain mitochondrial redox responses to direct cortical stimulation in vivo. *Neuropharmacology* **21,** 1051–1058.

Nuutinen E. M. (1984) Subcellular origin of the surface fluorescence of reduced nicotinamide nucleotides in the isolated perfused rat heart. *Basic Res. Cardiol.* **79,** 49–58.

O'Connor M. J. (1977) Origin of Labile NADH Tissue Fluorescence, in *Oxygen and Physiological Function* (Jobsis F. F., ed.) Professional Information Library, Dallas, Texas.

Partridge W. M., Crane P. D., Mietus L. J., and Oldendorf W. H. (1982) Normogram for 2-deoxyglucose lumped constant for rat brain cortex. *J. Cereb. Blood Flow Metab.* **2,** 197–202.

Passonneau J. V., Hawkins R. A., Lust D. W., and Welsh F. A. (Eds.) (1980) *Cerebral Metabolism and Neural Function.* Williams and Wilkins, Baltimore, Maryland.

Petroff O. A. C., Prichard J. W., Behar K. L., Alger J. R., den Hollander J. A., and Shulman R. G. (1985) Cerebral intracellular pH by $^{31}$P nuclear magnetic resonance spectroscopy. *Neurology* **35,** 781–788.

Phelps M. E., Kuhl D. E., and Mazziotta J. C. (1981) Metabolic mapping of the brain's response to visual stimulation: Studies in man. *Science* **211,** 1445–1448.

Piantadosi C. A. and Jobsis-Vandervliet F. F. (1984) Spectrophotometry of cerebral cytochrome a,a$_3$ in bloodless rats. *Brain Res.* **304,** 89–94.

Pikarsky S. M., LaManna J. C., Sick T. J., and Rosenthal M. (1985) A computer-assisted rapid scanning spectrophotometer with applications to tissues in vitro and in vivo. *Comp. Biomed. Res.* **18,** 408–421.

Plum F., Posner J. B., and Troy B. (1968) Cerebral metabolic and circulatory responses to induced convulsions in animals. *Arch. Neurol.* **19,** 1–13.

Ponten U., Ratcheson R. A., Salford L. G., and Siesjo B. K. (1973) Optimal freezing conditions for cerebral metabolites in rats. *J. Neurochem.* **21,** 1127–1138.

Powers W. J. and Raichle M. E. (1985) Positron emisson tomography and its application to the study of cerebrovascular disease in man. *Stroke* **16,** 361–376.

Prichard J. W. and Shulman R. G. (1986) NMR spectroscopy of brain metabolism in vivo. *Ann. Rev. Neurosci.* **9,** 61–85.

Prichard J. W., Alger J. R., Behar K. L., Petroff O. A. C., and Shulman R. G. (1983) Cerebral metabolic studies in vivo by $^{31}$P NMR. *Proc. Natl. Acad. Sci. USA* **80,** 2748–2751.

Quistorff B. (1975) A mechanical device for the rapid removal and freezing of liver or brain tissue from unanesthetized and nonparalyzed rats. *Anal. Biochem.* **68,** 102–118.

Quistorff B. and Chance B. (1977) Two- and Three-Dimentional Analysis on Brain Oxygen Delivery, in *Oxygen and Physiological Function* (Jobsis F. F., ed.) Professional Information Library, Dallas, Texas.

Quistorff B., Gjedde A., and Hansen A. J. (1979) Spatial analysis of the freeze trapped brain provides for temporal resolution of an event.

Metabolic-electrical and blood flow changes during spreading depression. *Acta Physiol. Scand.* **105**, 42A.

Radda G. K. and Seeley P. J. (1979) Recent studies on cellular metabolism by nuclear magnetic resonance. *Ann. Rev. Physiol.* **41**, 749–769.

Raichle M. E. (1979) Quantitative in vivo autoradiography with positron emission tomography. *Brain Res. Rev.* **1**, 47–68.

Ramsay E., Van Buren J. M., Sick T. J., Rosenthal M., and Kreisman N. R. (1984) Oxygen Sufficiency During Seizures and in Human Epileptic Foci, in *Advances in Epileptology,* vol. 15 (Porter R. J., Mattson R. H., Ward A. A., and Dam M., eds.) Raven, New York.

Reivich M., Alavi A., Gur R. C., and Greenberg J. (1985) Determination of Local Cerebral Glucose Metabolism in Humans: Methodology and Applications to the Study of Sensory and Cognitive Stimuli, in *Brain Imaging and Brain Function* (Sokoloff L., ed.) Raven, New York.

Reivich M., Kuhl D., Wolf A., Greenberg J., Phelps M., Ido T., Casella V., Fowler J., Hoffman E., Alani A., Som P., and Sokoloff L. (1979) The ($^{18}$F)-fluorodeoxyglucose method for the measurement of local cerebral glucose utilization in man. *Circ. Res.* **44**, 127–137.

Renault G., Raynal E., Sinet M., Berthier J. P., Godard B., and Cornillault J. (1982) A laser fluorometer for direct cardiac metabolism investigation. *Optics Laser Tech.* **14**, 143–148.

Renault G., Raynal E., Sinet M., Muffat-Joly M., Berthier J. P., Cornillault J., Godard B., and Pocidalo J. J. (1984) In situ double-beam NADH laser fluorometry: Choice of a reference wavelength. *Am. J. Physiol.* H491–499.

Richter D. and Dawson R. M. C. (1948) Brain metabolism in emotional excitement and in sleep. *Am. J. Physiol.* **154**, 73–79.

Rosenthal M. and Jobsis F. F. (1971) Intracellular redox changes in the functioning cerebral cortex. II. Effects of direct cortical stimulation. *J. Neurophysiol.* **34**, 750–761.

Rosenthal M. and LaManna J. C. (1981) Applications of Optical Techniques to Brain Physiology, in *Advances in Physiological Sciences* vol. 8, *Cardiovascular Physiology: Heart, Peripheral Circulation and Methodology* (Kovach A. G. B., Monos E., and Rubanyi G., eds.) Pergamon, New York.

Rosenthal M. and Somjen G. G. (1973) Spreading depression, sustained potential shifts and metabolic activity of cerebral cortex of cats. *J. Neurophysiol.* **36**, 739–749.

Rosenthal M., LaManna J. C., Jobsis F. F., Levasseur J. E., Kontos H., and Patterson J. L. Jr. (1976a) Effects of respiratory gases on cytochrome a in intact cerebral cortex: Is there a critical $pO_2$? *Brain Res.* **108**, 143–154.

Rosenthal M., Martel D. L., LaManna J. C., and Jobsis F. F. (1976b) Oxidative energy metabolism in situ during and following short periods of transient cortical ischemia in cats. *Exp. Neurol.* **50,** 477–494.

Rosenthal M., LaManna J. C., Yamada S., Younts B. W., and Somjen G. G. (1979) Oxidative metabolism, extracellular potassium and sustained potential shifts in cat spinal cord in situ. *Brain Res.* **162,** 113–127.

Schuette W. H., Whitehouse W. C., Lewis, D. V., O'Connor M. J., and Van Buren J. M. (1974) A television fluorometer for monitoring oxidative metabolism in intact tissue. *Med. Instrument.* **8,** 331–333.

Schwab H. and Sies H. (1978) A new organ spectrophotometer for sensitive dual-wavelength absorbance measurement and spectral scanning of intact perfused organs. *Hoppe-Seyler's Z. Physiol. Chem.* **359,** 385–392.

Shanes A. M. and Berman M. D. (1955) Kinetics of ion movement in the squid giant axon. *J. Gen. Physiol.* **39,** 279–300.

Shinohara M., Dollinger B., Brown G., Rapoport S., and Sokoloff L. (1979) Cerebral glucose utilization: Local changes during and after recovery from spreading cortical depression. *Science* **203,** 188–190.

Shoubridge E. A., Briggs R. W., and Radda G. (1982) $^{31}$P NMR saturation transfer measurements of the steady state rates of creatine kinase and ATP synthetase in the rat brain. *FEBS Lett.* **140,** 288–292.

Sick T. J. and Kreisman N. R. (1981) Potassium ion homeostasis in amphibian brain: Contribution of active transport and oxidative metabolism. *J. Neurophysiol.* **45,** 998–1012.

Sick T. J., Rosenthal M., LaManna J. C., and Lutz P. L. (1982) Brain potassium ion homeostasis during anoxia and metabolic inhibition in the turtle and rat. *Am. J. Physiol.* **243,** R281–288.

Siemkowicz E. and Hansen A. J. (1981) Brain extracellular ion composition and EEG activity following 10 minutes ischemia in normo- and hyperglycemic rats. *Stroke* **12,** 236–240.

Siesjo B. K. (1978) *Brain Energy Metabolism.* Wiley, New York.

Snow T. R., Kleinmann L. H., LaManna J. C., Wechsler A. S., and Jobsis F. F. (1981) Response of cyt a,a$_3$ in the in situ canine heart to transient ischemic episodes. *Basic Res. Cardiol.* **76,** 289–304.

Sokoloff L. (1981) Localization of functional activity in the central nervous system by measurements of glucose utilization with radioactive deoxyglucose. *J. Cereb. Blood Flow Metab.* **1,** 7–36.

Sokoloff L., Mangold R., Wechsler R. L., Kennedy C., and Kety S. S. (1955) The effect of mental arithmetic on cerebral circulation and metabolism. *J. Clin. Invest.* **34,** 1101–1108.

Sokoloff L., Reivich M., Kennedy C., Des Rosiers M. H., Patlak C. S., Pettigrew K. D., Sakurada O., and Shinohara M. (1977) The ($^{14}$C)deoxy-glucose method for the measurement of local cerebral glucose utilization: Theory, procedure and normal values in the conscious and anesesthetized albino rat. *J. Neurochem.* **28,** 897–916.

Somjen G. G., Rosenthal M., Cordingley G., LaManna J. C., and Lothman E. (1976) Potassium, neuroglia and oxidative metabolism in central gray matter. *Fed. Proc.* **35,** 1266–1271.

Sylvia A. L. and Rosenthal M. (1978) The effect of age and lung pathology on cytochrome a,$a_3$ redox levels in rat cerebral cortex. *Brain Res.* **146,** 109–122.

Sylvia A. L., Harik S. I., LaManna J. C., Wilkerson T., and Rosenthal M. (1983) Abnormalities of cerebral oxidative metabolism with aging and their relation to the central noradrenergic system. *Gerontology* **29,** 248–261.

Thulborn K. R., du Boulay G. H., Duchen L. W., and Radda G. (1982) A $^{31}$P nuclear magnetic resonance in vivo study of cerebral ischaemia in the gerbil. *J. Cereb. Blood Flow Metab.* **2,** 299–306.

Veech R. L., Harris R. L., Veloso D., and Veech E. H. (1973) Freeze-blowing: A new technique for the study of brain in vivo. *J. Neurochem.* **20,** 183–188.

Velick S. F. (1961) Spectra and Structure in Enzyme Complexes of Pyridine and Flavine Nucleotides, in *Light and Life* (McElroy W. D. and Glass B., eds.) Johns Hopkins, Baltimore, Maryland.

Vyskocil F., Kriz N., and Bures J. (1972) Potassium-selective microelectrodes used for measuring the extracellular brain potassium during spreading depression and anoxic depolarization in rats. *Brain Res.* **39,** 255–259.

Walker J. L. (1971) Ion specific liquid ion exchanger microelectrodes. *Anal. Chem.* **43,** 89A–93A.

Welsh F. A., O'Connor M. J., and Langfitt, T. W. (1977) Regions of cerebral ischemia located by pyridine nucleotide fluorescence. *Science* **198,** 951–953.

Whittam R. (1961) Active cation transport as a pace-maker of respiration. *Nature* **191,** 603–604.

Wodick R. and Lubbers D. W. (1974) Quantitative evaluation of reflexion spectra of living tissues. *Hoppe-Seyler's Z. Physiol. Chem.* **355,** 583–594.

Yang C. C. and Legallais V. (1954) A rapid and sensitive recording spectrophotometer for the visible and ultraviolet region. I. Description and performance. *Rev. Sci. Inst.* **25,** 801–807.

# Use of Ion-Selective Microelectrodes and Voltammetric Microsensors to Study Brain Cell Microenvironment

## Charles Nicholson and Margaret E. Rice

## 1. Overview of Brain Cell Microenvironment

### 1.1. Definition of the Microenvironment

The dense aggregation of cells that make up the brain and spinal cord has always prompted discussion about the nature of the interstitial region. This region was termed the "extracellular space" and for many years its extent, and even its existence, was widely disputed. Later the concept of the extracellular space became extended and deepened, primarily because of Schmitt and Samson (1969), who recognized that the space was not merely a saline-filled gap between cells, but rather that it was a complex region defined by the membranes of the bounding cells. These membranes were adorned with a rich variety of long-chain proteoglycans that extended into the extracellular space, potentially endowing it with unusual properties. In recognition of potential functional importance of this domain, the term "brain cell microenvironment," henceforth abbreviated to BCM, was introduced (Schmitt and Samson, 1969).

It has become clear that a ceaseless flux of ions is crossing the membranes and profoundly influencing the minute-by-minute behavior of the interstitial space (Nicholson, 1980b), and that transmitters and neuromodulators are also part of the transmembrane traffic. In parallel with these developments has come a vastly increased understanding of the glial cell and its potential role in modulating events in the BCM. Finally the ultimate reference point for this system is the cerebrospinal fluid (CSF) and blood supply. The interfaces between these systems, i.e., the ventricular surfaces and blood–brain barrier, constitute a major component of the BCM (Cserr, 1986).

## *1.2. Issues for Research*

The most immediate aspect of the BCM is that it is the conduit for metabolic substrates to and from cells of the brain. Thus oxygen and glucose migrate from the blood vessels to the cellular constituents, and carbon dioxide moves in the opposite direction. This is not all that is necessary. Ion homeostasis must be maintained in the interstitial space for neurons to generate and transmit signals; $K^+$, $Ca^{2+}$, and $H^+$, particularly, must be held within narrow limits. Failure to control these aspects leads, within minutes, to a plethora of pathologies, and these issues are central to the management of stroke, ischemia, and edema. The mechanisms of the movement of metabolites and ionic stability in the BCM are primary areas of research.

Metabolic and ion homeostasis is a prerequisite (Cserr, 1986) for the second major property of the BCM. Given that one of the most unique aspects of the brain is the dense collection of cells and their interaction, a fundamental issue for research on the microenvironment is to what extent this domain permits new forms of neuronal interaction. For example, one normally thinks of the brain as a vast system of interconnections mediated by the cable-like structures that comprise the axon–synapse–dendrite pathway. This "telephone network" topology has traditionally affected all our thinking about brain morphology and hodology. Yet the BCM provides a different channel for the intercommunication of neuronal aggregates. From this viewpoint the proximity of different elements (and possibly glia) has quite a new importance (Nicholson, 1979). A major issue, therefore, is, how a substance released into the BCM at one point migrates to another.

Beyond the roles of the BCM in the ongoing function of the mature brain are issues of the BCM in development. It has long been known that the developing nervous system possesses a larger extracellular space and greater abundance of the extracellular glycosaminoglycan, hyaluronate, than does the adult (Margolis et al., 1986). Moreover it is likely that aspects of the complex developmental process depend on the existence of diffusion gradients in the BCM to guide migrating processes.

Our understanding and identification of the proteoglycans of the central nervous system (CNS) is an ongoing process (Schmitt and Samson, 1969; Nicholson, 1980b; Margolis et al., 1986). There has been imaginative speculation about the possible role of these compounds; one concept that has been discussed is that the BCM and its long-chain molecules are capable of transducing certain forms of radio frequency energy (Adey, 1975). Despite some provocative experiments, this idea remains speculative and will not be discussed further here.

For concrete discussion we may vastly simplify the chemical makeup of the BCM by dividing all substances into two classes: informational and energetic. Informational substances (Schmitt, 1984) are usually present in small amounts (e.g., micromolar quantities), and there exist quite specific receptors and transduction mechanisms for them. These substances bring about changes in the behavior of the system. In contrast, energetic molecules are present in larger amounts and have a major role in providing metabolic energy. It is necessary to accommodate large fluxes of such substances.

## 2. Ion-Selective Microelectrodes

A major tool in the contemporary study of the BCM is the ion-selective microelectrode (ISM). Typically the modern ISM consists of two barrels, one of which is fabricated as a conventional microelectrode and filled with a saline solution isotonic with that of the BCM. The tip of the other barrel contains a liquid membrane that is selective for a specific ion and is back-filled with a solution of the ion being sensed. When the ISM is placed in a solution of an appropriate ion, a potential develops across the liquid membrane that is proportional to the logarithm of the ratio of the ion activities on either side of the membrane. As we shall see below, the ISMs are calibrated in such a way that the ISM responds to the concentration of the ion of interest. Two barrels are used to allow the local potential in the tissue to be measured and subtracted from the signal recorded on the ion barrel, which includes this potential contribution. Recent symposia and books that deal with ISMs include: Sykova et al. (1981), Zeuthen (1981), Lubbers et al. (1981), Kessler et al. (1985), and Ammann (1986).

### 2.1. History of Technique

ISMs were originally developed using glass membranes. One of the first designs was the Hinke type with an exposed sensing tip; later a recessed tip design was formulated by Thomas. These types of electrode are described in detail by Thomas (1978). Such designs were successful for sensing $Na^+$ and $H^+$ (pH) and to a lesser extent for sensing $K^+$. Their disadvantages were the difficulty of fabrication, the extent of the exposed tip (in the Hinke design) or slow response time (Thomas type), and the complexity of fabricating double-barreled designs (De Hemptinne, 1980).

Liquid membrane designs depend on the existence of a suitable liquid ion exchanger and the ability to insert and hold it in the electrode tip. The first suggestion for such a microelectrode seems to have been by Orme (1969), but that investigator failed to find a way of retaining the exchanger. Later Corning laboratories became interested in fabricating an exchanger for measuring tetraalkylammonium compounds. It was soon appreciated that such an exchanger would sense $K^+$ in plasma or similar fluids, and Walker (1971) devised a method of holding the exchanger in the electrode tip. The essential step consisted of silanizing the inside of the microelectrode tip so that the glass and exchanger would form a hydrophobic bond. This remains a crucial aspect of ISM fabrication and most electrode failures stem from inadequate silanization.

A typical exchanger consists of an organic solvent with a high dielectric constant to which are added a few percent by weight of a carrier molecule. Nitrobenzene is an effective solvent, but nitro-ortho-xylene is preferred. A typical carrier molecule is K tetra (*p*-chloro)phenylborate, which selectively transports tetralkylammonium ions. In the absence of a tetralkylammonium ion (i.e., in typical brain fluid), this type of exchanger transports $K^+$.

The exchanger described belongs to the class of charged-carrier exchangers and has only a limited selectivity, i.e., ability to discriminate between cations, and so senses $Na^+$ to some degree (typically about 40–80 times less effectively than $K^+$). Simon and coworkers in Zurich recognized that neutral carriers would offer much higher selectivity (*see* Ammann et al., 1983; Ammann, 1986, for an extensive survey of neutral carrier-based electrodes). One of the first neutral carriers available was the antibiotic valinomycin, which is a highly selective $K^+$ carrier (Frant and Ross, 1970; Oehme and Simon, 1976; Wuhrmann et al., 1979). Today many neutral carrier exchangers are available, some with excellent selectivities and sensitivities.

The introduction of neutral carriers removed some, but not all, technical problems. The original charged carriers, being dissolved in a high dielectric fluid, imparted considerable resistance to the ISM; typically for a $K^+$ ISM made with Corning exchanger, the resistance of the electrode will be 1000 M$\Omega$ or more. This placed considerable demands on the buffer amplifiers available at that time. But the exchanger made with valynomycin is several orders of magnitude more resistive. Because of this it could not be used in microelectrodes until formulations were found that enabled small amounts of additives (tetraphenylborate for example) to be incorporated into the exchanger. Other adjustments were also made to the solvent to enhance selectivity, and today's neutral carrier exchangers are carefully formulated "cocktails" (*see* Ammann et al., 1983; Ammann, 1986).

## 2.2. Principles of Operation

Typically a liquid membrane containing ion-selective carriers separates two ionic solutions of different concentration. The concentration gradient across the ion exchanger tends to drive ions from the region of high concentration to that of low concentration. Only the ions for which the carrier exists can be transported so that charge separation occurs and a potential develops across the liquid membrane. The electric field associated with this potential is such as to oppose the further movement of ions, and finally a steady state is reached in which the diffusional and electrical forces are equal and opposite in magnitude. Considerations of the thermodynamic work involved in setting up such an ideal system lead to the Nernst equation for the potential difference:

$$E_1 - E_2 = (RT/zF)\ln(a_1/a_2) \tag{1}$$

where $E_1 - E_2$ is the potential difference across the ion exchanger, $a_1$ is the ion activity at the ISM tip, $a_2$ is the ion activity in the backfill solution, $R$ is the universal gas constant, $T$ is absolute temperature, $F$ is Faraday's electrochemical equivalent, and $z$ is the valence of the ion.

In practice the membrane is subject to interference from other ions, which may be divalent, and an extended, and largely empirical, equation has been developed known as the extended Nicolsky equation (Ammann et al., 1983; Ammann, 1986)

$$E_1 - E_2 = (RT/z_iF)\ln(a_i + k_{ij}a_j^{z_i z_j}) \tag{2}$$

where $k_{ij}$ represents the interference of ion species $j$ on the ion being sensed, species $i$.

Recently a more fundamental equation has been provided by Simon and coworkers that can be reduced to the Nicolsky equation with suitable assumptions (Ammann et al., 1983). This equation provides the most complete description of the potential developed by an ISM at the present time.

Despite the impressive laboratory studies on ion exchangers, the resulting theories are of limited practical value to the experimental physiologist because of the complexity of the microelectrode/phyiological system under study. It is always advisable to test the ISM of choice under conditions that approximate its real application to determine interferences and other problems, rather than rely on published data.

### *2.3. Choice of Exchangers*

Table 1 lists commonly used ion exchangers and indicates their major properties. The most popular exchanger remains the Corning $K^+$ exchanger (type 477317), despite its rather poor selectivity against $Na^+$ and its sensitivity to numerous other compounds. It is popular because it is easy to work with, producing relatively low resistance ISMs, which implies low noise and lack of sensitivity to mechanical artifacts. Valinomycin-based exchangers were difficult to use, but the presently available cocktail (Fluka) is quite usable.

For measuring $[Ca^{2+}]_o$, the ETH 1001 ligand, in suitable cocktail form, is the best choice because of its excellent selectivities. Many other $Ca^{2+}$-sensing compounds are described in the literature, but their use is not recommended because they almost all have poor selectivities against some common compounds.

To measure extracellular $Na^+$, ETH 227-based cocktails are usually suitable. This cocktail does have a significant interference from $Ca^{2+}$. The quoted interferences (Steiner et al., 1979; O'Doherty et al., 1979; Dagostino and Lee, 1982) seem to refer to conditions that would be encountered intracellularly, where $[Ca^{2+}]_o$ is below 1 $\mu M$; in the BCM the concentration of free $Ca^{2+}$ is about 1.2 mM and we find that this can appear to look like as much as 20–40 mM of $Na^+$. Dietzel et al, (1982), in their studies of stimulus-evoked changes in $[Na^+]_o$, also found a high apparent interference. This indicates caution in the use of this exchanger for measuring small extracellular signals. The exchanger is also sensitive to $Li^+$, so this cation cannot be used as a sodium substitute in experiments employing it. An alternative $Na^+$ exchanger for extracellular work is based on a cocktail containing ETH 157 (*see* Coles and Orkand, 1985, for details).

Before ETH227 was available, a $Na^+$ exchanger based on the

Table 1
Some Available Liquid-Membrane Ion Exchangers
for Use in Microelectrodes[a]

| Ion | Exchanger | Manufacturer | Interferences[b] | Major references |
|---|---|---|---|---|
| $K^+$ | KTpClPB[c] | Corning (477317) | $Na^+$, TAA[+d] | Numerous |
| $K^+$ | Valinomycin | Fluka (60031) | $Rb^+$, $Cs^+$ | Oehme and Simon (1976), Wuhrmann et al. (1979) |
| $Na^+$ | ETH227 | Fluka (71176) | $Ca^{2+}$, $Li^+$ | Steiner et al. (1979) |
| $Ca^{2+}$ | ETH1001 | Fluka (21048) | $Mg^{2+}$ | Oehme et al. (1976) |
| $Cl^-$ | N/A | Corning (477315) | $HCO_3^-$, A[-e] | Numerous |
| $Cl^-$ | N/A | Corning (477913) | $HCO_3^-$, $A^-$ | Baumgarten (1981) |
| $H^+$ | Tridodecylamine | Fluka (82500) | DNP | Schulthess et al. (1981), Amman et al. (1981) |
| TAA[+d] | KTpClPB[c] | Corning (477317) | $K^+$, and so on | Nicholson and Phillips (1981) |
| $A^-$ | Aliquat 336S | *See* refs. | $Cl^-$, and so on | Nicholson and Phillips (1981) |
| $A^-$ | Crystal violet | *See* refs. | $Cl^-$, and so on | Nicholson and Phillips (1981) |

[a]This table is not exhaustive.

[b]Interferences are generally more extensive than shown; this table only indicates some of the more relevant ones. Original literature and laboratory tests should be used.

[c]KTpClPB = K Tetra (*p*-chloro)phenylborate.

[d]TAA = Tetra alkylammonium.

[e]$A^-$ = Anion.

antibiotic monensin was formulated (Kraig and Nicholson, 1976) that does not have a high $Ca^{2+}$ interference; unfortunately it does suffer from $K^+$ interference. For those interested in using this exchanger, an improved crystallization procedure was described later (Nicholson and Kraig, 1981).

Chloride can be sensed with the Corning 477315 exchanger. This charged-carrier exchanger is a quaternary ammonium compound, the precise formulation of which has not been revealed. A liquid membrane from the Orion company contains distearyldimethyl ammonium (Koryta, 1975, p. 133), but it is inferior to the Corning product. At the present time the Corning exchanger is the best available for $Cl^-$, and a new formulation (Baumgarten, 1981; Corning 477913) has somewhat improved its properties. Nevertheless, as with the charged-carrier $K^+$ sensor, this anion exchanger is sensitive to many anionic substances, most notably $HCO_3^-$, and the majority of $Cl^-$ substitutes, as well as drugs like penicillin (Speckmann et al., 1983). Recent work by Simon's group (Oesch et al., 1986) raises the hope that neutral carrier-based chloride sensors may become available soon.

It is worth mentioning that AgCl on an Ag wire is a chloride-specific electrode with selectivities as good as, if not better than, the liquid membrane (Neild and Thomas, 1973; Nicholson and Kraig, 1975; Saunders and Brown, 1977). Such microelectrodes may sometimes be used with advantage.

$H^+$ ions or pH can be sensed with great precision using a neutral carrier-based sensor incorporating tridodecylamine (Schulthess et al., 1981; Ammann et al., 1981). ISMs made with this ligand have excellent performance in the physiological range of pH. The original paper recommended equilibrating this exchanger with $CO_2$, and several investigators do this, but we have never found it necessary. This exchanger does not seem to show significant interferences, however it is rendered nonfunctional in the presence of the metabolic blocker dinitrophenol (a proton carrier), and caution should be exercised if similar compounds are used. Note that an earlier neutral carrier pH ligand from Simon's group (Erne et al., 1979) did not work well.

Alternatives to the liquid-membrane exchanger for $H^+$ include glass-membrane electrodes (Thomas 1978), which are excellent in selectivity, but have the drawbacks mentioned earlier, and metal-interface electrodes. The best known of the latter are antimony-based sensors. These have the merit of very low resistance and hence low noise and fast response time. Unfortunately they are affected by a wide variety of endogenous compounds (Quehenberger, 1977; Satake et al., 1980a,b), and validating the accuracy of the response is difficult and time-consuming.

For measuring tetraalkylammonium ions, the Corning $K^+$ exchanger can be used effectively. Examples will be described in detail in section 5.

For the measurement of various anions, other than $Cl^-$, the quaternary ammonium compound Aliquat 336S (methyltricapryl ammonium; Koryta, 1975, p. 133) is useful. This is dissolved in nitro-ortho-xylene to improve discrimination against $Cl^-$ (use of 1-decanol enhances selectivity in favor of $Cl^-$). An alternative carrier is the dye Crystal Violet (Senkyr and Petr, 1978). The use of these anion exchangers in ISMs has been described in detail (Phillips and Nicholson, 1979, 1981; Nicholson and Phillips, 1981) and will be discussed again in sections 4 and 5. Typical ions that can be sensed with a high selectivity over $Cl^-$ include $SbF_6^-$, $AsF_6^-$, $\alpha$-nathphalenesulfonate, $SCN^-$, $AuCl_4^-$, and salicylate.

A variety of other exchangers are described in the literature, including neutral carriers for $Li^+$ (Guggi et al., 1975; Zhukov et al., 1981; Metzger et al., 1984). A bicarbonate ISM was described by Khuri et al. (1974), but it has been hard to reproduce and some doubt exists as to what this type of charged carrier actually senses (Funck et al., 1982). A possible neutral-carrier electrode for $HCO_3^-$ has been discussed (Funck et al., 1982).

A neutral carrier-based ISM sensitive to $Mg^{2+}$ was described by Lantner et al. (1980), but since it was significantly more selective to $Ca^{2+}$ than to $Mg^{2+}$, it is only appropriate for intracellular recording. Recently new ligands for $Mg^{2+}$ have been described (Behm et al., 1985) and these have much better selectivities and may become available in a cocktail suitable for ISMs.

A number of exchangers are also available from companies other than those listed in Table 1, but frequently the composition or source of the exchanger is not precisely defined and such products should be avoided when possible, since this adds one more complication to an already difficult area.

A considerable literature exists on exchangers that have been formulated and used in macroelectrodes (Koryta, 1975; Lakshminarayanaiah, 1976; Ammann et al., 1983). Such exchangers may work in microelectrodes, but often exhibit different characteristics because of high impedance properties or some unexpected interaction with the glass surface of the microelectrode. Some papers have claimed that the leakage resistance of a fine glass capillary is comparable with that of the exchanger in some instances (Lewis and Wills, 1980; Coles et al., 1985), but this does not seem to be a general problem.

## *2.4. Methods of Fabrication*

The fundamental process is to put a thin layer of silane on the inside of a microelectrode tip so that it will subsequently retain the liquid ion exchanger. This process is sensitive to environmental factors, particularly local humidity, as well as the state of the glass and the shape of the electrode tip. Numerous methods have been devised, but some experimentation is usually necessary. Any investigator who is familiar with the fabrication of glass microelectrodes will be able to make ISMs relatively easily.

Two basic strategies exist for making ISMs: the use of a silane solution or the use of a silane vapor. The method used in our laboratory is a solution approach and will be described in detail. It is relatively simple, but can only make ISMs with a tip outer diameter (od) in excess of 1 $\mu$m (double-barreled theta tubing) and is more commonly employed to make tips of 2–6 $\mu$m. In contrast, vapor methods can produce tips below 1 $\mu$m as well as larger diameter electrodes. Vapor methods work well in many laboratories and should be tried (references will be provided below). We are only describing the solution method because that is the one with which we have extensive experience.

### *2.4.1. An ISM Fabrication Protocol*

This method originated in the laboratory of Prof. D. Lux at the Max-Planck Institute, Munich (Lux, 1974a), and was, in turn, derived from procedures given by Walker (1971), Vyskocil and Kriz (1972), and Lux and Neher (1973). We have modified the method slightly over the years.

We use double-barreled theta glass (R & D Scientific Glass, Box 198, Spencerville, MD 20868). To facilitate fabrication, we use 2–3 mm od glass, but if the available puller cannot accommodate such large pieces, then smaller tubing may be used. As an alternative to theta glass, one can twist two ordinary capillaries together in a puller before pulling in the normal way, or use prewelded double capillary tubing. In some experiments it may be feasible to use a single-barreled microelectrode to which a reference is glued after fabrication.

The theta tubing is cut into lengths of about 10 cm and, using an old pair of forceps and eye goggles for protection (essential), about 5 mm of the capillary is chipped away at either end (Fig. 1A). An alternative strategy is to use a diamond grinding wheel (*see* Brown and Flaming, 1977). After 20 or so pieces of glass have been cut they are cleaned by immersing in a mixture composed of equal parts of concentrated $H_2SO_4$ and 30% $H_2O_2$ (Vyskocil and Kriz, 1972). Needless to say, this mixture is very hazardous and must be made, handled, and used with appropriate precautions. After the glass has stood in this for several hours (overnight, for example), the mixture is diluted with water and discarded, and the pieces of capillary are washed in several changes of distilled water. Finally, each piece of glass is connected to a small hose and vacuum so that additional distilled water can be drawn through the tubing. This is followed by acetone and then air, or the washed capillaries are kept in a jar of acetone. In the former case the air-dried capillaries are sealed at each end with Parafilm (American Can Co.) until they are pulled; in the latter case the capillaries are blown dry with nitrogen (used in the silanization process) just prior to pulling. The cleaning procedure may be uneccessary and it may suffice to simply draw some acetone through the cut tubing, to remove glass fragments and dust, prior to fabrication. If, however, there is reason to think that the capillary is dirty or contaminated with oil or mold, for example, then this cleaning procedure is effective.

We pull electrodes in a vertical microelectrode puller. An instrument of this type is manufactured by Narishige (Type PE-2) or Kopf (Type 720). In fact, any electrode puller will function, but as noted above some types will necessitate the use of smaller-diameter tubing. We pull the tubing with a quite long shank (about 1 cm) and this seems to facilitate our filling method. For the vapor method, other geometries may be better. A long shank is also advantageous if the electrode is to be inserted deep into the brain or if other electrodes are to be glued to it.

As pulled, the electrode tip is usually too small in diameter and we routinely break it under a compound microscope (at $100\times$ total magnification) by bumping the tip against a microscope slide edge. If the electrode has a gradual taper, it can be broken quite precisely by this method. One may also use a micromanipulator to do this, but most people quickly learn to do it manually. We typically start with a tip od of 2–3 $\mu$m. To get reproducible sizing, it is best to have a reticule mounted in the eyepiece of the microscope. A monocular microscope is adequate, but objectives greater than $10\times$ usually have insufficient working distance to allow for electrode manipulation.

After breaking the tip, the side of the electrode that is to contain the exchanger is filled with an appropriate backfill solution (e.g., 100 m$M$ KCl for a $K^+$-sensing ISM) and the other side is filled with 150 m$M$ NaCl, for extracellular work. It is important that these

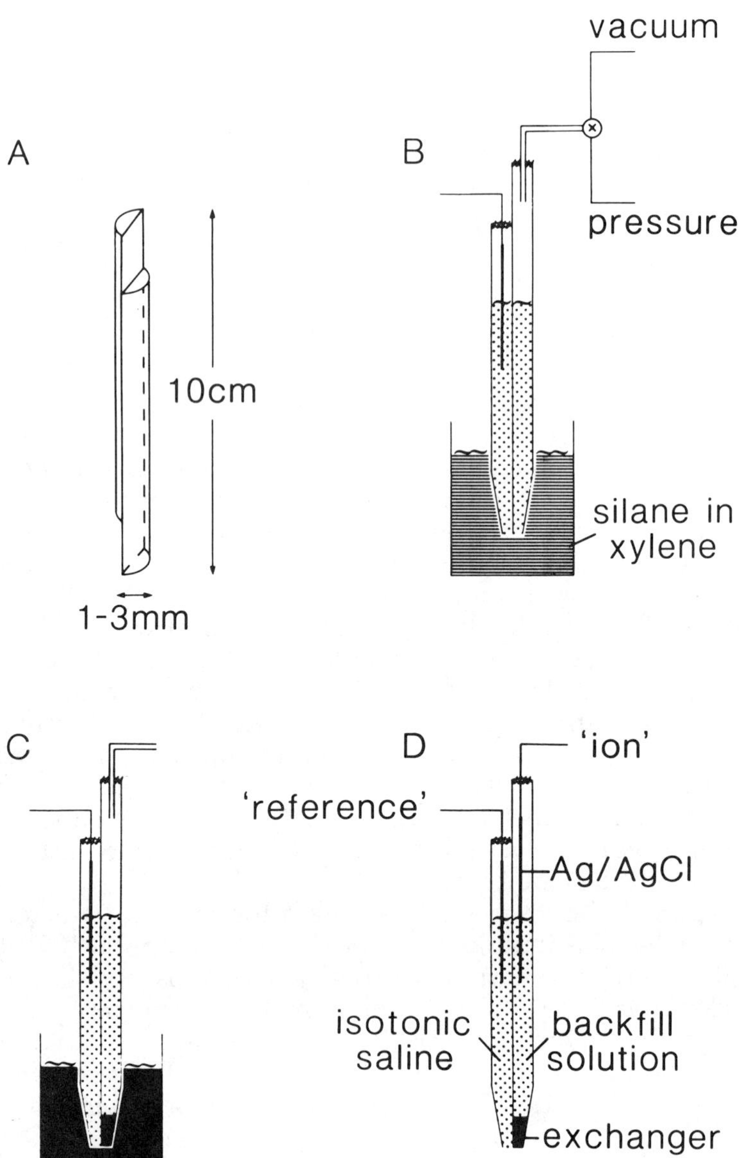


Fig. 1.   Fabrication of ISMs using silane liquid. A 10-cm length of theta-capillary glass (A) is chipped down at each end and cleaned. The tube is pulled in a conventional electrode puller and the tip broken to a few micrometers, then each barrel is filled with appropriate electrolyte, and an Ag/AgCl wire is inserted in the reference barrel and a Teflon tube

solutions do not contain bacteria (particularly the ion backfill), so we routinely inject them through a 0.22 μm filter attached to a long 30-gage needle. After filling, the back of the electrode is held in a Bunsen flame to completely dry that area and prevent a salt bridge from forming between the barrels. A chlorided Ag wire is inserted in the reference barrel and a fine Teflon tube (Small Parts Inc., 6901 NE Third Ave., Miami, FL 33138) inserted in the other side. Both wire and tube are sealed in with dental wax (Kerr "Sticky Wax," Emeryville, CA 94608). Because of the high melting point of the wax, Teflon tubing is used. The other end of the tube is connected (via a syringe needle) to a pressure/vacuum system and the electrode held vertically with a suitable micromanipulator. The electrode tip is viewed with a horizontally mounted dissection (stereo) microscope at magnifications between 10× and 40× in order to see the silane solution enter the electrode.

To silanize the ion side of the electrode, the tip is immersed in a small container of silane solution (Fig. 1B). A cuvet is ideal since the immersed tip must be viewed through the side. The silane solution consists of 4% trimethylchlorosilane (Sigma) dissolved in xylene (earlier formulations used dimethylchlorosilane and carbon tetrachloride as solvent). The solution should be made fresh each day as a stock solution. As the cuvet level falls because of evaporation of the solvent, the residual solution should be discarded and more stock solution used; otherwise the concentration of the silane rises. Both the organic solvent and silane are harmful if breathed, and the procedure should be carried out in a fume hood or at least with good ventilation. Note also that the silane vaporizes and releases HCl when it comes into contact with water. This can lead to rapid rusting of nearby ferrous objects.

Using a pressure of 20–40 psi from a tank of nitrogen, the backfill solution is expelled into the silane as a small bubble, then

---

in the ion barrel. Both are sealed with wax, the Teflon tube connected to a vacuum/pressure source and the tip placed in 4% trimethyldichlorosilane in xylene (B). Silane is drawn several hundred micrometers into the ion barrel and expelled a number of times, then the electrode is placed in the liquid ion-exchanger (C) and allowed to fill to the level of the silanization. Finally the Teflon tube is withdrawn from the ion barrel and another Ag/AgCl wire inserted and sealed into the ion barrel (D). Further details in text.

vacuum is applied and the silane is drawn into the ion barrel to a height of several tens or hundreds of μm and then blown out again (Fig. 1B). This procedure is repeated several times and then the silane expelled for a final time and the silane solution replaced with the exchanger (Fig. 1C). At this point the Teflon tube is left open (exposed to atmospheric pressure) and the exchanger spontaneously fills the electrode tip. If the filling is not spontaneous, a brief vacuum pulse can be applied. The exchanger should rise to the same level as the silanization.

The electrode is now removed from the filling setup, the Teflon tube pulled out, and a chlorided silver wire inserted into the ion barrel. Finally the wire is sealed in place with more dental wax (Fig. 1D). ISMs are either stored with their tips in an appropriate medium or with tips in the air. Sometimes the noise characteristics are improved by one of these procedures.

It is not essential to use a pressurized gas/vacuum system for this method, although it does speed up production. An alternative is simply to use a 5-mL disposable syringe and apply pressure and suction as appropriate.

A commonly encountered problem with this filling method is that the electrolyte blows out into the silane, but the silane will not enter into the electrode tip. If this occurs, one should persist with the pressure/vacuum cycle; usually the silane enters suddenly. If, after 20 or more attempts, the silane still resists entry, the tip of the electrode should be further broken against the microscope slide.

### 2.4.2. Vapor Silanization Methods

This approach was introduced by Coles and collaborators who have provided detailed descriptions (Munoz et al., 1983) as well as important discussions of the whole nature of silanization (Deyhimi and Coles, 1982). The original methods were quite complex and numerous attempts have been made to simplify them; resulting publications usually claim that the methods are very easy (e.g., Ronnau, 1984; Borrelli et al., 1985) and the investigator should certainly try these approaches. In our experience, the vapor methods can be capricious and may work well for one type of exchanger but not for another. Other investigators use them routinely without difficulty and it must be emphasized that they are essential for fine-tipped ISMs. These methods are covered elsewhere in this volume.

### 2.4.3. Older Methods

Some of the earliest methods involved the use of siloxane polymers rather than silane, e.g., "Siliclad" (Clay Adams; now discontinued) (Walker, 1971) or 1107 fluid (Dow Corning) (Futamachi, personal communication; Nicholson and Kraig, 1975). These methods involved baking the polymer and often resulted in plugged electrode tips. Such methods are generally inferior to those described above, but may be revived for special applications.

## 2.5. Instrumentation

Instrumentation is quite conventional, except that two requirements must be met. First the ion barrel of an ISM has an impedance of $10^9$ $\Omega$ (charged carrier) to as much as $10^{11}$ $\Omega$ (neutral carrier) (Dagostino and Lee, 1982) so that the buffer amplifier to which it is connected must have both a high input impedance and a low bias (leakage) current. Typical FET operational amplifiers have input impedances of $>10^{12}$ $\Omega$, which is adequate, but not all have a low bias current. A value of less than 0.1 pA is desirable. A second requirement is that the potential recorded on the reference barrel can be subtracted from the signal on the ion side. This can be accomplished with a differential amplifier. Usable circuits are shown in Fig. 2 (*see* also Tsien, 1980). It is useful to apply capacity neutralization to both barrels even though the ion signal does not have a very-high-frequency component, because it enhances the performance of the reference side and enables field potentials or spikes to be recorded from it.

Since ISMs do not have a high-frequency response, it is often advantageous to filter the output. Typically, a double-pole active filter with cut-off at a few Hertz (Hz) is adequate. For some applications it may be better to add filters between the headstages and the subtraction amplifiers so that the subtraction occurs after the filtering.

Commercial amplifiers may also be used, for example those manufactured by WPI (375 Quinnipiac Ave., New Haven, CT 06513) or Axon Instruments (1437 Rollins Rd., Burlingame, CA 94010), but the circuits shown in Fig. 2 can be fabricated relatively inexpensively. To record the data, a chart recorder is almost essential. This provides a continuous record of the experiment and enables electrode drift or dc shifts to be detected, as well as providing a convenient format for the experimental narrative. Two- or four-channel recorders manufactured by Gould are excellent, but less expensive instruments, or even EEG polygraphs, can be used. For quantitative work or the recording of evoked responses, a digital oscilloscope or appropriately programmed microcomputer is useful. We use either a Nicolet 3091 or 2090. If a computer

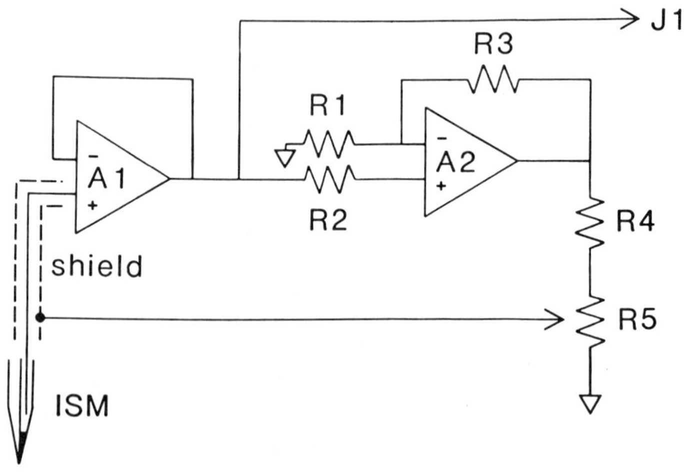



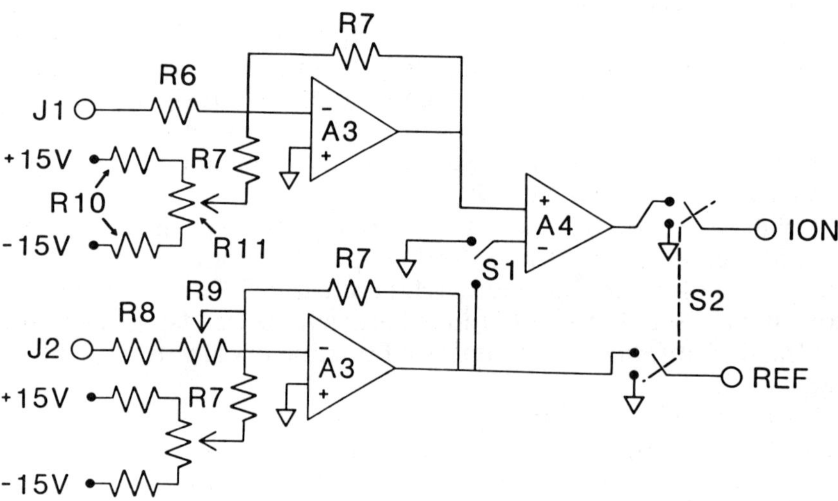



Fig. 2.   Circuits for ion recording. Two similar headstages are used, one for the ion barrel and another for the reference. Operational amplifier A1 is an FET device (Analog Devices 515L) and A2 is not critical (e.g., National LM301A or 741). Resistor values are: R1(3 KΩ), R2(1 KΩ), R3(12 KΩ), R4(2 KΩ), and R5(2.5 KΩ). A2 supplies the capacity neutralization to the shield of the wire that goes the ion or reference barrel. The output, J1, goes to the offset and substraction unit. The latter circuit consists of two

interface is not being used, an oscilloscope with disk storage would be advantageous. An alternative is to use a conventional laboratory instrumentation magnetic tape recorder. Records also can be photographed directly from the chart recorder paper. An appropriate red filter and panchromatic film combination can be used to remove the red chart paper grid on some types of paper; one should also avoid polygraphs with curved recording pen tracks if possible.

The recording environment is quite normal, but the high impedance of the ISMs can create problems. Small movements of the investigator can produce large electrical artifacts and it may be necessary to screen the preparation with grounded mesh or aluminum foil. The severity of this problem is related to the humidity of the environment and is most pronounced when the air is very dry; under these conditions static electricity buildup on the investigator can also damage the FET amplifiers, and personnel should make sure that they ground themselves before connecting the ISM to the headstage, or even during the entire experiment.

An important aspect of low-frequency recording is the indifferent electrode. An inadequately chloride-coated bare silver wire will tend to drift in its junction potential; moreover, AgCl itself forms an anion-selective interface so that experiments in which the anion content of the bathing media are altered can lead to unpredictable dc shifts on this type of electrode. The best electrodes are commercially available indifferent electrodes ("reference" electrodes), but these often obtain their superior characteristics by allowing a rapid flow of $4M$ KCl from the electrode. This is usually unacceptable in a neurobiological experiment. Consequently, we fabricate indifferent electrodes by drawing a 4% agar in $1M$ KCl

---

identical 10× gain stages with ability to add offset voltages (A3 and A4, LM301 or 741). The outputs go to an instrumentation amplifier (differential amplifier) for accurate subtraction (A4, Analog Devices AD520, or AD521 or similar). Resistor values are: R6(1 K$\Omega$), R7(10 K$\Omega$), R8(900 $\Omega$), R9(200 $\Omega$ for trimming gain), R10(5.6 K$\Omega$), R11(10 K$\Omega$, 10 turn). Ion headstage connects to J1, reference to J2. Switch S1 is used for test purposes and removes the subtraction. S2 is used to ground the output to zero the chart recorder. Filters with a few Hertz cutoff are either connected to the outputs or at J1 and J2.

solution into a Pasteur pipet and letting it set. Before the agar solidifies, we insert a chlorided Ag wire into the back of the pipet together with some cotton wool soaked in 1*M* KCl and seal the electrode with silicone rubber (e.g., aquarium sealant). These electrodes are kept with their open ends in 1*M* KCl until they are used. They typically last for 1–3 mo.

## *2.6. Calibration*

Calibration methods for ISMs have been discussed at length in the literature (Oakley et al., 1978; Lee, 1981). The basic requirement is to verify the sensitivity and selectivity of the ISM in solutions similar in composition to the medium likely to be encountered in practice. One can take the normal balanced physiological saline as the "interferent" and add to it known amounts of the ion to be measured, typically in doubling steps, if practical. If the ion of interest is likely to fall in concentration, then, of course, one must extend the range below that of the basic saline. The resulting voltage-versus-concentration curve is fitted with the Nicolsky equation (Eq. 2) using a sequence of values for interference until the least-squares regression is optimized. At this point the slope and interference are known. If 4–6 solutions are used, the estimate is not highly accurate, but is adequate to indicate if the ISM properties change during the experiment.

The ISMs are calibrated shortly after they are made and again at the end of the experiment. For a charged-carrier electrode, there will often be some loss in selectivity during an experiment (12–18 h), possibly increased by contact with the brain. Thus a fresh $K^+$ electrode (using Corning 477317 exchanger) may have a selectivity of 80:1 against $Na^+$, but by the end of the experiment it may only be 60:1. Generally, neutral-carrier ISMs remain stable longer. Frequently, however, they exhibit noise immediately after fabrication and may have to be stored with their tips in a suitable saline solution for a day or so before use.

Prior to the experiment, ISMs are also tested for drift, noise, and insensitivity. The latter problem is caused by (1) loss of exchanger, (2) faulty silanization, and/or (3) salt bridge between the two barrels at the filling end. Only the third problem can be corrected easily. This is done by removing the wax seal and silver wires, reflaming the end of the electrode, and replacing the wires and wax.

In the past a number of papers have displayed data in units of ion activity instead of concentration, on the basis that this is what ISMs actually measure. Although this statement is technically correct, calibration procedures such as that outlined above implicitly take into account the activity and actually calibrate the ISM in terms of dissolved concentration (Somjen, 1984b). This is simply because the calibrating solutions are formulated in a medium that closely approximates the extracellular ionic constituents and strength. If, for example, 6 m$M$ of $K^+$ are now dissolved in this artificial extracellular medium, the activity of this ion will indeed be reduced and the voltage measured on the ISM will reflect this, but we interpret the voltage as indicating the presence of 6 m$M$ of $K^+$, and this is an accurate and reproducible interpretation. If we actually wish to work in terms of activity, rather difficult and involved assumptions and calculations are needed (*see* Meier et al., 1982). For this reason, the majority of papers that label responses in terms of activity are actually dealing with concentration, as can be seen by looking at the calibration protocol. It is only when measurements are made in media that differ significantly from that of the calibrating solution that activities may have to considered. It should be emphasized again that it is always advisable to check a given ISM under working conditions for interference, rather than rely too much on published data for electrodes manufactured under other conditions and calibrated in other solutions.

## 2.7. Other Issues

The lifetime of ISMs varies. With the fabrication method described above, $K^+$ electrodes last about 1–2 d, whereas $Ca^{2+}$, $Na^+$, and $H^+$ ISMs are good for longer. It has been reported that the vapor method results in electrodes with much longer lifetimes.

Most ISMs' failures stem from problems with silanization. Humidity is a factor. When fabricated in conditions of excessive humidity, ISMs only work for an hour or two. In this case a dehumidifier will improve yield. In very dry conditions, electrodes also fail, so it seems that there is a critical range of values that is probably quite dependent on both the fabrication method and the exchanger being used.

The speed of response of an ISM is quite difficult to measure and must be obtained by rapidly changing the concentration of the ion while not generating any artifact. Typical "low"-resistance ISMs made from the Corning 477317 exchanger respond in 20–100 ms, whereas neutral carriers may take 50 ms–1 s to change. Much of this delay is related to the capacity neutralization capabilities of the amplifier. A method of reducing the response time of the ISM by using a coaxial design has been described (Ujec et al., 1981; *see also* Pumain et al., 1983).

Recently it has been suggested that when long columns of ion exchanger are used in ISMs, they may be very sensitive to temperature gradients and the level of the solution in which they are immersed (Vaughan-Jones and Kaila, 1986).

## 3. Voltammetric Microsensors

Voltammetry, like potentiometry (i.e., the technique used with ISMs), is a technique that is based on fundamental electrochemical principles; the similarity stops there, however. To a certain extent, the techniques are opposites. In voltammetry, it is voltage that is clamped and current flow that provides the analytical information. To avoid distortion of the monitored current, voltammetric microsensors (VMs) are constructed from low-impedance materials such as carbon or platinum, in contrast to the extremely high-impedance ISMs. When the potential applied ($E_{app}$) to a VM is sufficient to cause the oxidation (electron loss) or reduction (electron gain) of species in solution, the current measured is proportional to the concentration of electroactive species present. Because of the prevalence of antioxidants in the BCM, most electroactive substances in the brain are kept in the reduced state. Consequently, the most useful electrode reactions for studying neuroactive species *in situ* involve anodic oxidation, in which a positive potential is applied to the VM. Several recent reviews of in vivo voltammetric methods and results include those of Adams and Marsden (1982), Gonon et al. (1983b), and Stamford (1985).

### *3.1. History of In Vivo Voltammetry*

Although voltammetric principles had been studied since before the turn of the century, the first in vivo use of the technique was not reported until the mid-1950s when Clark introduced a membrane-covered Pt electrode to monitor oxygen levels in blood and various tissues (Clark, 1956). Platinum was found to be the most stable and least interactive metal tested for use in vivo. In addition to their utility for cathodic $O_2$ reduction, these electrodes were also used anodically to monitor the passage of a reductant (such as ascorbic acid or hydrogen) introduced into the circulation to assess blood flow (Clark, 1960; Frommer et al., 1961; Koryta et al., 1971). Nonmetallic materials used for oxygen electrodes included pyrolytic graphite (Clark and Clark, 1964) and glassy carbon (Clark and Lyons, 1965). Interestingly, when Clark and Lyons used glassy carbon to determine oxygen levels in the brain, they also examined the effect of locally injected ascorbic acid on the anodic current.

The first anodic voltammetry to study endogenous rather than infused substances in the brain was described by Kissenger et al. in 1973. Earlier, Adams (1958) had introduced graphite paste as an electrode material that has better characteristics for oxidation reactions than does Pt. The subsequent investigations of anodic reaction pathways of catecholamines that Adams and coworkers carried out with these electrodes (*see* Adams, 1969) led to the suggestion that voltammetric methods could be used to study these neurotransmitter substances in the brain. Although Adams and coworkers had hoped initially to measure dopamine in the rat striatum, the first observed in vivo current–voltage curves (voltammograms), as well as the known high ratio of ascorbic acid to catecholamine concentration in brain tissue, led to the conclusion that the major signal contributor was ascorbate. The initial disappointment of this discovery was compensated for by the recognition that VMs could be used to monitor the fate of intracerebrally injected drugs or neurotransmitter substances (McCreery et al., 1974). Later, graphite paste electrodes implanted in rat ventricles were used to detect and identify biogenic amine metabolite formation following neuronal stimulation (Wightman et al., 1976).

For several years, however, the ascorbate interference held back the progress of VM studies. Although investigators were able to induce voltammetric signals and identify possible components with the use of pharmacological agents (see Adams, 1976; Marsden, 1979), the possibility of interference was still there. To combat this, Lane and Hubbard (1976) introduced a halide-modified Pt electrode that gave separate oxidation peaks for ascorbate and dopamine. This VM, unfortunately, was unstable in brain tissue. The later electrochemically pretreated carbon fiber VMs introduced by Gonon et al. (1980) were found to effect the same separation, and to maintain it in the brain. The more recent Nafion (Gerhardt et al., 1984a) and stearic acid (Blaha and Lane, 1983) modifications have also improved the selectivity of VM measurements. These and other recent techniques to enhance the selectivity and applicability of voltammetric methods for studying the brain cell microenvironment have contributed to increasing interest in the field and will be described in the following sections.

### 3.2. Principles of Voltammetry

The simplest potential waveform that could be applied to a VM is a square-wave voltage step, as seen in Fig. 3A. The practical range of $E_{app}$ is limited at both the positive (anodic) and negative (cathodic) ends by solvent reactions. Potentials more positive than about +1.0 V vs. Ag/AgCl yield large background current from the oxidation of water, whereas potentials more negative than –0.3 V vs. Ag/AgCl lead to background current from the reduction of both hydrogen ions and dissolved oxygen. Within this potential window, when no electroactive species are present in solution, an initial current spike can be observed (Fig. 3A). This initial current flow is caused by the charging of the electrode surface as the potential is stepped to the new level. This is called capacitative or charging current. The residual current that persists after the charging current spike has ended is the result of a finite level of background oxidation observable at any positive $E_{app}$.

The presence of an electroactive substance in the solution produces an increase in current measured at any time during the potential step, as seen in Fig. 3A. The basic principle of this Faradaic current is illustrated in Fig. 3B for the oxidation of a catechol such as dopamine (DA) or norepinephrine (NE). Here, the electrode surface is charged to a positive potential, sufficient to oxidize the catechol. Two electrons per molecule are given up into the electrode, producing the concentration-dependent current flow as well as the quinone oxidation product. The decrease of current with time is caused by the depletion of oxidizable molecules at the electrode surface as the reaction continues; current flow is therefore diffusion-limited. The entire exponential decay with time can be described by the Cottrell equation as follows:

$$i_a = nFAC(D/\pi t)^{1/2} \tag{3}$$

where the anodic current $i_a$ at any time $t$ depends on: $n$, number of electrons transferred per molecule; $F$, the Faraday; $A$, electrode area; $C$, concentration; and $D$, the diffusion coefficient of the analyte. This simple voltage step or chronoamperometric technique has been used successfully in vivo for a variety of applications, as discussed below. The other techniques described in section 3.3,

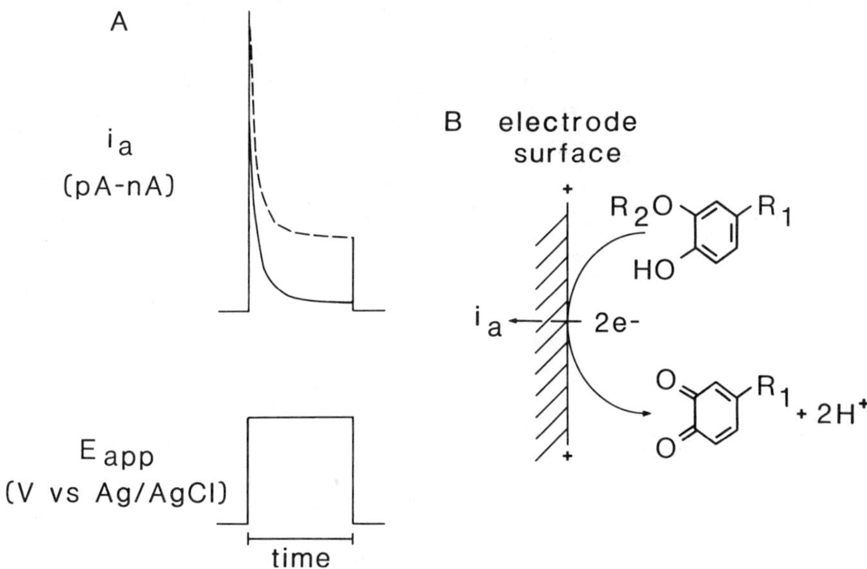



Fig. 3. Basic voltammetric events at a VM. (A) Current response (anodic current $i_a$) at a VM with square-wave voltage pulse application (chronoamprometry). The smooth line represents the capacitive and background current contibutions; the dashed line represents current flow from the oxidation of electroactive species in solution (Faradic current) plus background. (B) The oxidation reaction at the electrode surface, for catechol species ($R_1$ = H). Specifically, $R_1$ = $CH_2CH_2NH_2$: $R_2$ = H (DA), $R_2$ = $CH_3$ (3-methoxytyramine); $R_1$ = $CH_2CH_2COOH$: $R_2$ = H (DOPAC), $R_2$ = $CH_3$ (HVA); $R_1$ = $CH_2CHOHNH_2$: $R_2$ = H (NE), $R_2$ = $CH_3$ (nor-MET); $CH_2CHOHCH_2OH$: $R_2$ = H (DOPEG), $R_2$ = $CH_3$ (MOPEG); $R_1$ = $CH_2CHOHCH_2CH_3$: $R_2$ = H (epinephrine), $R_2$ = $CH_3$ (MET).

although admittedly more complicated than chronoamperometry, are based on the same principles.

What bioactive compounds are also electroactive? A chart of most of the known detectable substances and their oxidation potentials is given in Fig. 4. Catecholamine and indoleamine neurotransmitters and their metabolites are the species that have been studied most extensively. As can be seen, the products of monoamine oxidase (MAO) metabolism [DOPAC (3,4-dihydroxy-phenylacetic acid), DOPEG (3,4-dihydroxyphenylglycol), and 5-HIAA (5-hydroxyindolacetic acid)] oxidize at about the same potential as the respective parent molecules [DA, NE, and 5-HT (5-hydroxytryptamine)] because only the side chain is altered by

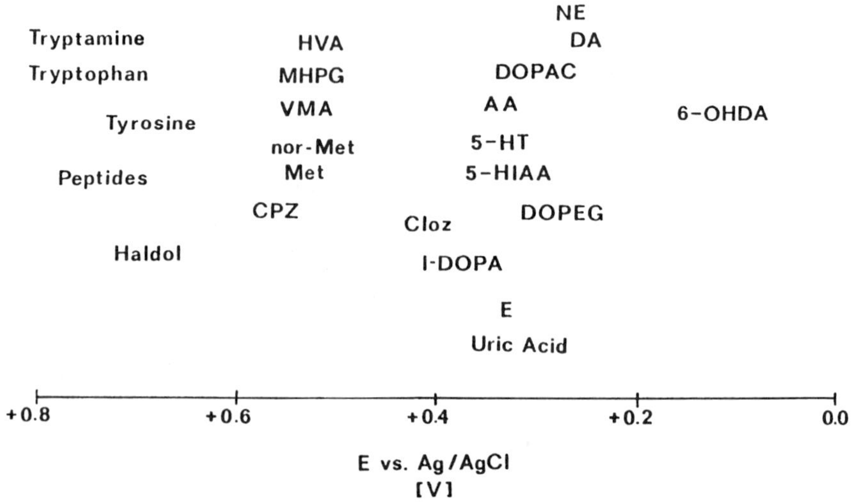


Fig. 4.   Oxidation potential ranges of selected endogenous compounds and pharmacological agents determined at graphite paste electrodes. Abbreviations: 6-OHDA, 6-hydroxydopamine; NE, norepinephrine; DA, dopamine; DOPAC, dihydroxyphenylacetic acid; AA, ascorbic acid; 5-HT, 5-hydroxytryptamine, serotonin; 5-HIAA, 5-hydroxyindoleactic acid; DOPEG, dihydroxyphenylglycol; E, epinephrine; Cloz, clozapine; *l*-DOPA, *l*-dihydroxyphenylalanine; HVA, homovanillic acid; MHPG, MOPEG, 3-methoxy-4-hydroxyphenylglycol; VMA, vanillylmandelic acid; nor-Met, normetanephrine; Met, metanephrine; CPZ, chlorpromazine; Haldol, haloperidol.

MAO. Methoxylated metabolites [HVA (homovanillic acid), MOPEG (3-methoxy-4-hydroxyphenylglycol), VMA (vanillylmandelic acid), MET (metanephrine), and nor-MET (normetanephrine)], however, oxidize at a higher potential because of the electron-donating methyl group that must be lost during oxidation (*see* Fig. 3B). Similarly, the amino acid precursors of these species, tyrosine and tryptophan, oxidize at even higher potentials because they lack the second hydroxyl group on the ring altogether. Peptides and proteins that contain these amino acids are also oxidizable at similar potentials. Two other endogenous compounds, ascorbic and uric acid, oxidize at potentials similar to that needed for catechol oxidation. Although these compounds are biologically relevant also, they constitute the major interference for in vivo measurements of biogenic amines. A few neuroactive drugs such as haloperidol and chlorpromazine have also been included in Fig.

4. Many related compounds are also electroactive. Consequently, prior to use, any pharmacological agent, antioxidant, or peptide (including enzymes) should be checked for electroactivity in the potential range to be examined.

The major difference between the voltammetric measurements just described and the potentiometric ISM measurements discussed earlier is that voltammetric detection necessarily alters the structure of the original substance examined. Consequently, voltammetry is a destructive technique. What this means for an in vivo experiment is that only the first measurement *in situ* samples the unperturbed extracellular concentration of electroactive species at the electrode tip. At an unmodified VM, the primary substance detected is ascorbic acid, which undergoes irreversible oxidation at the electrode surface. Subsequent measurements, therefore, reflect the depletion caused by each previous sampling. Eventually (generally after 10–20 measurements), a stable baseline can be attained wherein the number of electroactive molecules consumed during each sample (governed by the duration of the measurement and the area of the electrode) is matched by the replenishment of these species from the surrounding environment by diffusion or other processes during the interval between samples (Cheng et al., 1979; Lindsay et al., 1980a; Cheng, 1982; Albery et al., 1983). This "baselining" behavior is much more pronounced in vivo than in vitro because of the restricted diffusion in the brain compared to that in free solution, as discussed in section 5.

The amount of depletion of electroactive substances at the VM tip can be minimized in several ways. The most straightforward of these is to decrease the duration of the measurement and/or to increase the interval between measurements. Another method is to maintain an $E_{app}$ between measurements that is sufficient to re-reduce the oxidation product. Although this is ineffective when ascorbate is the major signal component, it can be used with modified electrodes (*see* section 3.5) that detect primarily the reversibly oxidized catecholamines and/or their acid metabolites. In fact, little "baselining" behavior is observed when ascorbate is eliminated, for example by ascorbic acid oxidase (Nagy et al., 1982; Schenk et al., 1983).

### 3.3. Voltammetric Techniques and Instrumentation

Voltammetric measurements involve two basic components: the application of a stable potential and the measurement of the resultant current flow. Both the waveform of the $E_{app}$ and the pattern of current sampling can be varied to provide different electrochemical information. Commonly used voltage waveforms include the square-wave pulse (for chronoamperometry), the linear ramp (linear sweep voltammetry), and the triangular wave (cyclic voltammetry). In the so-called pulsed techniques (such as differential and normal pulse voltammetry), a square-wave pulse is superimposed on a linear ramp or on another voltage pulse. The current output for each of these techniques can then be monitored for the duration of the $E_{app}$ or sampled periodically, depending on the information wanted.

The central element of a voltammetric instrument is the potentiostat. This device is used to set the potential difference between the working electrode and a reference electrode. The potential of the reference electrode, often Ag/AgCl, provides a stable benchmark for the $E_{app}$ so that comparisons among apparent oxidation potentials obtained under different conditions can be made. Most modern potentiostats are based on a three-electrode system in which a third so-called auxilliary (or counter) electrode is also used. This electrode (usually a silver or Pt wire) connected to the feedback loop of the voltage clamp operational amplifier (*see* Fig. 5) is used to sense any change in the $E_{app}$ and to correct for the change by current injection into the feedback loop. The major cause of drift in the $E_{app}$ is the voltage drop that occurs as current flows through the finite resistance of the solution. With large analytical electrodes that often pass milliamps of current, this IR drop can introduce a significant error in the $E_{app}$.

With the small current levels (often subnanoampere) measured at VMs used for neurochemical studies, however, a solution resistance of up to a M$\Omega$ (higher than a few k$\Omega$ is unlikely) would cause a voltage drop of only 1 mV. Consequently, a simpler two-electrode potentiostat is adequate for most in vivo applications. The potentiostat circuit presented schematically in Fig. 5 can be used in either a two- or three-electrode configuration, although we have routinely used only two electrodes. Also illustrated in Fig. 5 is a current monitor circuit. The low currents measured at in vivo electrodes dictate that high-quality, low-noise circuitry be used for the current measurement system. In the following discussion of the various voltammetric techniques that have been used in vivo, reference will be made to other circuit designs, as well as to commercially available instruments that have been used for each method. More detailed information about general circuit requirements as well as about various electroanalytical techniques can be found in the recent text edited by Kissinger and Heineman (1984).

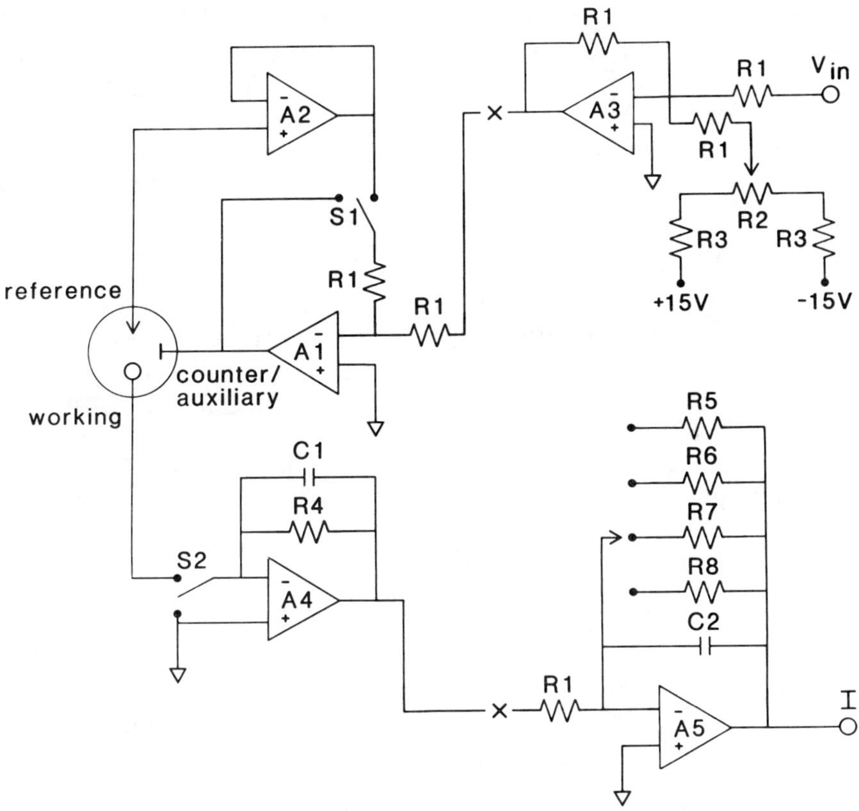



Fig. 5.    Potentiostat circuit for voltammetric measurements. Amplifier A3 supplies the required potential, or sum of potentials. Amplifier A1 acts as either buffer amplifier when system is used in two-electrode mode (no reference) or as a voltage clamp when amplifier A2 is switched in via S1. A2 is a follower or impedance buffer depending on the impedance of the reference electrode. A4 and A5 form the current measuring circuit via a virtual ground configuration. A1, A3, A4, and A5 are Burr-Brown 111 operational amplifiers; A2 is either an FET amplifier (headstage, cf. Fig. 1) or a general purpose operational amplifier. Resistor values: R1(10 k$\Omega$), R2(10 k$\Omega$), R3(10 k$\Omega$), R4(1 M$\Omega$), R5-R8(1 M$\Omega$–10 k$\Omega$ depending on gains required). C1 is 0.1 $\mu$F for filtering. S2 may be controlled by a relay.

### 3.3.1. Chronoamperometry

The voltage waveform and current response for a chronoamperometric measurement were described in section 3.1 and illustrated in Fig. 3A. In a typical *in situ* experiment, the voltage pulse is applied for no longer than 1 s, with an interval between pulses of 1 s–5 min. Current sampling is done during the last portion of the voltage pulse so that charging current contributions to the electrode background are avoided. The elimination of charging current makes chronoamperometry one of the most sensitive voltammetric techniques available, especially in comparison to the conventional scanning methods in which the electrode surface charging occurs continuously as the $E_{app}$ is ramped upward. Several methods of sampling the chronoamperometric current have been introduced. The chronoamperometric output can be sampled at a discrete time (Gerhardt and Adams, 1982; Schenk et al., 1983), integrated during the last portion of the pulse (Cheng et al., 1980; Keller et al., 1983) or sampled and averaged during this period (Lindsay et al., 1980b; Blakely and Duvarney, 1983).

Like the three-electrode potentiostat, the equation (Eq. 3) that describes the chronoamperometric current response was derived for large planar electrodes, for which all mass transfer to the electrode surface is by linear diffusion (*see* Adams, 1969, pp. 43–61). The VMs used in vivo, however, are small enough that hemispherical diffusion (or "edge effects") may also contribute to the mass transfer process and thus to the magnitude of the current measured. Consequently, a second term is added to the basic Cottrell equation that predicts the additional current contributions from such edge effects (for a complete discussion of the additional terms, *see* Dayton et al., 1980b; Galus et al., 1982; Schenk and Adams, 1984).

Three aspects of the modified equation are important for practical consideration. The first is that the contribution of the spherical term is inversely dependent on the radius of an electrode. Hence, the current measured at VMs of decreasing radius will have an increasing spherical component and a decreasing linear component. The second aspect is that the spherical term is time-independent. This means that, although the linear term may dominate at short times, at longer times while the linear diffusion current continues to decrease, the spherical contribution is constant, yielding a steady-state current. The third and most important practical aspect of the derived equation is that both terms indicate a direct dependence of the measured current on concentration, such that linear calibration curves can be obtained, just as with larger, conventional electrodes. As will be shown, these diffusion considerations are relevant for the other voltammetric techniques discussed below, as well as for chronoamperometry.

The application of the voltage pulse for a chronoamperometric measurement can be done in two instrumentally distinct ways. Fortunately, the resulting current flow can be described by the equations discussed above for either method. In the first, a switch (either manual or Teledyne relay) is placed between the working electrode and the current monitor, such that no current can flow when the switch is open. The $E_{app}$ is held at the step voltage. When the circuit is closed by activating the switch, the chronoamperometric measurement is initiated. Schenk et al. (1983) used a Princeton Applied Research (PAR) model 174A Polarographic Analyzer to provide the $E_{app}$, then the current output from the 174A was read from an oscilloscope. An external timing device sent a pulse to close a Teledyne relay switch for the desired duration of the chronoamp and also controlled the interval between pulses. Although even more elaborate and expensive instruments could be designed for these measurements, the use of a single $E_{app}$ necessitates only the simplest potentiostat circuitry, such as that illustrated in Fig. 5 (by activation of S2 in the current monitor) or the battery-powered apparatus described by Gerhardt and Adams (1982). In the latter, an inexpensive digital multimeter (Micronta, No. 22-197) was used to read out the sampled current.

In the second chronoamperometry method, the working electrode is initially at system ground, with a resting potential applied that is insufficient to cause either anodic or cathodic current to flow. Typically, the resting potential is –0.2–0.0 V vs. Ag/AgCl. The voltage pulse is then stepped from the resting potential to the desired voltage. Because current is allowed to flow at all times, when the voltage is stepped back to the resting $E_{app}$ a charging current spike will be observed with the opposite polarity of that seen with the positive step. In addition, if the measured electrochemical reaction is reversible (i.e., the oxidation product can be reduced again) cathodic current will flow as electrons are given back from the electrode to the oxidized molecules still at the electrode surface. As discussed in section 3.1, this "re-reduction" step can minimize depletion of substances (other than ascorbate) at the electrode tip in vivo to provide a stable baseline.

We use the basic circuit in Fig. 5 to apply a voltage step from a resting potential. For this method, the resting potential is set by the potentiostat voltage divider and the magnitude of the potential step is set by an external voltage source (such the WPI model 305 Isolation Unit). The duration of the pulse as well as the interval between pulses is controlled by a timing circuit. Current is sampled and recorded at the end of the pulse by a digital oscilloscope (e.g., the Nicolet 3091). The various timing sequences and $E_{app}$ values involved in these measurements can be readily computer-controlled, of course. Both microprocessor- (Cheng et al., 1980; Lindsay et al., 1980b) and microcomputer- (Apple II Plus, Blakely and Duvarney, 1983; Kuhr et al., 1984) controlled chronoamperometry systems have been introduced. Computer control also facilitates the use of multielectrode recordings. Each of the systems just mentioned are designed for use with at least two and up to 16 working electrodes. Data analysis programs are generally necessary to expedite the evaluation of the increased volume of information that can be collected under computer control.

The selection of pulse duration and repetition interval for a chronoamperometric experiment is governed primarily by the nature of the information sought. For example, a 1-s pulse with a 1–5 min sampling interval may be adequate to follow the in vivo effects of an administered drug that is known to have a time course of action that lasts for several hours (e.g., Huff and Adams, 1980; Blaha and Lane, 1984). In contrast, the stimulated release of a transmitter by $K^+$ (Rose et al., 1985) or by electrical stimulation (Kuhr et al., 1984) produces a signal that has returned to baseline in only 30 s to a few minutes. A 2–5 s sampling interval is necessary to monitor these responses accurately. To limit the total amount of oxidation that occurs with such frequent sampling, a short voltage pulse duration is recommended. We typically use pulses of 100 ms. Current measured at earlier times with shorter pulses contains a significant charging current component. The backstep-corrected pulse voltammetry used by Kuhr et al. (1984) allowed slightly shorter pulses to be used. This latter technique is discussed more thoroughly in the section on differential normal pulse voltammetry (section 3.3.6).

Although chronoamperometry has been criticized as being a less selective technique than a scanning or differential pulse technique, it is in fact only limited by the selectivity of the electrode used. For example, if an electrode has been modified to exhibit distinct oxidation potentials for ascorbate, catechols, and indoles (*see* section 3.5.1), then changes in catechol concentration can be monitored selectively when the resting $E_{app}$ is sufficient for ascorbate oxidation and the potential step is set between the catechol and indole oxidation potentials. A commercially available polarograph, the Solea Tacussel Biopulse (72 a 78 rue d'Alsace, F69100 Villeurbanne, France) does exactly this in its differential amperometry mode.

Although the scanning techniques discussed below permit the simultaneous monitoring of ECF levels of several species to be done, they lack the time resolution possible with chronoamperometric measurements. Consequently, the capability of chronoamperometry to monitor the time course of BCM events, as well as the inherent sensitivity of the technique and the simplicity of the instrumentation needed for its implementation, make it among the most useful of the available voltammetric methods for studying the dynamics of the brain cell microenvironment.

### 3.3.2. Linear Sweep and Cyclic Voltammetry

In linear sweep voltammetry, the waveform of the $E_{app}$ is a linear voltage ramp, as illustrated in Fig. 6. As the ramping potential approaches the oxidation potential of an electroactive substance, current begins to flow, increasing until the diffusional processes cannot keep up with the rate of oxidation. At this point, under linear diffusion conditions, the current flow decreases with an inverse dependence on $t^{1/2}$, exactly as in a chronoamperometric measurement (*see* Adams, 1969). With small electrodes, however, the predominance of spherical diffusion at slow scan rates ($< 20$ mV/s) often leads to voltammograms that exhibit an oxidation plateau rather than a peak because of the steady-state spherical contributions to the current response. At faster scan rates, hence shorter sampling times, oxidation peaks can be obtained because of the greater linear diffusion contribution to the current (*vide supra*). Both current responses are shown in Fig. 6. In this figure, all voltammograms are presented using standard electrochemical convention (*see* Adams, 1969) in which the voltage becomes increasingly more positive to the left of the voltage scale and anodic (oxidation) current is negative. In other figures taken from the literature and in Fig. 3A, the opposite convention is used so that peaks and concentration increases appear positive.

When a triangular wave is used instead of the linear sweep, the resulting current–voltage profile is referred to as a cyclic voltammogram, also illustrated in Fig. 6. In this technique, the rising portion of the triangular wave produces an anodic peak as described above, then the direction of the ramp is reversed and the potential scanned back to the resting level. This decreasing ramp produces a cathodic peak as the just-oxidized substance is again reduced. With larger electrodes, cyclic voltammetry is a valuable electroanalytical technique because the pattern of the oxidation and reduction peaks obtained for a given compound provide a "fingerprint" for its identification. With VMs, however, reverse

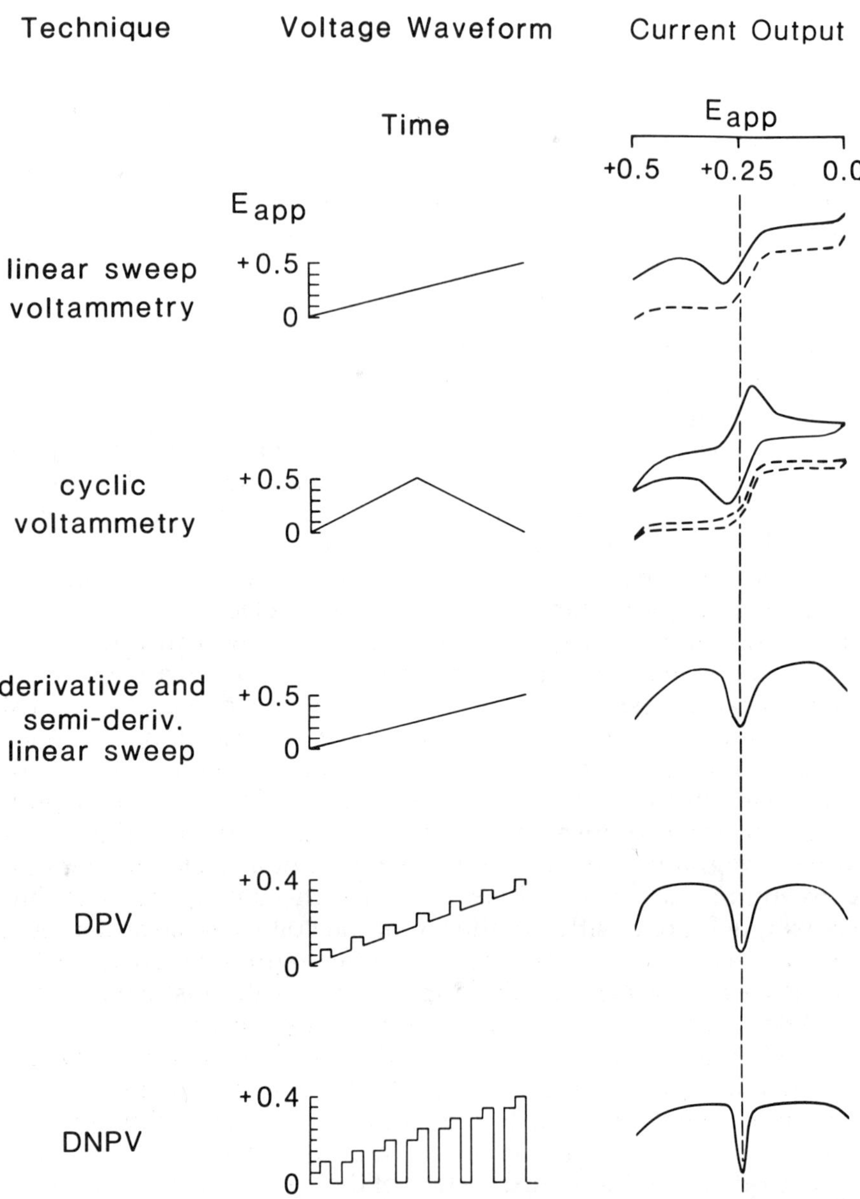



Fig. 6. Voltage and current waveforms for voltammetric techniques applied in vivo. The voltammograms (current output) are presented using standard electrochemical convention in which the voltage scale becomes more positive to the right and anodic current is negative. Each voltammogram represents the approximate current output expected

current is generally not seen for any species (except at faster scan rates) because spherical diffusion facilitates the removal of oxidized molecules from the diffusion layer before they can be reduced (Dayton et al., 1980a; Wightman, 1981; Galus et al., 1982). With the exception of the high-speed cyclic voltammetry discussed below, single-sweep methods have consequently found wider application than have cyclic methods for in vivo studies.

Several commercial polarographs have been used for in vivo studies. The most often cited are the PAR model 174 and 174A (EG&G Princeton Applied Research, P.O. Box 2565, Princeton, NJ, 08540). Also available are the DCV-5 cyclic voltammetry amplifier and CV37 *In Vivo* Voltammograph from BAS (Bioanalytical Systems, 2701 Kent Ave., Purdue Resarch Park, West Lafayette, IN 47906). The potentiostat shown in Fig. 5 can again be utilized, as well, by connecting a ramp or triangular wave generator as the input voltage. A suitable ramp generator circuit has been described by McCreery et al. (1974). O'Neill et al. (1983) have developed a microprocessor-based system that produces a ramp by a staircase of 2-mV steps. A useful feature of this system is that the digitally stored current can be used to construct difference voltammograms. This means that the baseline scans can be recorded and averaged, then subtracted from stimulus-evoked responses. Another advantage of a computer-controlled system is that the interval between scans can be precisely controlled. Timing is important in these scanning operations because of the need to replenish oxidized species at the electrode tip, as in chronoamperometry. Slow scan rates of 5–20 mV/s, which improve peak resolution, generally require 5–30-min intervals between scans for steady-state peak heights to be obtained. In addition, the appearance of the background response is also dependent on scanning interval.

←――――――――――――――――――――――――――――――――――――――――――――――――――――――――――――

for an electroactive substance (such as DA) that has an oxidation potential of +0.25 V vs. reference. The dashed-line voltammograms illustrate the oxidation plateaux seen with VMs. Oxidation peaks (smooth lines) are seen with larger electrodes and/or faster scan rates in cyclic and linear sweep voltammetry. With the derivative techniques, semiderivative, differential pulse (DPV), and differential normal pulse (DNPV) voltammetry, peak-shaped current maxima are obtained with both VMs and convential electrodes. Details of each technique are given in the text.

### 3.3.3. Semiderivative Voltammetry

Although slow scan rates improve the resolution of oxidation potentials and minimize charging current contributions to a linear sweep voltammogram, they also lead to oxidation plateaus rather than peaks, as discussed above. This conflict makes system optimization somewhat difficult. Consequently, several differential techniques have been used to enhance linear sweep peak shape. One of the first of these was semiderivative voltammetry, suggested by Lane et al. (1979). In this technique, the current output is electronically differentiated by a ladder network circuit. The manipulation produces a peak at the potential where the current output exhibits a maximal rate of change (*see* Fig. 6), as in any differentiation. Because the charging current is relatively constant with a constant scan rate, its contributions are minimized. Background currents at the solvent limits are, however,.enhanced.

The semiderivative circuit is fairly straightforward (*see* Lane et al., 1979 for discussion and additional references) and can be attached to the current output source of any potentiostat or polarograph. The microprocessor-based voltammetry system described by O'Neill et al. (1983), for example, included a semidifferentiated output. A semiderivative output option is also available with the BAS CV37 *In Vivo* Voltammograph mentioned earlier. Similar modulation of linear sweep output can be done with a simpler operational amplifier derivative circuit as well.

### 3.3.4. High-Speed Cyclic Voltammetry

As alluded to above, a notable exception to the preference for linear sweep over cyclic voltammetry for in vivo measurements is the work of Millar and coworkers who have introduced high speed cyclic voltammetry as a new and promising in vivo technique. This 100-Hz triangular wave voltammetry was initially introduced as a means of quantifying iontophoretic pulses of catecholamines (Armstrong-James et al., 1981). The VMs used for these studies were made from 7-$\mu$m carbon fibers (Armstrong-James and Millar, 1979), so a sensitive technique was needed. Because the peak current in a scanning measurement is proportional to scan rate, high-speed voltammetry is an appropriate choice. With the 40–1400 V/s scan rates used, 10–50 n$M$ iontophoresed catecholamines can be detected. Recently, these investigators capitalized on the irreversibility of ascorbate oxidation at solid electrodes to selectively measure endogenous ascorbic acid (AA) using this technique (Stamford et al., 1984a,b). In these studies, the difference between the anodic and cathodic currents is assumed to be proportional to ascorbate concentration.

As might be anticipated for this fast scanning operation, peak-shaped current maxima are observed even with semimicroelectrodes because of the extremely short measurement time. The charging current component of these scans is large as well so that the actual voltammograms are almost unrecognizable to those accustomed to more traditional electrochemistry. When the background current is subtracted from a stimulated release measurement, however, the resulting difference voltammogram appears quite classical. In fact, Millar et al. (1985) recently obtained elegant in vivo voltammograms of DA released by electrical stimulation. Both the actual and difference voltammogram are illustrated in Fig. 7. In addition to the "fingerprint" identification of substances possible with this technique, the speed of the measurement (hence minimal depletion) permits a very short interval between measurements to be used. A 25-ms interval for the monitoring of stimulated DA release has been reported (Stamford et al., 1986; Kuhr and Wightman, 1986). The circuitry for this technique has been described by Armstrong-James et al. (1981). Recently, an instrument based on this design has become commercially available (Ensman Instrumentation, 4151 Broadway, Bloomington, IN 47401). Because of the nature of the method, even the small fiber electrodes used can produce nanoamp current levels; consequently the waveform generator and current subtraction capabilities, rather than the current detection system, become the limiting factor. Present detection limits are submicromolar for both unmodified (Stamford et al., 1986) and Nafion-coated carbon fiber electrodes (Kuhr and Wightman, 1986).

### 3.3.5. Differential Pulse Voltammetry

In differential pulse voltammetry (DPV), the voltage waveform is again a linear ramp; however, superimposed on the ramp are small (typically 50 mV) square-wave pulses, as illustrated in Fig. 6. For a DPV measurement, current is instrumentally sampled just prior to and again at the end of each pulse. The difference between these sampled current levels is plotted against the ramp voltage. As in semiderivative voltammetry, a peak is obtained at the voltage at which the current exhibits the greatest rate of change. Electrode charging current is subtracted out of these DPV scans because it is virtually constant during both current sampling periods. Faradaic current output, on the other hand, can actually

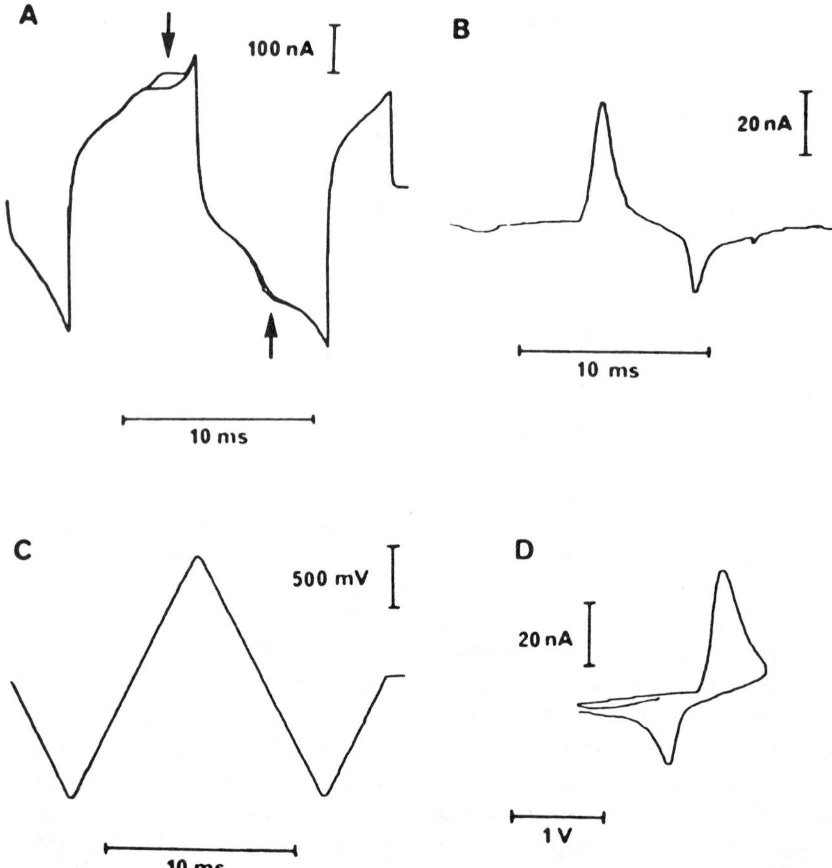

Fig. 7.    Fast cyclic voltammetry (FCV) waveforms. (A) Current waveforms seen in the caudate nucleus before and after stimulation of the MFB. Two waveforms are shown superimposed. The electroactive material released by the stimulation produces the current increment shown at arrows. (B) Current increments from (A) shown amplified with control current waveform subtracted. (C) Voltage waveform used for FCV. (D) Waveform in (B) plotted as a function of voltage rather than time. This is the cyclic voltammogram used for identification of electroactive material. The concentration of material is measured by the height of the oxidation peak [in (B)], Millar et al., 1985).

be enhanced with DPV, because the falling edge of one superimposed pulse can reduce the species just oxidized at the electrode surface, which increases the number of molecules available for oxidation by the next pulse.

As with any derivative method, DPV also enhances peak definition. Consequently, this technique is often used to increase the resolution of two voltammetric peaks with slightly overlapping potentials. For example, the peak separations facilitated at the electrochemically pretreated electrode surfaces discussed in section 3.5.1 can be maximized using DPV. This has led to the virtually exclusive use of DPV for in vivo studies with these electrodes. The advantages of the increased resolution possible with DPV have been realized with unmodified electrodes as well, such that greater peak separation, hence selectivity, can be obtained without a pretreatment step (Brazell and Marsden, 1982; Mos et al., 1981; Sharp et al., 1984a).

Several commercial polarographs that perform DPV are available. The PAR 174A and the Solea Taccussel Biopulse, both discussed earlier for other techniques, as well as the Taccussel PRG-5, have DPV capabilities. One other manufacturer is Metrohm, Ltd. (CH-9100 Herisau, Switzerland), which has a range of voltammetric units. The Taccussel Biopulse has the added feature that it will carry out the pretreatment protocol necessary to effect the separation of catechols, indoles, and ascorbic acid (*see* section 3.5.1).

The main disadvantage of DPV is that rather slow scan rates (often 5–10 mV/s) must be used to maintain good peak separation. Also, background oxidation currents are enhanced, such that at high-sensitivity settings, peaks must be measured on a sloping rather than flat baseline. This decreases the quantitative accuracy of DPV (similar problems can be met with other scanning techniques, as well).

### 3.3.6. Normal Pulse and Differential Normal Pulse Voltammetry

In normal pulse voltammetry (NPV), a voltage ramp is produced by a series of increasing voltage pulses, separated by a return to resting potential. Current flow is recorded during the last portion of each pulse. The similarity of this technique to the single potential step measurements discussed in section 3.3.1 led to the use of the descriptive term "scanning chronoamperometry" by Wightman and coworkers (Ewing et al., 1981b).

NPV offers several advantages over conventional scanning methods. One of these noted early by Lane and Hubbard (1976) and again more recently by Ewing et al. (1981a) is that the filming of solid electrodes by catecholamine oxidation products is minimized with this technique because the extent of oxidation is reduced with only periodic voltage application. Another advantage cited in a study using NPV at unmodified carbon fiber electrodes is that the sensitivity of the measurement is greater than that seen with linear sweep (Ponchon et al., 1979). The reason for this is two-fold; first, the charging current contribution is reduced as in chronoamperometry, because the current is sampled at the end of a pulse, and second, as in DPV, the Faradaic current can also be increased by the reduction of just-oxidized molecules by the step back to resting potential between pulses.

Recent modifications of NPV have further improved the sensitivity of the technique. Gonon et al. (1984b) reported that a detection limit for DA of 5 nM can be attained using differential NPV at a pretreated carbon fiber electrode. For this measurement, a small voltage pulse is superimposed at the end of each potential step, as illustrated in Fig. 6. Current sampling and differentiation are carried out as in DPV. The PAR 174A has NPV capabilities, whereas the Solea Taccussel Biopulse offers DNPV.

An even more sophisticated modification of NPV has been developed by Wightman and coworkers (Ewing et al., 1981a,b). To maximize Faradaic current flow with each NPV voltage step, very short (92 ms) pulses are used, with current integration during the last third of the pulse. At this time point, the charging current is still significant. To eliminate it, the equal and opposite capacitive current measured as the pulse ends and the $E_{app}$ returns to the resting level is added to the forward step current. Because of the primarily spherical diffusion at these electrodes, little Faradaic reduction current is seen with the return to the resting potential so these measurements are thus "backstep corrected" for charging current contributions.

The instrumentation originally developed for this backstep-corrected normal pulse voltammetry can also be used to produce differential normal pulse voltammograms by subtracting the current of the previous pulse from that of a given step (Ewing et al. 1981a,b). Subsequently introduced computer control of the system allowed the stored voltammogram data to be plotted as current versus time for a single potential or subtracted from a later scan to produce a difference voltammogram (Ewing et al., 1982). Unfortunately, the instrumentation for these procedures is not yet commercially available. This digitized technique, however, offers the combined advantages of the sensitivity time resolution of chronoamperometry and the fingerprint identification possible with a scanning technique.

### *3.4.  Electrode Fabrication*

As mentioned earlier, the materials used for voltammetric measurements are typically of low resistance to minimize distortion of the current output. Carbon, because of its inert surface and wide anodic potential window, is the material used most widely for anodic voltammetry. Platinum is still used successfully in Clark oxygen electrodes, but these cathodic electrodes will not be considered further in this chapter. The form of carbon usually used is graphite, so the terms carbon and graphite are used interchangeably in the voltammetric literature and in the following discussion.

In addition to the use of a low-resistance electrode material, small voltammetric current measurements also require good electrical contact at all interfaces. Graphite paste- and graphite epoxy-to-metal contacts appear to have sufficiently low impedance for most applications. The contact between the electrode wire and the potentiostat and current monitor must then be optimized as well. Rusty or loose alligator clips are one of the greatest potential sources of noise in these measurements. The best method for eliminating this noise source is to use matched gold amphenol connectors (part no. 220 p02 100 and 112 s02 100, Amphenol North America) soldered onto the bared end of the electrode contact wire and onto the wire leads from the potentiostat. This procedure will be the final common step for all the electrode fabrication schemes described below. Other considerations such as cable length and shielding have been discussed by Ewing et al. (1981a).

#### 3.4.1.  Graphite Paste

The simplest VMs to prepare for in vivo use are those made with carbon paste. This mixture of graphite powder and an inert water-immiscible pasting liquid provides an electrode surface that exhibits low charging and residual current levels. As originally introduced (Adams, 1958), carbon paste was made from an industrial grade lamp black, such as Acheson No. 38 graphite powder (which can still be purchased very inexpensively from Fisher Scientific). More recently, a purified graphite powder, Ultra Carbon (UCP-1-M, Ultra Carbon Corp., P.O. Box 747, Bay City, MI 48707) has been used for in vivo measurements (Conti et al., 1978; Plotsky et al., 1982). The main advantage of Ultra Carbon is that its advertised 1-$\mu$m average particle size facilitates the making of smaller electrodes than is possible with the >10-$\mu$m particle lamp black. The choice of pasting liquid depends on the nature of the experiment. For in vitro or acute in vivo studies, a straight-chain hydrocarbon such as hexadecane ($C_{16}H_{34}$, from Sigma) mixed in a 2:1 (by weight) graphite-to-liquid ratio shows the best electron transfer characteristics, i.e., produces sharper peaks in a voltammogram (Rice et al., 1983). For longer in vivo use, however, a more viscous liquid such as paraffin oil (Nujol, mineral oil; 0121-1, light paraffin oil, Fisher Scientific) or silicon oil (catalogue no. 17,563-3, Aldrich Chemical Co.) is recommended (Kissinger et al., 1973; Plotsky et al., 1982; O'Neill et al., 1982b). A 3:1 graphite-to-oil ratio is generally used.

To make a Nujol carbon paste, 1 g of Nujol is thoroughly mixed with 3 g of graphite. Thorough mixing usually requires 10–30 min of stirring with a glass stirring rod. When stored in a tightly closed container, graphite paste is stable for years. The electrode itself is made from a 4–5-cm length of Teflon-coated Ag wire (e.g., Ag 10T, Medwire Corp., Mt. Vernon, NY 10553) by sliding the outer sheath to 1 mm beyond the end of the flush-cut wire. A loop in the wire will prevent the Teflon from sliding farther. The resulting well can then be filled with carbon paste by gently tamping the tip of the electrode into a small amount of the mixture that has been transferred to a clean, flat surface (such as a glass microscope slide). These electrodes are usually 200–300 μm od. Smaller ones are harder to pack because of the flexibility of the Teflon sheath. Glass capillary tubing (3–5 mm od) pulled as described below for graphite epoxy electrodes can also be used to make electrodes of smaller diameter. One method is to insert a contact wire that is slightly smaller than the desired diameter of the electrode as far as it will go into the pulled glass, then to cement it in place. The tip of the capillary can then be sanded down with fine sand paper until only 1 mm extends beyond the wire. This well can then be packed as above.

### 3.4.2. Graphite Epoxy

To improve the durability of graphite paste for chronic implants, Conti et al. (1978) incorporated an epoxy resin into the paste formulation. For this modification, 0.45 g of triethylenetetramine thoroughly mixed with Shell Epon 815 resin (catalogue nos. 1251 and 0572, respectively; Polysciences, Inc., Warrington, PA 18976) is added to 1.35 g of graphite paste. This mixture remains fluid for 1–2 h and is sufficient for >50 electrodes. The electrode-making procedure has been presented schematically by Schenk and Adams (1984), so only a written description of the procedure is given here. Lengths (6 cm) of sodium glass (3–5 mm od) are pulled to a 2–3-cm taper with a standard glass puller such as the Narishige

PE-2. The glass shaft is then cut (if desired) to extend only 1 cm above the taper to facilitate force packing of the graphite mixture from this end. The sealed glass tip should also be opened. To pack these capillaries, the barrel is inverted and tamped into a portion of the graphite-epoxy until it is filled. The graphite can then be forced farther down into the shank with a wooden rod like that from a cotton-tipped applicator. When the paste is forced too far, however-er, the liquid will separate from the graphite, making the last segment nonconductive.

Once the shank is packed, it is cut from the shaft using small scissors. An insulated wire with the insulation removed from the last 2–3 mm is inserted as far as possible (without breaking the glass) into the graphite. The wire can then be sealed in place with a drop of epoxy. The electrodes cure in 24 h and have a shelf-life of about 2 wk. Before use, the unfilled glass at the tip of the electrode is carefully snipped back to the level of the graphite epoxy. If the exposed surface is not sensitive enough, the electrode can be cut back further into the graphite to provide a fresh surface. The practical tip diameter of these GEC electrodes is 100–200 μm, although 50-μm tips have been reported (Rose et al., 1985).

A commercially available heat-curing graphite-epoxy (PX-Graphpoxy; Dylon Industries, Berea, OH) has also been used successfully for in vivo applications (Lindsay et al., 1980b; Schenk and Adams, 1984; Rose et al., 1985). The use of Graphpoxy saves the time needed to prepare the carbon paste and the epoxy mixture (but care must be taken to buy batches that are fairly fluid in composition). Further, electrodes can be made and used the same day. The shelf-life of Graphpoxy-GECs is about 10 d. When kept refrigerated, the opened material is useful for 6 mo to a year. Electrodes are prepared as described above, except that after the contact wire is inserted, the electrodes are cured at 300°F for 4 h or at 350°F for 1 h, and then are ready for immediate use.

### 3.4.3. Carbon Fibers

Carbon fibers offer two main advantages over graphite paste and graphite epoxy as an electrode material. The first is increased mechanical stability. Even with an epoxy hardener, surface particles are lost from graphite powder-based electrodes with time. The second advantage is the size of the electrode tip. In the world of the microenvironment, smaller is better to minimize the amount of tissue damage caused by a measuring probe. In contrast to the 100–200-μm GECs described earlier, carbon fiber electrodes are 8–35 μm in outer diameter. The first in vivo measurements utilizing pyrolytic graphite fibers were reported by Gonon et al. (1978). This introduction was followed by more thorough characterization by Ponchon et al. (1979). Although the construction of fiber electrodes requires more delicacy and patience than is needed for their carbon paste counterparts, the advantages tend to overshadow the inconvenience, making them the electrode of choice for many investigators (cf., Gonon et al., 1984a).

As might be anticipated, an electrode whose surface area is only that defined by the diameter of a carbon fiber can create a serious problem for current sampling. Gonon and coworkers have circumvented this problem by extending an 8–12-μm fiber 500 μm beyond an insulating pulled-glass capillary tip, such that the surface area is increased 250-fold to give currents comparable to those seen with GEC electrodes (Ponchon et al., 1979; Adams and Marsden, 1982). A detailed description of the electrode construction procedure has recently been given (Gonon et al. 1984a). These electrodes are also commercially available from Solea Tacussel (MFC 1;72 a 78. rue d'Alsace, F 69100 Villeurbanne, France). Another method to increase the current output from the 8-μm fibers is to use multiple fibers in the electrode. The current output from electrodes made with 2–20 fibers can be increased by a factor of 10–100 (Crespi et al., 1982; Hahn et al., 1985).

Rather than lose spatial resolution by using an extended fiber or go to the trouble of gluing fibers together, we prefer to use a single 35-μm graphite fiber extended 50 μm below the glass insulation, similar to that used by Nagy et al. (1985a). The usual current output is in the low picoampere range. These larger fibers also offer the advantage of being easier to work with than the more fragile 8–12-μm fibers. The uncoated 35-μm carbon fibers can be obtained from Avco (Specialty Materials Division, 2 Industrial Ave., Lowell, MA 01815). We begin VM construction by cutting 10–12-cm lengths of fiber and 8–10-cm lengths of non-fiber-filled 1-mm od borosilicate glass. The various steps of the subsequent procedure are illustrated in Fig. 8. To improve the seal at the carbon-glass interface, we insert one carbon fiber in each length of tubing and pull the glass with the fiber inside (Fig. 8A). Normally the fiber is sealed to only one of the two shanks (Fig. 8B). The length of fiber extending beyond the glass should be broken back to about 1 cm to reduce the possibility of shattering the glass sheath by fiber movement. If the puller heating coil is too hot, the fiber can be burnt and will be brittle and exhibit high background currents. Fortunately, when this happens the fiber appears to be curved or wavy (rather than straight as usual) at the glass seal, so that the problem can be

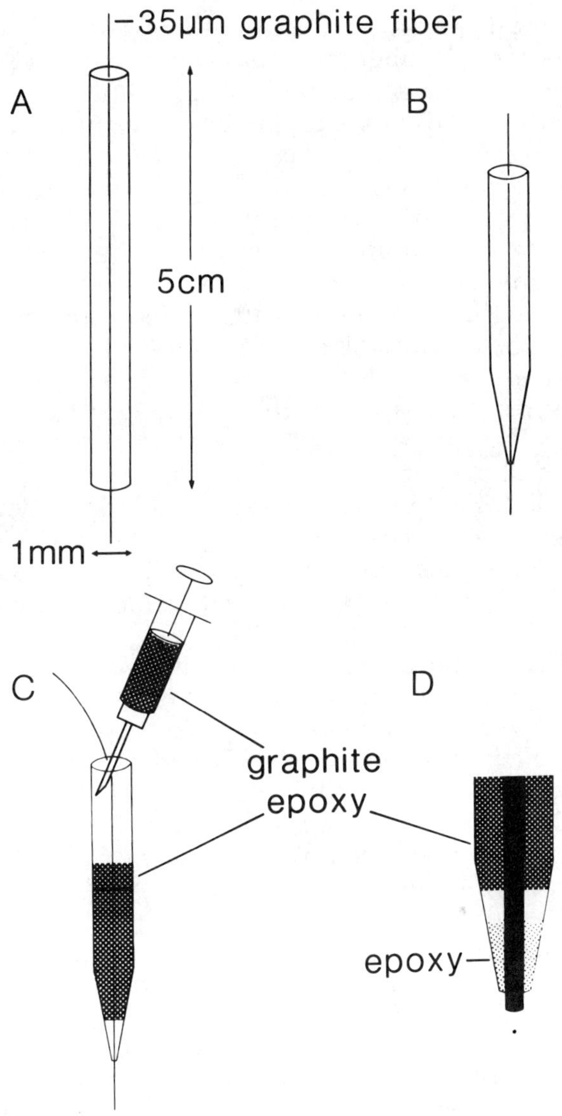



Fig. 8.    Carbon fiber VM fabrication procedure. (A) Carbon fiber (35 µm diameter) threaded through non-fiber-filled glass tubing. (B) Pulled capillary with carbon fiber sealed inside. (C) Graphite-epoxy mixture back-filled into electrode for electrical contact. (D) Close view of VM tip showing epoxy seal at the glass/carbon interface.

recognized and the heat setting can be adjusted before others are made. Similarly, if the fiber can be easily pulled out of the glass, it is an indication that the coil is not hot enough.

After pulling, the VM is backfilled with a mixture of graphite and epoxy (Fig. 8C). This serves two functions: (1) the graphite mixture provides electrical contact between the carbon fiber and the wire lead that is subsequently inserted and (2) the fluid epoxy separates from the graphite at the tip and by capillary action is drawn into any gaps in the glass-fiber interface to complete a water-tight seal (Fig. 8D). We mix 10 g of Shell Epon 815 resin with 2 g of triethylenetetramine then add 1.5–1.8 g of Ultra Carbon (the sources were given in the discussion of GEC electrodes). A 1-mL disposable syringe equipped with a 21-gage 1.5-in.-long needle is optimal for filling the electrode barrel with this conductive mixture. As suggested above, an insulated wire, bared at least 1 cm at the end, is inserted. Filled electrodes are stored tip-down, while the epoxy cures (12–24 h). The shelf-life of these VMs is at least several months. When an electrode is to be used, the extended fiber is cut back with iris scissors or scalpel blade under a dissection scope to a length of 50–60 μm.

In addition to the 35-μm carbon fibers just described and the 8-μm fibers discussed earlier, a few other fiber types have been used for in vivo measurements. The two most successful are the 5-μm Thornell P-55 fibers (Union Carbide, New York, NY) introduced by Wightman and coworkers (Dayton et al., 1980a,b) and the 7-μm Grafil fibers (Messrs. Courtaulds Ltd., Coventry England) first described by Armstrong-James and Millar (1979). In the Thornell fiber VMs, the fiber is cut flush with the outer glass and epoxy sheath, while in the Grafil VMs, the fiber protrudes only a few 10s of μm beyond the glass. Because of the small surface area of these VMs, rather sophisticated voltammetric techniques and instrumentation are needed to enhance their current output, as discussed in sections 3.3.4 and 3.3.6. A useful comparison of the in vitro properties of graphite paste, GEC, and both cylindrical and flush-cut (disk) carbon fiber VMs has recently been carried out by Kovach et al. (1984).

### 3.4.4. Calibration

In the early days of in vivo voltammetry, VMs were calibrated in a pH 7.4 buffer solution with linear increments of 4-methylcatechol, which is readily available and has oxidation properties similar to that of DA. This procedure tested the level of the residual current of an electrode as well as the sensitivity and linearity of its response to concentration changes. All of these characteristics are good indicators of electrode performance and must still be evaluated. They say nothing, however, about the selectivity of a VM for detecting a given substance in vivo, which is a concern of growing importance. Fortunately, the composition of the electroactive portion of the extracellular fluid (ECF) has become better known from both voltammetric (e.g., Schenk et al., 1983) and in vivo dialysis data (e.g., Zetterstrom et al., 1983), so that the approximate ECF concentration of both the substance(s) of interest and possible interferents can be incorporated into the calibration solution.

Specifically, for a complete evaluation of a VM, several measurements in the supporting electrolyte should first be made to determine the background current of the electrode, as well as to test for good electrical contact and minimal noise level. The same measurement protocol (potential range, sampling interval, and so on) to be used in vivo must, of course, be used in vitro. A standard electrolyte solution is $0.1M$ sodium phosphate buffer, adjusted to pH 7.4, with 0.9% NaCl added to provide chloride for the proper function of the Ag/AgCl reference electrode. (This reference can be made by anodizing a Teflon-insulated Ag wire, with the last centimeter of insulation removed, in $1M$ KCl for 30 s to 1 min until the white AgCl coating appears.) After the background current has been recorded, the response of the VM to various species electroactive in the potential range examined should be tested. For example, if studies of the DA system are to be made, the electrode response in 200-$\mu M$ ascorbate, 10–20-$\mu M$ uric acid, 20-$\mu M$ DOPAC, and 0.1–1 $\mu M$ DA would be evaluated. These concentrations can be obtained by adding incremental steps to the buffer solution to evaluate the sensitivity of the VM to each component. The ability of the VM to detect concentration changes of a single component in the presence of the others can be evaluated with subsequent increments.

Electrodes should be calibrated both before and after use in vivo. Generally, some loss in sensitivity is observed, but ideally not more than 20–25%. This loss is caused by filming of the VM surface by oxidation products as well as by proteins and other large molecules from the ECF. The loss from oxidation product filming can be minimized by decreasing the duration of each measurement (e.g., faster scan rates or shorter voltage pulse) and maintaining the electrode at a negative resting potential between measurements (e.g., –0.2 V vs. Ag/AgCl). Also, the VM must be transferred to the calibration buffer immediately upon its removal from the brain before any adhering substances can dry and further block the surface. The postcalibration is thought to be more reliable than the precalibration because it takes into account the effect of the brain on the electrode (Ewing et al., 1982). An interesting *in situ* calibration procedure has been proposed by Morgan and Freed (1981), who determine the magnitude of an in vivo signal by comparison with the response obtained for the appearance of electroactive acetaminophen in the BCM after a standard ip injection.

## *3.5. Electrode-Selectivity Enhancements*

Because of the variety of endogenous electroactive substances in the brain (*see* Fig. 4), the correct interpretation of an in vivo response is not a simple task. Fortunately, there are several experimental parameters that can be controlled to limit the number and nature of contributing species. The most straightforward way of eliminating signal components is to carefully select the potential step or range of a voltage scan applied to the VM. For example, as seen in Fig. 4, stepping or scanning to +0.4 V vs. Ag/AgCl allows for the monitoring of the primary catecholamines and their acid metabolites, but eliminates contributions from methoxylated metabolites, amino acid precursors, and peptides.

Further selectivity can be achieved when measurements are carried out in a brain region in which one transmitter system predominates. A large number of in vivo studies have been made in the primarily dopaminergic caudate nucleus of the rat for this reason. Also, activation of a given neurotransmitter system by electrically stimulating a specific pathway or using pharmacological agents with putative selectivity for the system helps to limit the source of a response. Other pharmacological agents such as synthesis or metabolism inhibitors, uptake blockers, enzymes, and selective neurotoxins can then be used to elucidate the major contributor(s) to a given signal by eliminating or enhancing the ECF concentration of possible candidates.

Unfortunately, in many types of in vivo experiments pharmacological treatments can alter the responsiveness of the brain to a given stimulus. Further, the lack of specificity of action of many drugs may confound the data, rather than clarify them. Consequently, it is better to improve the specificity of the measurement for a given substance by improving the selectivity of the probe electrode, rather than by altering the matrix in which the measurement is made. As a means to this end, several surface-modified or otherwise selective VMs have been introduced in the past few years.

### 3.5.1. Electrochemical Pretreatment

As alluded to in earlier sections, a major interferent in the in vivo detection of catecholamines and their metabolites is ascorbic acid (AA). Ascorbate not only oxidizes at the same potential as the catechols at unmodified VMs, but also is found in 10–1000 times greater concentration. Gonon and colleagues discovered, however, that after carbon fiber VMs (described earlier) were subjected to a high-frequency triangular voltage wave of a specific positive potential range for a fixed interval, individual oxidation peaks for DA and ascorbate could be resolved, as in Fig. 9 (Gonon et al., 1980, 1981). These parameters have since been optimized to improve the stability of the separation in vivo. The currently recommended protocol is a 70-Hz 0 to +3 V triangular wave applied to the VM for 20 s followed by a +1.5-V dc step applied for 5 s then a –0.8-V step for another 5 s (Gonon et al., 1984a). The oxidation peak for ascorbate appears at –50 mV vs Ag/AgCl and that of DA and DOPAC at +150 mV, with these electrochemically pretreated fiber VMs using differential pulse voltammetry.

Immediately after the pretreatment procedure, 5-HT and 5-HIAA are oxidized at +300 mV with these VMs, such that three distinct peaks are obtained. Serotonin (5-HT) oxidation products tend to film the electrode surface, however, so that the separation of the peaks is quickly lost. Consequently, Cespuglio et al. (1981) developed a variation on the above pretreatment paradigm strictly to measure the indole species. Recently, Crespi, working with Marsden's group, reported that when this same indole procedure is applied to a three-carbon-fiber VM, all three peaks, as well as a fourth corresponding to HVA, can be stably monitored in vivo for several hours (Crespi et al., 1984a). Similar results were later described for a monofiber VM after a slight modification of the indole procedure (Crespi et al., 1984b).

Although the exact function of each of the various pretreatment steps is not well understood, that such anodic treatment of graphite should improve electron transfer reactions (e.g., decrease oxidation potentials) is an accepted property of carbon electrodes (*see* Engstom, 1982; Rice et al., 1983). Presumably the graphite surface is activated in some way, either by surface clearing or

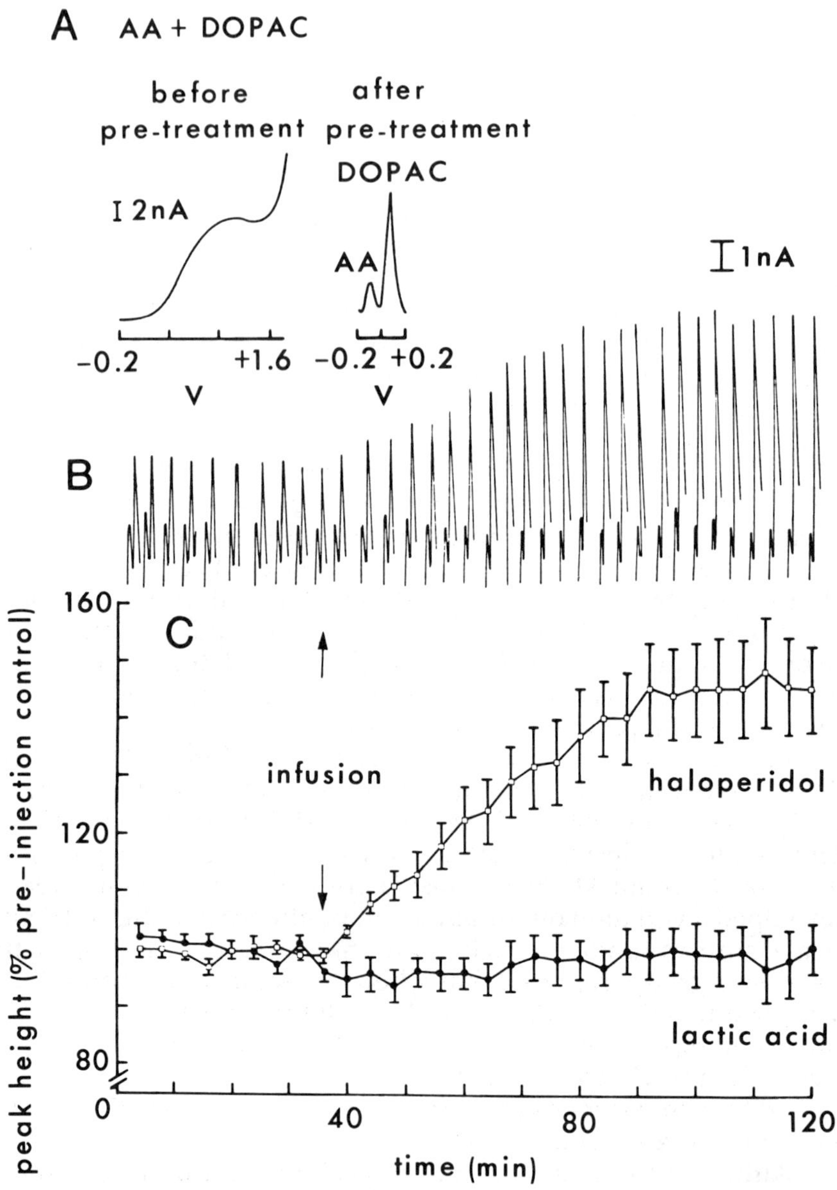

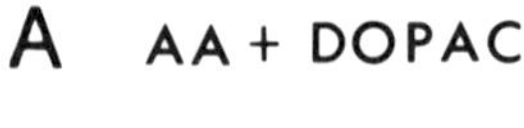


Fig. 9. In vitro and in vivo response of pretreated carbon fiber electrodes. (A) DPV voltammogram obtained in a solution of ascorbate ($5 \times 10^{-4}M$) and DOPAC ($5 \times 10^{-5}M$) before and after electrical pretreatment (modified from Sharp et al., 1984b). (B) A typical series of voltammograms recorded every 4 min from the nucleus accumbens of a chloral hydrate-

through the formation of oxygen-containing functionalities. Similar separate oxidation peaks for ascorbate and catechols have also been reported after chemical and electrochemical pretreatment of carbon fibers (from a different source than those used by Gonon) (Plotsky, 1982) and also after the electrochemical pretreatment of GEC (Graphpoxy) electrodes (Falat and Cheng, 1982).

The advantage of using VMs that can simultaneously monitor changes in the extracellular concentrations of four different classes of species is obvious. In addition, these electrodes are extremely sensitive for monoamines. Gonon et al. (1984b) recently reported 50-n$M$ changes in ECF DA levels in the striatum following amphetamine administration. The disadvantages of these probes are more subtle. First, the background currents of these pretreated VMs are elevated over untreated levels. This usually necessitates the use of pulsed voltammetric techniques (commonly DPV) to minimize charging current contributions. Second, the remarkable sensitivity of these VMs is in large part the result of the adsorption of catechols, especially DA, on the electrode surface. This adsorption causes a large increase in the time constant of the electrodes such that monitoring the actual time course of a stimulated response is not possible with these VMs (Gonon and Buda, 1985). The increased time constant will also affect the accuracy of concentration estimates. Also, although ascorbate can be separately monitored with these VMs, they lack sensitivity for it, so again more than qualitative observations of AA concentration changes are difficult.

Nonetheless, these modified electrodes have already proved to be a valuable tool for in vivo studies. Further selectivity enhancements may also be possible from adventitious pretreatment conditions. For example, Hahn et al. (1985) reported that the relative sensitivities of VMs to DOPAC and ascorbate could be altered by changing the pH of the pretreatment medium. The coupling of

---

anaesthetized rat. The increase in the DOPAC peak following haloperidol infusion (2.5 µg/0.5 µL) into the ipsilateral ventral tegmental (VTA) area is clearly demonstrated. (C) The effect of haloperidol (2.5 µg/0.5 µL, n = 6) and lactic acid vehicle (0.5 ΩL, 6 mg/mL, $n = 4$) infusion into the VTA on the DOPAC peak in the nucleus accumbens. The results are expressed as a percentage of a preinjection control period, taking the mean of the last ten readings before infusion as 100%. The error bars represent the standard error of the mean [(B) and (C) after Maidment and Marsden, 1985].

electrochemical pretreatment with the use of Nafion, as discussed below, also offers promise for improved selectivity (Brazell et al., 1986).

### 3.5.2. Stearic Acid Graphite Paste

In contrast to the effect of electrochemical modification in which the oxidation potential of AA is reduced to below that of DA, the incorporation of stearic acid into graphite paste (75 mg stearate/ g graphite) has the opposite effect of pushing its oxidation potential to the solvent limit (Blaha and Lane, 1983). The same effect is seen for the quasireversible DOPAC as well; however, the oxidation potentials of NE and DA are unchanged at these modified surfaces. Consequently, stearate-modified VMs can be used to selectively detect the catecholamines in the presence of ascorbate and DOPAC without interference. Further, the oxidation potential of 5-HT is distinct from NE and DA, so that it may be selectively eliminated as well (Blaha and Lane, 1983). Similar selectivity has also been seen with graphite epoxy VMs made with dodecylsulfate-modified paste (Keller et al., 1983).

The presence of the stearic acid chains at the VM surface serves two functions. The first is to make the electrode surface negatively charged (from the ionized carboxylic acid groups) so that the cationic amines will be attracted and the anionic acid metabolites, as well as ascorbate, will be repelled. The second is that the extended stearate chains probably block the electrode surface to an extent, which inhibits electron transfer from irreversible and quasireversible species like AA and DOPAC, respectively, more than from the more reversibly oxidized DA and NE. This latter idea is supported by the observation that similar electrode selectivity toward NE and DA is seen with unmodified, but tightly packed, Nujol carbon paste VMs (Echizen and Freed, 1983).

Although these VMs offer clear advantages for selective neurotransmitter detection, they are not without their limitations. The size of the VMs is restricted to that possible with carbon paste. Further, there have been reports that the measured current for DA can vary with AA concentration at these electrodes (Newell and Calhoun, 1982). This is a common feature of electrodes that exhibit a higher oxidation potential for AA than for DA (Dayton et al., 1980b). Because AA is thermodynamically easier to oxidize than DA (which is seen with pretreated VMs), DA oxidized in the presence of AA is reduced by charge transfer with ascorbate. This DA can then be reoxidized at the electrode surface, resulting in enhanced current that can depend on the concentrations of both AA and DA. If the AA concentration remains constant, this sensitivity enhancement is a bonus. However, as discussed later, AA as well as catechol levels in the BCM are altered with neuronal activity.

### 3.5.3. Nafion

Another technique developed to eliminate anion contributions to a voltammetric response is to coat a VM with the polysulfonated Teflon derivative Nafion (Gerhardt et al., 1984a). Like the negative surface charge obtained with stearic acid, this polyanionic coating repels like-charged AA, DOPAC, and 5-HIAA. In addition, this polymer extracts cationic amines (Nagy et al., 1985a), so that the sensitivity of Nafion-coated VMs for biogenic amine detection is enhanced as well. The selectivity of these electrodes for DA compared to AA and DOPAC (Gerhardt et al., 1984a) or uric acid (Rice and Nicholson, 1987) is usually at least 100:1.

Our procedure for coating 35-$\mu$m carbon fiber VMs with Nafion (5% solution, EW 1100; Solution Technology, Inc., P.O. Box 171, Mendenhall, PA 19357; or catalogue no. 27,470-4, Aldrich Chemical Co.) is a modified version of that described by Brazell et al. (1985). These workers reported that the application of a positive potential to the electrode when it was dipped in the negatively charged Nafion increased both the ease of coating fiber VMs and the resultant selectivity of the VMs against anions. We use a Ag/AgCl wire, which has a small loop at the end as the counter electrode, for this procedure to help ensure that a constant potential is applied. A drop of Nafion solution is placed at the looped end of this wire. The $E_{app}$ is continuously on (+0.5 V) as the fiber electrode is lowered into the drop for 10–20 s. This first coat is allowed to air dry for a few minutes, then a second coating is done. The sensitivity of the electrode to 50-$\mu M$ increments of AA and UA and 0.1–0.5-$\mu M$ increments of DA is then evaluated and more Nafion applied if necessary. Rinsing and hot air or oven drying (at low temperatures: 50–60°C) is recommended before recoating to eliminate the noise possible from trapped water layers.

Brazell et al. (1987) have recently found that a superior coat of Nafion can be obtained with a 3–4-s single exposure when an even higher potential (+3.1 V) is used for the "electroplating." For this procedure, the electrode is lowered into the Nafion drop, the potential is applied for 3–5 s after the electrode is in position, then the electrode is removed with the potential still on. The Nafion coat is then hardened (as per the manufacturer's recommendation) by heating in a 60°C oven for 15 s. With fiber VMs that have also been electrochemically pretreated, the sensitivity for DA over DOPAC can reach 400–500:1 (Brazell et al., 1987).

One of the main advantages of Nafion coating is that it can be applied to most in vivo VM types. It was originally introduced for use with GEC electrodes, but has recently been used to improve the selectivity of carbon paste electrodes (Mueller, 1986). It has also been used with both untreated and electrochemically pretreated carbon fibers (Kuhr and Wightman, 1986; Brazell et al., 1987). An important factor here is that not only does the Nafion not alter the effects of pretreatment or distort the resulting DPV voltammogram, it can also serve to protect a modified surface from being altered by exposure to brain tissue (Brazell et al., 1987). Further, Nafion-coating an electrode apparently does not increase the response time of an electrode, as indicated by the similarity of the time course of subsecond DA-release signals recorded with either coated (Kuhr and Wightman, 1986) or uncoated (Stamford et al., 1986) carbon fiber VMs.

### 3.5.4. Other Methods

One reason that various fiber electrode pretreatments were originally investigated was because unmodified fibers are relatively insensitive, possibly because of surface properties introduced during the manufacturing process. Wightman's group has taken advantage of their untreated fibers' lack of sensitivity for ascorbate and DOPAC to increase measurement selectivity without a pretreatment step (Dayton et al., 1980b; Ewing et al., 1982). In these studies, a voltammogram is recorded before and after a stimulus, then the difference between the two scans is generated electronically and compared to the in vitro voltammograms of possible contributors. Even though the oxidation of AA is inhibited at these VMs, little catalytic current is seen for DA because the spherical diffusion of oxidized products away from the surface is so efficient at these electrodes (5-$\mu$m carbon fiber disk described in section 3.4.3) (Dayton et al., 1980b).

Another electrode modification that completely eliminates AA contributions to both Faradaic and catalytic current is to immobilize the enzyme ascorbic acid oxidase (AAO, EC 10.1.3.3; catalogue no. 236 341; Boehringer-Mannheim, St. Louis, MO) (Nagy et al., 1982). This AAO reaction layer on a carbon paste or GEC electrode allows for the diffusion of various electroactive species to the electrode surface for voltammetric oxidation, except for ascorbate, which is oxidized by the enzyme. AA serves as the specific cofactor for the AA oxidase-catalyzed reduction of $O_2$ to $H_2O$. Although these electrodes have been successfully used in brain slice release studies (Rice and Adams, 1982), they lose activity in vivo, possibly because of local oxygen depletion that inhibits the enzyme.

## 4. Applications of ISMs and VMs to Measurement of Endogenous Compounds

One of the most exciting discoveries that has been made with ISMs is that the brain cell microenvironment is in a state of continuous flux with ion concentrations ebbing and flowing with brain activity. We are presently only able to glimpse a tiny part of this chemical interchange, because of our limited repertoire of sensors. Moreover, the logarithmic response of an ISM means that it is best suited to measuring changes in ions with low extracellular concentration. Not surprisingly, therefore, most work has been done on $[K^+]_o$, with $[Ca^{2+}]_o$ running second in popularity. Recently it has become possible to look at $[H^+]_o$ as well, and a few studies have looked at $[Na^+]_o$ and $[Cl^-]_o$, particularly in spreading depression and anoxia where large changes are evident.

In contrast to ISMs, VMs have been used most sucessfully to measure changes in the levels of electroactive species having the highest concentrations in the ECF. With the exception of DA, few actual neurotransmitters have been directly detected in vivo because their typically submicromolar levels (Zetterstrom et al., 1983) are below the detection limits for many VMs. More commonly, the metabolites of transmitters, such as DOPAC, 5-HIAA, and HVA, which are generally 10–20 $\mu M$ in the ECF, have been used as indicators of neurotransmitter release and/or turnover. In addition, the high concentration of AA has not only made it readily detectable, but has also led to an increasing interest in its possible role as a neuromodulator.

This section will not try to review the considerable literature that has accrued from ISM and VM methods, but rather take a few examples of work, including both early and recent studies, to indicate the experimental possibilities and provide access to the literature.

## *4.1. Stimulation*

When recordings are made from the extracellular space, we are usually dealing with a *population* response; that is the change that originates in a group of cells rather than a single one. In the case of chemical signals, as discussed here, the responses are modified both by the geometry of the neuronal population involved and the diffusion properties of the BCM. The latter will be detailed in section 5.

### *4.1.1. Natural Stimulation*

Comparatively few studies of ions in the CNS have used natural stimulation to evoke chemical responses. An early study by Singer and Lux used visual input to evoke $K^+$ responses in the lateral geniculate nucleus (Singer and Lux, 1973). Sykova and coworkers (Czeh et al., 1981; Sykova, 1983; Sykova et al., 1985) have demonstrated $K^+$ changes in the spinal cord of the frog following thermal and tactile stimulii to the skin. Heinemann et al. (1977) showed both $K^+$ and $Ca^{2+}$ responses in the cerebral cortex evoked by tactile stimuli.

By far the largest number of studies using natural stimuli have been on the retina where it is believed that some components of the electroretinogram (ERG) are generated by changes in $[K^+]_o$ (*see*, for example, Oakley and Green, 1976; Oakley and Steinberg, 1982; Kline et al., 1978; Karwoski et al., 1985; Fig. 10).

Although VM measurements in the retina have not yet been reported, retinal stimulation by light has been used to induce NE release in hippocampus-locus coeruleus tissue grafted into the anterior chamber of the eye (Gerhardt et al., 1984b). The grafted tissue receives acetylcholine input, which is activated by visual stimuli.

### *4.1.2. Behavioral Studies*

Related to the studies of natural stimuli discussed above are several types of behavior that have been found to elicit voltammetric signals in various brain regions. Because of the link between DA and motor activity, most of the studies have examined dopaminergic areas. For example, Yamamoto et al. (1982) found increases in presumed DOPAC in the striatum with turning behavior. O'Neill and Fillenz (1985) reported a higher correlation of spontaneous motor activity with DA release (indicated by HVA levels) in the nucleus accumbens than from release in the striatum. Broderick et al. (1983) observed that the inhibition of amphet-

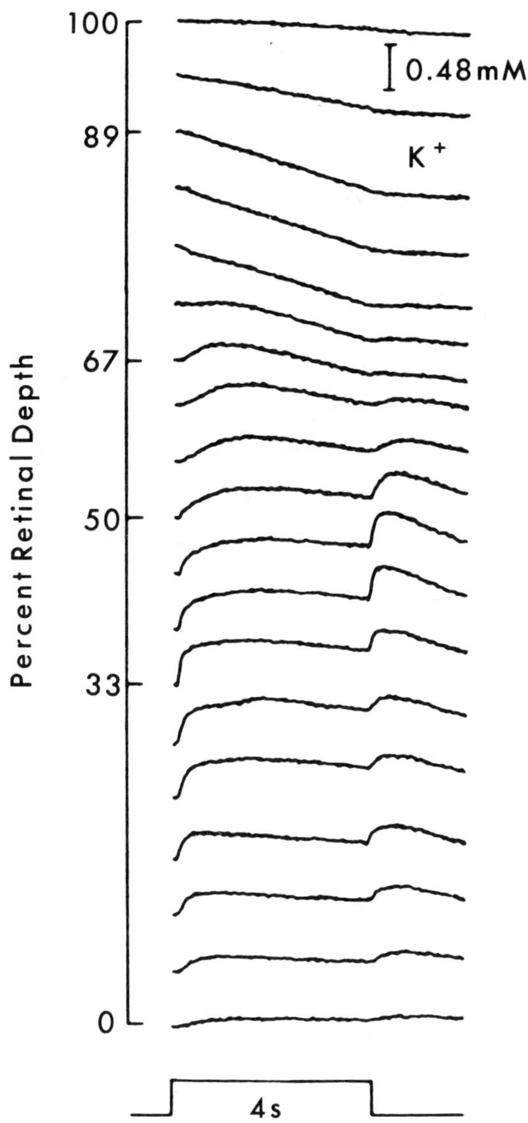



Fig. 10. Depth profile of light-evoked change in [K$^+$] in a frog retina. Inner limiting membrane corresponds to 0% depth and pigment epithelium to 100%. Resting [K$^+$]$_o$ is 2.9 m$M$. Light applied for 4 s (modified from Karwoski et al., 1985).

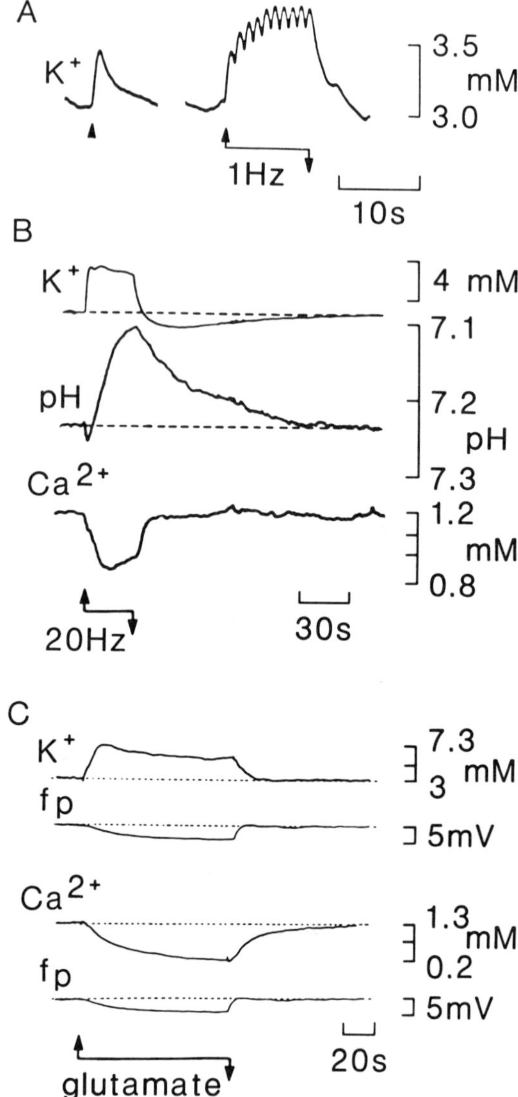



Fig. 11.  Comparison of various types of evoked ionic signals. (A) shows $[K^+]_o$ change evoked by a single stimulus to the surface of the cerebellar cortex of the rat. Note the long time course of the decay; this enables the evoked responses to summate at frequencies as low as 1 Hz. (B) shows the three ions that can be reliably measured in the BCM. At 20-Hz stimulation, $[K^+]_o$ increases by 4 m$M$ (from a baseline of 3 m$M$) and then undershoots the resting level after the cessation of the stimulus. In

amine-induced stereotypical behavior by enkephalin was accompanied by a reversal of the increase in VM signal usually seen with amphetamine.

In other behavioral studies with nonselective VMs, changes in ECF concentrations of electroactive substances (possibly ascorbate) were reported in the striatum of the awake rhesus monkey when the animal was presented with food or with a stressful stimulus (Lindsay et al., 1981). Knott et al. (1981) also examined the effects of various behaviors and stress on both catecholamine and indoleamine signals in the rat. More recently, Gonon et al. (1983a), using pretreated VMs, demonstrated an increase in DOPAC in the locus coeruleus as the result of immobilization stress. VMs have also been used to follow circadian changes in the levels of ascorbate and HVA (cf., O'Neill et al., 1982a).

### 4.1.3. Electrical Stimulation

The most common type of stimulation of the CNS is by means of electrical pulses. Such techniques are widely used and will not be described in detail here (e.g., Patterson and Kesner, 1981). The response of $[K^+]_o$ to a single local surface stimulus applied to the rat cerebellum is shown in Fig. 11A. It is evident that although the stimulus lasted only a few tens of microseconds, the evoked $[K^+]_o$ increase persisted for several seconds. Because of this feature, repetitive stimuli at frequencies of 1 Hz or more cause $K^+$ to accumulate (Fig. 11A). As the frequency or duration of the stimulus train increases, so does the magnitude of the $K^+$ response; more than a hundred papers attest to such data. When other types of ISM are used, changes in $[Ca^{2+}]_o$ and $[H^+]_o$ can be demonstrated (see below and Fig. 11B). Changes in $[Na^+]_o$, and $[Cl^-]_o$ are much more difficult to detect because of the high background. In a few instances, the ionic variations around single cells have been meas-

---

contrast pH ($[H^+]$) first shows a rapid alkaline-going transient and then exhibits a slow, long-lasting acid-going excursion that outlasts the stimulus train (modified from Kraig et al., 1983). Finally $[Ca^{2+}]_o$ decreases during stimulation (modified from Nicholson et al., 1978). (C) demonstrates the ionic changes that can be produced by local iontophoresis of glutamate in the rat neocortex (fp, field potential; positive up) (modified from Pumain and Heinemann, 1985).

ured (Neher and Lux, 1973; Heyer and Lux, 1976; Hounsgaard and Nicholson, 1983); such experiments graphically illustrate the relationship between the single cell and the population response (Fig. 12).

One of the earliest methods used to produce an in vivo VM response that could be attributed to a single chemical species was electrical stimulation. Wightman et al. (1976) stimulated the median forebrain bundle (MFB), then monitored metabolite formation in the lateral ventricle. Cyclic voltammetry confirmed that the main substance produced was HVA. In this same report, midline raphe nuclei were also stimulated, producing an increase in ventricular 5-HIAA. Marsden et al. (1979) later found the raphe stimulation caused a release peak in hippocampus, but not in striatum. More recently, Wightman and coworkers have returned to MFB stimulation to elicit DA release in the striatum, as illustrated in Fig. 13 (Kuhr et al., 1984; Ewing and Wightman, 1984; Millar et al., 1985). O'Neill et al. (1984b) reported dentate gyrus-specific release of ascorbate with stimulation of the perforant pathway.

### 4.1.4. Chemical Stimulation

Ionic changes in the brain cell microenvironment can be initiated by the application of chemical agents that depolarize neurons. This type of study has been quite limited and has usually involved excitatory amino acids. Heinemann and coworkers used glutamate, aspartate, and homocysteic acid to cause $[K^+]_o$, to increase and $[Ca^{2+}]_o$ to drop (Pumain and Heinemann, 1985; Fig. 11C). Potassium increases were seen in the cerebellar slice (Hounsgaard and Nicholson, 1983) in response to the iontophoresis of glutamate. Gamma-aminobutyric acid (GABA) (Deschenes and Feltz, 1976) has also been used.

As in studies using ISMs, the effect of amino acids on ECF levels of electroactive species has also been investigated. O'Neill et al. (1984b) examined the level of ascorbate in striatum following intraperitoneal injection of glutamate, aspartate, tyrosine, and glycine. Only the excitatory amino acids produced an increase. Microinfusion of glutamate in striatum also increased ECF ascorbate. Using both a $K^+$ ISM and a Nafion-coated VM, Nagy et al. (1985b) reported a release of DA and $K^+$ with pressure ejection of glutamate, but an increase only in the DA signal with low tyramine concentrations. A decrease in $Ca^{2+}$ was also seen with the glutamate-induced DA release (Moghaddam and Adams, 1986; Fig. 14).

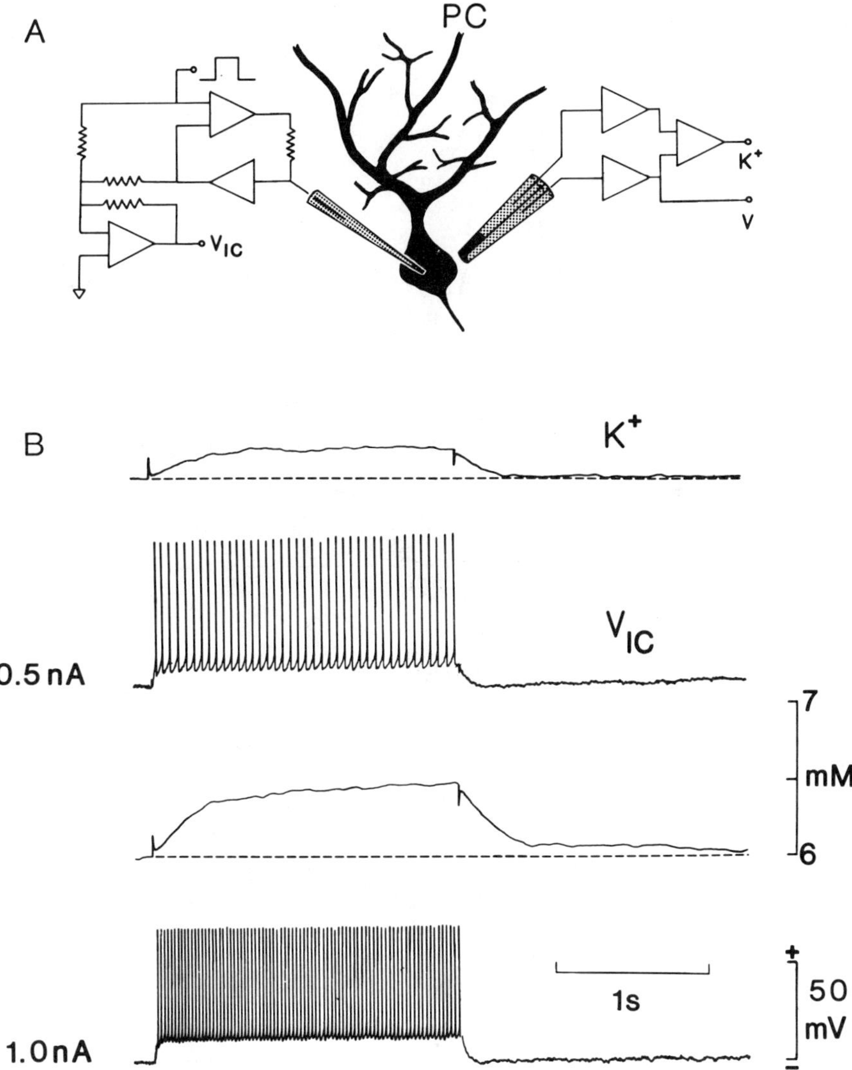



Fig. 12.   $[K^+]_o$ accumulation around the soma of a single Purkinje cell. An intracellular electrode and bridge amplifier is used to pass depolarizing current into the cell body in a slice preparation. An ISM just outside the membrane records the resulting ion signal (*see* upper figure; ground not shown). With 0.5 nA of applied current, a train of regular action potentials is elicited and a modest $K^+$ increase is seen. A doubling of current increases the firing frequency and the $K^+$ Hounsgaard and Nicholson, unpublished).

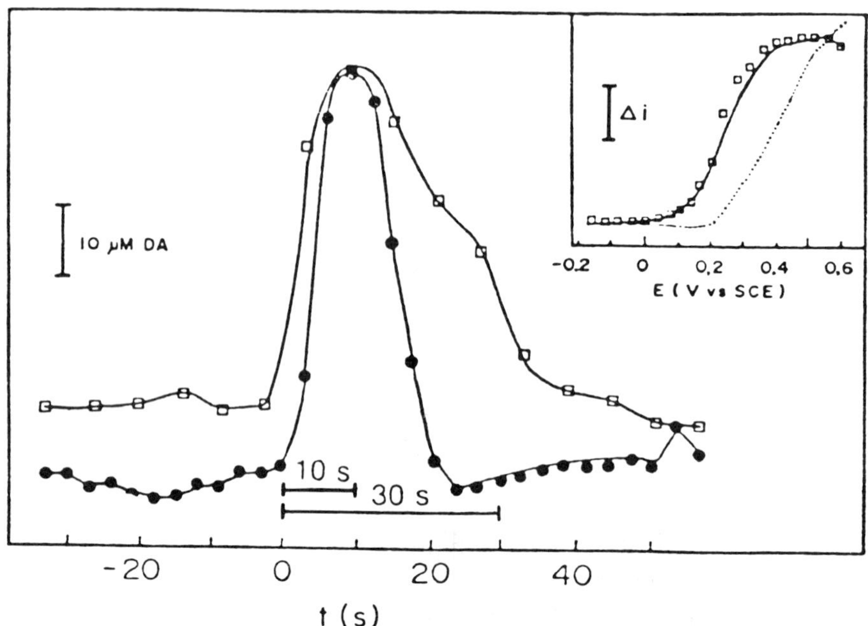

Fig. 13.   Effect of the duration of the electrical stimulation of the median forebrain bundle (MFB) on the response of microelectrodes. Response of a microvoltammetric electrode placed in the caudate nucleus during a 10-s (circles) or a 30-s (boxes), 80-μA, 60 Hz sinusoidal stimulation of the MFB (indicated by the horizontal bars). The time course data represent the concentration of DA as indicated by repetitive chronoamperometric measurements at 0.5 V vs. SCE (saturated calomel electrode) taken at 6-s (boxes) or 3-s (circles) intervals. Insert: Difference voltammograms. Solid and dotted lines, DA (40 μ*M*) and ascorbate (200 μ*M*) in pH 7.4 buffer after in vivo use, respectively; boxes, in vivo result obtained by subtraction of a scan before stimulation from that taken at the peak of stimulation. Current scales for DA, ascorbate and the in vivo response: i = 25, 25, and 20 p*A*, respectively (modified from Kuhr et al., 1984).

In addition, a number of voltammetry studies have also examined the effects of various pharmacological agents, especially amphetamine (Wilson and Wightman, 1985) and neuroleptics such as haloperidol (Maidment and Marsden, 1985; Fig. 9; Nagy et al., 1985b). Changes in ascorbate, DOPAC, and DA have been reported following either systemic or pressure ejection of these agents. Because of the long time course of action of many drugs,

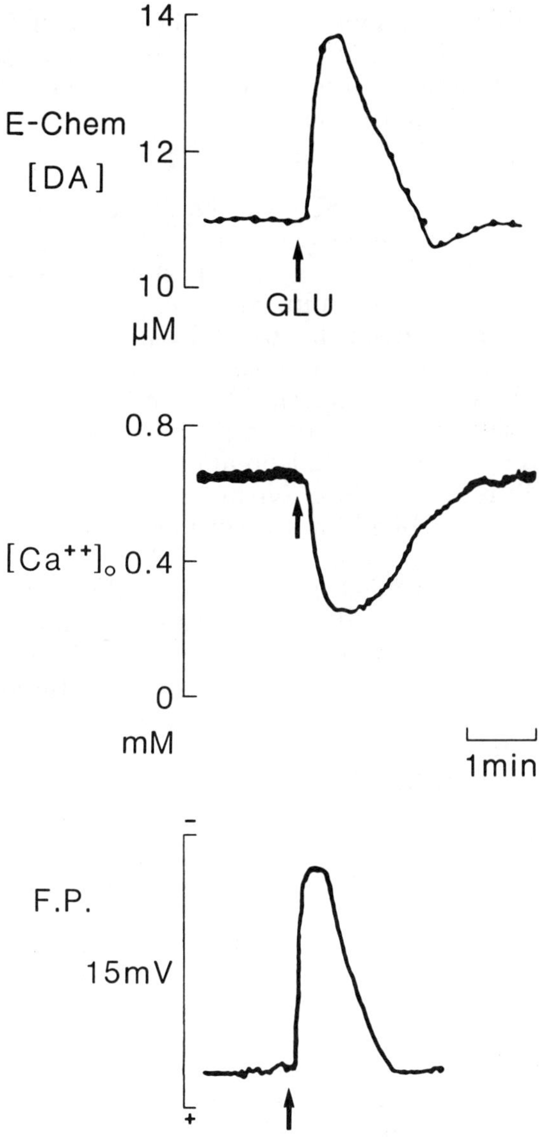



Fig. 14.   Simultaneous monitoring of VM, ISM, and field potential (FP) changes after pressure ejection of glutamate in the rat striatum. The Nafion-coated VM detects DA release as the ECF $Ca^{2+}$ concentration decreases and the FP becomes more negative. Note the similarity in onset time of each response after ejection of 10 m$M$ glutamate (100 nL) (after Moghaddam and Adams, 1986).

such studies are ideally suited for investigation by the slower-peak voltammetry techniques. Pressure-ejected $K^+$ has also been used to release DA, for which chronoamperometry offers advantages, as discussed in section 3.3.1.

### 4.1.5. Spreading Depression and Anoxia

Spreading depression (SD) was discovered by Leao (1944) and reveals itself as a transient extinction of all neuronal activity in a cortical region. This depression travels slowly across the area involved at speeds of 3–9 mm/min. An attractive hypothesis for SD is that it is primarily caused by the release of $K^+$ into the BCM (Grafstein, 1956). An alternative suggestion is that the release of glutamate is the primary agent in SD (Van Harreveld, 1959); both $K^+$ and glutamate may be involved (Van Harreveld, 1978) or massive transmitter release (Nicholson and Kraig, 1981). SD has been reviewed by Bures et al. (1974), Nicholson and Kraig (1981), and in a special issue of the *Annais da Academia Brazieleira de Ciencias* **56**(4), 371–531.

It is evident from this brief description that SD is a favorable phenomenon to explore with ISMs, and the earliest papers in the field did indeed report massive increases in $[K^+]_o$ during SD (Vyscocil et al., 1972; Prince et al., 1973). Numerous studies in cerebral cortex, cerebellum, and retina have now confirmed that $[K^+]_o$ rises to 20–40 m$M$ during SD (see Nicholson and Kraig, 1981).

In concert with the $K^+$ changes, other large ionic variation have been shown with ISMs to occur during SD, as will be detailed below. The ion changes during SD are summarized in Fig. 15. It is evident that ISMs have made a major contribution to defining the events of SD and producing evidence for and against competing hypotheses. SD is currently undergoing a revival of interest because of its possible link to migraine headache (Lashley, 1941; Gardner-Medwin and Mutch, 1984). As will be discussed in section 5, large decreases in extracellular volume also accompany SD. It seems that most measurable endogenous compounds change their extracellular concentration during SD and so this phenomenon provides an ideal proving ground for new ISM and VM technologies.

Under anoxic conditions, the BCM undergoes similar changes to those described for SD, with the important exception that the changes are not transient (*see* Hansen, 1985). In fact anoxic episodes often begin with SD-like rapidity.

Although it has been commonly observed that concentrations

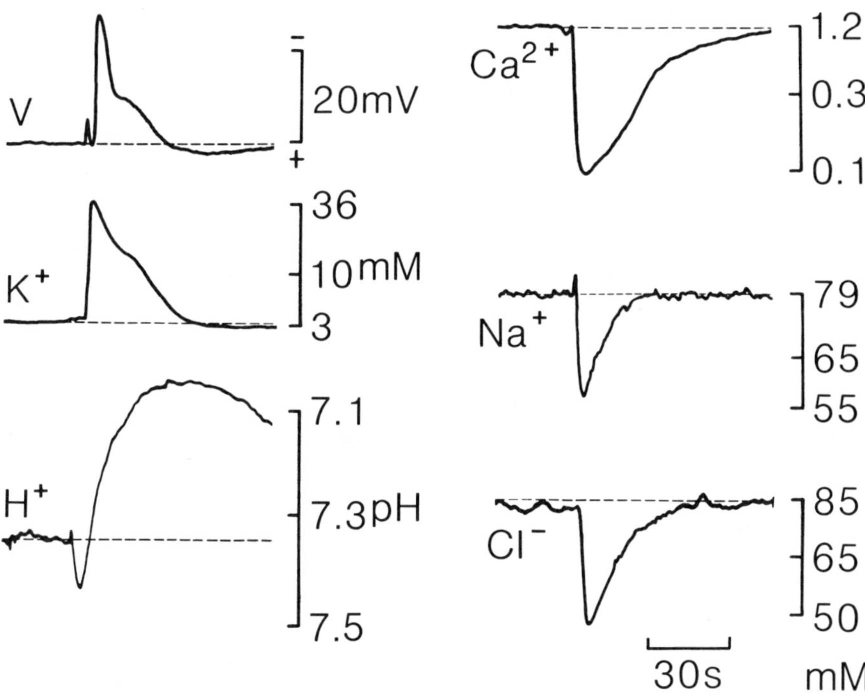



Fig. 15. Extracellular ionic changes during SD in rat cerebellum. Each record is taken from a separate experiment and the cerebellum was conditioned by reducing the NaCl content of the Ringer solution bathing the brain, which accounts for the low baseline levels of these ions (from Nicholson, 1984).

of most electroactive species increase dramatically with anoxia and death (cf. Marsden et al., 1979). The general explanation is that, as with ion changes, electroactive substances move down their concentration gradient, which is typically from intracellular to extracellular. Only recently has the release of specific electroactive substances during SD and/or anoxia been reported (Rice and Nicholson, 1987; Moghaddam et al., 1987). Using carbon fiber VMs, we monitored extracellular ascorbate during SD in the isolated turtle cerebellum. Not only was AA concentrated in the BCM because of extracellular volume decrease, but AA was also released from intracellular stores at the same time that the major ion shifts occurred (Rice and Nicholson, 1987). DA apparently is released at this same time point during SD in the rat striatum. It may precede the major ion shifts that occur during anoxia, however, indicating a fundamental difference between the two phenomena (Moghaddam et al., 1987).

### 4.1.6. Blocking Agents

An indispensible tool in the study of ionic variations in the BCM is the use of blocking agents. These may interact with ionic channels directly, for example tetrodotoxin and the voltage-dependent sodium channel, or aminopyridine and voltage-dependent potassium channels (e.g., Nicholson et al., 1976). The action may be on synaptic transmitters, when effective blockers exist, on neuromodulators, or on the uptake systems for these transmitters. Finally the agents may work on the metabolic aspects of the system, such as when ouabain and related compounds are used to inhibit the Na/K pump or blockers of glycolysis (Ullrich et al., 1982; *see* also chapter by Rosenthal and Sick, this volume).

## 4.2. Results with Specific Ions and ISMs

### 4.2.1. $[K^+]_o$

Almost all forms of stimulation lead to a rise in $[K^+]_o$ from a baseline level of about 3 m$M$ to as much as 10–12 m$M$. At this level, a ceiling is established (Heineman and Lux, 1977; Somjen, 1979; Nicholson, 1980b) that is only broken through by either pathological events, such as SD or anoxia, or in the immature CNS. The mechanisms by which a $K^+$ ceiling is maintained are not entirely clear, but active transport via the Na/K ATPase-mediated pump and passive spatial buffering (Nicholson, 1980b) are plausible candidates. How the ceiling level is "sensed" remains obscure.

Following cessation of the stimulus train, the $[K^+]_o$ level drops back to baseline and undershoots for a period. This feature is generally attributed to the action of the Na-K pump, which is responding to the entry of Na into cells (Heinemann and Lux, 1975). We also know that glia may be involved in removing $K^+$ from the stimulated brain region by a passive "spatial-buffer" mechanism (Gardner-Medwin, 1983, 1986), to be described in more detail below.

The characteristics described above for evoked $K^+$ responses are quite similar throughout the nervous system. Areas that have been studied in detail include the spinal cord (e.g., Krnjevic and Morris, 1972; Sykova, 1981; Somjen, 1979), cerebral cortex and hippocampus (Heinemann and Lux, 1975, 1977; Krnjevic et al., 1982a,b), and cerebellum (Nicholson et al., 1977, 1978; Hounsgaard and Nicholson, 1983). In general, these types of response also persist even in epileptic seizure, including adherence to the ceiling level, and may play a role in the pathology (Prince et al., 1973; Lux, 1974b).

A significant difference seems to occur in the immature CNS, in which the ceiling for $[K^+]_o$ may not exist (Mutani et al., 1974; Ransom et al., 1985). Ransom has studied this phenomenon in detail in the optic nerve of the rat and has linked the absence of a potassium ceiling to the paucity of glia in early development. Other conditions exist in which the ceiling value is broken, namely spreading depression and anoxia.

It is very likely that changes in $[K^+]_o$ bring about depolarization of glial cells, leading to circulating extracellular currents that generate the so-called "slow potentials" of the CNS (Somjen, 1973; Heinemann et al., 1979). Whether such potentials have a functional effect is still speculative, although under some conditions they may induce changes in neuronal behavior (Jefferys, 1981; Taylor and Dudek, 1984; Traub et al., 1985). Our own studies confirm that extracellular field strengths of about 20 mV/mm are adequate to change neuronal behavior (Chan and Nicholson, 1986).

It is still unclear if $[K^+]_o$ changes have direct effects on neurons, although several papers have produced intriguing evidence in some quite specific cases (Alkon and Grossman, 1978; Malenka et al., 1981; Spira et al., 1984). In this context an important issue is the possible role of $[K^+]_o$ in seizure. It has been established that this ionic variation very likely plays a role in this disorder, through cellular depolarization, but is not the sole causal agent (Fisher et al., 1976a; Somjen, 1984b; Rutecki et al., 1985). As regards SD, it is virtually certain that the rise in $[K^+]_o$ associated with SD is important in the development of the phenomenon and its propagation, as suggested originally by Grafstein, but here again other processes and mechanisms play a role in the triggering and subsequent evolution of the wave.

### 4.2.2. $[Ca^{2+}]_o$

In contrast to $K^+$, the usual response of $[Ca^{2+}]_o$ to stimulation is to fall. These falls are quite small in magnitude, 0.1–0.4 m$M$, but do represent an appreciable fraction of the baseline concentration. No clearly defined "floor" has been shown, and larger decreases in $[Ca^{2+}]_o$ can be shown with application of excitatory amino acids (Pumain and Heinemann, 1985) and during SD or anoxia (Nicholson et al., 1977, 1978; Somjen and Giacchino, 1985; Hansen, 1985).

The functional relevance of the falls in $[Ca^{2+}]_o$ have been discussed (Nicholson, 1980a); basically the changes are sufficient to diminish synaptic transmission and enhance neuronal excitability, but clear experimental evidence for this is lacking. Again a possible role in seizure has been postulated (Heinemann et al., 1977), and some experiments indicate that early fast $Ca^{2+}$ transients precede seizure (Pumain et al., 1983). These issues will be covered by Pumain in this volume.

### 4.2.3. $[H^+]_o$

The recently introduced liquid exchanger cocktail for detecting changes in $[H^+]_o$ has led to increased interest in this ion, i.e., extracellular pH, but relatively few studies have been done so far. The overall pattern to emerge is that all forms of stimulation lead to a biphasic change in $[H^+]_o$ consisting of a fast decrease followed by a longer slower increase, i.e., an alkaline-going transient followed by a prolonged acid-going phase. This pattern is seen both with stimulation and seizure (Kraig et al., 1983; Somjen, 1984a) and during SD (Kraig et al., 1983, 1985; Somjen, 1984; Mutch and Hansen, 1984). Such changes have in fact been seen on a number of occasions with older methods (*see* Kraig et al., 1983, for references), but the reliability of the other types of microelectrode in these conditions was always questioned.

The acid-going phase of the evoked change in $[H^+]_o$ is generally explicable in terms of metabolic recovery processes (Kraig et al., 1983), but the initial alkaline-going phase is more difficult to explain, although it is sensitive to $Mn^{2+}$, suggesting some involvement of $Ca^{2+}$ channels. An explanation in terms of increased blood-flow (Urbanics et al., 1978; Somjen, 1984a) is insufficient because the shift occurs in isolated brain tissues (Kraig et al., 1983; Nicholson et al., 1985).

### 4.2.4. $[Na^+]_o$ and $[Cl^-]_o$

Few studies have looked at $[Na^+]_o$ and $[Cl^-]_o$ changes under stimulated conditions because the relative magnitudes of the changes are so small and because the available exchangers are not ideal for extracellular work. Dietzel et al. (1982) observed that, during stimulation, $[Na^+]_o$ fell by 4–7 mM and later rose by 2–6 mM in some cortical areas, whereas $[Cl^-]_o$ always increased by an average of 7 mM. In contrast, during SD and anoxia, the decreases in extracellular $Na^+$ and $Cl^-$ can approach 100 mM (Nicholson and Kraig, 1981; Do Carmo and Martins-Ferreira, 1984; Hansen, 1985).

The fact that both these ions fall by a similar amount is a consequence of the need to maintain electroneutrality in large ion shifts; only $Na^+$ and $Cl^-$ are present in such large amounts.

Under normal conditions, changes in these two majority ions are unlikely to affect function. In pathological conditions they will be significant, but so many other events occur concomitantly that it is hard to assign a precise consequence to such ion shifts. One fact is that the ions almost certainly move inside cells and must be actively extruded if neuronal excitability is to be restored. A related set of studies involved $Li^+$ (Grafe et al., 1983).

## 4.3. Results with Electroactive Species and VMs

### 4.3.1. Catecholamines

As already noted, the catecholaminergic system that has been studied primarily is that of DA. Although the basal ECF levels of DA have been suggested to be below 50 n$M$ (Gonon et al., 1984b), levels following $K^+$ or electrical stimulation can reach up to 50 $\mu M$ (Kuhr et al., 1984; Gerhardt et al., 1984a; Gonon and Buda, 1985). Direct monitoring of DA released following neuroleptic administration (Blaha and Lane, 1984) and decreased following phencyclidine (Howard-Butcher et al., 1984) have also been reported in studies using stearic acid-modified VMs. Using pretreated carbon fiber VMs in the striatum of rats treated with the MAO inhibitor pargyline, Gonon and Buda (1985) have directly monitored DA levels after electrical stimulation and the administration of various dopaminergic agonists and antagonists. This study demonstrated that the release of DA is influenced by both the frequency of impulse flow and by striatal autoreceptors.

Somatodendritric DA autoreceptor control of the mesolimbic system as well has been demonstrated by Maidment and Marsden (1985). With VMs implanted in the nucleus accumbens, DOPAC concentration changes, used to indicate DA turnover, decreased transiently after DA, but increased following haloperidol infusion into the ventral tegmental area (Fig. 9). Similar results have been reported for DOPAC changes in the olfactory tubercle and accumbens (Louilot et al., 1985) and for HVA levels in striatum and accumbens, but not in frontal cortex (O'Neill and Fillenz, 1985).

An earlier study by Mos et al. (1981) also indicated that catechol concentrations decreased after administration of HA-966, which depresses DA neuron firing rate. This report was especially interesting in that the entry of the electroactive HA-966 into the brain was simultaneously monitored with the decrease in catechol peak using DPV. Changes in DOPAC/ascorbate DPV peak have also been reported from the local injection of a thyrotropin-releasing factor (TRH) analog (Sharp et al., 1984a). An increase was seen in the accumbens, but not the striatum. This finding is interesting in light of the apparently different VM responses observed for striatal and accumbens dopaminergic systems to certain neuroleptics (Huff and Adams, 1980).

There have been few VM studies of catecholamines in brain regions other than those just mentioned. In one such study in the hypothalamus, Crespi et al. (1985) reported that another peptide, growth hormone-releasing factor (GHRF), affected dopaminergic but not serotonergic activity in the arcuate nucleus; neither system, however, responded in the medial or lateral nuclei. Another recent investigation of a nondopaminergic area is also an extremely good example of the use of pharmacological tools to identify the major components of a voltammogram peak. In these reports, the locus coeruleus (a noradrenergic cell body region) was examined for its response to various drug and behavioral stimuli (Gonon et al., 1983a; Buda et al., 1983). Interestingly, the major contributor to the in vivo catechol peak again was found to be DOPAC. The major contributor to the $K^+$-stimulated release signal monitored in the locus coeruleus-innervated thalamus using Nafion-coated VMs, however, was reported to be NE (Gerhardt et al., 1984a).

Signal validation is still a critical part of VM measurements. Notable examples of signal misinterpretation were the early studies using amphetamine as a DA-releasing agent. Large, dose-dependent concentration increases attibuted to DA were seen using chronoamperometry with unmodified VMs (Conti et al., 1978; Gonon et al., 1978; Huff et al., 1979). Other in vivo and in vitro evidence supported this interpretation (*see* Stamford, 1985, for a summary). Gonon et al. (1980, 1981), however, found that with pretreated fibers, only the ascorbate peak increased after amphetamine administration. This has since been confirmed using intracerebral dialysis (Zetterstrom et al. 1983; Clemens and Phebus, 1984) (*see* section 4.3.3).

The coupling of techniques, such as VM studies with dialysis or the use of VMs with different selectivities, offers much promise for correct signal identification. Sharp et al. (1984b) used dialysis and pretreated fiber VMs in parallel studies to confirm, for example, the DOPAC increase following haloperidol administration. To confirm that the $K^+$-induced signal measured with Nafion-coated electrodes in striatal slices was caused only by DA release, Rice et al. (1985b) extracted catechols from the slice surface near the VM tip for later liquid chromatographic (LC) analysis. In both of these studies, the chemical analysis corresponded extremely well with the VM responses. Another means of confirmation that is becoming more popular (because an LC is not required) is to compare the responses at two different types of VMs; for example, a pretreated fiber in parallel with an electrode that is insensitive to DOPAC (Kuhr et al., 1984; Kovach et al., 1984); or an unmodified fiber compared to a Nafion-coated, and/or acid oxidase, modified VM (Rice and Nicholson, 1987).

### 4.3.2. Indoleamines

The monitoring of serotoninergic activity by VMs has been a popular use of the technique, primarily because a distinct voltammetric peak at the indole oxidation potential can be seen in many brain regions, even with unmodified VMs (cf., Marsden, 1979). Until the recent reports of interference from uric acid (Zetterstrom et al., 1982), the +0.3 V vs. Ag/AgCl peak was thought to be specifically due to 5-HIAA and 5-HT oxidation. Because pargyline treatment did not completely abolish this peak, it was suggested by some groups that 5-HT and 5-HIAA were in equal ECF concentrations (Kennett and Joseph, 1982), unlike the dopaminergic system in which DOPAC levels are at least 100-fold greater than DA (Zetterstrom et al., 1983; Gonon et al., 1984b). Others, however, reported that 5-HIAA was the primary substance detected as an indicator of serotoninergic activity in both the striatum (Cespuglio et al., 1981) and spinal cord (Rivot et al., 1983). Recent reports, in which the enzyme uricase was applied, however, suggest that 30% of the indole peak at electrochemically pretreated carbon fiber VMs is attributable to uric acid (Crespi et al., 1983), whereas the peak seen at some unmodified VMs may be entirely caused by oxidation of this adenosine metabolite (O'Neill et al., 1984a; Mueller et al., 1985). Fortunately, most of the indole studies have been made with modified fibers so that the reported changes in 5-HIAA levels are probably correct. Other studies in which the voltammetric response has been shown to be affected by pargyline should also be valid.

Early reports examined the increase in voltammetric signal after electrical stimulation of the serotoninergic raphe nuclei. Stimulation produced increases in ventricular 5-HIAA (Wightman et al., 1976) and in the ECF in hippocampus, but not in the striatum

(Marsden et al., 1979). More recently, VM methods have been used to measure 5-HT turnover in the dorsal raphe nucleus by monitoring the rate of 5-HIAA decrease after pargyline (Echizen and Freed, 1984). The close correlation of these in vivo data with other measurements of 5-HT turnover support the validity of using 5-HIAA levels to indicate serotoninergic activity.

Lamour et al. (1983) made good use of this assumption by using VMs to map serotoninergic activity in the somatosensory cortex, in which the highest levels were found to be in the most superficial layers. The reinervation of a denervated serotoninergic target area was also recently examined by voltammetric methods. McRae-Degueurce et al. (1984) reported that the 5-HIAA levels in the striatum of raphe-lesioned rats were restored following transplantation of neonatal mesencephalic raphe nuclei fragments into the lateral ventricle. In other studies that may be described as chemical neuroanatomy, Scatton and coworkers have examined the nature and localization of the GABA(gamma-aminobutyric acid)ergic influence on the 5-HT system (Scatton et al., 1984, 1985). Systemic and raphe infusion of GABA-mimetics were found to decrease 5-HIAA in both the dorsal raphe and striatum. GABA agonists prevented these decreases, as did lesions of the habenular nuclei. These results indicate that GABA depresses serotoninergic activity at both the cell bodies and nerve endings, and that its influence is through the habenulodorsal pathway.

In another area of active research, circadian changes in 5-HIAA levels in the ECF in various brain regions have been monitored to examine the putative role of 5-HT in sleep. Paradoxically, Cespuglio (1982) found that the highest levels of 5-HIAA in both cortex and striatum occurred during the awake stage, with lower levels occurring during sleep. In subsequent studies of the four raphe nuclei, greater increases in 5-HIAA were seen during active waking (e.g., eating, grooming) than during passive wake states (Crespi and Jouvet, 1984). The changes were also more pronounced when the animal maintained a given vigilance state for 15–20 min. An interesting aspect of both of these investigations is the simultaneous monitoring of EEG activity with the voltammetric information. Recently, circadian changes in serotoninergic activity in the pineal gland were also reported (Ikeda and Nagatsu, 1985). Similar increases were reported in indole peak height during the dark (wake) cycle. Because pargyline was found to have little effect on peak amplitude, however, the indole measured was concluded to be *N*-acetylserotonin.

### 4.3.3. Ascorbic Acid

Interest in the role of ascorbate in neurotransmission has increased since the early days of in vivo voltammetry when it was considered to be merely an interferent. The first demonstration of AA concentration changes in vivo was presented by Gonon et al. (1980), who found that at pretreated fiber VMs, the separate AA peak increased after amphetamine, whereas the catechol peak decreased. This increase was later attributed to dopaminergic transmission (Gonon et al., 1981). Clemens and Phebus (1984) carried out voltammetric and brain dialysis studies in parallel to more fully characterize the interaction of AA with the dopaminergic system. They found that in addition to the DA-releasing agent amphetamine, the DA agonist pergolide also increased AA levels. This suggests that the AA change is a DA-receptor-mediated event, which was also supported by their earlier finding that the amphetamine-induced increase could be prevented by the DA antagonist spiperone (Clemens and Phebus, 1983).

Wilson and Wightman (1985) have recently pointed out that the conditions under which AA levels in the ECF are altered is much more complex than the simple picture presented above. They reported that local infusion of amphetamine into the striatum did not alter DA levels (in agreement with Lane et al., 1976), but infusion into the substantia nigra did increase striatal concentration, although slowly. These and other data suggest that although DA activity can indeed increase AA levels, it may be the interaction of DA with other elements of the nigrostriatal pathway that causes the increase, rather than direct DA receptor stimulation. For example, the putative amino acid transmitter glutamate (O'Neill et al., 1984b) has been shown to increase ECF concentrations of AA. Fillenz and coworkers have further suggested that AA may serve as an index of excitatory amino acid release based on these and other in vivo and in vitro studies (*see* O'Neill et al., 1984b and Stamford, 1985 for a summary).

In addition to these data concerning stimulated AA levels, it has also been established by Schenk et al. (1982) that the resting ECF concentration of AA is homeostatically controlled in the BCM. In a study of incubated striatal slices, it was found that although 75% of the total tissue AA was lost after a 30-min incubation, the ECF ascorbate level of about 300 $\mu M$ from first point measurement was indistinguishable from the level seen for a fresh slice, as well as for a chronically implanted intact animal (anesthetized animals had somewhat lower levels; Schenk et al., 1983). We have found similar levels in the in vitro turtle cerebellum, as well (Rice and Nicholson, 1987). Regional differences in ECF levels of ascorbate have also been reported (Stamford et al., 1984b). The highest levels were found in corpus callosum and in CSF, with lower levels in the cortex and still lower levels in the striatum. These data are interesting in light of whole tissue analyses of AA content, in which striatal levels were the highest seen (Milby et al., 1982). This regulation of ECF may be important if AA acts as a neuromodulator, as suggested by Gardiner et al. (1985), who found that AA at varying physiological levels (monitored with fast cyclic voltammetry) can either suppress or enhance the activation of neostriatal neurons by glutamate.

## 5. Use of ISMs and VMs to Determine Biophysical Properties of BCM

A major issue that can be addressed in this area is how substances migrate in the narrow spaces between cells of the brain. A study of the nature of this process in turn leads to the estimation of what have become designated as the "volume fraction" and the "tortuosity" of the extracellular space. ISMs and VMs are well suited to this type of measurement because they measure at a point. By introducing a "probe" substance that can be sensed in the presence of the normal background, the diffusion properties of the probe, as modified by the BCM, can be determined.

### 5.1. Diffusion Theory

The emphasis in the following sections will be on practical measurements so that we shall not derive the theory but summarize and refer to the original literature.

#### 5.1.1. Diffusion from a Point Source in a Complex Medium

The diffusion equation and its solution for a free medium, e.g., an aqueous medium, are well known (Crank, 1975). They represent a macroscopic continuum description of the thermally driven collisions between the probe molecules and the solvent. Since the concentration of the solvent, water, is some $55M$ these collisions are exceedingly frequent and the probe executes a three-dimensional random walk.

Within a complex medium, that is, one composed of obstructions, the probe molecules occasionally encounter and are reflected from the obstructions. This leads to modifications of the diffusion equations. A basic issue is whether the macroscopic diffusion equations are still valid. By the use of suitable averaging procedures, developed originally for the study of porous media, it can be shown that the continuum equations remain applicable providing that two new factors, the volume fraction and the tortuosity, are introduced (Nicholson and Phillips, 1981, Nicholson and Rice, 1986). Volume fraction is the proportion of the brain volume that is extracellular and is synonomous with the extracellular space. Tortuosity is a measure of the increase in total path length that a randomly diffusing particle travels because of the presence of obstructions. In other words, if two similar particles are found at the same straight-line distance from a starting point, one in a simple medium and the other in a complex, then the particle in the second medium will have "walked" a greater distance.

Volume fraction is designated by $\alpha$ and is such that $1 \geqslant \alpha \geqslant 0$. It can be roughly thought of as "magnifying" the source density, $q$, of emitted particles in a complex medium and will appear in the diffusion equation in the term $q/\alpha$. Tortuosity is designated by $\lambda$ and is such that $\lambda \geqslant 1$. It appears as a factor that reduces the diffusion coefficient, $D$, to $D^*$ where $D^* = D/\lambda^2$.

The basic equation for diffusion of a substance in a complex medium is:

$$(D/\lambda^2)\nabla^2 C - q/\alpha = dC/dt \tag{4}$$

where $\nabla$ is the Laplacian (sum of second spatial derivatives), $C$ is concentration, and $t$ is time.

It remains to define the source term, $q$. We shall be concerned with two types of source: iontophoresis and pressure ejection. The type of experiment that is envisaged is depicted in Fig. 16 (upper panel).

5.1.1.1. IONTOPHORETIC SOURCE.    With iontophoresis a current is used to carry the probe molecule out of the tip of a micropipet; consequently the probe must be charged (in fact large currents may lead to sufficient water movement or local heating so that neutral molecules or even ions of inappropriate sign can be expelled, but such processes are hard to quantify and we assume that the currents to be used are sufficiently small to avoid these complications). The topic of iontophoresis in general has been covered recently (Hicks, 1984; Stone, 1985).

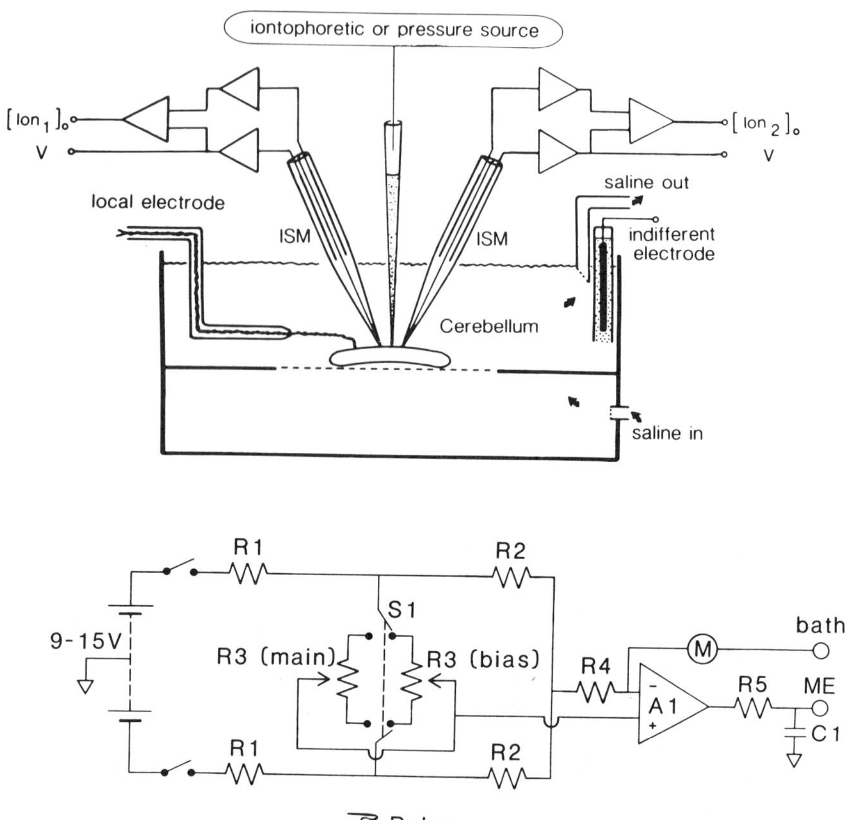



Fig. 16. (Top panel) Experimental configuration for diffusion measurements in isolated turtle brain. The brain rests on nylon mesh submerged in flowing Ringer solution. An array of two ISMs and a source electrode is lowered into the brain. The source is either connected to an iontophoresis unit or a pressure injection system. (Lower panel) Constant current iontophoresis circuit. An FET operational amplifier is used (A1: Analog Devices 515J). Values of resistors are: R1 (15 K$\Omega$), R2 (49.9 $\Omega$), R3 (2 K$\Omega$), R4 (100 K$\Omega$), R5 (14 K$\Omega$). Cl is 0.01 $\mu$F. Switch S1 is controlled by a relay to synchronise the unit with the data aquisition system; it switches between bias and main current. A nanoammeter, M, is placed in the lead to the bath; the other output connects with the iontophoresis electrode.

Only a certain fraction of the emitted ions will belong to the probe species because some ions of opposite charge will enter the iontophoresis electrode and because the small size and fixed charge distribution on the glass surface of the micropipet may discriminate in favor of ions of a particular charge. This means that if a current $I$ is applied, only a fraction $n$ will be carried by the probe ion. The parameter $n$ is called the transport number and is an empirical quantity that must be determined experimentally for a given iontophoretic electrode. The source term magnitude is defined by $In/F$, where $F$ is Faraday's electrochemical equivalent.

With iontophoresis, the probe ions are emitted at a constant rate from what is described mathematically as a point source. The actual concentration just outside the tip of the electrode is ill-defined, but may greatly exceed that of the filling solution within the electrode. Usually the iontophoretic current is stepped by an amount $I$ and the appropriate solution to Eq. 4 in an isotropic, complex medium of large extent is:

$$C = ((In\lambda^2)/(4\pi FDr\alpha))\mathrm{erfc}(r\lambda/(2Dt)^{1/2}) \tag{5}$$

where 'erfc' is the complementary error function. For an iontophoretic pulse of finite duration, an expression similar to Eq. 5, commencing at the end of the pulse, must be subtracted from it (Nicholson and Phillips, 1981).

5.1.1.2. PRESSURE SOURCE.    The second type of source that we shall consider is the pressure ejection source. Here a micropipet is filled with a concentration $C_f$ of the probe ion and pressure is used to eject a small bolus of the probe. In practice the concentration of the probe is either isotonic with the extracellular fluid (to avoid osmotically driven solute movement at the tip), or the probe is dissolved in a suitable medium to achieve isotonicity. In this way either a relatively low concentration of the probe can be released or more than one substance expelled at once. Furthermore the ejected probe need not be charged.

If the ejected volume of probe is assumed spherical and the radius is small compared to the distance to the ISM and the bolus is rapidly ejected (in relation to the time for diffusion), then it can be treated as an instantaneous point source and the solution to Eq. 5 will be:

$$C = (UC_f\lambda^3/\alpha(2\pi Dt)^{2/3})\,\exp(-r^2\lambda^2/4Dt) \tag{6}$$

where $U$ is the volume of ejected substance (liters).

With pressure, a finite amount of known concentration is ejected, but the rate of ejection need not be considered provided that it is rapid enough. Of course, an extended ejection time can be incorporated into the equation (resulting in error functions instead of exponentials), as can a bolus of finite extent. In the latter case consideration must be given to whether the bolus forms a probe-filled cavity in the brain or whether it infiltrates the extracellular space; these issues have been examined elsewhere (Nicholson, 1985).

### 5.1.2. Diffusion from a Planar Surface

Substances can also be introduced into the brain by exposing one of the surfaces to an elevated concentration. Typically, this would be the pial or ventricular surface. In such a case the concentration at the surface is fixed, $C_o$, (providing that the applied solution is well stirred), and the diffusate enters in such a way that the extracellular concentration is (e.g., Crank, 1975):

$$C = C_0 \, \mathrm{erfc}(x\lambda/2(Dt)^{1/2}) \tag{7}$$

where $x$ is the depth below the surface. Note that only $\lambda$ enters into this expression, so that $\alpha$ cannot be determined from measurements of concentration versus time. The volume fraction can be estimated, however, if the total amount of probe ion in the tissue can be determined.

In practice, Eq. 7 often has to be modified to take uptake into account because the time course of experiments employing this paradigm is frequently much longer than that based on point-source methods.

### 5.1.3. Uptake and Charge Interaction

The theory considered above assumes that the only impediments to diffusion are those offered by the geometry of the BCM. Over a sufficiently long time, however, any substance released into the BCM will be lost from that region either by uptake into cells or across the blood–brain barrier or ventricular surfaces, since the concentration gradient will always be directed out of the BCM for a foreign substance. If the probe is well chosen there will be little loss over quite long periods (e.g., 30 min) and removal need not be considered. Explicit studies of tetraalkylammonium ions have been made (Nicholson and Phillips, 1981).

Alternatively one may use compounds that are taken up readily, and by appropriate analysis of the modified diffusion curves obtain information about the uptake process. A few papers have attempted to do this for $K^+$ (Pape and Katzman, 1972; Lux and Neher, 1973; Krnjevic and Morris, 1974; Fisher et al., 1976b; Vern et al., 1977; Cordingly and Somjen, 1978); these studies are all somewhat flawed because the tortuosity, volume fraction, and spatial buffering were not explicitly recognized at that time. Studies have also been made on $Ca^{2+}$ (Nicholson, 1980a; Morris and Krnjevic, 1981; see below for discussion). Other studies have looked at tetralkylammonium penetration into the brain (Nicholson and Phillips, 1981, Hansen and Olsen, 1980). These studies considered only a very simple form of uptake in which the process was proportional to the concentration in the BCM. In these cases the point source solutions or plane surface solutions can easily be modified to incorporate the uptake term. More commonly, however, one waits until the diffusion process comes into a steady state with the uptake process, at which time the concentration is only a function of depth:

$$C = C_o \exp(-x\lambda(k/D)^{1/2}) \tag{8}$$

where $k$ is the uptake parameter. This method can be used also with radioisotopes and extended to obtain both $\alpha$ and $\lambda$, as noted above. This topic will not be described here, but the reader is referred to Fenstermacher and Patlak (1975).

In the case of many compounds of interest (e.g., transmitters), uptake is likely to be more complex and obey nonlinear kinetics. A Michaelis-Menten formulation is frequently considered in such cases, but two caveats need to be mentioned. First, such a kinetic description, although popular, may not represent the actual situation, and, second, the addition of this type of uptake term to the diffusion equation results in a new equation that cannot be solved analytically and is not trivial to evaluate numerically. It would therefore seem that accurate experimental and theoretical resources should be available before attempting to characterize in detail uptake from the BCM.

Another issue that is frequently raised is the possible role of charge interaction in the BCM. Given that membranes are endowed with numerous long-chain glycosaminoglycans and proteoglycans, all bearing negative charges, it is reasonable to discuss potential interactions between these fixed charges and migrating ions. This problem would be akin to diffusion through an ion-exchange matrix, but experimental evidence for this in the brain is inconclusive and our own recent studies on $Ca^{2+}$ (Nicholson and Rice, 1986; Nicholson and Rice, 1987), as well as other observations (Nicholson, 1980; Somjen, 1984b), do not require the presumption of such interactions. Studies with VMs and cationic transmitters by Rice et al. (1985a) (*see* section 5.2.4) do suggest the possibility of charge interactions.

## 5.2. Diffusion Experiments with ISMs

The most detailed work on diffusion in the BCM has relied on the use of point sources and ISMs. This type of study will now be described. Recent studies of this type with VMs are summarized in section 5.4.

### 5.2.1. Choice of Probes

Probe ions must satisfy several criteria: (1) there must be an exchanger that is highly selective for the probe over endogenous ions (2) the probe must remain in extracellular space for an appreciable time (unless uptake is being studied specifically—see above), and (3) the ion must be nontoxic at the concentrations used. These issues have been discussed (*see* Nicholson and Phillips, 1981).

We have concluded that the cation tetramethylammonium ($TMA^+$) is the best available probe; tetraethylammonium also functions well, but eventually enters cells and affects voltage-dependent potassium conductances. The exchanger for $TMA^+$ is the Corning $K^+$ exchanger [i.e, K tetra (*p*-chloro)phenylborate] (*see* Table 1). Selectivities against interfering ions are represented by K values where $K = k_{ij}a_j^{z_i/z_j}$ (*see* Eq. 2). In this case the primary interference is probably $K^+$, and $K = 0.09$ (Nicholson and Phillips, 1981) i.e, all interfering ions in normal Ringer only look like this amount of $TMA^+$.

With regard to anions, we found $\alpha$-naphthalenesulfonate ($\alpha$-NS) and $AsF_6$ to be most useful (Nicholson and Phillips, 1981, Phillips and Nicholson, 1981). The first is slightly preferable since it remains extracellular during SD, whereas $AsF_6^-$ does not (Phillips and Nicholson, 1979). $\alpha$-NS is not ideal, however, since it too appears to enter cells during the later phases of SD (Do Carmo and Martins-Ferreira, 1984) and certainly does so during prolonged anoxia (see later). This may reveal interesting data about the ion channels that open during these phenomena. Two iontophores are suitable for measuring these anions: methyltricaprylyl ammonium chloride (Aliquat 336S, Fluka) and the dye Crystal Violet (J. T. Baker Chemicals) (Nicholson and Phillips, 1981a; Phillips and Nicholson, 1981). When disolved in 3-nitro-*ortho*-xylene, the selectivity against typical mammalian Ringer is described by $K = 0.26$ ($\alpha$-NS) and $K = 0.004$ ($AsF_6$) (Nicholson and Phillips, 1981a). Since this Ringer contains 145 m$M$ $Cl^-$, the selectivities are adequate.

Other compounds such as 5-HT (Kriz and Sykova, 1981; Rice

and Nicholson, 1985) and potentially a variety of drugs and compounds with quaternary ammonium groups can be studied at a relatively high concentration. In these cases, the Corning exchanger 477317 is used, again making use of its rather poor selectivity. Of course, when using novel compounds the ISM/exchanger combination must be carefully evaluated to determine that the response still conforms to the Nicolsky equation and that the ISM does not change its characteristics after being immersed in the new ion for some time.

For determining the behavior of endogenous ions in the BCM by diffusion methods, one is limited to $K^+$ and $Ca^{2+}$, since other ions such as $Na^+$ and $Cl^-$ have too high a background.

### 5.2.2. Iontophoresis of Probe Ions

In this method a glass micropipet is pulled conventionally (using stock containing an internal fiber to facilitate filling) and may be bent if it is to be glued close to the ISM so that the shanks of the two electrodes are parallel. This reduces the possibility of a spacing change between the tips as the electrode pair is lowered through the tissue.

To bend the shank of the electrode, it is inserted horizontally in a coil of Nichrome wire, and current is passed through the wire until the glass softens and begins to droop slightly, then the current is switched off. If the bend is not more than about 30°, the electrode remains patent and the internal fiber does not break. The bending may sometimes be accomplished using a horizontal electrode puller with the pulling circuits disabled.

The tip of the pipet may be broken to about 1 $\mu$m and filled with an isotonic solution of the probe ion of choice. An Ag/AgCl wire is sealed in the open end of the electrode and then the iontophoretic electrode is glued to the ISM. At first sight it might seem that glueing is unnecessary, and, with adequate dual manipulators rigidly mounted, it may be. A major source of error in these experiments, however, is electrode spacing, and glueing is a simple and economical way to reduce this problem. Two manipulators may be used to position the microelectrodes, using the 100× (total magnification) compound microscope that was employed to break the tips.

An alternative method of positioning is to hold the electrodes in a piece of plasticine or similar pliable substance on a microscope slide and adjust the position of the electrodes manually under the microscope. An accuracy of 10 μm or better can be achieved by this simple method. Typical spacings range from 50 to 300 μm.

The electrodes are either glued with "5-min" epoxy or dental cement (using repair liquid as the catalyst rather than the normal solution). In both cases it is critical to put a bead of glue within a few hundred micrometers of the tip to ensure spacing stability. Other beads of glue are placed further up, and a glass cross-bridge may be applied using old capillary tubing if the shafts diverge considerably. With both types of glue, it is best to wait 20 min for curing. Note that the choice of glue is critical because it must set in such a way that it maintains dimensional stability; glues that set by evaporation of solvent or waxes that set by cooling are not suitable.

The iontophoresis of the ion may be accomplished by using a commercial iontophoresis unit, available from most companies that supply instruments for electrophysiology. Alternatively, the circuit shown in Fig. 16B may be used. This is adapted from the design published by Lux et al. (1970) and Globus (1973). This circuit is a constant-current system that requires one FET-operational amplifier.

The iontophoresis current is in one of two states, a low-amplitude mode (typically 20–40 nA), which we term the "bias current" and is applied as a background until a higher amplitude pulse (50–300 nA) is activated for 50–200 s. The bias current has the same polarity as the higher current and is to be distinguished from the backing or holding current often used in iontophoresis of drugs to retain the compound within the electrode.

The bias current is essential in these experiments since it keeps the tip of the iontophoresis electrode filled with the ion to be ejected; a backing current has the opposite effect and this means that when the higher current pulse is applied, the tip must first be replenished with the ion. This results in a "warm-up" phase of iontophoresis in which the transport number is increasing. This is difficult to treat theoretically (Dionne, 1976; Purves, 1979). A further advantage of the bias current is that it elevates the background level of the ion being sensed above that of the interfering background and so better defines the ISM response.

The high current pulse is initiated by a timing signal that originates in the data collection system. It is important that the iontophoresis unit is isolated from the rest of the system so that the current can be accurately measured in the return ground (Fig. 16, lower panel). Both the iontophoresis unit and nanoammeter are battery powered in our system to maintain the isolation. The nanoammeter only measures the current during the steady bias or the long pulse and does not have to respond rapidly. The meter also indicates problems with the iontophoresis electrode such as blocking.

We usually superfuse the exposed brain or isolated brain region with a physiological saline solution that contains a small quantity of the probe ion (e.g., 0.2 m$M$ TMA$^+$). This acts as a local calibration point for the ISM when it is raised into the fluid and also provides a finite background level in the brain in a similar manner to that generated by the bias current.

At the end of the experiment, and sometimes during the course of it, the transport number of the iontophoresis electrode is checked by making measurements in a dilute agar gel. This is typically made up of 0.3–0.4% agar (Difco or similar quality that remains transparent) dissolved in 150 m$M$ NaCl. This solution is allowed to gel in a small plastic sieve that sits on top of the brain (in vivo) or is immersed adjacent to the tissue (in vitro, *see* Hounsgaard and Nicholson, 1983). By curve fitting (see below), the iontophoretic curves in this medium can be used to determine $n$, the transport number, and $D$, the diffusion coefficient. The former value is quite variable, although for a given electrode fabrication procedure constant values will be obtained. In contrast, the estimated value of $D$ should closely agree with that obtained by other methods; significant deviations from the norm may indicate electrode spacing errors.

### 5.2.3. Pressure Ejection of Probe Ions

Injection electrodes are made in a similar manner to those used for iontophoresis. The filling solution can be a mixture of substances and need not be ionized. A Teflon tube, similar to that used in the ISM fabrication process, is waxed into the shaft of the electrode and connected to a pressure ejection system. We find the Picospritzer (General Valve, 202 Fairfield Rd., Fairfield, NJ 07006) to be very suitable for this purpose. Essentially this instrument is an electronically controlled gas valve that is connected to a standard gas cylinder. Pulses as brief as 2 ms can be given. Typically we use a range of 20–200 ms at a pressure of 20–40 psi nitrogen.

As with iontophoresis, measurements in agar serve to establish the system characteristics. Here they are used to determine the amount of substance ejected by the pressure pulse and $D$. The reproducibility of the volume ejected is still being assessed in our laboratory. One procedure that reduces the effects of possible variation is to eject a mixture of a well-characterized substance (such as TMA$^+$) and a substance to be studied (such as Ca$^{2+}$) in the brain and use established values of volume fraction and tortuosity together with the TMA$^+$ diffusion profile to estimate volume ejected, and then use this information to interpret the diffusion properties of the other compound (Nicholson and Rice, 1986).

### 5.2.4. Analysis of Data

In order to derive the parameters from the diffusion measurements, appropriate expressions (such as Eqs. 5 or 7) must fitted to the actual data. Space does not permit us to describe this in detail here; only to outline the salient points.

A typical set of data is shown in Fig. 17, where both agar and brain data are seen. They are presented on logarithmic scales, as the data is actually recorded, and linear scales, which show the relative magnitudes correctly, and normalized. The latter is obtained by dividing each linear curve by the asymptotic value, i.e., that which it would obtain at infinite time. This representation enables the time courses to be compared for brain and agar and shows the increased rate of rise in the agar.

Nicholson and Phillips (1981) described a detailed methodology for curve fitting using a linear regression approach and a programmable calculator. That description is still relevant, but we have improved the computational aspects of the analysis. In our original method, data was digitized by hand from chart records and keyed in to the calculator. This is economical but tedious. Subsequently we used a digital oscilloscope to capture the data and transferred the digitized waveform to a microcomputer for analysis (Hounsgaard and Nicholson, 1983). We continue to use this method, since it combines the flexibility of the oscilloscope with the arithmetic power of a computer.

The present program embodies a more versatile curve-fitting algorithm: the simplex algorithm (Nelder and Mead, 1965; Caecci and Cacheris, 1984), which enables virtually any expression to be fitted to the data, including expressions that incorporate uptake. Other popular nonlinear curve-fitting routines are based on the Marquardt algorithm (1963) and should be equally effective. The program is written in Pascal and is about 2000 lines in length and runs on an IBM PC. It is also evident that this type of analysis can be done with microcomputers equipped with analog-to-digital converter boards, particularly if scientific software packages are

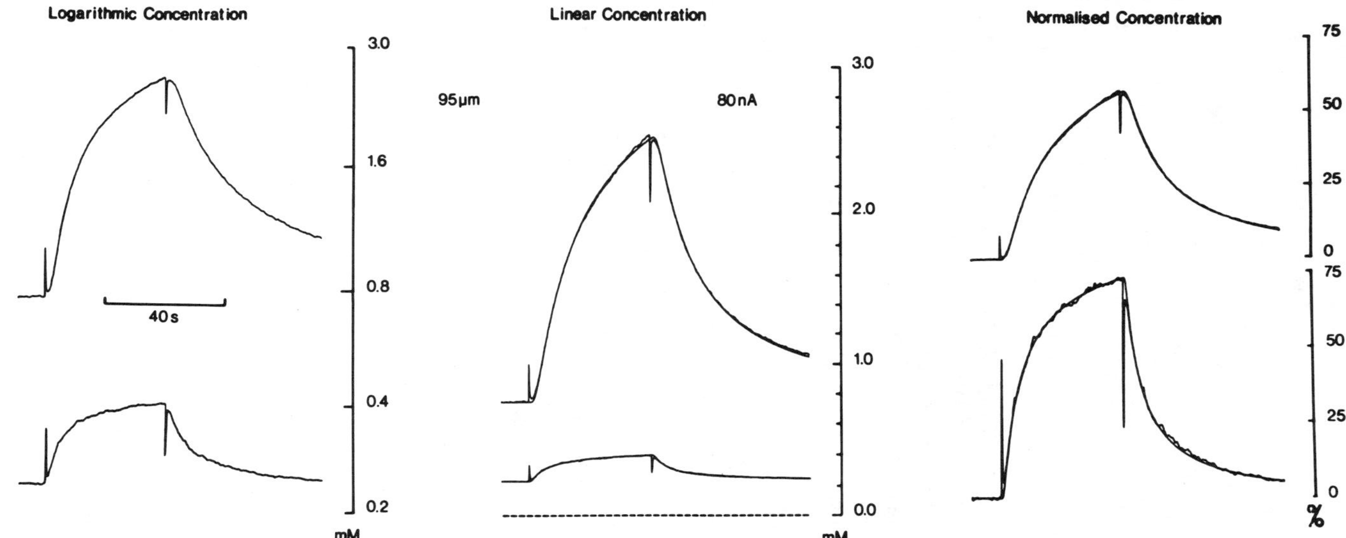



Fig. 17.   Diffusion curves generated by a TMA$^+$ source 95 μm from an ISM. In each panel the upper figure is from guinea pig cerebellar slice and the lower record is from a control agar block. The left set shows logarithmic data as recorded, the center set shows same data on linear scale together with a theoretical curve, and the right panel shows normalized data, i.e., linear data divided by asymptotic value (after Hounsgaard and Nicholson, 1983).

available to drive the boards and provide effective curve-fitting procedures.

### 5.2.5. Results Obtained For Volume Fraction and Tortuosity

Studies on the cerebellum using $TMA^+$ in a variety of species from fish to rat have revealed values of between 0.18 (18%) and 0.28 (28%) for the volume fraction, and values between 1.49 and 1.84 for tortuosity (Nicholson and Rice, 1986). Similar values have been found with other ions (Nicholson and Phillips, 1981) and in other brain regions (Lehmenkuhler et al., 1985).

The values quoted above can be obtained by both iontophoresis and pressure ejection (Fig. 18). These data also indicate the "amplification" effect of the signal obtained in the brain compared to that in the agar caused by the interplay of the volume fraction and tortuosity.

Use of VMs has revealed different values for cationic transmitters (Rice et al., 1985a) because of either charge interaction or some residual uptake (*see* section 5.4). In contrast we have not found evidence that $Ca^{2+}$ diffuses abnormally (Nicholson and Rice, 1986; Nicholson and Rice, 1987).

It is probably accurate to conclude that the "typical" extracellular space has a volume of about 20% and a tortuosity of 1.6. These values are in good agreement with those obtained by radiotracer methods (e.g., Levin et al., 1970; Fenstermacher and Patlak, 1975) and other techniques (Van Harreveld, 1972; Cragg, 1979). Knowledge of these parameters can be used to solve practical problems about $O_2$ diffusion and the entry of substances into slices (Nicholson and Hounsgaard, 1983).

## 5.3. Measurement of Changes in Volume of the Extracellular Space

There is growing evidence that the size of the extracellular space is a dynamic variable that can vary locally and in concert with ionic fluxes across membranes. The methods described above can be used to follow the changes, if they are not too rapid. Modifications of these methods allow for better time resolution.

### 5.3.1. Measurements with Iontophoretic Pulses

Dietzel et al. (1980) iontophoresed $TMA^+$ or $choline^+$ for periods of 0.5–5 s in the sensorimotor cortex and used the changes in amplitude of the probe signals, detected with an ISM 20–80 μm away, as a measure of changes in the volume of the extracellular

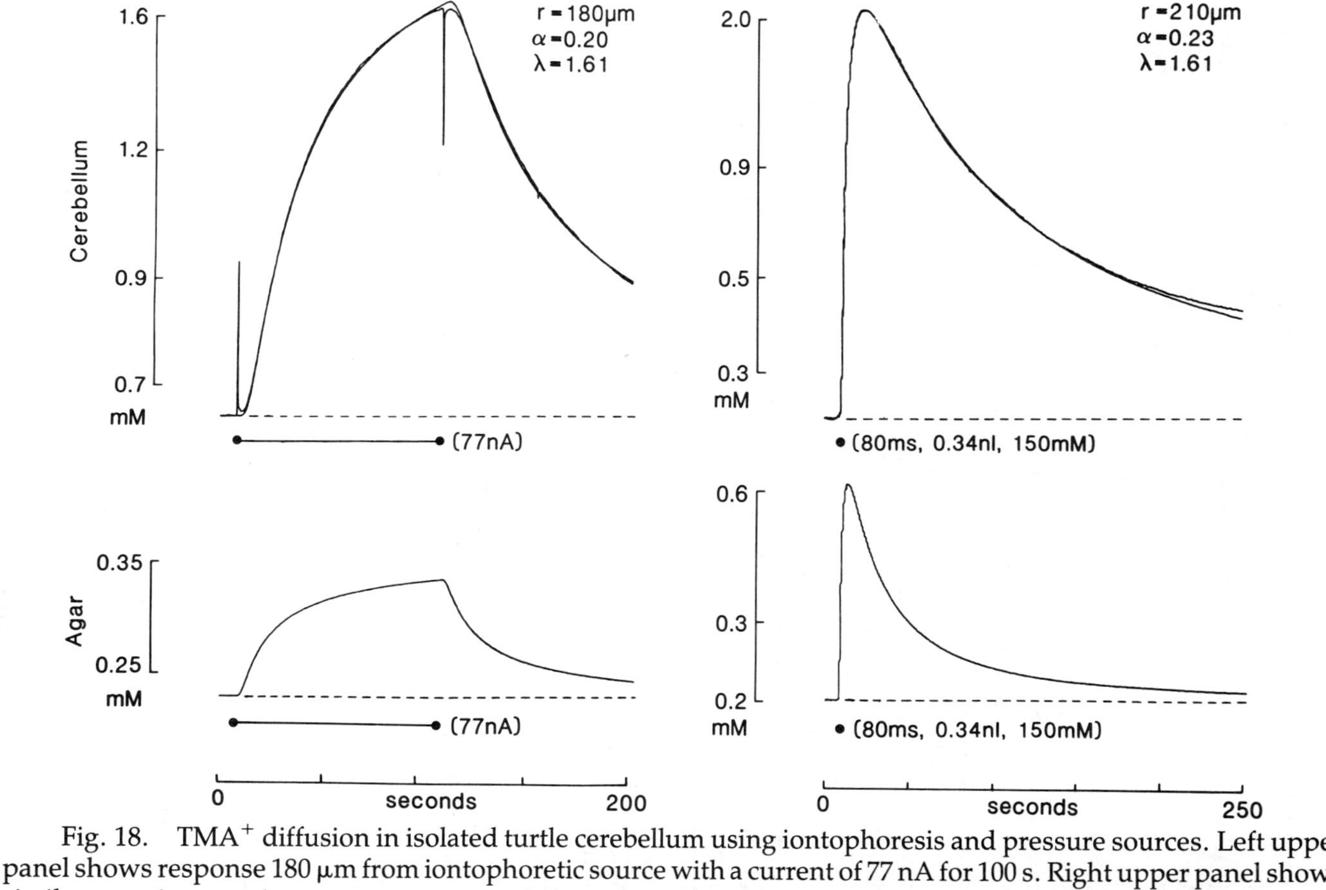



Fig. 18.    TMA$^+$ diffusion in isolated turtle cerebellum using iontophoresis and pressure sources. Left upper panel shows response 180 µm from iontophoretic source with a current of 77 nA for 100 s. Right upper panel shows similar experiment using pressure source at 210 µm from ISM. The source was 0.34 nL of 150 m$M$ TMA$^+$ ejected in 80 ms. In each case the lower figure is the corresponding theoretical agar control value; theoretical brain values have been superimposed on the data records (from Nicholson and Rice, 1986).

space. It was found that the size of the extracellular space apparently decreased by 30–50%, as inferred from the change in size of the probe signals, during electrical stimulation of the cortex. Later (Dietzel et al., 1982), these measurements were combined with measurements of changes in $Na^+$ and $Cl^-$ in the extracellular space to try to provide a complete account of the water movements associated with neuronal activity.

Further studies by Dietzel and Heinemann (1986) confirmed changes of up to 30% in the extracellular space of the cortex during seizure and enabled a detailed model of ion and water movements to be formulated.

Recently this type of technique has been used by Ransom et al. (1985) to describe changes in the extracellular volume with development in the optic nerve.

This technique is based on the assumption that the tortuosity of the BCM remains constant during stimulation and that the transport number of the iontophoretic electrode also remains unchanged. Both these assumptions appear reasonable at the present time. As noted by Dietzel et al. (1980), it is necessary to avoid the use of backing currents and make iontophoretic injections at constant intervals to reduce dilution of the probe in the iontophoretic electrode tip. Possibly a forward bias current would be more effective here.

### 5.3.2. Volume Measurements with Constant Probe Background

As noted above, it is possible to infiltrate a probe ion into the extracellular space by superfusing one of the brain surfaces with a suitable probe ion. If a suitable ISM is then placed in the brain and a disturbance initiated that causes a change in the size of the extracellular space, then the local probe concentration will change to reflect this, as water leaves or enters the extracellular space.

This approach was used by Phillips and Nicholson (1979) with the probe ions $TEA^+$ and $\alpha\text{-}NS^-$ to show that the size of the extracellular space decreased by a factor of about two during SD. Later studies confirmed these results, using other quaternary ammonium compounds (Hansen and Olsen, 1980; Nicholson et al., 1981) and $\alpha\text{-}NS^-$ during SD (Do Carmo and Martins-Ferreira, 1984) and administration of $CO_2$ (Kent et al., 1985).

This method is quite simple since iontophoresis is not required, but it depends on the probe ion remaining extracellular during the event of interest. For SD this is usually the case, but in anoxia, for example (Fig. 19), only $TMA^+$ remains outside cells in

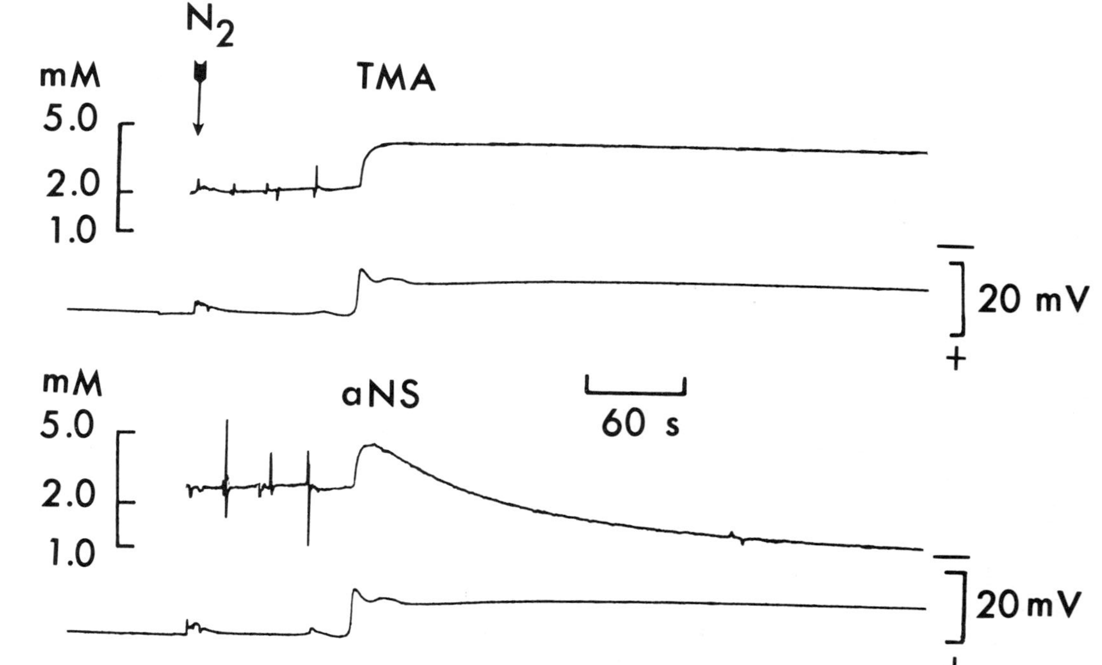



Fig. 19.   Behavior of extracellular probe ions during terminal anoxia. Brain was superfused with 10 m$M$ of TMA$^+$ and 10 m$M$ of $\alpha$-NS$^-$. After the ions had infiltrated the BCM, nitrogen was administered to the anesthetized rat and the heart stopped beating. At the time of rapid onset of the anoxia (indicated by the sudden negative extracellular potential), the TMA$^+$ and $\alpha$-NS$^-$ signals increased, indicating a reduction of extracellular volume, but only the TMA$^+$ remained high. This suggests that $\alpha$-NS$^-$ may enter cells after a short time and also indicates the superiority of TMA$^+$ as a long-term extracellular probe ion (Nicholson, unpublished).

the long term; $\alpha$-NS$^-$ enters (*see* also Do Carmo and Martin-Ferreira, 1984).

The fact that some ions enter cells during the stimulation or disturbance can be turned to advantage. In the original study of Phillips and Nicholson (1979), anions smaller than $\alpha$-NS$^-$ decreased their concentration in the extracellular space and so provided a clue to the size of the anion channel that must open during SD (between 0.6 and 1.12 nm in diameter).

A final caveat concerning this method is that if the probe is infiltrated from one surface of the brain only and there is active blood flow, then a standing gradient will occur with depth. If the disturbance causes any type of brain movement, the ISM will move through the gradient and possibly register a false signal. If a totally isolated preparation is available (such as a slice), then a uniform concentration can be achieved, but it may require a significant time (Nicholson and Hounsgaard, 1983).

### 5.3.3. Spatial Buffering of $K^+$

One of the mechanisms by which $K^+$ is removed is by induced current flow through extended core conductors, provisionally identified with glial cells. An extensive theoretical and experimental analysis of this problem has been provided by Gardner-Medwin (1983) and Gardner-Medwin and Nicholson (1983). $K^+$ is probably removed from the extracellular space by other mechanisms as well, notably uptake into cells (Gardner-Medwin, 1986). Thus the quantitative treatment of this problem is complex, and the interested investigator should consult the papers cited above.

## 5.4. Biophysical Applications of VMs

Althuh the use of ISMs and probe ions to evaluate the BCM has provided valuable information about its structural properties, the use of voltammetric electrodes to monitor neuroactive substances can provide information about its functions. The application of voltammetry to address such issues is still in its infancy, however. In fact, only recently has auxilliary chemical analysis confirmed that a parent transmitter (specifically DA) can escape the synaptic cleft and appear in the ECF at levels that are detectable by voltammetric methods (Zetterstrom et al., 1983; Rice et al., 1985b). With this ascertained, questions concerning the factors that could affect biogenic amine movement in the BCM become more relevant.

Dayton et al. (1983) examined the effect of the restricted diffusion in brain tissue on voltammetric measurements. This is important for quantitative work because of the dependence of the voltammetric current on the diffusion coefficient of the measured substance. They found that with 100-ms sample times, free solution diffusion in the presumed pool at the electrode tip was monitored, whereas at longer times, the restricted diffusion of the BCM dominated. In this study, Dayton et al. also determined in vivo diffusion coefficients in rat cortex for AA, DOPAC, and $\alpha$-methyldopamine using the point source equations with microsyringe injection of each substance. The values obtained corresponded to $\lambda$ values of about 1.6 for all three species. Lower concentrations than predicted were observed in each case, but $\alpha$ was not included in the equation used.

We later examined the diffusional properties of a variety of catecholamine-related species and metabolites in the rat striatum (Rice et al., 1985b; Fig. 20). Using a pressure-ejection pipet as the point source with simultaneous monitoring at two differently spaced GEC electrodes, we found that the anionic species measured (AA, DOPAC, ferrocyanide, HVA) all diffused through the brain apparently unimpeded except by tortuosity, much as Dayton et al. (1983) had reported. Cationic species (DA, NE, MET, 3-methoxytyramine, $\alpha$-methyldopamine), however, all apparently diffused through the BCM at a rate reduced by a factor of 10 from that in free solution. This apparent charge discrimination, as well as supporting in vitro evidence, suggested interactions with polyanionically charged extracellular matrix components were affecting the migration of these substances, as postulated by various investigators (e.g., Schmitt and Samson, 1969; Nicholson, 1980b). The differences between the concentration–time profiles for ascorbate and 3-methoxytyramine are illustrated in Fig. 20.

Another viable explanation for these data, however, is that some form of uptake is influencing the diffusion of these ions. This latter postulate could explain the discrepancy between our data for $\alpha$-methyldopamine and that of Dayton and coworkers, because the striatum has a much greater aminergic innervation than does the cortex, which may correlate with a more active uptake system. We have also recently found that the cation 5-HT apparently diffuses normally in the also minimally aminergic turtle cerebellum (Rice and Nicholson, 1985). That the cortex also has at least localized regions of efficient uptake, however, has been noted by Armstrong-James et al. (1981). They reported that the steady-state

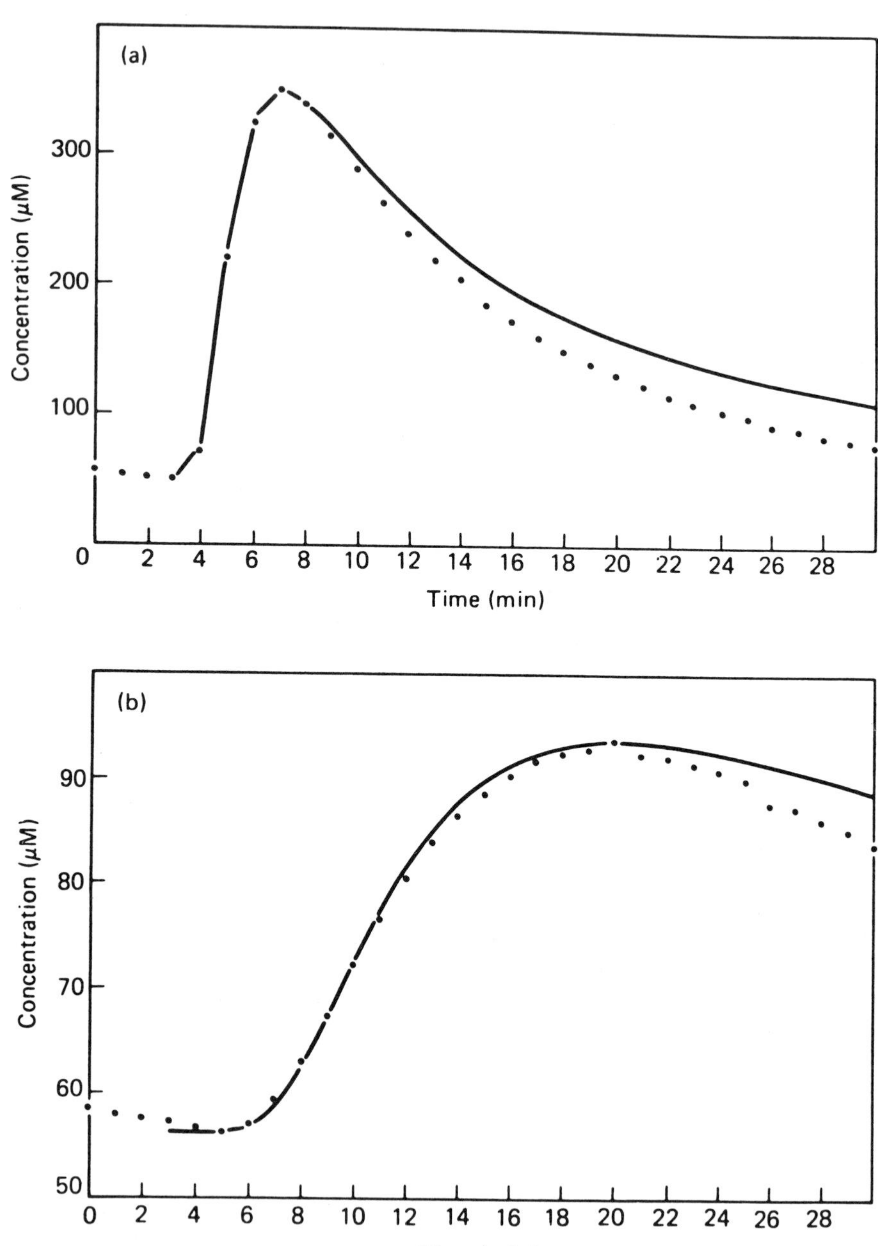


Fig. 20. Experimental and calculated concentration–time curves for ascorbic acid and 3-methoxytyramine for the same (700 μm) diffusion distance. (a) Ascorbate: unconnected points, experimental (chrono-

concentrations of iontophoresed NE and DA (monitored using high-speed cyclic voltammetry) were lower in some cortical areas than in saline, but higher in others. Because restricted diffusion in the BCM would lead to higher brain concentrations, localized uptake processes were suggested to cause the lower levels.

Although iontophoresis and pressure ejection are useful tools for studying the role of uptake processes in the control of ECF levels of transmitters and their metabolites, perhaps more information can be gained by studying the effect of uptake on endogenously released substances. Ewing and Wightman (1984) have suggested, from electrical stimulation studies examining DA release, that DA is not free to diffuse in the brain because an extraneuronal uptake mechanism rapidly clears DA from the ECF. This conclusion was based in part on the much more rapid decrease in the release signal than predicted for diffusion alone. In similar stimulated-release studies, Stamford et al. (1984c) were able to suggest a $K_m$ of 8 μM for this high-capacity uptake system. In a more qualitative approach, Gerhardt et al. (1985) have examined the decay time for a $K^+$-stimulated release response to evaluate the effects of MPTP damage to the dopaminergic system of the mouse. Such functional studies open yet another avenue for the application of voltammetric techniques. The coupling of the voltammetric measurement of the time course of neurotransmitter-related events with simultaneous ISM measurements of ion flux as demonstrated by Nagy et al. (1985b) should provide an even better analysis of BCM function.

## 6. Limitations and Prospects for the Techniques

Both ISMs and VMs have become respectable research tools and few question the rationale of their use. Consequently, present research is concerned with either using the techniques or further

---

amperometry) data taken every 1 min. Smooth line, computer simulated using an in vivo diffusion coefficient ($D^*$) of 0.34 × $10^{-5}$ cm$^2$/s with the time of the concentration maximum ($t$) occurring at 4 min. (b) 3-Methoxytyramine: unconnected points, experimental data taken every 1 min. Smooth line simulated using $D^* = 0.080 \times 10^{-5}$ cm$^2$/s with $t = 17$ min (from Rice et al., 1985a).

improving them rather than demonstrating that they function. Despite this evidence of maturity, it is fair to say that they are only used by a relatively small number of people. It is worthwhile, therefore, to examine some of the limitations presently inherent in these methods.

## 6.1. Limitations

### 6.1.1. Sensitivity

Both ISMs and VMs are capable of measuring a variety of substances that are intrinsic to the BCM. Thus ISMs respond accurately to $K^+$, $Ca^{2+}$, and $H^+$, and VMs can determine ascorbate and a variety of metabolites of transmitters. What neither technique appears able to record reliably are the "informational substances" of the brain; we are always compelled to infer their behavior from the more "metabolically" oriented substances. The underlying reason for this lack of sensitivity appears to be that informational substances are able to act in the nanomolar range, and ISMs and VMs are unable to function optimally in the hostile (to microsensors) environment of the brain.

### 6.1.2. Selectivity

The sensitivity issue is compounded by the problem of selectivity. It is here, however, that the most evident progress is being made. The introduction of neutral carrier cocktails for use in ISMs has eliminated much of the uncertainty that was previously associated with charged carriers, although the latter remain useful for quantitative work with probe ions. Likewise, new methods of surface treatment of VMs have greatly improved their discriminatory capabilities. Some price has to be paid for these enhancements, however. In the case of ISMs, the use of neutral carriers has increased electrode resistance, and with it, noise, as well as slowing down the speed of response somewhat. Again, surface treatments of VMs significantly lengthen the time that these sensors require to respond. Fortunately many of the chemical changes that occur in the brain do so rather slowly.

### 6.1.3. Size of Sensor

A recurring issue is how a sensor that is several micrometers in diameter can accurately measure events in the clefts between cells that are of the order of a hundred Angstroms across. This question has been largely answered by pointing to the reproducibility of results from one laboratory to the next. On a more quantitative basis, the excellent curve fits that can be achieved with point source diffusion attest to the fact that the sensor does couple effectively to the BCM and that the dead space around the tip of an electrode is about the same size as the electrode diameter.

An encouraging trend has been the continuing miniaturization of both ISMs and VMs. ISMs can now be fabricated, in standard double-barrel configuration, with tips below 1 μm in diameter. In fact this is probably too small for extracellular work because the tip tends to lodge inside a cell rather than create a small space around it. VMs have also decreased their size greatly, thanks to the introduction of carbon fibers. These sensors can be as small as 5 μm in diameter, but such fibers often have to have a fairly long exposed tip to obtain sufficient current for measurements to be practical. Even this limitation may be removed by using, for example, high-speed cyclic voltammetry, which considerably enhances the available current.

## 6.2. Prospects

### 6.2.1. Some Accomplishments

Both ISMs and VMs have revealed a previously unsuspected world of change and fluctuation in the chemical composition of the BCM. These variations hint at the potential for communication and information processing that exists within this channel. For the reasons described above, this vision remains a tantalizing conjecture rather that a proven fact.

At a more mundane level, ISMs have helped us to see the ramifications of $K^+$ homeostasis and its failure in SD and anoxia. We also have a much better idea of some of the roles of glial cells. VMs have brought ascorbate to our attention and confirmed that powerful and rapid uptake mechanisms exist for most neuroactive compounds. We have also gained a more quantitative picture of the biophysics of the BCM and are even able to apply those ideas to the understanding of the behavior of compounds that we presently cannot measure.

### 6.2.2. Future Directions

ISMs have reached a plateau as extracellular sensors and will probably be increasingly applied as intracellular probes (see other chapters in this book). The continued commitment to the development of neutral carriers by Simon and coworkers can be expected to produce better sensors for ions, particularly for anions, which have been rather innaccessible to date. ISMs will be increasingly used in areas that are interdisciplinery, for example acid–base balance, brain metabolism, and the exploration of the blood–brain barrier.

VMs are a rapidly expanding field in the context of the BCM. With the recent impressive demonstrations that a number of compounds can be reliably identified in the brain, a surge of interest in this technology is beginning. Many years of painstaking development by Adams and others have begun to bear fruit in the form of reliable electrochemical tools. New surface treatments and coatings as well as continuing reduction in size are likely to further expand the applicability of the methods.

## 6.3. Conclusions

Three points are worth emphasizing. First, these techniques have significant limitations. To overcome these one must understand the limitations and then design well thought out and original experiments. We do not need further demonstrations that $[K^+]_o$ rises during stimulation or that an insult of sufficient magnitude releases enough DA to reach the level of detectability of a VM. The second point is that both ISMs and VMs are point sensors that provide accurate information about only a tiny area of the brain at one time. There is inevitably a lot of variation in such measurements and they are not always representative of the larger picture. Third, the major strengths and weaknesses of both the methods discussed here are in the sensor itself. No amount of elaborate data processing can overcome that fact, although it obviously helps to extract information that is there. But the real improvements will come about at the sensor and sensor/brain interface and this continues to be the most productive area for technical development.

## Acknowledgments

Christine Cronin made the illustrations and did the photography. The work was supported by USPHS grants NS-13742 and NS-07745 (MER).

# References


Adams R. N. (1958) Carbon paste electrodes. *Anal. Chem.* **30,** 1576.

Adams R. N. (1969) *Electrochemistry at Solid Electrodes.* Marcel Dekker, New York.

Adams R. N. (1976) Probing the brain with electroanalytical techniques. *Anal. Chem.* **48,** 1126A–1138A.

Adams R. N. and Marsden C. A. (1982) Electrochemical Methods for Monoamine Measurements *In Vitro* and *In Vivo,* in *Handbook of Psychopharmacology* vol. 15 (Iversen L. L., Iversen S. D., and Snyder S. H., eds.) Plenum, New York.

Adey W. R. (1975). Evidence for Cooperative Mechanisms in the Susceptibility of Cerebral Tissue to Environmental and Intrinsic Electric Fields, in: *Functional Linkage in Biomolecular Systems* (Schmitt F. O., Schneider D. M., and Crothers D. M., eds.) pp. 325–342, Raven, New York.

Albery W. S., Fillenz M., and O'Neill R. D. (1983) The compartment model for chronically implanted voltammetric electrodes in the rat brain. *Neurosci. Lett.* **38,** 175–180.

Alkon D. L. and Grossman Y. (1978) Evidence for nonsynaptic neuronal interaction. *J. Neurophysiol.* **41,** 640–653.

Ammann D. (1986) *Ion-Selective Microelectrodes* Springer-Verlag, Berlin.

Ammann D., Lanter F., Steiner R. A., Schulthess P., Shijo Y., and Simon W. (1981) Neutral carrier based hydrogen ion selective microelectrode for extra- and intracellular studies. *Anal. Chem.* **53,** 2267–2269.

Ammann D., Morf W. E., Anker P., Meier P. C., Pretsch E., and Simon W. (1983) Neutral carrier based ion-selective electrodes. *Ion-Selective Electrode Rev.* **5,** 3–92.

Armstrong-James M. and Millar J. (1979) Carbon fibre microelectrodes. *J. Neurosci. Meth.* **1,** 279–287.

Armstrong-James M., Fox K., Kruk Z. L., and Millar J. (1981) Quantitative iontophoresis of catecholamines using multibarrel carbon fibre electrodes. *J. Neurosci. Meth.* **4,** 385–406.

Baumgarten C. M. (1981) An improved liquid ion exchanger for chloride ion-selective microelectrodes. *Am. J. Physiol.* **241,** C258–C263.

Behm F., Ammann D., Simon W., Brunfeldt K., and Halstrom J. (1985). Cyclic octa- and decapeptides as iontophores for magnesium. *Helv. Chim. Acta* **68,** 110–118.

Blaha C. D. and Lane R. F. (1983) Chemically modified electrode for *in vivo* monitoring of brain catecholamines. *Brain Res. Bull.* **10,** 861–864.

Blaha C. D. and Lane R. F. (1984) Direct *in vivo* electrochemical monitoring of dopamine release in response to neuroleptic drugs. *Eur. J. Pharmacol.* **98,** 113–117.

Blakely R. D. and Duvarney R. C. (1983) A microcomputer-controlled system for monitoring multiple voltammetric electrodes *in vivo. Brain Res. Bull.* **10,** 315–320.

Borrelli M. J., Carlini W. G., Dewey W. C., and Ransom B. R. (1985) A simple method for making ion-selective microelectrodes suitable for intracellular recording in vertebrate cells. *J. Neurosci. Meth.* **15,** 141–154.

Brazell M. P. and Marsden C. A. (1982) Differential pulse voltammetry in the anaesthetized rat: Indentification of ascorbic acid, catechol and indoleamine peaks in the striatum and frontal cortex. *Br. J. Pharmacol.* **75,** 539–547.

Brazell M. P., Feng J., Kasser R. J., Renner K. J., Moghaddam B., and Adams R. N. (1987) An improved method for Nafion coating carbon fiber electrodes for *in vivo* electrochemistry. *J. Neurosci. Meth.* (in press).

Brazell M. P., Moghaddam B., and Adams R. N. (1985) A new technique for coating carbon fiber electrodes for *in vivo* electrochemistry. *Soc. Neurosci. Abst.* **15,** 353.1.

Broderick P. A., Blaha C. D., and Lane R. F. (1983) *In vivo* electrochemical evidence for an enkephalinergic modulation underlying stereotyped behavior: Reversibility by naloxone. *Brain Res.* **269,** 378–381.

Brown K. T. and Flaming D. G. (1977) New microelectrode techniques for intracellular work in small cells. *Neuroscience* **2,** 813–827.

Buda M., De Simoni G., Gonon F., and Pujol J.-F. (1983) Catecholamine metabolism in the rat locus coeruleus as studied by *in vivo* differential pulse voltammetry. I. Nature and origin of the contributors to the oxidation current at +0.1 V. *Brain Res.* **273,** 197–206.

Bures J., Buresova O., and Krivanek J. (1974) *The Mechanism and Application of Leao's Spreading Depression of Electroencephalographic Activity* Academic, New York.

Caecci M. S. and Cacheris W. P. (1984) Fitting curves to data. *Byte* **9,** 340–362.

Cespuglio R. (1982) Voltammetric Detection of Brain 5-Hydroxyindoleacetic Acid Fluctuations During the Sleep-Waking Cycle, in *Advances in Pharmacology and Therapeutics* II vol. 1 (Yoshida H., Hagihara J., and Ebaslu S., eds.) Pergamon, Oxford. 241–246.

Cespuglio R., Faradji H., Ponchon J. L., Buda M., Riou F., Gonon F., Pujol J.-F., and Jouvet M. (1981) Differential pulse voltammetry in brain tissue. I. Detection of 5-hydroxyindoles in the rat striatum. *Brain Res.* **223,** 287–298.

Chan C. Y. and Nicholson C. (1986). Modulation by applied electric fields of Purkinje and stellate cell activity in the isolated turtle cerebellum. *J. Physiol.* (Lond.), **371,** 89–114.

Cheng H.-Y. (1982) Compartment model for chronoamperometric measurements *in vivo*. *J. Electroanal. Chem.* **135**, 145–151.

Cheng H.-Y., Schenk J., Huff R., and Adams R. N. (1979) *In vivo* electrochemistry: Behavior of microelectrodes in brain tissue. *J. Electroanal. Chem.* **100**, 23–31.

Cheng H.-Y., White W., and Adams R. N. (1980) Microprocessor-controlled apparatus for *in vivo* voltammetry. *Anal. Chem.* **52**, 2445–2448.

Clark L. C. (1956) Monitor and control of blood and tissue oxygen tensions. *Trans. Am. Soc. Artif. Intern. Organs* **2**, 41–45.

Clark L. C. (1960) Intravascular polarographic and potentiometric electrodes for the study of the circulation. *Trans. Am. Soc. Artif. Intern. Organs* **6**, 348–354.

Clark L. C. and Clark E. W. (1964) Epicardial oxygen measured with a pyrolytic graphite electrode. *Ala. J. Med. Sci.* **1**, 142–148.

Clark L. C. and Lyons C. (1965) Studies of a glassy carbon electrode for brain polaroghrphy with observations of the effect of carbonic anhydrase inhibition. *Ala. J. Med. Sci.* **2**, 253–259.

Clemens J. A. and Phebus L. A. (1984) Brain dialysis in conscious rats confirms *in vivo* electrochemical evidence that dopaminergic stimulation releases ascorbate. *Life Sci.* **35**, 671–677.

Coles J. A. and Orkand R. K. (1985) Changes in sodium activity during light stimulation in photoreceptors, glia and extracellular space in drone retina. *J. Physiol.* **362**, 415–435.

Coles J. A., Munoz J. L., and Deyhimi F. (1985) Surface and Volume Resistivity of Pyrex Glass Used for Liquid Membrane Ion-Selective Microelectrodes, in *Ion Measurements in Physiology and Medicine* (Kessler M., Harrison D. K., and Hoper J., eds.) pp. 67–73, Springer-Verlag Berlin.

Conti J. C., Strope E., Adams R. N., and Marsden C. A. (1978) Voltammetry in brain tissue: Chronic recording of stimulated dopamine and 5-hydroxytryptamine release. *Life Sci.* **23**, 2705–2716.

Cordingly G. E. and Somjen G. G. (1978) The clearing of excess potassium from extracellular space in spinal cord and cerebral cortex. *Brain Res.* **151**, 291–306.

Cragg B. (1979) Brain extracellular space fixed for electron microscopy. *Neurosci. Lett.* **15**, 301–306.

Crank J. (1975) *The Mathematics of Diffusion* 2nd Ed. Clarendon, Oxford.

Crespi F. and Jouvet M. (1984) Differential pulse voltammetric determination of 5-hydroxyindoles in four raphe nuclei of chronic freely moving rats simultaneously recorded by polarographic technique: Physiological changes with vigilance states. *Brain Res.* **299**, 113–119.

Crespi F., Cespuglio R., and Jouvet M. (1982) Differential pulse voltammetry in brain tissue. III. Mapping of the rat serontonergic raphe nuclei by electrochemical detection of 5-HIAA. *Brain Res.* **270,** 45–54.

Crespi F., Paret J., Keane P. E., and Morre M. (1984a) An improved differential pulse voltammetry technique allows the simultaneous analysis of dopaminergic and serotonergic activities *in vivo* with a single carbon-fibre electrode. *Neurosci. Lett.* **52,** 159–164.

Crespi F., Sharp T., Maidment N., and Marsden C. A. (1984b) Differential pulse voltammetry: Simultaneous *in vivo* measurement of ascorbic acid, catechols and 5-hydroxyindoles in the rat striatum. *Brain Res.* **322,** 135–138.

Crespi F., Paret J., Keane P. E., Morre M., Coude F. X., and Roncucci R. (1985) Growth hormone-releasing factor modified dopaminergic but not serotonergic activity in the arcuate nucleus of hypothalamus in the rat, as recorded by differential pulse voltammetry. *Brain Res.* **348,** 367–370.

Crespi F., Sharp T., Maidment N., and Marsden C. A. (1983) Differential pulse voltammetry *in vivo*—evidence that uric acid contributes to the indole oxidation peak. *Neurosci. Lett.* **43,** 203–207.

Cserr H. F. (1986) The neuronal microenvironment *Ann. NY Acad. Sci.* **481,** 1–391.

Czeh G., Kriz N., and Sykova E. (1981) Extracellular potassium accumulation in the frog spinal cord induced by stimulation of the skin and ventrolateral columns. *J. Physiol.* **320,** 57–72.

Dagostino M. and Lee C. O. (1982) Neutral carrier $Na^+$- and $Ca^{2+}$-selective microelectrodes for intracellular application. *Biophys. J.* **40,** 199–207.

Dayton M. A., Brown J. C., Stutts K. J., and Wightman R. M. (1980a) Faradaic electrochemistry with microvoltammetric electrodes. *Anal. Chem.* **52,** 1842–1847.

Dayton M. A., Ewing A. G., and Wightman R. M. (1980b) Response of microvoltammetric electrodes to homogeneous catalytic and slow heterogeneous charge-transfer reactions. *Anal. Chem.* **52,** 2392–2396.

Dayton M. A., Ewing A. G., and Wightman R. M. (1983) Diffusion processes measured at microvoltammetric electrodes in brain tissue. *J. Electroanal. Chem.* **146,** 189–200.

De Hemptinne A. (1980) Intracellular pH and surface pH in skeletal and cardiac muscle measured with a double-barrelled pH microelectrode. *Pflugers Arch.* **386,** 121–126.

Deschenes M. and Feltz P. (1976) GABA-induced rise of extracellular potassium in rat dorsal root ganglia: An electrophysiological study in vivo. *Brain Res.* **118,** 494–499.

Deyhimi F. and Coles J. A. (1982) Rapid silylation of a glass surface: Choice of reagent and effect of experimental parameters on hydrophobicity. *Helv. Chim. Acta* **65,** 1752–1759.

Dietzel I. and Heinemann U. (1986) Dynamic variations of the brain cell microenvironment in relation to neuronal hyperactivity. *Ann. NY Acad. Sci.* **481,** 72–84.

Dietzel I. Heinemann U., Hofmeier G., and Lux H. D. (1980) Transient changes in the size of the extracellular space in the sensorimotor cortex of cats in relation to stimulus-induced changes in potassium concentration. *Exp. Brain Res.* **40,** 432–439.

Dietzel I., Heinemann U., Hofmeier G., and Lux H. D. (1982) Stimulus-induced changes in extracellular $Na^+$ and $Cl^-$ concentration in relation to changes in the size of the extracellular space. *Exp. Brain Res.* **46,** 73–84.

Dionne V. E. (1976) Characterization of drug iontophoresis with a fast microassay technique. *Biophys. J.* **16,** 705–717.

Do Carmo R. J. and Martins-Ferreira H. (1984) Spreading depression of Leao probed with ion-selective microelectrodes in isolated chick retina. *Ann. Acad. Brasil Cienc.* **56,** 401–421.

Echizen H. and Freed C. R. (1983) *In vivo* electrochemical detection of extraneuronal 5-hydroxyindole acetic acid and norepinephrine in the dorsal raphe of urethane-anaesthetized rats. *Brain Res.* **277,** 55–62.

Echizen H. and Freed C. R. (1984) Measurement of serotonin turnover in rat dorsal raphe nucleus by *in vivo* electrochemistry. *J. Neurochem.* **42,** 1483–1486.

Engstrom R. C. (1982) Electrochemical pretreatment of glassy carbon electrodes. *Anal. Chem.* **54,** 2310–2314.

Erne D., Ammann D., and Simon W. (1979) Liquid membrane pH electrode based on a synthetic proton carrier. *Chimia* **33,** 88–90.

Ewing A. G. and Wightman R. M. (1984) Monitoring stimulated release of dopamine with *in vivo* voltammetry. II. Clearance of released dopamine from extracellular fluid. *J. Neurochem.* **43,** 570–577.

Ewing A. G., Wightman R. M., and Dayton M. A. (1982) *In vivo* voltammetry with electrodes that discriminate between dopamine and ascorbate. *Brain Res.* **249,** 361–370.

Ewing A. G., Dayton M. A., and Wightman R. M. (1981a) Pulse voltammetry with microvoltammetric electrodes. *Anal. Chem.* **53,** 1842–1847.

Ewing A. G., Withnell R., and Wightman R. M. (1981b) Instrument design for pulse voltammetry with microvoltammetric electrodes. *Rev. Sci. Instrum.* **52,** 454–458.

Falat L. and Cheng H.-Y. (1982) Voltammetric differentiation of ascorbic acid and dopamine at an electrochemically treated graphite/epoxy electrode. *Anal. Chem.* **54,** 2108–2111.

Fenstermacher J. D. and Patlak C. S. (1975) The Exchange of Material Between Cerebrospinal Fluid and Brain, in *Fluid Environment of the*

*Brain* (Cserr H. F., Fenstermacher J. D., and Fencl V., eds.) Academic, New York. 201–214.

Fisher R. S., Pedley T. A., Moody W. J. Jr., and Prince D. A. (1976a) The role of extracellular potassium in hippocampal epilepsy. *Arch. Neurol.* **33**, 76–83.

Fisher R. S., Pedley T. A., and Prince D. A. (1976b) Kinetics of potassium movement in normal cortex. *Brain Res.* **101**, 223–237.

Frant M. S. and Ross J. W. Jr. (1970) Potassium ion specific electrode with high selectivity for potassium over sodium. *Science* **167**, 987–988.

Frommer P. L., Pfaff W. W., and Braunwald E. (1961) The use of ascorbate dilution curves in cardiovascular diagnosis. *Circulation* **24**, 1227–1234.

Fujimoto M., Matsumura Y., and Satake N. (1980) General properties of antimony microelectrode in comparison with glass microelectrode for pH measurement. *Jpn. J. Physiol.* **30**, 491–508.

Funck R. J. J., Morf W. E., Schulthess P., Ammann D., and Simon W. (1982) Bicarbonate-sensitive liquid membrane electrodes based on neutral carriers for hydrogen ions. *Anal. Chem.* **54**, 423–429.

Galus Z., Schenk J. O., and Adams R. N. (1982) Electrochemical behavior of very small electrodes in solution. Double potential step, cyclic voltammetry and chronopotentiometry with current reversal. *J. Electroanal. Chem.* **135**, 1–11

Gardiner T. W., Armstrong-James M., Caan A. W., Wightman R. M., and Rebec G. V. (1985) Modulation of neostriatal activity by iontophoresis of ascorbic acid. *Brain Res.* **344**, 181–185.

Gardner-Medwin A. R. (1983) Analysis of potassium dynamics in brain tissue. *J. Physiol.* (Lond.) **335**, 393–426.

Gardner-Medwin A. R. (1986) A new framework for assessment of potassium buffering mechanisms. *Ann. NY Acad. Sci.* **481**, 287–302.

Gardner-Medwin A. R. and Mutch W. A. C. (1984) Experiments on spreading depression in relation to migraine and neurosurgery. *Ann. Acad. Brasil. Cienc.* **56**, 423–430.

Gardner-Medwin A. R. and Nicholson C. (1983) Changes of extracellular potassium activity produced by electric current through brain tissue in the rat. *J. Physiol.* **335**, 375–392.

Gerhardt G. A. and Adams R. N. (1982) Battery-powered apparatus for chronoamperometric measurements. *Anal. Chem.* **54**, 1888–1889.

Gerhardt G. A., Oke A. F., Nagy G., Moghaddam B., and Adams R. N. (1984a) Nafion-coated electrodes with high selectivity for CNS electrochemistry. *Brain Res.* **290**, 390–394.

Gerhardt G. A., Palmer M. R., Seiger A., Adams R. N., Olson L., and Hoffer B. J. (1984b) Adrenergic transmission in hippocampus-locus coeruleus double grafts in oculo: Demonstration by in vivo electrochemical detection. *Brain Res.* **306**, 319–325.

Gerhardt G. A., Rose G., Stromberg I., Conboy G., Olson L., Jonsson G., and Hoffer B. (1985) Dopaminergic neurotoxicity of 1-methyl-4-phenyl-1,2,3,6-tetrahydropyridine (MPTP) in the mouse: An *in vivo* electrochemical study. *J. Pharmacol. Exp. Ther.* **235,** 259–265.

Globus A. (1973) Iontophoretic Injection Techniques, in *Bioelectric Recording Techniques* vol. 1A, (Thompson R. F. and Patterson M. M., eds.) pp. 23–38, Academic, New York.

Gonon F. G. and Buda M. J. (1985) Regulation of dopamine release by impulse flow and by autoreceptors as studied by *in vivo* voltammetry in the rat striatum. *Neuroscience* **14,** 765–774.

Gonon F., Buda M., Cespuglio R., Jouvet M., and Pujol J.-F. (1980) *In vivo* electrochemical detection of catechols in the neostriatum of anaesthetized rats: Dopamine or DOPAC. *Nature* **286,** 902–904.

Gonon F., Buda M., De Simoni G., and Pujol J.-F. (1983a) Catecholamine metabolism in the rat locus coeruleus as studied by *in vivo* differential pulse voltammetry. II. Pharmacological and behavioral study. *Brain Res.* **273,** 207–216.

Gonon F., Cespuglio R., Buda M., and Pujol J.-F. (1983b) *In Vivo* Electrochemical Detection of Monoamine Derivatives, in *Methods in Biogenic Amine Research* (Parvez A., Nagatsu T., Nagatsu I., and Parvez H., eds.) pp. 165–188, Elsevier, New York.

Gonon F., Buda M., and Pujol J.-F. (1984a) Treated Carbon Fiber Electrodes for Measuring Catechols and Ascorbic Acid, in *Measurement of Neurotransmitter Release In Vivo* (Marsden C. A., ed.) pp. 153–171, John Wiley, Chichester.

Gonon F. G., Navarre F., and Buda M. J. (1984b) *In vivo* monitoring of dopamine release in the rat brain with differential normal pulse voltammetry. *Anal. Chem.* **56,** 573–575.

Gonon F., Cespuglio R., Ponchon J.-L., Buda M., Jouvet M., Adams R. N., and Pujol J.-F. (1978) Mésure électrochimique continué de la liberation de dopamine realisée in vivo dans le neostriatum du rat. *CR Acad. Sci. Paris* **286,** 1203–1206.

Gonon F. G., Fombarlet C. M., Buda M. J., and Pujol J. F. (1981) Electrochemical treatment of pyrolytic carbon fiber electrodes. *Anal. Chem.* **53,** 1386–1389.

Grafe P., Reddy M. M., Emmert H., and ten Bruggencate G. (1983) Effects of lithium on electrical activity and potassium ion distribution in the vertebrate central nervous system. *Brain Res.* **279,** 65–76.

Grafstein B. (1956) Mechanism of spreading cortical depression. *J. Neurophysiol.* **19,** 154–171.

Guggi M., Fiedler U., Pretsch E., and Simon W. (1975) A lithium ion-selective electrode based on a neutral carrier. *Anal. Lett.* **8,** 857–866.

Hahn Z., Cespuglio R., Faradji H., and Jouvet M. (1985) Factors influencing the properties of voltammetric carbon fibre electrodes: The importance of the pH of the medium used for the electrical treatment and of the resin coating of the fibres. *J. Biochem. Biophys. Meth.* **11,** 265–275.

Hansen A. J. (1985) Effect of anoxia on ion distribution in the brain. *Physiol. Rev.* **65,** 101–148.

Hansen A. J. and Olsen C. E. (1980) Brain extracellular space during spreading depression and ischemia. *Acta. Physiol. Scand.* **108,** 355–365.

Heinemann U. and Lux H. D. (1975) Undershoots following stimulus-induced rises of extracellular potassium concentration in cerebral cortex of cat. *Brain Res.* **93,** 63–76.

Heinemann U. and Lux H. D. (1977) Ceiling of stimulus induced rises in extracellular potassium concentration in the cerebral cortex of cat. *Brain Res.* **120,** 231–249.

Heinemann U., Lux H. D., and Gutnick M. J. (1977) Extracellular free calcium and potassium during paroxysmal activity in the cerebral cortex of cat. *Exp. Brain Res.* **27,** 237–243.

Heinemann U., Lux H. D., Marciani M. G., and Hofmeier G. (1979). Slow Potentials in Relation to Changes in Extracellular Potassium Activity in the Cortex of Cats, in *Origin of Cerebral Field Potentials* (Speckmann E.-J. and Caspers H., eds.) pp. 33–48. Thieme, Stuttgart.

Heyer C. B. and Lux H. D. (1976) Properties of a facilitating calcium current in pace-maker neurones of the snail, *Helix pomatia*. *J. Physiol.* **262,** 319–348.

Hicks T. P. (1984) The history and development of microiontophoresis in experimental neurobiology. *Prog. Neurobiol.* **22,** 185–240.

Hounsgaard J. and Nicholson C. (1983) Potassium accumulation around individual Purkinje cells in cerebellar slices from guinea-pig. *J. Physiol.* **340,** 359–388.

Howard-Butcher S., Blaha C. D., and Lane R. F. (1984) A comparison of CNS stimulants with phencyclidine on dopamine release using *in vivo* voltammetry. *Brain Res. Bull.* **13,** 497–501.

Huff R. M. and Adams R. N. (1980) Dopamine release in N. accumbens and striatum by clozapine: Simultaneous monitoring by *in vivo* electrochemistry. *Neuropharmacology* **19,** 587–590.

Huff R., and Adams R. N., and Rutledge C. O. (1979) Amphetamine dose-dependent changes of *in vivo* electrochemical signals in rat caudate. *Brain Res.* **173,** 369–372.

Ikeda M. and Nagatsu T. (1985) Monitoring of circadian fluctuations of N-acetylserotonin in the rat pineal body by differential pulse voltammetry. *J. Neural Transm.* **62,** 321–329.

Jefferys J. G. R. (1981) Influence of electric fields on the excitability of granule cells in guinea-pig hippocampal slices. *J. Physiol.* **319,** 143–152.

Karwoski C. J., Newman E. A., Shimazaki H., and Proenza L. M. (1985) Light-evoked increases in extracellular $K^+$ in the plexiform layers of amphibian retinas. *J. Gen. Physiol.* **86,** 189–213.

Keller R. W., Stricker E. M., and Zigmond M. J. (1983) Environmental stimuli but not homeostatic challenges produce apparent increases in dopaminergic activity in the striatum: An analysis by *in vivo* voltammetry. *Brain Res.* **279,** 159–170.

Kennett G. A. and Joseph M. H. (1982) Does *in vivo* voltammetry in the hippocampus measure 5-HT release? *Brain Res.* **236,** 305–316.

Kent T. A., Nagy G., Oke A., Preskorn S. H., and Adams R. N. (1985) Effect of $CO_2$ on a brain extracellular space marker and evidence of its neuronal modulation. *Brain Res.* **342,** 141–144.

Kessler M., Harrison D. K., and Hoper J., eds. (1985) *Ion Measurements in Physiology and Medicine* Springer-Verlag, Berlin.

Khuri R. N., Bogharian K. K., and Agulian S. K. (1974) Intracellular bicarbonate in single skeletal muscle fibers. *Pflug. Arch.* **349,** 285–294.

Kissinger P. T. and Heineman W. R., eds. (1984) *Laboratory Techniques in Electroanalytical Chemistry* Marcel Dekker, New York.

Kissinger P. T., Hart J. B., and Adams R. N. (1973) Voltammetry in brain tissue—a new neurophysiological measurement. *Brain Res.* **55,** 209–213.

Kline R. P., Ripps H., and Dowling J. E. (1978) Generation of b-wave currents in the skate retina. *Proc. Natl. Acad. Sci. USA* **75,** 5727–5731.

Knott P. J., Hutson P. H., Scraggs R. P., and Curzon G. (1981) Electrochemical Recording of Brain Catecholamine and Serotonin Release During Behavioral Changes, in *Function and Regulation of Monoamine Enzymes: Basic and Clinical Applications* (Usdin E., Weiner N., and Youdim M. B. H., eds.) pp. 771–780, Macmillan, New York.

Koryta J. (1975) *Ion-Selective Electrodes* Cambridge University Press, Cambridge.

Koryta J., Pradac J., Pradacova J., and Ossendorfova N. (1971) Organic oxidation reduction systems as electrochemical indicators for monitoring in organs *in vivo. Experientia* (suppl.) **18,** 367–373.

Kovach P. M., Ewing A. G., Wilson R. L., and Wightman R. M. (1984) *In vitro* comparison of the selectivity of electrodes for *in vivo* electrochemistry. *J. Neurosci. Meth.* **10,** 215–227.

Kraig R. P. and Nicholson C. (1976) Sodium liquid ion exchanger microelectrode used to measure large extracellular sodium transients. *Science* **194,** 725–726.

Kraig R. P., Ferreira-Filho C. R., and Nicholson C. (1983) Alkaline and acid transients in cerebellar microenvironment. *J. Neurophysiol.* **49,** 831–850.

Kraig R. P., Pulsinelli W. A., and Plum F. (1985) Hydrogen ion buffering during complete brain ischemia. *Brain Res.* **342,** 281–290.

Kriz N. and Sykova E. (1981) Sensitivity of $K^+$-Selective Microelectrodes to pH and Some Biologically Active Substances, in *Ion-Selective Microelectrodes and Their Use in Excitable Tissues.* (Sykova E., Hnik P., and Vyklicky L., eds.) pp. 25–45. Plenum, New York

Krnjevic K. and Morris M. E. (1972) Extracellular $K^+$ activity and slow potential changes in spinal cord and medulla. *Can. J. Physiol. Pharmacol.* **12,** 1214–1217.

Krnjevic K. and Morris M. E. (1974) Extracellular accumulation of $K^+$ evoked by activity of primary afferent fibers in the cuneate nucleus and dorsal horn of cats. *Can. J. Physiol. Pharmacol.* **52,** 852–871.

Krnjevic K., Morris M. E., and Reiffenstein R. J. (1982a) Stimulus-evoked changes in extracellular $K^+$ and $Ca^{2+}$ in pyramidal layers of the rat's hippocampus. *Can. J. Physiol. Pharmacol.* **60,** 1643–1657.

Krnjevic K., Morris M. E., Reiffenstein R. J., and Ropert N. (1982b) Depth distribution and mechanism of changes in extracellular $K^+$ and $Ca^{2+}$ concentrations in the hippocampus. *Can. J. Physiol. Pharmacol.* **60,** 1658–1671.

Kuhr W. G. and Wightman R. M. (1986) Real-time measurement of dopamine release in rat brain. *Brain Res.* **381,** 168–171.

Kuhr W. G., Ewing A. G., Caudill W. L., and Wightman R. M. (1984) Monitoring of stimulated release of dopamine with *in vivo* voltammetry. I. Characterization of the response observed in the caudate nucleus of the rat. *J. Neurochem.* **43,** 560–569.

Lakshminarayanaiah N. (1976) *Membrane Electrodes* Academic, New York.

Lamour Y., Rivot J. P., Pointis D., and Ory-Lavollee L. (1983) Laminar distribution of serotonergic innervation in rat somatosensory cortex, as determined by *in vivo* electrochemical detection. *Brain Res.* **259,** 163–166.

Lane R. F. and Hubbard A. T. (1976) Differential double pulse at chemically modified platinum electrodes for *in vivo* determination of catecholamines. *Anal. Chem.* **48,** 1287–1293.

Lane R. F., Hubbard A. T., and Blaha C. D. (1979) Application of semidifferential electroanalysis to studies of neurotransmitters in the central nervous system. *J. Electroanal. Chem.* **95,** 117–122.

Lane R. F., Hubbard A. T., Fukunaga K., and Blanchard R. J. (1976) Brain catecholamines: Detection *in vivo* by means of differential pulse voltammetry at surface modified platinum electrodes. *Brain Res.* **114,** 346–352.

Lantner F., Erne D., Ammann D., and Simon W. (1980) Neutral carrier based ion-selective electrodes for intracellular magnesium activity studies. *Anal. Chem.* **52,** 2400–2402.

Lashley K. S. (1941) Patterns of cerebral integration indicated by the scotomas of migraine. *Arch. Psychiat.* **46,** 331–339.

Leao A. A. P. (1944) Spreading depression of activity in the cerebral cortex. *J. Neurophysiol.* **7,** 359–390.

Lee C. O. (1981) Ionic activities in cardiac muscle cells and application of ion-selective microelectrodes. *Am. J. Physiol.* **241**, H459–H478.

Lehmenkuhler A., Caspers H., and Kersting U. (1985). Relations Between DC Potentials, Extracellular Ion Activities, and Extracellular Volume Fraction in the Cerebral Cortex with Changes in $PCO_2$, in *Ion Measurements in Physiology and Medicine* (Kessler M., Harrison D. K., and Hoper J., eds.) pp. 199–213, Springer-Verlag, Berlin.

Levin V. A., Fenstermacher J. D., and Patlak C. S. (1970) Sucrose and insulin space measurements of cerebral cortex in four mammalian species. *Am. J. Physiol.* **219**, 1528–1533.

Lewis S. A. and Wills N. K. (1980) Resistive artifacts in liquid-ion exchanger microelectrodes. Estimates of $Na^+$ activity in epithelial cells. *Biophys. J.* **31**, 127–138.

Lindsay W. S., Herndon J. G., Blakely R. D., Justice J. B., and Neill D. B. (1981) Voltammetric recording from neostriatum of behaving rhesus monkey. *Brain Res.* **220**, 391–396.

Lindsay W. S., Justice J. B., and Salamone J. (1980a) Simulation studies of *in vivo* electrochemistry. *Computers Chem.* **4**, 19–26.

Lindsay W. S., Kizzort B. L., Justice J. B., Salamone J. D., and Neill D. B. (1980b) Microcomputer controlled multielectrode system for *in vivo* electrochemistry. *Chem. Biomed. Environ. Instr.* **10**, 311–330.

Louilot A., Buda M., Gonon F., Simon H., le Moal M., and Pujol J. F. (1985) Effect of haloperidol and sulpiride on dopamine metabolism in nucleus accumbens and olfactory tubercle: A study by *in vivo* voltammetry. *Neuroscience* **14**, 775–782.

Lubbers D. W., Acker H., Buck R. P., Eisenman G., Kessler M., and Simon W., eds. (1981) *Progress in Enzyme and Ion-Selective Electrodes* Springer-Verlag, Berlin.

Lux H. D. (1974a) Fast recording ion specific microelectrodes: Their use in pharmacological studies in the CNS. *Neuropharmocology* **13**, 509–517.

Lux H. D. (1974b) The kinetics of extracellular potassium: Relation to epileptogenesis. *Epilepsia* **15**, 375–393.

Lux H. D. and Neher E. (1973) The equilibration time course of $[K^+]_o$ in cat cortex. *Exp. Brain Res.* **17**, 190–205.

Lux H. D., Loracher C., and Neher E. (1970) The action of ammonium on postsynaptic inhibition of cat spinal motoneurons. *Exp. Brain Res.* **11**, 431–447.

Maidment N. T. and Marsden C. A. (1985) *In vivo* voltammetric and behavioral evidence for somatodendritic autoreceptor control of mesolimbic dopamine neurones. *Brain Res.* **338**, 317–325.

Malenka R. C., Kocis J. D., Ransom B. R., and Waxman S. G. (1981). Modulation of parallel fiber excitability by postsynaptically mediated changes in extracellular potassium. *Science* **214**, 339–341.

Margolis R. U., Aquino M. M., Klinger J. A., Ripellino J. A., and Margolis

R. K. (1986) Structure and localization of nervous tissue proteoglycans. *Ann. NY Acad. Sci.* **481,** 46–52.

Marquardt D. W. (1963) An algorithm for least-squares estimation of nonlinear parameters. *J. Soc. Indust. Appl. Math.* **11,** 431–441.

Marsden C. A. (1979) Functional aspects of 5-hydroxytryptamine neurones: Application of electrochemical monitoring *in vivo. Trends Neurosci.* **1,** 230–234.

Marsden C. A., Conti J., Strope E., Curzon G., and Adams R. N. (1979) Monitoring 5-hydroxytryptamine release in the brain of the freely moving unanaesthetized rat using *in vivo* voltammetry. *Brain Res.* **171,** 85–99.

McCreery R. L., Dreiling R., and Adams R. N. (1974) Quantitative studies of drug interactions. *Brain Res.* **73,** 23–33.

McRae-Degueurce A., Serrano A., Sandillon F., Privat A., and Scatton B. (1984) *In vivo* voltammetric measurement of extracellular 5-hydroxyindoleacetic acid in the denervated striatum after transplantation of mesencephalic raphe neurones. *Neurosci. Lett.* **48,** 97–102.

Meier P. C., Lanter F., Ammann D., Steimer R. A., and Simon W. (1982) Applicability of available ion-selective liquid-membrane microelectrodes to intracellular ion-activity measurements. *Pflugers Arch.* **393,** 23–30.

Metzger E., Ammann D., Schefer U., Pretsch E., and Simon S. (1984) Lipophilic neutral carriers for lithium-selective liquid membrane electrodes. *Chimia* **38,** 440–442.

Milby K., Oke A., and Adams R. N. (1982) Detailed mapping of ascorbate distribution on rat brain. *Neurosci. Lett.* **28,** 15–20.

Millar J., Stamford J. A., Kruk Z. L., and Wightman R. M. (1985) Electrochemical, pharmacological and physiological evidence of rapid dopamine release in the rat caudate nucleus following electrical stimulation of the median forebrain bundle. *Eur. J. Pharmocol.* **109,** 341–348.

Moghaddam B. and Adams R. N. (1986) Recent developments in *in vivo* voltammetry: Applications to studies of chemical dynamics in the neuronal microenvironment. *Ann. NY Acad. Sci.* **481,** 106–114.

Moghaddam B., Schenk J. O., Stewart W. D., and Hansen A. J. (1987) Temporal relationship between neurotransmitter release and ion flux during spreading depression and anoxia. *Can. J. Physiol. Pharmacol.* (in press).

Morgan M. E. and Freed C. R. (1981) Acetaminophen as an internal standard for calibrating *in vivo* electrochemical electrodes. *J. Pharmacol. Exp. Ther.* **219,** 49–53.

Morris M. E. and Krnjevic K. (1981) Slow diffusion of $Ca^{2+}$ in the rat's hippocampus. *Can. J. Physiol. Pharmacol.* **59,** 1022–1025.

Mos J., Broxterman H. J., and Van Bennekom W. P. (1981) *in vivo* voltammetric investigations into the action of HA-966 on central dopaminergic neurons. *Brain Res.* **207**, 465–470.

Mueller K. (1986) In vivo voltammetric recording with Nafion-coated carbon paste electrodes: Additional evidence that ascorbic acid release is monitored. *Pharmacol. Biochem. Behav.* **25**, 325–328.

Mueller K., Palmour R., Andrews C. D., and Knott P. J. (1985) *In vivo* voltammetric evidence of production of uric acid by rat caudate. *Brain Res.* **335**, 231–235.

Munoz J-L., Deyhimi F., and Coles J. A. (1983) Silanization of glass in the making of ion-sensitive microelectrodes. *J. Neurosci. Meth.* **8**, 231–247.

Mutani R., Futamachi K. J., and Prince D. A. (1974) Potassium activity in immature cortex. *Brain Res.* **75**, 27–39.

Mutch W. A. C. and Hansen A. J. (1984) Extracellular pH changes during spreading depression and cerebral ischemia: Mechanisms of brain pH regulation. *J. Cereb. Blood Flow Metab.* **4**, 17–27.

Nagy G., Gerhardt G. A., Oke A. F., Rice M. E., Adams R. N., Moore R. B., III, Szentirmay M. N., and Martin C. R. (1985a) Ion exchange and transport of neurotransmitters in Nafion films on conventional and microelectrode surfaces. *J. Electroanal. Chem.* **188**, 85–94.

Nagy G., Moghaddam B., Oke A., and Adams R. N. (1985b) Simultaneous monitoring of voltammetric and ion-selective electrodes in mammalian brain. *Neurosci. Lett.* **55**, 119–124.

Nagy G., Rice M. E., and Adams R. N. (1982) A new type of enzyme electrode: The ascorbic acid eliminator electrode. *Life Sci.* **31**, 2611–2616.

Neher E. and Lux H. D. (1973) Rapid changes of potassium concentration at the outer surface of exposed single neurones during membrane current flow. *J. Gen Physiol.* **61**, 385–399.

Neild T. O. and Thomas R. C. (1973) New design for a chloride-sensitive microelectrode. *J. Physiol.* **231**, 7P–8P.

Nelder J. A. and Mead R. (1965) A simplex method for function minimization. *Comput. J.* **7**, 308–313.

Newell G. A. and Calhoun E. H. (1982) Comparison of modified carbon paste and carbon fibre electrodes for the electrochemical measurement of catecholamines in the presence of ascorbic acid. *Soc. Neurosci. Abst.* **8**, 889.

Nicholson C. (1979) Brain Cell Microenvironment as a Communication Channel, in *The Neurosciences Fourth Study Program* (Schmitt F. O. and Warden F. G., eds.) pp. 457–476, Massachusetts Institute of Technology, Cambridge.

Nicholson C. (1980a) Modulation of extracellular calcium and its functional implications. *Fed. Proc.* **39**, 1519–1523.

Nicholson C. (1980b) Dynamics of the brain cell microenvironment. *Neurosci. Res. Prog. Bull.* **18**, 183–322.

Nicholson C. (1984) Comparative neurophysiology of spreading depression in the cerebellum. *Anais Acad. Brasil Cienc.* **56**, 481–494.

Nicholson C. (1985) Diffusion from an injected volume of a substance in brain tissue with arbitrary volume fraction and tortuosity. *Brain Res.* **333**, 325–329.

Nicholson C. and Hounsgaard J. (1983) Diffusion in the slice microenvironment and implications for physiological studies. *Fed. Proc.* **42**, 2865–2868.

Nicholson C. and Kraig R. P. (1975) Chloride and potassium changes measured during spreading depression in catfish cerebellum. *Brain Res.* **96**, 384–389.

Nicholson C. and Kraig R. P. (1981) The Behavior of Extracellular Ions During Spreading Depression, in *The Application of Ion-Selective Microelectrodes* (Zeuthen T., ed.) pp. 217–238, Amsterdam, Elsevier, North Holland.

Nicholson C. and Phillips J. M. (1975) Chloride and potassium changes measured during spreading depression in catfish cerebellum. *Brain Res.* **96**, 384–389.

Nicholson C. and Phillips J. M. (1981) Ion diffusion modified by tortuosity and volume fraction in the extracellular microenvironment of the rat cerebellum. *J. Physiol.* **321**, 225–257.

Nicholson C. and Rice M. E. (1987) Calcium diffusion in the brain cell microenvironment. *Can. J. Physiol. Pharmacol.*, in press.

Nicholson C. and Rice M. E. (1986) The migration of substances in the neuronal microenvironment. *Ann. NY Acad. Sci.* **481**, 55–66.

Nicholson C., Phillips J. M., Tobias C., and Kraig R. P. (1981) Extracellular Potassium, Calcium and Volume Profiles During Spreading Depression, in (Vyklicky L., eds. and Sykova E., Hnik P.,) pp. 25–45, *Ion-Selective Microelectrodes and Their Use in Excitable Tissues* Plenum, New York.

Nicholson C., Kraig R. P., Ferreira-Filho C. R, and Thompson P. (1985) Hydrogen Ion Variations and Their Interpretation in the Microenvironment of the Vertebrate Brain, in *Ion Measurements in Physiology and Medicine* (Kessler M., Harrison D. K., and Hoper J., eds.). pp. 229–235, Springer-Verlag, Berlin.

Nicholson C., Steinberg R., Stockle H., and ten Bruggencate G. (1976) Calcium decrease associated with aminopyridine-induced potassium increase in cat cerebellum. *Neurosci. Lett.* **3**, 315–319.

Nicholson C., ten Bruggencate G., Steinberg R., and Stockle H. (1977) Calcium modulation in brain extracellular microenvironment demonstrated with ion-selective micropipette. *Proc. Natl. Acad. Sci. USA* **74**, 1287–1290.

Nicholson C., ten Bruggencate G., Stockle H., and Steinberg R. (1978) Calcium and potassium changes in extracellular microenvironment of cat cerebellar cortex. *J. Neurophysiol.* **41,** 1026–1039.

Oakley B., II. and Green D. G. (1976) Correlation of light-induced changes in retinal extracellular potassium concentration with c-wave of the electroretinogram. *J. Neurophysiol.* **39,** 1117–1133.

Oakley B., II. and Steinberg R. H. (1982) Effects of maintained illumination upon $[K^+]_o$ in the subretinal space of the frog retina. *Vision Res.* **22,** 767–773.

Oakley B., II., Miller S. S., and Steinberg R. H. (1978) Effect of intracellular potassium upon the electrogenic pump of frog retinal pigment epithelium. *J. Membrane Biol.* **44,** 281–307.

O'Doherty J., Garcia-Diaz J. F., and Armstrong W. McD. (1979) Sodium-selective liquid ion-exchanger microelectrodes for intracellular measurements. *Science* **203,** 1349–1351.

Oehme M. and Simon W. (1976) Microelectrode for $K^+$ based on a neutral carrier and comparison of its characteristics with a cation exchanger sensor. *Anal. Chim. Acta.* **86,** 21–25.

Oehme M., Kessler M., and Simon W. (1976) Neutral carrier $Ca^{2+}$-microelectrode. *Chimia* **30,** 204–206.

Oesch U., Ammann D., Pham H. V., Wuthier R. Z., and Simon W. (1986) Design of anion-selective membranes for clinically relevant sensors. *J. Chem. Soc.* **82,** 1179–1186.

O'Neill R. D. and Fillenz M. (1985) Simultaneous monitoring of dopamine release in rat frontal cortex, nucleus accumbens and striatum: Effect of drugs, circadian changes and correlations with motor activity. *Neuroscience* **16,** 49–55.

O'Neill, R. D., Fillenz M., and Albery, W. J. (1982a) Circadian changes in homovanillic acid and ascorbate levels in the rat striatum using microprocessor-controlled voltammetry. *Neurosci. Lett.* **34,** 189–193.

O'Neill R. D., Grunewald R. A., Fillenz M., and Albery W. J. (1982b) Linear sweep voltammetry with carbon paste electrodes in the rat striatum. *Neuroscience* **7,** 1945–1954.

O'Neill R. D., Fillenz M., Albery W. J., and Goddard N. J. (1983) The monitoring of ascorbate and monoamine transmitter metabolites in the striatum of unanaesthetized rats using microprocessor-based voltammetry. *Neuroscience* **9,** 87–93.

O'Neill R. D., Fillenz M., Grunewald R. A., Blomfield M. R., Albery W. J., Jamieson C. M., Williams J. H., and Gray J. A. (1984a) Voltammetric carbon paste electrodes monitor uric acid and not 5-HIAA at the 5-hydroxyindole potential in the rat brain. *Neurosci. Lett.* **45,** 39–46.

O'Neill R. D., Fillenz, M., Sundstrom L., and Rawlins J. N. P. (1984b) Voltammetrically monitored brain ascorbate as an index of excitatory amino acid release in the unrestrained rat. *Neurosci. Lett.* **52,** 227–233.

Orme F. W. (1969) Liquid Ion-Exchanger Microelectrodes, in *Glass Microelectrodes* (Lavallee M., Schanne O. F., and Hebert N. C., eds.) pp. 376–395, Wiley, New York.

Pape L. G. and Katzman R. (1972) $K^{42}$ distribution in brain during simultaneous ventriculocisternal and subarachnoid perfusion. *Brain Res.* **38,** 49–69.

Patterson M. M. and Kesner R. P., eds. (1981) *Electrical Stimulation Research Techniques* Academic, New York.

Phillips J. M. and Nicholson C. (1981) Microelectrodes for Novel Anions and Their Application to Some Neurophysiological Problems, in *Progress in Enzyme and Ion-Selective Electrodes* (Lubbers D. W., Acker H., Buck R. P., Eisenman G., Kessler M., and Simon W., eds.) pp. 15–20, Springer-Verlag, Berlin.

Phillips J. M. and Nicholson C. (1979) Anion permeability in spreading depression investigated with ion-selective microelectrodes. *Brain Res.* **173,** 567–571.

Plotsky P. M. (1982) Differential voltammetric measurement of catecholamines and ascorbic acid at surface-modified carbon filament microelectrodes. *Brain Res.* **235,** 179–184.

Plotsky P. M., de Greef W. J., and Neill J. D. (1982) *In situ* voltammetric microelectrodes: Application to the measurement of median eminence catecholamine release during simulated suckling. *Brain Res.* **250,** 251–262.

Ponchon J.-L., Cespuglio R., Gonon F., Jouvet M., and Pujol J.-F. (1979) Normal pulse polarography with carbon fiber electrodes for *in vitro* and *in vivo* determination of catecholamines. *Anal. Chem.* **51,** 1483–1486.

Prince D. A., Lux H. D., and Neher E. (1973) Measurement of extracellular potassium activity in cat cortex. *Brain Res.* **50,** 489–495.

Pumain P. and Heinemann U. (1985) Stimulus- and amino acid-induced calcium and potassium changes in rat neocortex. *J. Neurophysiol.* **53,** 1–16.

Pumain R., Kurcewicz I., and Louvel J. (1983) Fast extracellular calcium transients: Involvement in epileptic processes. *Science* **222,** 177–179.

Purves R. D. (1979) The physics of iontophoretic pipettes. *J. Neurosci. Meth.* **1,** 165–178.

Quehenberger P. (1977) The influence of carbon dioxide, bicarbonate and other buffers on the potential of antimony microelectrodes. *Pflug. Arch.* **368,** 141–146.

Ransom B. R., Yamate C. L., and Connors B. W. (1985) Activity-dependent shrinkage of extracellular space in rat optic nerve: A developmental study. *J. Neurosci.* **5,** 532–535.

Rice M. E. and Adams R. N. (1982) Electrochemical monitoring of endogenous neurotransmitter release from rat thalamic slices. *Soc. Neurosci. Abst.* **8,** 475.

Rice M. E. and Nicholson C. (1985) Serotonin migration in brain cell microenvironment. *Soc. Neurosci. Abst.* **11,** 42.

Rice M. E., Galus Z., and Adams R. N. (1983) Graphite paste electrodes: Effects of paste composition and surface states on electron-transfer rates. *J. Electroanal. Chem.* **143,** 89–102.

Rice M. E., Gerhardt G. A., Nagy G., Hierl P. M., and Adams R. N. (1985a) Diffusion coefficients of neurotransmitters and their metabolites in brain extracellular fluid space. *Neuroscience* **15,** 891–902.

Rice M. E., Oke A. F., Bradberry C. W., and Adams R. N. (1985b) Simultaneous voltammetric and chemical monitoring of dopamine release *in situ. Brain Res.* **340,** 151–155.

Rice M. E. and Nicholson C. (1987) Both release and extracellular volume change regulate interstitial ascorbic acid in turtle brain, submitted.

Rivot J. P., Ory-Lavollee L., and Chiang C. Y. (1983) Differential pulse voltammetry in the dorsal horn of the spinal cord of the anaesthetized rat: Are the voltammograms related to 5-HT and/or to 5-HIAA? *Brain Res.* **274,** 311–319.

Ronnau K. (1984) A simplified method for silanisation of double barrelled ion-sensitive microelectrodes. *Experientia* **40,** 1019–1020.

Rose G., Gerhardt G., Stromberg I., Olson L., and Hoffer B. (1985) Monoamine release from dopamine-depleted rat caudate nucleus reinnervated by substantia nigra transplants. An *in vivo* electrochemical study. *Brain Res.* **341,** 92–100.

Rutecki P. A., Lebeda F. J., and Johnston D. (1985) Epileptiform activity induced by changes in extracellular potassium in hippocampus. *J. Neurophysiol.* **54,** 1363–1374.

Satake N., Matsumura Y., and Fujimoto M. (1980a) Temperature coefficient of and oxygen effect on the antimony microelectrode. *Jpn. J. Physiol.* **30,** 671–687.

Satake N., Matsumura Y., and Fujimoto M. (1980b) Protein effect on the antimony microelectrode in application to biological fluid. *Jpn. J. Physiol.* **30,** 689–700.

Saunders J. H. and Brown H. M. (1977) Liquid and solid-state $Cl^-$-sensitive microelectrodes. *J. Gen. Physiol.* **79,** 507–530.

Scatton B., Serrano A., and Nishikawa T. (1985) GABAmimetics decrease extracellular concentrations of 5-HIAA (as measured by *in vivo* voltammetry) in the dorsal raphe of the rat. *Brain Res.* **341,** 372–376.

Scatton B., Serrano A., Rivot J. P., and Nishikawa T. (1984) Inhibitory GABAergic influence on striatal serotonergic transmission exerted in the dorsal raphe as revealed by *in vivo* voltammetry. *Brain Res.* **305,** 343–352.

Schenk J. O. and Adams R. N. (1984) Chronoamperometric Measurements in the Central Nervous System, in *Measurement of Neurotransmitter Release In Vivo* (Marsden C. A., ed.) pp. 193–208, John Wiley, Chichester.

Schenk J. O., Miller E., Gaddis R., and Adams R. N. (1982) Homeostatic control of ascorbate concentration in CNS extracellular fluid. *Brain Res.* **253**, 353–356.

Schenk J. O., Miller E., Rice M. E., and Adams R. N. (1983) Chronoamperometry in brain slices: Quantitative evaluation of *in vivo* electrochemistry. *Brain Res.* **277**, 1–8.

Schmitt F. O. (1984) Molecular regulators of brain function: A new view. *Neuroscience* **13**, 991–1001.

Schmitt F. O. and Samson F. E. (1969) Brain cell microenvironment. *Neurosci. Res. Prog. Bull.* **7**, 277–417.

Schulthess P., Shijo Y., Pham H. V., Pretsch E., Ammann D., and Simon W. (1981) A hydrogen ion-selective liquid-membrane electrode based on tri-*n*-dodecylamine as neutral carrier. *Anal. Chim. Acta* **131**, 111–116.

Senkyr J. and Petr J. (1978) Liquid Ion-Selective Electrodes Based on Basic Dyes, in *Ion Selective Electrodes* (Pungor E., ed.) pp. 559–565, Elsevier, Amsterdam.

Sharp T., Brazell M. P., Bennett G. W., and Marsden C. A. (1984a) The TRH analogue CG 3509 increases *in vivo* catechol/ascorbate oxidation in the nucleus accumbens but not in the striatum of the rat. *Neuropharmacology* **23**, 617–623.

Sharp T., Maidment N. T., Brazell M. P., Zetterstrom T., Ungerstedt U., Bennett G. W., and Marsden C. A. (1984ab) Changes in monoamine metabolites measured by simultaneous *in vivo* differential pulse voltammetry and introcerebral dialysis. *Neuroscience* **12**, 1213–1221.

Singer W. and Lux H. D. (1973) Presynaptic depolarization and extracellular potassium in the cat lateral geniculate nucleus. *Brain Res.* **64**, 17–33.

Somjen G. G. (1973) Electrogenesis of sustained potentials. *Prog. Neurobiol.* **1**, 201–237.

Somjen G. G. (1979) Extracellular potassium in the mammalian central nervous system. *Ann. Rev. Physiol.* **41**, 159–177.

Somjen G. G. (1984a) Acidification of interstitial fluid in hippocampal formation caused by seizures and by spreading depression. *Brain Res.* **311**, 186–188.

Somjen G. G. (1984b) Interstitial Ion Concentration and the Role of Neuroglia in Seizures, in *Electrophysiology of Epilepsy* (Schwartzkroin P. A. and Wheal M., eds.) pp. 304–341, Academic, London.

Somjen G. G. and Giacchino J. L. (1985) Potassium and calcium concentrations in interstitial fluid of hippocampus formation during paroxysmal responses. *J. Neurophysiol.* **53,** 1098–1108.

Speckmann E.-J., Elger C. E., and Lehmenkuehler A. (1983) Penicillin activity in brain tissue: A method for continuous measurement. *Electroencephalog. Clin. Neurophysiol.* **56,** 664–667.

Spira M. E., Yarom Y., and Zeldes D. (1984) Neuronal interactions mediated by neurally evoked changes in the extracellular potassium concentration. *J. Exp. Biol.* **112,** 179–197.

Stamford J. A. (1985) *In vivo* voltammetry: Promise and perspective. *Brain Res. Rev.* **10,** 119–135.

Stamford J. A., Kruk Z. L., and Millar J. (1984a) A double-cycle high-speed voltammetric technique allowing direct measurement of irreversibly oxidised species: Characterization and application to the temporal measurement of ascorbate in the rat central nervous system. *J. Neurosci. Meth.* **10,** 107–118.

Stamford J. A., Kruk Z. L., and Millar J. (1984b) Regional differences in extracellular ascorbic acid levels in the rat brain determined by high speed voltammetry. *Brain Res.* **299,** 289–295.

Stamford J. A., Kruk Z. L., and Millar J. (1986) Sub-second striatal dopamine release measured by in vivo voltammetry. *Brain Res.* **381,** 351–355.

Steiner R. A., Oehme M., Ammann D., and Simon W. (1979) Neutral carrier sodium ion-selective microelectrode for intracellular studies. *Anal. Chem.* **51,** 351–353.

Stone T. W. (1985) *Microiontophoresis and Pressure Ejection* Wiley, Chichester.

Sykova E. (1981) $K^+$ changes in the extracellular space of the spinal cord and their physiological role. *J. Exp. Biol.* **95,** 93–109.

Sykova E. (1983) Extracellular $K^+$ accumulation in the central nervous system. *Prog. Biophys. Mol. Biol.* **42,** 135–189.

Sykova E., Hnik P. and Vyklicky L., eds. (1981) *Ion-Selective Microelectrodes and Their Use in Excitable Tissues* Plenum, New York.

Sykova E., Kriz N., and Hajek I. (1985) Extracellular $K^+$ Accumulation in the Spinal Cord and Its Role in Primary Afferent Depolarization and Poststimulation Analgesia, in *Ion Measurements in Physiology and Medicine* (Kessler M., Harrison D. K., and Hoper J., eds.) pp. 214–220, Springer-Verlag, Berlin.

Taylor C. P. and Dudek F. E. (1984) Excitation of hippocampal pyramidal cells by an electrical field effect. *J. Neurophysiol.* **52,** 126–142

Thomas R. C. (1978) *Ion-Sensitive Intracellular Microelectrodes* Academic, London.

Traub R. D., Dudek F. E., Snow R. W., and Knowles W. D. (1985)

Computer simulations indicate that electrical field effects contribute to the shape of the epileptiform field potential. *Neuroscience* **15,** 947–958.

Tsien R. Y. (1980) Low-cost active-probe electrometer with special features for testing ion-selective micro-electrodes. *J. Physiol.* **308,** 6P–7P.

Ujec E., Keller O., Kriz N., Pavlik V., and Machek J. (1981) Double-Barrel Ion Selective [$K^+$, $Ca^{2+}$, $Cl^-$] Coaxial Microelectrodes (ISCM) for Measurements of Small and Rapid Changes in Ion Activities, in *Ion-Selective Microelectrodes and Their Use in Excitable Tissues* (Sykova E., Hnik P., and Vyklicky L., eds.) pp. 41–45, Plenum, New York.

Ullrich A., Steinberg R., Baierl P., and ten Bruggencate G. (1982) Changes in extracellular potassium and calcium in rat cerebellar cortex related to local inhibition of the sodium pump. *Pflug. Arch.* **395,** 108–114.

Urbanics R., Leninger-Follert E., and Lubbers D. W. (1978) Time course of changes of extracellular $H^+$ and $K^+$ activities during and after direct electrical stimulation of the brain cortex. *Pflug. Arch.* **378,** 47–53.

Van Harreveld A. (1959) Compounds in brain extracts causing spreading depression of cerebral cortical activity and contraction of crustacean muscle. *J. Neurochem.* **3,** 300–315.

Van Harreveld A. (1972) The Extracellular Space in the Vertebrate Central Nervous System, in *The Structure and Function of Nervous Tissue* vol. IV (Bourne G. H., ed.) pp. 447–511, Academic, New York.

Van Harreveld A. (1978) Two mechanisms of spreading depression in the chick retina. *J. Neurobiol.* **9,** 419–431.

Vaughan-Jones R. D. and Kaila K. (1986) The sensitivity of liquid sensor, ion-selective microelectrodes to changes in temperature and solution level. *Pflugers Arch.* **406,** 641–644.

Vern B. A., Schutte W. H., and Thibalt L. E. (1977) $[K^+]_o$ clearance in cortex: A new analytical model. *J. Neurophysiol.* **40,** 1015–1023.

Vyskocil F. and Kriz N. (1972) Modifications of single- and double-barrel potassium specific microelectrodes for physiological experiments. *Pflug. Arch. Ges. Physiol.* **337,** 265–276.

Vyskocil F., Kriz N., and Bures J. (1972) Potassium-selective microelectrodes used for measuring the extracellular brain potassium during spreading depression and anoxic depolarization in rats. *Brain Res.* **39,** 255–259.

Walker J. L. (1971) Ion specific liquid ion exchanger microelectrodes. *Anal. Chem.* **43,** 89A–92A.

Wightman R. M. (1981) Microvoltammetric electrodes. *Anal. Chem.* **53,** 1125–1130A.

Wightman R. M., Strope E., Plotsky P. M., and Adams R. N. (1976) Monitoring of transmitter metabolites by voltammetry in cerebrospinal fluid following neural pathway stimulation. *Nature* **262,** 145–146.

Wilson R. L. and Wightman R. M. (1985) Systemic and nigral application of amphetamine both cause an increase in extracellular concentration of ascorbate in the caudate nucleus of the rat. *Brain Res.* **339,** 219–226.

Wuhrmann P., Ineichen H., Riesen-Willi U., and Letzi M. (1979) Change in nuclear potassium electrochemical activity and puffing of potassium-sensitive salivary chromosome regions during *Chironomus* development. *Proc. Natl. Acad. Sci. USA* **76,** 806–808.

Yamamoto B. K., Lane R. F., and Freed C. R. (1982) Normal rats trained to circle show asymmetric caudate dopamine release. *Life Sci.* **30,** 2155–2162.

Zetterstrom T., Sharp T., Marsden C. A., and Ungerstedt U. (1983) *In vivo* measurement of dopamine and its metabolites by intracerebral dialysis: Changes after *d*-amphetamine. *J. Neurochem.* **41,** 1769–1773.

Zetterstrom T., Vernet L., Ungerstedt U., Tossman U., Jonzon B., and Fredholm B. B. (1982) Purine levels in the intact rat brain. Studies with an implanted perfused hollow fiber. *Neurosci. Lett.* **29,** 111–115.

Zeuthen, T., ed. (1981) *The Application of Ion-Selective Microelectrodes* Elsevier/North Holland, Amsterdam.

Zhukov A. F., Erne D., Ammann D., Guggi M., Pretsch E., and Simon W. (1981) Improved lithium ion-selective electrode based on a lipophilic diamide as neutral carrier. *Anal. Chim. Acta* **131,** 117–122.

# Patch Clamp Recording Methodology

## Jerry M. Farley

## 1. Introduction

One of the most exciting recent advances in the field of neurobiology has been the development of the technique called single-channel or patch clamp recording. The utility and power of the method are now being exploited by neuroscientists using a wide variety of tissue preparations to answer questions that previously had been, at best, difficult to fathom. The purpose of this chapter is to provide basic information on the development and usefulness of the technique and practical information on how to set up and perform basic experiments. The reader will be referred to books and articles for a more thorough discussion of methods when those methods go beyond the scope of this chapter. An excellent book, entitled *Single Channel Recording*, edited by B. Sakmann and E. Neher (1983), covers thoroughly most aspects of this technique. In addition, several reviews about patch clamping have been written (Neher et al., 1978; Hamill et al., 1981; Sachs and Auerbach, 1983; Auerbach and Sachs, 1984; Rae and Levis, 1984; Rae, 1985; Dwyer, 1985). Finally this chapter is not designed to be a complete review of the literature, since there are over 250 articles and reviews currently available on this subject.

Recording from a discrete area of membrane, one of the primary purposes behind the development of the patch clamp, was first shown to be feasible by Fatt and Katz (1952). They demonstrated that the current flowing through the area of membrane at endplates in frog muscle could be discretely recorded by positioning a microelectrode very near the synaptic junction. The currents were measured by the voltage drop they caused across resistance developed by the close aposition of the microelectrode tip with the membrane. This is the focal recording technique.

A variant of this technique was developed by Strickholm (1961) in which a larger-tipped glass pipet was pressed against the synaptic connection. Using this method it is possible to obtain lower noise and higher resolution recordings of postsynaptic events such as miniature endplate currents. The resistance between the pipet lumen and the bath solution was high enough to allow the potential of the cell membrane under the lumen of the pipet tip to be changed. Fishman (1975) improved the seal resistance of a focal current recording electrode with squid axon membrane using sucrose as an insulating medium around the tip of the pipet. The increase in seal resistance allowed for better spatial control of the voltage gradient across the membrane, as well as increasing the signal-to-noise ratio. These features (1) recording from a discrete area of membrane; (2) decreased electrical noise; and (3) the ability to change the membrane potential of the patch (or cell) are all similar to those of the patch clamp or single-channel recording technique.

The patch clamp recording technique was first used by Neher and Sakmann (1976). They recorded single acetylcholine-activated currents from the membrane of denervated frog muscle. Basically the technique consisted of pressing a glass pipet up against the membrane to electrically isolate the membrane patch within the lumen of the pipet. The pipets used had a tip diameter of less than 1 $\mu$m. This meant that a few channels could be isolated in the extra-synaptic membrane of denervated frog muscle enclosed within the lumen of the pipet tip. The resolution of this recording configuration allowed these authors to measure the current flowing through the individual channels that were induced to open by the binding of acetylcholine. This is very impressive when one considers that the amplitudes of the currents were on the order of only 1–4 pA ($10^{-12}$ A).

In the initial experiments of Neher and Sakmann (1976) the seal resistance was less than 100 M$\Omega$. One great improvement in the method occurred when it was found that gentle suction of the pipet greatly increased the resistance of the seal between the pipet tip and the cell membrane (Hamill et al., 1981). This method has been termed the "gigohm seal" technique, a name that reflects the magnitude of the resistance of the seal between the glass pipet tip and the membrane. Seal resistances of 10 to 100 G$\Omega$ ($10^9$ $\Omega$) can be readily obtained. With this improvement one could accurately control the potential over a wide voltage range across the small membrane patch within the lumen of the pipet. In addition the mechanical stability of the seal allowed the patch of membrane to be pulled off of the cell and yet remain attached to the pipet tip. This permits the recording of currents under conditions in which both the potential and ionic gradient across the membrane can be controlled. Other recording configurations will be discussed below.

The utility of the patch clamp/single-channel recording technique has not been as yet fully realized. Our understanding of the physiological and pharmacological properties of ionic channels is being expanded daily by use of this method. Some of the uses to which the method has been put are given in Table 1. As can be seen, the list of possibilities is lengthy.

## 2. Fundamental Concepts

This section will deal with a brief discussion of the fundamental concepts, theory, advantages, and disadvantages of patch clamp/single-channel recording methodology.

### 2.1. Theory

If a glass pipet (connected to the appropriate electronics) is sealed tightly against a cell, then any current flowing through the patch of membrane enclosed by the pipet will flow through the electronic circuit attached to the pipet rather than leak out around the area of its attachment to the membrane. In addition, any current applied to the membrane patch will flow through the patch rather than the seal. These represent the fundamental principles behind the development of the patch clamp/single-channel recording technique. The very high seal resistances that can be attained with this method mean that the current flowing across the patch of membrane attached to the pipet can be accurately measured, and the potential gradient can be accurately controlled. Thus, two requirements for voltage clamp experiments are fulfilled. Finally, because of the extremely high resistance of the seal, the current noise of the system is very low. This means that with the advent of the extremely low-noise electronic circuitry now available, the currents flowing through single channels can be resolved. The development of "gigaseal" recording technique has increased the fidelity of recordings and the usefulness of patch clamp methodology to the point that many, if not most, electrophysiologists use or contemplate the use of the technique.

### 2.2. Advantages and Disadvantages

There are a number of advantages to the use of this method. In many ways patch clamp experiments are much easier to perform

Table 1

Uses of the Patch Clamp Technique

| Technique | Reference[a] |
| --- | --- |
| Whole-cell recording | |
|   Membrane potential | Nakayama et al., 1984 |
|   Internal dialysis | |
|   Channel kinetics | Bean, 1985 |
|   Channel selectivity | Hagiwara and Ohmori, 1982 |
|   Second messenger effects on channels | Trautmann and Marty, 1984; Forscher and Oxford, 1985 |
|   Pharmacology of channels | |
|   Transmitter release | Neher and Marty, 1983 |
| Recording from membrane patches | |
|   Control of ionic gradient across the membrane | Dani and Eisenman, 1984 |
|   Channel gating mechanisms (e.g., by neurotransmitters) | Jackson, 1984; Ogden and Colquhoun, 1983; Sine and Steinbach, 1984 |
|   Channel kinetics | Horn et al., 1984; Dionne and Leibowitz, 1982; Aldrich et al., 1983 |
|   Channel selectivity | Dwyer and Farley, 1984 |
|   Pharmacology of channels | Neher and Steinbach, 1978; Quandt and Narahashi, 1982; Patlak and Horn, 1982 |
|   Second messenger effects | Abrams et al., 1984; Greger et al., 1985; Ewald et al., 1985 |
|   Insertion of channels into lipid bilayers | Montal et al., 1984; French et al., 1984 |
| Loose patch clamp | |
|   Distribution of channels in the membrane | Almers et al., 1983 |
| Focal recording from very small nerve fibers | Forda et al., 1982 |

[a]The references given above are meant only as examples of the possible uses of the patch clamp technique and are not the only references. These references may or may not be included in the text.

than those that use microelectrodes or other types of recording configurations. Once an appropriate cell preparation has been found that will allow for the formation of a tight seal of the pipet with the membrane, making the seals and obtaining data is fairly straightforward. Indeed, using chick myotubes we find that gigaseals can be obtained almost 100% of the time. This however does not indicate the rate of successful experiments, since the patch may not contain the channel we are trying to study, a vesicle may form at the pipet tip during the manipulation of the patch, too many different channel types may be present, and/or any one of sundry other problems may arise. The technique also permits the unambiguous determination of channel properties, as well as the effect an experimental manipulation has on these properties. For example, the hypothesis that local anesthetics in low concentrations blocked the response of striated muscle to acetylcholine by shortening the lifetime of the channel rather than by reducing single-channel current amplitude was shown to be true by Neher and Steinbach (1978).

In addition to the ease of use, the conditions to which the membrane is exposed can be accurately controlled. In most recording configurations the ionic gradient across the cell membrane can be determined. This permits the physiology and pharmacology of the channel to be examined at an essentially molecular level and has provided us with a better understanding of the functioning of ionic channels. The development of this method has also provided a means of studying the electrophysiological properties of cells from which it had previously been difficult or impossible to record. In fact, if a cell can be isolated such that there are areas of naked membrane exposed, then it will most likely be possible to record from that cell using this method. For example, it has always been difficult to record membrane properties of smooth muscle cells because of their small size (i.e., 3–20 $\mu$m diameter). It is now possible to isolate these cells from several tissues and perform voltage clamp experiments on the single cells (Benham and Bolton, 1983; Bolton et al., 1985; Kuriyama and Kitamura, 1985). This ability should rapidly improve our understanding of the physiology and pharmacology of this very important group of tissues.

The above are just a few of the advantages of the technique. However, there are some disadvantages or limitations to the method. The first disadvantage has to do with obtaining cells that are suitable for use in this technique. The cells from which you wish to record must not have any overlying connective tissue or cell layer (e.g., glial cells, fibroblasts, and so on). This generally means that the tissue must be enzymatically treated and the cells dispersed. This process obviously disrupts interesting cell-to-cell connections. The possibility also exists that the enzyme treatment of the tissue has deleterious effects on the property that you wish to study. Primary culture of cells dispersed from tissues is a useful technique that permits the effects of enzyme dispersion to be reversed. Even this has its problems in that many cells de-differentiate in culture (e.g., *see* review by Chamley-Campbell et al., 1979). Many investigators use clonal cell lines. This has the great advantage of providing a ready source of cells. However, if you are interested in the physiology of "normal" cells, then this may not be the route you wish to follow. Second, the recording technique can disrupt the normal metabolism and cytoskeleton of the cell and thus may alter the activity of channels. When recording from unattached patches or the whole cell, soluble components necessary for normal metabolic function are lost. Very little is known of the effect changes in metabolic activity or cytoskeletal structure have on the activity of channels. A final difficulty with the technique is that of data acquisition and analysis. The amount of data you must acquire in order to meaningfully analyze single-channel recordings is massive. Hundreds of single-channel events need to be analyzed to generate open and closed time histograms. The use of a computer in the analysis of kinetic properties of channels is almost a necessity. If the channel you are studying has a low frequency of occurrence or declines in frequency with time (i.e., is not stationary), then analysis of open and closed time histograms may be difficult or impossible. Generally, these problems can be overcome or minimized.

## 2.3. Recording Configurations

Obviously the major advantage of the technique is the ability to record currents under voltage clamp from the whole cell or from an isolated membrane patch. The various recording configurations and their uses will be considered below and are illustrated in Fig. 1.

### 2.3.1. Cell-Attached Patch

Initially upon making a gigohm seal the pipet is attached to the cell, isolating a small patch of membrane on the cell surface. Single-channel currents are recorded in this configuration. Since the membrane patch is still connected to the cell, the membrane potential across the patch is determined by the potential at which the pipet interior is being held and the potential of the cell interior. Thus, in order to determine the potential across the patch, the cell's membrane potential must be determined and the pipet potential must be subtracted from it. This configuration can be one of the most stable with regard to the activity of channels within the patch since the cytoskeleton of the cell is relatively undisturbed and the metabolic processes that may be involved in the maintenance of channel activity are still intact. Recently, however, it has been suggested that large vesicles may form at the tip of patch pipets after gigaseal formation. The size of the vesicles suggests that alterations in the cytoskeleton have occurred (Milton and Caldwell, 1986). The configuration has the disadvantage of not permitting the control of the internal ionic content of the cell. In addition, if any treatment of the cell changes the cell's membrane potential, this will be reflected in a change of the potential across the membrane patch. Combinations of a standard two-microelectrode voltage clamp to control membrane potential and the patch clamp have been utilized to eliminate this problem (Neher and Sakmann, 1976; Lux and Brown, 1984).

The uses of this configuration include the recording of single-channel currents under relatively normal physiological conditions (either neurotransmitter or voltage gated). Another use of attached patches is to determine whether a neurotransmitter alters channel activity via a second messenger-mediated process. These possibilities are illustrated in Fig. 2. For example, a second messenger-mediated change in channel activity could be tested by forming an attached patch and then applying a neurohormone to the cell membrane outside of the lumen of the patch pipet. If the neurohormone causes the activation or inactivation of channels by the intracellular release of second messengers, then an increase or decrease in channel activity of single channels within the patch would be expected (*see* Camardo et al., 1983; Reuter, 1983).

One variant of this technique, developed by Almers et al. (1983), has been termed the "loose patch clamp." The method is similar to the patch clamp except that the pipet tip is 5–20 μm in diameter. The large membrane patch is voltage clamped and the currents flowing through the membrane are measured via a pipet loosely pressed against the cell membrane. This technique is useful for mapping the density of channels in the cell membrane. Another interesting use of this type of configuration is to allow for high-

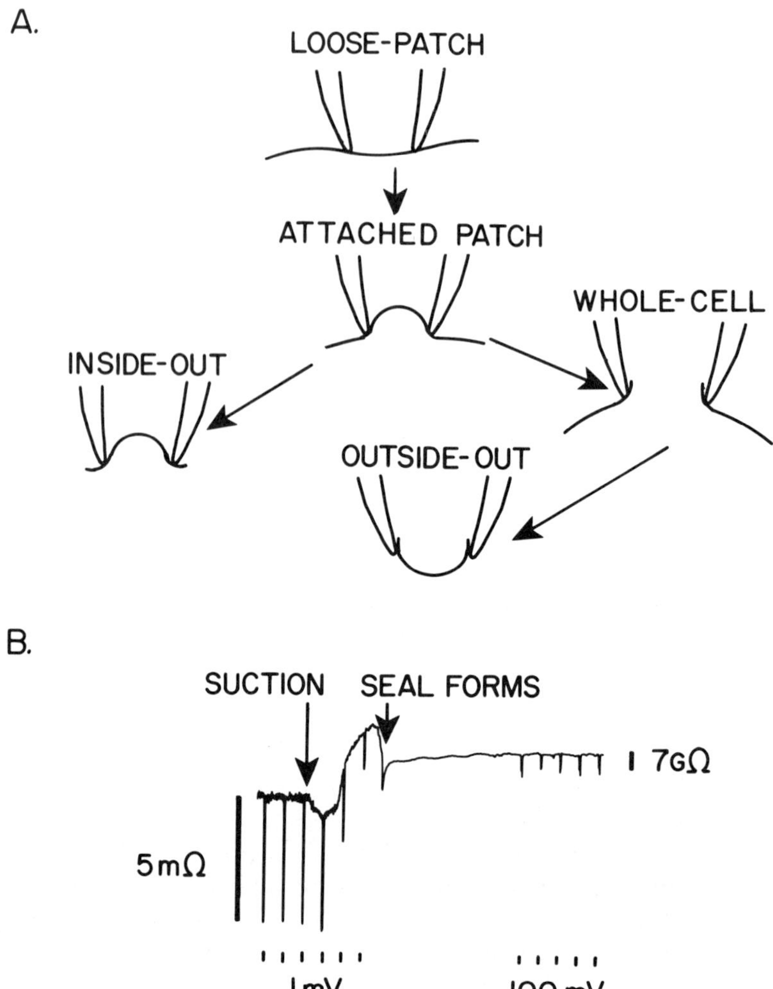



Fig. 1.   Recording configurations possible using the patch clamp technique. (A) The initial contact of the pipet to the cell is low resistance (>20 MΩ). This arrangement is useful in the loose patch clamp and for focal recording. A small suction on the pipet pulls the membrane into the pipet, creating a tight, very-high-resistance seal with the membrane. This is the attached or on-cell patch. Moving the pipet tip away from the cell tears the patch of membrane off the cell, but leaves it attached to the pipet. This is the inside-out configuration. If during the on-cell configuration the suction on the pipet is briefly increased, the patch of membrane within the lumen of the pipet can be disrupted. The interior of the pipet is now in

resolution focal recording from extremely small nerve endings (Forda et al., 1982).

### 2.3.2. Inside-Out Patch

One of the amazing properties of the gigohm seal of the pipet with the membrane is that of mechanical stability. If the pipet is withdrawn from the cell after formation of the attached patch, the membrane remains attached to the pipet tip. Under the appropriate conditions it is possible to obtain a single layer of membrane covering the tip of the pipet. The potential of the patch of membrane attached to the tip of the pipet is now entirely determined by the patch clamp circuitry and is the bath potential minus the pipet potential. In addition this recording configuration has the advantage of allowing for complete control of the ionic gradients across the membrane patch and permits the administration of drugs, enzymes, or other agents directly to the internal membrane surface. However, the membrane patch has its normal cytoskeleton disrupted and is removed from most if not all of the metabolic control mechanisms that might affect channel activity. These two factors should be considered before using this patch configuration. One other problem frequently encountered is that when the patch is pulled off, a membrane vesicle forms at the tip of the pipet. Some techniques to reduce the formation of vesicles will be discussed later. Inside-out patches are useful in a number of ways, for example in the study of ionic channel permeability (by substitution of ions in the intracellular and extracellular solution; Dwyer and Farley, 1984), the effects of the addition of drugs intracellularly on ionic channel properties (Patlak and Horn, 1982), or second messenger-mediated processes (Camardo et al., 1983). Most of the advantages deal with the ease with which the ionic environment bathing the intracellular membrane surface can be changed.

---

communication with the cell interior. This is the whole-cell configuration. If the pipet is slowly withdrawn from the cell, the membrane reseals over the tip with the extracellular surface facing the bath, forming an outside-out patch (from Hamill et al., 1981). (B) A current record showing the formation of a "gigaseal" using chick myotubes. The noisy trace is the current measured with the pipet pressed against the membrane (1 mV pulses; 5 M$\Omega$ resistance). Suction on the pipet (at the arrow) causes a rapid decrease in noise as the seal forms. This patch had a seal resistance of 7 G$\Omega$ (100 mV pulses).

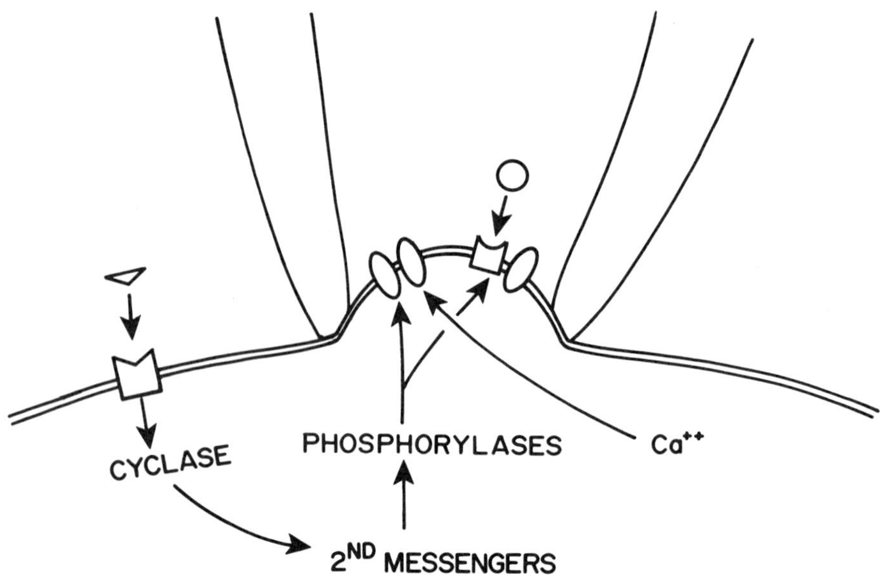



Fig. 2.   Possible uses of the patch clamp technique. Some very interesting uses of the patch clamp technique involve the effects of neurotransmitters or second messengers on ionic channel properties. This schematic representation illustrates possible ways that channels in the membrane isolated within the pipet lumen could be affected by neurotransmitters or intracellular messengers. A neurotransmitter(circle) could be applied to the membrane within the pipet to directly activate the channels. Alternatively, a neurotransmitter (triangle) could be applied to the cell and the effects measured on channel activity within the patch. Presumably these effects would be mediated via intracellular messengers (cAMP, cGMP, $Ca^{2+}$, and so on).

### 2.3.3. Outside-Out Patch

The outside-out patch is made by first forming a cell-attached patch, then rupturing the membrane within the lumen of the pipet. The membrane within the lumen is broken by very briefly increasing the suction on the back of the pipet and thus mechanically disrupting the patch. Another method is to apply a brief large voltage change to the pipet. The membrane within the pipet can be disrupted with little or no effect on the resistance of the seal. This procedure results in the solution within the pipet being contiguous with the cytoplasm. This is the whole-cell patch clamp configuration that we will discuss in the next section. If the pipet is then slowly pulled away from the cell, the membrane is also pulled away from the cell, forming an isthmus that narrows down until it finally pinches off and seals over the tip of the pipet. In this configuration the outside cell surface is facing the bath, and the cytoplasmic surface is bathed by the solution within the pipet. Single-channel currents are recorded in this configuration. The utility of the outside-out patch is the ease with which the external membrane surface can be superfused with drugs, neurotransmitters, or other agents. Vesicles can form in this configuration also, except that now the membrane reseals within the pipet lumen. It has been our observation that the occurrence and stability of channels when recording in this configuration are much lower than in the attached patch. This may just be a function of the tissue (chick myotubes) or the receptors we were studying (acetylcholine receptor/channel), but it is a possible problem that should be considered when designing experiments. This problem may arise because the normal membrane cytoskeleton and metabolic processes are disrupted by this procedure.

### 2.3.4. Whole-Cell Recording

The whole-cell arrangement is obtained as discussed in the last section. The internal environment of the cell is now contiguous with the solution within the pipet. This means that the internal ionic environment of the cell will equilibrate with the pipet solution. Since the volume of the pipet solution will be many times greater than that of the cell, the ionic environment within the cell will approximate that of the pipet. Thus, in the whole cell arrangement the intracellular and extracellular ionic environment can be controlled. Loss of soluble material important in the maintenance of a normal metabolic environment occurs (Fenwick et al., 1982).

The cytoskeleton of the cell may be altered little during whole-cell recording. Therefore if the internal ionic composition is designed to support normal metabolic function, the cell should operate as if it were physiologically normal. Whole-cell recording permits the recording of currents through the entire cell membrane to be made. The data obtained in this procedure are similar to those that have been obtained by other voltage clamp methods (e.g., two-microelectrode, axial wire, sucrose, vaseline gap, and so on). Now it is possible to obtain this data from very small cells. There is a limitation to the size of cells that can be voltage clamped using this method. Some cells are too large. The capacity transient of large cells and the pipet voltage errors induced by large whole cell currents flowing through the series resistance between the pipet and the interior of the cell limit the size of the cell that can be used. In addition, the larger the cell the slower will be the exchange of the cytoplasmic diffusable components with those of the pipet. Cells that are not spherical but elongate should be much less than a space constant in length so that during voltage clamp experiments the membrane voltage can be held constant and uniform across all parts of the cell membrane. Very small cells can be voltage clamped using this recording configuration (e.g., smooth muscle, Benham and Bolton, 1983; chromaffin cells, Fenwick et al., 1982; macrophages, Ypey and Clapham, 1984). This configuration can also be used to record membrane potentials by connecting the patch pipet to a standard high-impedance microelectrode amplifier or to the patch clamp amplifier used in the current clamp mode. Therefore, it is possible to determine the membrane potential of the cell after recording from an attached patch on the same cell. The whole-cell recordings will be the response of a population of channels on the cell to whatever stimulus is provided (i.e., voltage step, neurotransmitter application, and so on). This means that the effects of treatments on the population of channels is more easily determined with this method, although the recording of single-channel currents is also possible (Fenwick et al., 1982). Since the cell is reasonably intact, the effect of internal application of second messengers or enzymes (e.g., catalytic subunit of adenyl cyclase) on ionic channel properties or number can be determined.


## 3. Getting Started

In this section we will consider the basic equipment and accessories needed to begin patch clamping. We will first consider the minimal electronic equipment required, then the other equipment needed to set up. We will then look at the design of the pipet holder and the chamber.

### 3.1. Basic Equipment

The following electronics are necessary for the basic patch clamp setup. First a patch clamp amplifier is required for the measurement of single-channel or whole-cell currents. Some type of oscilloscope is also required to display the signal being amplified by the patch clamp amplifier. Finally there must be some method of recording the data for subsequent analysis. The recording device could be an FM recorder, a computer, or a video recorder interfaced for recording analog data. Initially even a chart recorder could be used. Some other pieces of equipment that are very useful include a pulse generator to supply voltage pulses to the patch clamp amplifier, an electronic filter, and a circuit for analog leakage subtraction. Much of this equipment may be readily available in many electrophysiological laboratories.

Any oscilloscope will suffice to display the current or voltage being recorded during an experiment, although a digital oscilloscope is suitable for this purpose since the display can easily be manipulated to allow for examination of the rising or falling phases of the single-channel current. This permits a close examination of the currents to be made to see if they are rounded, a good indication of the formation of a vesicle or a large series resistance at the tip of the pipet. We use both analog and digital oscilloscopes. The digital oscilloscopes are Nicolet models 3091 or 4094 (Nicolet Instrument Corp., Madison, WI) and the analog oscilloscope is a Tektronix model 5113 (Tektronix, Beaverton, OR).

The patch clamp amplifier decided upon should have several characteristics. Commercially available amplifiers have a very-low-noise headstage with a frequency boost circuit, a circuit to supply the holding potential to the pipet, capacity cancellation capabilities, some type of display for voltage and current, an electronic filter, and probably a circuit to allow for the measurement of membrane potential. Those amplifiers capable of whole cell recording will have a circuit for compensation of series resistance. In addition you may want to be able to perform both single-channel and whole-cell recording with the same circuit. Some amplifiers, such as the model EPC 7 (List Electronics, Darmstadt/Eberstadt, West Germany), have a special headstage that can be switched between whole-cell and single-channel recording, whereas other amplifiers require an exchange of the headstage. The commercially available amplifiers differ primarily in the accessory circuits that are supplied with, or can be added to, the amplifier. For example, the amplifier may have variable frequency cutoff filters, a pulse generator, or leakage subtraction capabilities. It is the accessories or the price that will most likely determine the brand purchased. If the researcher is adventuresome and plans to build a patch clamp amplifier, Sigworth (1983) has analyzed the characteristics of the circuit in detail. We have built two such amplifiers, but now use either an EPC 7 or an EPC 5 (List Electronics, Darmstadt/Eberstadt, West Germany).

In addition to the electronic equipment, some type of table or bench on which to set up the equipment is required. A vibration isolation table is useful for this purpose. We use a Technical Mfg. Corp. model 61 air suspension table to eliminate vibration (Technical Mfg. Corp., Woburn, MA). Enclosing the table with a Faraday cage made of screen wire (aluminum or copper) is also sometimes necessary to decrease electrical interference. A microscope will be required to visualize the cells. Many laboratories use inverted microscopes since most of the preparations that are used are in culture or are dispersed from the whole tissue. This makes visualization from below feasible. The primary advantage of the inverted microscope is that there is a fair amount of space above the chamber for manipulation of the pipet holder and the headstage. We have used several models of inverted microscopes; a Zeiss Invertoscope Model D phase contrast microscope (Carl Zeiss Inc., Thornwood, NY), a Nikon Diaphot (Nikon Inc., Garden City, CA) outfitted with Hoffman modulation contrast optics (Modulation Optics Inc., Greenvale, NY), and a Swift model PF 100 phase contrast microscope (Swift Inst. Inc., San Jose, CA). The microscope, whether inverted or upright, should be outfitted with phase contrast, Hoffman modulation contrast, or Nomarski optics. The increase in contrast is needed so that the cells and the pipet tip can be easily visualized. Stable micromanipulators are necessary to allow fine positioning of the pipet to be done. Any stable, good-quality mechanical, motor-driven or, other type of manipulator could be used. We use Narashige model MO-103 hydraulic micromanipulators (Narashige Scientific Inst. Lab., Tokyo, Japan). These have the advantage of permitting the pipet to be moved without having to touch anything on the vibration-free table. The MO-103 manipulators we have drift slightly, thus on-cell patches are difficult to maintain for any length of time. A chamber that can be mounted on the microscope stage or frame is necessary. The chamber should have a means of temperature control to maintain a constant bath temperature during recording. Finally the equipment necessary for making the patch pipets is required. This equipment consists of a pipet puller, a microscope modified for fire polishing the tips, and a heated coil for curing the coating on the tip of the pipet (if a Sylgard coating is used).

## 3.2. Cell-Isolation Procedures

The formation of gigaseals requires that the cell membrane to which the pipet is sealed be free of all overlying cells and connective tissue. This usually requires that the cells be isolated from the tissue. One source of cells is to develop or purchase a cell line that can then be maintained in cell culture. There are many clonal cell lines available (American Type Culture Collection, Rockville, MD) that may be useful. For example, N1E-115 neuroblastoma cells have been used in the study of sodium channels using the patch clamp technique (Quandt and Narahashi, 1982). Clonal lines can be very useful, and the ease with which they can be maintained in culture makes them attractive as an alternative to using freshly isolated cells. Many experiments may, however, require, or be better performed by, using freshly isolated cells maintained in a physiological state as close to normal as possible.

Mammalian cells invariably have to be treated with enzymes to isolate them from the tissue. When dispersing cells from a primary tissue source, the best procedures will be gentle and reasonably rapid. The cells that are isolated should be calcium-tolerant, which basically means that the cell membrane is intact. For example, cells such as smooth muscle, striated muscle, and heart should be elongate and relaxed. Epithelia should have beating cilia when cilia are present. The cells should not be swollen or have blebbing of the membrane and have a reasonable oxygen consumption. In addition, living cells will be able to attach to the substrate that has been provided, whereas dead or dying cells generally will not.

Protocols have been developed by several investigators that attempt to fulfill these requirements by providing a very gentle means of dissociating tissues. Those investigators examining the properties of smooth muscle and heart have developed some of the gentlest procedures for the isolation of cells (Fay and Singer, 1977; Haworth et al., 1980; Mitra and Morad, 1985). These procedures have been necessary because of the fragile nature of the isolated cells. It would seem logical that the gentler procedure is always to be desired for the isolation of any cell type, in order to minimize the damage to the cells. There are some basic similarities that exist among the various methods. (1) The isolated cells are not exposed to medium containing zero calcium, although the tissue may be exposed to this medium. (2) The tissue and isolated cells are subjected to only gentle agitation or trituration. (3) Centrifugation of the cells, if performed at all, is at low speed. Many enzymes have been utilized in isolation procedures, such as, typsin, collagenase, hyalouronidase, elastase, and protease. Most of the procedures that seem gentlest use collagenase as the major enzyme, alone or in concert with other enzymes (e.g., protease, Dionne and Leibowitz,

1982; Mitra and Morad, 1985; elastase, Bechem and Pott, 1985).
Deoxyribonuclease (DNAase) is sometimes added to the dissocia-
tion medium to hydrolyze DNA released from damaged cells.
DNA is very sticky and may cause cells to clump.

Described below are the isolation procedures used for the
isolation of tracheal smooth muscle cells from adult animals. These
protocols have been modified from those reported by the authors
listed above. There are three major steps to the isolation procedure,
including incubating of the tissue in two different enzyme solutions, then harvesting and plating the cells.

The tissue is removed, minced, and then exposed to protease
(Type XIV, Sigma Chemical Co., St. Louis, MO) at a concentration
of 0.2–1 mg/mL. Any standard physiological medium can be used
to make up this solution, but the solution should be prepared
without calcium since, as noted by Mitra and Morad (1985), protease contains about 15% calcium acetate by weight. Unlike the
report of Mitra and Morad (1985), we do not find that the tissue
must be exposed to zero calcium solution prior to exposure to the
enzymes to obtain good dispersion of cells. In addition, we use two
separate enzyme treatments instead of combining protease with
the other enzymes. It has been our observation that the protease
rapidly digests other enzymes (most noticeably collagenase) that
are included with it in solution. The protease breaks down many of
the cell–cell connections and tends to make the tissue pieces become very soft and flocculent. The minced tissue is placed in 5–10
mL of enzyme solution in a 15-mL, conical-bottom centrifuge tube
that is then incubated at 37°C. The tube is slowly rotated at 1–3 rpm
end-over-end to gently agitate the tissue. The tissue remains in this
solution for 15–45 min, after which the tissue pieces are removed
from the enzyme solution. This is accomplished by allowing the
tissue fragments to settle, low-speed centrifugation, or filtration
through nylon mesh (100–200 μm, Tetko Inc., Elmsford, NY). The
tissue pieces are then placed in 5–10 mL of a physiological solution
containing approximately 0.2 m$M$ calcium and collagenase (Type
1, 1–2 mg/mL), elastase (10–20 units per mL), and DNAase (Type II,
0.01%). The calcium is necessary for collagenase activity, and inhibits the residual tryptic activity that is present in most collagenase preparations. This solution should not contain magnesium, which would inhibit the collagenase activity. The tube containing the enzymes and tissue is slowly rotated, as mentioned
above. The cells are then harvested from this solution at 30-min
intervals for 1–1.5 h. The tissue pieces are removed from the

enzyme solution by allowing them to settle or by low-speed centrifugation (<100$g$). The cells that have been released are removed from the enzyme solution by low-speed centrifugation (100–200$g$ for 2–5 min), and washed with Dulbeccos minimum essential medium. The centrifugation procedure is then repeated. The loose cell pellet at the bottom of the tube is broken apart by very gentle trituration using a 10-mL pipet. The trituration is performed as few times as needed to disperse the cells. The cells obtained during each collection period can be pooled and are now ready for plating.

For more information on the isolation of cell types, there are several books available on these procedures and on cell culture techniques (Nelson and Leiberman, 1981; Jakoby and Pastan, 1979; Paul, 1975; Freshney, 1983).

We have isolated tracheal smooth muscle, gland, and epithelial cells using this procedure or a slight modification of it. Once the cells are isolated they should be allowed to attach to some substrate at least loosely. We allow our cells to attach to collagen-coated plastic coverslips or to cleaned glass. Cultured cells will require some type of matrix on which to grow. The tissue culture books mentioned earlier will provide information on the best type of matrix for many cell types. Collagen is probably the most common substrate used for this purpose. Although soluble collagen is available and can be made from rat tails, we find that the most convenient and inexpensive way to make a collagen solution is from freeze-dried collagen (either rat tail, human placental, or calf skin from Sigma Chem. Co., Saint Louis, MO). The freeze-dried collagen is dissolved in glacial acetic acid (1:1000 diluted with Hank's balanced salt solution) by stirring in the cold for several hours. The final concentration of the collagen is 1 mg/mL. This solution should be stored in the refrigerator, not frozen. This solution can be sterile filtered. Coverslips are coated by placing a very small amount, about 8 $\mu$L, onto a 22-mm square coverslip and then spreading the collagen into a very thin layer using a glass rod bent at a 90° angle at the tip. The coverslips are placed in Petri dishes and allowed to air dry. If desired, the coverslips can be sterilized by UV irradiation under a germicidal lamp. We use 22-mm square plastic coverslips for most of our culture work. These are convenient to use since they are easy to cut into any desired shape using scissors. This may be important if the chamber has a limitation on the size or shape of the coverslips or there are limitations in the number of cells that may be used. We routinely use only one coverslip of cells each day. Plastic coverslips cannot, however, be used with Nomarski optics.

The cells can be carefully pipeted onto the coated coverslips and allowed to incubate at 37°C for 1–4 h. This is generally enough time for most mammalian cells to attach loosely to the coverslip. Cells will also sometimes attach rapidly to acid-cleaned glass (0.1–1N HCl for 1–2 h) or to glass that has been coated with poly-*l*-lysine (Fisher, 1975). For whole-cell recording studies, the researcher may want the cells to only lightly attach so that they can be pulled free of the substrate and raised near the surface of the solution. This procedure will reduce the capacitance of the pipet with the bath.

Not all cells must be isolated from the tissue in which they reside. Two preparations that have recently been used in patch clamp studies are isolated kidney tubules (Gogelein and Greger, 1984; Hunter et al., 1984) and brain slices (Gray and Johnston, 1985). Both preparations were treated with enzymes, but the cells were not disrupted from the tissue matrix. These successes raise the possibility of using the patch clamp/single-channel recording technique in cells with reasonably intact cell-to-cell contacts.

### 3.3. The Chamber

The chamber used to maintain the cells during patch clamp studies can be as simple or as complex as you like. For example, recording may be done from cells in the culture dishes in which they are grown. A chamber should have certain features. (1) A half cell must be connected to the chamber and to the patch clamp circuit. (2) It is very helpful if some means of changing the bath solution is provided. (3) It is best that the chamber be designed to permit accurate control of the bath temperature. (4) Finally, if the cells are to be viewed from the bottom, the chamber should have a glass bottom to allow for an undistorted view of the cells. Obviously the chamber must fit on the microscope stage. Also, it is useful to enclose the chamber in a solid aluminum or brass sheet that covers the entire microscope stage. This will help prevent overflows of the chamber from damaging your microscope (suggestion of Dr. Clay Armstrong).

The chamber should ideally be constructed from a thermally conductive material that is nonporous and has a high electrical resistance. Acrylic and other types of plastics have been used in the past for chambers, but are on the whole extremely poor thermal conductors. Corning (Corning Glass Works, Corning, NY) has developed a machinable glass ceramic called Macor®. This ceramic is handled like hard brass as far as machining goes and has a much higher thermal conductivity than plastics, almost zero porosity, and a very high dielectric constant. The chamber illustrated in Fig. 3 is made from this ceramic (Dr. G. Oxford suggested the use of Macor®). A solid-state temperature control module (thermoelectric module 801-2003-01, Cambion Thermionic Corp., Cambridge, MA) is in direct contact with the ceramic block. This module has a very high cooling/heating capacity (~22 W). The thermoelectric module can be connected to a temperature control circuit or it can be powered directly (as we do) from a variable voltage dc power supply. We use an Acopian (Acopian Corp., Easton, PA) power supply for this purpose. The chamber in Fig. 3 has only one thermoelectric module, but another could be added if additional heating/cooling capacity were desired. The other surface of the module is in contact with an aluminum block. The block and the aluminum plate surrounding the chamber form a heat sink for the thermoelectric module. Water is not circulated through the heat sink in this design, but this could be easily incorporated by drilling a U-shaped tube in the aluminum heat sink and circulating water through it.

The ceramic block in which the bath is incorporated has horizontal holes drilled through it that are directly under the thermoelectric module. This is a means of preheating/cooling the solution perfusing the bath. The bath consists of a slot milled through the ceramic with stepped sides on the bottom. The steps in the sides provide a place to glue the glass coverslips that will form the double bottom of the chamber. The coverslips have the good optical qualities necessary when using an inverted microscope (especially if differential interference contrast optics are used). Dry air should be trapped between the coverslips to keep condensation from occurring. The air forms a good insulation layer to help maintain the bath at a uniform temperature. The distance between the coverslips should be kept small when using an inverted microscope, since microscope objectives generally have working distances of about 2 mm (at 20×). There are some extremely long working distance objectives available that have working distances of 6–10 mm at 20–40×.

The ceramic block is surrounded by an insulating layer of acrylic to which it is glued. The acrylic is surrounded by and glued to the aluminum heat sink as well. This whole assemblage forms a cover for the top of the microscope stage and is connected to the

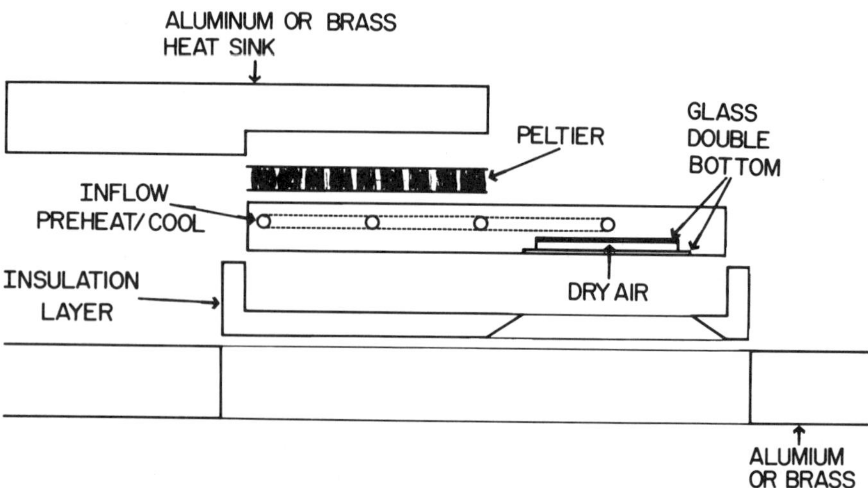


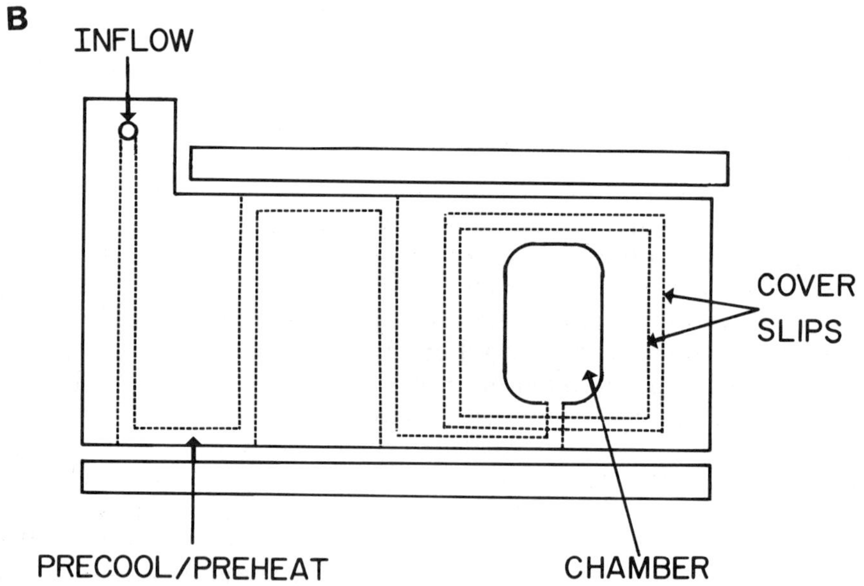


Fig. 3. An exploded view of a chamber. (A) Side view. The temperature-controlled chamber is made of a machinable ceramic. The slot milled into this block is the actual chamber into which the cells will be placed. The bottom of the chamber is made from two glass coverslips glued to steps milled in the ceramic. Between the coverslips (0.080 in., ~1 mm) is an insulation barrier of dry air. The inflow into the chamber is via

stage manipulator so that the chamber can be positioned over the objective. If the stage on the microscope does not have a lock-down mechanism, it may be useful to install a locking screw to keep the stage from drifting. This is especially necessary if recording from attached patches.

Chabala et al. (1985) have designed a chamber that has features similar to the one described above. In addition they give a circuit for making a digital thermometer/temperature control unit. This circuit design is similar to that of Caldwell (1977). Another chamber design is described by Kakei and Noma (1984).

Changing the bath solution can be performed by continuously perfusing the bath and removing the solution by suction. We do not use this method because of the increased electrical noise that it seems to generate. Instead, a push/pull syringe arrangement was designed that permits the rapid exchange of bath solution to be made with little or no increase in noise during solution changes. This device is shown in Fig. 4. The syringes are mounted pointing in opposite directions side by side in a frame. Disposable or glass syringes (6 or 10 mL) can be used. One syringe is for perfusing the chamber, and the other is to remove solution from the chamber. Both syringes are fitted with three-way polypropylene valves (Drummond Scientific Co., Broomall, Pa; #590-92) like those found on repeating pipetors. The valves permit the syringes to be filled or emptied without having to manually switch them. The syringe for

---

holes (~0.060 in.) drilled through the ceramic underneath the thermoelectric device. The solution is precooled as it flows through this S-shaped pathway. The thin ceramic strips (1/8 in.) on each side of the block are glued on the sides to cover slots that form connections between the holes for precooling (shown in the top view in B). The outflow from the chamber occurs by suction via a bent glass capillary tube (not shown). A small hole (not shown) is drilled at an angle into the chamber to permit insertion of a small thermocouple for temperature measurement. The ceramic block is surrounded with at least 1/8 in. of plastic insulator. The Peltier module used for temperature control is 1/8 in. thick and is in direct contact with the ceramic below and an aluminum heat sink above. The aluminum heat sink is coupled by 4–40 screws (not shown) to, and is in direct contact with, the aluminum plate, which surrounds the insulator and in which the chamber is imbedded. The aluminum plate surrounding the chamber forms a large heat sink for the Peltier device.

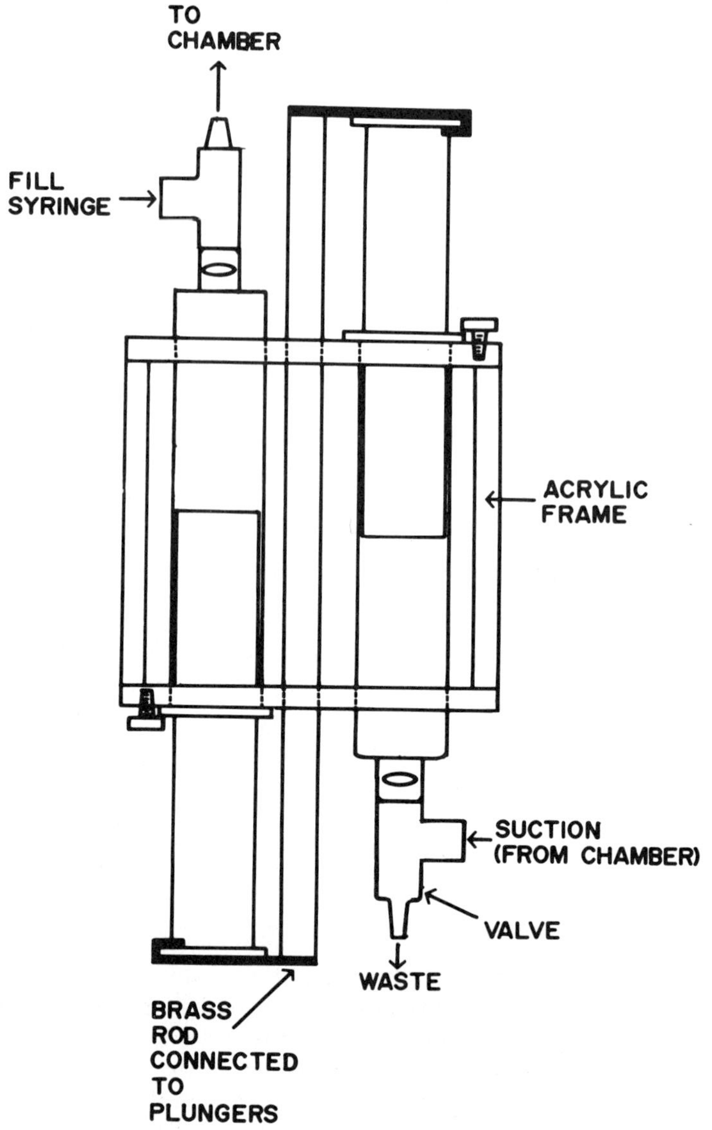


Fig. 4.    Push–pull syringe. The push–pull syringe for changing bath solutions consists of two 6- or 10-mL plastic or glass syringes mounted in a box frame made of 3/8-in. acrylic plastic. The syringes are secured by 8–32 nylon screws. A 1/4-in. brass rod is inserted through the frame between the syringes and couples the plungers together. The luer-lock tips of the syringes are connected to plastic valves, such as those on a repeating pipetor. The valve ports are connected to the perfusion bottles and bath such that as one syringe is emptied to perfuse the bath, the other syringe removes solution from the bath. Conversely, when the syringe that perfused the bath is filled, the syringe that removed the bath solution empties into a waste reservoir.

perfusing the bath should be connected to a reservoir from which it can be filled. In this arrangement, when one syringe plunger is depressed, the other plunger is pulled out the same amount. If the syringes are reasonably well matched in volume, then the amount of solution put into the chamber and the amount withdrawn will be equal and the bath level will remain constant. Since the entire system is filled with solution, the tip of the tube for removing the solution should be at the bottom of the bath. This prevents air from entering the system.

A method of rapidly changing the solution bathing the membrane was described by Yellen (1982). In this method several inflow tubes are mounted side by side in the chamber. Three different solutions are introduced simultaneously into the chamber through these tubes. The pipet tip is put directly in the inflow stream coming from one of the tubes. Switching solutions is achieved by moving the pipet tip to the solution issuing from an adjacent tube. Another method for the rapid change of solutions is the microflow system developed by Krishtal and Pidoplichko (1980).

## 4. The Patch Pipet

### 4.1. Glass and Shape

The pipet can be fabricated from flint, borosilicate, aluminasilicate, or other types of glass tubing. Soft glasses such as flint glass tubing can be pulled at relatively low heat and are easy to fire polish. These are also reported to seal with membranes easily. However, soft glass does not have as great a dielectric constant as harder glass, which means it must be coated or the baseline noise of the system will be much greater than with the harder glasses. Harder glasses, such as borosilicate or aluminasilicate, have lower noise properties. The sealing properties of the glass with the cell will have to be determined for each cell type used. However, we have found that Corning glass 8161 and Kimax 52 glass will seal with the cells we have tried, such as chick myotubes, tracheal epithelium, and vascular and tracheal smooth muscle. Kimax 52 glass is very inexpensive and readily available as melting point tubing from most major laboratory suppliers. Corning 8161 glass is available from specialty glass suppliers such as Garner Glass Co. (Claremont, CA), who will make glass tubing to your specifications. Corning 8161 glass is reported to be one of the best glasses for forming seals (Rae, 1985). This has been our observation as well.

Both glass types can be pulled to provide very blunt pipets. The diameter of the glass tubing used will depend in part on the requirements of your pipet holder. The glass best suited for your experiment can be determined only when considering cell type, the type of pipet holder, and the experiment to be performed (particularly whether whole-cell or single-channel recordings are to be made). A much more complete discussion of the properties of various glasses is given by Corey and Stevens (1983) and Rae and Levis (1984).

The shape of the pipet tip is important to consider. The primary characteristics of the tip that need to be controlled are the diameter of the orifice at the tip of the pipet, the length of the shank, and the wall thickness. The wall thickness at the tip will be controlled in part by the wall thickness of the glass tubing with which you start. Thicker walls at the tip of the pipet result in a lower-noise pipet and often better stability of the patch (Sakmann and Neher, 1983). The diameter of the tip and the length of the shank are controlled during the pulling process. The size of the tip will normally be in the range of 0.5–5 $\mu$m in diameter. The diameter of the tip used depends on the experiment being performed. Larger tip diameter pipets are generally used for whole cell recording since larger tip diameters provide for a lower series or access resistance to the interior of the cell. This improves the fidelity of the recording and increases the rate of dialysis of the intracellular fluid. Smaller tip diameters are used for single-channel recording. When the density of ionic channels in the membrane is very high, the smallest tip diameters will be needed to reduce the number of channels within the membrane patch at the tip of the pipet. We generally have seal resistances in chick myotubes of 10–40 G$\Omega$ using pipets made of Kimax 52 glass with tips around 1–2 $\mu$m in diameter. The length of the shank is not a great consideration for single-channel recording unless the researcher will be trying to perfuse the tip of the pipet. Shorter shanked "bullet-shaped" pipets are more suitable for whole-cell recording to reduce the series resistance and allow for rapid exchange of the pipet solution.

## 4.2. Fabrication

Patch pipets can be pulled using several standard microelectrode pullers. The one that we use is the Kopf model 700C (David Kopt Inst., Tajunga, CA) vertical puller. This puller is commonly found in many laboratories. Several manufacturers make pullers designed specifically to pull patch pipets. Since I am most familiar with the Kopf 700C vertical puller, I will describe in detail the method for pulling patch pipets using it. The glass tubing we use is 1 mm diameter Kimax 52 melting point tubing (Kimble Glass Co., Toledo, OH). We have modified the puller by removing the rod that trips the fast pull and by shielding the coil from drafts with a plastic cover. A patch pipet is fabricated using the two-stage pulling process described by Hamill et al. (1981). The glass is clamped into the puller and the heat setting on the puller is adjusted for a relatively high heat (on the Kopf this is about 20 A). The pipet begins to elongate when heated, and the pull is stopped after the pipet has been lengthened by about 6 mm. The amount of narrowing of the glass during the first pull is very important, since this will have great effects on tip size and shank length and will be a determinant of the heat required during the second pull. In order to keep the length of the first pull consistent, it is stopped mechanically. We use a very simple stop made from 1/4-in. thick acrylic, as illustrated in Fig. 5A. Once the first pull is completed, the heat should be turned off. The rod that turns off the heater coil can be set so as to trip the heater coil microswitch once the first pull is completed. The upper clamp on the pipet is then loosened and the pipet is raised such that the thinned portion is now in the center of the coil. In order to make the second positioning consistent, a ramp is incorporated into the top of the Lucite stop (Fig. 5A). When the stop is pushed under the lower clamp, it rides up the ramp to the second level. The upper clamp is tightened and the pipet is ready to be pulled again. The second pull is made at a much lower heat setting than the first (usually around 14.5 A on the Kopf). In order to increase the ease with which the second heat can be consistently set, we have the pipet puller plugged into a switch box that allows it to be switched from line voltage to a variable voltage transformer. The second lower heat setting is obtained by switching the puller to be powered from the transformer, which is set to between 60 and 70 V. A simple schematic illustrating this is shown in Fig. 5B. This simple method allows for reduction of the heat of the coil without having to adjust the heat settings on the puller. Adjustments in coil heat are made by varying the input voltage to the puller. This method of adjusting the second heat setting will not work and should not be used on pullers that have circuits that actively control the heat of the coil. After the second pull has been made, the pipet is removed from the upper clamp. The lower pipet can be used, but has a much more elongated shank and thus may have the wrong characteristics for a particular experiment.

**A**

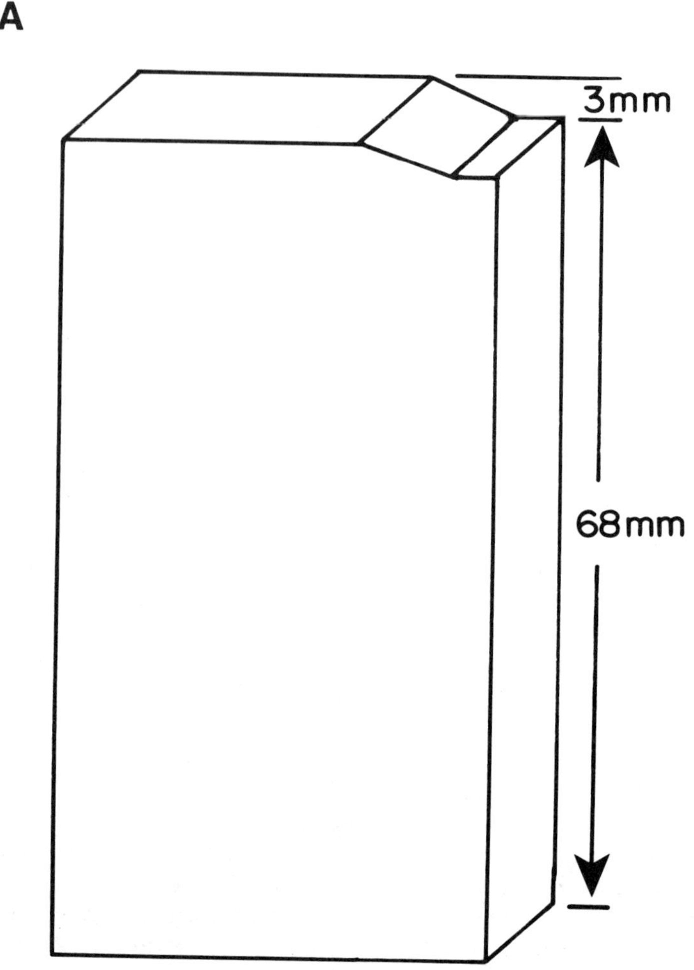



After the pipets are pulled they should routinely be coated with Sylgard® 182 (Dow Corning Corp., Midland, MI), polystyrene Q-dope (GC Electronics, Rockford, IL), or other hydrophobic substance (such as varnish). Sylgard® is one of the most common coatings used. The coating reduces the capacitance of the pipet and keeps solution from climbing up the pipet. Recordings will have less drift and lower noise with coated pipets. We have used primarily polystyrene Q-dope, since we find it is easier to use. To coat the tips only requires painting the tip of the pipet with the dope using a fine brush (an old nail polish brush works well). The pipets are then allowed to air dry. If the tips are inadvertently covered

**B**

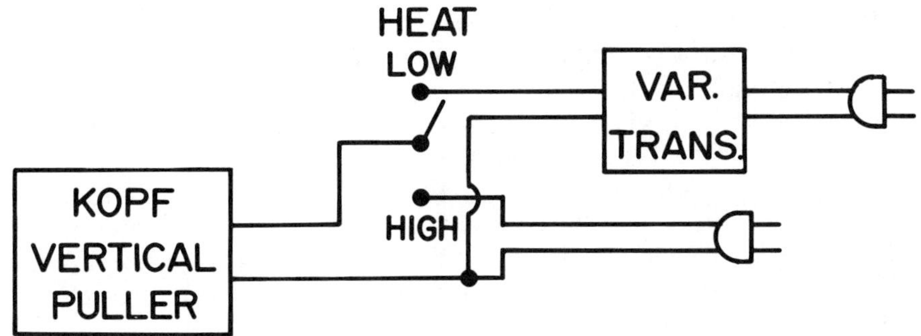



Fig. 5.    (A) A simple device made from 1/4- or 3/8-in. acrylic plastic to provide a stop for the first-stage pull of the electrode. The ramp permits easy, consistent repositioning of the glass pipet for the second-stage pull. (B) A diagram of a simple method to provide the high and low heat settings during pipet fabrication using the Kopf 800 C puller. The puller is plugged into a box with a switch wired as shown. The puller can be switched between line voltage for high heat and the reduced power supplied by a variable transformer for low heat.

with Q-Dope, they can still be used since during fire polishing the coating will be burned off. In Mississippi the humidity is very high, so after the pipets are pulled we must place them (either coated or uncoated) in a micropipet storage jar that contains a drying agent such as silica gel. Pipets stored in this way have been successfully used weeks after they were pulled. If the Q-dope is kept relatively thick, you can put a heavy coat on the pipets that will further improve the noise characteristics of the pipet. The Q-dope should be allowed to dry completely before the pipet is used. A Sylgard® coating has been shown to provide the best noise characteristics (Corey and Stevens, 1983). Sylgard® is wound around the shank of the pipet down to near the tip. Once the coating has been completed, the Sylgard® is cured by gently heating the pipet tip. Additional coats may be applied to increase the thickness of the coating. Curing the Sylgard® may be performed using the pipet puller heating coil on a very low heat setting or, more conveniently, by making a separate electrically heated coil (Corey and Stevens, 1983).

We fire polish the tip of the pipet under an inexpensive, upright microscope, the stage of which has been modified to hold an electrically heated wire and a small manipulator to move the pipet. The microscope does not have to be fancy. We use a Bausch and Lomb student microscope with 4, 10, and 45× objectives and 20× oculars (Bausch and Lomb, Rochester, NY). The working distance of the 45× objective is about 0.6 mm. The wire that we use for fire polishing the tips is a 100-μm platinum-iridium (90–10%) wire coated with glass. The glass coating is made by inserting the wire into a fine piece of glass tubing and heating the wire and melting the glass or, alternatively, by using a glass pipet and "painting" the heated wire with glass. A glass coating is required to keep the lens used for fire polishing from being platinum coated. The wire is linear and stretched between two electrical connectors separated by an insulator of Teflon®. This is illustrated in Fig. 6A. The insulator block is then connected to the movable microscope stage to allow for the positioning of the wire underneath the lens. In order to be able to heat this wire, it is attached to a filament transformer that is in turn connected to a variable transformer, as illustrated in Fig. 6B. We have mounted a small Narashige model C1 manipulator directly on the microscope to position the pipet. The pipet is laid in a small groove in a block attached to the manipulator. The pipet should be angled upward slightly so that the lens, when focused, does not hit the shank of the pipet. Fire polishing of the pipet is performed by moving the pipet tip into the microscope field (magnification 600–900×) near the wire. The wire is then heated by briefly closing the connection to the filament transformer. The wire should be heated so that it is a dull red, just hot enough to fire polish the tip. If the wire is too hot, the fire polishing will proceed at a rate too rapid to control. The heating of the wire will produce a change in the diffraction pattern of the light produced by the glass at the pipet tip, generally after 1–2 s. This is sometimes observed as a darkening at the extreme tip of the pipet. The change in appearance is the signal to stop heating and remove the pipet tip from the vicinity of the wire. It is desirable that only the tip of the pipet be fire polished such that the shank be narrowed as little as possible. Controlling the heat and duration of fire polishing will generally take care of this problem. However, if there is a problem, heating can be limited to the extreme tip of the pipet by blowing a stream of air along the tip. Fire polishing is performed just before the pipet is to be filled and used.

The pipette can now be filled with solution. The solutions should be filtered at 0.2 μm using a syringe fitted with a filter unit.

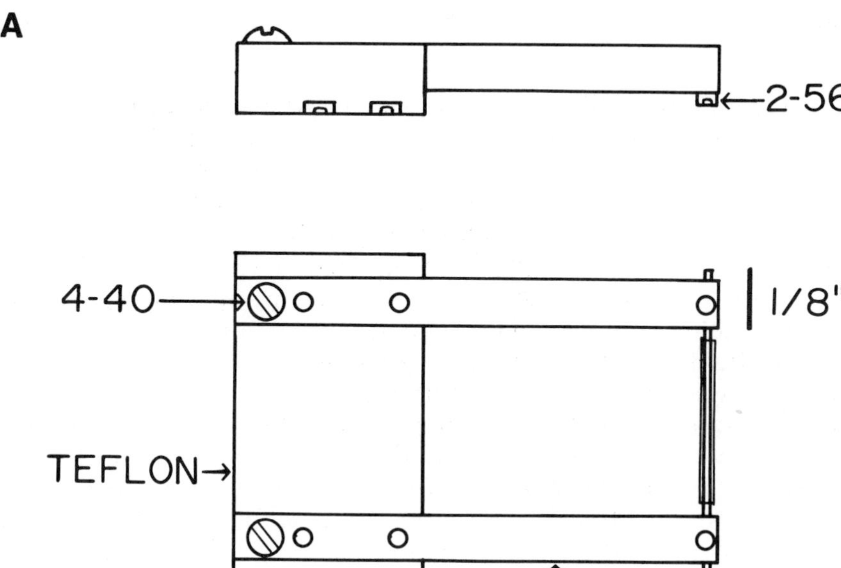


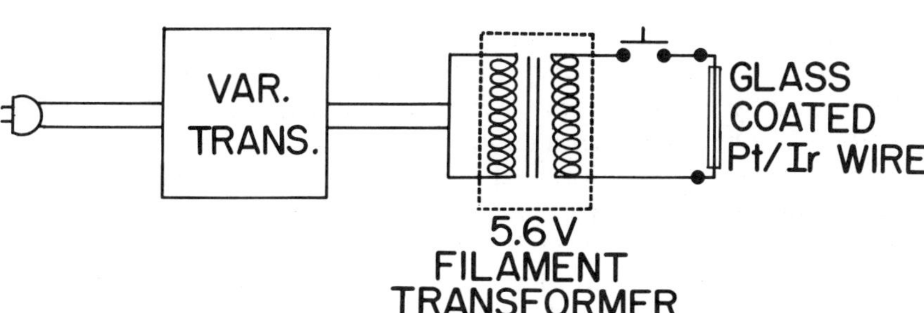


Fig. 6. (A) A drawing of the wire assembly used for fire polishing. The platinum wire is tightly held between two brass rods by 2-56 screws. The brass rods are separated by a block of Teflon®, Delrin®, or other appropriate insulation material. The wire is held straight in this design, since this reduces the movement of the wire as it is heated. The 4-40 screws are used to make electrical connections to the platinum wire. The block can be attached to the stage manipulator or mounted on a slide that is held by the manipulator. (B) The wire is electrically heated by connecting it to a filament transformer (4–6 V), which is in turn connected to a variable transformer. The push button switch is of the momentary-contact type and is normally open.

A small volume of filtered solution is placed in small beaker or vial. The tip is filled with this solution by suction. A 6- or 12-mL syringe is used to produce the suction. The syringe is fitted with a tubing adaptor for syringes (Clay Adams, Div. of Becton, Dickson and Co., Parsippany, NJ), chosen such that the pipet will fit into the hole in the end of it. A small O-ring, 1/16 in. (Small Parts, Miami, FL), is included in the adapter that will seal around the pipet when the adapter is tightened. The male threaded portion of the adapter should be flattened to provide a good surface for compression of the O-ring. With the pipet in place the tip can be dipped into filtered solution and the syringe plunger pulled back. This fills the tip and the shank. The pipet can then be backfilled to the desired level. The syringe used for backfilling the pipet can be the same as the one used for filtering solution, but now fitted with a 30-gage needle on the output of the filter unit. If solutions being used should not come into contact with stainless steel (e.g., chelation reactions with EGTA, Corey and Stevens, 1983), a special syringe can be made for filling by heating the body of a 1-cc disposable plastic syringe and pulling it out to a very fine tube. The pipet is now ready to be inserted into the holder.

### 4.3. Pipet Holder

In buying or building a pipet holder, consideration should be given to characteristics required for your particular application. The characteristics you may wish to take into account are (1) the ability to exchange the pipet solution, (2) the capacity to change the anion in the pipet solution without large changes in voltage offsets, and (3) the electrical noise that arises because of the pipet holder. In addition, a decision must be made as to whether an "inline," straight design, or an angled design is to be used. Some features should be standard on all holders. The holder should plug directly into the headstage. Also, a port for connecting a suction line to the holder needs to be provided, as well as an airtight coupling of the glass pipet to the holder. The coupling should permit the pipet to be changed easily, but still be firmly held. Finally, a half-cell must be provided for measurement of voltage and currents.

We use the 90° angled holder (shown in Fig. 7) in which the pipet solution is electrically coupled to the silver/silver chloride half-cell via a salt bridge. Since the connection to the half-cell is via a salt bridge, large changes in the ionic composition of the solution can be made (especially in chloride concentration) without large

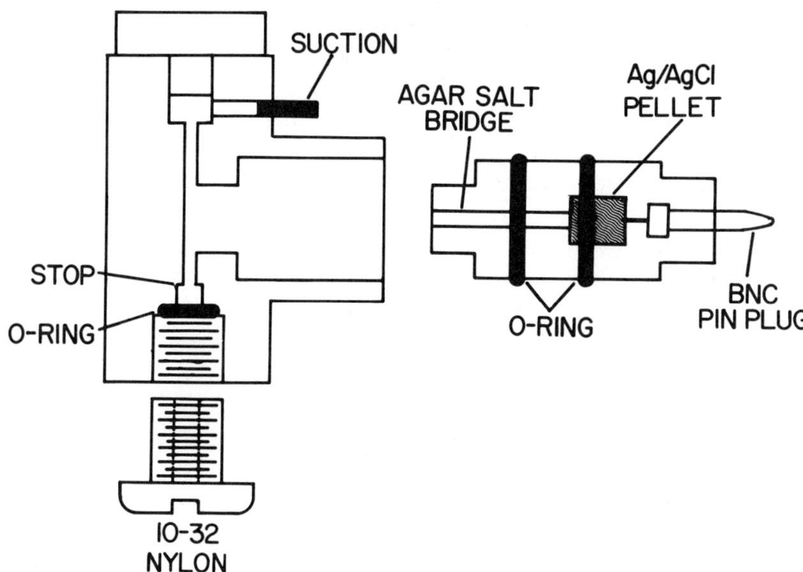


Fig. 7. Pipet holders. This pipet holder consists of two main parts, the body, made of acrylic plastic, and the side arm, cast from a plastic casting resin. The body is made from 3/8 in. sheet plastic and has a threaded hole to accept a 10-32 nylon screw. The nylon screw has a centered hole drilled through it that is only slightly larger than the glass tubing to be used for the pipets. Tubing up to about 2 mm in diameter can be used. The pipet is inserted through the screw and O-ring up to the stop. When the screw is tightened it compresses a 3/16-in. od O-ring. The stop provides mechanical stability during suction. The top cap on the body is removable to permit backfilling of the pipet and cleaning of the body. The side arm (0.25 in. diameter) has a Ag/AgCl pellet connected to a gold-plated pin plug. The pellet is in contact with a 5% agar salt bridge. The side arm O-rings are 1/4 in. od and 1/8 in. id and are set in grooves 0.030 in. deep. The O-rings are compressed upon insertion of the side arm into the body of the holder (0.265 in. diameter hole). The fit should be tight to prevent any movement of the pipet tip during experiments. The holder plugs directly into the BNC connector of the head stage. The BNC connector has been fitted with an insert (not shown) made of Teflon®, which will form a tight stable connection with the side arm.

shifts in the half-cell potential. The removable side arm contains the silver/silver chloride pellet, which is in contact with a 5% agar bridge. The agar is equilibrated with 0.1–1$M$ KCl or other chloride-containing solution. The side arm is made from a polyester casting resin molded around the pellet. The grooves for the O-rings are cut into the side-arm. The side arm is made removable so that the bath and pipet holder half-cells can be shorted together by storing them with their salt bridges connected by adding solutions of 0.1–1*M* KCl and connecting the pin plugs. This procedure equalizes the offsets between the half-cells to within a few microvolts after 24 h. There is a positive stop for the pipet in the body of the holder to prevent movement of the pipet during suction. By tightening the screw, an O-ring is compressed and provides an airtight seal around the pipet. We have made the body of the holder from acrylic plastic so that we could see any bubbles in the solution. However, it would be better from the standpoint of noise to make the body from Teflon®, Delrin®, or Kel-F®. Kel-F® has the added advantage of being translucent. Both the pipet and the holder contain solution in this design. Thus, one disadvantage of the design is the increase in noise (when compared with the amplifier alone) that will occur because of the added capacitance of the holder. The holder is plugged directly into the BNC connector found on most patch clamp headstages. An insert made of Teflon® and thoroughly cleaned by boiling in acetone is inserted in the BNC connector. This insert should provide a tight and mechanically stable fit for the pipet holder.

A pipet holder with lower current noise than the one described above is designed to eliminate the salt bridge in the side arm. This holder has a small 700–750 μm diameter silver/silver chloride pellet that can be inserted directly into the pipet down to the shank. The pellet is connected to a male BNC connector pin press fit into a Delrin® plug that takes the place of the side arm. The pipet is filled only to the level of the pellet and no further. This design has the advantage that the capacitance of the pipet holder and pipet is reduced to that at the tip of the pipet. Great care should be taken to keep the inside of this holder dry or the current noise will increase. The body of the holder is made of Delrin®. If desired, chlorided silver wire can be used in place of the pellet. Hamill et al. (1981) provide the design for an inline pipet holder with features similar to this one.

Another type of pipet holder permits perfusion of the tip of the pipet. Methods for exchanging the solution at the tip of the electrode have been described by Cull-Candy et al. (1981), Soejima and Noma (1984), and Kaneko and Tachibana (1985). These holders are generally designed for use in whole cell recording. The basic features include a very small inflow tube that extends to as near the tip as possible and an outflow tube that extends part way down the pipet. Small inflow tubes can be made from polyethylene tubing or a piece of coating stripped from a Teflon®-coated wire (idea from Dr. Clay Armstrong's laboratory). A coated wire of 60–100 μm will yield a tube slightly smaller in diameter because of stretching. The tube can be glued to a small stainless (30-gage) Teflon® or polyethylene tube with a cyanoacrylate glue. The half-cell electrode in this system consists of a small Teflon®-coated silver wire that has been chlorided at the tip, where the coating has been stripped away. The wire from which the Teflon tubing was obtained may be used for this purpose. The wire is inserted along with the inflow tube through the outflow tube made of stainless steel. The inflow tube can be connected to the perfusion solution and the outflow tube to a trap for collecting the waste solution. The inflow and outflow tubing should be kept as short as practical.

## 4.4. Pellets

We prefer to make electrical connections to the pipet and bath solutions by using silver/silver chloride half-cells in the form of pellets rather than chloride silver wire. The advantages of pellets are that: (1) they are stable, showing little drift in offset voltages during an experiment, (2) they require a minimum of care (i.e., they do not have to be kept shielded from light or rechlorided and are mechanically stable), and (3) they have a low electrical resistance. The pellets are made according to the method of Martin et al. (1970). Pellets can also be purchased from companies such as WPI (W-P Instruments, New Haven, CT).

We make our own pellets to suit our needs. The pellets are made from a fine mixture of silver powder (1 part) and silver chloride (2 parts) compressed around a silver wire. The pellet in the side arm of the pipet holder in Fig. 7A is about 1/8 in. in length and 1/8 in. in diameter. Pellets of this type have been in use and have remained stable for years. The pellet press for making this type of pellet is shown in Fig. 8A. This particular press is made entirely from stainless steel. A much smaller diameter pellet is needed if it is to be inserted into the glass pipet. The press illustrated in Fig. 8B was made from aluminum (clamp and piston) and brass (press) and can be used to make pellets of about 750 μm in diameter.

To make a pellet using a press you first clamp the halves (or quarters) of the body of the press together. An appropriate length of cleaned silver wire is then centered in the cylinder of the press. A

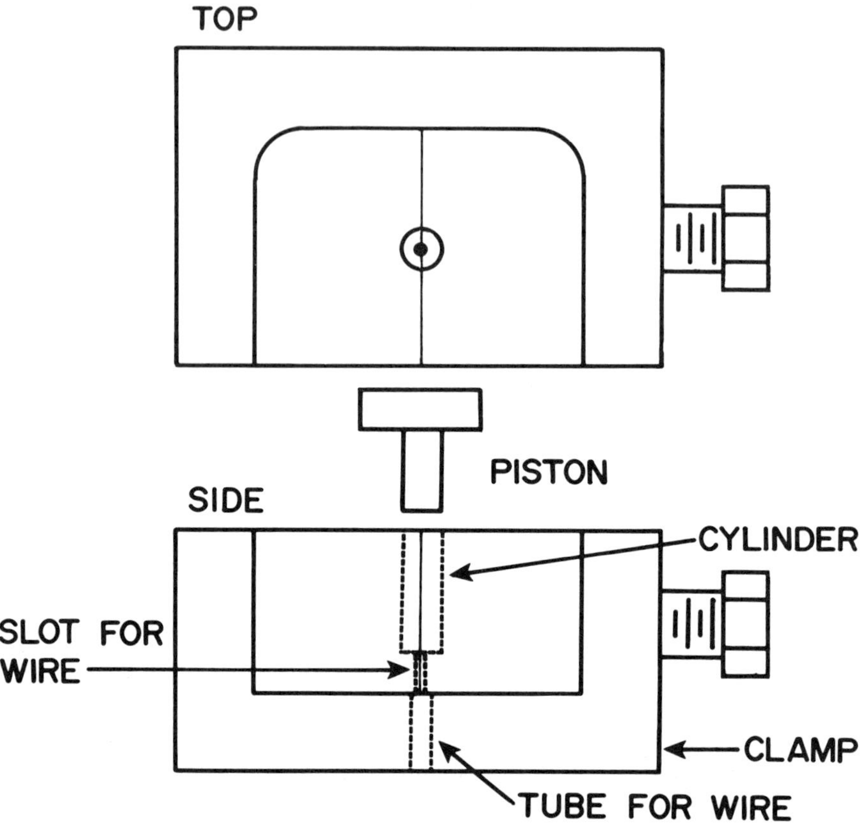



Fig. 8.  Pellet presses for Ag/AgCl half cells. (Top panel) Schematic representation of a press for producing 1/8-in. diameter pellets. The press is made of stainless steel and consists of four parts. A clamp holds the two halves of the press together by tightening the bolt. The two halves of the body of the press have a 1/8-in. diameter hole between them when clamped together. The hole has a very small slot at the bottom for centering the silver wire in the cylinder. Finally a piston inserts tightly into the cylinder and stops approximately 1/8 in. from the bottom of the cylinder. (Bottom panel) This press can be used to make 750 $\mu$m diameter pellets (0.030 in.) approximately 6 mm in length. The press is shown cut in half in the side view. There is a clamp (two, 3/8-in. bolts) that holds the four quarters of the body of the press together. The body of the press has concentric cylinders of two diameters centered between the four quarters. The piston has two corresponding diameters, the largest (0.185 in., 4.6

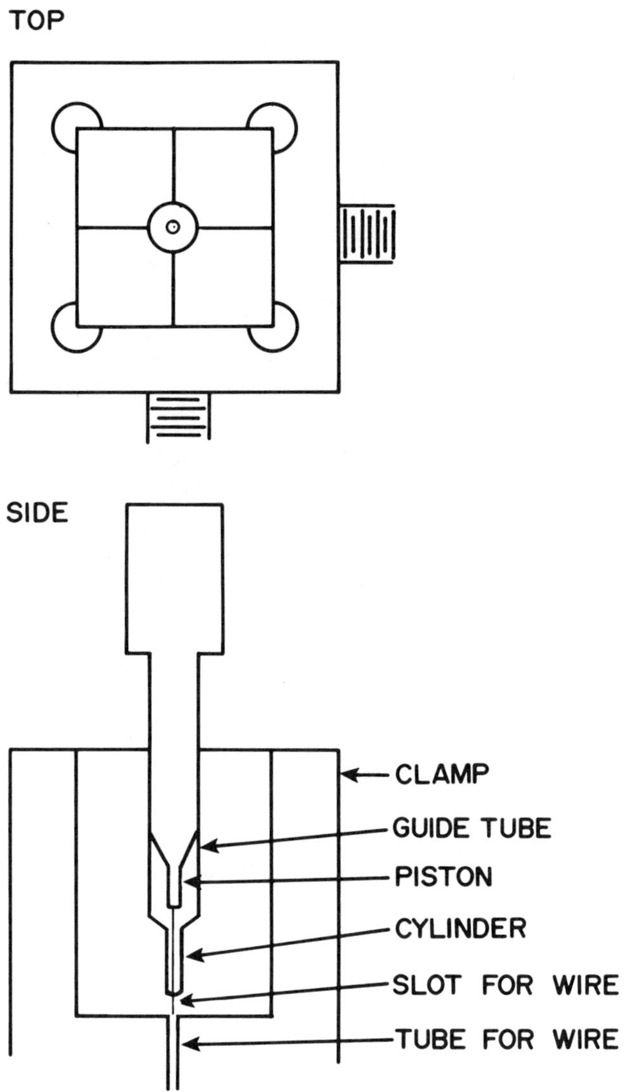



mm) for centering a support and the smaller (0.030 in., 0.75 mm) to provide compression when the pellet is made. The small piston is approximately 0.045 in. (~1 mm) in length. A small notch is provided at the bottom of the smallest cylinder for the silver wire to pass through. This notch can be made by filing down a corner of one of the quarters of the body. The silver wire extends into and is centered by a small hole drilled in the clamp block.

small slot is provided at the bottom of the press through which the wire is inserted. The diameter of the wire is not critical, but obviously the wire around which the small pellet will be formed should have a diameter of much less than 750 μm (e.g., 50 μm). The wire for the larger pellets should have a small hook or loop at the tip to improve the mechanical stability of the pellet. The silver/silver chloride mixture is then poured into the cylinder and the mixture compressed gently by hand. This step is repeated until it is difficult to compress the powder. The piston is then struck with a hammer until the head of the piston rests on the top of the cylinder block. This final compression tightly packs the powder around the wire. The cylinder block and piston are then removed from the clamp. The larger pellet is removed by rolling it out of the cylinder using a straight-edged piece of metal, a spatula for example. The small pellets generally fall out of the press once the cylinder blocks are removed from the clamp. The pellet is now ready to be trimmed and mounted. The pellets are mechanically stable and can be drilled into if needed. Since they will be in contact with the bath or pipet solution, it is important to electrically isolate the solder connection to the silver wire. We have found that the pellets can be imbedded in a polyester casting resin (the kind found in hobby shops), which provides a good, stable isolation of this solder connection. The resin is poured around the pellet mounted in a Teflon® mold. Once the resin has hardened (by baking at ~140°F), the casting is removed from the mold and a hole is drilled through the plastic and into the pellet. The hole can then be filled with agar to form a salt bridge. A picture of the mold and a description of the casting procedure is provided by Dwyer (1985).


## 5.  Solutions

One of the challenges in the study of ionic channels has always been to isolate the current flowing across the membrane to one ion channel type without altering the characteristics of the channel being studied. The use of the patch clamp technique has not changed this problem. The difficulty has been compounded in a way by the single-channel recording technique, in that the activity of low-density channels can now be observed. Thus, because of the plethora of channels that are now known to exist, the success of an experiment in large part depends on the ability to design solutions that eliminate the currents flowing through unwanted channels.

The isolation procedure can be performed in one of two ways: (1) by careful design of the ionic composition of the solution and (2) by the addition of pharmacologic agents.

## 5.1. Ionic Composition

The primary method of isolating a single type of current uses the selectivity of channels for passing specific types of ions. This property allows one to design solutions in which the major ion will only pass through one channel type. An example of such a procedure is the one that we have used to study nicotinic acetylcholine-activated ionic channels (Dwyer and Farley, 1984). The buffered extracellular solution contains (in m$M$): Cs$^+$, 120; Cl$^-$, 124; and Ba$^{2+}$, 2. The intracellular solution contains (in m$M$): Cs$^+$, 120 and Cl$^-$, 120. These experiments were performed using an inside-out patch. The logic behind the design of the solutions in the study of this channel is as follows. Cesium does not go through sodium channels (Hille, 1972) and it blocks many potassium channels (Hagiwara et al., 1976). Barium is used to block calcium-activated potassium channels (Meech, 1978) and to provide a divalent cation extracellularly. Barium will go through calcium channels that have a small conductance and are labile, so they generally are not a problem during these experiments. Thus, these channels are eliminated from the record, whereas the acetylcholine-activated channel, which is permeant to cesium, can be observed. One other type of channel that is observed when utilizing this solution is the chloride channel present in muscle (Blatz and Magleby, 1985). Replacing the intracellular chloride with glutamate or gluconate helps to eliminate current through this channel since these anions are less permeant than chloride. In addition, since the channel that we were studying was activated by acetylcholine, it was simple to show that the occurrence of the channel depended on the presence of this neurotransmitter.

Table 2 gives a list of general types of channels and the solutions that may be useful for their isolation. The acetylcholine-activated ionic channel would be in the category of nonselective cation channels. Concentrations are not given in the table since these will vary depending on the preparation used and experiment to be performed. When several cations or anions are given for a single type of channel, the ions may be equally useful depending on your experimental protocol in the study of that channel. Although it is not shown in the table, we use mannitol to maintain

Table 2

Solution Components for Isolation of Currents[a]

| Channel | | Na$^+$ | K$^+$ | Cs$^+$ | Tris$^{+b}$ NMG$^+$ | Ca$^{2+}$ Ba$^{2+c}$ | Cl$^{-d}$ | F$^{-e}$ | Gluc$^f$ Glut Asp | MgATP$^g$ CPK, CP GTP | EGTA$^h$ |
|---|---|---|---|---|---|---|---|---|---|---|---|
| | | | | | | Cations | | Anions | | Other | |
| Na$^+$ | Ext | ✓ | | | | ✓ | ✓ | | | | |
| | Int | ✓ | | | | | | ✓ | ✓ | | ✓ |
| K$^+$ | Ext | | ✓ | | | ✓ | ✓ | | | | |
| | Int | | ✓ | | | | | ✓ | ✓ | | ✓ |
| Nonspecifi-cation | Ext | | | ✓ | | ✓ | ✓ | | | | |
| | Int | | | ✓ | | | | ✓ | ✓ | | ✓ |
| Ca$^{2+}$ | Ext | ✓ | | ✓ | ✓ | ✓ | ✓ | | | | |
| | Int | ✓ | | ✓ | ✓ | | | ✓ | ✓ | ✓ | ✓ |
| Cl$^-$ | Ext | ✓ | | ✓ | ✓ | ✓ | ✓ | | | | |
| | Int | ✓ | | ✓ | ✓ | | ✓ | | | | ✓ |

[a]The general composition of extracellular (Ext) and intracellular (Int) solutions are given in this table.
[b]Used as poorly permeant cation substitutes—Tris-(hydroxymethyl)-aminomethane, $N$-methylglucosamine.
[c]Used to block calcium-activated potassium channels and as a permeant ion for calcium channels.
[d]Cl$^-$ (4–10 m$M$) may need to be added internally to all solutions for Ag/AgCl half cell if a salt bridge is not used.
[e]F$^-$ is a metabolic poison and may affect channel activity.
[f]Gluconate, glutamate, aspartate are poorly permeant anions.
[g]Added to internal solution to stabilize calcium channel activity. Creatinine phosphokinase (CPK), creatinine phosphate (CP).
[h]Added to internal solution to chelate calcium, may help prevent vesicle formation.

the osmotic pressure of solutions if we dilute the major permeant ion. Ideally, if the permeant ion is diluted, it is desirable to replace it with a compound of the same charge, but that is inert. Unfortunately, at physiological concentrations many "inert" ions become channel-blocking agents (e.g., *see* Dwyer and Farley, 1984).

One other consideration in designing solutions is to provide components that maintain a constant activity of the ionic channel you are interested in studying. Most analyses of the kinetic properties of the channel require a stationary system; that is, one that has properties that are stable with time. Also, in any study of the action of a drug on a channel it is important to know that any changes in channel frequency or properties are caused by the action of the drug and not instability in the preparation. An example of non-stationary channel activity is shown in Fig. 9, for the acetylcholine-activated channel of chick myotube. Note that there are three active channels at the beginning of the recording, but with time the frequency of occurrence of the channels declines, presumably because of desensitization. Calcium currents are notoriously difficult to maintain whether they are recorded from excised patches or during whole-cell recording. The decline in channel number with time is apparently caused by the loss of some soluble intracellular component (Kostyuk et al., 1981; Fenwick et al., 1982). It has been shown that perfusion of the cell with solutions containing MgATP and EGTA (Kostyuk et al., 1981; Forscher and Oxford, 1984) and creatinine phosphokinase plus creatinine phosphate (Forscher and Oxford, 1985) reduces or abolishes the decline in calcium channel activity normally observed. Apparently the calcium channel requires an active metabolic process to maintain its excitability. Addition of the high-energy phosphates and creatinine phosphokinase seems to maintain the appropriate metabolic components necessary for channel activity.

If one is attempting to characterize an unknown type of channel, probably the first thing to do is determine the selectivity of the channel. This is performed by changing the concentration gradient of the major permeant ion across the membrane. This is generally done by using mixtures of ion types. For example, one might examine the selectivity of the sodium channel by its ability to pass ions such as sodium, potassium, cesium, rubidium, and small organic ions (Hodgkin and Huxley, 1952; Hille, 1971, 1972). Selectivity is measured relative to a particular ion (e.g., sodium) by the shifts caused in the reversal potential upon dilution or replace-

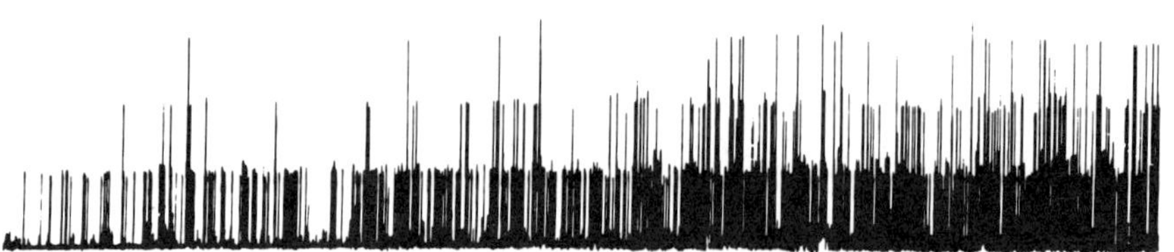

Fig. 9.   Nonstationary acetylcholine-activated ionic channel activity. This record was obtained from an inside-out patch of chick myotube membrane. The patch was bathed in 120 m$M$ cesium chloride on both sides and both solutions were buffered to pH 7.2 with 5 m$M$ Hepes buffer. The pipet or external solution also contained 2 m$M$ barium and 100 n$M$ acetylcholine chloride. The membrane potential is –100 mV and the temperature is 10°C. Note that channel activity decreases with time. The vertical and horizontal scales are 4 pA and 30 s, respectively.

ment of the major ion with the test cation. For further information on the determination of selectivity, see Hille (1984). This procedure still may not be adequate to identify a channel any further than to say it is a sodium, potassium, or chloride channel. The actual identification of a very specific channel type is performed by determining (1) the conductance and ion selectivity of the channel, (2) the kinetic properties and voltage sensitivity of the channel, (3) the pharmacology of the channel, and (4) the channel's regulation by neurotransmitters. These are not necessarily listed in the order of their importance.

## 5.2. Pharmacology

Pharmacological agents that are useful in the elimination or identification of ionic channels can be classed into three broad categories: those that block channel activity, those that activate channels, and those that modulate channel activity. When using pharmacological agents it is very important to know the specificity of those agents for the type of channel you wish to eliminate or study, with consideration given to the channels that exist within the cell preparation you are using. We will first consider the blocking agents.

Probably the most specific ion channel blocking agent is tetrodotoxin. It is known to block only the voltage-gated sodium channel of nerve and muscle (Narahashi et al., 1964). At concentrations up to 1 $\mu M$ it does not appear to have effects on other channels. Another example of a very specific blocking agent is that of $\alpha$-bungarotoxin, which blocks the receptor to which acetylcholine binds to activate the nicotinic cholinergic receptors of striated muscle (Lee, 1970). Calcium channels can be eliminated by use of divalent cations such as $Co^{2+}$, $Ni^{2+}$, and so on or the organic calcium channel antagonists verapamil or nifedipine (Fleckenstein, 1977). The organic agents may have actions on sodium channels as well (Yatani and Brown, 1985). Calcium-activated potassium channels can be blocked by replacing external calcium with barium, as already noted. Many potassium channels can be eliminated by replacing potassium with cesium or even sodium (Hille, 1973; Fukushima, 1982), although other agents such as tetraethylammonium, 4-diaminopyridine, or 2,4-diaminopyridine are commonly used (Yeh et al., 1976; Kirsch and Narahashi, 1978). A review of many agents that may be useful for channel identification or as probes of ionic channels has been written by Narahashi (1974) and covered by Hille (1984).

Another method of identifying a specific type of channel is to activate that channel by a drug or neurotransmitter. As mentioned previously, we activate the nicotinic receptors in chick myotubes with acetylcholine. Channels that may be identified in this way include channels coupled directly to neurotransmitter receptors. Channels not directly coupled to the neurotransmitter receptors may have their properties indirectly modified via transmitter-activated second messenger processes. For example identification of calcium-activated potassium channels is performed by examining the sensitivity of the currents to intracellular calcium concentration (Meech and Standen, 1975). The use of cyclic AMP- and cyclic GMP-mediated processes to identify the properties of channels is only beginning, but in the future will no doubt provide other criteria for channel classification. Certainly this avenue for the modulation of channel activity is known to be important for several channels (Ca channels, Reuter, 1983; Ca-dependent K-channels, de Peyer et al., 1982 and Ewald et al., 1985; K-channel, Camardo et al., 1983; other channels, Green and Gillette, 1983).

An additional important criterion in the identification of specific channels is the voltage sensitivity of channel kinetics and conductance. The use of this property goes well beyond the intent of this chapter. The reader is referred to two books that deal in detail with this subject (Hille, 1984; Adelman, 1971).

## 6. Starting the Experiment

The cells should now be in the chamber. Once the pipet is filled and in the pipet holder, the holder is plugged into the headstage. At this point, if this has not already been done, a cell should be found. The tip of the pipet is lowered into the bath very near the cell from which recording will take place.

### 6.1. Voltage Offsets

When the tip of the pipet enters the bath, the noise of the output from the patch clamp amplifier will greatly increase and the amplifier output voltage may saturate (*see* Fig. 1B). The offset is caused by the currents flowing in response to potential differences at solution interfaces and between the half cells. It can be balanced by adjusting the offset control on the patch clamp amplifier until the current monitor reads zero (a meter on the EPC 5 and 7 amplifiers). Once the offset has been nulled, it is best to leave this control alone. Some drift in the baseline is normal at this time. Many amplifiers incorporate a sample-and-hold circuit to provide a means of keeping the headstage balanced while making a patch. Voltage offsets can be minimized by shorting the bath and the pipet half-cell electrodes together when not in use. A very small voltage pulse (0.01–1 mV) applied to the pipet can be used to determine the pipet resistance. The current amplitude obtained in response to the voltage pulse divided into the test voltage amplitude is a measure of the pipet resistance.

### 6.2. Making the Gigohm Seal

The tip is now lowered until it gently touches the cell. This should be visible as an increase in the pipet resistance of 2–3 times (e.g., a 1/2–2/3 decrease in the amplitude of the current recorded in response to the test pulse). Gentle suction on the pipet will increase the resistance being measured, and the noise of the current output should decrease (this is illustrated in Fig. 1B at the arrow). Suction is performed by mouth via a small light polyethylene tube attached to the pipet holder. This tube should be securely fastened at several places (e.g., the headstage, the manipulator, and the table top) to prevent any movement of the tube from reaching the pipet. During the suction a gigohm seal may form, usually rather abruptly. Occasionally gigaseals will form spontaneously with no suction, although this does not occur often using our preparations. Some cells (tissue-cultured vascular smooth muscle and tracheal epithelium) seem to require a longer time and more suction to form a seal than do chick myotubes when using Kimax 52 tubing. We find that after attaching smooth muscle and epithelium to the pipet tip with suction, alternately increasing and decreasing the pressure in the pipet sometimes results in the formation of a high-resistance seal. Another method that is sometimes useful is to attach the cell to the pipet and then let it sit at atmospheric pressure for a minute or so. Gentle suction after this rest period sometimes results in the formation of a seal.

The resistance of the seal can be determined by applying a small voltage pulse to the pipet (10–100 mV). The seal resistance should be determined in the absence of any channel activity. Seal resistances of greater than 10 G$\Omega$ should be obtained. The leakage and capacity currents can be balanced now if an attached patch is to be used. Leakage currents can be eliminated from the membrane current by subtracting the scaled pipet voltage from the current. A simple circuit to provide this capability is shown in Fig. 10. Ampli-

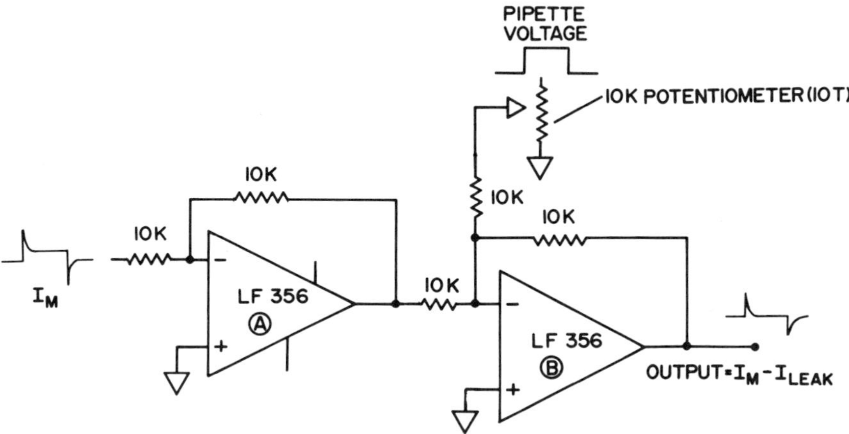


Fig. 10. Leakage subtraction circuit. This is a simple circuit to provide a means of analog subtraction of leakage current. The circuit subtracts the inverted membrane current from the scaled pipet voltage. It is assumed that the current monitor output should not be inverted and that the pipet voltage output and current signal output are of the same sign. If the pipet voltage change and current signal are opposite in sign, then amplifier A is unnecessary and the current monitor output can be directly connected to the 10-K resistor on amplifier B. The amplifiers used are LF 356 (National Semiconductor, Santa Clara, CA).

fier A (an inverting amplifier) may or may not be needed depending on the polarity of the current monitor output from the patch clamp amplifier. The leakage current is subtracted by adjusting the potentiometer until the dc level of the current trace does not shift in response to a voltage pulse. Only the capacity currents flowing at the beginning and end of the pulse should be evident after leakage subtraction (illustrated schematically in Fig. 10).

The capacity currents flowing at the beginning and end of the pulse should also be eliminated by adjusting the capacity cancellation controls on the patch clamp amplifier. The manual for a particular amplifier should be consulted for details. Once the transients have been neutralized, the bath should be maintained at a relatively constant level. The reason for this is that the capacitance of the pipet to ground is a function of the surface area of the pipet exposed to the solution. Thus, if it is practical, lowering the bath level or raising the pipet tip near the solution surface (for detached membrane patches or during whole-cell recording) before adjustments are made can greatly decrease the capacity transient. In addition, in the whole-cell recording configuration the series resistance should be compensated. The exact procedure for this can be found in the manual for a particular amplifier. Series resistance is the electrical resistance between the pipet lumen and the cell interior. Any current that flows across the membrane must also flow through this resistance. The resulting voltage drop across the series resistance causes the pipet voltage to be different from the cell's actual potential. The larger the currents, the greater the need for series resistance compensation. A thorough discussion of the theory behind capacity transient cancellation and series resistance compensation is given by Sigworth (1983) and Auerbach and Sachs (1984). These procedures should be performed after the final patch configuration and tip position have been attained.

### 6.2.1. Inside-Out Patch

An inside-out patch is formed by briskly pulling the pipet tip away from the cell. Both extracellular (pipet) and intracellular (bath) solutions are controlled in this configuration. The membrane potential across the patch is controlled by the holding potential and voltage pulses supplied to the patch clamp amplifier. The membrane potential is equal to the negative of the pipet potential if the bath is at ground potential. The output will be equal to the negative of the membrane current if the amplifier's current monitor output is noninverting. Generally the output voltage of the current monitor is adjustable in the range of 10–1000 mV/pA. The actual gain setting used will depend on the size of the currents being recorded and the requirements of the other components of the researcher's setup. For example, one would not want to record 12 pA currents at a gain of 1000 mV/pA since this would require the patch amplifier to output 12 V.

### 6.2.2. Outside-Out Patch

The formation of the outside-out patch configuration is initiated by rupturing the patch of membrane isolated by the pipet tip after the gigaseal is formed. The pipet should contain intracellular solution. One method of rupturing the membrane is to apply a very brief suction to the pipet. A second method is to apply a very brief, large voltage change to the pipet (<1 ms, 300–400 mV) or, if all else fails, to turn the patch clamp amplifier off and on again. Once the membrane is ruptured, the pipet is slowly withdrawn vertically from the cell. After formation, the resistance of the patch should be measured and the leakage and capacity currents cancelled. In addition, careful checks should be made throughout the experiment to determine whether or not a vesicle has formed. The membrane potential is equal to the pipet potential if the bath is maintained at ground potential. The output of the patch clamp amplifier in this configuration is directly proportional to the membrane current if the amplifier output is noninverting.

### 6.2.3. Whole-Cell Recording

During the formation of an outside-out patch, the whole–cell configuration is passed through. Either the membrane potential of the cell or the current flowing across the cell membrane can be recorded. Membrane potential is recorded in the current clamp mode available on most patch clamp amplifiers or by attaching the pipet to a standard high-impedance microelectrode amplifier. The whole cell currents recorded under voltage clamp conditions can be in the nanoampere range for larger cells. This means that the headstage of the patch amplifier must have a lower gain (e.g., changed from one with a 10–100 GΩ feedback registor to one with a 100–500 MΩ feedback resistor). Depending on the amplifier available, it may be necessary to change the headstage or switch the headstage to a lower gain (as with the EPC-7) in order to obtain the lower sensitivity of the headstage.

There are several things that must be taken into consideration when performing the whole-cell voltage clamp. First, the cell must be reasonably small to (1) have a cell capacitance within the range of the transient cancellation circuitry (a maximum of around 100 pF) and (2) keep the peak whole-cell currents within the range that the amplifier is capable of measuring (25–50 nA usually). Second, in this recording configuration it is also important to minimize the effects of the series resistance of the pipet tip. The effects of series resistance can be reduced in several ways. The tip of the pipet should be made larger than tips for single-channel recording (3–5 μm). The increased tip diameter will also increase the rate of dialysis and decrease the probability that the membrane will reseal over the tip. Series resistance compensation circuitry is built into those patch clamp amplifiers capable of whole-cell recording. It may be possible to compensate for over 90% of the series resistance. The importance of series resistance compensation can be seen in a very simple example. If the series resistance during an experiment is 3 MΩ, then a membrane current of 1 nA will result in a 3-mV voltage drop across the tip and a 10-nA current will result in a 30-mV voltage drop at the tip. This means that the pipet voltage and the actual cell voltage will differ by this much. Compensating for 90% of the series resistance means that the 10-nA current will result in only a 3-mV error in the cell potential. Clearly, smaller currents or series resistance compensation will reduce the error in the cell potential and permit a more accurate determination of the time course and voltage dependence of channel properties.

### 6.2.4. Vesicle Formation

A problem that will be faced in using excised patches or in whole-cell recording is the formation of a vesicle or reformation of the membrane across the pipet tip. The formation of vesicles can be a particularly vexing problem under certain conditions and with some cell types. Several procedures given below may be of some use to reduce the incidence of vesicle formation. Most of the procedures given below are for inside-out patches and have been tried on chick myotubes, which seem to have a penchant (in our hands at least) for forming vesicles.

We prefer to prevent the formation of vesicles if possible. Inside-out patches are formed by a rapid lateral motion of the tip of the pipet away from the cell. Our observation has been that pulling the pipet away slowly and vertically increases the chance of forming a vesicle, particularly in chick myotubes. Also we find that excessive movement of the tip after detaching the patch enhances the chances of vesicle formation. Movement of the solution (such as occurs during solution changes) does not seem to greatly increase the incidence of vesicles. One of the most important ways to reduce the formation of vesicles is to perform the experiments at a low bath temperature. A temperature of 10°C greatly decreases the incidence of vesicle formation when compared with higher temperatures (e.g., 24°C), where vesicles form almost every time. Removal of calcium from the internal solution by EGTA, fluoride, or other calcium chelating agents seems to be important in preventing the membrane from resealing over the tip. Horn and Patlak (1980) suggest that fluoride may be particularly well suited to prevent vesicle formation. Another method that appears to prevent the formation of vesicles is to raise the ionic strength of the internal solution. Raising the internal (bath solution for inside-out patches) cation concentration from 120 m$M$ to 240 or 360 m$M$ prevents vesicle formation. Conversely, with 40–60 m$M$ cation concentrations internally, vesicles form very frequently (>90% of the time). If a low ionic strength solution is used intracellularly, then balancing the osmotic pressure of the pipet and bath solutions with mannitol will decrease the incidence of vesicle formation. Mannitol only needs to be added to increase the intracellular solution osmolarity. The extracellular solutions may have a low osmolarity without increasing the incidence of vesicles (T. Dwyer, in press). Vesicles form when using small diameter pipets more regularly than with larger tipped pipets, so adjusting the size of the tip or its shape may reduce the formation of vesicles. If a vesicle does form, raising the tip of the pipet briefly out of the solution can rupture the vesicle (Hamill et al., 1981). The time required out of the solution to disrupt the vesicle will be 1–2 s. This procedure is particularly useful for steepening the rise time of currents (*see* Fig. 11C).

## 7. Data Collection and Storage

### 7.1. Filtering Data

The amplitudes of single-channel currents are very small, and even with the lowest-noise headstage circuitry it is necessary to filter the output of the patch clamp amplifier to improve the signal-to-noise ratio so that the currents can be seen.

Most patch clamp amplifiers have some type of integral filter circuit. However, a secondary circuit is sometimes preferable. This

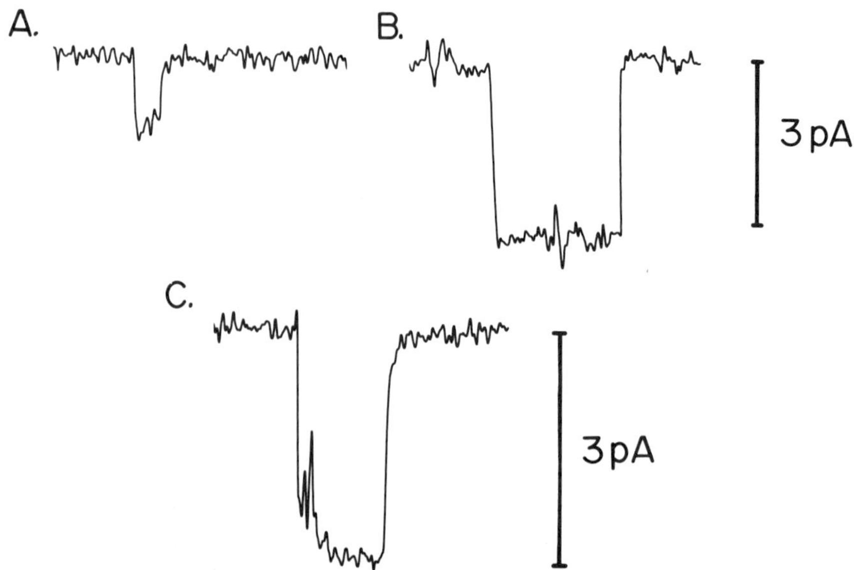



Fig. 11. Vesicle formation. (A) A typical current trace observed after a vesicle has formed. The acetylcholine-activated current rises rapidly but decays slowly. This current was obtained from a patch bathed on both sides with 120 mM CsCl and held at –80 mV. (B) The same patch as in (A) after it was briefly pulled out of the bath. The current shown now has steep rising and falling edges and a conductance of 42 pS. (C) Another indication of vesicle formation is the rounded rising and falling phase of the currents. The current amplitude will sometimes be very near to that in the absence of the vesicle as shown here. However, the outward currents may be very small or nonexistent. This current was obtained at –70 mV.

should be a Bessel filter with a frequency response of dc to a variable cut-off frequency. The roll-off characteristics of this filter should be reasonably steep, 4–8 dB per decade. We do not filter our records before storage below the bandwidth of the FM (frequency modulation) tape recorder. This means that the currents can be replayed and refiltered later at a lower bandwidth for analysis. Capacity cancellation should be performed prior to filtering, near the full bandwidth of the amplifier, since capacity currents are distorted by filtering. We use an Ithaco model 4302 dual filter (Ithaco Filter Corp., Ithaca, NY). If the data is stored in a computer, it can be filtered digitally as well. Brown et al. (1984) have described the implementation of a zero-phase lag digital filter that may be useful.

## 7.2. Storage

A mass storage device is a necessity when measuring single-channel current characteristics because of the large number of events that must be accumulated in order to determine accurately the kinetic properties of the channel. There are several manufacturers of FM tape recorders starting in price at about $5000. We have used a Vetter Model B recorder and a Racal Store 4DS recorder. The FM recorders have the advantage of permitting the recording of data at one speed and the replaying of it at slower speed for analysis. These recorders generally have four or more tracks as well, which permits several types of data to be stored simultaneously. The disadvantages are their initial cost, which can be quite high, and the cost of good-quality magnetic tape. An 1800-ft roll of audio mastering tape costs ~$10. An 1800-ft tape can be used to record about 48 min of data at 7.5 ips. Thus, it is easy to use large numbers of tapes very quickly. A recording speed of 7.5 ips provides a bandwidth of 2.5–5 KHz, depending on the FM recorder being used.

Another mass storage device is a digital VCR audio recorder modified to permit recording of dc signals. The recorder consists of a digital audio processor (most often a SONY, PCM 501 ES) that converts the analog signal to a signal that can be stored on a standard VCR. The signal can be converted back to an analog signal by the same audio processor. The advantages of this device are its relatively low cost, $2000–3000, and the capability it provides for recording of data for 2–8 h on a single tape at a bandwidth of ~20 KHz. The disadvantages of this recorder are that it is not possible to change the playback speed of the VCR recorders currently available or to precisely or rapidly start and stop the tape. In addition, only two tracks of dc data and two of ac-coupled data can be recorded. However, for most patch clamp experiments, two channels of dc-coupled data are sufficient, one for current and the other for voltage.

Finally, a computer may be used for mass storage. It is possible to sample and store data on a hard disk or streaming digital magnetic tape at sampling rates of over 10,000 data points/s. At this sampling rate, a 10-megabyte hard disk or tape can store no more than 16 2/3 min of data. A 100-megabyte disk or tape can store up to 2.8 h of data. It is useful to have the data on a medium easily accessed by the computer. However, tape drives and hard disks of >20 megabyte storage capacity can be expensive. Sachs et al. (1982) and Sigworth (1983) have provided good explanations of the requirements for using the computer to sample and analyze data. One of the newest innovations in computer storage is the write-once read-mostly (WORM) drive, which is a laser disk recording system. It offers the possibility of large amounts of data storage (>100 megabyte) on a removable disk. The data is permanently stored on the disk.

Whatever mass storage device is chosen, we find it is very helpful to put a short series of calibration pulses onto the storage medium. It is important to provide a ready means of calibration of the recording system and to provide an internal check on the gain of the recording system at the beginning of the experiments. The second reason is important because it is possible to inadvertently change the gain at several places in the recording process. The simplest means of calibration is to connect a 10-G$\Omega$ resistor (1 or 2% precision resistor, Eltech Inst. Inc., Daytona Beach, FL or Victoreen Inc., Cleveland, OH) in the BNC input of the headstage and the bath ground. A 10-mV signal applied to the pipet voltage input will provide a 1-pA signal at the current monitor output. It is important not to touch the body of this resistor since skin oils and sweat can cause a reduction in the resistance between the leads. The resistors can be boiled in reagent grade acetone to clean them.

### 7.3. Data Analysis

The analysis of single-channel or whole-cell data is beyond the intent of this chapter. There are a number of articles and reviews that are concerned with this subject and these are listed here: Colquhoun and Hawkes, 1981, 1982; Dionne and Leibowitz, 1982; Horn and Lange, 1983; Fitzhugh, 1983; Horn et al., 1984; Leibowitz and Dionne, 1984; Yellen, 1984; Roux and Sauve, 1985; Jackson, 1985; Liebovitch and Fischbarg, 1985; Labarca et al., 1985. In addition, the book *Single Channel Recording* (edited by B. Sakmann and E. Neher, 1983) contains a good discussion of this subject, as do some of the reviews listed in the introduction to this chapter.

## Acknowledgments

Jerry M. Farley was supported by National Institutes of Health (NS17789), Department of Defense Contract DAMD-83-C-3248, and the Cystic Fibrosis Foundation. The chamber in Fig. 3, push-pull syringe in Fig. 4, stop in Fig. 5A, firepolishing assembly in Fig.

6A, pipet holder in Fig. 7, and pellet presses in Fig. 8 were built by Mr. Joe Ed Smith, who also assisted in the design of these items. I wish to thank him for his excellent assistance. I would also like to thank Martha Wood, who drew the figures and proofread the text. I am also very grateful to Dr. Terry Dwyer who provided Fig. 1B and who has been of great source of help and suggestions during the writing of this chapter. I also thank Dr. Clay Armstrong for the idea of using Teflon wire coating and Dr. Gerry Oxford for the suggestion of using Macor®. My appreciation to Gloria Griffis for typing this chapter.

# References


Abrams T. W., Castellucci V. F., Camardo J. S., Kandel E. R., and Lloyd P. E. (1984) Two endogenous neuropeptides modulate the gill and siphon withdrawal reflex in *Aplysia* by presynaptic facilitation involving cAMP-dependent closure of a serotonin-sensitive potassium channel. *Proc. Natl. Acad. Sci. USA* **81,** 7956–7960.

Adelman W. J., Jr. (1971) *Biophysics and Physiology of Excitable Membranes* Van Nostrand Reinhold, New York.

Aldrich R. W., Corey D. P., and Stevens C. F. (1983) A reinterpretation of mammalian sodium channel gating based on single channel recording. *Nature* **306,** 436–441.

Almers W., Stanfield P. R., and Stuhmer W. (1983) Lateral distribution of sodium and potassium channels in frog skeletal muscle: Measurements with a patch clamp technique. *J. Physiol.* **336,** 261–284.

Auerbach A. and Sachs F. (1984) Patch clamp studies of single ionic channels. *Ann. Rev. Biophys. Bioeng.* **13,** 269–302.

Bechem M. and Pott L. (1985) Removal of calcium current inactivation in dialysed guinea-pig atrial cardioballs by Ca chelators. *Pflugers Arch.* **404,** 10–20.

Bean B. (1985) Two kinds of calcium channels in canine atrial cells. *J. Gen. Physiol.* **86,** 1–30.

Benham C. D. and Bolton T. B. (1983) Patch-clamp studies of slow potential sensitive potassium channels in longitudinal smooth muscle cells of rabbit jejunum. *J. Physiol.* **340,** 469–486.

Blatz A. L. and Magleby K. L. (1985) Single chloride-selective channels active at resting membrane potentials in cultured rat skeletal muscle. *Biophys. J.* **47,** 119–123.

Bolton T. B., Lang R. J., Takewaki T., and Benham C. D. (1985) Patch and whole-cell voltage clamp of single mammalian visceral and vascular smooth muscle cells. *Experientia* **41,** 887–894.

Brown A. M., Lux H. D., and Wilson D. L. (1984) Activation and inactivation of single calcium channels in snail neurons. *J. Gen. Physiol.* **83,** 751–769.

Caldwell W. M. (1977) DC proportional heater-cooler temperature controller for physiological research. *J. Appl. Physiol.* **43,** 160–163.

Camardo J. S., Shuster M. J., Siegelbaum S. A., and Kandel E. R. (1983) Modulation of a specific potassium channel in sensory neurons of *Aplysia* by serotonin and cAMP-dependent protein phosphorylation. *Cold Harbor Symposia on Quantitative Biology* **48** (1), 213–220.

Chabala L. D., Sheridan R. E., Hodge D. C., Power J. N., and Walsh M. P. (1985) A microscope stage temperature controller for the study of whole-cell or single-channel currents. *Pflugers Arch.* **404,** 374–377.

Chamley-Campbell J., Campbell G. R., and Ross R. (1979) The smooth muscle cell in culture. *Physiol. Rev.* **59,** 1–61.

Colquhoun D. and Hawkes A. G. (1981) On the stochastic properties of single ion channels. *Proc. Roy. Soc. Lond.* **B211,** 205–235.

Colquhoun D. and Hawkes A. G. (1982) On the stochastic properties of bursts of single channel openings and of clusters of bursts. *Phil. Trans. Roy. Soc. Lond.* **B300,** 1–59.

Corey D. P. and Stevens C. F. (1983) Science and Technology of Patch-Recording Electrodes, in *Single-Channel Recording* (Bert Sakmann and Erwin Neher, eds.) Plenum, New York.

Cull-Candy S. G., Miledi R., and Parker, I. (1981) Single glutamate-activated channels recorded from locust muscle fibers with perfused patch clamp electrodes. *J. Physiol.* **321,** 195–210.

Dani J. A. and Eisenman G. (1984) Acetylcholine-activated channel current-voltage relations in symmetrical $Na^+$ solutions. *Biophys. J.* **45,** 10–12.

de Peyer J. E., Cachelin A. B., Levitan I. B., and Reuter H. (1982) $Ca^{2+}$-activated $K^+$ conductance in internally perfused snail neurons is enhanced by protein phosphorylation. *Proc. Natl. Acad. Sci. USA* **79,** 4207–4211.

Dione V. E. and Leibowitz M. D. (1982) Acetylcholine receptor kinetics. A description from single-channel currents at snake neuromuscular junctions. *Biophys. J.* **39,** 253–261.

Dwyer T. (1985) A patch clamp primer. *J. Electrophysiol. Tech.* **12,** 15–29.

Dwyer T. (1986) Guanidine block of single channel currents activated by acetylcholine. *J. Gen. Physiol.,* **88,** 635–650.

Dwyer T. M. and Farley J. M. (1984) Permeability properties of chick myotube acetylcholine-activated channels. *Biophys. J.* **45,** 529–539.

Ewald D. A., Williams A., Levitan I. B. (1985) Modulation of single $Ca^{2+}$-dependent $K^+$-channel activity by protein phosphorylation. *Nature* **315,** 503–506.

Fatt P. and Katz B. (1952) Spontaneous subthreshold activity at motor nerve endings. *J. Physiol.* **117,** 109–128.

Fay F. S. and Singer J. J. (1977) Characteristics of response of isolated smooth muscle cells to cholinergic drugs. *Am. J. Physiol.* **232,** C144–154.

Fenwick E. M., Marty A., and Neher E. (1982) A patch clamp study of bovine chromaffin cells and of their sensitivity to acetylcholine. *J. Physiol.* **331,** 577–597.

Fisher K. A. (1975) "Half" membrane enrichment: Verification by electron microscopy. *Science* **190,** 983–85.

Fishman H. M. (1975) Patch voltage clamp of squid axon membrane. *J. Memb. Biol.* **24,** 265–279.

FitzHugh R. (1983) Statistical properties of the asymmetric random telegraph signal, with applications to single-channel analysis. *Math. Biosci.* **64,** 75–89.

Fleckenstein A. (1977) Specific pharmacology of calcium in myocardium, cardiac pacemakers, and vascular smooth muscle. *Ann. Rev. Pharmacol. Toxicol.* **17,** 149–166.

Forda S. R., Jessell T. M., Kelly J. S., and Rand R. P. (1982) Use of the patch electrode for sensitive high resolution extracellular recording. *Brain Res.* **249,** 371–378.

Forscher P. and Oxford G. S. (1984) Norepinephrine modulation of Ca-channels in internally dialysed sensory neurons. *Biophys. J.* **45,** 181a.

Forscher P. and Oxford G. S. (1985) Modulation of calcium channels by norepinephrine in internally dialyzed avian sensory neurons. *J. Gen. Physiol.* **85,** 743–763.

French R. J., Worley J. F. III, and Krueger B. K. (1984) Voltage-dependent block by saxitoxin of sodium channels incorporated into planar lipid bilayers. *Biophys. J.* **45,** 301–310.

Freshney R. I. (1983) *Culture of Animal Cells. A Manual of Basic Technique* Alan R. Liss, New York.

Fukushima Y. (1982) Blocking kinetics of the anomalous potassium rectifier of tunicate egg studied by single channel recording. *J. Physiol.* **331,** 311–331.

Gogelein H. and Greger R. (1984) Single channel recordings from basolateral and apical membranes of renal proximal tubules. *Pflugers Arch.* **401,** 424–426.

Gray R. and Johnston D. (1985) Rectification of single GABA-gated chloride channels in adult hippocampal neurons. *J. Neurophysiol.* **54,** 134–142.

Green D. J. and Gillette R. (1983) Patch- and voltage-clamp analysis of cyclic AMP-stimulated inward current underlying neurone bursting. *Nature* **306,** 784–785.

Greger R., Schlatter E., and Gogelein H. (1985) Cl⁻-channels in the apical cell membrane of the rectal gland "induced" by cAMP. *Pflugers Arch.* **403**, 446–448.

Hagiwara S. and Ohmori H. (1982) Studies of calcium channels in rat clonal pituitary cells with patch electrode voltage clamp. *J. Physiol.* **331**, 231–252.

Hagiwara S., Miyazaki S., and Rosenthal N. P. (1976) Potassium current and the effect of cesium on this current during anomalous rectification of the egg membrane of a starfish. *J. Gen. Physiol.* **67**, 621–638.

Hamill O. P., Marty A., Neher E., Sakmann B., and Sigworth F. J. (1981) Improved patch-clamp techniques for high-resolution current recording from cells and cell-free membrane patches. *Pflugers Arch.* **391**, 85–100.

Haworth R. A., Hunter D. R., and Bukoff H. A. (1980) The isolation of $Ca^{2+}$-resistant myocytes from the adult rat. *J. Mol. Cell Cardiol.* **12**, 715–723.

Hille B. (1971) The permeability of the sodium channels to organic cations in myelinated nerve. *J. Gen. Physiol.* **58**, 599–619.

Hille B. (1972) The permeability of the sodium channel to metal cations in myelinated nerve. *J. Gen. Physiol.* **59**, 637–658.

Hille B. (1973) Potassium channels in myelinated nerve. Selective permeability to small cations. *J. Gen. Physiol.* **61**, 669–686.

Hille B. (1984) *Ionic Channels of Excitable Membranes* Sinauer Associates, Sunderland, Massachusetts.

Hodgkin A. L. and Huxley A. F. (1952) Currents carried by sodium and potassium ions through the membrane of the giant axon *Loligo*. *J. Physiol.* **116**, 449–472.

Horn R. and Lange K. (1983) Estimating kinetic constants from single channel data. *Biophys. J.* **43**, 207–223.

Horn R. and Patlak J. (1980) Single channel currents from excised patches of muscle membrane. *Proc. Natl. Acad. Sci. USA* **77**, 6930–6934.

Horn R., Vandenberg C. A., and Lange K. (1984) Statistical analysis of single sodium channels. Effects of N-bromoacetamide. *Biophys. J.* **45**, 323–335.

Hunter M., Lopes A., Boulpaep E. L., and Giebisch G. (1984) Single channel recordings of calcium-activated potassium channels in the apical membrane of rabbit cortical collecting tubules. *Proc. Natl. Acad. Sci. USA* **81**, 4237–4239.

Jackson M. B. (1984) Chemically activated channels in muscle and spinal cord. *Ann. Neurol.* **16** (suppl.), S52–S58.

Jackson M. B. (1985) Stochastic behavior of a many-channel membrane system. *Biophys. J.* **47**, 129–137.

Jakoby W. B. and Pastan I. H. (1979) *Methods in Enzymology* vol. 58 *Cell Culture* Academic, New York.

Kakei M. and Noma A. (1984) Adenosine-5'-triphosphate-sensitive single potassium channel in the atrioventricular node cell of the rabbit heart. *J. Physiol.* **352,** 265–284.

Kaneko A. and Tachibana M. (1985) Effects of L-glutamate on the anomalous rectifier potassium current in horizontal cells of *Carassius auratus* retina. *J. Physiol.* **358,** 169–182.

Kirsch G. E. and Narahashi T. (1978) 3,4-diaminopyridine: A potent new potassium channel blocker. *Biophys. J.* **22,** 507–512.

Kostyuk P. G., Veslosky N. S., and Fedulova S. A. (1981) Ionic currents in the somatic membrane of rat DRG Neurons. II. Calcium currents. *Neuroscience* **6,** 2431–2437.

Krishtal D. A. and Pidoplichko, V. I. (1980) A receptor for protons in the nerve cell membrane. *Neuroscience* **5,** 2325–2327.

Kuriyama H. and Kitamura, K. (1985) Electrophysiological aspects of regulation of precapillary vessel tone in smooth muscles of vascular tissues. *J. Cardiovasc. Pharmacol.* **7** (suppl. 3), S119–S128.

Labarca P., Rice J. A., Fredkin D. R., and Montal M. (1985) Kinetic analysis of channel gating. Application to the cholinergic receptor channel and the chloride channel from *Torpedo californica. Biophys. J.* **47,** 469–478.

Lee C. Y. (1970) Elapid neurotoxins and their mode of action. *Clin. Toxicol.* **3,** 457–472.

Leibowitz M. D. and Dionne V. E. (1984) Single-channel acetylcholine receptor kinetics. *Biophys. J.* **45,** 153–163.

Liebovitch L. S. and Fischbarg J. (1985) Determining the kinetics of membrane pores from patch clamp data without measuring the open and closed times. *Biochim. Biophys. Acta* **813,** 132–136.

Lux H. D. and Brown A. M. (1984) Patch and whole cell calcium currents recorded simultaneously in snail neurons. *J. Gen. Physiol.* **83,** 727–750.

Martin A., Wickelgren O., and Beranek R. (1970) Effects of iontophoretically applied drugs on spinal interneurones of the lamprey. *J. Physiol.* **207,** 653–665.

Meech R. W. (1978) Calcium-dependent potassium activation in nervous tissue. *Ann. Rev. Biophys. Bioeng.* **7,** 1–18.

Meech R. W. and Standen N. B. (1975) Potassium activation in *Helix aspersa* neurones under voltage clamp: A component mediated by calcium influx. *J. Physiol.* **249,** 211–239.

Milton R. L. and Caldwell J. H. (1986) Vesicles extracted from skeletal muscle fiber via suction: Possible implications for loose and tight patch voltage clamping. *Biophys. J.* **49**(2, part 2), 172a.

Mitra R. and Morad M. (1985) A uniform enzymatic method for dissociation of myocytes from hearts and stomachs of vertebrates. *Am. J. Physiol.* **24,** H1056–H1060.

Montal M., Labarca P., Fredkin D. R., and Suarez-Isla B. A. (1984) Channel properties of the purified acetylcholine receptor from *Torpedo californica* reconstituted in planar lipid bilayer membranes. *Biophys. J.* **45,** 165–174.

Nakayama T., Kurachi Y., Noma A., and Irisawa, H. (1984) Action potential and membrane currents of single pacemaker cells of the rabbit heart. *Pflugers Arch.* **402,** 248–257.

Narahashi T. (1974) Chemicals as tools in the study of excitable membrane. *Physiol. Rev.* **54,** 813–889.

Narahashi T., Moore J. W., and Scott W. R. (1964) Tetrodotoxin blockage of sodium conductance increase in lobster giant axons. *J. Gen. Physiol.* **47,** 965–974.

Neher E. and Marty A. (1983) Discrete changes of cell membrane capacitance observed under conditions of enhanced secretion in bovine adrenal chromaffin cells. *Proc. Natl. Acad. Sci. USA* **79,** 6712–6716.

Neher E., and Sakmann B. (1976) Single-channel currents recorded from membrane of denervated frog muscle fibers. *Nature* **260,** 799–802.

Neher E. and Steinbach J. H. (1978) Local anaesthetics transiently block currents through single acetylcholine channels. *J. Physiol.* **277,** 153–176.

Neher E., Sakmann B., and Steinbach J. H. (1978) The extracellular patch clamp: A method for resolving currents through individual open channels in biological membranes. *Pflugers Arch.* **375,** 219–228.

Nelson P. G. and Lieberman M. (1981) *Exitable Cells in Tissue Culture* Plenum, New York, London.

Ogden D. C. and Colquhoun D. (1983) The efficacy of agonists at the frog neuromuscular junction studied with single channel recording. *Pflugers Arch.* **3,** 246–248.

Patlak J. and Horn R. (1982) Effect of *n*-bromoacetamide on single sodium channel currents in excised membrane patches. *J. Gen. Physiol.* **79,** 333–351.

Paul J. (1975) *Cell and Tissue Culture* (5th Ed.) Churchill Livingston, Edinburgh.

Quandt F. N. and Narahashi T. (1982) Modification of single Na$^+$ channels by batrachotoxin. *Proc. Natl. Acad. Sci. USA* **79,** 6732–6736.

Rae J. L. (1985) The application of patch clamp methods to ocular epithelia. *Curr. Eye Res.* **4,** 409–419.

Rae J. and Levis R. (1984) Patch voltage clamp of lens epithelial cells: Theory and practice. *Mol. Physiol.* **6,** 115.

Reuter H. (1983) Calcium channel modulation by neurotransmitters, enzymes and drugs. *Nature* **301,** 569–574.

Roux B. and Sauve R. (1985) A general solution to the time interval omission problem applied to single channel analysis. *Biophys. J.* **48,** 149–158.

Sachs F. and Auerbach A. (1983) Single-channel electrophysiology: Use of the patch clamp. *Meth. Enzymol.* **103,** 147–177.

Sachs F., Neil J., and Barkakati N. (1982) The automated analysis of data from single ionic channels. *Pflugers Arch.* **395,** 331–340.

Sakmann B. and Neher E., eds. (1983). *Single-channel Recording* Plenum, New York.

Sigworth F. J. (1983) Electronic Design of the Patch Clamp, in *Single Channel Recording* (Sakmann B. and Neher E., eds.) Plenum, New York.

Sine S. M., and Steinbach, J. H. (1984) Activation of a nicotinic acetylcholine receptor. *Biophys. J.* **45,** 175–185.

Soejima M. and Noma A. (1984) Mode of regulation of the ACh-sensitive K-Channel by the muscarinic receptor in rabbit atrial cells. *Pflugers Arch.* **400,** 424–431.

Strickholm A. (1961) Impedance of a small electrically isolated area of the muscle cell surface. *J. Gen. Physiol.* **44,** 1073–1088.

Trautmann A. and Marty A. (1984) Activation of Ca-dependent K channels by carbamoylcholine in rat lacrimal glands. *Proc. Natl. Acad. Sci. USA* **81,** 611–615.

Yatani A. and Brown A. (1985) The calcium channel blocker nitrendipine blocks sodium channels in neonatal rat cardiac myocytes. *Circ. Res.* **56,** 868–875.

Yeh J. S., Oxford G. S., Wu C. H., and Narahashi T. (1976) Dynamics of aminopyridine block of potassium channels in squid axon membrane. *J. Gen. Physiol.* **68,** 519–535.

Yellen G. (1982) Single $Ca^{2+}$-activated nonselective cation channels in neuroblastoma. *Nature* **296,** 357–359.

Yellen G. (1984) Ionic permeation and blockade in $Ca^{2+}$-activated $K^+$ channels of bovine chromaffin cells. *J. Gen. Physiol.* **84,** 157–186.

Ypey D. L. and Clapham D. E. (1984) Development of a delayed outward-rectifying $K^+$ conductance in cultured mouse peritoneal macrophages. *Proc. Natl. Acad. Sci. USA* **81,** 3083–3087.

# Analysis of Ion Fluxes and Fluid Compartmentation in Brain Slices

## Wolfgang Walz

## 1. Introduction

Brain slices from cerebral cortex represent the classical model with which the first studies of cellular mechanisms of water and ion homeostasis were conducted. These experiments involved the analysis of ion and water movements with radiotracers. The amount of information gained with this approach has been summarized in reviews by Marchbanks (1970), Katzman and Pappius (1973), and Hertz and Schousboe (1975). The difficulty of this method is, however, clearly apparent: Compromised energy metabolism of the cells leads to swelling and changes in the sodium:potassium ratio. In addition even the most healthy slices contain 25% dead cells. The radiotracer methods give information about different compartments in the *whole* tissue; therefore one can not correct readily for the compromised metabolic function. In contrast, sophisticated electrophysiological methods are now available for the measurement of extracellular and intracellular functions in brain slices (*see* chapters by Nicholson and Rice and by MacVicar and O'Beirne in this volume). Here, *individual* cells can be tested before the experiment if they conform to certain standards. The inevitable cell swelling that creates certain difficulties for the interpretation of compartmental fluxes seems not to compromise the electrophysiological function of individually tested neurons. With the use of homogeneous primary cultures of neural cells, an alternative system with fewer limitations became available. Brain slices have, however, gained new popularity in measuring energy metabolism and electrical transmission in certain pathological states induced by anoxia, hypoglycemia, and ischemia (*see* Lipton and Whittingham, 1984). In this context, analyses of fluid compartments and ion fluxes are quite frequently used to give a more complete picture about events involved in such states.

A very good introduction to the use of brain slices and their electrical stimulation is available in this series (Lipton, 1985). Therefore, for practical aspects of procedures concerning preparation of slices, solutions and oxygenation, chambers, electrical stimulation, and so on, the reader may refer to this publication. MacVicar and O'Beirne (this volume) covers aspects of electrophysiological work with neurons in brain slices. Finally, the theoretical background for kinetic analysis involving radiotracers is dealt with by Kimelberg and Walz (this volume) and is not repeated here.

## 2. Properties of the Preparation

### 2.1. Slicing Technique

There are three main methods available to make slices of fresh brain tissue. (1) Hand-cutting with or without a guide is the most simple and has been thoroughly described by McIlwain and Rodnight (1962). A mechanical guided tissue slicer is commercially available (e.g., from Stoelting Co., Chicago, IL, USA). (2) Mechanical chopping is described by Alger et al. (1984); commercially used tissue choppers are the Sorval and McIlwain choppers. (3) Cutting with a vibratome is widely used, although more expensive. Currently, two vibrating knives are available commercially: the Oxford Vibratome (Oxford Laboratories, USA) and the Vibroslice Oscillating Tissue Slicer designed by Jefferys (1981) (available from Campden Instruments, Oxford, UK and Frederick Haer, Brunswick, ME, USA). Currently, no studies indicate conclusively that one method is superior to any other in yielding healthier slices. It is clearly evident that much depends upon the proper and careful handling of the tissue. It is recommended that slices from younger, male animals are used and that care is taken to ensure proper oxygenation. More details can be obtained in Lipton (1985). Holding chambers for the storage of tissue slices are available commercially (e.g., from Medical Systems Corp., Greenvale, NY) or may readily be built in a workshop. It must be emphasized that proper oxygenation is probably the most important factor.

### 2.2. Surface Effects

The area directly adjacent to the cut surface of a slice represents damaged tissue and thus may be a source of erroreous measurements. Total tissue water and inulin spaces of second slices of guinea pig cerebral cortex, which have two cut surfaces, are larger than those of first slices, which have one cut and one intact surface (Varon and McIlwain, 1961). Cohen (1974) distinguishes two tissue compartments that form the surface space. First, a superficial layer consisting of damaged cells that may be easily permeated by water and various marker substances and that have an abnormal electrolyte content (e.g., high sodium concentration). Second, cutting increases the contribution of adherent films of medium on the cut surface to the total tissue water, tissue electrolytes, and tissue compartment defined by markers. The surfaces of brain slices have been studied microscopically (Pappius et al., 1962; Pappius; 1982; Moller et al., 1974) and were found to have damaged, swollen neurons and glial cells in the outer layers. Figure 1 shows the regional differences of fluid content in cortical brain slices of 0.5–0.6 mm thickness as estimated by Moller et al. (1974) from biochemical experiments and verified by electron microscopy. Moller et al. found increased swelling at the uncut pial

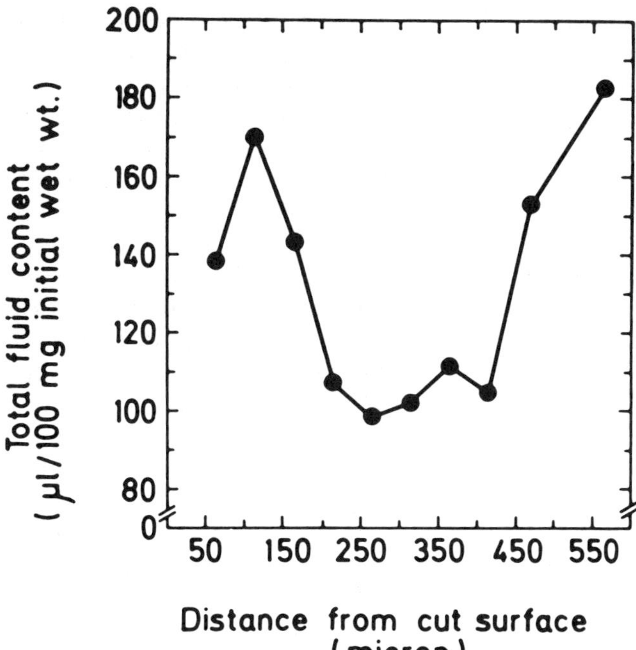


Fig. 1.   Regional differences of fluid content in 0.5–0.6 mm thick cortical brain slices incubated for 1 h in a potassium-rich medium (from Moller et al., 1974).

surface, indicating that the swelling cannot solely be caused by a traumatic treatment of the tissue during the slicing. From these data it seems apparent that a tissue slice has a "sandwich" structure with destroyed outer layers of about 50 μm on both cut and uncut surfaces (Garthwaite et al., 1979; Bak et al., 1980). The disturbance and swelling in these layers can be present on both neurons and glial cells. The adherent film of medium on cut surfaces at 37°C adds 5% to the fluid content of a 0.4-mm slice.

### *2.3. Slice Thickness*

The maximal slice thickness is limited mainly by the development of an anoxic core, and the minimal slice thickness should, of course, be much larger than the total "edge damage" of the sandwich structure. Fujii et al. (1981, 1982) found U-shaped depth profiles for $pO_2$ in guinea pig olfactory cortex slices maintained in an interface chamber at 37°C. Their slices (320–480 μm thick) exhibited a similar $pO_2$ at both surfaces; in slices thicker than 430 μm, there was an anoxic central core. Their slices were, however, superfused very slowly so that the $pO_2$ at the surface was only half that in the original buffer. Lipton and Whittingham (1984) calculated (on the basis of Warburg's data with liver slices) that brain slices of up to 700 μm could be regared as nonanoxic. Assuming that the diameter of the damaged outer parts remains constant at about 100 μm regardless of the brain slice thickness, 20% of a 500-μm slice would consist of damaged cells, whereas the equivalent value for a 700-μm slice would be only 14%. Bak et al. (1980), however, compared the morphology of 300 and 700 μm neostriatal slices and found that the 300-μm slices exhibited a characteristic three-layered sandwich structure. The middle layer was 100 μm thick and had 88% of the cells intact. This ratio did not change during a 5-h incubation. In contrast to thin slices, thick slices displayed no layering of intact cells in cross-section. Moreover, after 2 h of incubation, the majority of cells throughout these thick slices exhibited swelling (98%), with only small patches of intact cells. The reason for this dramatic effect is unclear. Taking all these observations into account, a slice thickness of 400–500 μm can be recommended. It should be noted, however, that a slice with about 450 μm thickness consists of at least 20–25% damaged tissue resulting from the cut surface, even under optimal conditions.

## 3. Experimental Chambers

Most of these experiments involve preincubation with a tracer or marker with the subsequent monitoring of the time course of washout or efflux. For these experiments, two different strategies can be used. The cells can be kept in a holding chamber or holder and at certain time intervals transferred in their holder from one reservoir of washing fluid to the other. The efflux of tracer or marker in each reservoir is then determined, as well as the remainder in the slice at the end of the series. Alternatively, the slice may be kept in a chamber with fast-flowing solution, loaded by a tracer and marker for some time, and then perfused by switching to washing fluid. This perfusate is then collected at certain time intervals for determining the quantity of marker or tracer. At the end of the experiment the remainder of the substance in the slice is determined.

Arnfred et al. (1970) found that, by comparing extent of cellular swelling with the potassium content, transferred brain slices seem only able to maintain an optimum potassium concentration if their free mobility is preserved. They described a new type of tissue holder in which the slices are freely floating (Fig. 2A). White et al. (1978) constructed a perfusion chamber that allows for the collection of perfusion fluid and rapid exchange of solutions (Fig. 2B). The slice is kept fixed by the downward flow of the solutions that press it onto the fitted tissue support. This arrangement has the advantage that the slice can be electrically stimulated, and this effect can be measured in terms of the efflux of substances. This is not possible with the holder designed by Arnfred et al. (1970).

## 4. Delineation of Fluid Compartments

It is of interest to have detailed information about the extent of intracellular water space in order to trace intracellular water space changes and to calculate ion concentrations from the measured ion contents. Changes in extracellular space can, of course, also be measured.

### 4.1. Choice of Marker Substance

Usually a slice is exposed to a marker, which is assumed to equilibrate with the extracellular space and not to enter the in-

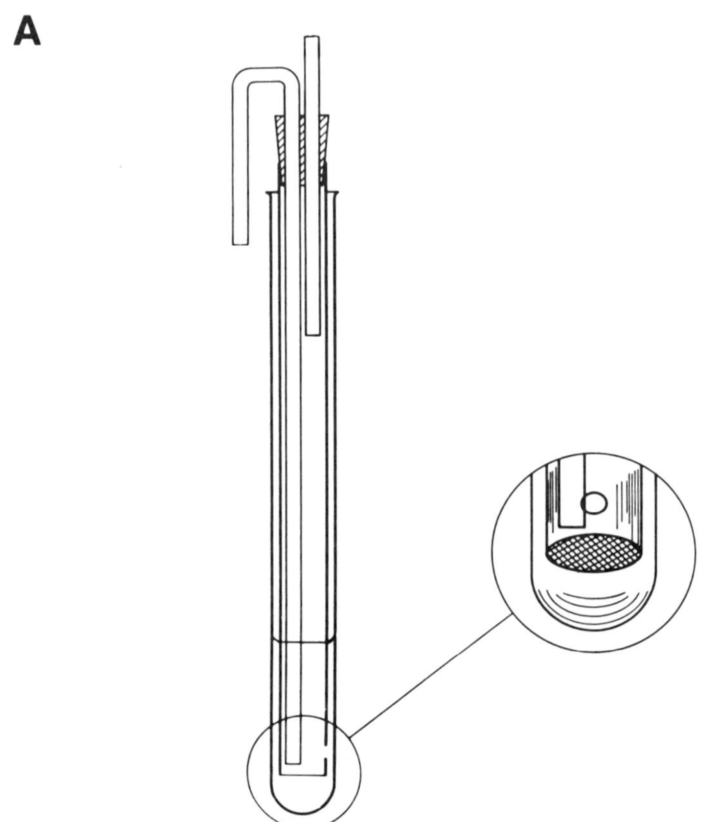



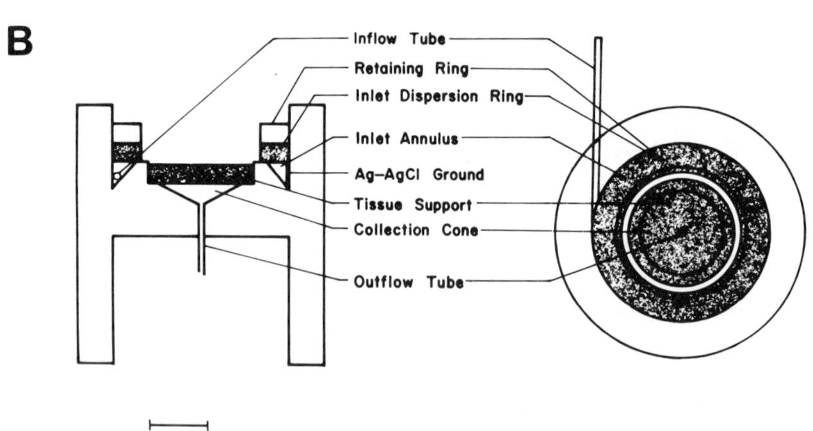

Fig. 2.   (A) Tissue holder for brain slices. For details, *see* Arnfred et al., (1970). (B) Schematic diagram of a tissue chamber. For details, *see* White et al., (1978).

426

tracellular compartment. No ideal marker exists and it is therefore advisable to name the measured extracellular space the marker space, assigning to it the name of the markers used. A detailed discussion of marker spaces in brain slices can be found in Cohen (1972), Katzman and Pappius (1973), and Pappius (1982). In the present chapter, the most commonly used markers are critically discussed.

### 4.1.1. Ions

Chloride, thiocyanate, iodide, and sulfate have been used to determine the marker space. Chloride is an extremely unsuitable marker because it is reported to accumulate actively in glial cells via pumps (Kimelberg, 1981; Walz and Hertz, 1983). Thiocyanate penetrates an intracellular compartment, usually yielding a space much higher than the inulin space. This compound is also extremely sensitive to details of preparation (Cohen, 1972) and binds to proteins (Pollay et al., 1966). Iodide and sulfate have none of these disadvantages, but compared with other markers recommended (below), they also have no real advantages. Both the sulfate and iodide space give constantly higher marker spaces than those measured with the inulin method.

### 4.1.2. Nonmetabolized Carbohydrates

Inulin, sucrose, D-sorbitol, and D-mannitol are used as extracellular markers. Amtorp (1979) found that efflux rates for mannitol and sucrose were considerably slower than from agar sheets. There was some space in brain slices that was inaccessible to inulin, but accessible to mannitol and sucrose; this probably represented damaged tissue cells. Sorbitol is not recommended because it is a minor constituent of neural tissue (Stewart et al., 1967). Inulin has long been considered an ideal marker. Lund-Andersen and Moller (1977) found, however, that inulin slowly penetrates cells in brain slices, seriously compromising its function as an extracellular marker. Nevertheless, in other respects inulin is almost ideal. Thus, when the inulin space is measured by kinetic analysis of efflux data under steady-state conditions, the cellular penetration of inulin does not play a role (see below). The average molecular weight of commercial inulin is most often 5000—5500. This polymer is stable in neutral and moderately basic solutions, but it is degraded by acid and its solubility in water increases markedly with temperature (Phelps, 1965). It is difficult to dissolve at room temperature, but solutions containing 1–2% by weight can be easily prepared by gentle heating and can be subsequently chilled without precipitation (Cohen, 1972). In view of the high molecular weight of inulin, solutions have to be incubated for a longer time to equilibrate completely with all extracellular spaces. Peroxidase and ferritin penetrate cells (Becker et al., 1968; Selwood, 1970), and these proteins must therefore be considered a poor choice.

### 4.1.3. Evaluation

A critical evaluation of these markers leaves only one reasonable choice—inulin, but only if this marker is used in efflux experiments in which the inulin space is calculated from the kinetic analysis of the efflux data (see below) and *not* from equilibrium distribution within a brain slice.

## 4.2. Determination of the Extracellular Space with $^{14}$C-Inulin

Traditionally, the inulin space, obtained after incubation of brain slices for 60 min with inulin, has been considered the extracellular space. The method described below, based on that described by Lund-Andersen (1974), represents a much more accurate efflux kinetics analysis.

After preparation, each slice was incubated in 4 mL of medium containing 0.9 Ci/mg. In some experiments the inulin was present from the start of the incubation, and in some cases it was added after a certain period of preincubation. After incubating for 60 min, the slice was removed with the aid of forceps, dipped quickly in nonradioactive medium, and placed in a rapid transfer holder (see above). While tissue was freely floating in the holder, the wash-out was performed by transferring the holder through a series of test tubes, each containing 4 mL of inulin-free medium. Initially the holder was transferred every 2 min and finally every 30 min (*see* Fig. 3). The frequent change of medium ensured that practically no back-diffusion of inulin from the wash-out medium to the slice would occur. After wash-out, the wet and dry weights of the slice were determined as described by Lund-Andersen and Hertz (1970) (*see* also, chapter by Go this volume), and the inulin remaining in the tissue was determined as follows (Schousboe and Hertz, 1971): one slice was treated with 100 $\mu$L of 1.0$N$ HNO$_3$ for 30 min at 100°C, and, after dilution, 1.5 mL of the samples added to scintillation fluid. The scintillation fluid was ethanol/toluene (10:7 vol/vol) containing 4 g PPO (2.5-diphenyloxazole) and 0.1 g dimethyl-

POPOP (1,4-*bis*-2-(5-phenyloxazolyl)-benzene) per 700 mL toluene. The radioactivity found in each test tube was added to that remaining in the tissue at the end of the experiment to give the total radioactivity of the slice immediately before the washout started. Wash-out curves were constructed showing the fractions (expressed as percentages) of the initial radioactivity remaining in the tissue at different times during the wash-out (Fig. 3).

Cohen (1972) recommends heating the inulin-containing medium samples for 30 min at 80–90°C or allowing them to stand at room temperature for 4 h. Under these conditions, the inulin hydrolyzes to fructose; otherwise inulin may gradually precipitate from the scintillation mixture, reducing counting efficiency. Inulin precipitates more readily from the standards because of their higher concentrations, yielding high values for the inulin space. A method to measure unlabeled inulin colorimetrically is given by Varon and McIlwain (1961), and a method based on measuring the fructose formed by the acid hydrolysis of inulin is described by Kulka (1956).

Figure 3 shows a typical efflux curve for inulin. (For an in

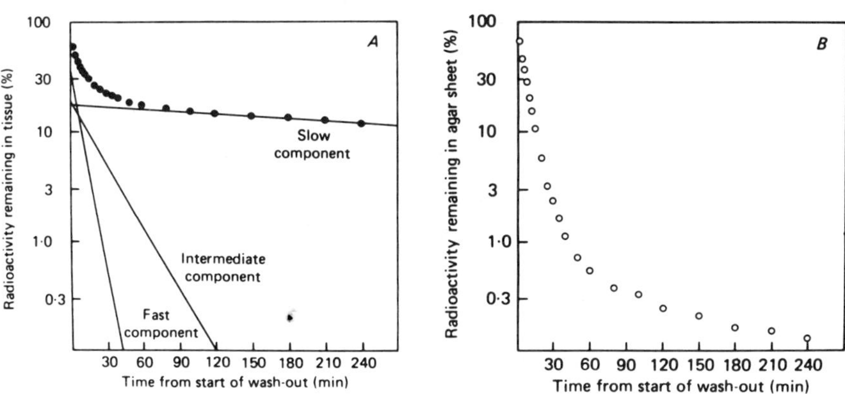


Fig. 3. (A) Wash-out of [14]C-inulin from 0.6-mm thick cortical brain slices following 60 min of incubation plotted in a semilogarithmic diagram. The ordinate is the amount of radioactivity remaining in the slice at specified times and the abscissa is the time in min from start of wash-out. The experimental curve is represented by solid circles and the resolution into three exponential functions designated slow component, intermediate component, and fast component was performed as described in the text. (B) A semilogarithmic plot of the efflux of [14]C-inulin from 0.6 to 0.8 mm thick agar sheets incubated for several hours with the inulin tracer before wash-out (from Amtorp, 1979).

troduction into efflux kinetics, *see* chapter by Kimelberg and Walz, this volume). Three time constants can easily be calculated. The slow component has a half-time of 426 min and probably represents the wash-out of inulin from an intracellular compartment. The intermediate half-time is 15 min and its origin is probably inulin penetration into damaged cells of the outer portions of the sandwich structure and its subsequent wash-out. The fast component is 4.3 min, as compared with 3.8 min for efflux of inulin from an agar sheet of the same thickness, and is therefore extracellular equilibrated inulin.

The extracellular space can be calculated by extrapolating the slope of the fast component to time 0 (beginning of wash-out) and using the dpm of $^{14}$C-inulin of this compartment at time 0:

$$\mu L \text{ of water} = \frac{\text{dpm inulin at time 0 (fast component)}}{\text{dpm inulin per } \mu L \text{ incubation fluid}}$$

The $\mu$L of extracellular water can now be expressed as per mg final wet weight, initial wet weight, or dry weight. The dry weight of brain slices does not change during incubations, because there is only negligible loss of soluble materials (Bourke and Tower, 1966). If the volume of extracellular space is simply required to enable calculation of the concentrations of ions from their total content in slices, it is acceptable to express the fluid space as per mg final wet weight (or dry weight). If, however, absolute changes during the incubation are of interest, e.g., swelling as it occurs during anaerobic conditions (see below), volume should be expressed as per initial weight of slice (*see* Fig. 4). Then the increased swelling is accompanied by an increase in noninulin space, whereas the dry weight and inulin space remain unchanged. The same data calculated on the basis of final weight imply that a decrease in both dry weight and the inulin space has occurred and suggest that the noninulin space increased to a smaller extent than was actually the case. If it is not possible to measure initial wet weight, this has to be measured in parallel preparations. Instead of a comparison between wet and dry weight, tritiated water content may be used to estimate the total water space of a slice. Equilibration takes place with the total water content of the slices within 2 min (Cohen et al., 1968). Total water content and inulin space can be measured at the same time with a double-labeling program for $^3$H/$^{14}$C.

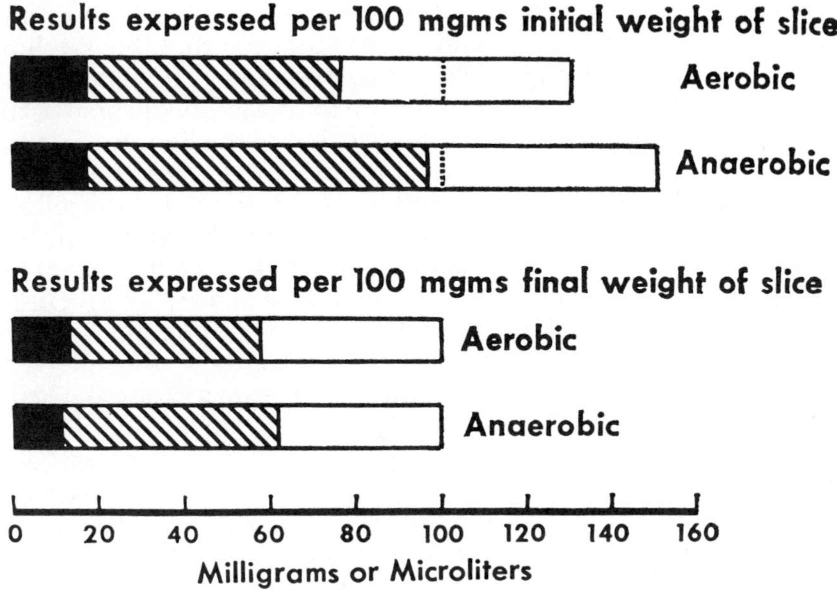


Fig. 4. Two methods of expressing fluid distribution in vitro. A slice of assumed weight 80 mg after 60 min of aerobic incubation weighs 104 mg, has a dry weight of 16 mg, and contains inulin equivalent to 40 μL of inulin space. The corresponding figures after 60 min of anaerobic incubation are 120, 16, and 40. When expressed per 100 mg of the initial weight of the slice, the total length of the bar represents the weight or volume of the slice after incubation, and the portion to the right of the vertical broken line indicates the extent of swelling. The black portion represents dry weight and the open portion the tracer space (in this case, the inulin space). The shaded portion, representing the difference, represents the water not equilibrated with the tracer (from Katzman and Pappius, 1973).

## 5. Compromised Function in the Brain Slice

Unfortunately, there is a major problem involved with this method, and that is considered in this section. Immediately (within 10 s) after incubation, a brain slice increases its wet weight or tritiated water content by about 10% (*see,* e.g., Katzman and Pappius, 1973; Hertz and Schousboe, 1975). This swelling continues in the next 1–2 h and reaches, depending on the incubation conditions an increase of up to 70% over an unincubated slice. Figure 5 illustrates such increases under different conditions. This phenomenon occurs in slices from all brain regions (Lipton and Whittingham, 1984). Originally, it was thought that this swelling consists of intra- and extracellular components. However, these interpretations did not account for the permeability of some of the intracellular spaces to inulin. A reappraisal of the literature, in the light of this inulin permeability of the cell membrane (with damaged and undamaged cells), points toward a complete intracellular localization of this additional water space during brain slice incubation (*see,* e.g., Lipton and Whittingham, 1984). Swelling is not the only phenomenon that occurs during the first phase (1–2 h) of brain slice incubation: the intracellular potassium/sodium ratio is significantly lower than that of the *in situ* preparation. Swelling has caused considerable difficulties since a prerequisite for the discussion of factors such as ion distribution and transport is that the content of the substance in the incubated tissue is expressed in relation to a constant weight. Accordingly, concentrations of ions or other compounds have often been expressed in relation to the initial wet weight with a correction made for a presumed extracellular swelling. The demonstration that the swelling is not an extracellular phenomenon implies that such a correction can no longer be regarded as permissible. The swelling may be reduced by cooling and aggravated by anaerobic incubation (Fig. 5). Section 2.2 suggests that about 25–30% of the cells in a slice are damaged because of surface effects of slicing. These cells probably lose their metabolites and their ability to engage in biochemical reactions, but have retained their protein content and dry weight. Thus, their presence would lead to an approximate 25% decrease in the concentrations of energy-related metabolites, respiration, and associated parameters. There is, however, a 50% reduction of high-energy phosphates in brain slices during the incubation (Lipton and Whittingham, 1984), and there is a high ratio of glycolysis to oxidation of glucose for which damaged or dead cells do not account. It appears, therefore, that the metabolism of the "non-leaky" cells must be compromised. Lipton and Whittingham (1984) made an interesting calculation to estimate the extent of the compromised function: 25–30% of the cells are effectively dead. The remaining 75% of the tissue is composed of functioning cells and they are associated with metabolites whose overall levels are reduced by 25% (the other 25% of the reduction is caused by the dead cells). Thus, the intact cells in the slice contain about two thirds of their normal content of high-energy phosphates. Because the cells are swollen, the actual concentrations of these compounds are

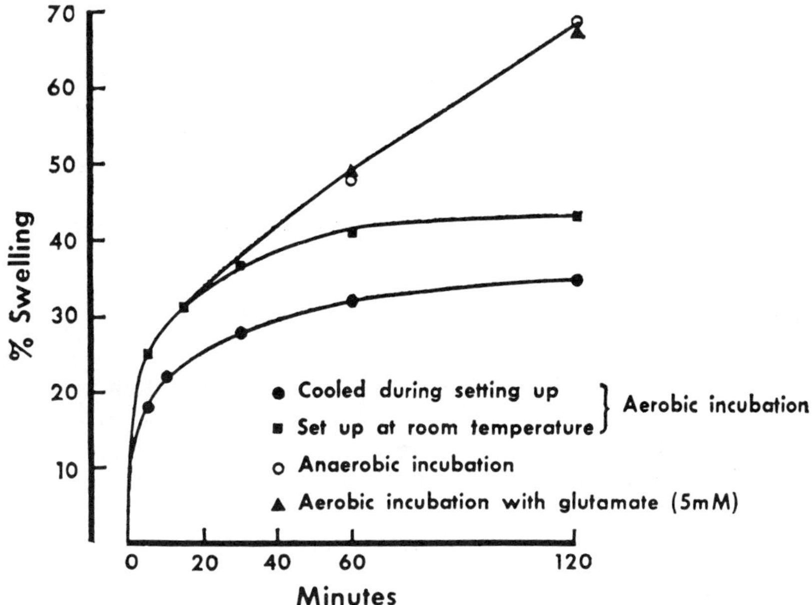


Fig. 5. The time course of the swelling of brain slices incubated in bicarbonate-buffered Ringer-type medium under different experimental conditions (from Katzman and Pappius, 1973).

about 50% of their *in situ* values. Since the cells are swollen by about 40–50%, their potassium concentration is reduced by about 25%, and their sodium is increased by about the same amount. This is an indication of a decrease in the ratio of pumping rate to the permeability for sodium ions. Nevertheless, cells in brain slices maintain an almost steady metabolic state for many hours and show quite normal electrophysiological responses.

## 5.1. Possible Protective Measures

It is very important to use gentle cutting techniques. Lipton and Whittingham (1984) suggest that barbiturate anesthetics could be administered to the animals in vivo to protect against damage from ischemia. Most of the irreversible damage seems, however, to be caused by accumulation of calcium in cells of the slice (Lipton et al., 1985). It may therefore be helpful to use during the preparation of the slices a modified salt solution, which has some protective effects. Calcium-free salt solution and the addition of 2 m*M* cobalt (a calcium channel blocker) might be helpful. Lipton et al. (1985) reported that preincubation with creatine strongly protects the tissue against anoxic damage, maintaining ATP, potassium, and sodium levels.

## 6. Kinetic Analysis of Compartments for Potassium, Sodium, and Chloride Ions

There are generally three ways to investigate movements of these ions. (1) The uptake of these ions is studied by exposing the slices for the time of rectilinear uptake to the radioactive ion, terminating the uptake by washing in nonradioactive solution, and subsequently determining the remainder of radioactivity in the tissue. The procedure can be done in controls and compared with samples exposed to altered experimental conditions. This procedure gives information about unidirectional influx of ions. The values of ion uptake have to be expressed on a common basis. These could be: dry weight of tissue; initial wet weight; or weight of protein or DNA content. Calculating ion concentrations from the intracellular water content is possible, but should only be done using compartmentation analysis (see below). (2) The radiotracer is equilibrated with the tissue, then the experimental manipulations are introduced. If these require a change in the concentration of the ion under investigation in the solution, the specific radioactivity (KBq per g) should be left constant. Again, the experiment is terminated by rapid washing. This analysis gives the total intracellular content of the ions at any time. (3) The most elegant procedure is compartmental analysis. It represents a way to not only eliminate the compartment of the dead cells (which is a problem with the sodium ion), but also to distinguish between glial cells and neurons. It is essentially the same method as used to determine inulin efflux by previous equilibration with inulin and subsequent sampling of the washing fluid (see above). If these experiments are done in dual-labeling experiments with inulin or in parallel samples with inulin efflux, the intracellular concentration of the ion can be calculated. Thus this method gives an estimate concerning the compartments involved; the content of ions and their concentration in the intracellular and other compartments; and the rate of efflux out of these different compartments. Figure 6 gives an example of a potassium compartmental analysis. Four compartments are analyzed: Compartment I with a half-life of 30 s and 4.9% of the total potassium, probably associated with the extracellular space (Section 2.2). Compartment II, with a half-time of 5 min, contains 14% of the total content and is probably the glial compartment. Compartment III has a half-time of 17 min with 74.3% of the potassium and is the neuronal compartment. Compartment IV has a half-time of 43 min and contains 6.7% of the potassium. It is not clear what compartment IV represents. Hertz and Frank (1978) showed a dramatic increase of compartment II after exposure to high potassium (Fig. 6). Care has to be taken if sodium uptake or content (as above) is measured. Intracellular sodium concentration is probably as low as 10 m$M$, whereas in the portion of damaged cells, the concentration will approach the extracellular value (about 140 m$M$). A similar problem exists for the chloride ion. Nevertheless, if only changes in equilibrated contents as a reaction to external stimuli (drugs, electrical pulses, and so on) are measured, these considerations might be less important.

Techniques for the electrical stimulation of brain slices can be found, for example, in a chapter of this series (Lipton, 1985). When slices are electrically stimulated, there is a loss of potassium from the non-inulin space and an uptake of sodium, but little change in the chloride distribution (Bachelard et al., 1962; Cummins and McIllwain, 1961; Keesey et al., 1965; Keesey and Wallgren, 1965). On the cessation of stimulation there is a reuptake of potassium and extrusion of sodium. Measurements of the flux of potassium and sodium during stimulation show increases by a factor of 2.5 for potassium and of 5 for sodium.

## 7. Kinetic Analysis of Compartments for Calcium Ions

Slices inclubated with $^{45}$calcium reach equilibrium within 30 min, and efflux experiments can then be performed. Efflux kinetics from cortical brain slices of calcium show two efflux constants, one for extracellular and one for intracellular calcium (Rubiales de Barioglia and Orrego, 1982). In order to gain more information about the considerable amount of calcium sequestered, by, for example, mitochondria and endoplasmic reticulum, one has to modify these methods somewhat.

The efflux should be presented as the fractional rate of loss per unit time (Hopkins and Neal, 1971):

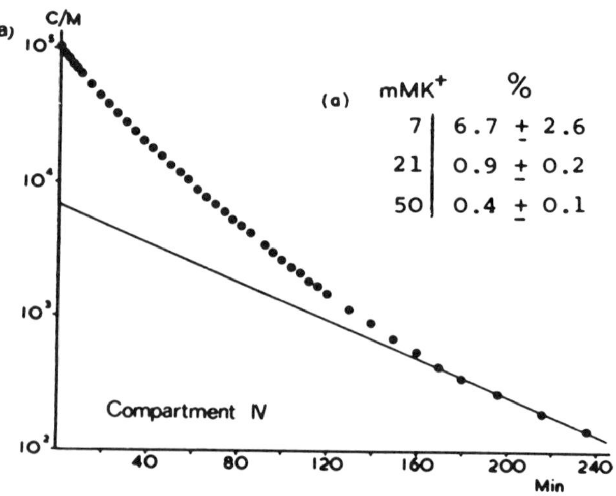

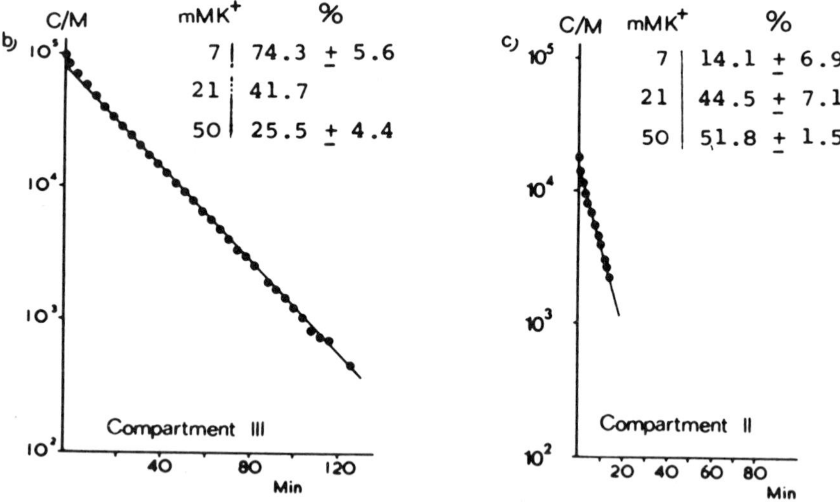

Fig. 6. An example of kinetic analysis of compartments with different exchange rate for $^{42}$K in a rat brain slice. Inserted in a, b, and c are the relative magnitudes of each of the compartments II–IV after incubation in media with 7.21, 21, or 50 m$M$ potassium (from Hertz and Franck, 1978).

$$f = \frac{\Delta A}{\Delta t \times A_t}$$

where $\Delta A$ represents the amount of calcium lost in the time interval $\Delta t$, and $A_t$ is the amount of radioactive calcium in the tissue at the midpoint of the interval $\Delta t$. In order to determine $A_t$, the calcium released from the slice during the whole experiment is added to the amount of calcium extracted from the tissue at the end of the experiment. This gives the value $A_0$. $A_t$ is the difference between $A_0$ and the total calcium lost up to the middle of $\Delta t$. The expression of calcium efflux as log % of calcium remaining in the tissue against time is too insensitive for detecting small changes in efflux. This problem is overcome by expressing efflux as the fractional rate constant against time. The cellular efflux has then to be analyzed for its different intracellular compartments with the use of drugs and removal of sodium (e.g., carbonyl cyanide, *m*-chlorophenylhydrazone, veratridine, caffeine; for a more detailed discussion, *see* Bygrave, 1977). Some authors (Cooke and Robinson, 1971) use subcellular fractionation to locate the calcium content of the intracellular compartments.

## 8. Conclusions

With careful handling, brain slices provide a useful model for compartmental analysis of ion and water movements at the cellular level. The fact that they include about 25% dead cells is inevitable and can be overcome by using efflux analysis. The intracellular swelling that causes so many problems and is not apparent during intracellular studies of neurons in slices is a severe limitation. It might be only partly attributable to the fact that this method studies the properties of the whole tissue, whereas electrophysiology is the study of single cells that can be tested before an experiment is started. The intracellular swelling may well represent only the glial compartment (MacVicar and Walz, unpublished results) and this might explain why it is not manifest in electrophysiological recordings of neurons in slices up to 12 h old. The swelling of glial cells may therefore represent not pathological manifestations, but rather may reflect one of the homeostatic mechanisms with which glial cells react to changes in their environment. Thus, further experiments concerning glial responses in brain slices are necessary and it is quite possible that in this area of the work with brain slices important advances may be forthcoming.

## Acknowledgments

The author is a scholar of the Medical Research Council of Canada. Dr. Andrew J. Greenshaw (University of Alberta) is thanked for helpful suggestions in improving the text.

## References

Alger B. E., Dhanjal S. S., Dingledine R., Garthwaite J., Henderson G., King G. L., Lipton P., North A., Schwartzkroin P. A., Sears T. A., Segal M., Whittingham T. S., and Williams J. (1984) Brain Slice Methods, in *Brain Slices* (Dingledine R., ed.) Plenum, New York.

Amtorp O. (1979) Distribution of inulin, sucrose and mannitol in rat brain cortex slices following *in vivo* or *in vitro* equilibration. *J. Physiol.* **294,** 81–90.

Arnfred T., Hertz L., Lolle L., and Lund-Andersen H. (1970) An improved holder for transfer of brain slices during *in vitro* incubation. *Exp. Brain Res.* **11,** 373–375.

Bachelard H. S., Campbell W. J., and McIlwain H. (1962) The sodium and other ions of mammalian cerebral tissues, maintained and electrically stimulated *in vitro. Biochem. J.* **84,** 225–232.

Bak I. J., Misgeld U., Weiler M., and Morgan E. (1980) The preservation of nerve cells in rat neostriatal slices maintained *in vitro:* A morphological study. *Brain Res.* **197,** 341–353.

Becker N. H., Hirano A., and Zimmerman H. M. (1968) Observations of the distribution of exogenous peroxidase in the rat cerebrum. *J. Neuropathol. Exp. Neurol.* **27,** 439–452.

Bourke R. S. and Tower D. B. (1966) Fluid compartmentation and electrolytes of cat cerebral cortex *in vitro.* I. Swelling and solute distribution in mature cerebral cortex. *J. Neurochem.* **13,** 1071–1097.

Bygrave F. L. (1977) Mitochondrial Calcium Transport, in *Current Topics in Bioenergetics* vol. 6 (Sanadi D., ed.) Academic, New York.

Cohen S. R. (1972) The Estimation of Extracellular Space of Brain Tissue *In Vitro,* in *Research Methods in Neurochemistry* vol. 1 (Marks N. and Rodnight R., eds.) Plenum, New York.

Cohen S. R. (1974) The contribution of adherent films of liquid on the cut surfaces of mouse brain slices to tissue water and "extracellular" marker spaces. *Exp. Brain Res.* **20,** 421–434.

Cohen S. R., Blasberg R., Levi G., and Lajtha A. (1968) Compartmentation of the inulin space in mouse brain slices. *J. Neurochem.* **15,** 707–720.

Cooke W. J. and Robinson J. D. (1971) Factors influencing calcium movements in rat brain slices. *Am. J. Physiol.* **221,** 218–225.

Cummins Y. T. and McIlwain H. (1961) Electrical pulses and the potassium and other ions of isolated cerebral tissues. *Biochem. J.* **79,** 330–341.

Fujii T., Baumgartl H., and Lubbers D. W. (1982) Limiting section thickness of guinea pig olfactory cortical slices studied from tissue $pO_2$ values and electrical activities. *Pflugers Arch.* **393,** 83–87.

Fujii T., Buerk D. G., and Whalen W. J. (1981) Activation energy in the mammalian brain slice as determined by oxygen microelectrode measurement. *Jpn. J. Physiol.* **31,** 279–283.

Garthwaite J., Woodhams P. L., Collins M. J., and Balazs R. (1979). On the preparation of brain slices: Morphology and cyclic nucleotides. *Brain Res.* **173,** 373–377.

Hertz L. and Frank G. (1978) Effect of Increased Potassium Concentrations on Potassium Fluxes in Brain Slices and in Glial Cells, in *Dynamic Properties of Glial Cells* (Schoffeniels E. G. Franck, D. B. Tower, and L. Hertz, eds.) Pergamon, Oxford.

Hertz L. and Schousboe A. (1975) Ion and Energy Metabolism of the Brain at the Cellular Level, in *International Review of Neurobiology* vol. 18 (Pfeiffer C. C. and Smythies J. R., eds.) Academic, New York.

Hopkins J. and Neal M. J. (1971) Effect of electrical stimulation and high potassium concentrations on the efflux of [$^{14}$C] glycine from slices of spinal cord. *Br. J. Pharmacol.* **42,** 215–223.

Jefferys J. G. R. (1981) The Vibroslice, a new vibrating blade tissue slicer. *J. Physiol.* **324,** 2P.

Katzman R. and Pappius H. M. (1973) *Brain Electrolytes and Fluid Metabolism.* Williams and Wilkins, Baltimore.

Keesey J. C. and Wallgren H. (1965) Movements of radioactive sodium in cerebral-cortex slices in response to electrical stimulation. *Biochem. J.* **95,** 301–310.

Keesey J. C., Wallgren H., and McIlwain H. (1965) The sodium, potassium and chloride of cerebral tissues: Maintenance, change on stimulation and subsequent recovery. *Biochem. J.* **95,** 289–300.

Kimelberg H. K. (1981) Active accumulation and exchange transport of chloride in astroglial cells in culture. *Biochim. Biophys. Acta* **646,** 179–184.

Kulka R. G. (1956) Colorimetric estimation of ketopentoses and ketohexoses. *Biochem. J.* **63,** 542–548.

Lipton P. (1985) Brain Slices, in *Neuromethods* vol. 1 (Boulton A. A. and Baker G. B., eds.) Humana, Clifton, New Jersey.

Lipton P. and Whittingham T. S. (1984) Energy Metabolism and Brain Slice Function, in *Brain Slices* (Dingledine R., ed.) Plenum, New York.

Lipton P., Hurtenbach C., and Kass I. R. (1985) Ca uptake during anoxia in the rat hippocampal slice. *Soc. Neurosci. Abstr.* **11** (1), 394.

Lund-Andersen H. (1974) Extracellular and intracellular distribution of inulin in rat brain cortex slices. *Brain Res.* **65,** 239–254.

Lund-Andersen H. and Hertz L. (1970) Effect of potassium and of glutamate on swelling and on sodium and potassium content in brain cortex slices from adult rats. *Exp. Brain Res.* **11,** 199–212.

Lund-Andersen H. and Moller M. (1977) Uptake of inulin by cells in rat brain cortex. *Exp. Brain Res.* **28,** 37–50.

Marchbanks R. M. (1970) Ion Transport and Metabolism in Brain, in *Membranes and Ion Transport* vol. 2 (Bittar E. E., ed.) Wiley-Interscience, London.

McIlwain H. and Rodnight R. (1962) *Practical Neurochemistry,* Churchill, London.

Moller M., Hertz L., Molgaard K., and Lund-Andersen H. (1974) Concordance between morphological and biochemical estimates of fluid spaces in rat brain cortex. *Exp. Brain Res.* **22,** 299–314.

Pappius H. M. (1982) Water Spaces, in *Handbook of Neurochemistry* vol. 1 (Lajtha A., ed.) Plenum, New York.

Pappius H. M., Klatzo I., and Elliott K. A. C. (1962) Further studies on swelling of brain slices. *Can. J. Biochem. Physiol.* **40,** 885–898.

Phelps C. F. (1965) The physical properties of inulin solutions. *Biochem. J.* **95,** 41–47.

Pollay M., Stevens A., and Davis C. (1966) Determination of plasma-thiocyanate binding and the Donnan ratio under simulated physiological conditions. *Anal. Biochem.* **17,** 192–200.

Rubiales de Barioglio and Orrego F. (1982) A study of calcium compartments in rat brain cortex thin slices: Effects of veratridine, lithium and of a mitochondrial uncoupler. *Neurochem. Res.* **11,** 1427–1435.

Schousboe A. and Hertz L. (1971) Effects of potassium on indicator spaces and fluxes in slices of brain cortex from adult and new-born rats. *J. Neurochem.* **18,** 67–77.

Selwood L. (1970) Electron microscopy of the fate of exogenous ferritin in the feline visual cortex. *Z. Zellforsch.* **107,** 6–14.

Stewart M. A., Sherman W. R., Kurien M. M., Moonsammy G. I., and Wisgerhof M. (1967) Polyol accumulations in nervous tissue of rats with experimental diabetes and galactosaemia. *J. Neurochem.* **14,** 1057–1066.

Varon S. and McIlwain H. (1961) Fluid content and compartments in isolated cerebral tissues. *J. Neurochem.* **8,** 262–275.

Walz W. and Hertz L. (1983) Comparison between fluxes of potassium and of chloride in astrocytes in primary cultures. *Brain Res.* **277,** 321–328.

White W. F., Nadler J. V., and Cotman C. W. (1978) A perfusion chamber for the study of CNS physiology and pharmacology in vitro. *Brain Res.* **152,** 591–596.

# Ion Transport and Volume Measurements in Cell Cultures

H. K. Kimelberg and Wolfgang Walz

## 1. Introduction

Studies on transmembrane ion movements in brain cells are of fundamental importance to understanding brain function. Thus, studies on the content and fluxes of major ions such as $K^+$, $Na^+$, $Cl^-$, and $Ca^{2+}$ in neurons are critical to understanding the effects of conductance changes during excitatory or inhibitory events (Hille, 1984; Katz, 1966), and changes in $Ca^{2+}$ conductance are critical for transmitter release (Hille, 1984; Douglas, 1978). Changes in pH also affect membrane conductances (Moody, 1983). The complexities of the ion transport processes present in the major nonneuronal cells of the brain (glia, endothelia, and ependyma) are also now beginning to be appreciated and studied, and many of these processes appear to be electrically silent. Such processes appear likely to be involved in control of extracellular ion concentrations and pH and thus will also be important for neuronal function (Varon and Somjen, 1979; Kimelberg and Bourke, 1982; Kimelberg and Ransom, 1986). Exaggeration of such processes may underly the swelling of astroglia frequently seen in various pathological states (Kimelberg and Ransom, 1986). The maintenance of low intracellular sodium concentrations ($[Na^+]_i$) is also important for maintaining inward $Na^+$ gradients for secondary active cotransport of transmitters and other substances into both neurons and glia (Fonnum et al., 1980).

In order to understand these processes at both the cellular and molecular level, it is important to be able to study individual cells. Apart from various microelectrode techniques, however, including the measurement of ion activities using ion-specific microelectrodes (ISMs) with which recordings from individual cells in intact tissue can be made, investigators studying ion transport have had to treat nervous tissue as a two-compartment system in which one can only clearly delineate the total extracellular space (ECS) from the total intracellular space. This is clearly a serious barrier to understanding ion transport at the cell level. An obvious approach to studying the properties of individual classes of brain cells is to study them as separate populations in culture. In addition, this also appears to have advantages over trying to isolate individual cells from brain tissue, since the latter methods are time-consuming and often result in a poor cell yield or a high proportion of damaged cells and preparations of varying purity. Nervous system cell cultures can be grown as primary cultures or cell lines. The latter also include transformed (malignant) cells that have the advantage of growing rapidly and can be passaged for long periods of time, but may well have altered properties. The different types of cultures and their relative merits and usefulness for ion transport studies will be discussed in the next section. Also see Hertz et al. (1985a,b) for recent reviews of neuronal and glial cultures.

## 2. Types of Cell Cultures

Cell cultures can be classified according to a variety of criteria, such as whether the cells are normal or transformed to a malignant state. Both these types can be grown directly from tissue (primary cultures) and can then be successively passaged. In this way a line can be set up that can be passaged up to a maximum of 50 times for normal cells and indefinitely for transformed cells (Hayflick and Moorhead, 1961). Usually cells from such lines are frozen at an early passage and then thawed and grown for a fixed number of passages when required. When such lines are established from a single cell by dilution of the original suspension into single-cell samples, the line is a clone. Many cultures are grown substrate-attached in different kinds of plastic vessels. This type of culture makes uptake studies and subsequent washing techniques relatively simple, and cultures of normal cells from the nervous system are invariably of this type. Some cell types, e.g., lymphocytes or ascites tumor cells, can be grown as suspension cultures, and in this case ion transport studies require filtration and sedimentation techniques. Such procedures have been well worked out for red blood cells, which have been extensively used in transport studies, and many of these procedures are well-suited for measurements of rapid changes. Cells in suspension are very convenient for studies using spectroscopic and fluorescent techniques and are particularly suited to Coulter counter or light-scattering techniques to measure cell volume. As will be discussed in section 4.3, indirect tracer studies can be used for studying cell volume in attached cells, but these introduce important methodological problems of their own. Detachment of such cells, however, usually with trypsin, may affect the leakiness and physiological functioning of cell membranes, introducing uncertainty with regard to physiological function. Attached cells are, however, very convenient for electrophysiology. Cultures can also be started from excised blocks of tissue, and these are referred to as organotypic or explant cultures. For ion-transport studies, such cultures have the same interpretative limitations as a tissue slice, but like slices they can also be useful for electrophysiology. Thus a number of different cultures suited to particular purposes are available.

Apart from such methodological questions, there is the equally important question of interpretation; how closely do the cells in the different culture types correspond to cells *in situ?* For example, it is often assumed that cells in explant or organotypic cultures will represent cells *in situ* most closely, since they reproduce to some extent the three-dimensional relations of tissue. It is unlikely that transformed cell cultures are representative of normal cells in culture from which they have been derived. A transformed cell line can be very useful, however, if it expresses a particular transport mechanism. In contrast, primary cultures are likely to correspond more closely to normal cell types present in the brain than transformed cells and provide an accessibility and homogeneity that is not provided by the explant cultures. There are now methods for preparing astrocyte cultures (e.g., *see* Hertz et al., 1985a,b), and these can be easily applied to different brain regions (Schousboe and Divac, 1979; Kimelberg and Katz, 1986; Hansson et al., 1984). Oligodendrocytes can be prepared from neonatal brain tissue, or oligodendrocytes isolated from lamb brain can be maintained in culture (Hertz et al., 1985a). Of these preparations, the astrocytes are easiest to grow; like transformed cells they proliferate and can also be passaged, although the total number of passages that they can undergo has not been examined. Neurons and oligodendrocytes do not or slowly proliferate, and thus the amount of material obtained is small. In addition, neurons especially need precise conditions in which to grow, and these include a background carpet of astrocytes or factors derived from such cultures or brain tissue (Hertz et al., 1985b; Varon and Somjen, 1979). Cell cultures are usually grown in a suitable growth medium containing serum, although defined media in which serum is replaced by a number of defined compounds are being used increasingly, especially for studies emphasizing cell growth and development (Bottenstein, 1985; Morrison and De Vellis, 1981, 1983).

Table 1 represents a list illustrating some of the types of cultures used for studies on ion transport, cell volume, and electrophysiology and some of the results obtained in these studies. See also the chapter by Hertz et al. (1985b) in volume 1 of this series for a review of neuronal and glial cultures and references to the original literature.

## 3. Ion Measurements

### 3.1. Nonisotopic Methods

These methods include a diverse variety of techniques such as measurement of the ion content of extracts of the cell contents by flame photometry or atomic absorption spectroscopy (AAS). They also include use of fluorescent dyes or other probes sensitive to ions such as $Ca^{2+}$. In the latter technique the probes are incorporated into the cells and have the advantage that dynamic changes in real time of viable cells can be continuously measured. Nonisotopic methods also include the use of ion-sensitive microelectrodes, which will be dealt with in other chapters in this volume (*see* chapters by Nicholson/Rice and Schlue/Dietmar).

#### 3.1.1. Spectroscopic and Photometric Measurements of Ion Content

For monolayer cultures, the external medium has to be first removed, and the cells washed several times. The cells are then lysed, and the ion contents of the solution determined and referred to some measure of total cell content, usually milligrams of total cell protein. The washing media can vary, but Sanui and Rubin (1979) have reported that $CO_2$-free $0.25M$ sucrose is better than physiological electrolyte solutions such as Ringer's type media for this purpose, since the latter, but not the former, appear to remove protein and surface exchangeable cations, particularly $Ca^{2+}$. Either the ion content of the cells as they grow in growth media or the ion content of the cells after experimental manipulations are usually measured. In the latter case the following protocol is followed: The growth medium is first poured off, and the cells are then washed in the reaction medium. The basic reaction medium generally used in our laboratories consists of the following components (in mmol/L); NaCl, 122; KCl, 3; $CaCl_2$, 1.3; $MgSO_4$, 0.4; $KH_2PO_4$, 1.2; $NaHCO_3$ or HEPES, 25; and glucose, 10. The solution pH is adjusted to pH 7.4 with $1N$ NaOH in the case of HEPES and by bubbling with 5%

$CO_2$/95% air for the bicarbonate solution. The total osmolality is approximately 300 osmol/kg. The cells will then be incubated in this medium for varying times at 37°C in a 5% $CO_2$/95% air incubator for $HCO_3^-$-buffered solutions, with different additions depending on the experiment. If a $CO_2$ incubator is not available, then 25 m$M$ HEPES or some other fixed buffer can be substituted for the $HCO_3^-$/$CO_2$ buffer. Use of such buffers may, however, have unknown physiological effects, since such buffers are obviously unphysiological. Thus with bicarbonate buffer, $H_2CO_3$ or especially $CO_2$ can readily cross cell membranes, and thus also buffer the cells intracellularly (Roos and Boron, 1981). In relation to this, it should be noted that the $H^+$ initially produced intracellularly by dissociation of $H_2CO_3$ is usually extruded, e.g., by $Na^+$/$H^+$ exchange (Roos and Boron, 1981). This is then buffered by the excess $HCO_3^-$ in the outside medium that, because of the relative volumes of usually at least 1000:1 extra- to intracellular volume, acts as a reservoir. *In situ* the extracellular space is usually about 20% of the intracellular, so here neutralization of $CO_2$ is likely to be principally by intracellular buffers. In contrast, it is not clear if the unionized forms of HEPES or other buffers are always permeable.

After appropriate experimental treatments, the cells are washed rapidly using a repipet (Labindustries) or automatic pipet, with each addition of wash being rapidly and completely aspirated. The total time for washing should be 10–30 s. We use ice-cold 0.29M sucrose buffered with 10 m$M$ Tris-nitrate, pH 7.4, and also containing 0.5 m$M$ Ca nitrate to maintain cell adhesion to the substrate, to wash our cells (see *also* Sanui and Rubin, 1979). For 12-well plates we wash three to seven times with 1.0 mL per well. *See* Fig. 1 for effects of washing on $^{14}$C-urea and $^{14}$C-3-*O*-methyl-D-glucose retention by a primary astrocyte monolayer culture. After the final wash, the cells are treated with 1mL of 5% perchloric acid and brought to a final volume of 2.5 mL with distilled water. After centrifugation, the $Na^+$, $K^+$, $Mg^{2+}$, or $Ca^{2+}$ content of the supernatant can be determined using atomic absorption spectrophotometry (AAS). Determinations of how much material is required will have to be made in each case; for AAS, the total protein from a 60-mm diameter culture dish with contents diluted to 2–3 mL is found to be adequate. This volume can be further adjusted if necessary. An example of the use of AAS to measure the ion content of primary rat astrocyte cultures is shown in Fig. 2, in which the effects of addition and then removal of ouabain on $K^+$ and $Na^+$ content were followed. As shown in Fig.

Table 1
Representative Neural Cell Cultures Used in Ion Transport, Electrophysiology, and Cell Volume Studies

| Type | Used for studies on | | | Properties found | Representative references |
| --- | --- | --- | --- | --- | --- |
| | Ion transport | Electro physiology | Cell volume | | |
| Glioma $C_6$ (rat) | Yes | Yes | Yes | Membrane potentials of −36 and −90mV have been reported. (Na + K) pump- and furosemide-sensitive $K^+$ + $Cl^-$ cotransport present. Intracellular water content of 4.0 and 6.4 µL/mg protein. | Erecinska and Silver, 1986; Kukes et al., 1976a,b.; Kimelberg, 1974; Johnson et al., 1982 |
| Neuroblastoma (mouse) | Yes | Yes | No | (Na + K) pump- but no furosemide-sensitive cotransport present. | Johnson et al., 1982; Kimelberg, 1974 |
| Glioma LRM (rat spinal cord) | Yes | Yes | Yes | Anion exchange- and furosemide-sensitive cotransport. $E_m$ −43 mV. | Wolpaw and Martin, 1984 |
| Primary astrocyte cultures (rat and mouse) | Yes | Yes | Yes | Anion exchange- and furosemide-sensitive cotransport. $Na^+/H^+$ exchange. Large negative membrane potentials (−60 to −90 | Kimelberg et al., 1979a,b; Kimelberg and Hirata, 1981; Hirata et al., 1983; Bowman and Kimelberg, 1984; Walz et |

| | | | | | |
|---|---|---|---|---|---|
| | | | | mV). Depolarization in response to some transmitters. Shows regulatory volume decrease. Volumes of 3–5 µL/mg protein. | al., 1984; Kettenman and Schachner, 1985; Kimelberg and Frangakis, 1985 |
| Primary neuronal cultures (various types) | No | Yes | No | Membrane potentials of around −40 to −60 mV. Potentials responsive to transmitters. Spontaneous spike activity. | Ransom et al., 1977; Peacock, 1979. Greatly increasing number of electrophysiological and patch clamp studies since 1980. Consult recent journal issues, especially *J. Neurosci.* and *Nature.* |
| Primary oligodendrocyte cultures (embryonic mouse spinal cord) | No | Yes | No | Large negative membrane potentials. Respond to neurotransmitters. | Kettenman et al., 1983b, 1984 |
| Explant cultures (spinal cord and cerebellum) | No | Yes | No | Studies of both neurons and glia. Depolarization of both to neurotransmitters. Responses of glia thought to be caused by $K^+$ released from neurons. | Hosli and Hosli, 1983; Hosli et al., 1982 |
| Glioma hybrids | No | Yes | No | Depolarization to different agents. | Heumann et al., 1982 |

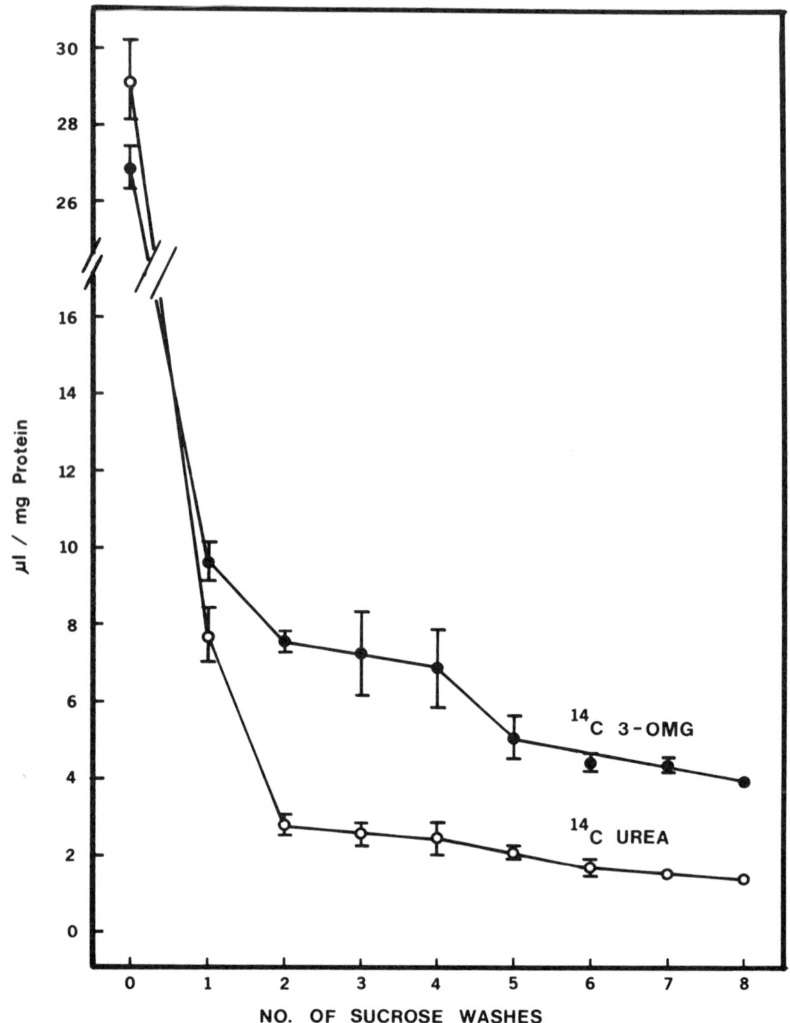


Fig. 1. Effect of successive washings on $^{14}$C-3-$O$-methyl glucose (3-OMG) and $^{14}$C-urea content of rat astrocyte cultures after exposure to the labeled compounds for 10 min. The growth media was first poured off and the cells, growing in 12-well trays (Costar), were washed with $HCO_3^-$-buffered, balanced salt solution (NaCl, 122; KCl, 3.3; $MgSO_4$, 0.4; $CaCl_2$, 1.3; K phosphate, 1.2; $NaHCO_3$, 25; glucose, 10; pH 7.4, all concentrations in m$M$) and then allowed to incubate in 0.5 mL of this medium for 15 min in a 5% $CO_2$/95% air atmosphere. $^{14}$C-3-OMG (0.3 µCi) or $^{14}$C-urea (1 µCi) were then added. After 10 min, the cells were washed with ice-cold 290 m$M$ sucrose, 10 mM Tris-$NO_3$, 0.5 m$M$ Ca($NO_3$)$_2$, pH 7.4 (1 mL/wash) by

2B, the reaccumulation of $K^+$ after removal of ouabain corresponded closely to $^{86}Rb^+$ uptake.

For determination of ion content by flame photometry, the method of preparation is essentially identical to that described above for AAS. In this case, however, the cells are scraped into 2.5 mL distilled water containing 1.5 m$M$ LiCl as a standard rather than precipitated with perchloric acid. This is then sonicated, and the ion and protein content of aliquots are measured.

### 3.1.2. Measurements Using Fluorescent Probes

These can be very useful for continuous measurements and to show the immediate effects of various additions. For example, Grinstein and colleagues (Grinstein and Furuya, 1983; Grinstein et al., 1984a,b,c), using lymphocytes in suspension culture, have shown that the use of fluorescent dyes to measure $pH_i$, $E_m$, and $[Ca^{2+}]_i$, together with a Coulter counter to measure cell volume, can be a very fruitful experimental approach to study the changes involved in cell volume regulation. One problem with nervous system cultures, however, is that they are grown, in almost all cases, as monolayers, and it is unlikely that they can be made to grow as suspension cultures. Thus one has to detach the cells to do such measurements. Cells can be detached by (1) scraping with a rubber policeman, (2) incubation in $Ca^{2+}$-free medium containing

---

rapid addition and removal of each wash with an automatic pipet and aspiration using a suction pump and trap. In the case of $^{14}C$-3-OMG, the wash medium also contained 0.2 m$M$ phloretin to block the equilibratory glucose carrier (Kletzien et al., 1975). After the final wash the cells were dissolved in 1 mL of 1$N$ NaOH and left for 15 min at room temperature. An aliquot (0.1 mL) was taken for cell protein determination (Kimelberg et al., 1979a), and the remainder was added to 1 mL 1$N$ HCl. This together with 14 mL of Liquiscint was then transferred to a 20–mL scintillation vial, and $^{14}C$ radioactivity was then counted in a Packard Tricarb scintillation counter (Model #3330). Cell volume was calculated based on the specific activity CPM/$\mu$L of the $^{14}C$-labeled compound in the medium and assuming equilibration of the compound inside and outside the cell. For the $^{14}C$-urea experiments, 28-d primary astrocyte cultures prepared from 1-d-old rat cerebral cortex were used, whereas for the $^{14}C$-3-OMG experiments the cultures were 23-d secondary cultures passaged from primary astrocyte cultures.

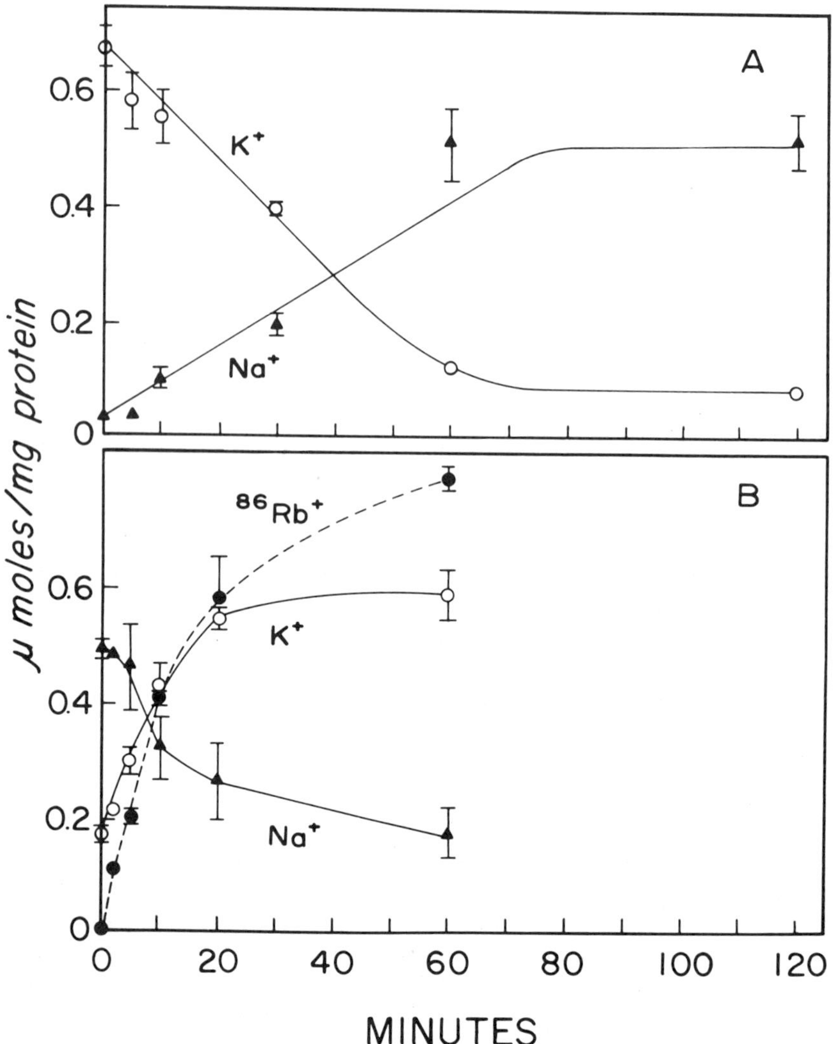


Fig. 2. Ion contents of primary rat astrocyte cultures after addition (A) and removal (B) of ouabain. (A) Complete growth media was poured off from 3-wk-old cultures growing in 60-mm culture dishes and started from the cerebral hemispheres of 1-d-old rats. The cells were then washed twice in 3 mL of HEPES buffered medium: NaCl, 122; KCl, 3; CaCl$_2$, 1.3; MgSO$_4$, 1.2; KH$_2$PO$_4$, 1.2; HEPES (titrated to give a final pH of 7.4 with NaOH at 37°C), 20; glucose, 10. All values represent mmol/L. A 2.5-mL portion of this medium at a temperature of 37°C and containing 1 m$M$ ouabain was then added. For some dishes, 2 μCi $^{14}$C-sucrose was also

1.0 m$M$ EDTA or EGTA, or (3) the use of 0.25% trypsin $\pm$ 1 m$M$ EDTA or EGTA. One simply has to test which treatment gives the best cell viability and dispersion as a single-cell suspension. It is also important to make sure that detachment of the cells does not alter the properties one is interested in. Thus it will be important to correlate these studies with studies on attached cells using the isotope techniques described later in this chapter.

For the different techniques used, the original references cited above should be referred to, and, because of space limitations, a complete description will not be given here. In brief, when the cells are properly in suspension they are then loaded with the probes. Cells are incubated with the probes for 30–60 min in the solution to be used for experiments. Unincorporated probes can then be separated from incorporated probes by centrifuging the cells and removing the supernatant. The cells can be washed once or twice more and the cells resuspended for final experimentation. Examples of such probes are (1) intracellular pH (pH$_i$)—5,6-dicarboxyfluorescein diacetate or *bis*-carboxyethyl carboxyfluorescein acetoxymethyl ester, (2) membrane potential ($E_m$)—3,3'-dipropylthiadicarbocyanine [di-S-C$_3$-(5)], and (3) intracellular calcium [Ca$^{2+}_i$]—quin-2-acetoxymethyl ester. These probes are lipophilic and thus rapidly permeate cell membranes. In the case of

$\longleftarrow$

added. At the times indicated the medium was poured off from groups of wells and the cells were rapidly washed seven times (15–20 s) in 3 mL ice-cold 0.32$M$ sucrose solution and assayed for K$^+$ and Na$^+$ levels by AAS (see text). The extracellular adherent medium remaining after washing the cells was determined from the amount of $^{14}$C-sucrose present, and was found to average 0.17 $\mu$L/mg protein. The Na$^+$ values were corrected for this, which represented 0.023 $\mu$mol Na$^+$/mg protein. For K$^+$, the correction required was insignificant. Values shown are the means $\pm$SD with $n = 3$. (B) Three-week cultures were treated with 1 m$M$ ouabain in growth medium without fetal calf serum containing 142 m$M$ Na$^+$ and 4.9 m$M$ K$^+$, for 2 h at 37°C. The dishes were then washed twice in serum-free growth media. Two mL of the serum-free growth media at 37°C was then added, and the Na$^+$ and K$^+$ content of the cells determined as for (A) above. In addition, 2 $\mu$Ci $^{86}$Rb$^+$ was added to some of the cells after the same treatment as above, and the uptake of K$^+$ was calculated based on the uptake of $^{86}$Rb$^+$. The Na$^+$ content was corrected for remaining adherent medium as in (A) (from Kimelberg et al., 1979b, with permission).

dyes used for intracellular pH (1) and $Ca^{2+}$ (3), once inside the cell these compounds are cleaved by specific esterases to form a less lipophilic and therefore relatively membrane-impermeant product (Grinstein et al., 1984a,b,c). For $Ca^{2+}$, new probes known as Fura-1 and Fura-2 have recently become available and show superior fluorescence characteristics and better selectivity for $Ca^{2+}$ (Grynskiewicz et al., 1985). Corrections and standardizations for these methods are very important and will be completely described in the references to Grinstein and colleagues cited above. Also, since these compounds are lipophilic, corrections should be made for any binding to membranes.

## 3.2. Isotopic Methods

The methods differ depending on whether cells in monolayers attached to the culture dishes or cells in suspensions are used, and, as pointed out in section 3.1.2 above, nervous system cultures are usually monolayers. The techniques will also differ according to whether one uses beta or gamma emitters as radiotracers and also if one studies ions that are free in the cytoplasm or are significantly surface-membrane or intracellularly bound, as in the case of $Ca^{2+}$. The different approaches that have been used will now be considered.

### 3.2.1. Experimental Chambers

When cells are growing in suspension they are first washed and then collected by centrifugation. Unincorporated isotopes in the extracellular space can be removed by several centrifugations and washings. This poses the problem, however, that fast kinetics cannot be studied; there is also a problem of loss of diffusible intracellular contents caused by several washings. Alternatively, centrifugation through silicone, or another suitable substance, removes most of the extracellular medium. The cells can then be collected as a pellet or centrifuged into a lower layer of trichloroacetic acid that serves both to precipitate cell protein and retain the cell contents (*see* Table 2 and legend for methodology).

Because most neural cultures grow as monolayers, they can be manipulated while still attached to their culture dish, making rapid washing procedures simpler. Cells can be grown in multiwell, 12- or 24-well plastic trays or single 35- or 60-mm diameter dishes. In addition, for fast kinetic determinations, cells can be grown in special transport measurement chambers, as, for example, described by Pearce et al. (1981). Figure 3 shows the perfusion chamber used by Pearce and coworkers for studying cerebellar cultures grown on 22 × 22 mm coverslips.

Table 2
Intracellular Spaces of Glioma Cells (Rat LRM55 Cells)

| Space marker | Method | $n^a$ | Space ($\mu$L/mg prot)[b] |
|---|---|---|---|
| [14]C-Urea | Attached cells[c] | 9 | 2.47 ± 0.58 |
| [14]C-Urea | Suspended cells, centrifugation[d] | 4 | 3.07 ± 0.51 |
| [14]C-3-O-methyl-glucose | Suspended cells, centrifugation[d] | 4 | 1.97 ±0.24 |
| Tritiated Water | Suspended cells, centrifugation[d] | 4 | 3.93 ± 0.43 |
| — | Suspended cells, microscopic[e] | 2 | 5.72 ± 0.68 |

[a]Number of experiments.

[b]Values are mean ± SD.

[c]Cells remained attached to the substrate in multiwell plates. Urea space was determined from measurements of the radioactivity ([14]C content) of the medium and the cells by methods used for transport experiments on the same cells. (Martin and Shain, 1979).

[d]Cells were removed from culture dishes with enzymes (mixture of trypsin, collagenase, and chick serum). Cells were allowed to equilibrate with space markers and then collected by centrifugation through silicone oil. Intracellular space was calculated from radioactivity in the pellet and supernatant. Correction for entrained extracellular fluid was made with [14]C-sucrose.

[e]Cells were removed from culture dishes with enzymes and photographed on a hemocytometer grid under phase contrast optics. Volumes were calculated from cell diameters measured on projected images by using hemocytometer grid lines as the reference distance (unpublished work of D. L. Martin).

## 3.2.2. Radiotracers and Experimental Solutions

$K^{42}$ and $Na^{22}$ are gamma emitters, so that they can be easily counted in a gamma-counting system and the same sample then used afterward for other determinations such as those for proteins, DNA, and so on. Alternatively, their beta emission is of sufficient energy that they can be counted suspended in water in a liquid scintillation counter by Cerenkov radiation. $Ca^{45}$, $Cl^{36}$, and $Na^{24}$ are weaker beta emitters and must be counted in liquid scintillation fluids. This means that the counting sample can no longer be used afterward for other measurements, and the samples must be first divided for radioactivity determination and any additional measurements. Scintillation counting is com-

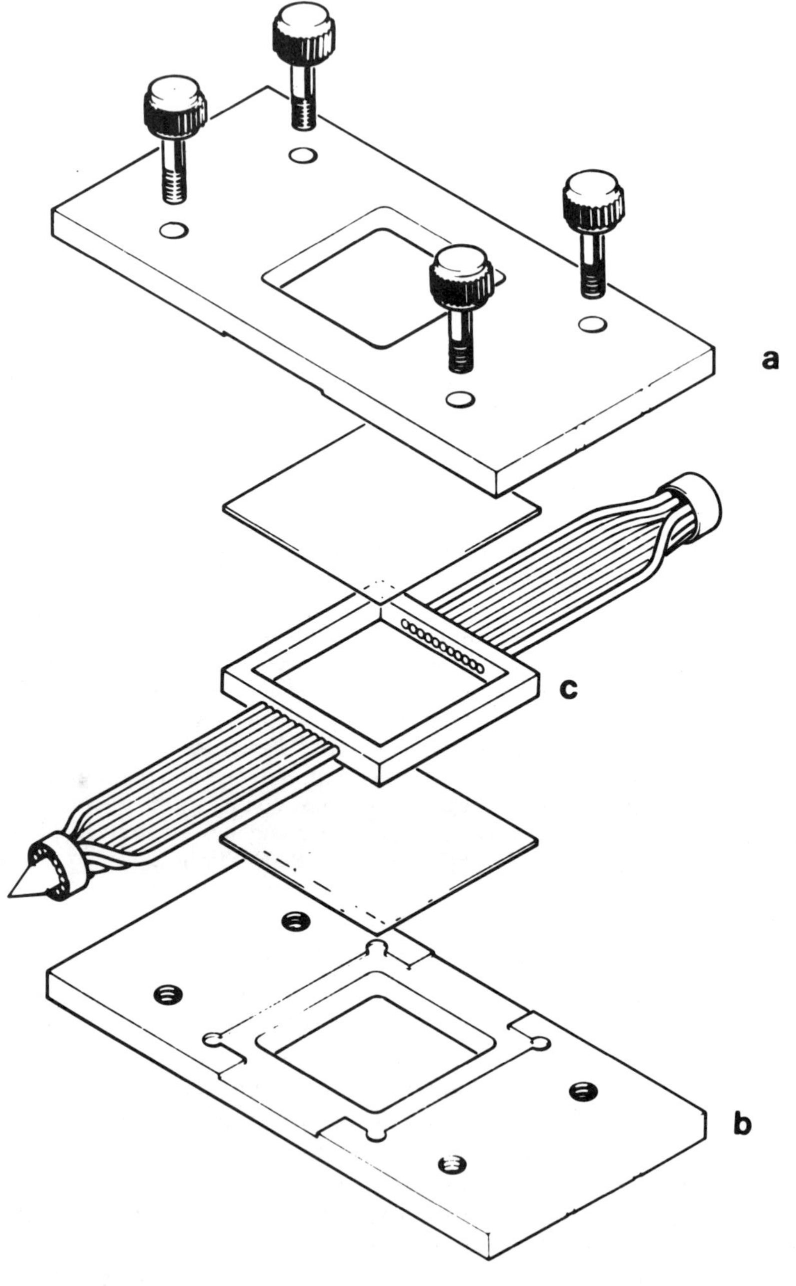

a
b
c
454

plicated by the possibility of quenching caused by absorption of the light signal emitted from the phosphors in the scintillation fluid excited by the beta radiation. Such quenching can be chemical quenching or caused by coloration of the sample. Standard correction procedures are easily applied and in newer instruments can be programmed into the readout. With cell cultures, unlike many tissues, quenching does not seem to be a serious problem. Using manually operated or programmed discriminators or "windows," counts from two radioactive labels can be separated, allowing double label experiments to be readily done. Scintillation fluid should be selected to optimize the counts, and one can prepare one's own formula or use one of many commercial brands that are available. We use a scintillation fluid that gives optimal results for $Ca^{45}$, $Cl^{36}$, and double-labeling experiments for $H^3$ and $C^{14}$ (used for volume measurements of attached cells); it has the following composition: 10 g 2,5-diphenylaxazole and 1 g dimethyl POPOP mixed well in 2 L of toluene. Thereafter a liter of Triton X-100 is added and mixed.

As discussed above for nonisotopic methods of determining ion content (section 3.1.1), the ion radiotracer experiments should be done in a physiological salt solution with the same ions present as in the growth medium, plus glucose and preferably a bicarbonate buffer system, which has to be kept in equilibrium with the correct $CO_2$ partial pressure in the gas phase. Other buffers, such as HEPES, can be used to replace bicarbonate (see above), but such substitutes should be checked for any effects this may cause. Also, incubation with salt solution could itself change the physiological parameters of the cultured cells, perhaps because of serum withdrawal. Therefore care has to be taken to incubate controls for the same amount of time as the experimental samples to determine any changes that may be caused simply by incubation with buffered salt solutions. The washing procedure could also be a problem. The exact duration of the washing procedure or the amount of washes needed should be determined experimentally and depend on the

---

Fig. 3. An exploded view of the perfusion chamber used by Pearce et al. (1981) showing the three components of the chamber and two 22 × 22 mm coverslips. The two coverslips were inserted, with the cell-bearing surfaces pointing inward, into the recesses of the outer elements a and b. For details, *see* Pearce et al. (1981).

relation of extra- to intracellular concentration of the ion in question. The temperature of the washing solution should be around 4°C (ice-cold). Salt solution could be used if the amount of tracer lost during washing is negligible. Otherwise buffered isotonic sucrose can be used. Calcium-free washes can often prevent loss of ions by blocking of ion channels, but may decrease adhesion of cells to the substrate. The addition of transport inhibitors to the washing medium for blocking efflux is an alternative approach.

### 3.2.3. Experimental Protocol: Efflux and Influx Measurements

Two efflux methods for measuring efflux into the medium can be used; a sampling and a washing method as reported by Boonstra et al. (1981a) for neuroblastoma cells. In addition, one can also measure the decrease in cell content. These three efflux methods and an influx method (Boonstra et al., 1981a) are described below:

3.2.3.1. SAMPLING EFFLUX METHOD.    Medium from cells growing in 35-mm diameter culture dishes is replaced by 1 mL of growth medium at 37°C. The cells are loaded with $^{42}$KCl or $^{86}$RbCl, added to a final concentration of approximately 2–3 n$M$ (specific activity usually around 1 Ci/g K$^+$ at time of shipment) by incubating for at least 3 h at 37°C. At the end of the incubation period, the medium containing the radionuclides is removed, and the cells are washed five times in 1 mL phosphate-buffered saline at room temperature, the whole procedure being completed within 20 s. The final addition is the reaction medium. Sequential samples of 0.05 mL of this medium are taken at various time intervals and assayed for radioactivity by measuring Cerenkov radiation using a liquid scintillation counter. After the final sample is taken, the cells are lysed with ice-cold 10% trichloroacetic acid. The lysate is then transferred to a scintillation vial, and its radioactivity is determined. The initial intracellular content of the isotope at zero time is determined from the sum of the counts leaving the cells during the time of the experiment and those remaining at the end of the experiment, as obtained from the trichloroacetic acid lysate.

3.2.3.2. WASHING EFFLUX METHOD.    Cells are equilibrated with isotope as above. Labeled medium is removed from the cells, and the cells washed with 1 mL phosphate-buffered saline at room temperature at various intervals over a 10-min period, the washing fluid being removed each time and is taken for counting. The radioactivity remaining in the cells at the end of the experiment is again determined from a trichloroacetic acid lysate. In order to distinguish between radioactivity in the washing fluid that is caused by efflux from the cells and extracellular radioactivity remaining after removal of the high-specific-activity incubation medium, a correction procedure was developed. For this correction, some of the cells are exposed to 15 $\mu M$ gramicidin at 37°C for 15 min. Gramicidin increases the membrane permeability to various ions, including $K^+$ and $Na^+$, and thus the intra- and extracellular concentrations become equal in its presence. After this preincubation in gramicidin, the cells are washed at intervals and the washing fluid is counted for radioactivity. In the case of the gramicidin-treated cells, all the $K^+$ is taken as extracellular, since the intracellular $K^+$ has equilibrated with the extracellular $K^+$ and is thus negligible, since $[K^+]_i$ is $>> [K^+]_o$. The radioactivity in the washes from experiments is then corrected for initial extracellular activity using the data from gramicidin-treated cells (Walz and Hertz, 1983). That these data indeed represent removal of the remainder of the initial radioactivity was confirmed by experiments in which inulin was used as an extracellular marker. A check for possible cell loss during the washing procedure was made by measuring the loss of protein-incorporated $^3$H-alanine (specific activity, 5.5 mCi/mmol) during the procedure from cells previously incubated with the amino acid (1.8 $\mu M$ final concentration) for 24 h. This method is only really applicable to $K^+$ where $[K^+]_i >> [K^+]_o$.

3.2.3.3. MEASUREMENT OF EFFLUX BY MEASURING CELL CONTENT.     After loading cells with isotope to equilibrium, the cells are then washed three to seven times as described in section 3.1.1 so that the isotope left is predominantly that present in the cells. A small residual fraction (10%) may be present in the extracellular medium, but this can be ignored since the efflux curve should identify this as a rapid initial loss. The cells are then washed again as before at varying intervals of time for a group of cells for each time period. The cell isotope content and cell protein can then be determined for each time. Cell radioactivity content expressed in cpm/mg protein can then be plotted as a function of time. In this method, however, each group of cells (usually three to four wells or dishes) can only be used for one time point. Plotting the log of cpm/mg protein as a function of time allows one to measure the slope that represents

$$\frac{\log \text{cpm}_1 - \log \text{cpm}_2}{(t_1 - t_2)} \times 2.304 = k \tag{1}$$

where $k$ is the first-order rate constant expressed in units of reciprocal time (*see* also, section 5.4), and $cpm_1$ and $cpm_2$ represent the cpm at two time points, $t_1$ and $t_2$, respectively. Multiplication of the initial ion content in nmol/mg protein by this rate will give the efflux in nmol/mg protein × time. In this case one can measure unidirectional efflux under steady-state conditions, i.e., no net change in ion content is occurring, but with the isotope one is tracing the unidirectional efflux, or net loss.

Figure 4 represents an example of a $^{36}Cl^-$ efflux experiment with primary astrocyte cultures, in which the efflux was measured by measuring cell content. The effect of transtimulation of $^{36}Cl^-$ efflux by addition of external $Cl^-$ or $HCO_3^-$ ion is shown. In this case a single-rate process was observed, and the values for $k$ for this process, calculated as described above, are given on the graph.

3.2.3.4. INFLUX MEASUREMENTS.    In this method one usually measures the unidirectional influx of isotope. As usual, growth medium is removed and replaced by fresh growth or balanced salt medium at 37°C, usually with one or two washes with the same medium to ensure complete removal of growth media. A short period of time is then usually allowed for equilibration (15–30 min), and this can include addition of an inhibitor if such an agent requires some time for complete interaction. In this case, however, one has to take into account possible rearrangement of ion gradients caused by the inhibitor. Tracer radioisotope is then added and the cells left at 37°C. At various time intervals the labeled medium is removed and the cells washed three to seven times in 1 mL phosphate-buffered saline (pH 7.3) or sucrose medium, at room or ice-cold temperature. Such washing can be completed within 10–20 s (*see* section 3.1.1). In order to rapidly measure the contribution of extracellular isotope, the label can be added and the cells immediately washed before the isotope has time to permeate the cells. If, however, the influx rate is very fast, this procedure cannot be used. This question of the extracellular, adherent medium space is a difficult one. Sequestration of material in a slowly exchangeable extracellular region, e.g., below the monolayer, would not be corrected for by such a zero-time value. Also, extracellular markers such as mannitol or inulin may not rapidly enter this region (*see* also Rindler et al., 1979), or may bind to the cells. These possibilities are difficult to evaluate. As shown in Fig. 1, extensive washing may be the best correction, since the extracellular space is preferentially diluted. Care should be taken to see that the isotope is pure, i.e., there are no membrane-permeable breakdown products and, second, that a compound used as an extracellular marker is definitely not taken up. The isotope content is determined from the trichloroacetic acid lysate as described above, or the cells can be solubilized in a convenient volume of $1N$ NaOH.

### 3.2.4. Calcium and Magnesium

These two ions bind to cell surfaces. Sanui and Rubin (1979) developed a method to measure total, intracellular, and surface-bound calcium and magnesium in cell cultures. They used five rinses with $CO_2$-free isotonic sucrose solution (pH near neutrality), which removes essentially no cellular proteins or cations. This has an advantage over salt solutions and buffer in that the protein loss during washing is minimized. A sixth 10-s wash with pH 4 ($CO_2$-saturated) sucrose solution displaces essentially all extracellularly bound $Mg^{2+}$ and $Ca^{2+}$ (*see* Fig. 5).

Kurzinger et al. (1980) found three rate constants for calcium efflux from neuroblastoma–glioma cell hybrids and glioma cells. They are thought to be associated with surface-bound, free intracellular $Ca^{2+}$ and intracellular vesicular calcium stores, respectively. A low concentration (0.5 $\mu M$) of the ionophore A23187 blocks the calcium exchange in the intracellular stores without influencing the calcium permeability of the cell membrane significantly. This treatment with low A23187 makes studies of calcium efflux mechanisms through cell membranes feasible.

### 3.2.5. Methodological Considerations of Extracellular Space

One very important consideration with regard to use of isotopes is correction for adherent extracellular medium, as has been mentioned in the two preceding sections. Because of the small intracellular volume of the cell monolayer, any adherent medium will be significant. This can even be a problem after exhaustive washing (six to seven times), if the cell content of the substance is low, for example for $Na^+$. With such extensive washing, however, one has the problem of loss of intracellular material or of cells themselves. Thus a compromise protocol must be used. Corrections for intracellular isotope include adding isotope and washing immediately before the isotope has time to enter the cell, as mentioned above; this is referred to as a zero-time correction. Alternatively time-dependent entrapment may be a problem and one can use a marker that is known not to be taken up by the cells. Examples of this are radiolabeled sucrose, mannitol, or inulin (but see potential problems noted in 3.2.3.4).

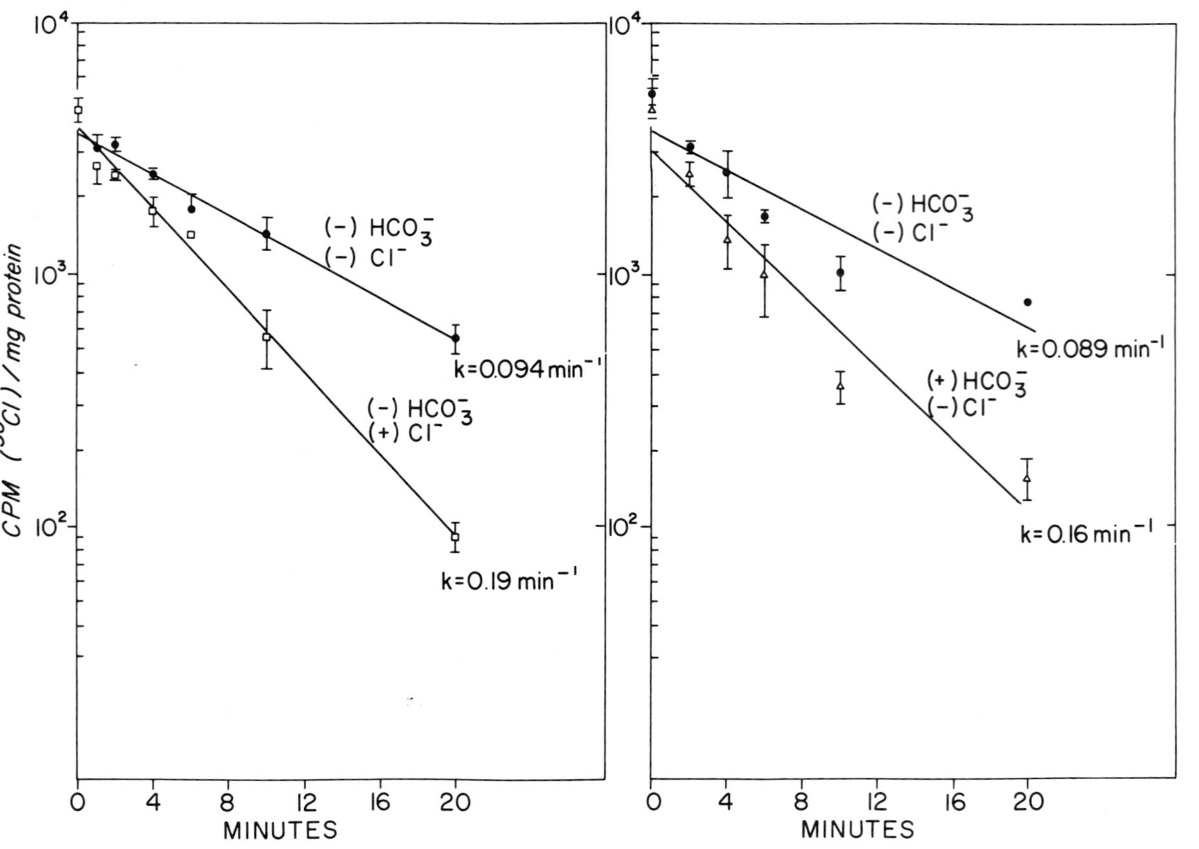
$10^4$
$10^3$
$CPM$ ($^{36}Cl$) / mg protein
(−) HCO₃⁻
(−) Cl⁻
k=0.094 min⁻¹
(−) HCO₃⁻
(+) Cl⁻
$10^2$
k=0.19 min⁻¹
0  4  8  12  16  20
MINUTES
$10^4$
(−) HCO₃⁻
(−) Cl⁻
$10^3$
k=0.089 min⁻¹
(+) HCO₃⁻
(−) Cl⁻
$10^2$
k=0.16 min⁻¹
0  4  8  12  16  20
MINUTES

Fig. 4. Efflux of $^{36}Cl^-$ from preloaded cells into $Cl^-$- and $HCO_3^-$-free media and effects of added $Cl^-$ or $HCO_3^-$. Cells were equilibrated with $^{36}Cl^-$ in a HEPES-buffered medium (*see* Fig. 2). After 1 h at 37°C, the cells were washed seven times with 3 mL of ice-cold $0.32M$ sucrose by rapid addition using a Labindustries repipet and immediate removal by aspiration using a suction pump (total wash time about 20 s). Then warm media of the same composition, but with $Cl^-$ replaced by isethionate, or media containing normal $Cl^-$, or isethionate plus 10 m$M$ $NaHCO_3^-$, was added. At various times the cells were again washed as before, and the cpm/mg protein remaining in a group of 3 different cell cultures in culture dishes was determined. The different media conditions are shown on the graphs. Each point represents the mean of three values ±SD. The lines are the best fit from linear least squares regression analysis, and the values for $k$ represent the first-order rate constants calculated from the slopes as explained in the text. Eighteen-day primary astrocyte cultures were used (from Kimelberg, 1981, with permission.)

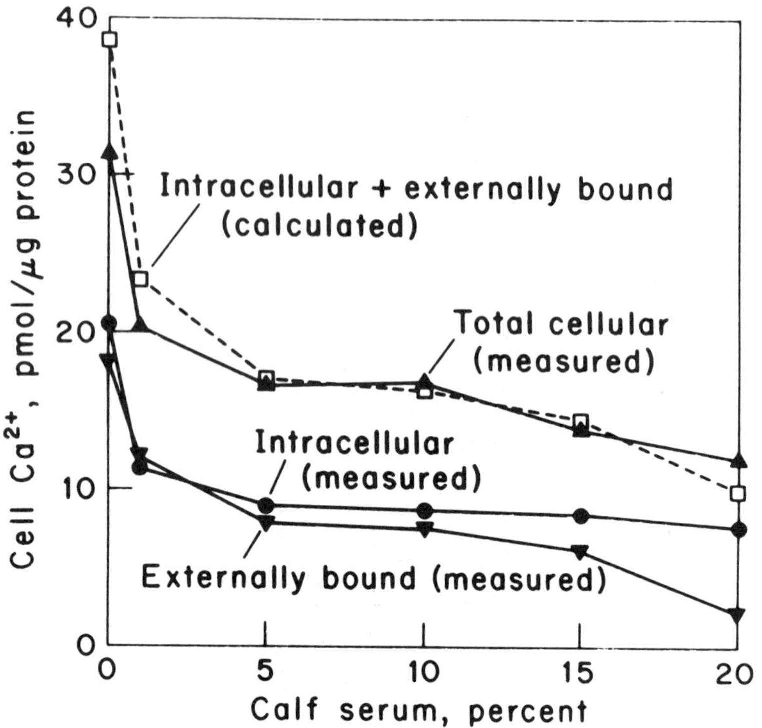



Fig. 5. Comparison of total cell calcium (measured) with total cell calcium as calculated from measured intracellular plus externally bound calcium of mouse Balb/c 3T3 cells. Cells were grown for 17 h in DME with the different serum concentrations shown. Cells were then washed five times with $CO_2$-free (pH, approximately 7) 0.25$M$ sucrose solutions. Half were then washed once in $CO_2$-free (pH, approximately 7) or $CO_2$-saturated (pH 4) sucrose solutions. Final washes were saved and $Ca^{2+}$ was determined, using AAS, in $CO_2$-free and $CO_2$-saturated solutions. The difference in these was taken as representing the externally bound $Ca^{2+}$. Intracellular $Ca^{2+}$ levels were then determined in all cells after they were given a final rinse with $CO_2$-saturated sucrose solution (from Sanui and Rubin, 1979, reproduced with permission).

### 3.3. Relation of Ion Fluxes to Intracellular Ion-Selective Microelectrode and Conductance Measurements

The intracellular content ($C_i$) of $K^+$, $Cl^-$, and $Na^+$ as measured for a certain number of cells is simply related to intracellular ion activities ($a_i$) as follows;

$$a_i = \frac{\gamma C_i}{V} \tag{2}$$

where $\gamma$ is the intracellular activity coefficient, and $V$ is the intracellular volume or water content of the cells. A comparison of radiotracer measurements of potassium and sodium contents and measurement of the volume of cultured astrocytes by Walz and Hertz (1984) and Walz and Hinks (1985) showed a relatively close approximation of these values with the directly measured activities of other authors in glial cells. The situation is different for calcium (and probably magnesium). Probably less than 1% of the total intracellular calcium exists as intracellular free calcium, with the remainder bound to proteins and in intracellular stores (Requena et al., 1977).

If the efflux, $j$, of an ion is entirely passive, e.g., when it is channel-mediated, a unidirectional flux rate can be measured and a permeability coefficient can be calculated. Thus for unidirectional efflux;

$$j_{eff} = \frac{\text{flux in mol/s}}{A} \tag{3}$$

where $A$ is the area of the cell surface and has to be known in all circumstances in which one is interested in calculating the permeability coefficient, $P$. $P$ in cm/s can then be calculated for the efflux of a positive ion from its concentration determined by specific microelectrodes as;

$$P = \frac{j_{eff}}{\gamma C_i} \times \frac{(1 - e^{-EF/RT})}{EF/RT} \tag{4}$$

where $C_i$ is the intracellular ion concentration and $E$ is the membrane potential. The sign of $E$ changes in relation to the direction of the ion movement, depending on whether the movement is being assisted $(+)$ or opposed $(-)$ by the electric field (Katz, 1966).

Let us say the ion is $K^+$. Then the $K^+$ resistance $R_K$ $\Omega$ in cm$^2$ can be calculated as follows:

$$R_K = \frac{1}{g_K} = \frac{RT}{P_K F^2} \times \frac{E - E_K}{E} \times \frac{(e^{EF/RT} - 1)}{\gamma C_i e^{EF/RT}} - \gamma C_o \tag{5}$$

where $g_K$ is the $K^+$ membrane conductance, and $E_K$ is the potassium equilibrium potential. With this equation, Boonstra et al.

(1981b) were able to calculate changes in the $K^+$ resistance of neuroblastoma cells during the cell cycle. Relative changes of the membrane resistance can probably be calculated in a reliable way using this method. Since it is very difficult to accurately determine the cell surface area in cell culture, however, absolute calculations of isotope flux measurements are problematic. Some reasonable estimates, however, of calculating surface area and volume for monolayer astrocyte cultures are discussed in the following section.

### 3.4. Cell Surface Area

As mentioned above, knowing the surface area of a cell is critical to obtaining accurate estimates of the true flux of substances, since the flux is clearly a function of the number of transporting sites embedded in the surface membranes or simply the total membrane surface area if the substance is diffusing freely through the membrane, which is, however, only likely to pertain to lipid-soluble substances. This approach, however, assumes homogeneous distribution of transporting sites for all cells, which may not be the case. Thus in epithelial cells it is well-established that transport sites are asymmetrically distributed on the apical and basolateral surfaces (Warnock et al., 1983). In the nervous system this has recently been shown to occur for $K^+$ channels in Muller cells of the retina (Newman, 1984). Alternatively, if one has a highly specific, tightly bound ligand to measure the number of the transporting sites, then the flux can also be expressed as moles per site per unit time. Failing this, it is usual to express fluxes per cell or per milligram total cell protein. Fluxes based on diffusion coefficients $(D)$ are usually given in units of, $cm^2/s$, and permeability coefficients $(P)$, in cm/s (divide $D$ by width of the cell membrane), i.e., $P$ has the dimensions of a velocity (Stein, 1967).

$$P = \frac{\text{flux}}{\text{concentration gradient} \times \text{area}}$$

which in the cgs system is equivalent to:

$$\frac{\text{mol/s}}{\text{mol/cm}^3 \times \text{cm}^2} = \text{cm/s}$$

$D$ is then equivalent to $P \times \text{cm} = \text{cm}^2/s$

As an example of an estimate of cell surface area for monolayer cultures, we have estimated the cell surface of primary astrocyte cultures (Kimelberg et al., 1979b) by assuming that the water content represented 80% of the cell mass. The volume of 4.8 $\mu$L cell water was obtained from a measured $K^+$ content of 0.61 $\mu$mol/mg cell protein and an intracellular $K^+$ concentration of 127 m$M$ estimated from Nernst plots of membrane potential vs $[K^+]_o$. We made the simplifying assumption that the shape of the cells was spherical. From this we obtained a value of $6.7 \times 10^{-6}$ cm$^2$ for the average area per cell. This gave reasonable values for pK and pNa using an equation that uses concentrations rather than activities (Eq. 4). For comparison, a value of 6.4 $\mu$L water/mg protein has been measured for $C_6$ glioma cells (Kukes et al., 1976a). In Table 2 are shown different values for the volume of monolayer LRM55 glioma cells based on a variety of different methods, including morphometric measures of detached LRM cells assuming a spherical shape. Astrocytes in culture, when not treated with dibutyryl cyclic AMP or other agents, often assume a flat shape. If we approximate this to a circle and use an average diameter of 20 $\mu$m, we then get a radius of $10^{-3}$ cm or a surface area for one face of the flat, disk-like cells of $3.14 \times 10^{-6}$ cm$^2$. Since the cells are approximately flat, we can assume that the total area is double this, or 6.28 $\times 10^{-6}$ cm$^2$. This value is actually quite close to the value of 6.7 $\times$ $10^{-6}$ cm$^2$ calculated above based on an estimate of cell volume and assuming the cell was spherical. Thus a value of around $6 \times 10^{-6}$ cm$^2$/cell may not be too inaccurate. Although this approach seems to hold for untreated astrocyte cultures, assumptions of a spherical shape may be at variance with the actual shape of the cell. Infoldings of the cell membrane or microvilli can lead to underestimations of the true surface area, and this problem together with the existence of cell processes of complex shape in vitro or in vivo can make the question of accurately measuring cell surface area very difficult.

We have found input impedances of about 2–10 M$\Omega$ for primary cultures of rat astrocytes (Hirata and Kimelberg, unpublished work; *see* Fig. 6). If this was a single cell, the area-specific resistance (resistance $\times$ area) for the larger value would be $10 \times 10^6$ $\Omega \times 6.28 \times$ $10^{-6}$ cm$^2$ = 62.8 $\times$ cm$^2\Omega$, taking the value of 6.28 cm$^2$ as the area of the cell. This is very low for cells, and since astrocytes in culture are known to form syncytia (Kettenmann et al., 1983a), we should multiply this value by the number of cells likely to form the syncytium. If this number is 10, the area-specific resistance will be 628 $\Omega \times$ cm$^2$; if 20, it will be 1256 $\Omega \times$ cm$^2$ (a reasonable value for cell membrane resistance), and so on. Also for electrophysiology we

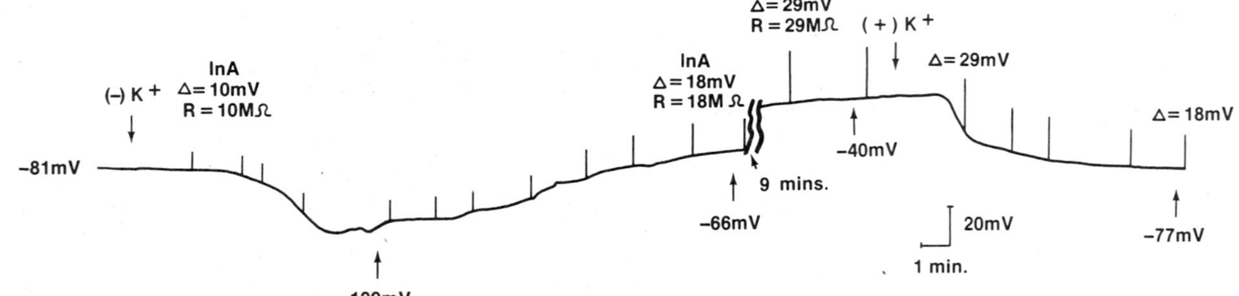



Fig. 6. Effect of changing to K$^+$-free medium on membrane potentials and resistance of a cell from a rat primary astrocyte culture. The medium was of the same composition as in Fig. 1, and was changed to K$^+$-free medium at the point marked by the arrow, (–) K$^+$. Na$^+$ concentration was simply increased to compensate for the small decrease in K$^+$. K$^+$ was reintroduced at (+) K$^+$. Vertical pulses are voltage responses to 1 namp (1 nA) injected current. The changes in potential and calculated resistances are shown in the figure. Eighteen-day cultures were treated with 1 m$M$ dibutyryl cyclic AMP 1 h prior to the experiment. This caused many of the cells to round up, facilitating impalement. Recording and injection of current was done with a single electrode, temperature was around 35°C (Hirata and Kimelberg, unpublished work).

may select for larger cells (Kimelberg et al., 1979a), so perhaps we can multiply further by a factor of 2 or 3. Flux measurements do not present these problems, since fluxes are referred to and are proportional to the number of cells or milligram of cell protein, although the area per cell needs to be known to calculate absolute permeabilities, as discussed above. For electrical measurements, however, one impales a single cell, but an unknown number of cells can be joined by low-electrical-resistance junctions (gap junctions), reducing the effective resistance.

We have applied these measurements of surface area and measured fluxes to primary astrocyte cultures (Kimelberg et al., 1979b). A $K^+$ efflux of $1.08 \times 10^{-11}$ mol/cm$^2$/s gave a $P_K$ of $3.70 \times 10^{-7}$ cm/s with $[K^+]_i = 127$ mM or $127 \times 10^{-6}$ mol/cm$^3$. Applying a simplification of the Eq. (5) from section 3.2.6 for $K^+$, namely

$$g = z\, C_i P F^2 / RT \tag{6}$$

then

$$
\begin{aligned}
g &= 1 \times 127 \times 10^{-6} \times 3.70 \times 10^{-7} \times (96{,}500)^2/8.314 \times 310 \\
&= 1.7 \times 10^{-4}\ \text{Cs/s/V/cm}^2 \\
&= 1.7 \times 10^{-4}\ \text{S/cm}^2
\end{aligned}
$$

or

$$R = 5882\ \Omega \times \text{cm}^2$$

This is probably too high for total cell resistance. It, of course, does not include conductances to other ions such as $Cl^-$, $Na^+$, or $HCO_3^-$, but the electrical properties of these cells are mainly those of a $K^+$-permselective membrane so that we may conclude that at least 90% of the current is being carried by $K^+$. In that case, the probable source of error would be estimates of $P$ based on surface area. The surface area used per cell was $6.7 \times 10^{-6}$ cm$^2$, the efflux was measured in mol/mg protein, and we found that there were $3.7 \times 10^6$ cells per mg protein. As estimated above, $6.7 \times 10^{-6}$ cm$^2$ seems a reasonable area, even for a flat cell. If we assume flux is only through the upper face, however, then the effective surface area will be only $3.35 \times 10^{-6}$ cm$^2$, and $R$ will also be about half the value calculated above, namely 2941 $\Omega$ cm$^2$. Clearly if the average area per cell is slightly less than the value used, the resistance could easily be within the "normal" range of 1000–2000 $\Omega \times$ cm$^2$ quoted for many cells (Hille, 1984). A further complication is that we assume that all the fluxes measured are conductive. We used $K^+$

efflux since this will not include ouabain-sensitive $(Na^+ + K^+)$ pump-driven $K^+$ uptake. We cannot, however, exclude that some of the efflux is an electrically neutral process such as furosemide-sensitive $K^+ + Na^+ + 2Cl^-$ efflux. Based on the large presumed conductance (channel-mediated) fluxes for $K^+$ in mouse, but not rat astrocytes (*see* Table 3), the resistance of the mouse cells may be very low indeed, say 50–100 $\Omega \times cm^2$. In this case fluxes and the electrically measured input impedance ($5 \times 10^6 \, \Omega \times 10^{-5} \, cm^2 = 50 \, \Omega \times cm^2$) may be quite close indeed. Thus the relation of ion flux measurements to measured electrical resistance is complicated by uncertainty about values for effective surface area and the proportion of the fluxes that occur via conductance and electrically neutral processes—these relationships have probably only been really solved for the red blood cell, in which, for example, the $Cl^-$ conductance flux is about six orders of magnitude less than the total flux measured (Lassen et al., 1978).

Table 3

Components of $K^+$ Unidirectional Influx in Different Cell Culture Systems (in nmol $\times$ mg/protein/min)[a]

| | Ouabain-sensitive | Furosemide-sensitive | Other (probably channel-mediated) | Total |
|---|---|---|---|---|
| Glioma cells | 7 | 7 | 0 | 14 |
| Neuroblastoma cells | 6 | 0 | 0 | 6 |
| Rat astrocytes in primary culture | 33 | 12 | 10 | 55 |
| Mouse astrocytes in primary culture | 33 | 12 | 1486 | 1541 |
| Cortical neurons in primary culture (mouse) | — | — | — | 88* |

[a]Values from Johnson et al. (1982), Walz et al. (1983), Walz and Kimelberg (1985), and Walz (unpublished results for cortical neurons; separate components were not determined).*

## 4. Volume and pH Measurements

### *4.1. Coulter Counter Methods*

In most cases when studying changes in ion transport, it is important to know if there are accompanying volume changes. This can be done simultaneously with probe methods for $pH_i$, $E_m$, or $Ca^{2+}$ using detached cells and a Coulter counter equipped with a channelyzer. This instrument sizes the population, giving a read-out of mean volume and distribution (Cheung et al., 1982). Fluorescence measurements can be made on the same preparation using a separate fluorescent spectrophotometer. Aliquots can be removed from the sample in the fluorescent spectrophotometer cuvet to measure volume at varying intervals in the Coulter counter.

### *4.2. Measurement of the $^3H_2O$ Space*

For this procedure, cells growing in a monolayer will first have to be detached as discussed above, since water movements are so rapid that even a brief washing will remove all the isotope.

Experimental manipulations are done on dissociated cell suspensions in the presence of $^3H_2O$ and $^{14}C$-sucrose in a suitable volume with shaking to prevent settling of the cells. After completion of the incubation period, the cell suspension is added to the centrifuge tube of an Eppendorf high-speed bench centrifuge containing silicone oil (*see* Table 2) or a corn oil/dibutyryl phthalate mixture (Cheung et al., 1982). The cells are then spun for 1 min, and the incubation medium and silicone are removed and the pellet suspended in distilled water by sonication or dissolved in a suitable volume of $1N$ NaOH. Alternatively, the tip of the plastic centrifuge tube can be cut off and the pellet dissolved in detergent and then counted (Cheung et al., 1982), or the cells can be centrifuged into a lower layer of trichloroacetic acid. Examples of values obtained by the silicone method for LRM55 glioma cells are given in Table 2.

### *4.3. Isotopic Methods for Cell Volume and Intracellular pH of Attached Cells*

#### 4.3.1. Cell Volume

For attached cells, intracellular volume can be measured using equilibration with $^{14}C$-3-$O$-methyl glucose (3-OMG), as originally described for hepatocytes (Kletzien et al., 1975). We have already used this method for primary astrocyte cultures, and it seems to work quite well (Kimelberg and Frangakis, 1985). We usually measure uptake of $^{14}$C-3-OMG for 20 min and express the content per mg protein. We also subtract a $^3$H-sucrose, -mannitol, or -inulin space to correct for extracellular, adherent medium. This can be done simultaneously with $^{14}$C-3-OMG on the same sample, as a double-label experiment. The $^3$H-sucrose uptake is measured for 10 or 20 min. Typically we use 1 μCi of $^3$H-labeled extracellular marker and about 0.3 μCi of [$^{14}$C]-3-OMG per 0.5 mL reaction volume per well in a 12-well plate. As shown in Fig. 7, 0.5 m$M$ phlorizin did not inhibit either $^{14}$C-3-O-OMG or $^{14}$C-D-glucose uptake, whereas the initial rate of uptake of $^{14}$C-3-O-MG was 80% inhibited and that for glucose was 95% inhibited by 0.2 m$M$ phloretin. These values are close to those found for chick intestine epithelial cells and suggest that uptake is indeed of the equilibratory rather than the Na$^+$-dependent concentrating type (Randles and Kimmich, 1978). We found that use of $^3$H-rather than $^{14}$C-labeled 3-O-MG gave us anomalously high values, presumably because of concentration of tritiated impurities, as reported by Lubin (1980).

Figure 1 shows the effect of successive washings on the $^{14}$C-3-OMG and $^{14}$C-urea space of astrocyte cultures. As can be seen, most of the radioactivity is removed in the first one or two washes. Thereafter a reasonably constant volume of 2 μL for $^{14}$C-urea and 4 μL for $^{14}$C-3-OMG per mg cell protein was obtained. The reasons for such differences are not clear, although the values of 2–4 μL are within the range of those reported for other cells (Kletzien et al., 1975; Rindler et al., 1979; Lubin, 1980; Bakker-Grunwald and Sinensky, 1979; and *see* Table 2).

### 4.3.2. Intracellular pH

By using the weak acid (p$K_a$ = 6.13) $^{14}$C-dimethyl oxazolidine-2,4-dione (DMO), it is possible to measure the intracellular pH based on equilibration of this compound (Waddell and Butler, 1959) and on measurements of intracellular volume using $^{14}$C-3-OMG, as described above. We have found that this technique seems to work quite well in primary astrocyte cultures (*see* Table 4). We apply the equation from Waddell and Butler (1959).

$$\text{pH}_i = 6.13 + \log\left[\left(\frac{C_t}{C_e}\left(1 + \frac{V_e}{V_i}\right) - \frac{V_e}{V_i}\right) \times \left(10^{(\text{pH}_e - 6.13)} + 1\right) - 1\right] \quad (7)$$

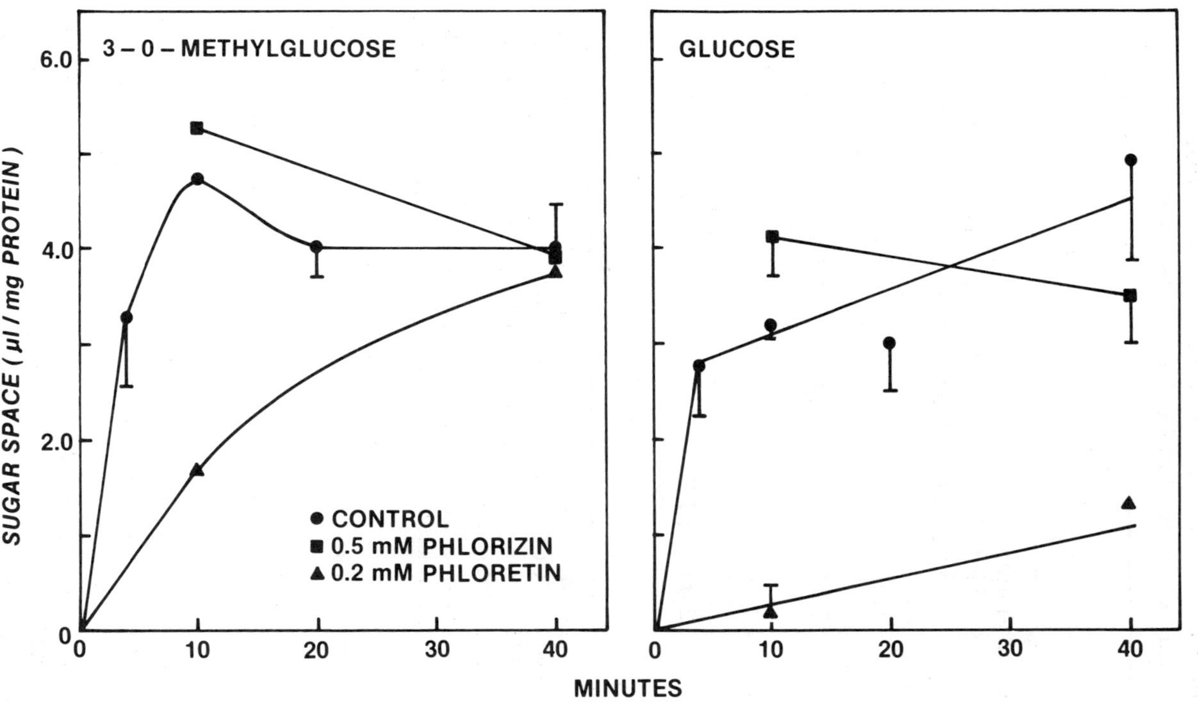



Fig. 7. Uptake of $^{14}$C-labeled 3-O-methyl glucose or glucose by primary astrocyte cultures in $HCO_3^-$-buffered medium. Inhibitors were added at the concentrations shown at 0 time. A $^{14}$C-sucrose space of 1.26 µL/mg protein determined concurrently in separate cultures was subtracted. The value of $n$ for each point = 3 – 6, ±SD (from Kimelberg and Frangakis, 1985, reproduced with permission).

471

Table 4
Effect of Na$^+$, Ouabain, and Furosemide on Internal pH
and Cell Volume of Cultured Astrocytes[a]

| | $V_i^b$ | pH$_o$ | pH$_I^e$ | $\Delta$pH (pH$_o$ – pH$_i$)[d] |
|---|---|---|---|---|
| HCO$_3^-$-buffered medium | | | | |
| Control | 3.11 ± 0.28 | 7.29 | 7.06 ± 0.07 | 0.23 |
| Na$^+$-free (K$^+$ = 50 m$M$) | 2.84 ± 0.23 | 7.39 | 6.76 ± 0.04 | 0.63[†] |
| + Ouabain (1 m$M$) | 3.01 ± 0.62 | 7.28 | 6.91 ± 0.04 | 0.37* |
| + Furosemide (5 m$M$) | 2.54 ± 0.18* | 7.44 | 7.03 ± 0.04 | 0.42** |
| HEPES-buffered medium | | | | |
| Control | 2.51 ± 0.10 | 7.34 | 7.34 ± 0.05 | 0 |
| Na$^+$-free (K$^+$ = 4.5 m$M$) | 2.09 ± 0.18* | 7.25 | 6.93 ± 0.14 | 0.33** |

[a]Cells 18–20 d old. Medium same composition as in Fig. 1. In Na$^+$-free HCO$_3^-$-buffered medium, K$_2$CO$_3$ was used, neutralized with KOH. Medium [K$^+$] was 50 m$M$. In HEPES-buffered medium, 10 m$M$ HEPES was used, neutralized with KOH. Medium [K$^+$] was kept at 4.5 m$M$.

[b]Intracellular volume in $\mu$L/mg protein using apparent equilibrium values for $^{14}$C-3-$O$-methyl-$D$-glucose after 40 min incubation and subtracting $^{14}$C-sucrose space.

[c]Intracellular pH determined by $^{14}$C-5,5-dimethyloxazolidine-2,4-dione ($^{14}$C-DMO) uptake also after 40 min. Extracellular space determined using $^{14}$C-sucrose equilibrated for 10 min.

[d]Values shown are means ± SD ($n$ = three separate dishes). *$p < 0.05$; **$p < 0.02$; [†] $p < 0.002$; Student's $t$-test for differences from appropriate controls (from Kimelberg et al., 1982).

where $C_t$ = total concentration of $^{14}$C-DMO in extra- + intracellular space and $C_e$ = concentration in the extracellular volume, which is usually 1 m$M$ in our experiments. $V_e$ = the extracellular volume and is equivalent to the $^{14}$C-sucrose space and $V_i$ = the intracellular volume. pH$_e$ = pH of medium. 6.13 = p$K_a$ of DMO. More recently $^{14}$C-labeled benzoic acid has been used to measure pH$_i$ in monolayers of cultured fibroblasts. Because of the lower p$K_a$ the concentration of undissociated benzoic acid can be ignored making it simpler to calculate pH$_i$ (L'Allemain et al., 1984).

# 5. Kinetic Aspects of Ion Movement Across Cell Membranes

Ions, like $Na^+$, $K^+$, and $Cl^-$, are extremely insoluble in lipids. The energy required for transfer from a medium with a high dielectric constant (water) to a medium with a low dielectric constant (lipid) is relatively high. Cell membranes are basically proteins embedded in lipid bilayers; therefore diffusion of ions across the lipid bilayer component of the plasma membrane is negligible. Ion transport is, therefore, restricted to movements mediated by specific ion channels and ion carriers (Stein, 1967; Hille, 1984). Exceptions to this are forms of bulk transport caused by invaginations of the cell membrane, namely exocytosis and pinocytosis, which cannot be treated by the normal mathematics of membrane transport. In general, channels are specific pores that mediate a "downhill" movement of ions that follow electrodiffusional criteria (Stein, 1967; Hille, 1984). Ion carriers usually use coupled co- or exchange transport systems and might be driven by facilitated diffusion, secondary active transport ("downhill" electrochemical gradient of another ion involved), or primary active transport (pump driven by direct input of metabolic energy, as in the hydrolysis of ATP). The criteria of selectivity, saturation, and stoichiometry are no longer the best for distinguishing channel and carrier modes, and the best criterion available is the rate of passage of ion. This is about $10^6$ ions/s for a channel and $10^3$ turnovers/s for an ion carrier. Basic transport rates for $K^+$ and the separate components revealed by use of different inhibitors for several neural systems in culture can be found in Table 3. As can be seen, the overall rate as well as the channel- and carrier-mediated components can vary considerably between different cultures.

## 5.1. Compartments

A physiological compartment (Solomon, 1960) is a spatial part of a biological system showing a certain homogeneity with respect to the substance considered, i.e., there are no important barriers inside such a compartment that would influence the distribution of the substance, and its concentration is therefore essentially the same throughout the compartment (Kotyk and Janacek, 1975). The total amount of substance leaving the compartment per unit time may be called the total efflux of the substance; the amount entering it during the same time as the total influx. When the area of the surface surrounding the compartment is known, the flow per unit area of the membrane (the unidirectional influx or efflux) may be calculated, and the properties of different membranes, regardless of their dimensions, may be compared (*also see* section 3.4). In the context of neurobiological problems and in homogeneous tissue culture there are two compartments (extracellular medium and cellular compartment) that are of importance for the main ions. In the case of calcium and magnesium, however, several compartments may exist (see below).

## 5.2. Unidirectional Flux

The unidirectional flux is defined as the number of moles of the substance that, being originally present in the first compartment, penetrates per unit time across a unit area of the membrane into the other compartment. If the area is not known, the unidirectional flux is quite often expressed per cell, per mg protein, or per DNA (*also see* section 3.4).

## 5.3. Advantages of the Use of Radiotracers

The use of radiotracers makes it possible to follow unidirectional fluxes of a substance under conditions in which the substance moves across the membrane in both directions. The cells can therefore be kept under physiological conditions. This is very important if the fluxes of ions, which are influenced by or affect membrane potential directly or indirectly (i.e., potassium and calcium), are being analyzed. To represent satisfactorily the behavior of unlabeled substances, the labeled substance must, of course, resemble them as much as possible, i.e., they have to be kinetically indistinguishable isotopes. This is the case for all biologically used isotopes, with the exception of only the very simple compounds of the lightest elements (ordinary and tritiated water) in which small isotope effects do exist.

Thus, if $C_i^*$ is the labeled form of an ion (i), and the labeled form is present in a negligible concentration, so that $C_i >> C_i^*$, the following equation

$$\frac{C_i + C_i^*}{C_i^*} \tag{8}$$

can be used for calculating total fluxes of the ion by simply multiplying the rate obtained from the radiotracer flux by this equation.

## 5.4. Rate Constants

The simplest kinetic case in a cell culture system is the concept of a compartment (the cells) and a reservoir (extracellular medium). A reservoir implies that the concentration of an ion in this compartment remains constant, even in the case in which it is accumulated in the cells, which is always the case for attached cell monolayers, since the volume of the reservoir is usually at least three orders of magnitude greater than the cell volume.

The efflux of a labeled substance from a compartment into the reservoir can be analyzed by keeping the concentration of the label in the reservoir at zero. This is obtained in practice by washing-out experiments (*see* section 3.2.3 and below). Under steady-state conditions, the concentration of the labeled ion $C_i$ in the (cellular) compartment varies with time ($t$) as follows;

$$C_i = C_{i(t=0)}\, e^{-kt} \tag{9}$$

This means that the concentration of the labeled ion inside the compartment decreases exponentially with time. $C_i$ = concentration in cells at time $t$, and $C_{i(t=0)}$ = initial concentration at time zero; $e$ is the base of natural logarithm = 2.71828 . . .). Figure 8 is a graphical representation of both the loss of label (dashed line) and the rate of appearance of label outside the cell (full line). Time ($t$) is represented in Fig. 8 as $x$. For practical purposes it is convenient to plot the values of $C_i/C_{i(t=0)}$ against time semilogarithmically, since the relationship is then a straight line, as shown in Fig. 9. The slope of this straight line (i.e., the change of the variable plotted on the vertical axis per unit time) corresponds to the rate constant ($k$) in units of time (*see also* Fig. 4 in section 3.2.3 for an experimental example of this). The total amount of the ion in the compartment can then be multiplied by the rate constant, $k$, and the total initial unidirectional efflux, $j$, then is equivalent to $kC_i$. If the area of the membrane across which the efflux takes place is known, the fluxes per unit area can be calculated. The time ($t_{0.5}$) in which the amount of $C_i$ decreases by one-half in the compartment can also be determined. The rate constant, $k$, can then be calculated the following way:

$$t_{0.5} = \frac{\ln 2}{k} = \frac{0.693}{k} \tag{10}$$

To study the unidirectional influx, the reservoir contains at the beginning all of the labeled ion; the cell compartment contains no labeled ion. The amount of the labeled ion ($C^*_i$) that then progressively appears in the cellular compartment is;

$$C^*_i = C^*_x (1 - e^{-kt}) \qquad (11)$$

where $C^*_x$ is the concentration in the reservoir. The function $(1 - e^{kt})$ has already been illustrated graphically in Fig. 8. The concentration $C^*_x$ is approached by the compartment if the species transported is electroneutral and completely passively distributed. If the species transported is charged, then the intracellular equilibrium concentration will be in electrochemical equilibrium with both the concentration and the potential, as discussed in the next section.

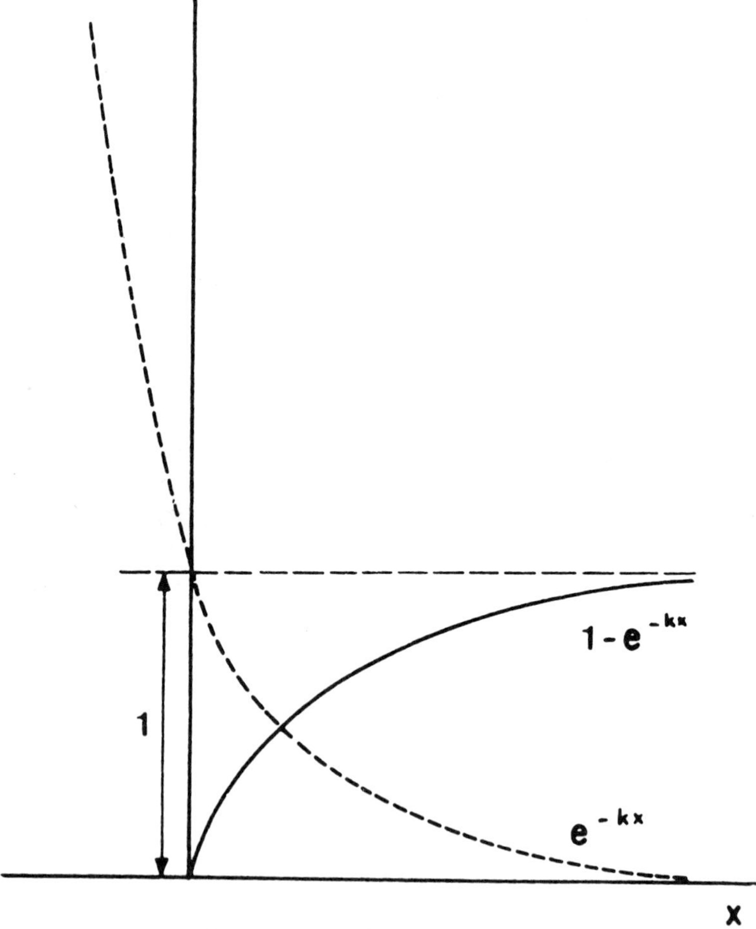


Fig. 8. Plot of $e^{-kx}$ and $1 - e^{-kx}$. See text for details.

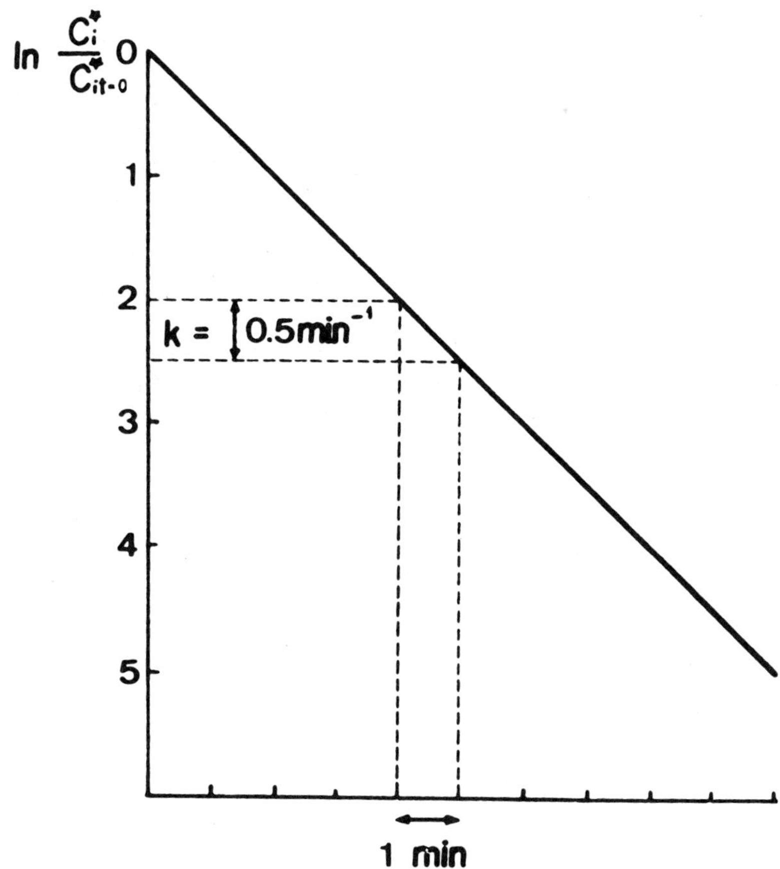



Fig. 9. Semilogarithmic plot of the exponential time course of $\ln C_i/C_i$ $(t = 0) = -kt$ (from Kotyk and Janacek, 1975).

It is important to note that the rate constant as determined by this analysis tells us nothing about the mechanisms of ion movement across the barrier. Different approaches must be chosen to determine the mechanisms. Thus one could examine how the changes in the initial unidirectional fluxes (or rate constants) vary with the concentration of the ion and see whether they correspond to Michaelis-Menten kinetics, see how the fluxes change with the membrane potential and/or chemical driving forces, add specific carrier, ion channel, or metabolic inhibitors, or calculate the flux ratio. Some of these approaches will be discussed below.

### 5.5. *Flux Ratio Analysis*

This method is used to determine if an ion is transported purely passively because of its electrochemical gradient (i.e., if it is only channel-mediated). It should be noted that the flux ratio analysis, although one of the most powerful methods in radioisotope transport measurements, is not sufficient by itself to decide if one is measuring a purely diffusional flux; it could be passively transported by a facilitated transport mechanism involving a carrier. Thus inhibitors or kinetic studies should also be used in additional experiments.

To derive the equation for the flux ratio, one can derive a relationship for the transmembrane unidirectional flux of a charged ion, which will be dependent on both the concentration gradient of the ion across the membrane and the membrane potential (Stein, 1967). The flux of the ion from inside to outside the flux ($j_i$) is given by;

$$ j_i = PC_i \, e^{\frac{zFE_i}{RT}} \tag{12} $$

where $P$ is the permeability coefficient, $C_i$ is the intracellular concentration, and $E_i$ is the intracellular potential. The other letters have their usual meanings. Similarly;

$$ j_o = PC_o \, e^{\frac{zFE_o}{RT}} \tag{13} $$

Therefore dividing Eq. (12) by (13),

$$ \frac{j_i}{j_o} = \frac{C_i}{C_o} \, e^{\frac{zF(E_i - E_o)}{RT}} \tag{14} $$

Since $E_i - E_o$ is the membrane potential $E_m$, then;

$$ \frac{j_i}{j_o} = \frac{C_i}{C_o} \, e^{\frac{zFE_m}{RT}} \tag{15} $$

### 5.6. *Compartmental Analysis*

If it is not known how many compartments are involved in a system, the simplest way to approach the problem is by measuring the efflux of the ion and then analyze the number of rate constants obtained. Figure 10 gives an example of the graphical curve-peeling technique for doing this (Perl, 1960; Feuerzeig and Tyler,

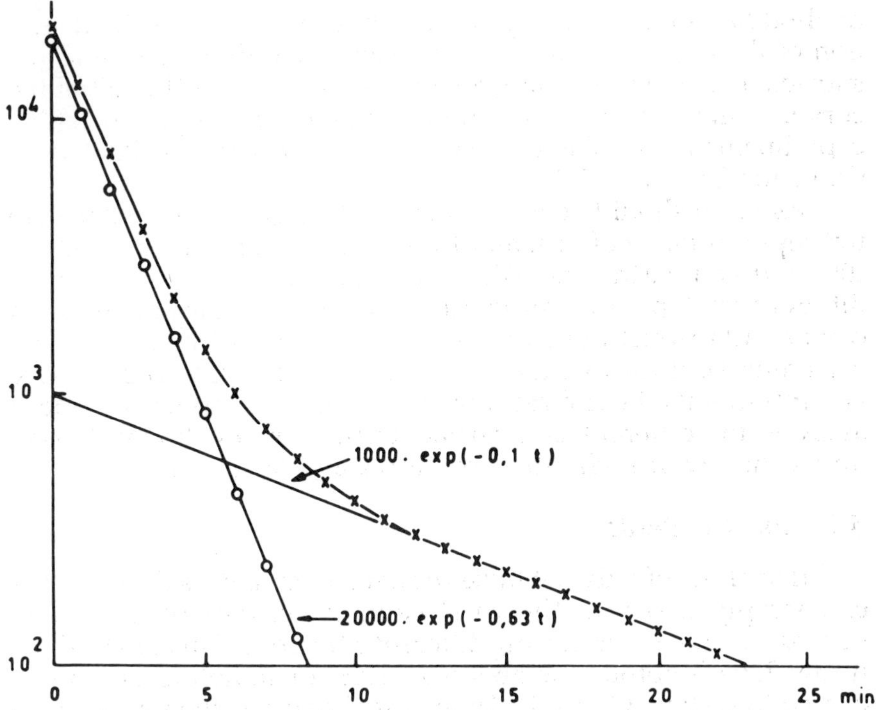


Fig. 10. Illustration of the graphical peeling techniques: the crosses *(X)* represent the original data points. An asymptote is drawn through the tail of the curve and is extrapolated to zero time. The value of this asymptote at zero time corresponds to $X_i$, whereas $\gamma i$, which corresponds to the negative slope of this line, is calculated from the half-time $t_{1/2}$ by means of the equation $\gamma = 0.693/t_1$, where 0.693 corresponds to 1n 2. The first asymptote can be represented by 1000 exp(–0.1$t$). The ordinates of the asymptote are subtracted from the original data points and the new ordinate values (0) are plotted. These values can be fitted by a straight line that follows the equation 20,000 exp(–0.63$t$). The original data points fit the equation 1000 exp(–0.1$t$) + 20,000 exp(–0.63$t$) (from Casteels and Droogmans, 1975).

1950). A straight line is fitted by eye through the tail of the semilogarithmic plot of the experimental data and extrapolated to the ordinate. The antilogarithm of the ordinate values are then subtracted from the earlier values until the values obtained are of the same order of magnitude as the experimental error. A straight line can then be fitted to the values obtained after subtraction. This method has some serious problems, however, such as the distortion of the experimental error on a semilogarithmic plot and the subjective element in fitting a straight line through the tail of the curve. Some programs for computer-assisted analysis in efflux experiments are available (Rossing, 1966; Mancini and Pilo, 1970; Cook and Taylor, 1971).

Again, it should be noted that the different rate constants are not representative of different transport mechanisms, but only of different compartments. Different compartments could include different cell types in a cell culture or a more complex distribution of ions within a cell, as in the case of calcium, in which different rate constants exist for exchange of surface-bound, intracellular free and intracellular bound calcium stores. For an overview of kinetic analysis for different compartments that are in different communication with each other, *see* Kotyk and Janacek (1975).

## 5.7. Ion Contents

If analysis of only net fluxes or net movements is desired, one can simply measure the total cellular content of an ion in homogenous systems during different situations. This can be done using flame photometric analysis, AAS, or radiotracers, as discussed in section 3.1. In the latter case, however, the tracer has to be in equilibrium with the compartments. This method will not measure homoexchange of ions, but only net change in ion content, such as accumulation of KCl by furosemide-sensitive cotransport or changes caused by Donnan forces (Stein, 1967; Katz, 1966). Also, net loss of KCl during shrinkage of cells during volume regulatory decrease (Grinstein et al., 1984a) can also be measured.

## 6. Use of Ion Replacement Studies and Inhibitors to Define Transport Processes

It is important to bear in mind that isotopic flux measurements trace total movement of ions, either as unidirectional influx or efflux, as described in the previous sections. Under certain conditions net fluxes may be occurring, and this will exclude certain processes such as homoexchange transport. In the steady state there is, by definition, no change in ion content, so that only net unidirectional fluxes of tracers are being measured, which initially, when isotope is only on one side, corresponds to the total unidirectional flux of this ion. As isotope accumulates inside the cell when measuring influx, isotope begins to leave at an increasing rate as its specific activity inside the cell increases. Influx remains constant since the specific activity in the large reservoir of medium outside the cell remains constant. Thus the apparent rate of increase of the intracellular isotope content becomes slower, until when the specific activity inside the cell reaches that on the outside, isotopic flux in is equal to isotopic flux out, and the isotope content of the cell remains constant with time. This time course and its equation have already been shown in Fig. 8 for a single first-order rate constant as the full line for $(1 - e^{-kx})$. Using the external specific activity of isotope, one can calculate the total content corresponding to the plateau value, and this is equivalent to the cell content of the total exchangeable ion.

Any ion may traverse the membrane by a number of different processes and routes, i.e., diffusional transport through membrane channels, primary active pump processes driven by metabolic energy, or secondary active processes linked to the movements of other ions that are themselves out of equilibrium. These different routes can be dissected out either by omitting ions if transport is linked to movement of other ions, changing the membrane potential if the ion is moving through a channel under the influence of the electrochemical gradient, or using specific inhibitors. It is also important to be careful about secondary effects of such changes; for example, does replacement of an ion influence the membrane potential so that any effect might be a secondary consequence of this rather than because of cotransport? Also, it should be borne in mind that inhibitors may not be specific, especially at higher concentrations, and may affect more than one process.

## 6.1. Ion-Replacement Studies

This approach can be used to examine the dependence of the transport of one ion on another ion. Thus if one wishes to examine whether part of the influx of $K^+$ (using either isotopic $^{42}K^+$ or the longer lived isotope $^{86}Rb^+$) is transported on the $K^+ + Na^+ + 2Cl^-$ cotransport system, one can replace $Na^+$ by another cationic species (e.g., choline$^+$ or Tris$^+$). Also, $Cl^-$ could be replaced by another anionic species (isethionate$^-$ or $SO_4^{2-}$). One problem with substitution for $Na^+$ is that there will also be a progressive inhibition of the $(Na + K)$ pump because of depletion of intracellular $Na^+$. If one omits $K^+$, there is the problem of changes in the membrane potential. This is shown for a primary astrocyte culture in Fig. 6, in which omission of $K^+$ (at the first arrow) leads to an initial hyperpolarization and then progressive marked depolarization, presumably caused by loss of $K^+$ from the cell. There is also a progressive increase in membrane resistance, in agreements with the concept that $K^+$ is the major transmembrane ion conductance in these cells.

In addition to identification of cotransport systems, ion-substitution studies can also be used to distinguish ion-exchange systems. Thus if a $Cl^-/Cl$ or $Cl^-/HCO_3^-$ exchange system is thought to be present, then the cells can be loaded with $^{36}Cl^-$, and the efflux rate of $^{36}Cl^-$ into isotope-free medium can be measured in the absence of $Cl^-$ or $HCO_3^-$ or in their presence. Such a study was originally used to show the existence of the anion-exchange system in primary astrocyte cultures, as shown previously in Fig. 4.

## 6.2. Inhibitors

The use of inhibitors constitutes a very convenient method of identifying the existence of particular transport processes. Major problems to be considered are to show that the inhibitors are specific for a single-transport process, and that they are being used within the concentration range at which they are known to be specific. In addition it is important to add the inhibitor in such a way as to ensure that it has fully reacted with its binding site, which is presumably the same as or close to the ion-transporting site. On the other hand, the inhibitor should not be present for so long that there is a significant rearrangement of ion gradients because of inhibition of transport processes, since this could indirectly change the electrochemical force on the ion or alter the kinetics of an electrically neutral ion-transport process involving more than one ion. Thus in the case of ouabain, prolonged inhibition will lead to loss of cell $K^+$ and gain of $Na^+$, resulting in plasma membrane depolarization reducing the passive influx of a positively charged species or the uptake of compounds that depend on the cotransport of $Na^+$ down an inwardly directed $Na^+$ electrochemical gradient. If possible it is best to add the inhibitor at the same time as an isotope or a change in condition that stimulates a net flux and to measure the change in the initial rate. Under these conditions the problem of indirect effects caused by ion-gradient rearrangement are minimized, with the only potential problem being an incomplete binding of the inhibitor. A listing of useful carrier or pump-mediated ion transport inhibitors with their sites of action

Table 5
Some Useful Inhibitors of Carrier- or Pump-Mediated
Ion-Transport Processes

| Inhibitor | Major target and/or secondary targets | Concentration range | Representative references |
|---|---|---|---|
| Ouabain | $(Na + K)$ ATPase-mediated $K^+$ influx or $Na^+$ efflux | Variable since sensitivity is species-dependent; $IC_{50}$ 0.1 μM in dogfish to 0.1 m$M$ in rat | Bonting, 1970 |
| Furosemide | $K^+ + Na^+ + 2Cl^-$ cotransport $Cl^-$/$HCO_3^-$ anion exchange | 0.1–5m$M$ | Brazy and Gunn, 1976; Warnock et al., 1983 |
| Bumetanide | $K^+ + Na^+ + 2Cl^-$ cotransport | 0.001–0.1 m$M$ | Warnock et al., 1983 |
| SITS; DIDS[a] | $Cl^-$/$HCO_3^-$ anion exchange | 0.01 – 1 m$M$ | Cabantchik and Rothstein, 1972 |
| Amiloride | $Na^+$/$H^+$ exchange | ~1 m$M$ | Rindler et al., 1979 |

[a]SITS, 4-acetamido-4'-isothiocyanostilbene-2,2'-disulfonic acid; DIDS, 4,4'-diisothiocyano-2,2'-stilbenedisulfonic acid.

and appropriate concentration ranges is given in Table 5. There is also a considerable number of inhibitors of diffusional or channel-mediated ion-transport processes, such as tetrodotoxin for voltage-dependent $Na^+$ channels or tetraethylammonium and $Ba^{2+}$ and 4-aminopyridine for different types of $K^+$ channels. Table 1 in both chapters 5 and 12 of the recent book by Hille (1984) should be consulted for a more complete listing of such inhibitors.

With this pharmacopoeia of inhibitors, used correctly, the experimenter is in a good position to begin to define the types of ion-transport processes present in a particular cell type. One last concept to be mentioned is that of additivity, whereby if the effects of maximally effective concentrations of inhibitors of different transport processes are additive, then it is assumed that the cell

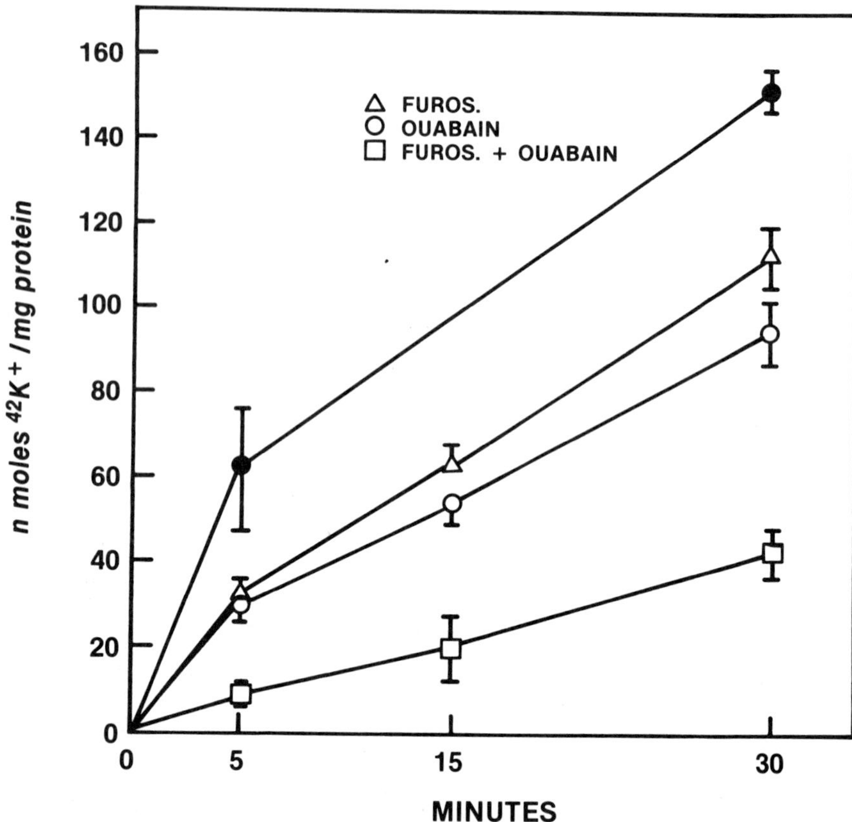


Fig. 11. Time course showing $^{42}K^+$ uptake and effects of ouabain and furosemide. Uptake of $^{42}K^+$ was measured under steady-state conditions in $HCO_3^-$-buffered medium (*see* Fig. 1), as described in section 3.2.3. Primary astrocyte cultures from neonatal rat cerebral cortex were pre-incubated for 15 min with furosemide, ouabain, or ouabain + furosemide before adding $^{42}K^+$ (10 original μCi per well per 0.5 mL were added, which represents only 2.5 μCi, since the $^{42}K^+$ had decayed approximately 75% by the time of the experiment). Each point represents the mean for three separate wells ± SEM. Cells were 27 d old. (●), control; (Δ), + 1 m*M* furosemide; (○), + 1 m*M* ouabain; (□), + 1 m*M* furosemide + 1 m*M* ouabain (from Kimelberg and Frangakis, 1985, with permission).

contains all these different transport processes. This is shown for inhibition of $^{42}K^+$ uptake by primary astrocyte cultures by both ouabain and furosemide (*see* Fig. 11), which was interpreted as showing that in these cells $K^+$ influx occurs partly via the (Na + K) pump and partly on the $K^+ + Na^+ + 2Cl^-$ cotransport system. The remaining ouabain plus furosemide-insensitive influx may be diffusional (Kimelberg and Frangakis, 1985).

## 7. Conclusions

Cultured cells constitute the simplest way of studying the ion-transport properties of individual cell types from the nervous system. Such studies *in situ* can only be done by electrical techniques using microelectrodes and extrapolating from measurements of conductance or changes in potential to ion movements. Such methods are technically difficult, especially to those inexperienced in such matters, and are often not easily interpretable. Also these techniques will not measure electrically silent or nonconductive ion-transport processes that seem to constitute a significant proportion of ion fluxes in the CNS, and only a few types of cells, usually large cells and those located in precise arrays, can be impaled. Ion-specific microelectrodes are also very useful for *in situ* studies, but again are difficult to use, and the interpretation of results has marked limitations. In contrast, flux studies are relatively easy to do with cell cultures, as we hope has been illustrated in this chapter. Complex kinetics, including the use of numerous inhibitors, can be studied, allowing for a sophisticated level of analysis. Since electrical measurements can also be done on these cell cultures, conductive versus nonconductive fluxes can be measured, and a relatively precise understanding of the driving forces can be potentially achieved.

The major methodological problems for a full interpretation of the data remain that of accurately determining surface area and, to a lesser extent, cell volume changes. The major interpretative problem is how the properties of cells in culture correspond to those of the same cells *in situ*. This problem is an inevitable consequence of the use of cell cultures and offsets, to some degree, the significant methodological advantages of this approach for analyzing basic mechanisms. Thus one has to guard against overinterpretation, especially that quantitative results obtained in culture necessarily represent what is occurring *in situ*, since as well as removing influences impinging on the cell *in situ* by growing cells in culture, the culture environment itself imposes its own characteristics. Information gained, however, from studies of cultures of different cell types and of different classes within each cell type are very important in alerting the researcher to potential and unexpected properties of these cells, and this applies to ion transport properties as much as any other. Also the results obtained in vitro may correspond closely to there existing in vivo. In most cases primary cultures of such cells from different regions of the brain are the most useful in this regard. Cell cultures are also very good for analyzing basic mechanisms, and in this case transformed or normal cell lines can be equally useful. Coculture is also becoming a powerful technique for studying the developmental and physiological interactions between different cell types, and may begin to overcome the lack of the normal interactions between cells that exist in culture; transport studies could be designed for such systems. It seems fair to conclude that nervous system cultures have proven themselves extremely useful thus far for addressing a large number of questions in neurobiology, including ion-transport experiments, because of the wide variety of experiments that can be done with them and the large number of different culture or co-culture combinations that are possible. It is to be expected that they will continue to be a powerful tool in neurobiology studies in general and transport studies in particular.

## Acknowledgments

Work quoted from one of the authors' (HKK) laboratories, as well as time spent writing this chapter, was partially supported by grant BNS 8213873 from NSF and NS 23750 from NINCDS. We thank Mrs. E. P. Graham for typing the manuscript and Dr. L. Martin for permission to quote his unpublished work in Table 2. We are grateful to the various authors and publishers for permission to reproduce published work, and these are separately acknowledged in the appropriate legends.

## References

Bakker-Grunwald T. and Sinensky M. (1979) $^{86}$Rb$^+$ fluxes in Chinese hamster ovary cells as a function of membrane cholesterol content. *Biochim. Biophys. Acta* **558,** 296–306.

Bonting S. L. (1970) Sodium-potassium activated adenosinetriphosphotose and cation transport, in *Membranes and Ion Transport* vol. 1, 257–363 (Bittar, E. E., ed.) Wiley, New York.

Boonstra J., Mumery C. L., Tertoolen G. L. G., Van der Saag P. T., and De Laat S. W. (1981a) Characterization of $^{42}$K$^+$ and $^{42}$Rb$^+$ transport and electrical membrane properties in exponentially growing neuroblastoma cells. *Biochim. Biophys. Acta* **643,** 89–100.

Boonstra J., Mummery C. L., Tertoolen L. G. J., Van der Saag P. T., and De Laat S. W. (1981b) Cation transport and growth regulation in neuroblastoma cells. *J. Cell. Physiol.* **107,** 75–83.

Bottenstein J. E. (1985) Growth and Differentiation of Neural Cells in Defined Media, in *Cell Culture in the Neurosciences* (Bottenstein J. E. and Sato G., eds.) Plenum, New York.

Bowman C. L. and Kimelberg H. K. (1984) Excitatory amino acids directly depolarize rat brain astrocytes in primary culture. *Nature* **311,** 656–659.

Brazy P. C. and Gunn R. B. (1976) Furosemide inhibition of chloride transport in human red blood cells. *J. Gen. Physiol.* **68,** 583–599.

Cabantchik Z. I. and Rothstein A. (1972) The nature of the membrane sites controlling anion permeability of human red blood cells as determined by studies with disulfonic stilbene derivatives. *J. Membrane Biol.* **10,** 311–330.

Casteels R. and Droogmans G. (1975) Compartmental Analysis of Ion Movements, in *Methods in Pharmacology* vol. 3 (Daniel E. E. and Paton D. M., eds.) Plenum, New York.

Cheung R. K., Grinstein S., Dosch H-M., and Gelfand E. W. (1982) Volume regulation by human lymphocytes: Characterization of the ionic basis for regulatory volume decrease. *J. Cell Physiol.* **112,** 189–196.

Cook D. A. and Taylor G. S. (1971) The use of the APL/360 system in pharmacology. A computer assisted analysis of efflux data. *Comput. Biomed. Res.* **4,** 157–166.

Douglas W. W. (1978) Stimulus-Secretion Coupling: Variations on the Theme of Calcium-Activated Exocytosis Involving Cellular and Extracellular Sources of Calcium, in *Respiratory Tract Mucosus* CIBA Foundation Symposium 54 New Series, Elsevier North-Holland, Amsterdam.

Erecinska M. and Silver I. A. (1986) The role of glial cells in regulation of neurotransmitter amino acids in the external environment. I. Transmembrane electrical and ion gradients and energy parameters in cultured glial-derived cell lines. *Brain Res.* **369,** 193–202.

Feuerzeig W. and Tyler S. A. (1950) A note on exponential fitting of empirical curves. *Argonne Nat. Lab. Quart. Rep., ANL* **4401,** 14–29.

Fonnum F., Karlsen R. L., Malthe-Sorenssen D., Sterri S., and Walaas I. (1980) High Affinity Transport Systems and Their Role in Transmitter Action, in *The Cell Surface and Neuronal Function* (Cotman C. W., Poste G., and Nicolson G. L., eds.) North-Holland, Amsterdam, New York, Oxford.

Grinstein S. and Furuya W. (1983) The electrochemical $H^+$ gradient of platelet secretory alpha-granules. Contribution of a $H^+$ pump and a Donnan potential. *J. Biol. Chem.* **258,** 7876–7882.

Grinstein S., Rothstein A., Sarkadi B., and Gelfand E. W. (1984a) Responses of lymphocytes to anisotonic media: Volume-regulating behavior. *Am. J. Physiol.* **246,** C204–C215.

Grinstein S., Elder B., Clarke C. A., and Buchwald M. (1984b) Is cytoplasmic $Ca^{2+}$ in lymphocytes elevated in cystic fibrosis? *Biochim. Biophys. Acta* **769,** 270–274.

Grinstein S., Cohen S., and Rothstein A. (1984c) Cytoplasmic pH regulation in thymic lymphocytes by an amiloride-sensitive $Na^+/H^+$ antiport. *J. Gen. Physiol.* **83,** 341–369.

Grynskiewicz G., Poenie M., and Tsien R. Y. (1985) A new generation of $Ca^{2+}$ indicators with greatly improved fluorescence properties. *J. Biol. Chem.* **260,** 3440–3450.

Hansson E., Ronnback L., Persson L. I., Lowenthal A., Noppe M., Alling C., and Karlsson B. (1984) Cellular composition of primary cultures from cerebral cortex, striatum, hippocampus, brainstem and cerebellum. *Brain Res.* **300,** 9–18.

Hayflick L. and Moorhead P. S. (1961) The serial cultivation of human diploid cell strains. *Exp. Cell Res.* **25,** 585–621.

Hertz L., Juurlink B. H. J., and Szuchet S. (1985a) Cell Cultures, in *Handbook of Neurochemistry* vol. 8 (Lajtha A., ed.) Plenum, New York.

Hertz L., Juurlink B. H. J., Szuchet S., and Walz W. (1985b) Cell and Tissue Cultures, in *Neuromethods* vol. 1 (Boulton A. A. and Baker G. B., eds.) Humana, Clifton, N.J.

Heumann R., Reiser G., Van Calker D., and Hamprecht B. (1982) Polyploid rat glioma cells. Production, oscillations of membrane potential and response to neurohormones. *Exp. Cell Res.* **139,** 117–126.

Hille B. (1984) *Ionic Channels of Excitable Membranes,* Sinauer, Sunderland, Massachusetts.

Hirata H., Slater N. T., and Kimelberg H. K. (1983) Alpha-adrenergic receptor-mediated depolarization of rat neocortical astrocytes in primary culture. *Brain Res.* **270,** 358–362.

Hosli L. and Hosli E. (1983) Localization and physiological properties of glycine and GABA receptors in cultures of rat CNS. *Adv. Biochem. Psychopharmacol.* **37,** 36–46.

Hosli L., Hosli E., Zehntner C., Lehmann R., and Lutz T. W. (1982) Evidence for the existence of alpha- and beta-adrenoceptors on cultured glial cells—an electrophysiological study. *Neuroscience* **7,** 2867–2872.

Johnson J. H., Dunn D. P., and Rosenberg R. N. (1982) Furosemide-sensitive $K^+$ channel in glioma cells but not neuroblastoma cells in culture. *Biochem. Biophys. Res. Commun.* **109,** 100–105.

Katz B. (1966) *Nerve, Muscle and Synapse* McGraw-Hill, New York.

Kettenmann H. and Schachner M. (1985) Pharmacological properties of gamma-aminobutyric acid-, glutamate-, and aspartate-induced depolarizations in cultured astrocytes. *J. Neurosci.* **5,** 3295–3301.

Kettenmann H., Orkand R. K., and Schachner M. (1983a) Coupling among identified cells in mammalian nervous system cultures. *J. Neurosci.* **3,** 506–516.

Kettenmann H., Sonnhof U., and Schachner M. (1983b) Exclusive potassium dependence of the membrane potential in cultured mouse oligodendrocytes. *J. Neurosci.* **3,** 500–505.

Kettenmann H., Gilbert P., and Schachner M. (1984) Depolarization of cultured oligodendrocytes by glutamate and GABA. *Neurosci. Lett.* **47,** 271–276.

Kimelberg H. K. (1974) Active potassium transport and ($Na^+$ + $K^+$)ATPase activity in cultured glioma and neuroblastoma cells. *J. Neurochem.* **22,** 971–976.

Kimelberg H. K. (1981) Active accumulation and exchange transport of chloride in astroglial cells in culture. *Biochim. Biophys. Acta* **646,** 179–184.

Kimelberg H. K. and Bourke R. S. (1982) Anion Transport in the Nervous System, in *Handbook of Neurochemistry* 2nd Ed., vol. 1 (Lajtha A., ed.) Plenum, New York.

Kimelberg H. K. and Frangakis M. V. (1985) Furosemide- and bumetanide-sensitive ion transport and volume control in primary astrocyte cultures from rat brain. *Brain Res.* **361,** 125–134.

Kimelberg H. K. and Hirata H. (1981) Electrophysiology of and sensitivity to furosemide and MK473 of Cl⁻ transport in primary astrocyte cultures. *Soc. Neurosci. Abst.* **7,** 698.

Kimelberg H. K. and Katz D. M. (1986) Regional differences in 5-hydroxytryptamine and catecholamine uptake in primary astrocyte cultures. *J. Neurochem.,* **47,** 1647–1652.

Kimelberg H. K. and Ransom B. R. (1986) Physiological and Pathological Aspects of Astrocytic Swelling, in *Astrocytes* (Fedoroff S. and Vernadakis A., eds.) Academic, Florida.

Kimelberg H. K., Biddlecome S., and Bourke R. S. (1979a) SITS-inhibitable Cl⁻ transport and Na⁺-dependent H⁺ production in primary astroglial cultures. *Brain Res.* **173,** 111–124.

Kimelberg H. K., Bowman C., Biddlecome S., and Bourke R. S. (1979b) Cation transport and membrane potential properties of primary astroglial cultures from neonatal rat brains. *Brain Res.* **177,** 533–550.

Kimelberg H. K., Bourke R. S., Stieg P. E., Barron K. D., Hirata H., Pelton E. W., and Nelson L. R. (1982) Swelling of Astroglia After Injury to the Central Nervous System: Mechanisms and Consequences, in *Head Injury: Basic and Clinical Aspects* (Grossman R. G. and Gildenberg P. L., eds.) Raven, New York.

Kletzien R. F., Pariza M. W., Becker J. E., and Potter V. R. (1975) A method using 3-*O*-methyl-D-glucose and phloretin for the determination of intracellular water space of cells in monolayer culture. *Anal. Biochem.* **68**, 537–544.

Kotyk A. and Janacek K. (1975) *Cell Membrane Transport. Principles and Techniques* 2nd Ed., Plenum, New York.

Kukes G., Elul R., and De Vellis J. (1976a) The ionic basis of the membrane potential in a rat glial cell line. *Brain Res.* **104**, 71–92.

Kukes G., De Vellis J., and Elul R. (1976b) A linked active transport system for $Na^+$ and $K^+$ in a glial cell line. *Brain Res.* **104**, 93–105.

Kurzinger K., Stadtkus C., and Hamprecht B. (1980) Uptake and energy-dependent extrusion of calcium in neural cells in culture. *Eur. J. Biochem.* **103**, 597–611.

L'Allemain G., Paris S., and Pouyssegur J. (1984) Growth factor action and intracellular pH regulation in fibroblasts. Evidence for a major role of the $Na^+/H^+$ antiport. *J. Biol. Chem.* **259**, 5809–5815.

Lassen U. V., Pape L., and Vestergaard-Bogind B. (1978) Chloride conductance of the Amphiuma red cell membrane. *J. Membrane Biol.* **39**, 27–48.

Lubin M. (1980) Control of growth by intracellular potassium and sodium concentrations is relaxed in transformed 3T3 cells. *Biochem. Biophys. Res. Commun.* **97**, 1060–1067.

Mancini P. and Pilo A. (1970) A computer program for multiexponential fitting by the peeling method. *Comput. Biomed. Res.* **3**, 1–14.

Martin D. L. and Shain W. (1979) High affinity transport of taurine and beta-alanine and low-affinity transport of gamma-aminobutyric acid by a single transport system in cultured glioma cells. *J. Biol. Chem.* **254**, 7076–7084.

Moody Jr. W. J. (1983) Intracellular pH regulation and cell excitability. *Neurol. Neurobiol.* **2**, 451–473.

Morrison R. S. and De Vellis J. (1981) Growth of purified astrocytes in a chemically defined medium. *Proc. Natl. Acad. Sci. USA* **78**, 7205–7209.

Morrison R. S. and De Vellis J. (1983) Differentiation of purified astrocytes in a chemically defined medium. *Dev. Brain Res.* **9**, 337–345.

Newman E. A. (1984) Regional specialization of retinal glial cell membrane. *Nature* **309**, 155–157.

Peacock J. H. (1979) Electrophysiology of dissociated hippocampal cultures from fetal mice. *Brain Res.* **169**, 247–260.

Pearce B. R., Currie D. N., Dutton G. R., Hussey R. E. G., Beale R., and Pigott R. (1981) A simple perfusion chamber for studying neurotransmitter release from cells maintained in monolayer culture. *J. Neurosci. Meth.* **3**, 255–259.

Perl W. (1960) A method for curve-fitting for exponential functions. *Int. J. Appl. Radiat.* **8,** 222.

Randles J. and Kimmich G. A. (1978) Effects of phloretin and theophylline on 3-O-methylglucose transport by intestinal epithelial cells. *Am. J. Physiol.* **234,** C64–C72.

Ransom B. R., Neale E., Henkart M., Bullock P. N., and Nelson P. G. (1977) Mouse spinal cord in cell culture. I. Morphology and intrinsic neuronal electrophysiologic properties. *J. Neurophysiol.* **40,** 1132–1150.

Requena J., Dipolo R., Brinley E. J., and Mullins L. J. (1977) The control of ionized calcium in squid axons. *J. Gen. Physiol.* **70,** 329–353.

Rindler M. J., Taub M., and Saier Jr. M. H. (1979) Uptake of $^{22}$Na$^+$ by cultured dog kidney cells (MDCK). *J. Biol. Chem.* **254,** 11431–11439.

Roos A. and Boron W. F. (1981) Intracellular pH. *Physiol. Rev.* **61,** 296–434.

Rossing R. B. (1966) Evaluation of a computer solution of exponential decay for washout curves. *J. Appl. Physiol.* **21,** 1907–1910.

Sanui H. and Rubin A. H. (1979) Measurement of total, intracellular and surface bound cations in animal cells grown in culture. *J. Cell. Physiol.* **100,** 215–225.

Schousboe A. and Divac I. (1979) Differences in glutamate uptake in astrocytes cultured from different brain regions. *Brain Res.* **177,** 407–409.

Solomon A. K. (1960) Compartmental Methods of Kinetic Analysis, in *Mineral Metabolism, An Advanced Treatise* (Comar C. L. and Bronner F., eds.) Academic, New York.

Stein W. D. (1967) *The Movement of Molecules Across Cell Membranes* Academic, New York.

Varon S. S. and Somjen G. G. (1979) Neuron-glia interactions. *Neurosci. Res. Prog. Bull.* **17,** 131–146.

Waddell W. J. and Butler T. C. (1959) Calculation of intracellular pH from the distribution of 5,5-dimethyl-2,4-oxazolidinedione (DMO). Application to skeletal muscle of the dog. *J. Clin. Invest.* **38,** 720–729.

Walz W. and Hertz L. (1983) Comparison between fluxes of potassium and of chloride in astrocytes in primary cultures. *Brain Res.* **277,** 321–328.

Walz W. and Hertz L. (1984) Sodium transport in astrocytes. *J. Neurosci. Res.* **11,** 231–239.

Walz W. and Hinks E. C. (1985) Carrier-mediated KCl accumulation accompanied by water movements is involved in the control of physiological K$^+$ levels by astrocytes. *Brain Res.* **343,** 44–51.

Walz W. and Kimelberg H. K. (1985) Differences in cation transport properties of primary cultures from mouse and rat brain. *Brain Res.* **340,** 333–340.

Walz W., Hertz E., and Hertz L. (1983) Lithium-potassium interaction in acutely treated cortical neurons and astrocytes. *Prog. Neuro-Psychopharmacol. Biol. Psychiat.* **7,** 697–702.

Walz W., Wuttke W., and Hertz L. (1984) Astrocytes in primary cultures: Membrane potential characteristics reveal exclusive potassium conductance and potassium accumulator properties. *Brain Res.* **292,** 367–374.

Warnock D. G., Greger R., Dunham P. B., Benjamin M. A., Frizzell R. A., Field M., Spring K. R., Ives H. E., Aronson P. S., and Seifter J. (1983) Ion transport processes in apical membrane of epithelia. *Fed. Proc.* **43,** 2473–2487.

Wolpaw E. W. and Martin D. L. (1984) Cl⁻ transport in a glioma cell line: Evidence for two transport mechanisms. *Brain Res.* **297,** 317–327.

# Electrophysiological Methods Applied in Nervous System Cultures

## H. Kettenmann

## 1. Introduction

This chapter describes the application of electrophysiological techniques in nervous system cultures. The characterization of the membrane properties of cultured neural cells with electrophysiological techniques had already started in 1956 on explant cultures of dorsal root ganglion cells (Crain, 1956) and on glial cells from the cerebellum and midbrain of the kitten (Hild et al., 1958). Subsequently, membrane properties of neurons and glial cells from different areas of the brain were characterized with a combination of electrophysiological, morphological, and immunocytochemical methods.

In cultures of mouse spinal cord, large neurons were found that exhibited membrane properties similar to motoneurons of the intact animal (Nelson et al., 1977; Ransom et al., 1977a,b,c). Chemical synapses between dorsal root ganglion cells and spinal neurons could be identified (Crain and Peterson, 1967). Sympathetic neurons from rat superior cervical ganglion were shown to exhibit their basic membrane properties in cell culture (Mains and Patterson, 1973; O'Lague et al., 1978a,b,c). By selecting appropriate culture conditions, these neurons could be triggered to produce and release either norepinephrine or acetylcholine, demonstrating that the culture environment determines the developmental fate of the cultured cell (Furshpan et al., 1976; Patterson, 1978). Large neurons of the cerebellum were maintained in explant (Calvet, 1974; Calvet and Calvet, 1979; Gahwiler, 1976) or microexplant cultures (Moonen et al., 1982), and these neurons had similar basic membrane and chemosensitivity properties as were known for Purkinje cells in vivo. The dendritic tree of these cultured neurons was reduced in size compared to Purkinje cells in the intact animal mostly by a reduction in the total number of segments, but it still showed similarities to Purkinje cells in vivo (Calvet et al., 1985). These cultures were therefore used as model systems for the study of cerebellar Purkinje cells. Recently, a Purkinje cell-specific antibody has been characterized (Weber and Schachner, 1984) that can be used for unequivocal identification of these cells in culture. Cortical neurons could also be cultured and impaled (Dichter, 1977; Godfrey et al., 1975), and all neurons were found to be sensitive to the inhibitory neurotransmitter $\gamma$-aminobutyric acid (GABA) (Dichter, 1980). Similarly, cultured neurons from olfactory bulb responded to GABA by an increase of cell membrane conductance (Frosch and Dichter, 1984). The electrophysiological properties of dissociated cultures of hippocampus are described by Peacock (1979).

Besides neurons, cell cultures from the nervous system contain many glial cells. It was noted recently that glial cells contain voltage- and transmitter-activated ion channels (Bevan and Raff, 1985; Chiu et al., 1984; Kettenmann et al., 1987; Shrager et al., 1985; Sonnhof and Schachner, 1986), and many studies have recently been undertaken to investigate the membrane properties of glial cells. The study of cultured glial cells was first carried out on presumed astrocytes (Hild et al., 1958), since enriched astrocytic cultures can be easily obtained (Moonen and Nelson, 1978; Walz et al., 1984; Wardell, 1966). The development of cell-type-specific immunological markers enabled electrophysiologists to identify the cell under investigation and to correlate physiological data with a subtype of the glial population (Kettenmann et al., 1983b). With the aid of this identification procedure, the membrane properties of oligodendrocytes and astrocytes could be analyzed, and it was found that these glial cells possess properties previously thought to be unique to neurons (Bevan and Raff, 1985; Gilbert et al., 1984; Hirata et al., 1983; Kettenmann et al., 1984a; MacVicar, 1984).

Cell cultures of neurons and glial cells were therefore used as model systems for investigating single cells or small groups of cells from a defined area of the brain in a defined environment. Properties of cultured cells cannot be assumed to match those of cells in the intact animal, however. First, the spatial orientation of the cell in the tissue has been lost during the dissociation of the tissue, and new connections and contacts are reestablished to a variable degree when the cells are maintained in culture. Second, many intrinsic properties of cultured cells, such as the expression of enzymes and receptors, are determined by environmental factors (Fischer and Wieser, 1983). Cellular properties can depend on developmental stage, solute environment, and cellular composition of the culture. All these parameters can be altered experimentally in culture, which, on one hand, allows for a study of cellular properties under different environmental conditions, and on the other hand leaves some uncertainty as to which of these properties is expressed in the intact animal. Thus, investigations in cell culture can only help to define the properties that defined cell types can aquire and the conditions under which these properties are expressed. Models of cellular properties can be developed together with simple tests to prove the validity of the model's basic assumptions. These models can then be tested in vivo using these simple tests to answer the question, under which developmental, physiological, or pathological conditions cells can express the properties characterized in culture. Investigations in cell culture are therefore not a substitute for in vivo experiments, but should be designed to circumvent complicated experiments in vivo and to accelerate the understanding of cellular properties in a complex nervous system.

## 2.  Culture Methods

The methods employed to prepare and maintain cultures of neural cells are described in several review articles (Fischbach and Nelson, 1977; Kruse and Patterson, 1973; Nelson, 1975; Patrick et al., 1978; Pfeiffer, 1982; Schlapfer, 1978). Basically, any cell in any type of culture can be characterized by electrophysiologists. Some electrophysiological techniques, however, are restricted to larger cells, e.g., intracellular ion-sensitive measurements. Cell cultures can be established using different procedures. When cells are cultured as explants, the topological relations between cells can partially be preserved (Corrigall et al., 1976; Crain and Peterson, 1967, 1974; Dreyfus et al., 1979), whereas purified populations of cells contain predominantly one cell type (Keilhauer et al., 1985; McCarthy and de Vellis, 1980; Meier et al., 1982; Podulso and Norton, 1972). Both approaches are helpful to electrophysiologists, one for the study of connections between populations of cells, the other for the characterization of the intrinsic membrane properties of a specific cell type.

### 2.1.  Explant Cultures

Explant cultures are obtained by cutting nervous tissue into small pieces and plating these tissue fragments into a culture dish. The surface of the dish is coated, e.g., with poly-L-lysine, laminin, or collagen (Bornstein, 1958; Manthorpe et al., 1983; Yavin and Yavin, 1974). Within a few hours the explant attaches to the culture dish, and glial and neuronal processes grow out. Glial and, to a smaller extent, neuronal cell bodies migrate into the outgrowth zone. Within a few days the outgrowth zone appears to be a few cell layers thick, and single cells can be distinguished under the microscope. The survival of neurons is generally improved as compared to dissociated cultures. Figure 1 shows explant cultures from mouse cerebellar cortex.

The flattening of the explant and the migration of neurons can be enhanced by the roller tube technique (Hogue, 1947). By continuous rotation of the culture grown on glass slides, explants flatten out and form a thin layer of cells. This culture system was used to maintain Purkinje cells that did not survive in dissociated cultures for periods of longer than few days. The basic membrane properties of cultured Purkinje cells could therefore be analyzed (Gahwiler, 1975, 1976, 1978). Another method to obtain cultured Purkinje cells of the cerebellum is with the use of microexplants. From these small pieces of cerebellar tissue of 200–400 μm in diameter, Purkinje cells are also reported to migrate out and survive for more than 1 mo (Moonen et al., 1982; Weber and Schachner, 1984).

By placing two explant pieces from defined, different areas of the brain into a culture dish, the formation of connections can be studied and analyzed with electrophysiological methods. Connections between dorsal root ganglia and spinal cord (Crain and Peterson, 1974) and between cerebellum and inferior olive (Gahwiler 1978), for example, were studied by simultaneous recording from both cultured brain regions.

## 2.2. Cell Monolayer Cultures

To obtain dissociated cultures, brain tissue is treated mechanically and/or enzymatically to obtain a single cell suspension. These cells are then plated into a culture dish (Fig. 2; Schlapfer, 1978). Intercellular connections have to be reestablished; the formation of chemical synapses between neurons (Ransom et al., 1977b) and electrical junctions between glial cells (Fischer and Kettenmann, 1985) occurs within a few days. Therefore, dissociated cultures can be used as a model to study the development of electrical activity and synapse formation (Barker and Ransom, 1978; Nelson, 1975; Nelson and Peacock, 1973; Peacock et al., 1973). Figure 2 displays a monolayer culture from mouse cerebellum after 1 and 3 d in vitro.

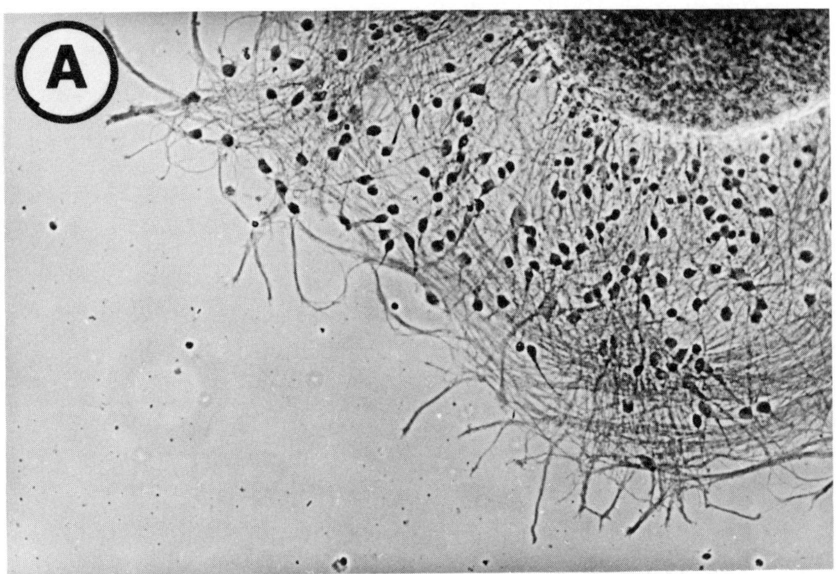

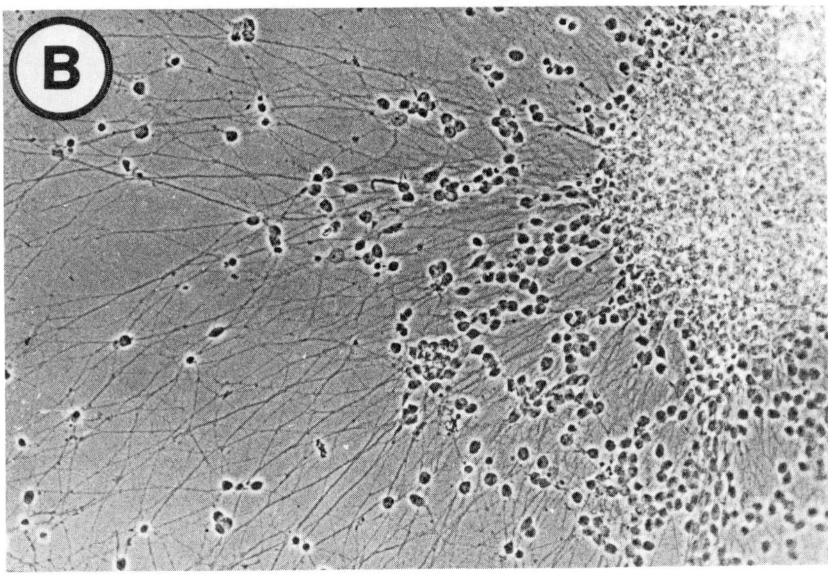

Fig. 1.    Explant culture from 6-d-old mouse cerebellar cortex. The explants were prepared as described (Fischer et al., 1986) and plated either on poly-D-lysine- (A, 3 d in vitro) or on laminin- (B, 2 d in vitro) coated glass coverslips. Neurites grow out from the cerebellar explant core and show a strong fasciculation on poly-D-lysine, but not on laminin. Mainly neuronal cell bodies, but also astrocytes migrate out from the explant core (for method and detailed description, *see* Fischer et al., 1986).

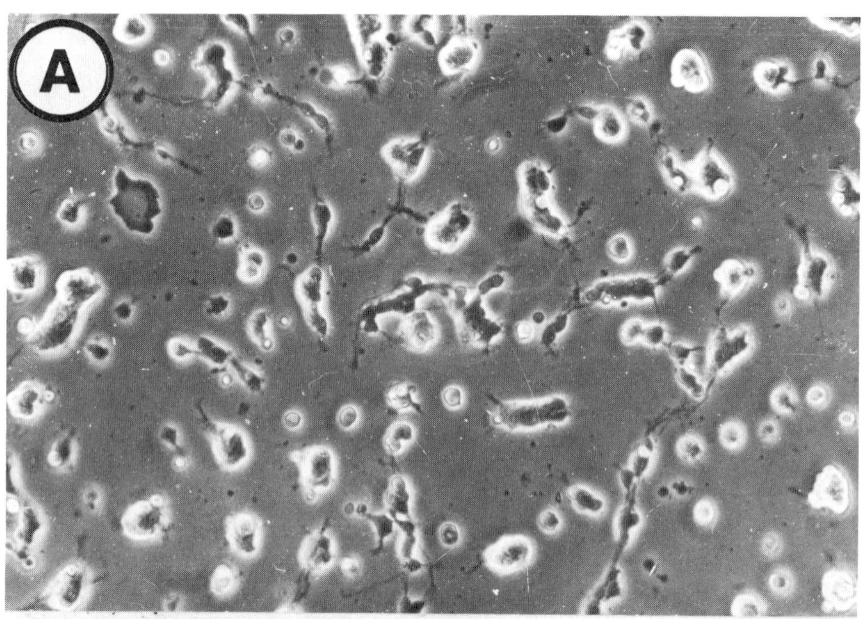

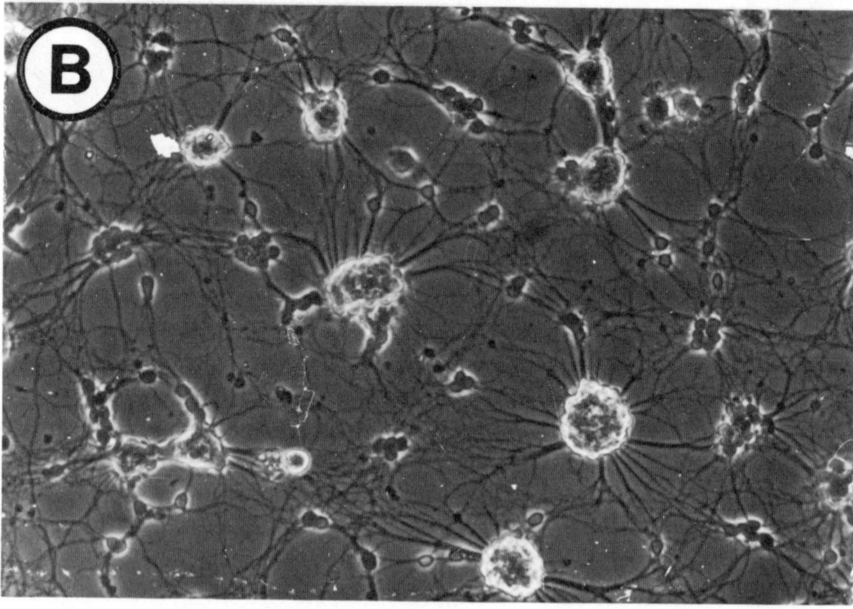

Fig. 2. Monolayer culture of mouse cerebellum. A,B: Cells from mouse cerebellum of postnatal d 6 are plated on poly-L-lysine as culture substrate. A is a phase contrast micrograph of the culture after 24 h in vitro; B after 3 d. (for methods, *see* Werz and Schachner, 1986).

## 2.3. Purified Cell Populations

Purified cell cultures ideally contain only one type of cell. The purification procedure is usually laborious and in some cases requires several purification steps (Fischer and Wieser, 1983). This type of culture offers the advantage that physiological data obtained from any cell in the culture can be attributed to one cell type. In these cultures, information mediated by cell contact or released factors from other cell types is lost. The basic membrane properties of a cell population can therefore be studied under the exclusion of influences from other cell types. The influence of single factors such as hormones or growth factors on membrane properties can be analyzed. From the central nervous system, enriched cultures of neurons, astrocytes (Fig. 3), and oligodendrocytes (Keilhauer et al., 1985; McCarthy and de Vellis, 1980; Meier et al., 1982; Podulso and Norton, 1972) can be obtained. It is even possible to maintain purified cultures of astrocytes at different developmental stages (Fischer et al., 1982). This allows for the study of the development of ion channels and carriers.

# 3. Recording Set-Up

Figure 4 shows the essential constituents of an electrophysiological set-up. The recording site has to be adapted to the requirements of the cultured cells. This implies that during electrophysiological recording, cultures should be maintained under conditions of controlled temperature, pH, and osmolarity. Most physiological experiments last only a few hours; thus sterility at the recording site is not necessary since after the experiment cultures are usually either discarded or preserved for morphological or immunocytochemical evaluation.

For long-term investigations of electrical properties, cultures can be kept under sterile conditions and penetrated with remote-controlled micromanipulators (Baer and Crain, 1971; Crain, 1973). When cultures are grown directly on an electrode array, long-term bioelectric activity from spontaneously active neurons can be monitored (Gross, 1979; Gross and Lucas, 1982; Gross et al., 1977).

## 3.1. Recording Chamber

The recording chamber is designed to maintain cultured cells for several hours during electrophysiological characterization. Con-

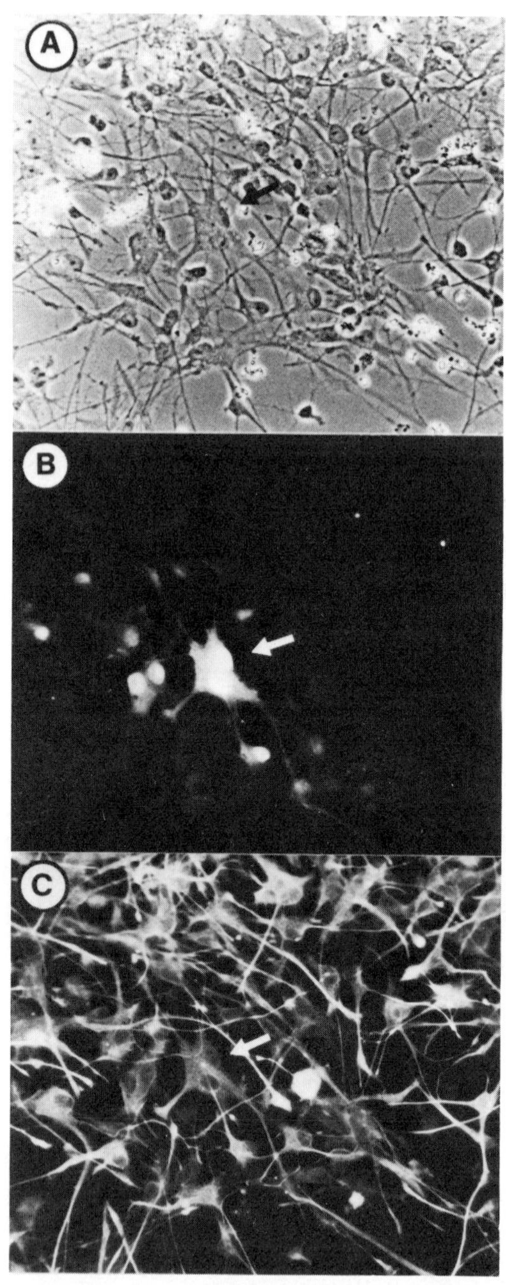

Fig. 3. Identification of astrocytes. After recording from a cultured cell marked by the arrow (A, phase contrast micrograph) of rat cortex,

trol devices for temperature and pH can be built in. The chamber is mounted directly on the stage of the inverted observation microscope.

Cultured cells are generally grown either on plastic dishes or glass coverslips, and the chamber is therefore constructed for reception of either dishes or coverslips. Recording from cells grown directly on the bottom of a culture dish has, however, the following disadvantages:

1. The recording chamber has to be adapted to the form of the culture dish. When other dishes are used, the recording chamber must be redesigned.
2. A rapid exchange of the bathing solution in a round culture dish is not possible. First, the volume of the solution in the dish is between 1 and 5 mL (depending on the diameter of the dish). The exchange of the solution with a flow rate that does not interfere with stable recording therefore lasts up to minutes. Second, a laminar flow of the solution in the round dish cannot be achieved, which in many cases leads to variable exchange times. The rate of exchange of the bathing solution is also dependent on the location of the cell in the dish. The solution in the center is replaced more rapidly than at the rim of the culture.

When cells are grown on glass coverslips, the form of the recording chamber can be optimized for rapid exchange. The time course of solution exchange can be determined by changing the $K^+$ concentration of the perfusate and measuring the concentration changes at different points in the recording chamber with a $K^+$-sensitive microelectrode (*see* section 4.4). The form of the chamber can then be optimized to ensure rapid exchange at the location of the cells.

---

Lucifer Yellow was injected. B. Subsequently the culture was immunolabeled with an antibody directed against glial fibrillary acidic protein, a marker for mature astrocytes (C). Rhodamine was used as fluorochrome. The injected cell and the surrounding electrically coupled cells were identified as astrocytes (Kettenmann et al., 1984a).

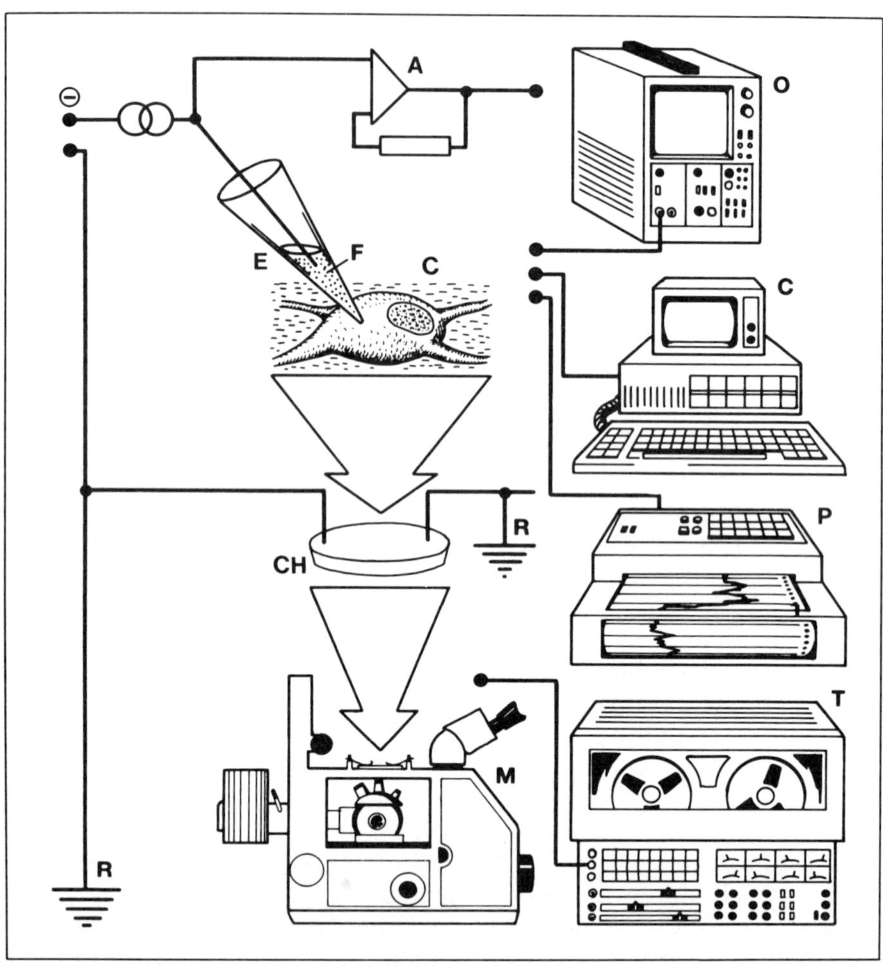


Fig. 4A.   Electrophysiological recording set-up for cell cultures. A Schematic representation. The cultured cells (C) are inside the recording chamber (CH) on the stage of an inverted microscope (M). Cells are penetrated with an electrode (E) filled with an electrolyte (F). The recorded potential compared to the reference potential (R) in the bathing solution is amplified (A) and stored and displayed by means of conventional electrophysiological devices like an oscilloscope (O), computer (C), chart recorder (P), or tape recorder (T) (from Wienrich and Kettenmann, 1984).

Fig. 4B.

Schematic drawings of recording chambers are shown in Fig. 5 (for reception of culture dishes) and Fig. 6 (for reception of glass coverslips). A built-in perfusion system can be used to exchange the bathing fluid. The control devices for temperature and pH are part of the chamber, and it is possible to alter the temperature in the bath during the experiment as described below (Gahwiler et al., 1972; Fig. 7).

## 3.2. The Microscope

An inverted microscope must be used to visualize the cells at the recording site. It is firmly connected to the recording table to avoid movement of the stage relative to the micromanipulators. The optical system is underneath the recording chamber, which allows for free access to the cultured cells. The illumination system is above the chamber, and a large working distance of that system is desirable for access of electrodes. For selection of cells, phase contrast optics are helpful. When the electrodes are in the recording position, however, phase contrast is usually lost. A camera coupled to a television monitor can be adapted to the microscope and is especially helpful in discussing the choice of the cell or the

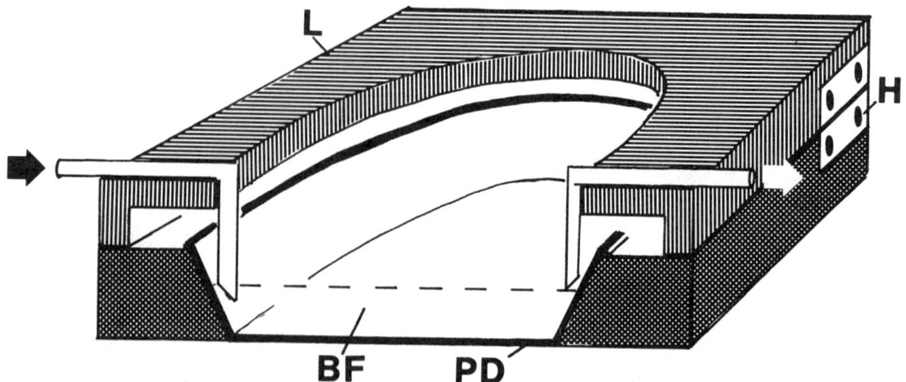



Fig. 5.   Schematic diagram of a recording chamber for reception of Petri dishes. A Petri dish (PD) is inserted into the central opening of the recording chamber (dotted pattern). A lid (L, lined pattern) connected with hinges (H) to the recording chamber contains the tubing for the perfusion system (arrow). The level of the bathing fluid (BF) in the Petri dish is controlled by the position of the outflow tube.

positioning of the electrode. It also minimizes possible damage of electrodes when used in combination with a remotely controlled electrode and stage placement system.

## 3.3.  Control of Temperature and pH

The temperature of the bathing solution can be controlled by heating or cooling either the recording chamber (Gahwiler and Bauer, 1975) or the perfusate. When the recording chamber is continuously perfused, the solution can be temperature controlled by passage through a heat-exchange system (Fig. 7). The pH is maintained either by stabilizing the pH of the perfusate or the solution in a nonperfused recording chamber with a pH buffer system, e. g., $CO_2/HCO_3^-$ or HEPES.

## 3.4.  Cell Penetration

The penetration of the cell membrane with microelectrodes is the most crucial step in recording from cells in culture. Cells can be penetrated either by a rapid movement of the recording electrode or by a current pulse after contact of the electrode with the cell membrane. The culture system offers the advantage that both the cell and the recording electrode can be visualized. This allows the

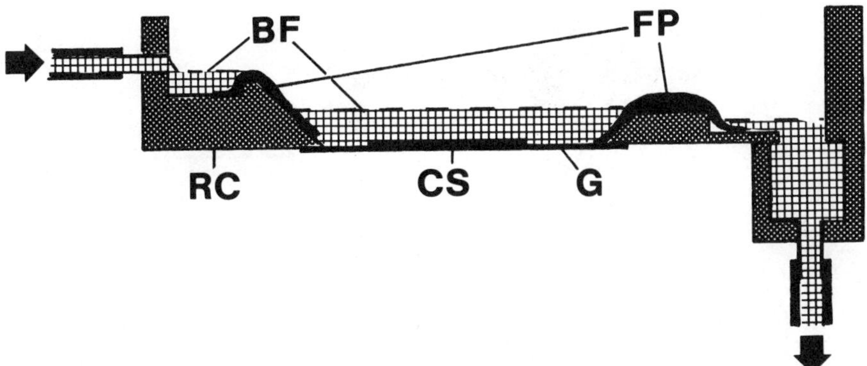



Fig. 6A.   Recording chamber for glass coverslips. A Schematic representation. This recording chamber is optimized for rapid exchange of the bathing fluid (BF). The perfusion fluid first flows into a small compartment that is connected with the main chamber by a strip of filter paper (FP). This arrangement prevents fluctuations in the bathing level. The glass coverslip with the cultured cells (CS) is placed on a thin glass surface (G). A second filter paper draws the fluid drop-free into another compartment, whereby a constant fluid level is maintained. The third chamber is emptied by gravity.

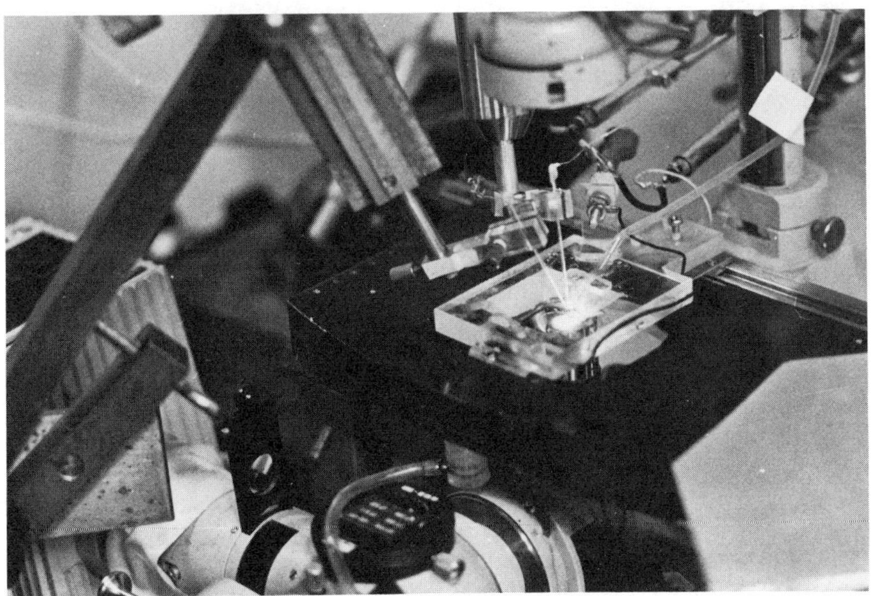

Fig. 6B.

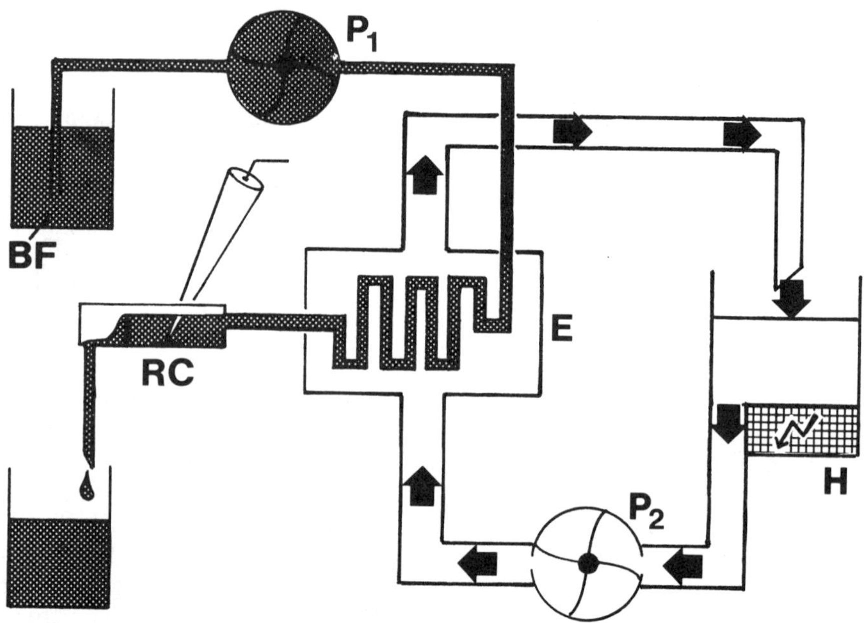


Fig. 7. Variation of temperature at the recording site. Cells are penetrated in the recording chamber (RC), which is continuously perfused with the bathing fluid (BF). A pump ($P_1$) regulates the inflow to the chamber. The bathing fluid passes a fluid filled heat exchange system, which is driven by a pump ($P_2$) and temperature controlled by an electrical heater (H) or a Peltier element. Rapid changes of temperature can be achieved by adding cold or hot water to the reservoir. This system is designed to control the temperature between 10 and 40°C.

electrode to be positioned at different locations in the cell. The advancement of the electrode must be precisely controlled since the cells are only a few micrometers thick, and the glass or plastic surface is directly underneath the cells. Therefore precise manipulators are required. For flat cells such as fibrous astrocytes, a remote-controlled micromanipulator with a step motor-driven (Sonnhof et al., 1982) or piezoelectric (Fromm et al., 1980) advancement is almost mandatory. A combination of advancement of the electrode and a current pulse was found to be optimal for the penetration of most neural cells. The size of the advancing step depends on the cell type and can vary between 0.6 μm for astrocytes and 5 μm for large neurons (Sonnhof et al., 1982). The best position for penetration with a step motor-driven micromanipulator is at a right angle with the surface of the cell soma (Fig. 8). The electrodes are usually fixed by the micromanipulator at an angle (10–20°) to the vertical, and the microscope picture yields the projection of cell and electrode position. Therefore electrodes are not positioned at the center of the soma.

## 3.5. Application of Drugs

Alterations of the ionic composition of the bathing fluid or application of drugs to the cell membrane can be achieved either by exchange of the recording bath, by pressure ejection from a micropipet, or by iontophoresis.

In culture, iontophoresis (Hicks, 1984) can only be used for applying substances in low concentrations (e.g., neurotransmitters in concentrations $\geq 10^{-4}M$), since the extracellular space around cultured cells is large compared to that in vivo (Macdonald and Barker, 1981; Ransom et al., 1977c). Pressure ejection from a micropipet positioned in the vicinity of the tested cell can be used to change the ionic composition of the extracellular fluid or to apply drugs (Macdonald and Barker, 1981; *see* Fig. 9). A pressure unit is connected to the back of a microelectrode, and by applying pulses of pressure in the range of 0.2 to 5 bar the test solution is pushed out of the electrode. The ejected volume depends on the tip diameter of the electrode, time of application, and applied pressure. The tip diameter should not exceed 10 μm to minimize leakage by diffusion. In contrast to iontophoretic application, large molecules such as antibodies can be easily ejected. The final concentration of the tested substance at the cell membrane, however, cannot be determined since the ejected solution is mixed with the bathing fluid. Figure 10 displays the decrease of $K^+$ concentration with distance from a pipet during ejection of elevated $K^+$ concentration. As an artifact of this method, electrically coupled cells can be transiently uncoupled by the bulk flow from the pressure ejection pipet. This results in an increase in cell input resistance in cultured mouse oligodendrocytes. The membrane potential is not affected (Kettenmann, 1985).

Drugs can also be applied by continuously perfusing the recording chamber and changing the perfusate. This method ensures that the concentration of the drug or the altered ionic composition of the bathing fluid can be controlled at the cell membrane (Fig. 11). The major disadvantages of applying drugs by exchange of the bath are that comparably large amounts of the test solution are

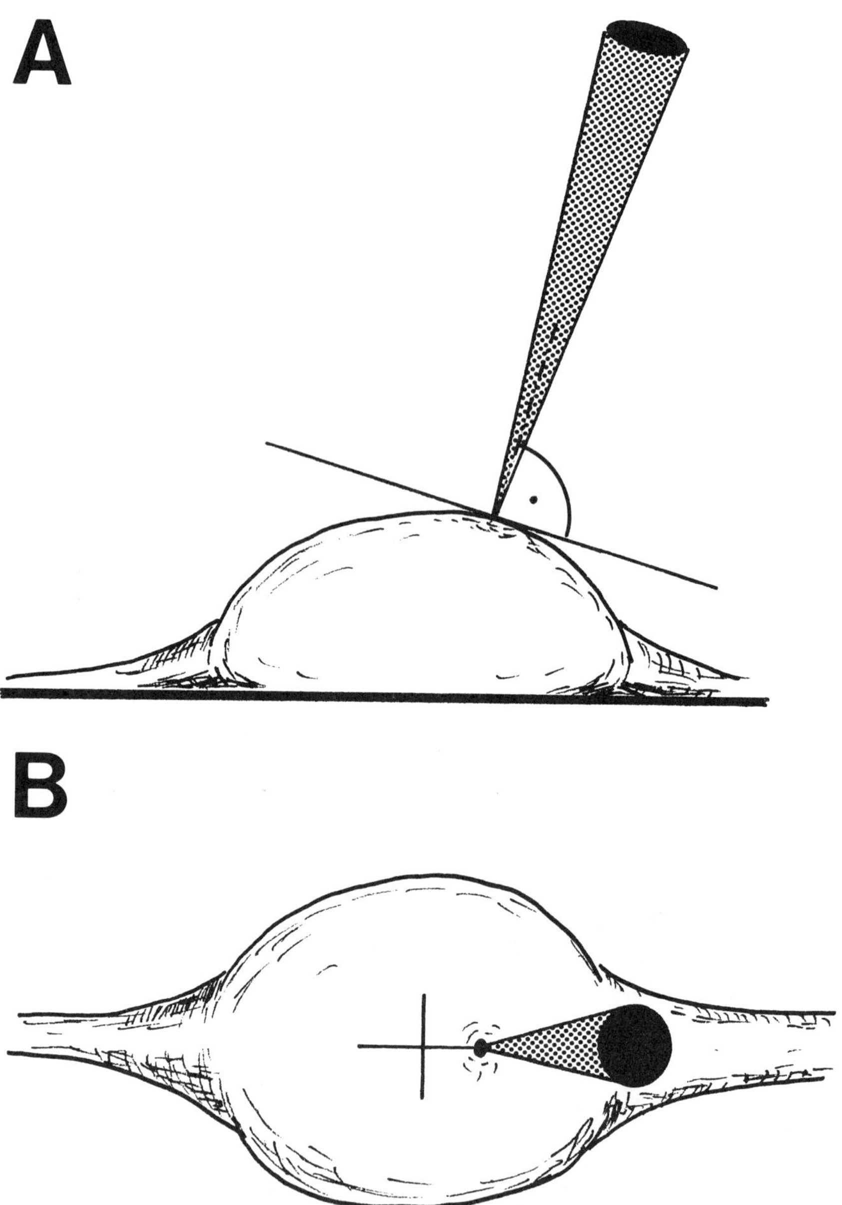

Fig. 8. Position of the electrode for cell penetration with a step motor-driven manipulator. Recording electrodes in culture are placed at a 10–20° angle to the vertical of the culture plane to permit an illumination of the culture for inspection in the microscope. For an optimal penetration,

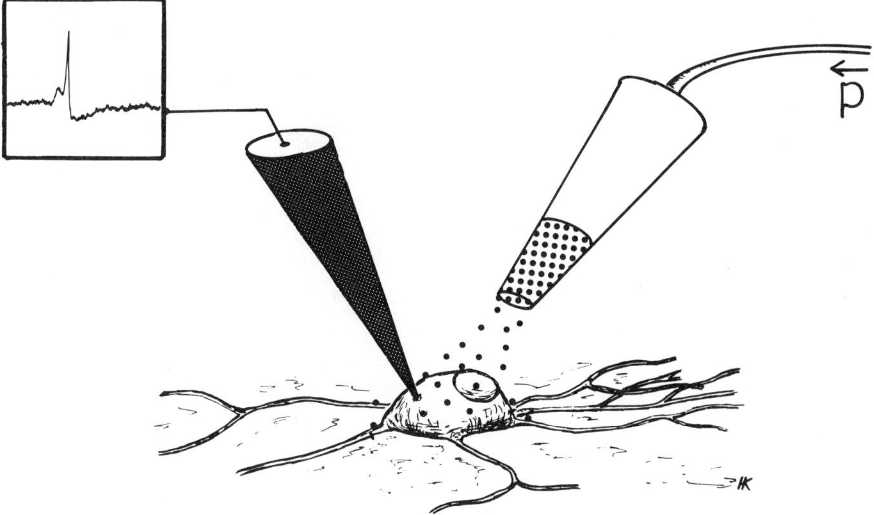


Fig. 9. Pressure ejection from a micropipet. A cell is penetrated with a microelectrode to monitor the electrical activity. Substances can be applied from a micropipet filled with the test solution. When pressure (p) is applied to the rear end of the pipet, the solution is ejected. To avoid leakage, the diameter of the tip should not exceed 10 μm.

required (about 10 mL for each test), and that the substance cannot be applied as rapidly as with pressure ejection from a micropipet. Moreover, the entire culture is exposed to the test solution, which usually limits the number of studies that can be carried out with a culture. Figure 11 shows the time course of exchange in a recording chamber (90% of the solution exchanged within 1 min).

The recording chamber must be optimized for rapid exchange (Fig. 6). The perfusate can be moved by either gravity or pumps. The flow rate into the chamber must be kept constant during the experiment to avoid variation in the responses. Therefore the

the electrode should be placed at a right angle to the cell membrane, as shown in A (vertical cut through the plane of the culture). B shows the same experimental arrangement as it is visualized in the inverted microscope. It can be seen that the point of penetration is not in the center of the cell.

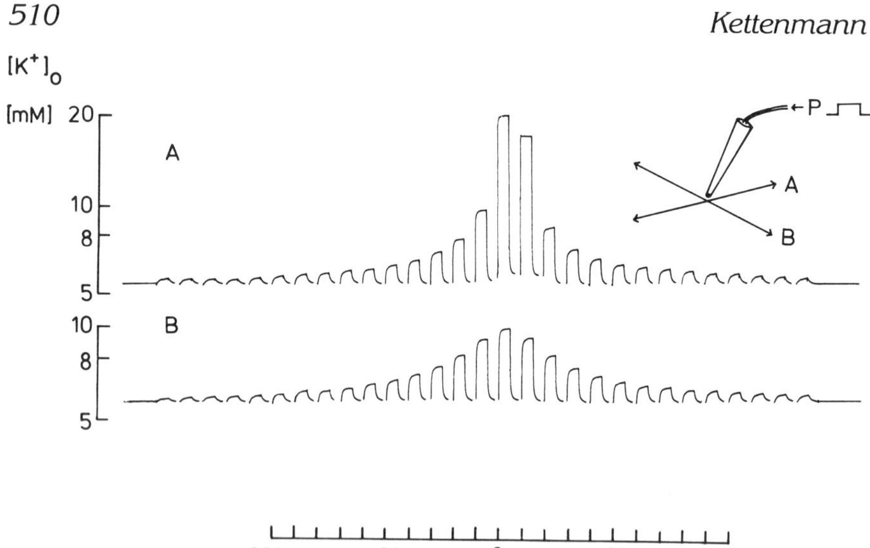



Fig. 10. Ejected pulses of elevated $K^+$ concentration (100 m$M$, $[K^+]$) measured at different distances from the ejection electrode. Solution with elevated $[K^+]$ was ejected from a fine micropipet (tip diameter, 1 $\mu$m) by pressure pulses of 5 bar and 20 s duration. After each pulse, the sensitive electrode was moved by 20 $\mu$m vertical to the axis of the pressure electrode. The responses of the $K^+$-sensitive electrode (A and B are two sets of movements) show a dependence of the $[K^+]$ on the distance from the micropipet. The concentration rapidly decreases within a few micrometers.

inflow should be driven by a speed-controlled pump rather than by gravity, since the exchange rate of a gravity-driven perfusion can be strongly affected by debris or salt crystals in the tubing system. The result of drug application with different perfusion rates is demonstrated in Fig. 12. The outflow must regulate a constant fluid level in the recording chamber and can be driven by gravity. The ionic composition and osmolarity of the perfusate must be identical with the culture medium. Table 1 compares perfusates used in different preparations.

## 3.6. Remote Control of the Recording Site

Manual positioning of more than one electrode at the recording site needs experienced and skillful experimenters. In many cases electrodes are broken during their positioning and the selection of the cell. A remote control of the recording site dramatically

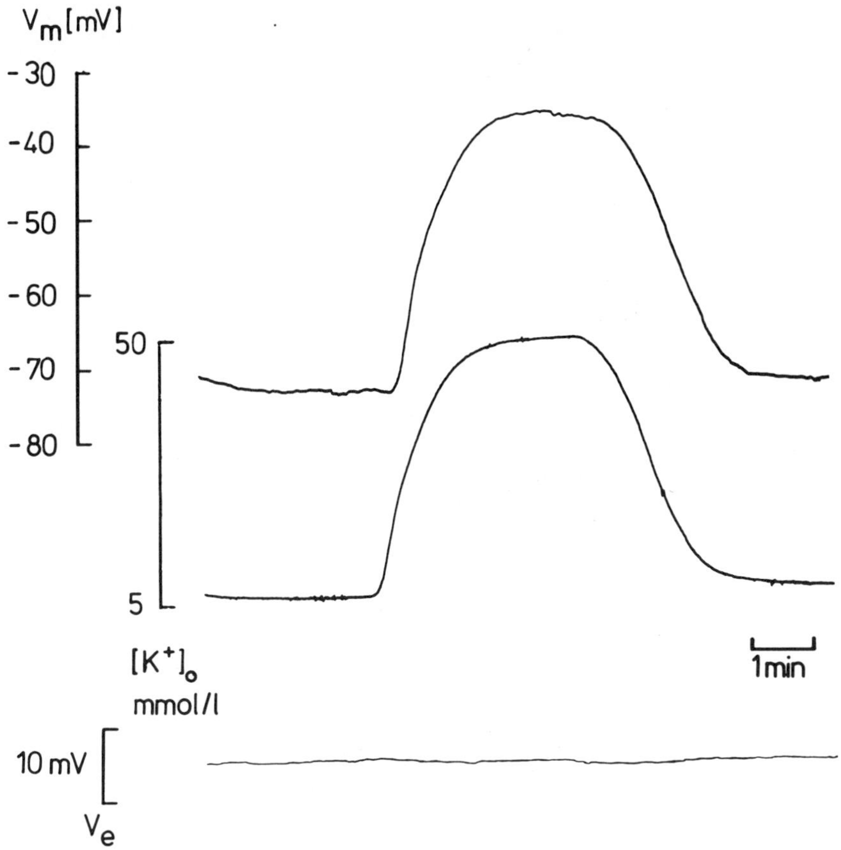



Fig. 11.   Response of the membrane potential to perfusion of the recording chamber with elevated $K^+$ concentration ($[K^+]_o$). During recording of membrane potential ($V_m$, upper trace) from a cultured oligodendrocyte of mouse spinal cord (Kettenmann et al., 1983b), the $[K^+]_o$ was raised in the bath by changing of the perfusate. A $K^+$-sensitive microelectrode placed close to the cell served to monitor the change in $[K^+]_o$ (middle trace). The recording of the reference channel of the $K^+$-sensitive electrode is shown on the lowest trace.

reduces the probability of damage to the electrodes. In a remote control system the movement of the electrodes, as well as the operation of the microscope, is driven by motors. The motors are controlled by a unit connected to a joystick panel. In combination with a television system the experimenter can position electrodes, focus the microscope image, and move the stage of the microscope

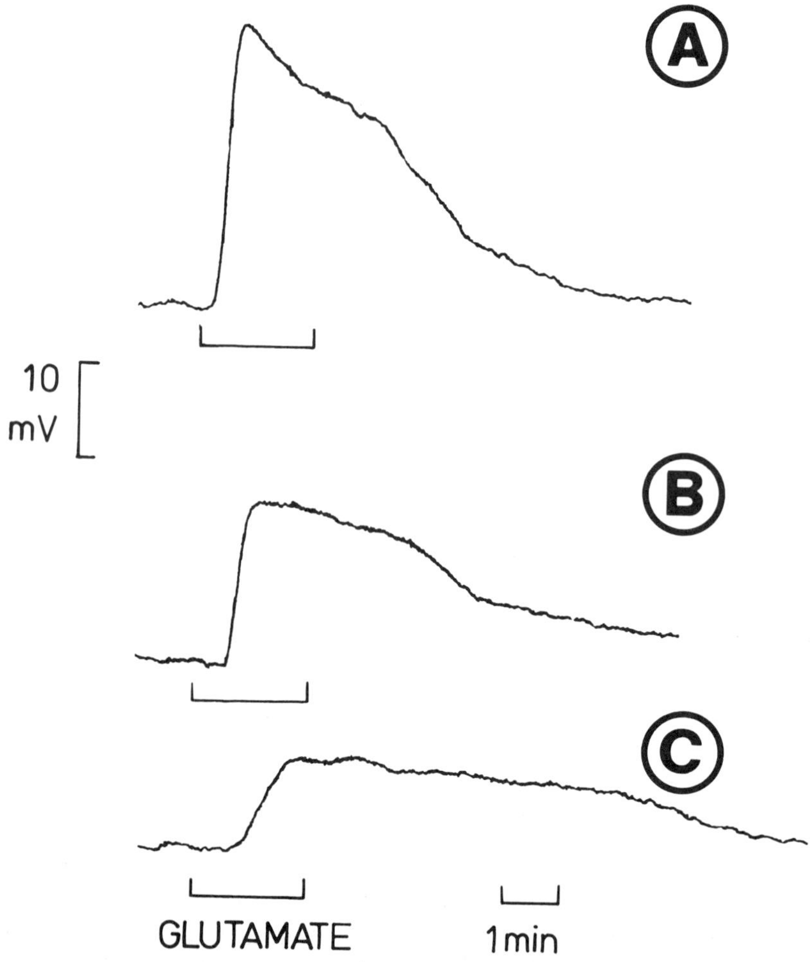



Fig. 12.   Drug application with variable perfusion speed. Glutamate was applied to a cultured astrocyte from rat cerebral hemispheres (Kettenmann et al., 1984a) as indicated by bar. The membrane potential depolarizes and subsequently repolarizes in the presence of glutamate (1 m$M$) because of desensitization of the depolarizing response. The speed of bath exchange was decreased from A (10 s for 90% exchange) to C (1 min for 90% exchange). Compared with A, the onset of the depolarization in B and C is delayed, the depolarizing phase prolonged, and the size of the depolarization reduced. To compare responses of membrane potential, the rate of bath exchange must be kept constant during the experiment to compare repetitive substance applications.

Table 1

Major Salt Contents of Perfusates Used in Different Preparations[a]

| Animal | Brain region | NaCl | KCl | CaCl$_2$ | MgSO$_4$ | MgCl$_2$ | NaH$_2$PO$_4$ | KH$_2$PO$_4$ | NaHCO$_3$ | Reference |
|---|---|---|---|---|---|---|---|---|---|---|
| Rat | Brain | 120 | 4.5 | 2 | 1 | | 1 | | 26 | Dimfel et al., 1979 |
| Rat | Spinal cord | 137 | 5 | 2.4 | | 2.2 | 1 | 0.18 | 2.9 | Hosli et al., 1978 |
| Mouse | Spinal cord | 143.2 | 5.4 | 1.8 | 0.8 | | 1 | | 26.2 | Kettenmann et al., 1983b |
| Rat | Brain | 122 | 3 | 1.3 | 1.2 | | | 1.2 | 10 | Kimelberg et al., 1979 |
| Rat | Cervical ganglion | 140 | 5.4 | 2.8 | | 0.18 | 0.56 | | 26 | O'Lague et al., 1978a |
| Rabbit | Brain | 137 | 5.4 | 1.3 | 0.4 | 7 | 0.5 | 0.4 | 4.2 | Wardell, 1966 |

[a]Numbers correspond to concentrations in m$M$.

513

at a safe distance from the recording site. Commercial systems are available that fulfill these requirements. This method introduces a large safety factor in the use of ion-sensitive microelectrodes that are laborious to fabricate and for experiments in which multiple electrodes are used.

## 4. Applications

### 4.1. Intracellular Dye Injection

Intracellular labeling of cells can be used to mark the location of the cell within the culture, visualize structures related to the cell, or identify the cell type in combination with a cell-type specific antibody. The structure and shape of the physiologically characterized cell can be documented by injecting a dye that spreads within the entire cell. As shown in Fig. 13, many fine processes of a cell are not visible in the phase contrast micrograph (Fig. 13A), but can be visualized after injection of a dye (Fig. 13B). When pairs of cells are investigated, these cells can be injected with different dyes. This method enables identification of the morphology of both cells and visualization of the possible contact areas between the cells.

All staining procedures generally used by electrophysiologists for intracellular staining of cells can also be applied in culture. Although many dyes require histological processing after the injection (e.g., Cobalt-staining; Pitman et al., 1972), some fluorescent dyes can be visualized immediately after injection. It is even possible to observe the event of injection if the recording microscope is equipped with appropriate excitation light and filter combinations. Lucifer Yellow is a highly fluorescent, nontoxic dye and can be easily injected into cells (Stewart, 1981). It is therefore commonly used in culture (Gahwiler, 1981; Kettenmann et al., 1983a,b). The dye is negatively charged, can be dissolved in water, and is used as an electrolyte in the recording electrode. Application of 5 nA of negative dc current for 5 min to the electrode is sufficient to stain an oligodendrocyte with a cell diameter of 30–50 μm. Since the resistance of a Lucifer Yellow-filled microelectrode is about fivefold that of a KCl-filled electrode (generally about 50 MΩ for penetration of cultured neural cells) with comparable tip diameter ($<0.5$ μm), the recordings of membrane potential are not as stable. It is therefore feasable to record the physiological signals with conventional electrodes and subsequently penetrate the cell for dye injection a second time with a Lucifer Yellow-filled electrode (Kettenmann et

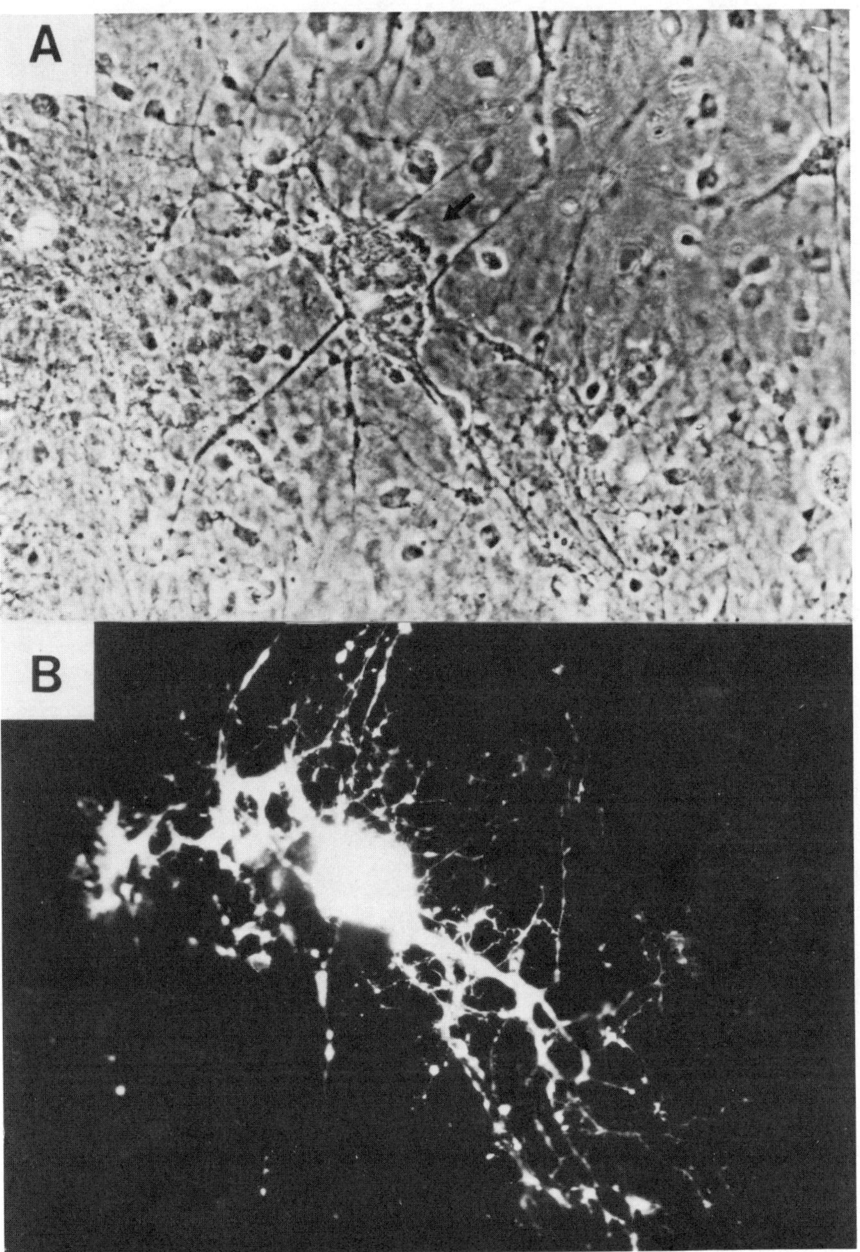

Fig. 13.    Lucifer Yellow-injected glial cell. A glial cell from mouse spinal cord was injected with the fluorescent dye Lucifer Yellow (Kettenmann et al., 1983b). A is the phase contrast micrograph (cell indicated by arrow) and B is the corresponding field inspected with fluorescence optics to visualize Lucifer Yellow staining.

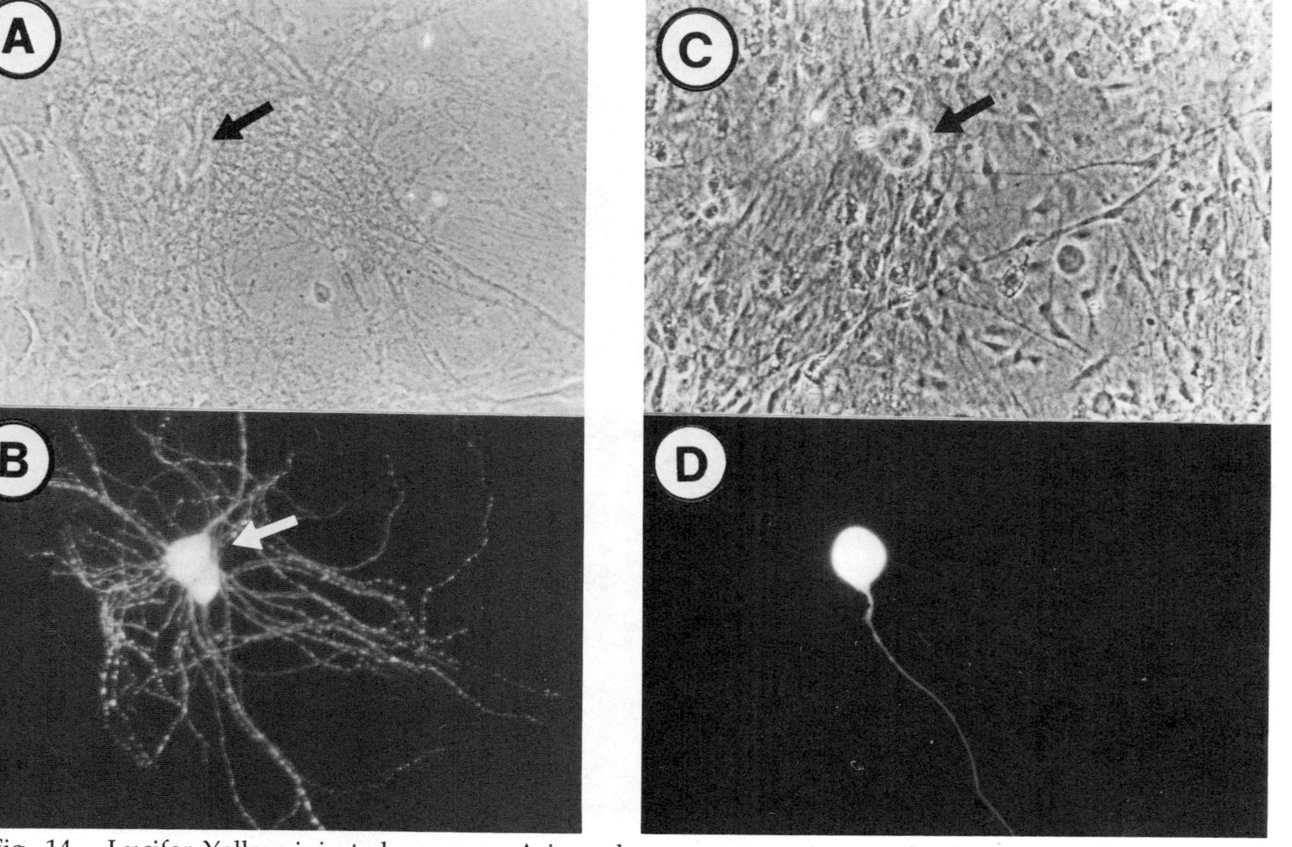

Fig. 14. Lucifer Yellow-injected neurons. A is a phase contrast micrograph of a rat spinal cord culture (Kettenmann et al., 1983c). Two neurons marked by an arrow are injected with Lucifer Yellow and the fluorescence is visualized in B. Similarly, a neuron from cultures of mouse dorsal root ganglion was injected (C and D). Neurons were identified by their ability to generate action potentials.

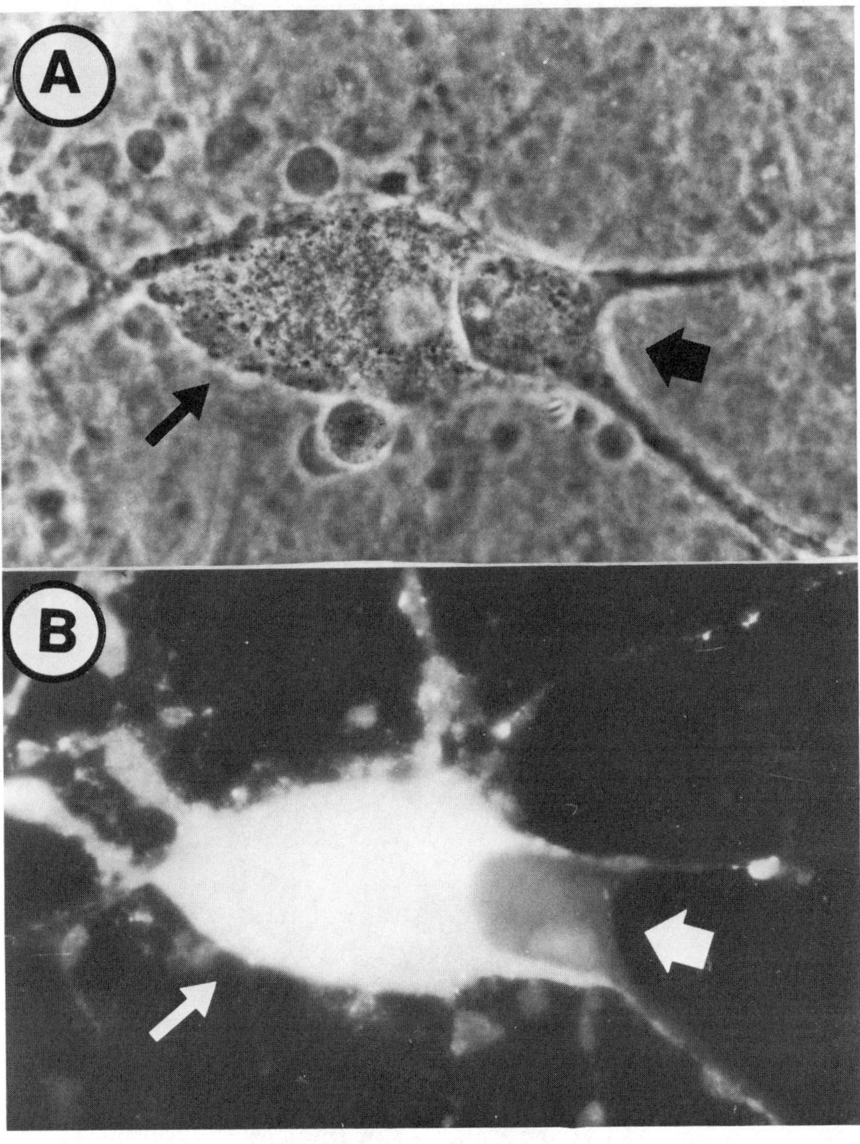

Fig. 15. Electrical coupling between glial cells. A glial cell was injected with Lucifer Yellow (Kettenmann et al., 1983a, narrow arrow). The dye can be seen to spread to a neighboring cell (wide arrow). A is the phase contrast, whereas B is the corresponding fluorescence micrograph.

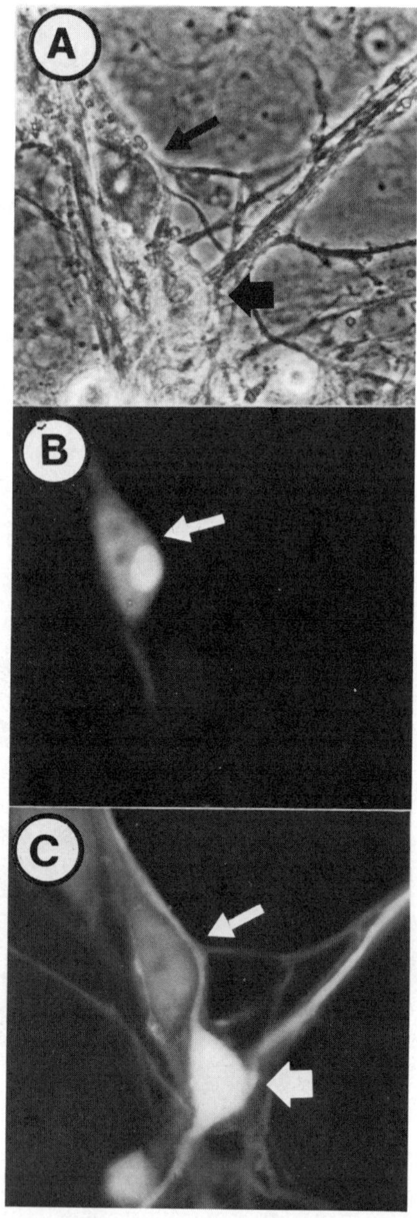

Fig. 16. SITS- and Lucifer Yellow-injected cells. Two glial cells (arrows) shown in the phase contrast micrograph (A) were impaled with a Lucifer Yellow- and SITS-filled microelectrode (Gilbert et al., 1982). One cell was injected with Lucifer Yellow (wide arrow), the other with SITS

al., 1983b). After injection of the dye, the culture can be fixed in 4% paraformaldehyde, and the fluorescence can be preserved for several weeks (Figs. 13 and 14). Lucifer Yellow passes gap junctions (Stewart, 1981) and can therefore be used to demonstrate electrical synapses between cells (Figs. 3 and 15). For investigating intercellular connections between pairs of cells, a second dye with a different fluorescence emission serves to label and discriminate between both cells. The stilbene-derivative SITS emits a blue light when appropriately excited. With the use of selected filter combinations (Table 2), Lucifer Yellow and SITS staining can be distinguished. One cell can therefore be stained with Lucifer Yellow and a second one with SITS (Fig. 16) (Gilbert et al., 1982). Injection of SITS requires more charge than Lucifer Yellow to yield a comparable staining intensity.

To correlate physiological data with ultrastructure, the characterized cell must be marked with a dye that can be visualized in the electron microscope. The staining procedure with horseradish peroxidase is commonly used to correlate physiological data with electron microscopic structures and has been applied to cell cultures (Calvet and Calvet, 1979; Neale et al., 1978).

## 4.2. Identification of Cell Types with Immunocytochemical Methods

The use of cell type-specific antibodies permits identification of individual cells in a heterogeneous population. In the nervous system, various antibodies are available to characterize the major cell types (Raff et al., 1979; Schachner, 1982a,b, 1984; Schachner et al., 1981a). It is therefore possible to distinguish between different types of glial cells and neurons. Among glial cells, astrocytes can be identified, e.g., by antibodies directed against glial fibrillary acidic protein, which is a unique characteristic of astrocytes (Bignami and Dahl, 1974; Eng et al., 1971). Oligodendrocytes are identified by antibodies against the antigens 01 to 04 (Schachner et al., 1981b; Sommer and Schachner, 1981) and Schwann cells in the peripheral

---

(narrow arrow). B shows the fluorescence of SITS staining; C shows Lucifer Yellow. Appropriate fluorescence optics are used to discriminate between SITS and Lucifer Yellow fluorescence. Lucifer Yellow crossed the gap junction between the two cells, whereas SITS did not permeate the electrical synapse.

Table 2

Characteristics of Dyes and Filter Combinations for Fluorescence
Microscopy Using the Intracellular Dyes Lucifer Yellow (LY)
and SITS, and the Fluorochromes Fluoresceine (FITC) and
Rhodamine (TRITC) for Visualization of Antibody Binding[a,b,c]

| Dye | LY | SITS | FITC | TRITC |
|---|---|---|---|---|
| Absorption max. | 430 | 350 | 490 | 555 |
| Emission max. | 540 | 430 | 530 | 580 |
| Excitation | 450–490 BP | 365/11 BP | 450–490 BP | 546/12 BP |
| Mirror | 510 | 395 | 510 | 580 |
| Barrier | 520 LP | 395–440 BP +490KP | 520 LP | 590 LP |

[a]Numbers show the wavelength in nm (peak values for absorption, emission,
and properties of filters).
[b]Abbreviations: BP, bandpass; KP, short pass.
[c]From Gilbert et al., 1982.

nervous system by anti-Ran1 antibodies (Fields, 1980). The properties of these neural cells can be compared to those of cells identified
as nonneural, such as macrophages and fibroblasts that also occur
in cultures of the central nervous system and often possess similar
morphological features. With appropriate immunological markers
it is even possible to discriminate different developmental stages of
astrocytes and oligodendrocytes (Schachner et al., 1982, 1983;
Sommer and Schachner, 1982). Antibodies are also available to
identify neurons (e.g., Rathjen and Schachner, 1984). Some antibodies recognize subtypes of neurons, e.g., Purkinje cells of the
cerebellum (Weber and Schachner, 1984). Table 3 displays a list of
antibodies that can be used to identify cell types in cultures of the
nervous system.

To correlate physiological data with a cell population, the
recorded cell must be identified with a cell type-specific antibody
(Fig. 17). The first step after physiological characterization is to
mark the recorded cell by injecting a dye such as Lucifer Yellow or
SITS. Subsequently the entire culture is stained with a cell type-
specific antibody. The binding pattern of the antibody is visualized
by a fluorochrome bound directly to the cell type-specific antibody
(direct immunofluorescence) or to a second antibody directed

Table 3
Cell Type-Specific Antibodies for Identification of Major Cell Classes in
Cultures of the Nervous System

| Antigen | Cell type | Reference |
| --- | --- | --- |
| GFAP | Astrocyte | Eng et al., 1971 |
| Fibronectin | Fibroblast | Schnitzer and Schachner, 1981 |
| L1 | Neuron (CNS) | Rathjen and Schachner, 1984 |
| 01–04 | Oligodendrocyte | Sommer and Schachner, 1981 Schachner et al., 1981a |
| Ran-1 | Schwann cell | Fields, 1980 |

against the first (indirect immunofluorescence) (Schnitzer and
Schachner, 1981a,b; Weir, 1978). The antibody-bound fluorochrome emits light at a different wavelength than the injected dye
(Fig. 3; Kettenmann et al., 1983a,b). Rhodamine-coupled antibodies can be used in combination with Lucifer Yellow, whereas SITS
can be combined with two additional fluorochromes, rhodamine
and fluorescein (Gilbert et al., 1982; *see* Table 2).

Antibodies can also be added prior to penetration of the cells,
and the labeled cells can then be selected by their fluorescence. The
limitation of this procedure is that the antibody must be directed
against a surface antigen, and it has to be proven that the binding of
the antibody does not influence the membrane properties of the
target cells. For oligodendrocytes, this has been shown for the
antibodies 01 to 012 (Kettenmann et al., 1985), and for neurons the
L1 antibody was found not to alter the gross electrophysiological
parameters such as action potential and spontaneous synaptic
activity (Kettenmann et al., 1983c). This method offers the advantage that the selected cell is from the desired population. As a
possible disadvantage, cells have to be exposed to fluorescent
light, which might cause cell damage.

## 4.3. Recording Membrane Potential

The resting membrane potential of a cell as the difference
between the intra- and extracellular potential can be measured
after penetration of the membrane with an electrolyte-filled glass

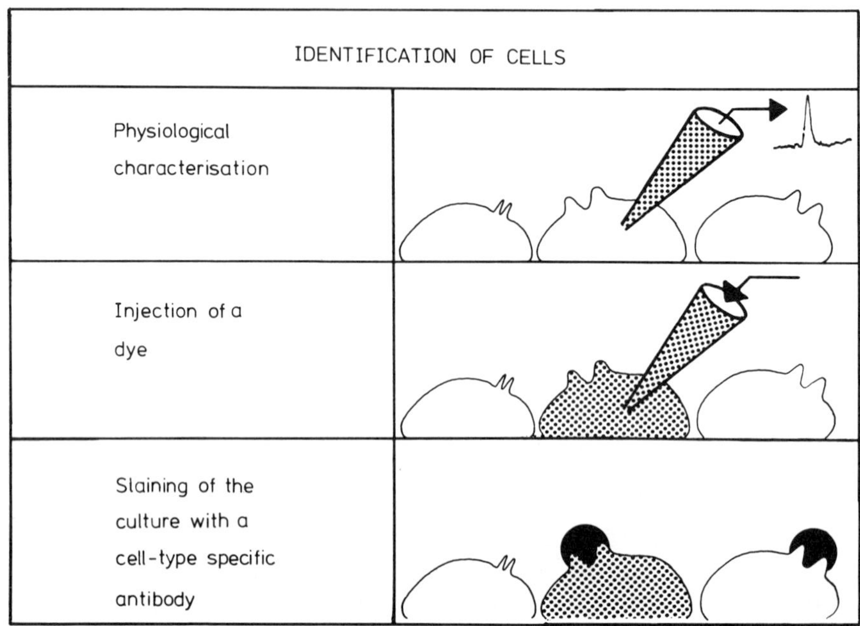



Fig. 17.  Identification of physiologically characterized cells by intracellular staining and subsequent application of a cell type-specific antibody.

microelectrode. The resting membrane potential of cultured cells is similar to that found in vivo; namely $-40$ to $-60$ mV for neurons and $-60$ to $-80$ mV for glial cells (*see,* e.g., Moonen and Nelson, 1978; Ransom et al., 1977a). The culture system offers the possibility of controlling or modifying the extracellular solution. The ionic dependence of the resting membrane potential can be studied independently of influences from other cells. The concentration of extracellular ions can be varied under controlled conditions, and the effect on the resting membrane potential can be studied. Figure 11 shows the membrane potential response of a cultured oligodendrocyte from mouse spinal cord to a change of the extracellular $K^+$ concentration. Similar measurements could never be carried out in the intact central nervous system since the extracellular ionic composition cannot be precisely controlled.

Cell membrane potential can also be recorded by applying the patch-clamp technique in the "whole cell recording configuration" (Hamill et al., 1981; Single-Channel Recording: Methology and Application, in this volume). After a gigaseal with the cell membrane is formed, the patched membrane area is disrupted by a current pulse or by suction applied to the recording pipet. The pipet then faces the cell interior, and the intracellular fluid of the cell is replaced by the pipet solution. This method permits the recording from small cells that cannot be successfully impaled with conventional microelectrodes.

## 4.4. Recording Ion Concentrations

Ion-sensitive microelectrodes can be used to record intra- and extracellular ion concentrations (*see* Intracellular Ion-Sensitive Microelectrodes, in this volume). The ionic concentration of the recording bath in culture is assumed to be identical to that of the fluid in the extracellular space around the cells. Only in deeper layers of explants have changes of ion concentration induced by neuronal activity been found to differ from the composition of the bathing solution (Hosli et al., 1981). In monolayer cultures, the ion activity of the space around the cells is identical to the rest of the bath solution. Therefore measurable changes in the extracellular ion composition induced by the activity of cells are not expected. Extracellular ion-sensitive measurements in culture can therefore be used only to measure and adjust the exchange speed of the perfusion (section 3.5; Fig. 12) or the effectiveness of a pressure ejection electrode (section 3.5, Fig. 10, Ransom and Holz, 1977).

The successful measurement of intracellular ion activities in cultured neural cells depends mainly on the quality of cell penetration and the properties of the ion-sensitive microelectrode. Double-barreled microelectrodes with fine-tip diameters ($< 0.5$ $\mu$m) must be used. Measurements of intracellular $[K^+]$ have been obtained from oligodendrocytes of the mouse spinal cord (Kettenmann et al., 1983b). The necessary silanization procedure and/or the filling of the electrode with the ion-sensitive resin changes the properties of the electrode tip and thereby decreases the probability of successful cell penetration. An additional problem is that current pulses can hardly be used without decreasing the ion sensitivity of the electrode. Therefore cells can only be penetrated with a rapid advancement of the microelectrode. This requires a step motor- or piezo-driven micromanipulator (Fromm et al., 1980; Sonnhof et al., 1982). Figure 18 displays the registration of membrane potential and $K^+$ concentration during the event of penetration of a cultured oligodendrocyte. For a more detailed description of ion-sensitive measurements, *see* Intracellular Ion-Sensitive Microelectrodes, in this volume.

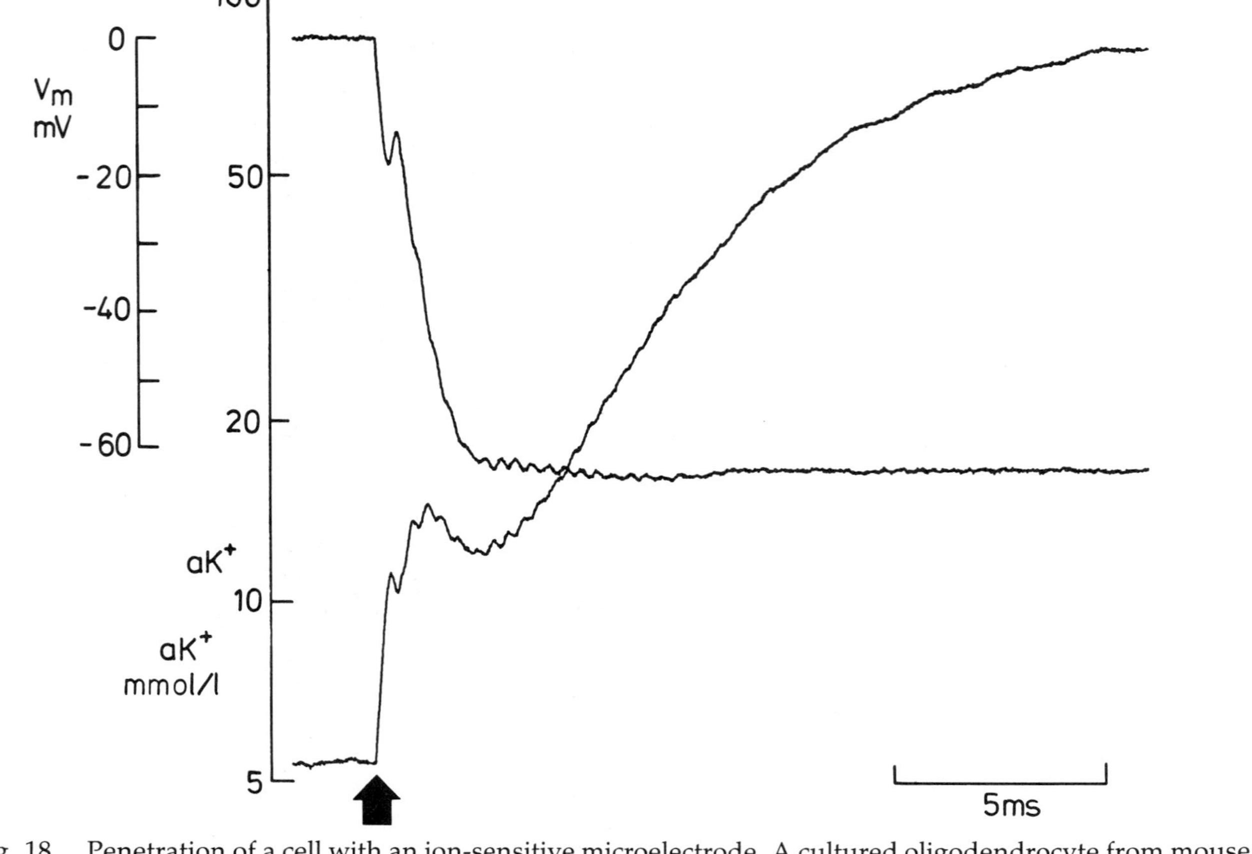



Fig. 18.   Penetration of a cell with an ion-sensitive microelectrode. A cultured oligodendrocyte from mouse spinal cord (Kettenmann et al., 1983a) was impaled with a $K^+$-sensitive microelectrode. After penetration of the cell (arrow) the potential $(V_m)$ changes from 0 to –60 mV, and the $K^+$-activity ($aK^+$) increases from 5.4 to about 70 m$M$. The response time of the $K^+$-sensitive channel is markedly slower than that of the potential recording channel.

### *4.5. Penetrating Cells With Two Electrodes*

Cell cultures enable the experimenter to position two electrodes on the surface of one selected cell and to double-penetrate it. Figure 19 shows that glial cells as well as neurons can be successfully penetrated with two electrodes, and a stable membrane potential can be recorded for many hours. Another application of pairs of electrodes is the penetration of two adjacent cells. By injecting current pulses in one of the penetrated pair, the connection between the two cells can be studied. Double penetrations therefore yield a powerful tool with which to study membrane properties and connections of cultured cells.

### *4.6. Measurement of Passive Membrane Properties*

The current-voltage response and the time constant of the membrane are important parameters for characterizing the passive membrane properties of cells (Rall, 1977). Cells must therefore be penetrated with two electrodes, one to record membrane potential and the other to inject current (Kettenmann et al., 1984c). Measurement of membrane potential via the current-injecting electrode using a bridge circuit can only be used in cells with relatively high cell input resistance ($>20$ M$\Omega$), such as small neurons; this method cannot be applied to large or extensively electrically coupled cells like astrocytes. In these cells with low input resistance (1–15 M$\Omega$) it is impossible to pass sufficient current to alter membrane potential without causing the electrodes to rectify (Kettenmann et al., 1984c). In culture, passive membrane properties have been determined for neurons of the dorsal root ganglion (Brown et al., 1981), spinal cord (Ransom et al., 1977c), astrocytes (Black and Kornblith, 1980), and oligodendrocytes (Kettenmann et al., 1984c). The current-voltage response of a cultured oligodendrocyte is shown in Fig. 20.

### *4.7. ·Recording Membrane Currents*

The voltage clamp technique is used to identify and characterize membrane currents. These membrane currents can be triggered by voltage steps or neurotransmitters. Conventional voltage clamp techniques can be applied in culture (e.g., Barker and McBurney, 1979; Barker and Mathers, 1981; Choi and Fischbach, 1981; Freschi, 1983; Mathers and Barker, 1981; Nelson and Lieberman, 1981). The patch-clamp technique in the whole cell configura-

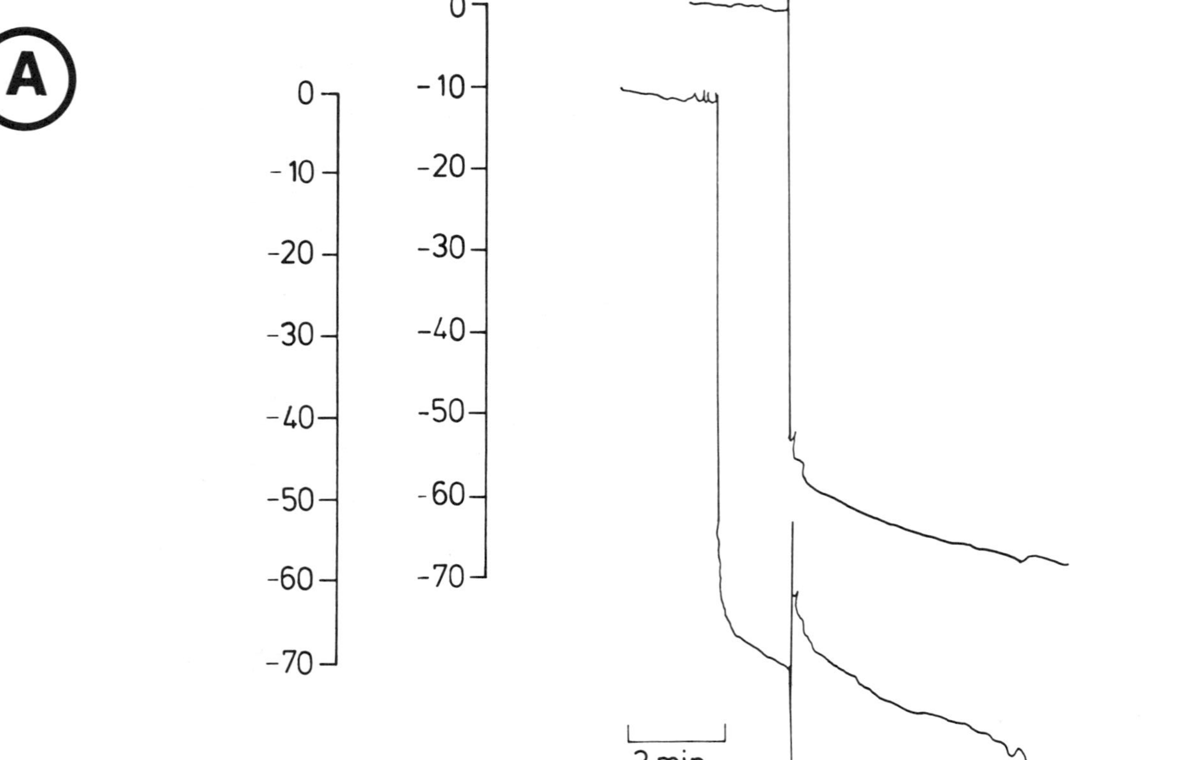


Fig. 19.   Penetration of cells with two electrodes. A: An oligodendrocyte from mouse spinal cord (Kettenmann et al., 1983a) was subsequently penetrated with two electrodes. During the event of penetration with the second electrode the membrane potential transiently decreases to more positive values because of the additional leakage current. B: During recording of membrane potential ($E_1$) from a cultured neuron of mouse dorsal root ganglion (Kettenmann et al., 1983a), a second electrode ($E_2$) was inserted into the cell. The depolarization of the cell caused by the penetration triggers an action potential that can be recorded with both electrodes ($E_1$, $E_2$).

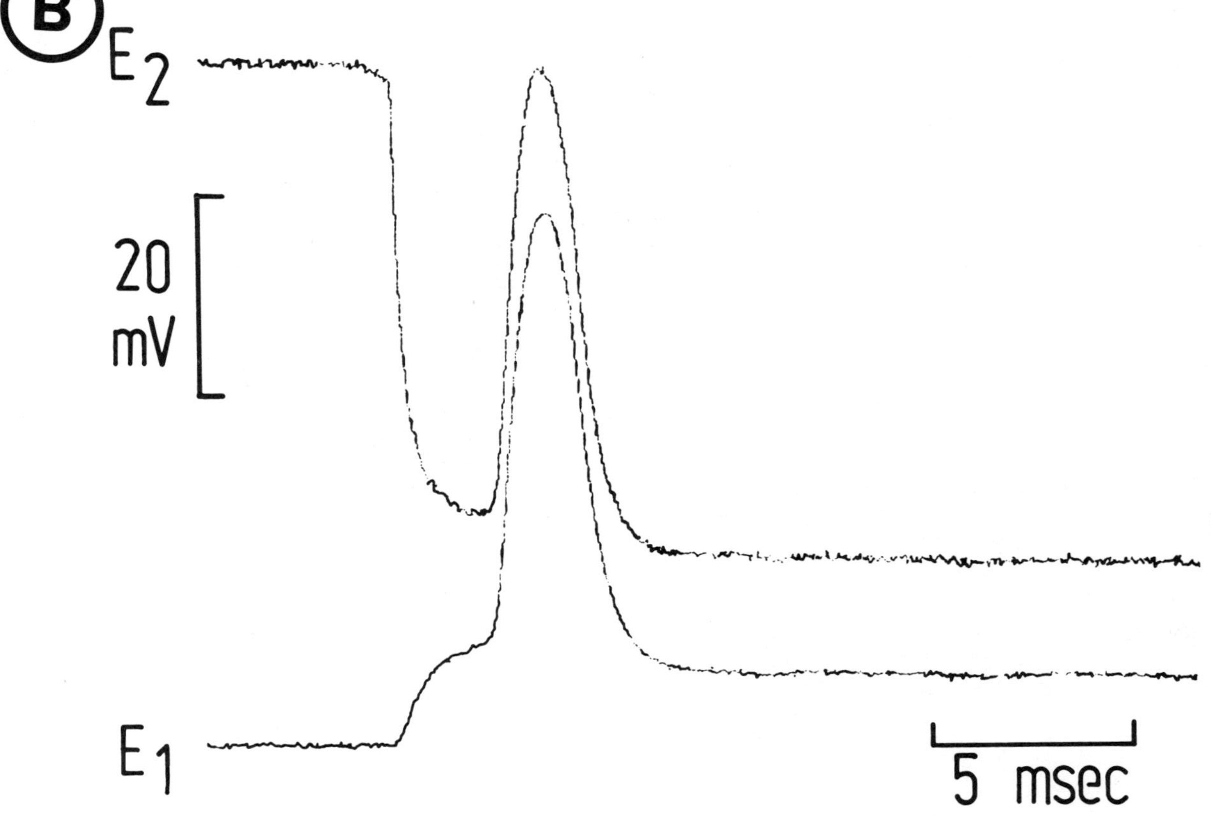


527

                                                                        

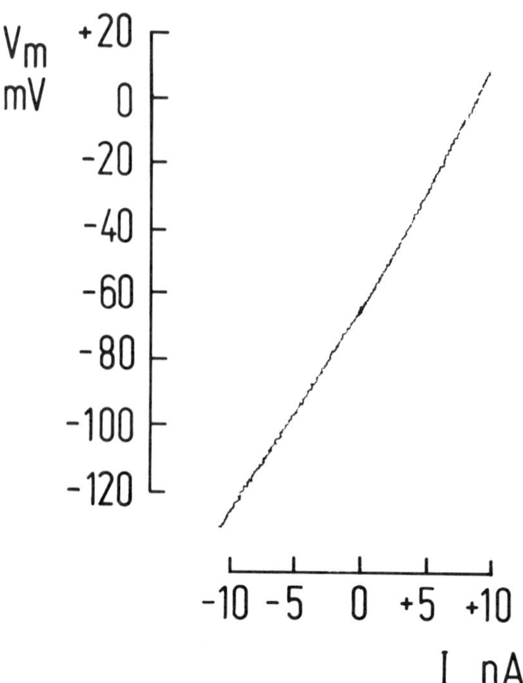


Fig. 20. Current–voltage curve of the cell membrane. An oligodendrocyte from mouse spinal cord (Kettenmann et al., 1984c) was penetrated with two electrodes, one serving for injection of current, the other for recording the membrane potential. Membrane potential $(V_m)$ is plotted against injected current (I).

tion can also be used to characterize membrane currents (Hamill et al., 1981). Figure 21 shows an example of membrane current induced by the neurotransmitter GABA in cultured astrocytes measured by means of the voltage clamp technique (Kettenmann et al., 1986).

### 4.8. Single-Channel Recording

The activity of single ion channels can be resolved with the patch-clamp technique (Hamill et al., 1981). A prerequisite of this technique is that the patch-clamp electrode has access to the clean surface of a cell. Therefore cells in intact tissue are largely not amenable, and in most investigations cultured cells have been utilized (Sakmann and Neher, 1983, 1984). Figure 22

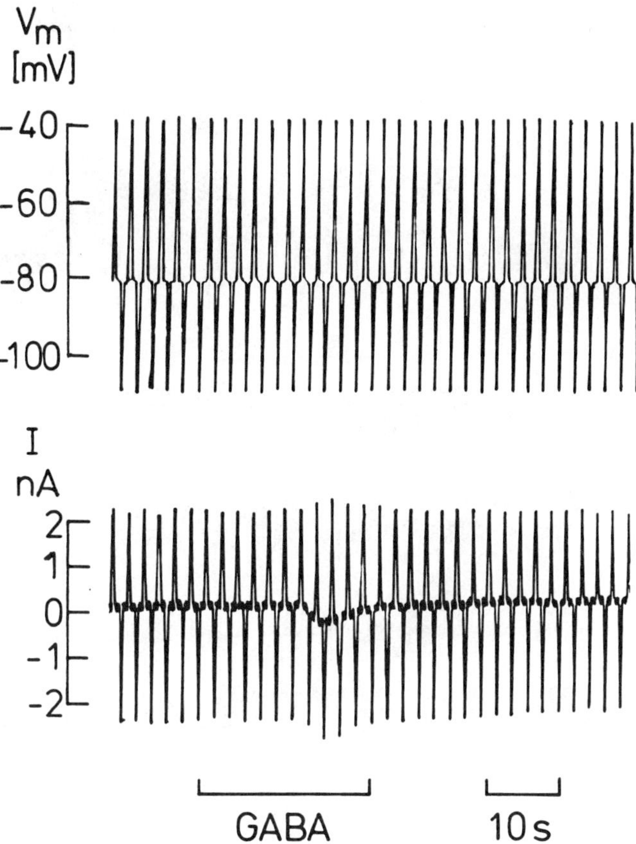


Fig. 21.   Voltage clamp recording of a GABA response. A cultured astrocyte from rat cerebral hemispheres (Kettenmann et al., 1984a) was impaled by two electrodes. The membrane potential $(V_m)$ was clamped at the resting value (–80 mV). When GABA (1 m$M$) was added to the bath, a transient current change (I) was observed. The potential was clamped in 2-s intervals for 400 ms at –40 and –110 mV. The polarity of the GABA-induced current is reversed at –40 mV with respect to the current at –80 and –110 mV, indicating that the reversal potential of the GABA response lies between –40 and –80 mV.

shows the activity of single $K^+$ channels from oligodendrocytes of mouse spinal cord (Kettenmann et al., 1982, 1984b). (For more detailed description of single-ion channel recording, *see* Single-Channel Recording: Methology and Application, in this volume.)

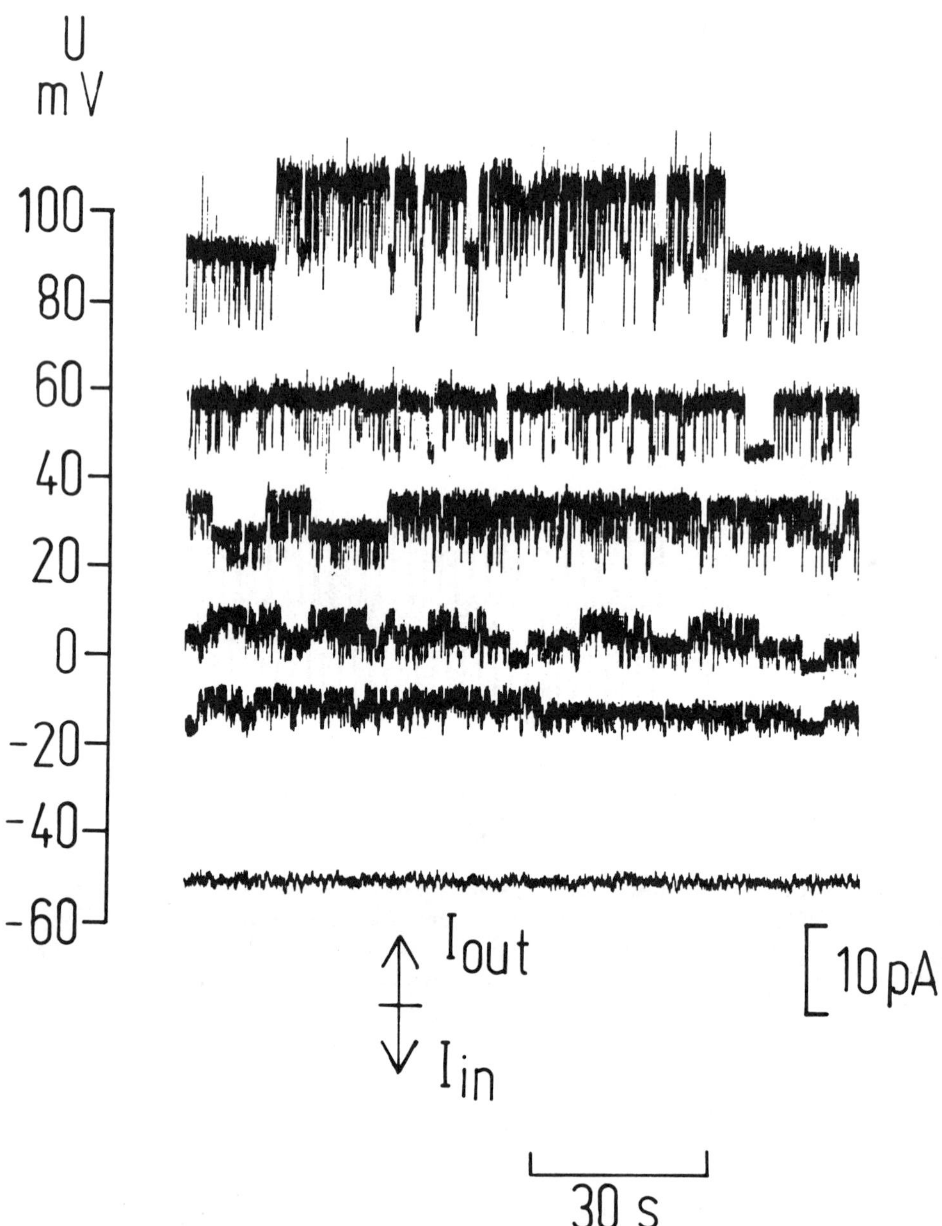


Fig. 22.   Activity of ion channels resolved with the patch-clamp technique. A membrane patch was electrically isolated from a cultured oligodendrocyte of mouse spinal cord (Kettenmann et al., 1982) and the

## 4.9. Intercellular Connections

The culture facilitates the study of cellular connections between identified cells. Pairs of cells can be impaled with microelectrodes. Current can be injected in one of the pair, and the resulting membrane response can be registered in the second cell (Fig. 23). Subsequently cells can be marked by injection of dye and cell type-specific antibodies can be used to identify the cells. This method allows for the characterization of electrical (gap junctions) and chemical synapses.

Gap junctions that allow current and small molecules to pass between cells are cytoplasmatic connections. Depending on the length constant of the membrane, the distance between the two electrodes, and the electrical resistance of the gap junction, a current pulse injected into one of the pair results in a voltage deflection in the second cell (Fig. 23). The voltage deflection is proportional to the injected current and can usually be reversed by changing the polarity of the injected current. Electrical synapses in culture have been found between astrocytes (Moonen and Nelson, 1978) and oligodendrocytes (Kettenmann et al., 1983a), but rarely between neurons (Ransom et al., 1977b), and in no case between Schwann cells (Kettenmann et al., 1983a).

Neurons form and maintain chemical synapses in culture (Crain and Peterson, 1967; Ransom et al., 1977b). Some of the properties of these structures can be studied with electrophysiological methods. As in the study of gap junctions, pairs of neurons are impaled. Stimulating one of the pair with depolarizing current sufficient to elicit an action potential can trigger a postsynaptic potential in the second cell. The postsynaptic potential (psp) can be either a depolarizing event (excitatory or inhibitory psp) or a hyperpolarizing event (inhibitory psp). Psps are blocked in solution with low extracellular $Ca^{2+}$ concentration and can therefore be easily distinguished from current spread through gap junctions. Characterized cells can subsequently be marked in culture by in-

---

current (I) was registered at different membrane potentials (U). Two current levels can be identified, indicating the presence of at least two channels or two conductance levels of one channel. The current amplitude increases in a depolarizing direction and is about 0 at the resting membrane potential.

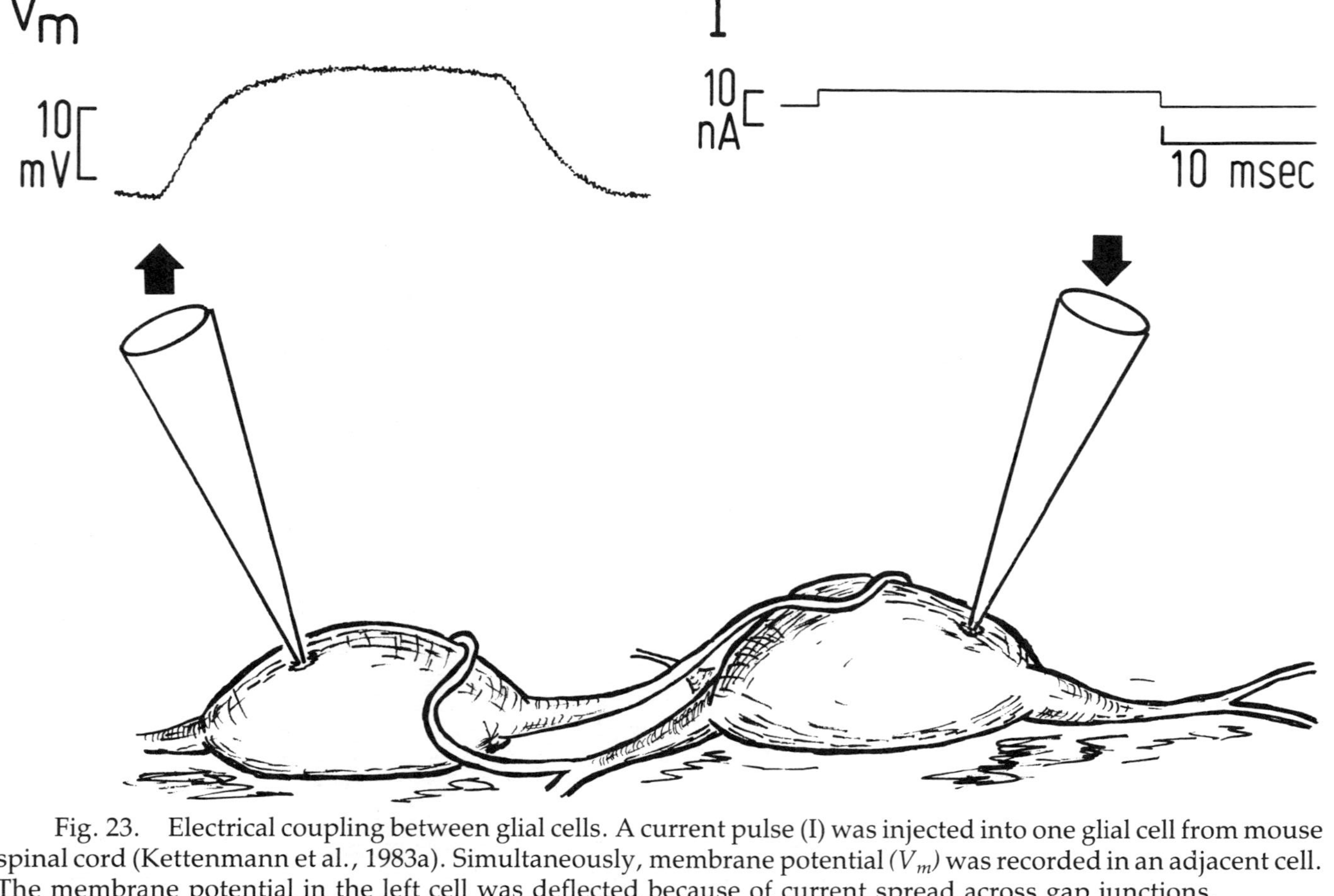



Fig. 23.    Electrical coupling between glial cells. A current pulse (I) was injected into one glial cell from mouse spinal cord (Kettenmann et al., 1983a). Simultaneously, membrane potential ($V_m$) was recorded in an adjacent cell. The membrane potential in the left cell was deflected because of current spread across gap junctions.

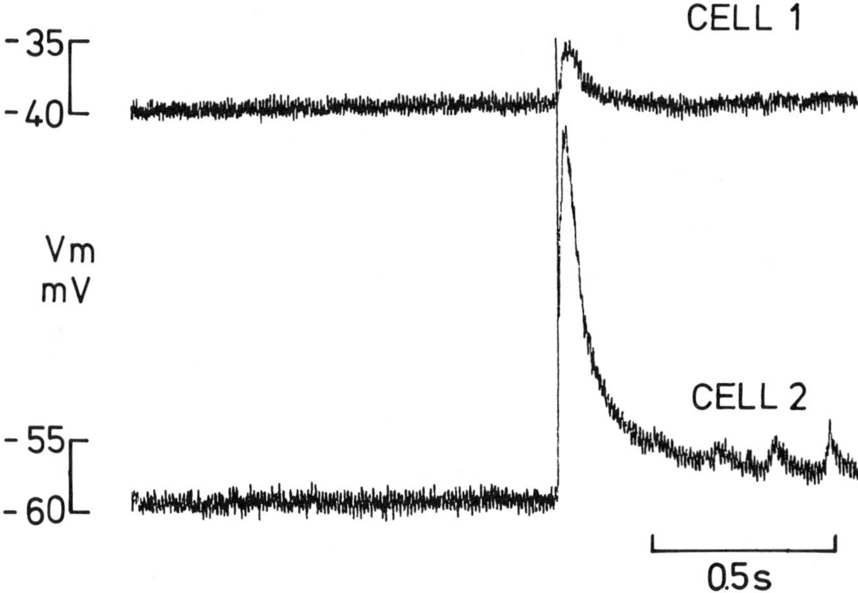


Fig. 24.   Simultaneous recording from two neurons. Two adjacent cultured neurons from rat brain (Kettenmann et al., 1983c) were impaled. The membrane potential $(V_m)$ was synchronously and spontaneously depolarized in both cells. This can either be the result of a direct connection between the two cells or of the activity of a third neuron that has synaptic contact to both recorded cells.

tracellular staining to reveal the extension of the processes and the possible areas of connection between the cells. Figure 24 shows an example of two synchronously active neurons.

### 4.10. *Microcomputer-Controlled Screening of Drugs*

The effect of drugs on membrane potential can be automatically tested if long-term stable recordings can be achieved. A microcomputer system controls the perfusion system, the sequence of drug application, and the storage of data. The culture is continuously perfused and substances are applied by changing the perfusate. The computer system controls the perfusion system by opening and closing magnetic valves or by placing the perfusion tube in different test bottles with the aid of a robot. The duration and sequence of the drug application is programmed. The microcomputer system simultaneously digitizes the recorded data

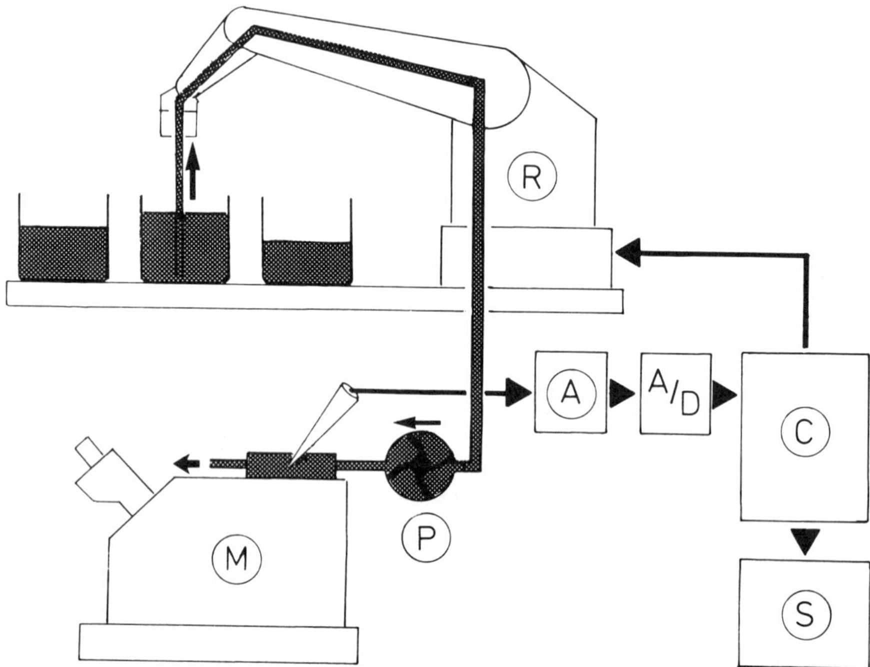

Fig. 25. Computer-controlled test of substances. A microcomputer system (C) can be used to control the application of drugs and to record electrophysiological properties. Cells are maintained in a perfused recording chamber on the stage of a microscope (M). The perfusate is moved by a pump (P). Electrophysiological data from a cell are amplified (A), digitalized by an A/D-converter, and stored by the computer (C) on a mass storage (S). Simultaneously the computer controls a robot (R), which changes the perfusate by moving the perfusion tube in different pots. A series of movements at defined times can be programmed, and drugs can be applied in fixed patterns.

and permanently stores the recorded traces with all necessary parameters such as voltage or time calibration (Fig. 25).

As an example, the membrane potential of cultured rat astrocytes can be registered for many hours. These cells have been found to react with membrane responses to the neurotransmitters glutamate, aspartate, and GABA (Bowman and Kimelberg, 1984; Kettenmann et al., 1984a; Kettenmann and Schachner, 1985). After penetration of the cell, a programmed pattern of agonists and antagonists can be tested on these cells. Similarly, cultured neurons can be used to screen for new agonists or antagonists of neurotransmitters.

## 5. Conclusions

In the past few years research on cultured brain cells, both neurons and glia, has yielded new information on the membrane properties of these cells and the interactions between them. Cells from many different areas of the brain have been cultured, and with the use of cell type-specific antibodies, cell populations can be distinguished and characterized. The application of novel electrophysiological techniques such as patch-clamp and ion-sensitive microelectrodes to cultured cells provides powerful tools for the characterization of membrane properties and cellular connections. With the use of these techniques, the membrane channels, transport systems for ions and other charged molecules, and properties of intercellular connections can be identified in clearly defined cellular populations. Advancements in the characterization of the cellular composition of different culture systems have been made, and electrophysiologists can compare cellular properties under different culture conditions.

In the future, research will focus on defined cell populations, either purified or identified with cell type-specific antibodies. The membrane properties, ion channels, and cellular connections of purified cell populations that lack influences from other cells can now be analyzed and compared to conditions in which defined factors such as hormones or neurotransmitters are added. In a next step the influence of the presence of other cell types on the membrane properties and formation of connections can be determined. By adding new, but clearly defined, influences, the developmental capacities of cell types can be defined. Cell cultures can therefore be used as model systems with which to describe the properties of defined cell types under different environmental conditions.

This precise control of the environment can never be achieved in the intact animal. The models of cellular behavior developed from in vitro experiments can be tested in the intact animal, howeverer. This procedure, development of models in culture and testing in the intact tissue, will accelerate the understanding of the properties of cells and connections in the central nervous system, and will eliminate many experiments on living animals.

## References

Baer S. C. and Crain S. M. (1971) Magnetically coupled micromanipulator for use within a sealed chamber. *J. Appl. Physiol.* **31,** 926–929.

Barker J. L. and Mathers D. A. (1981) GABA analogues activate channels of different duration on cultured mouse spinal neurons. *Science* **212,** 358–360.

Barker J. L. and McBurney R. N. (1979) GABA and glycine may share the same conductance channel on cultured mammalian neurones. *Nature* **277,** 234–236.

Barker J. L. and Ransom B. R. (1978) Amino acid pharmacology of mammalian central neurones grown in tissue culture. *J. Physiol.* **280,** 331–354.

Bevan S. and Raff M. (1985) Voltage-dependent potassium currents in cultured astrocytes. *Nature* **315,** 229–232.

Bignami A. and Dahl D. (1974) Astrocyte-specific protein and neuroglial differentiation. An immunofluorescence study with antibodies to the glial fibrillary acidic protein. *J. Comp. Neurol.* **153,** 27–37.

Black P. and Kornblith P. L. (1980) Biophysical properties of human astrocytic brain tumor cells in cell culture. *J. Cell. Physiol.* **105,** 565–570.

Bornstein M. B. (1958) Reconstituted rat-tail collagen used as a substrate for tissue cultures on coverslips in Maximow slides and roller tubes. *Lab. Invest.* **7,** 134–137.

Bowman C. L. and Kimelberg H. K. (1984) Excitatory amino acids directly depolarize rat brain astrocytes in primary culture. *Nature* **311,** 656–659.

Brown T. H., Perkel D. H., Norris J. C., and Peacock J. H. (1981) Electrotonic structure and specific membrane properties of mouse dorsal root ganglion neurons. *J. Neurophysiol.* **45,** 1–15.

Calvet M.-C. (1974) Patterns of spontaneous electrical activity in tissue cultures of mammalian cerebral cortex vs. cerebellum. *Brain Res.* **69,** 281–295.

Calvet M.-C. and Calvet J. (1979) Horseradish peroxidase iontophoretic intracellular labelling of cultured Purkinje cells. *Brain Res.* **173,** 527–531.

Calvet M.-C., Calvet J., and Camacho-Garcia R. (1985) The Purkinje cell dendritic tree: A computer-aided study of its development in the cat and in culture. *Brain Res.* **331,** 235–250.

Chiu S. Y., Schrager P., and Ritchie J. M. (1984) Neuronal-type $Na^+$ and $K^+$ channels in rabbit cultured Schwann cells. *Nature* **311,** 156–157.

Choi D. W. and Fischbach G. D. (1981) GABA conductance of chick spinal cord and dorsal root ganglion neurons in cell culture. *J. Neurophysiol.* **45,** 605–620.

Corrigall W. A., Crain S. M., and Bornstein M. B. (1976) Electrophysiological studies of fetal mouse olfactory bulb explants during development of synaptic functions in culture. *J. Neurobiol.* **7,** 521–536.

Crain S. M. (1956) Resting and action potentials of cultured embryo chick spinal ganglion cells. *J. Comp. Neurol.* **104,** 285–329.

Crain S. M. (1973) Microelectrode Recording in Brain Tissue Cultures, in *Methods in Physiological Psychology: Bioelectric Recording Techniques: Cellular Processes and Brain Potentials* (Thompson R. and Patterson M. M., eds.) Academic, New York.

Crain S. C. and Peterson E. R. (1967) Onset and development of functional interneuronal connections in explants of rat spinal cord-ganglia during maturation in culture. *Brain Res.* **6,** 750–762.

Crain S. M. and Peterson E. R. (1974) Development of specific sensory-evoked synaptic networks in fetal mouse spinal cord-brainstem cultures. *Science* **188,** 275–278.

Dichter M. A. (1977) Rat cortical neurons in cell culture: Culture methods, cell morphology, electrophysiology, and synapse formation. *Brain Res.* **149,** 279–293.

Dichter M. A. (1980) Physiological identification of GABA as the inhibitory transmitter for mammalian cortical neurons in cell culture. *Brain Res.* **190,** 111–121.

Dimpfel W., Pierau F.-K., and Haider S. G. (1979) Electrophysiological Studies on the Effects of the Neurotoxin Apamin on Cultured Neurons, in *Advances in Cytopharmacology* (B. Ceccarelli and F. Clementi, eds.) Raven, New York.

Dreyfus C. F., Gershon M. D., and Crain S. M. (1979) Innervation of hippocampal explants by central catecholaminergic neurons in co-cultured fetal mouse brain stem explants. *Brain Res.* **161,** 431–445.

Eng L. F., Vanderhagen J. J., Bignami A., and Gertl B. (1971) An acidic protein isolated from fibrous astrocytes. *Brain Res.* **28,** 351–354.

Fields K. L. (1980) The study of Schwann cells using antigenic markers. *Trends Neurosci.* **3,** 236–238.

Fischbach G. D. and Nelson P. G. (1977) Cell Culture in Neurobiology, in *Cellular Biology of Neurons, The Handbook of Physiology* (Kandel E. R., ed.) Williams & Wilkins, Baltimore, Maryland.

Fischer G. and Kettenmann H. (1985) Cultured astrocytes form a syncytium after maturation. *Exp. Cell Res.* **159,** 73–279.

Fischer G. and Wieser R. J. (1983) *Hormonally Defined Media: A Tool in Cell Biology* Springer Verlag, Heidelberg.

Fischer G., Leutz A., and Schachner M. (1982) Cultivation of immature astrocytes in a serum-free, hormonally defined medium. *Neurosci. Lett.* **29,** 97–302.

Fischer G., Künemund V., and Schachner M. (1986) Neurite outgrowth pattern in cerebellar microexplant cultures are affected by antibodies to the cell surface glycoprotein L1. *J. Neurosci.* **6,** 605–612.

Freschi J. E. (1983) Membrane currents of cultured rat sympathetic neurons under voltage clamp. *J. Neurophysiol.* **50,** 460–478.

Fromm M., Weskamp P. and Hegel U. (1980) Versatile piezoelectric driver for cell puncture. *Pflugers Arch.* **384,** 69–73.

Frosch M. P. and Dichter M. A. (1984) Physiology and pharmacology of olfactory bulb neurons in dissociated cell culture. *Brain Res.* **290,** 321–332.

Furshpan E. J., MacLeish P. R., O'Lague P. H., and Potter D. D. (1976) Chemical transmission between rat sympathetic neurons and cardiac myocytes developing in microcultures: Evidence for cholinergic, adrenergic, and dual-function neurones. *Proc. Natl. Acad. Sci. USA* **73,** 4225–4229.

Gahwiler B. H. (1975) The effects of GABA, picrotoxin and bicucullin on the spontaneous bioelectric activity of cultured cerebellar Purkinje cells. *Brain Res.* **99,** 85–95.

Gahwiler B. H. (1976) Spontaneous bioelectric activity of cultured Purkinje cells during exposure to glutamate, glycine, and strychnine. *J. Neurobiol.* **2,** 97–107.

Gahwiler B. H. (1978) Mixed cultures of cerebellum and inferior olive: Generation of complex spikes in Purkinje cells. *Brain Res.* **145,** 168–172.

Gahwiler B. H. (1981) Labeling of neurons within CNS explants by intracellular injection of Lucifer Yellow. *J. Neurobiol.* **2,** 187–191.

Gahwiler B. H. and Bauer W. (1975) Design of a temperature controlled microchamber for electrophysiological experiments in vitro. *Separatum Experientia* **31,** 369.

Gahwiler B. H., Mamoon A. M., Schlapfer W. T., and Tobias C. A. (1972) Effects of temperature on spontaneous bioelectric activity of cultured nerve cells. *Brain Res.* **40,** 527–533.

Gilbert P., Kettenmann H., Orkand R. K., and Schachner M. (1982) Immunocytochemical cell identification in nervous system culture combined with intracellular injection of a blue fluorescing dye (SITS). *Neurosci. Lett.* **334,** 123–128.

Gilbert P., Kettenmann H., and Schachner M. (1984) Gamma-aminobutyric acid directly depolarizes cultured oligodendrocytes. *J. Neurosci.* **4,** 561–569.

Godfrey E. W., Nelson P. G., Schrier B. K., Breuer A. C., and Ransom B. R. (1975) Neurons from fetal rat brain in a new cell culture system: A multidisciplinary analysis. *Brain Res.* **90,** 1–21.

Gross G. W. (1979) Simultaneous single unit recording in vitro with a photoetched laser deinsulated gold multimicroelectrode surface. *IEEE Transact. Biomed. Eng.* **26,** 273–279.

Gross G. W. and Lucas J. H. (1982) Long-term monitoring of spontaneous single unit activity from neuronal monolayer networks cultured on photoetched multielectrode surfaces. *J. Electrophysiol. Tech.* **9,** 55–67.

Gross G. W., Rieske E., Kreutzberg G. W., and Meyer A. (1977) A new fixed-array multi-microelectrode system designed for long-term monitoring of extracellular single unit neuronal activity in vitro. *Neurosci. Lett.* **6,** 101–105.

Hamill O. P., Marty A., Neher E., Sakmann B., and Sigworth F. J. (1981) Improved patch-clamp techniques for high-resolution current recording from cells and cell-free membrane patches. *Pflugers Arch.* **391,** 85–100.

Hicks T. P. (1984) The history and development of microiontophoresis in experimental neurobiology. *Prog. Neurobiol.* **22,** 185–240.

Hild W., Chang J. J., and Tasaki I. (1958) Electrical responses of astrocytic glia from the mammalian central nervous system cultivated in vitro. *Experientia* **14,** 220–221.

Hirata H., Slater N. T., and Kimelberg H. K. (1983) Alpha-adrenergic receptor-mediated depolarization of rat neocortical astrocytes in primary culture. *Brain Res.* **270,** 358–362.

Hogue M. J. (1947) Human fetal brain cells in tissue culture: Their identification and motility. *J. Exp. Zool.* **106,** 85–107.

Hosli L., Andres P. F., and Hösli E. (1978) Neuron-glia interactions: Indirect effect of GABA on cultured glial cells. *Exp. Brain Res.* **33,** 425–434.

Hosli L., Hosli E., Landolt H., and Zehntner C. (1981) Efflux of potassium from neurones excited by glutamate and aspartate causes a depolarization of cultured glial cells. *Neurosci. Lett.* **21,** 83–86.

Keilhauer G., Meier D. H., Kuhlmann-Krieg S., Nieke J., and Schachner M. (1985) Astrocyte support incomplete differentiation of an oligodendrocyte precursor cell. *EMBO J.* **44,** 2499–2504.

Kettenmann H. (1985) A reversible decrease in electrical coupling of cultured mouse glial cells induced by superfusion from a micropipette. *Neurosci. Lett.* **59,** 85–88.

Kettenmann H. and Orkand R. K. (1983) Intracellular SITS injection dye-uncouples mammalian oligodendrocytes in culture. *Neurosci. Lett.* **39,** 21–26.

Kettenmann H. and Schachner M. (1985) Pharmacological properties of GABA, glutamate and aspartate induced depolarizations in cultured astrocytes. *J. Neurosci.* **5,** 3295–3301.

Kettenmann H., Orkand R. K., Lux H. D., and Schachner M. (1982) Single potassium channel currents in cultured mouse oligodendrocytes. *Neurosci. Lett.* **32,** 41–46.

Kettenmann H., Backus K. H., and Schachner M. (1984a) Aspartate, glutamate and gamma-aminobutyric acid depolarize cultured astrocytes. *Neurosci. Lett.* **52,** 25–29.

Kettenmann H., Orkand R. K., and Lux H. D. (1984b) Some properties of single potassium channels in cultured oligodendrocytes. *Pflugers Arch.* **400,** 215–221.

Kettenmann H., Sonnhof U., Camerer H., Kuhlmann S., Orkand R. K., and Schachner M. (1984c) Electrical properties of oligodendrocytes in culture. *Pflugers Arch.* **401,** 324–332.

Kettenmann H., Sommer I., and Schachner M. (1985) Monoclonal cell surface antibodies do not produce short-term effects on electrical properties of mouse oligodendrocytes in culture. *Neurosci. Lett.* **54,** 195–199.

Kettenmann H., Backus K. H., and Schachner M. (1987) GABA opens chloride channels in cultured astrocytes. *Brain Res.* **404,** 1–9.

Kettenmann H., Orkand R. K., and Schachner M. (1983a) Coupling among identified cells in mammalian nervous system cultures. *J. Neurosci.* **3,** 506–516.

Kettenmann H., Sonnhof U., and Schachner M. (1983b) Exclusive potassium dependence of the membrane potential in cultured mouse oligodendrocytes. *J. Neurosci.* **3,** 500–505.

Kettenmann H., Wienrich M., and Schachner M. (1983c) Antibody L1 ejected from a micropipette identifies neurons without altering electrical activity. *Neurosci. Lett.* **41,** 85–90.

Kimelberg H. K., Bowman C., Biddlecome S., and Bourke R. S. (1979) Cation transport and membrane potential properties of primary astroglial cultures from neonatal rat brains. *Brain Res.* **177,** 533–550.

Kruse P. F. and Patterson M. K. (1973) *Tissue Culture: Methods and Application* Academic, New York.

Macdonald R. L. and J. L. Barker (1981) Neuropharmacology of Spinal Cord Neurons in Primary Dissociated Cell Culture, in *Excitable Cells in Tissue Culture* (Nelson P. D. and Lieberman M., eds.) Plenum, New York.

MacVicar B. A. (1984) Voltage-dependent calcium channels in glial cells. *Science* **226,** 1345–1347.

Mains R. E. and Patterson P. H. (1973) Primary cultures of dissociated sympathetic neurons. *J. Cell Biol.* **59,** 329–345.

Manthorpe M., Engvall E., Ruoslathi E., Longo F., Davis G. E., and Varon S. (1983) Laminin promotes neuritic regeneration from cultured peripheral and central neurons. *J. Cell Biol.* **97,** 1882–1890.

Mathers D. A. and Barker J. L. (1981) GABA and muscimol open ion channels of different lifetimes on cultured mouse spinal cord cells. *Brain Res.* **204,** 242–247.

McCarthy K. D. and de Vellis J. (1980) Preparation of separate astroglial and oligodendroglial cell cultures from rat cerebral tissue. *J. Cell. Biol.* **85,** 890–902.

Meier D. H., Lagenaur C., and Schachner M. (1982) Immunoselection of oligodendrocytes by magnetic beads. I. Determination of antibody coupling parameters and cell binding conditions. *J. Neuroscience Res.* **7,** 119–134.

Moonen G. and Nelson P. G. (1978) Some Physiological Properties of Astrocytes in Primary Cultures, in *Dynamic Properties of Glial Cells* (Schoffeniels E., Frank G., Hertz L., and Tower D. B. L., eds.) Pergamon, London.

Moonen G., Neale E. A., Macdonald R. L., Gibbs W., and Nelson P. G. (1982) Cerebellar macroneurons in microexplant cell culture. Methodology, basic electrophysiology, and morphology after horseradish peroxidase injection. *Dev. Brain Res.* **5,** 59–73.

Neale E. A., Macdonald R. L., and Nelson P. G. (1978) Intracellular horseradish peroxidase injection for correlation of light and electron microscopic anatomy with synaptic physiology of cultured mouse spinal cord neurons. *Brain Res.* **152,** 265–282.

Nelson P. G. (1975) Nerve and muscle cells in culture. *Physiol. Rev.,* **55,** 1–61.

Nelson P. G. and Lieberman M. (1981) *Excitable Cells in Tissue Culture* Plenum, New York.

Nelson P. G. and Peacock J. H. (1973) Electrical activity in dissociated cell cultures from fetal mouse cerebellum. *Brain Res.* **61,** 163–174.

Nelson P. G., Ransom B. R., Henkart M., and Bulloch P. N. (1977) Mouse spinal cord in cell culture. IV. Modulation of inhibitory synaptic function. *J. Neurophysiol.* **40,** 1178–1186.

O'Lague P. H., Potter D. D., and Furshpan E. J. (1978a) Studies on rat sympathetic neurons developing in cell culture. I. Growth characteristics and electrophysiological properties. *Dev. Biol.* **67,** 384–403.

O'Lague P. H., Furshpan E. J., and Potter D. D. (1978b) Studies on rat sympathetic neurons developing in cell culture. *Dev. Biol.* **67,** 404–423.

O'Lague P. H., Potter D. D., and Furshpan E. J. (1978c) Studies on rat sympathetic neurons developing in cell culture. III. Cholinergic transmission. *Dev. Biol.* **67,** 424–443.

Orkand R. K. (1977) Glial Cells, in Cellular *Biology of Neurons, The Handbook of Physiology* (Kandel E. R., ed.) Williams & Wilkins, Baltimore, Maryland.

Patrick J., Heinemann S., and Schubert D. (1978) Biology of cultured nerve and muscle. *Ann. Rev. Neurosci.* **1,** 417–443.

Patterson P. H. (1978) Environmental determination of autonomic neurotransmitter functions. *Ann. Rev. Neurosci.* **1,** 1–17.

Peacock J. H. (1979) Electrophysiology of dissociated hippocampal cultures from fetal mice. *Brain Res.* **169,** 247–260.

Peacock J. H., Nelson P. G., and Goldstone M. W. (1973) Electrophysiologic study of cultured neurons dissociated from spinal cords and dorsal root ganglia of fetal mice. *Dev. Biol.* **30,** 137–152.

Pfeiffer S. E. (1982) *Neuroscience Approach to Cell Culture,* CRC, Boca Raton, Florida.

Pitman R. M., Tweedle C. D., and Cohen M. J. (1972) Branching of central neurons: Intracellular cobalt injection for light and electron microscopy. *Science* **176,** 412–414.

Podulso S. E. and Norton W. T. (1972) Isolation and some chemical properties of oligodendroglia from calf brain. *J. Neurochem.* **19,** 727–736.

Rall W. (1977) Core Conductor Theory and Cable Properties of Neurons, in *Cellular Biology of Neurons, The Handbook of Physiology* (Kandel E. R., ed.) Williams & Wilkins, Baltimore.

Raff M. C., Field K. L., Hakomori S-I, Mirsky R., Pruss R. N., and Winter J. (1979) Cell type specific markers for distinguishing and studying neurons and major classes of glial cells in culture. *Brain Res.* **174,** 283–308.

Ransom B. R., and Holz R. W. (1977) Ionic determinants of excitability in cultured mouse dorsal root ganglion and spinal cord cells. *Brain Res.* **136,** 445–453.

Ransom B. R., Neale E., Henkart M., Bullock P. N., and Nelson P. G. (1977a) Mouse spinal cord in cell culture. I. Morphology and intrinsic neuronal electrophysiologic properties. *J. Neurophysiol.* **5,** 1132–1150.

Ransom B. R., Christian C. N., Bullock P. N., and Nelson P. G. (1977b) Mouse spinal cord in cell culture. II. Synaptic activity and circuit behaviour. *J. Neurophysiol.* **5,** 1151–1162.

Ransom B. R., Bullock P. N., and Nelson P. G. (1977c) Mouse spinal cord in cell culture. III. Neuronal chemosensitivity and its relationship to synaptic activity. *J. Neurophysiology* **5,** 1163–1177.

Rathjen F. G. and Schachner M. (1984) Immunocytological and biochemical characterization of a new neuronal cell surface component (L1 antigen) which is involved in cell adhesion. *EMBO J.* **3,** 1–10.

Sakmann B. and Neher E. (1983) *Single Channel Recording* Plenum, New York.

Sakman B. and Neher E. (1984) Patch clamp techniques for studying ionic channels in excitable membranes. *Ann. Rev. Physiol.* **46,** 455–472.

Schachner M. (1982a) Cell type-specific surface antigens in the mammalian nervous system. *J. Neurochem.* **39,** 1–8.

Schachner M. (1982b) Glial antigens and the expression of neuroglial phenotypes. *Trends Neurosci.* **5,** 225–229.

Schachner M. (1984) Cell Type-Specific Antigens in the Mammalian Nervous System, in *Molecular Biology Approach to the Neurosciences* (H. Soreq, ed.) John Wiley, New York.

Schachner M., Kim S. U., and Zehnle R. (1981a) Developmental expression in central and peripheral nervous system of oligodendrocyte cell surface antigens (O antigens) recognized by monoclonal antibodies. *Dev. Biol.* **83,** 328–338.

Schachner M., Sommer I., Lagenaur C., and Schnitzer J. (1981b) Monoclonal Antibodies Recognizing Subpopulations of Glial Cells in Mouse Cerebellum, in *Monoclonal Antibodies Against Neural Antigens* Cold Spring Harbor Reports in the Neurosciences (R. McKay, M. Raff, and L. Reichardt, eds.) Cold Spring Harbor, Long Island.

Schachner M., Sommer I., Lagenaur C., and Schnitzer J. (1982) Developmental Expression of Antigenic Markers, in Glial Subclasses, in *Neuronal-Glial Cell Interrelationships* (T. Sears, ed.) Life Sciences Research Report, vol. 20, Springer Verlag, Berlin.

Schachner M., Sommer I., Lagenaur C., Schnitzer J., and Berg G. (1983) Antigenic Markers of Glia and Glial Subclasses, in *Current Frontiers in Neurobiology Approached Through Cell Culture* (S. E. Pfeiffer, ed.) Vol. 2, CRC, Boca Rotan, Florida.

Schlapfer W. T. (1978) Tissue and Organ Culture, in *Methods in Neurobiology* (R. Lahue, ed.) Plenum, New York.

Schnitzer J. and Schachner M. (1981a) Characterization of isolated mouse cerebellar cell populations in vitro. *J. Neuroimmunol.* **1,** 457–470.

Schnitzer J. and Schachner M. (1981b) Developmental expression of cell type-specific markers in mouse cerebellar cells in vitro. *J. Neuroimmunol.* **1,** 471–487.

Shrager P., Chiu S. Y., and Ritchie J. M. (1985) Voltage-dependent sodium and potassium channels in mammalian cultured Schwann cells. *Proc. Natl. Acad. Sci. USA* **82,** 948–952.

Sommer I. and Schachner M. (1981) Monoclonal antibodies (01 to 04) to oligodendrocyte cell surfaces: An immunocytological study in the central nervous system. *Dev. Biol.* **83,** 311–327.

Sommer I. and Schachner M. (1982) Cells that are 04 antigen-positive and 01 antigen-negative differentiate into 01 antigen-positive oligodendrocytes. *Neurosci. Lett.* **29,** 183–188.

Sonnhof U. and Schachner M. (1986) Single voltage dependent $K^+$-channels in cultured astrocytes. *Neurosci. Lett.* **64,** 241–246.

Sonnhof U., Forderer R., Schneider W., and Kettenmann H. (1982) Cell puncturing with a step motor driven manipulator with simultaneous measurement of displacement. *Pflugers Arch.* **392,** 295–300.

Stewart W. W. (1981) Lucifer dyes—highly fluorescent dyes for biological tracing. *Nature* **292,** 17–24.

Walz W., Wuttke W., and Hertz L. (1984) Astrocytes in primary cultures: Membrane potential characteristics reveal exclusive potassium conductance and potassium accumulator properties. *Brain Res.* **292,** 367–374.

Wardell W. M. (1966) Electrical and pharmacological properties of mammalian neuroglial cells in tissue-culture. *Proc. Roy. Soc. Lond. (Biol.)* **165,** 326–361.

Weber A. and Schachner M. (1984) Maintenance of immunocytologically identified Purkinje cells from mouse cerebellum in monolayer culture. *Brain Res.* **311,** 119–130.

Weir D. M. (1978) *Handbook of Experiemental Immunology* vol. I, *Immunochemistry* Blackwell Scientific, Oxford.

Werz W. and Schachner M. (1987) Extracellular matrix molecules promote neuronal outgrowth of cerebelar cells (in preparation).

Wienrich M. and Kettenmann H. (1984) Intrazelluläre Farbstoffe zur Markierung einzelner neuraler Zellen in Kultur. *Kontakte* (Darmstadt) **3,** 30–44.

Yavin E. and Yavin Z. (1974) Attachment and culture of dissociated cells from rat embryo cerebral hemispheres on poly-L-lysine coated surface. *J. Cell Biol.* **62,** 540–546.

# Electrophysiological Methods for Studying Ionic Currents in Brain Slices and Cell Cultures

Brian A. MacVicar and Maeve O'Beirne

## 1. Introduction

The last two decades have seen a tremendous explosion in our knowledge concerning the properties of neurons of the central nervous system (CNS). The basis of this explosion is twofold: one is the development of the methodology to maintain isolated slices of brain tissue alive, and the second is the discovery of pharmacological tools to alter specific ionic channels of neurons. Some pioneering work describing neuronal electrophysiology has also been done on neurons cultured in slice preparations or after dissociation. The purpose of this chapter is to briefly describe the methodologies involved in maintaining isolated brain slices and cultures of both slices and dissociated cells and to describe the experimental paradigms that have been developed in the last few years to determine the ionic currents underlying the electrophysiological properties of the mammalian CNS. Also, another important area of research will be discussed: the measurement of extracellular ion concentrations in the CNS and the dynamic changes that occur during neuronal activity. Much of this work has been done in vivo, but extracellular and intracellular ion measurements have also been made in brain slice preparations. Because of the complexity of the intact CNS, there are few experiments correlating extracellular ionic changes with intracellular ionic currents. We will attempt to bridge this gap in this chapter by reviewing what is known to date about different ionic currents in mammalian CNS neurons and postulating what contribution each current may make to generating changes in extracellular ionic concentrations.

## 2.  Tissue Preparation

### *2.1.  Brain Slices*

The isolated brain slice preparation was first developed for biochemical studies by McIlwain in the early 1960s (McIlwain, 1961). Later, Yamamoto (Yamamoto and McIlwain, 1966) utilized this technique to obtain the first electrophysiological recordings from in vitro brain slices. By the mid-to-late 1970s, the use of isolated brain slices had become very widespread. Analysis of electrophysiological properties of neurons of the mammalian CNS has been greatly facilitated by dissemination of this technique. Several comprehensive reviews have been published on slice preparation techniques to which the reader is referred (Hatton et al., 1980; Kerkut and Wheal, 1981; Jahnsen and Laursen, 1983; Dingledine et al., 1980; Dingledine, 1984). It is therefore not necessary to do a complete detailed analysis of different slice methods. Some key factors will be described, however, that are necessary to keep in mind when slices are prepared.

The advantages of brain slice techniques are manyfold, and also apply to brain slice and dissociated cell cultures. For one, the extracellular ionic environment around neurons can be rigorously controlled and altered. This is extremely important for electrophysiological analysis because it allows for determination of the ionic basis of currents and the rapid introduction of pharmacological agents. Another factor is the ease of obtaining intracellular impalements. There are no vibrations or pulsations caused by the vascular or respiratory systems. It is possible to arrange several micromanipulators around the preparation to introduce several microelectrodes into the brain slice. For instance, recordings have been obtained from two adjacent neurons (MacVicar and Dudek, 1980a, 1981, 1982) and from different parts of the same cell in isolated brain slices (Masukawa and Prince, 1984; Llinas and Sugimori, 1980b). By judicious blocking and sectioning of brain slices, different parts of the cell, such as the soma or dendritic regions, can be cut away and intracellular recordings obtained from the remaining portions so that the electrophysiology of subcellular areas can be analyzed. Another recent advance is the partial enzymatic digestion of brain slices by collagenase or papain to expose cell bodies that are still viable (Gray and Johnston, 1985b). This allows for the use of patch clamp technology to obtain high-resistance seals from the exposed cells for whole cell voltage clamping and for patch clamping analysis of ionic currents. Synaptic connections with the dendritic tree are still viable, even though the cell body is free from its original cellular matrix. Therefore slice techniques can be combined with the most recent advances in electrophysiological techniques to permit sophisticated analysis of voltage-activated and ligand-activated ionic currents.

Viable brain slices have been prepared from many areas of the brain, including the hippocampus, hypothalamus, cortex, thalamus, striatum, substantia nigra, cerebellum, inferior olive, and brain stem. It is usually straightforward to determine anatomical structures within a slice because when transilluminated, cell body areas are more translucent than dendritic areas, and nerve fiber tracts are the most opaque. Therefore it is possible to visually place microelectrodes in the desired anatomically defined location. Even in the brain stem (Llinas and Yarom, 1981a) or thalamus (Jahnsen and Llinas, 1984), different nuclei can be reliably located. It is difficult to see individual cells by using normal optical techniques. Three techniques have been developed, however, to visualize neurons in intact brain slices. Each of these has certain advantages and disadvantages, the most important disadvantage being the limitation of the thickness of the slice. The use of Nomarski optics was first applied to 80-$\mu$m cerebellar slices by Yamamoto and Chujo (1978). This technique allows for excellent visualization of individual neurons and the microelectrode. It cannot be used, however, in slices that are over 100 $\mu$m thick. Llinas and Sugimori (1980a) have developed a Hoffman modulation contrast system to visualize cerebellar neurons in slices 100–200 $\mu$m thick. This is particularly useful when working with cerebellar slices, because the Purkinje cell dendritic field is restricted to a two-dimensional plane. Therefore Purkinje neurons and dendrites will be intact in a 200-$\mu$m slice. Most CNS neurons do not have such a convenient morphology and would have most of their dendrites shorn off in such a thin slice. MacVicar (1984a) has developed a system using an inverted microscope and infrared video microscopy to visualize neurons in slices up to 400 $\mu$m thick (Fig. 1). Although infrared optical systems do not have the resolution that shorter wavelength systems have, they can be used to visualize neurons in thick brain slices because tissue is more transparent in the infrared. Cell bodies as well as microelectrodes can be seen in slices 400 $\mu$m thick (Fig. 2). This system is particularly useful in observing neurons in which the cell bodies are not densely packed, such as interneurons in dendritic areas of the hippocampus (Fig. 3).

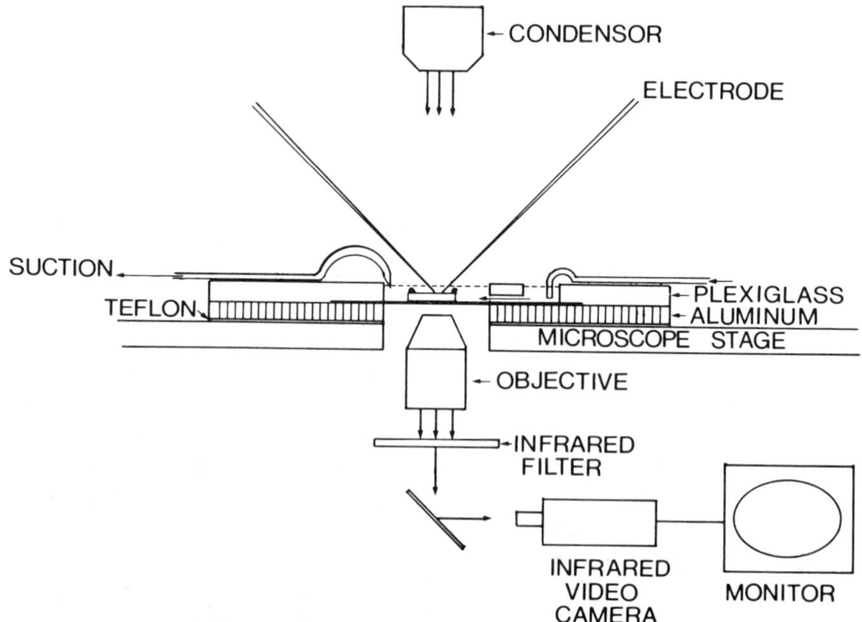



Fig. 1.  Schematic cross-section of the chamber for perfusion of brain slices and observation of cells using an infrared camera. The chamber is insulated from the microscope stage by a Teflon sheet. The plexiglass chamber is attached to the aluminum heating plate with Teflon screws. The brain slice rests on a coverslip, which is sealed to the chamber with inert high-vacuum grease. The perfusate is drawn off by applying suction to a hypodermic needle. A long working distance condensor is used for illumination. An infrared filter (e.g., Wratten 87C) is placed in the light path before the infrared video camera (IKegami with N-214 tube), and the cells are observed using a monitor (from MacVicar, 1984a).

The integrity of the extracellular space in isolated brain slices is an important factor, in the context of considering ion homeostasis and ionic concentrations in the extracellular space. This problem has been studied by Hounsgaard and Nicholson (1983), who have found that the extracellular space in an isolated brain slice has many of the same properties and dynamic capabilities as the tissue has in vivo. The percentage total tissue volume and tortuosity is approximately the same both in vivo and in vitro.

Following is a brief review of the methods for preparing brain slices, incorporating critical factors that we have found to be important in maintaining the viability of the brain slice.

The brain slice technique is applicable to a wide range of species, including rats, guinea pigs, rabbits, cats, monkeys, and even humans. It is not an important factor whether the animal is anesthetized or not. In our lab, similar results are obtained from animals in either case. A benefit of the preparation of brain slices, however, is that tissue that has not been exposed to anesthetics can be recorded from. This is not usually possible in vivo, where the animal is typically anesthesized. Critical factors in preparing slices to maintain viability are the speed of the dissection and, in particular, the avoidance of mechanical trauma to the tissue. It is usually more important to not bend, crush, or deform the tissue than it is to make a speedy dissection.

The brain is removed after the animal has been sacrificed by either decapitation or stopping the heart with a crushing blow to the back of the spinal cord (Figs. 4 and 5). The scalp is split, and the overlying skull is removed. The area of the brain under study dictates the care to be taken at this time. For instance, if slices are to be prepared from structures on the dorsal surface, such as the cortex or the cerebellum, great care should be taken to avoid applying pressure onto these areas while removing the skull. If the hypothalamus or areas close to the bottom of the skull are to be used, however, particular care has to be taken in the extraction of the brain from the skull case so that no pressure is applied to areas overlying the area of interest. It is very important to avoid compression, bending, or other mechanical trauma during the removal of the skull or removal of the brain from the skull case. After the brain is extracted, it is quickly cooled by immersion in ice-cold oxygenated saline. This slows down the metabolic rate of the tissue and prolongs tissue survival.

The next step is the blocking or dissection of the area to be sectioned, and again a critical factor is to avoid compression or mechanical trauma. Some areas, such as the hippocampus, can be gently dissected free of the surrounding tissue. We often use small plastic knives that have been filed to make the edges blunt. Other structures are not so easily dissected, and a single-edge razor blade is usually used to block out an appropriate area of the brain. The size of the tissue to be blocked obviously depends upon the brain area to be studied, but also determines the method of slicing. Large chunks of tissue or tissue containing a large number of myelinated fibers are best sectioned using a vibratome. If the area is smaller, however, such as the hippocampus, a tissue chopper works well. When a vibratome is used, the tissue (after blocking) is glued to the

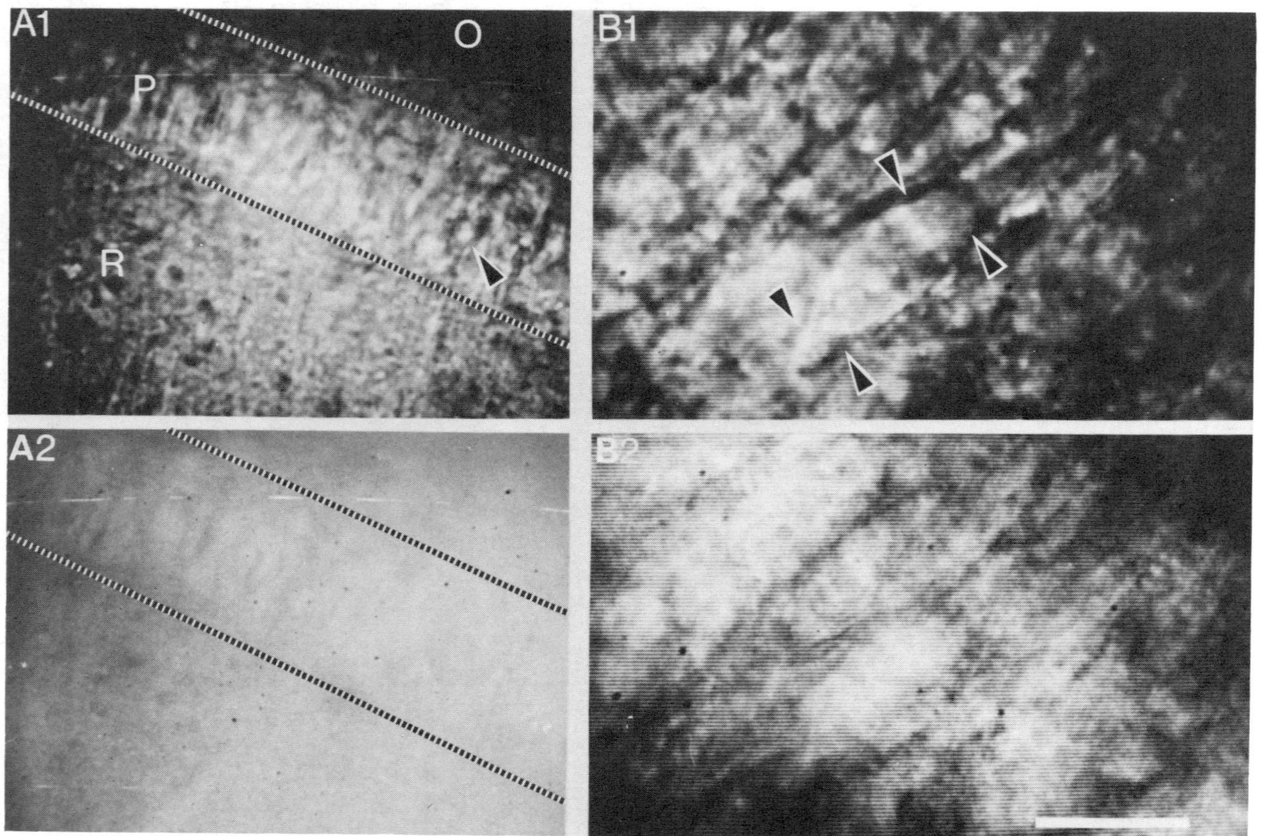


A1
O
P
R
B1
A2
B2

Fig. 2. CA1 pyramidal cells observed in 400-μm slices in the infrared. (A1) When the infrared filter was in place, individual pyramidal neurons (examples noted with arrowheads) were observed in the pyramidal cell body layer (P; delineated with dotted lines). The dense packing of cell bodies in this layer obscured some of the cell bodies. It was usually necessary to refocus up and down to determine the extent of the cell. The thick apical dendrites of some cells were visible extending into stratum radiatum (R). Stratum oriens (O) was relatively free of cell bodies, but contained scattered interneurons, possibly basket cells. (A2) The infrared filter was removed and the illuminating beam was filtered above 800 n$M$. Although the cell body layer was discernable, individual cells could not be observed when the visible spectrum was used. (B1) An individual neuronal cell body is illustrated at higher magnification. The outline of the cell is indicated by arrows. This cell was surrounded by other densely packed neurons and dendrites. (B2) In the visible spectrum with diminished contrast, individual cells were very hazy. Calibration bar represents 130 μm in A and 30 μm in B. (from MacVicar, 1984a).

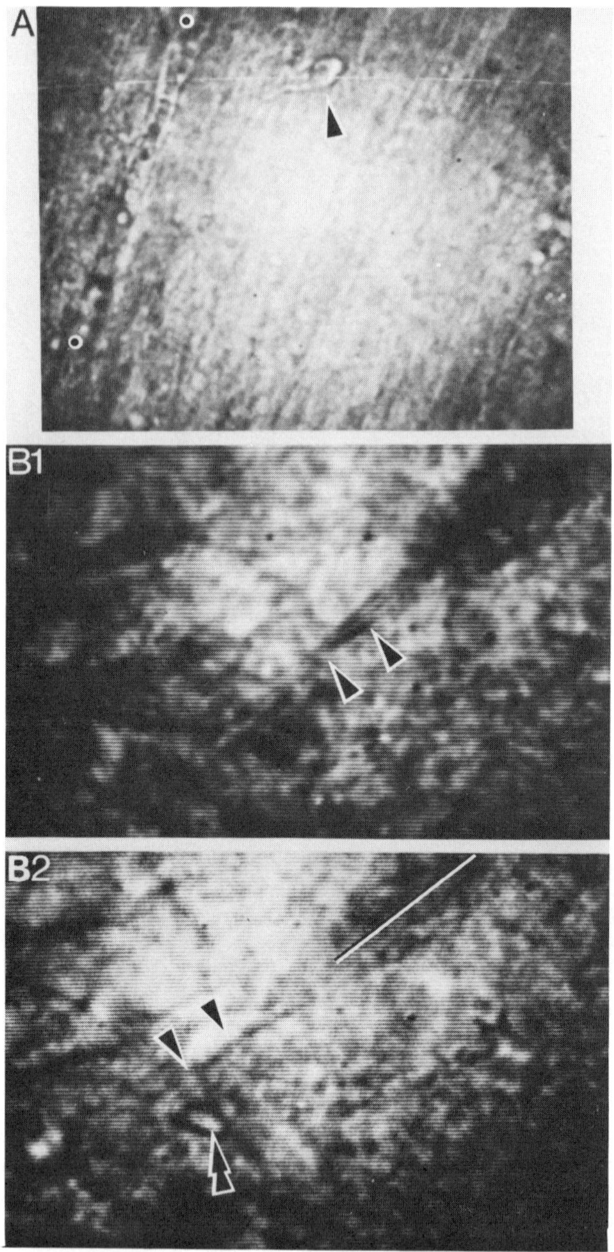

Fig. 3.  Observation of interneurons and microelectrodes in the infrared region. (A) In 400-μm slices of hippocampus, interneurons were
visible in the stratum radiatum (arrowhead). Capillaries (extending be-

cutting dish using Krazy glue, and blocks of agar are glued to the dish around it for support. To make slices on a tissue chopper, tissue to be sectioned is placed on a piece of filter paper on a movable stage and is kept moist by applying oxygenated artificial cerebrospinal fluid (CSF). After sectioning, the slices are transferred using a brush or a wide-mouth pipet to either a perfusion recording chamber or an incubation chamber for later transfer to a recording chamber.

There are several published designs of recording chambers (Figs. 6 and 7) that can basically be divided into two groups: interface (e.g., Schwartzkroin, 1975; Haas et al., 1979; Hatton et al., 1980) and submerged (White et al., 1978; Llinas and Sugimori, 1980; MacVicar, 1984a). In the interface chamber designs, perfusate reaches the top surface of the slices, which are covered with a thin layer of solution. Drying out of the slices is prevented by blowing a warmed and humidified gas mixture (95% $O_2$:5% $CO_2$) over the surface. In submerged designs, the slices are submerged under a rapidly flowing solution that is warmed and oxygenated. Some labs use lower temperatures (29 versus 35–37°C) when dealing with submerged slices (Nicholl and Alger, 1981) to increase viability, but this does not seem to be critical.

## 2.2. Cultured Slices and Explants

Principally because of the pioneering work of Gahwiler (1981, 1984a,b), it is now possible to maintain slices of hippocampus, hypothalamus (Armstrong and Sladek, 1982), or cerebellum for several weeks in tissue culture (Fig. 8). Many advantages of cultured slices are similar to those of other slice preparations, such as stability and the ability to quickly change extracellular solution. Also, like the dissociated cell culture (see below), cell membranes

---

tween dots) and the enclosed red blood cells were also particularly distinct. The striations are from the apical dendrites of the pyramidal cells. (B1 and 2) Microelectrodes could be seen within the slice by refocusing along the shaft. (B1) A thicker portion of the microelectrode was in focus (between arrows). The main shaft extends to the upper right (dark shadow indicated by line in B2). (B2) The tip (between arrows) was adjacent to a cell body (double arrowheads). Calibration bar in B represents 150 m$\mu$ in A and 30 m$\mu$ in B. (from MacVicar, 1984a).

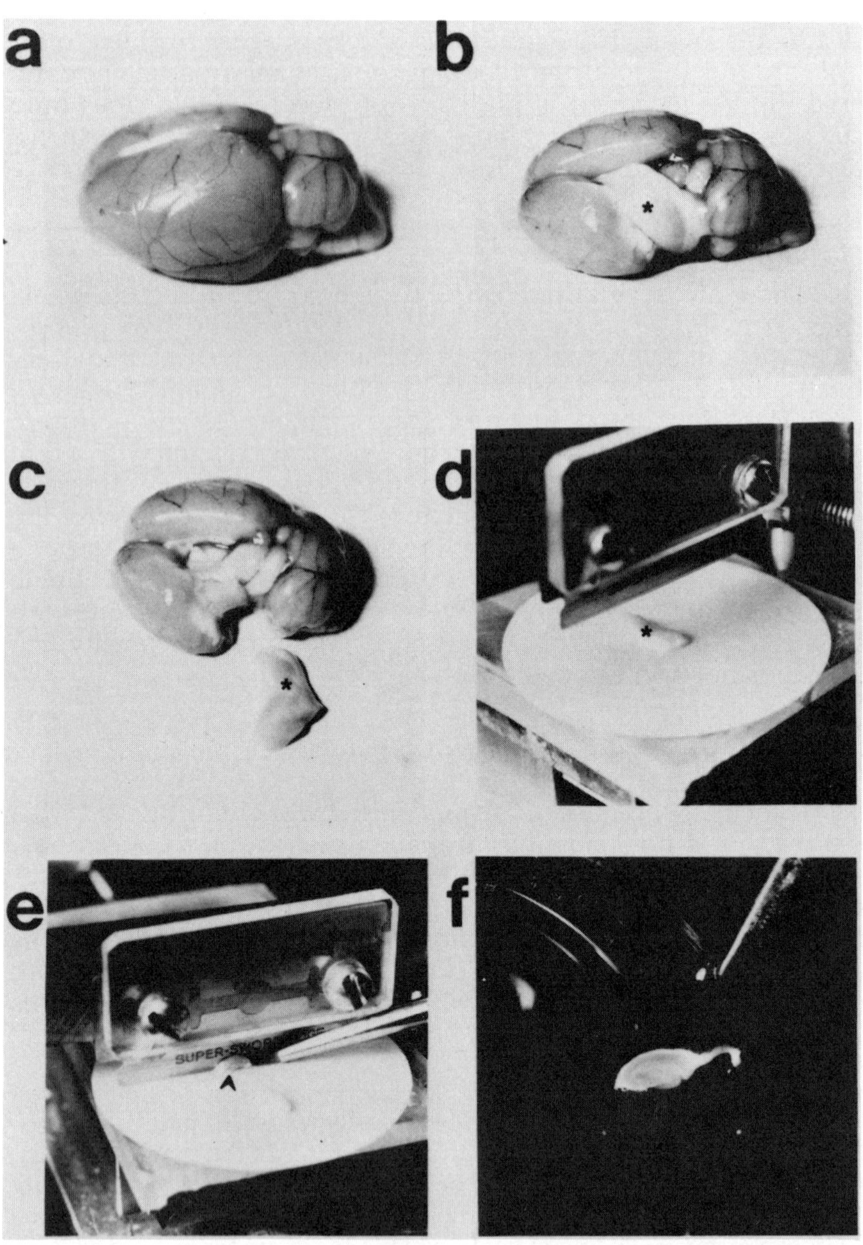

Fig. 4. Dissection of hippocampus and preparation of brain slices. (a) View of rat brain. The rat is decapitated and the scalp split. After the cranium and dura mater are removed, the brain is scooped out.

are "bare" and accessible for patch clamp experiments. The main advantage is that synaptic connectivity is maintained within slices, and that slices from different areas can be co-cultured and synaptic connections will form between them (e.g., Bird, 1985; Gahwiler and Brown, 1985). Hippocampal slices can be co-cultured with septal explants. Cholinergic axons will grow from the septal explant and innervate hippocampal pyramidal cells. The connections are functional, and cholinergic responses can be evoked in hippocampal neurons by stimulating the septal explant (Gahwiler and Brown, 1985). A detailed description of brain slice culture methods is given by Gahwiler (1981, 1984b).

For more acute experiments, explants can be kept viable for electrophysiological studies for up to 12 h without culturing (Bourque and Renaud, 1983, Llinas et al., 1981a; Muhlethaler and Llinas, 1984). Tissue is perfused with oxygenated artificial CSF through either the carotid artery in hypothalamic explants or through a vertebral artery in the case of brain stem explants (Fig. 9). The main advantage of this type of preparation is that synaptic connectivity that would be severed by tissue slicing can be maintained.

## 2.3. Dissociated Cultures

Dissociated cell cultures provide a good model in which to study ionic currents and cellular connections between neurons. Since the tissue is only 1–2 cells in thickness, the concentration of solutes bathing cells under study can be easily controlled. Also, cultured neurons can be used in patch clamp experiments because the cell surface is bare and high-resistance seals can be obtained. Therefore many patch clamp studies of single-channel currents

←—————————————————————————————————————————————

(b) Peeling away the overlying cortex reveals the hippocampus (indicated by asterisk). (c) The left hippocampus is carefully extracted. The relative size of the hippocampus compared to the brain is apparent here. (d) The hippocampus is positioned on a tissue chopper so that its longitudinal axis is at right angles to the razor blade. (e) Slices (indicated by arrow) are cut 400–500 μm thick from the hippocampus. (f) Slices are transferred to a Petri dish containing physiological solution at 4°C. It is possible to incubate slices at room temperature until they are transferred to a recording chamber, or they can be transferred immediately.

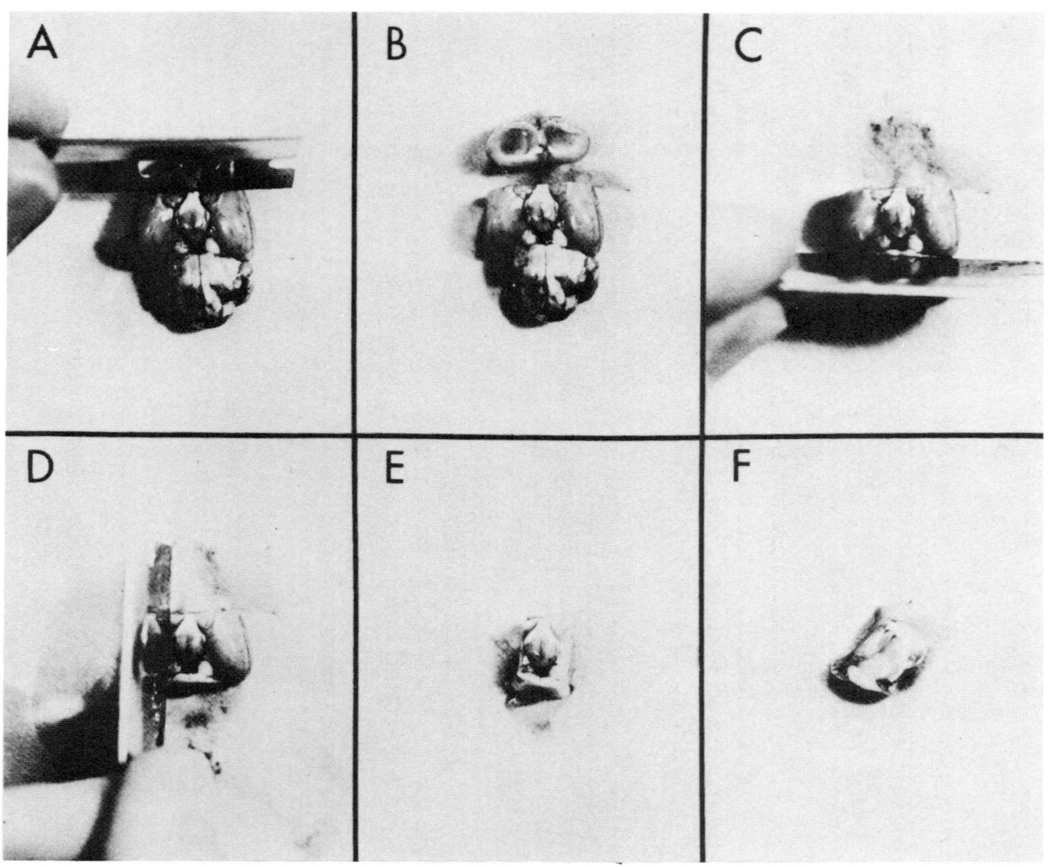

Fig. 5. Dissection of hypothalamus for preparation of brain slices. (A–H) Photographs of sequence of cuts used in blocking the hypothalamus. In F, the block shown in E has been placed on its side (ventral is to the right), so that horizontal cut (shown in G) at the level of the anterior commissure can be made. The tissue may then be picked up on the blocking blade (as in H) and transferred to a tissue chopper or a vibratome. (J) A slightly magnified view of the anterior face of the block showing anterior commissure (AC) and the optic chiasm (OC) (from Hatton et al., 1980).

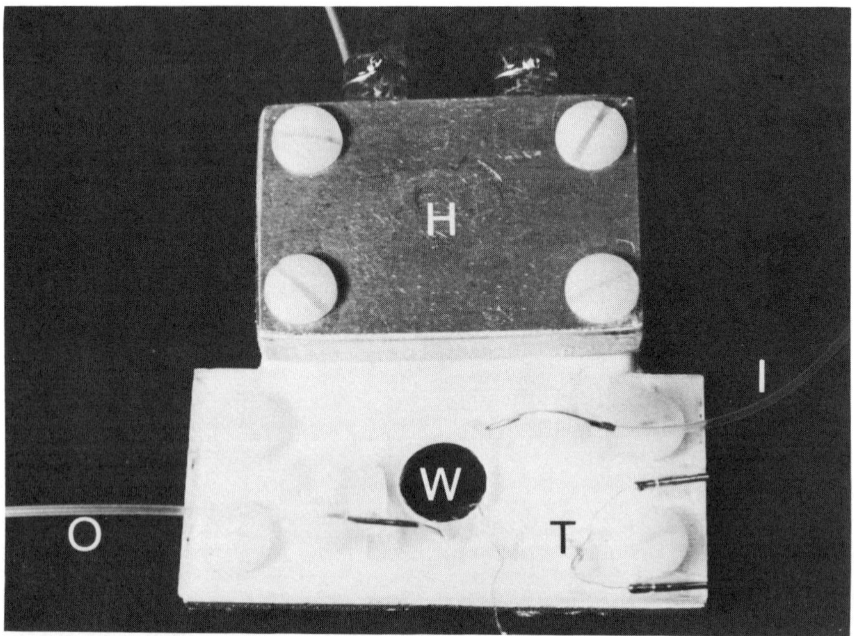


Fig. 6. Temperature-controlled perfusion chamber for brain slice preparations. The well (W) in which the slice is placed is drilled through a piece of plexiglass, which is attached to the aluminum heating plate (obscured) by Teflon screws. The bottom of the well is made by sealing a coverslip to the plexiglass with inert high-vacuum grease. Warmed and oxygenated physiological solution flows into the chamber through the inlet line (I), and the superfusate is removed by suction in the outlet line (O). A Peltier thermoelectric device is sandwiched between the heating plate and the heatsink (H), which is a water-cooled aluminum block. The temperature of the heating plate is measured with a thermistor (T) and controlled with a feedback circuit. Calibration bar, 3 mm (from MacVicar, 1984a).

have used cultured neurons. The composition of the growth medium can be manipulated to augment certain characteristics of the cells. For example, Higgins and Burton (1982) discovered that electrotonic coupling between cultured superior cervical ganglion neurons was dramatically increased in serum-free chemically defined medium. Subsequent studies determined that the increased insulin concentration in serum-free medium was responsible for the increased development of electrotonic coupling (Kessler et al., 1984; Wolinsky et al., 1985).

Various methods of cell dissociation have been developed by a variety of laboratories. The method used in our lab for the dissociation of hippocampal neurons is as follows:

The womb of an anesthetized, 18-d pregnant Sprague-Dawley rat is removed aseptically. The womb is washed in 70% ethanol for 10 s and then rinsed twice in separate washings of sterilized Hepes buffer. After removing the feti from the uterus, the fetal spinal cords are transected with a pair of jeweler's forceps. The scalp and skull are pealed away with the forceps, exposing the brain. After removing the brain with a blunt spatula, the hippocampi are dissected free and placed in chilled Hepes buffer.

Before dissociation, the tissue is transferred to plating medium containing serum. The presence of serum during the dissociation helps to prevent damage to the neurons. Mechanical dissociation is performed by drawing the tissue and plating medium through a fire-polished Pasteur pipet until the medium appears milky and few clumps of tissue remain. The cells are then diluted with more medium and plated on poly-L-lysine coated glass coverslips in Falcon 3000 tissue culture dishes. After 2 d, the plating medium is exchanged for growth media. The cells are then fed three times a week.

Enzyme treatment has been used by some investigators prior to mechanical dissociation (e.g., Bartlett and Banker, 1984). Using this protocol, the tissue is incubated in either trypsin or a mixture of enzymes for anywhere from 2 min to 2 h and then mechanically dissociated through a fire-polished Pasteur pipet. This is sometimes followed by centrifugation of the cells and resuspension of the cell pellet.

Many surface coatings have been used for neuronal cultures. These include poly-L-lysine, collagen, gelatin, or a layer of non-neuronal cells. The type of surface coating seems to depend more on personal preference than on cell type. Poly-L-lysine is the most common plating substance used. It has been shown to promote neurite extension and cell adhesion.


## 3. Analysis of Ionic Currents

### 3.1. Sodium

#### 3.1.1. Overview

The ubiquitous and most commonly studied current in neurons of both vertebrates and invertebrates is the fast sodium

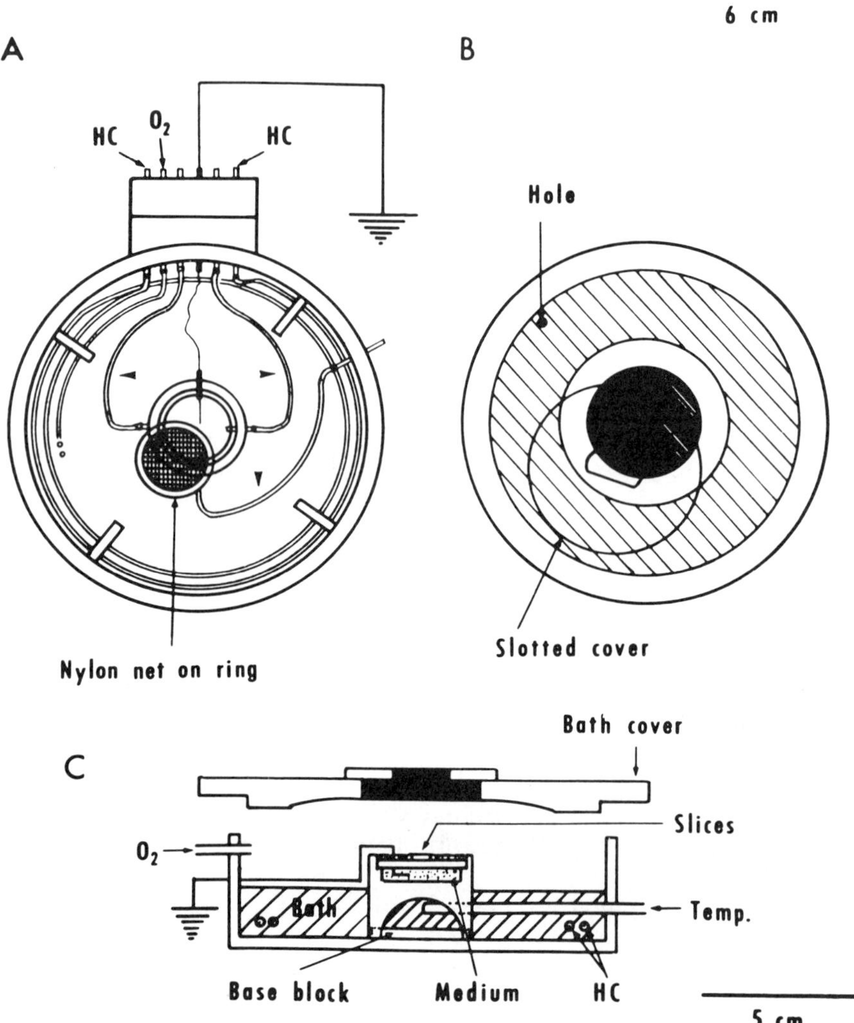


Fig. 7. Diagrams of simple plastic interface chamber for in vitro studies of brain slices. (A) Top view of the inside of a chamber showing the spatial relationships between the heating coils (HC), the oxygen input port ($O_2$), the ground wire, and the input ports with their connecting tubes (arrowheads) for infusing or drawing off medium. Entry to the outer bath and/or to the inner well is via 21-gage stainless steel tubing. All connecting tubes in the outer bath are PE 60 polyethylene. Three ports lead to the inner well; the two on opposite sides of the well are used for push–pull exchanges, and the third is used for constant infusion. The net

current underlying action potentials. The fast sodium current is typically a large current that has very fast activation and inactivation kinetics. There are technical problems in studying the sodium current in CNS neurons because of the difficulty in adequately voltage clamping such a large and fast current. Therefore this current has mainly been studied using voltage clamp techniques in model systems such as the squid giant axon (Hodgkin and Huxley, 1952a,b), neuroblastoma cells (Quandt and Narahashi, 1984), and invertebrate neurons (Adams et al., 1980). Most studies on CNS neurons have utilized current clamp techniques and intracellular recordings. The conclusion that responses in CNS neurons are caused by sodium currents rests mainly upon two lines of evidence: (1) the sensitivity of a response to tetrodotoxin (TTX), which has been shown to block the sodium channel in almost all neurons (Narahashi, 1974) and (2) the sensitivity of responses to replacing extracellular sodium with an impermeant cation, such as choline. Some experiments have also utilized the intracellular injection of a local anesthetic to reduce sodium currents (e.g., Connors and Prince, 1982). It should be noted, however, that some local anesthetics may have nonspecific actions on other currents. Although the large sodium current underlying the action potential in the mammalian CNS has not been rigorously studied, CNS neurons have another sodium current that has been studied to a greater extent. This other sodium-dependent current is of lower amplitude and slower kinetics, but is also voltage-activated (e.g., Llinas and Sugimori, 1980; MacVicar, 1985). Single-electrode voltage clamping as well as current clamp recordings have been used to study the properties of this current.

---

on the ring has been moved aside in order to show the inner well; its normal position is seated in the well, and as in (C). (B) Bath cover with slotted disk that covers the inner well. Slot allows access to slices. Hole in bath cover is for temperature probe. (C) Side view of chamber slightly modified from (A). In this model, $O_2$ enters above the level of the bath and must, therefore, be warmed and humidified prior to its entrance into the chamber. Also in (C) are shown the base block over which the inner well is press-fit and the heating coil wires as they course through the bath encased in polyethylene tubing. The coils in these chambers are made of enameled nichrome, are 0.005 in. diameter and have a resistance of ~27 ohm/ft (from Hatton et al., 1980).

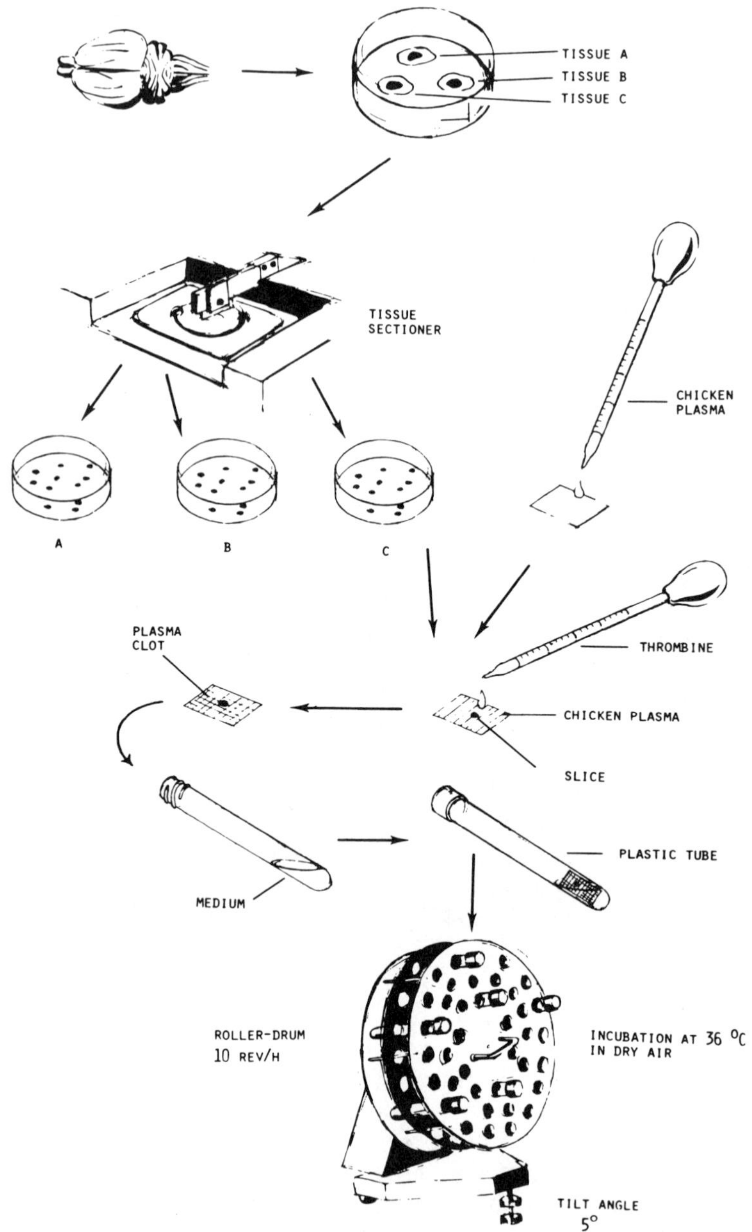



Fig. 8. Schematic illustrating method for preparing neuronal slice cultures. The brain is asceptically removed from 1- to 20-d-old rats and the tissue from the desired areas is dissected free. The explants are kept at

562

### 3.1.2. Voltage-Activated Sodium Channels

Most action potentials in CNS neurons are predominantly caused by a sodium current that is similar to that described in squid giant axon. That is, it has fast activation and inactivation kinetics and is blocked by TTX. This channel is important in the propagation of action potentials along axons and in generating somatic action potentials in neurons. Another sodium current that has a lower threshold and is noninactivating was first described in cerebellar Purkinje cells by Llinas and Sugimori (1980). Subsequent studies have reported this current in widespread areas, including neurons of the hippocampus (Hotson et al., 1979; MacVicar, 1985), thalamus (Jahnsen and Llinas, 1984a), cortex (Stafstrom et al., 1982; Connors et al., 1982), and deep cerebellar nuclei (Jahnsen, 1986). It is thought to play an important role in controlling repetitive firing behavior of neurons (Lanthorn et al., 1984). This current can be examined in the presence of agents that block calcium entry, such as divalent cations (0.5–2.0 m$M$), cobalt (1–5 m$M$), or manganese (1–5 m$M$) (Fig. 10). In Purkinje cells after calcium currents were abolished, action potentials and prolonged plateau potentials were evoked in solutions containing tetraethylammonium, which reduces potassium permeability. Lidocaine, a local anesthetic, preferentially blocked action potentials and not the plateau potentials (Sugimori and Llinas, 1983), therefore indicating a pharmacological difference. This low threshold, slowly inactivating sodium current is blocked by TTX and by removing extracellular sodium. In several neurons it is a slowly developing inward current that is activated by depolarizations that are below the threshold for the action potential and can act in concert with a low threshold inward calcium current to produce inward-going rectification (Hotson et al., 1979). Inward currents that are below threshold for action

← ——————————————————————————————————————————

room temperature in Geys balanced salt solution. Slices (300–500 $\mu M$ thick) are cut using a tissue chopper, then are placed in a drop of heparinized chicken plasma on a glass coverslip. The plasma is coagulated with a drop of thrombin to keep the tissue in place. Glass coverslips with tissue are transferred to plastic test tubes with growth media containing horse serum (25%), basal medium (Eagle) (50%), and Hanks' or Earles' balanced salt solution (25%) supplemented with glucose (6.5 mg/mL). The test tubes are transferred to a roller drum (10 rev/h) to ensure proper feeding and aeration of the cultures (from Gahwiler, 1981, 1984b).

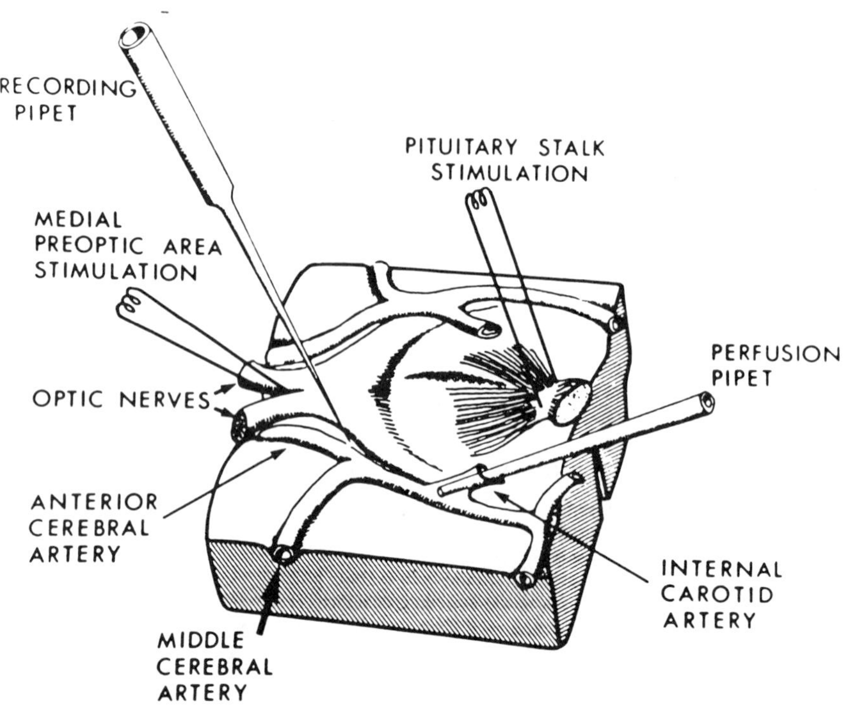



Fig. 9. Schematic illustration of the basal diencephalic explant utilized for the studies of Bourque and Renaud (1983). In this example, the pituitary has been removed from its stalk. The tissue block, measuring approximately 8 × 8 × 2 mm, is perfused with artificial media intravascularly through one of the anterior cerebral arteries. SON neurosecretory neurons may be activated antidromically via stimulation applied to the pituitary stalk and orthodromically via stimulation to the medial preoptic area (from Bourque and Renaud, 1983).

potentials can be important in controlling firing rates of neurons and can induce repetitive spiking. The threshold for this slow sodium current in hippocampal neurons is regulated by intracellular calcium concentration (MacVicar et al., 1986). In isolated slices of the hippocampus, repetitive slow wave activity can be recorded even after calcium currents and synaptic potentials have been blocked (Jefferys and Haas, 1982; Taylor and Dudek, 1982). It is possible that the slow inward sodium current underlies the prolonged depolarizations of these neurons during slow wave activity. Synchronization and spread of this activity through the hippocampus are most likely caused by extracellular potassium accumulation (Konnerth et al., 1984), electrotonic coupling (Mac-

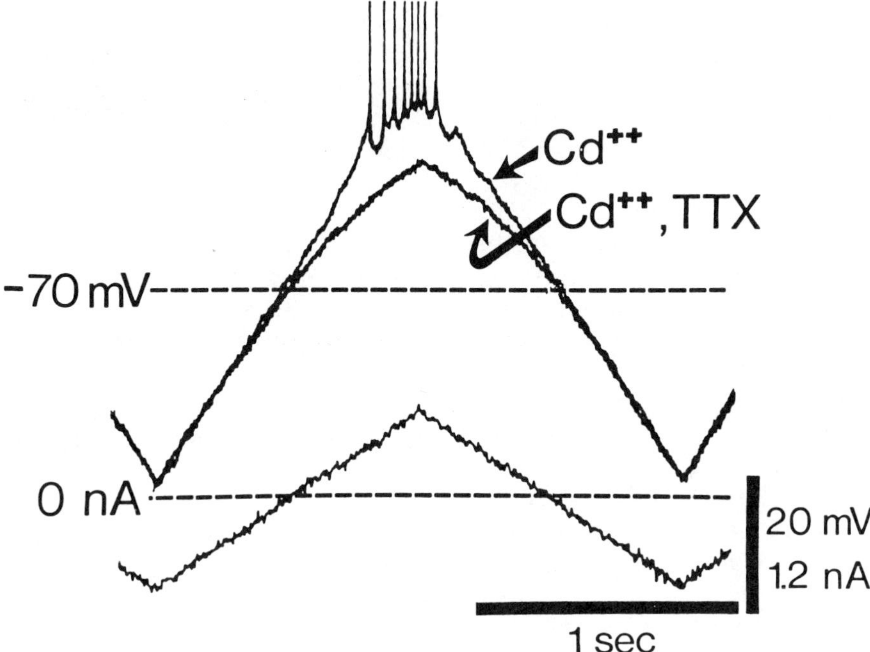


Fig. 10. Superimposed recordings from the same neuron illustrate the contribution of a persistent sodium current in depolarizing CA1 pyramidal cells. Oscillations of membrane potential were induced by injecting current with a triangular waveform (lower trace). The membrane potential fluctuations were initially recorded in the presence of cadmium to block any contributions from calcium currents. Identical current pulses were injected following the application of TTX. Traces were photographically superimposed to illustrate the contribution of a persistent inward sodium current in depolarizing the cell at different levels of membrane potential. It can be seen that as the cell is depolarized past –65 mV, there is a substantial contribution from the TTX-sensitive current. As the cell reaches approximately –50 mV, slow depolarizing potentials are evoked as well as action potentials. Action potentials are clipped (from MacVicar, 1985).

Vicar and Dudek, 1981), and ephaptic interactions (Taylor and Dudek, 1982).

### 3.1.3. Synaptic Currents and Nonspecific Cation Channels

Postsynaptic excitation at most synapses is mediated by ligand-operated channels that are permeable to sodium and potassium (Eccles, 1964). Because the resting potential of most cells is close to the equilibrium potential of potassium, the result of synaptic activation is predominantly a net influx of sodium leading to a postsynaptic depolarization. Presently there are no pharmacological tools for the selective block of this ionic channel; substitution of external sodium is useful, however, in determining the permeability of the channel to sodium ions. Substantial changes in extracellular sodium can occur during intense synaptic activation. It is postulated that some of the substantial changes in extracellular concentrations of sodium and potassium during spreading depression (Nicholson and Kraig, 1981) are caused by glutamate-activated sodium-potassium channels (Van Harreveld, 1978). Recently, analysis of glutamate-activated currents and channels has confirmed that there is a substantial sodium permeability to this channel (Cull-Candy and Ogden, 1985); in fact it seems to be equally permeable to both sodium and potassium. A subtype of the glutamate channel that is preferentially activated by *N*-methyl-D-aspartate (NMDA) is also significantly permeable to calcium (MacDermott et al., 1986) (discussed in section 3.2).

Single-channel analysis of calcium-activated channels has led to the discovery of another channel type that is permeable to sodium as well as other cations (Yellen, 1982). This nonspecific cation channel is activated by intracellular calcium and has been reported in cultured cardiac muscle fibers (Colquhoun et al., 1981), neuroblastoma cells (Yellen, 1982), and pancreatic acinar cells (Maruyama and Peterson, 1982). The channel has a conductance of approximately 30 pS and is equally permeable to potassium and sodium. This channel could underly the slow inward current that is critical in generating rhythmic activity in heart cells and bursting pacemaker neurons of *Aplysia* (Kramer and Zucker, 1985) and *Helix* (Swandulla and Lux, 1985). Although it is possible that it plays a role in generating pacemaker activity in mammalian CNS neurons (e.g., Andrew and Dudek, 1983), this has yet to be examined.

### 3.1.4. Sodium Exchange Mechanisms

In CNS neurons there are important exchange mechanisms for sodium-potassium and sodium-calcium. Intracellular loading of hippocampal neurons with sodium by glutamate application activates the sodium-potassium pump and causes a hyperpolarization (Thompson and Prince, 1986). Na-K-ATPase, the enzyme underlying the pump, extrudes three sodium ions for two potassium ions taken up, resulting in a net hyperpolarizing current. This helps restore sodium to the extracellular space and may be very important in restoring normal ion levels after spreading depression. Astrocytes also have a sodium-potassium exchange system that is activated by increasing extracellular potassium (Walz and Hertz, 1983). This is believed to be an important mechanism by which increases in extracellular potassium are reduced by glial cells. Sodium-potassium exchange mechanisms are tested by their sensitivity to strophanthidin or ouabain or by sensitivity to extracellular potassium and intracellular sodium. The sodium-calcium pump is another exchange mechanism that is electrogenic and could play a role in regulating sodium and calcium concentrations (Requena, 1983). Increasing intracellular calcium activates this pump mechanism in which one calcium is exchanged for three extracellular sodium molecules, thus creating a net electrogenic depolarization (Yau and Nakatani, 1984; Kimura et al., 1986; Mechmann and Pott, 1986). This is hypothesized to play a role in generating rhythmic depolarizations in certain forms of cardiac pacemaking. Sodium-calcium exchange could conceivably be a factor in depolarizing CNS neurons, but this has not been investigated to date. The sodium-calcium exchange current is defined by sensitivity to intracellular calcium and extracellular sodium. It can also be inhibited by external lanthanum (e.g., Yau and Nakatani, 1984).

## 3.2. Calcium

### 3.2.1. Overview

Dramatic decreases in extracellular calcium have been observed during intense synaptic activation and seizure-like activity in areas such as the hippocampus and neocortex (for review, *see* Nicholson, 1980). The loss of extracellular calcium probably represents calcium entry into neurons through voltage-activated calcium channels. Recent work, however, has also indicated two other possible sinks for extracellular calcium: (1) the glial network and (2) ionic channels activated by $N$-methyl-D-aspartate. The analysis of calcium currents in the CNS has been based mainly on intracellular recordings from neurons in intact slices. This has allowed for either voltage recordings or the use of single electrode voltage clamps to deduce the presence and the properties of calcium currents. Recently, patch clamp techniques have been applied to acutely dissociated neurons or to neurons partially dissociated, where the cell body is free of the slice, but the dendritic tree is still embedded within the slice (e.g., Gray and Johnston,

1985a). Therefore most studies have entailed the analysis of currents in dendrites or cell bodies. There are very few studies that analyze the properties of presynaptic calcium currents. The properties of presynaptic calcium currents are extrapolated from either properties of postsynaptic currents or properties of presynaptic currents in model systems.

### 3.2.2. Postsynaptic Calcium Spikes

The discovery of calcium electrogenesis in dendrites of CNS neurons (Llinas and Hess, 1976) was a major contribution to understanding CNS electrophysiology. This was first studied in the cerebellum by Llinas and coworkers (Llinas et al., 1977; Llinas and Sugimori, 1978, 1980), and since then dendritic calcium spikes have been found to be very widespread (Fig. 11). Similar calcium dependent dendritic spikes have been reported in the hippocampus (Schwartzkroin and Slawsky, 1977; Wong et al, 1979), inferior olive (Llinas and Yarom, 1981a,b), neocortex (Connors et al., 1982; Stafstrom et al., 1985), and thalamus (Jahnsen and Llinas, 1984). In fact, dendritic calcium spikes appear to be common features of CNS neurons. The evidence for calcium spiking is based upon the following criteria: that the regenerative potential is insensitive to TTX or to reductions of extracellular sodium; that it is dependent upon extracellular calcium concentrations; and that divalent cations, such as cadmium, manganese, or cobalt, block the spike. The properties of calcium spikes are often studied in the presence of tetraethlyammonium or intracellular cesium in order to reduce potassium conductances. Following the initial reports of calcium spiking, further studies indicated that calcium electrogenesis had several components. For example, Jahnsen and Llinas (1984) reported two types of calcium responses in thalamic neurons (Fig. 12). Multiple types of calcium channels have been described in many CNS neurons, including those of the inferior olive (Llinas and Yarom, 1981a,b) and hippocampus (Halliwell, 1983; Gray and Johnston, 1986). One is a high-threshold noninactivating calcium-dependent response and another is a low-threshold inactivating calcium-dependent response. Single-channel analyses of calcium currents with similar properties have been described in cultured neurons of the chick dorsal root ganglion (DRG) by Tsien's group (Nowycky et al., 1985). The high-threshold noninactivating response of CNS neurons is presumably generated by the same channel that is activated in DRG neurons at membrane potentials above −20 mV; this channel has a conductance of 25 pS (in 110 m$M$

barium). This has been called the L-type channel because of its large conductance and long-lasting opening. The low-threshold inactivating response is generated by a channel that is activated at membrane potentials between –50 and –60 mV and has a conductance of 8 pS (in 110 m$M$ barium). This channel is the T-type calcium channel because of its transient opening and tiny conductance. In chick DRG cells, these channels have different pharmacological sensitivities, where the T-type channel is preferentially blocked by nickel and the L-type channel is preferentially sensitive to cadmium and nifedipine. Another type of channel, called the N-type channel, has been described in DRG cells, but has not yet been described in the CNS.

The correlation of extracellular calcium changes with activation of the different intracellular currents is difficult and tenuous at present. Recent experiments have indicated, however, that a fairly large extracellular decrease can be attributed to entry of calcium into dendrites (Pumain and Heinemann, 1985). The largest signal should presumably be caused by activation of the L-type channel or high-threshold noninactivating calcium currents. This is the calcium current that presumably leads to calcium spikes and prolonged calcium entry. The determination of both spatial location of calcium currents and their different pharmacological sensitivities in CNS neurons will facilitate correlation of channel type with extracellular ionic signal.

### 3.2.3. Presynaptic Calcium Entry

Analysis of the entry of calcium into presynaptic terminals has not been successfully studied in the mammalian CNS because of the problems of electrophysiologically recording from presynaptic terminals. The only system in which this has been rigorously studied is the squid giant synapse (e.g., Llinas et al., 1981b). A few other systems, such as the sinus X-organ (e.g., Lemos et al., 1986) (a neuroendocrine organ in crustaceans), have been used as model systems to analyze calcium entry and its relationship to secretion. Therefore, the characteristics of calcium currents are not directly known, but they have been indirectly inferred from other experiments. For example, analysis of extracellular calcium transients during synaptic transmission in the hippocampus has shown that a substantial loss of extracellular calcium is caused by calcium entry into presynaptic terminals, and this can be differentiated from postsynaptic calcium entry (Schubert et al., 1986; Kon-

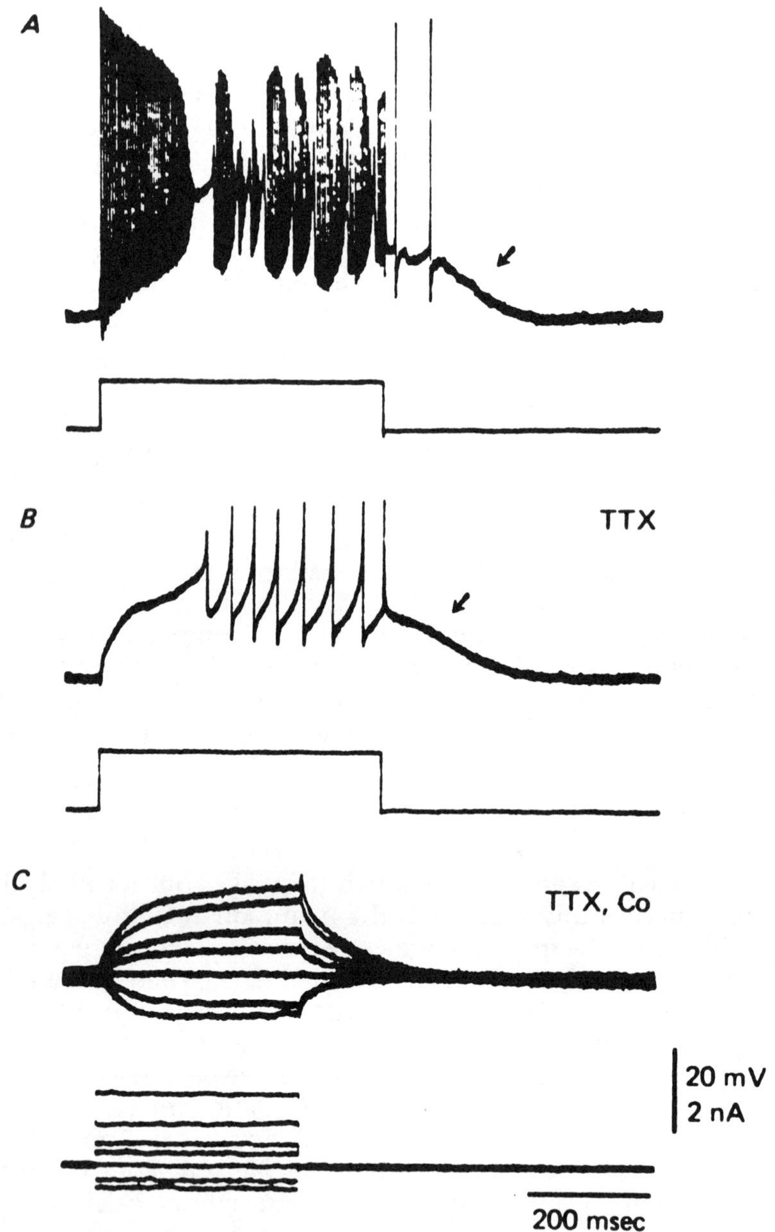


Fig. 11.  Calcium electrogenesis recorded intracellularly in cerebellar Purkinje cells. (A) Injection of depolarizing current evoked a complex pattern of spiking with fast and slow action potentials and an after-

nerth and Heinemann, 1983). Synaptic transmission is also blocked by several agents that block calcium entry and calcium currents, such as extracellular cadmium, cobalt, and manganese. The type of calcium channel that is associated with synaptic release of neurotransmitters is unclear at present. At the squid giant synapse there is evidence that the N-type calcium channel is the site of calcium entry that is associated with the release of vesicles. The rigorous experiment correlating the calcium channel type with secretion has not yet been performed, however.

### 3.2.4. Synaptic Calcium Currents

Most synaptically activated membrane currents have been shown to be caused by changes in conductances to sodium, potassium, or chloride, or a combination of these ions. Recently, however, the subtype of glutamate channel for which NMDA is an agonist has been shown to have a significant calcium permeability (MacDermott et al., 1986). This channel is blocked by magnesium in a voltage-dependent manner (Mayer et al., 1984; Nowak et al., 1984). Activation of the NMDA channel also increases intracellular calcium (MacDermott et al., 1986). This may be because of increased membrane calcium permeability or intracellular calcium release evoked by an intracellular messenger such as inositol triphosphate. It is possible, in areas such as the hippocampus, which have a high density of NMDA receptors (Monaghan et al., 1983), that some of the decrease in extracellular calcium during synaptic transmission could be caused by calcium entry through these channels. This would be most pronounced when postsynaptic cells were depolarized, resulting in a loss of the voltage-

←———————————————————————————————

depolarization (arrow). (B) Tetrodotoxin (TTX), which blocks sodium channels, was applied to differentiate between sodium- and calcium-dependent potentials. TTX blocked the fast action potentials, but did not affect the slow spikes or the after-depolarization. The same results was obtained when extracellular sodium was replaced with choline. (C) Cobalt, which blocks calcium channels, was then applied along with TTX and blocked both the slow spikes and the after-depolarization. This experiment demonstrates that fast action potentials in Purkinje cells are sodium-dependent, and that slow spikes and after-depolarizations represent calcium electrogenesis (from Llinas and Sugimori, 1980).

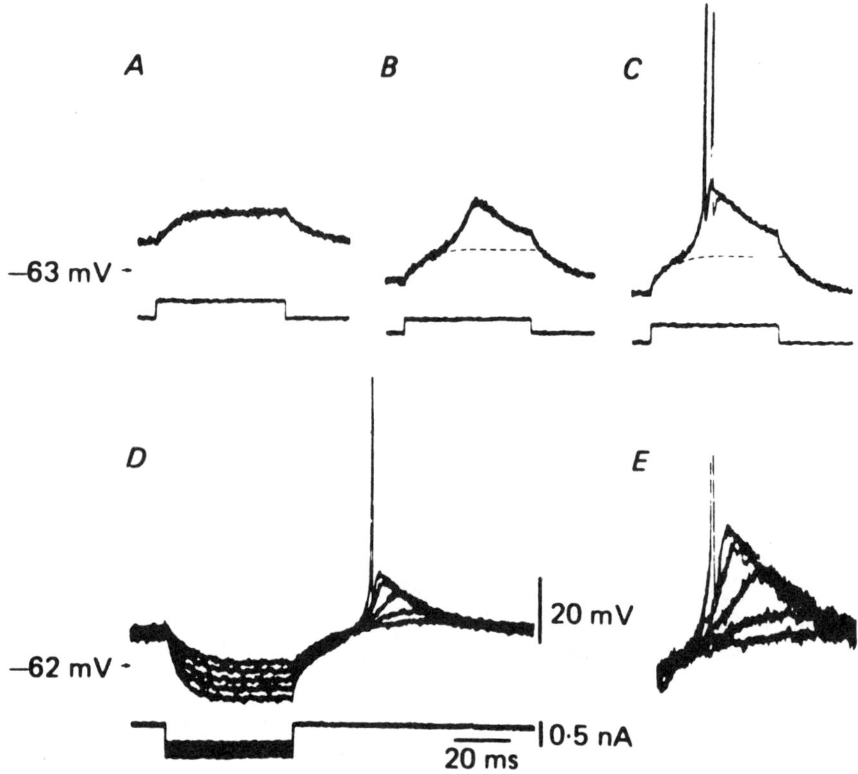


Fig. 12. The low threshold spike (LTS) recorded intracellularly in thalamic neurons. (A, B and C) Constant amplitude depolarizing current pulses superimposed on three different levels of membrane potential. In (A), the cell is depolarized above resting potential and no LTS is evoked. When the cell is hyperpolarized (B), the LTS alone is first evoked, and with further hyperpolarization (C), the LTS is larger and evokes fast action potentials. (D, E) Hyperpolarizing current pulses evoke a rebound LTS that increases in amplitude with greater hyperpolarization. These experiments illustrate a method of altering membrane potential to examine the activation properties of a response. The LTS is inactivated at resting potential, and the cell must be hyperpolarized for the LTS to be evoked. Other experiments, using TTX and calcium blockers (as illustrated in Fig. 11) demonstrate that the LTS is calcium-dependent and could play a role in generating rhythmic activity in the thalamus (from Jahnsen and Llinas, 1984).

dependent magnesium block of the NMDA channel. Hippocampal neurons have been shown to be further depolarized during intense synaptic activity by voltage-dependent disinhibition of NMDA responses (Herron et al., 1986). Although this has not been demonstrated, substantial calcium entry through the NMDA-activated channel could then cause a decrease in extracellular calcium. The NMDA receptor and calcium entry may have a role in the formation of long-term potentiation in the hippocampus.

### 3.2.5. Glial Calcium Current

Experiments on astrocytes in primary tissue culture have indicated that these cells have voltage-activated calcium channels (MacVicar, 1984b). Barium-dependent action potentials were observed after the potassium permeability of the glial cells was reduced by extracellular tetraethylammonium and barium. These action potentials are insensitive to TTX, are blocked by cadmium or manganese, and are enhanced by increasing concentrations of barium. The type of calcium channels involved in these responses has not yet been described. Glial calcium channels are modulated by neurotransmitters such as norepinephrine (noradrenaline) and adenosine and by intracellular cyclic adenosine monophosphate (cAMP) (MacVicar, 1986a,b; Chun et al., 1986). Cultured glial cells have a calcium-activated potassium conductance (Quandt and MacVicar, 1986). These currents have not yet been described in intact glial cells from the CNS. Muller cells, however, a type of glial cell in the retina, have been shown to have calcium channels in both the intact preparation and after dissociation (Newman, 1985). The presence of voltage-activated calcium channels in glial cells suggests another possible way in which extracellular calcium levels could be modulated in the CNS. At times, during intense neural activity, there will be an increase in extracellular potassium, which will depolarize glial cells. This in turn may cause activation of the voltage-activated calcium channels and thereby lead to calcium entry into glial cells. Calcium entry may further depolarize the cells and activate the calcium-activated potassium current, leading to a potassium efflux from the glial cell. In such a way, extracellular potassium signals may be enhanced by potassium efflux from the glial cell. Calcium entry into glial cells may also be a cause of substantial loss of calcium from the extracellular space that is observed during seizure activity.

## *3.3. Potassium*

### *3.3.1. Overview*

The analysis of potassium currents in the CNS is very complex at present. The cataloging of potassium currents has revealed at least seven different currents, and this is likely by no means complete at present. The separation of the different currents is determined by the pharmacological sensitivity and the voltage activation and inactivation properties. Potassium currents can be broadly defined as voltage-activated, calcium-activated, or ligand-activated.

### *3.3.2. Voltage-Activated Potassium Channels*

The original studies of Hodgkin and Huxley (1952a,b) described a potassium conductance, termed the delayed rectifier, that repolarizes the neuron during the action potential. This is a very widespread current that has been found in almost every neuron investigated. It is activated by depolarization and can be blocked by tetraethylammonium (Hermann and Gorman, 1981b; Thompson, 1977; Zbicz and Weight, 1985). Another slowly activating voltage-dependent potassium current is called the M-current. It was first discovered in frog sympathetic neurons (Adams et al., 1982a) and has since been described in cortical neurons (Constanti and Galvan, 1983; Halliwell and Adams, 1982), and even in human neurons (Halliwell, 1986). The M-current is so named because the current is blocked by acetylcholine acting through the muscarinic receptor.

Another voltage-sensitive potassium current in CNS neurons is called the A-current (Gustafsson et al., 1982; Yarom et al., 1986). This was first described in molluscan neurons (Connor and Stevens, 1971). It is characterized as a transient potassium current that is partially or completely inactivated at resting potential and is activated by depolarization. The A-current is preferentially sensitive to 4-aminopyridine (Hermann and Gorman, 1981a; Thompson, 1977; Zbicz and Weight, 1985; Gustaffson et al., 1982).

Many CNS neurons also have a potassium conductance activated by membrane hyperpolarization (e.g., Halliwell and Adams, 1982). This has been termed $I_Q$ or the anomalous rectifier. This current is preferentially blocked by 1–2 m$M$ extracellular cesium or by barium.

### 3.3.3. Calcium-Activated Potassium Channels

When CNS neurons are depolarized, a potassium current that is dependent upon calcium entry is slowly activated (e.g., Zbicz and Weight, 1985; Lancaster and Adams, 1986). Calcium entry during intense spiking or bursting causes an after-hyperpolarization that is caused by this current, the calcium-activated potassium current (e.g., Llinas and Sugimori, 1980; Hotson and Prince, 1980; Connors et al., 1982; Brown and Griffith, 1983) (Fig. 13). This potassium current is important in controlling calcium spiking because it is activated by calcium entry and helps to repolarize membrane potential to prevent prolonged calcium entry. It also causes

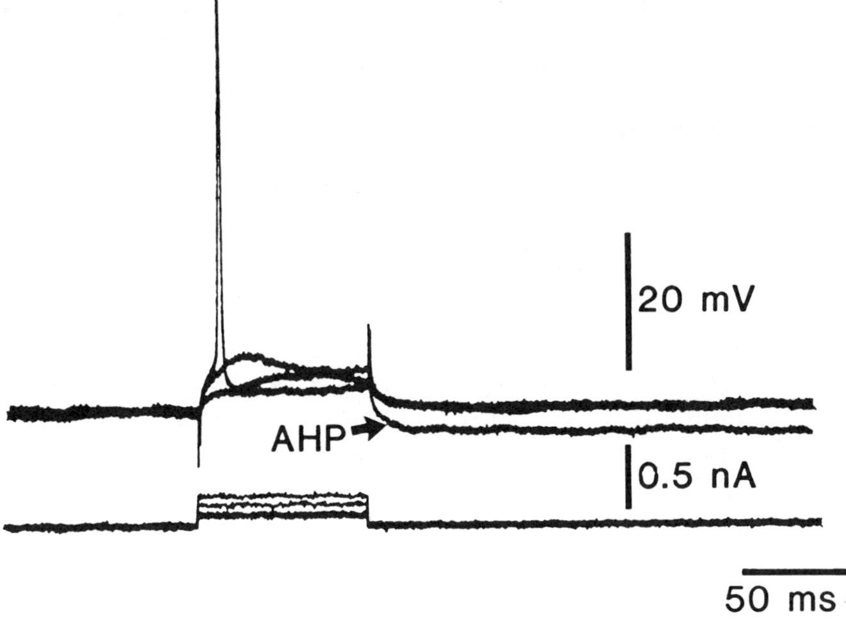


Fig. 13. Intracellular recording from a human cortical neuron with a pronounced after-hyperpolarization (AHP). Slices were prepared from cortical tissue surgically removed from patients for the treatment of epilepsy. The large AHP is most likely caused by a calcium-activated potassium conductance. This could be tested, as described in the text, by examining the sensitivity to tetraethylammonium or apamin. The AHP should be blocked by agents that block calcium entry, such as cobalt or cadmium or by injections of a calcium chelator such as EGTA.

spike accommodation (Madison and Nicoll, 1986). Calcium-activated potassium currents are blocked by tetraethylammonium (Hermann and Gorman, 1981b; Lancaster and Adams, 1986), by intracellular injection of EGTA, a calcium chelator (Alger and Nicoll, 1980), or by cobalt or cadmium, which block calcium entry. Cultured astrocytes also have a calcium-activated potassium channel (Quandt and MacVicar, 1986).

Recent work has demonstrated another calcium-activated potassium current (Ic) that is highly voltage sensitive and has faster activation kinetics (Lancaster and Adams, 1986). This current could play a role in spike repolarization in CNS neurons, as has been shown for sympathetic neurons (Adams et al., 1982b; MacDermott and Weight, 1982).

It is clear that CNS neurons have a bewildering complexity of voltage- and calcium-activated potassium conductances. As discussed by Lancaster and Adams (1986), it is difficult to envisage significant roles for all the currents described. Yet they are present and do contribute to the regulation of neuronal excitability.

### 3.3.4. Ligand-Activated Potassium Channels

Most excitatory neurotransmitters activate a channel that is permeable to both potassium and sodium (e.g., Eccles, 1964). However, there are recently described actions of acetylcholine and γ-aminobutyric acid (GABA) that may be caused by a specific transmitter-activated potassium conductance. Thalamic neurons of the nucleus reticularis are hyperpolarized by acetylcholine acting through an $M_2$ subclass of muscarinic receptors (McCormack and Prince, 1986). The hyperpolarization is caused by an increase in potassium conductance. This is in contrast to the usual acetylcholine actions of depolarizing by inactivation of the M-current.

GABA can also hyperpolarize hippocampal neurons independent of its action on chloride conductance. This has been shown to be mediated by a GABA-B receptor-linked increase in potassium conductance (Newberry and Nicoll, 1984).

## 3.4. Chloride

### 3.4.1. Overview

Chloride currents that are predominantly caused by synaptic activation of inhibitory synapses have been described in the CNS. GABA-activated inhibitory postsynaptic potentials (IPSPs) are caused by an increase in chloride permeability, which leads to a somatic hyperpolarization. Inhibitory synapses characteristically have their synaptic sites very close to the cell body. Therefore it has been possible to study IPSPs using voltage clamp techniques and single-electrode voltage clamps (e.g., Griffith et al., 1986). The predominant body of information, however, is from studies using intracellular recording techniques and replacement of chloride by impermeant anions to determine the chloride dependency of a response.

A calcium-activated chloride channel was first described in egg cells (Barish, 1983; Miledi, 1982) and has been seen in some neuron types (Owen et al., 1984, 1986; Mayer, 1985). This has also been observed in cultured mouse spinal neurons (Owen et al., 1986). It is possible that this chloride channel is present in the CNS.

A voltage-activated chloride channel has also just been described in cultured astrocytes using voltage clamp techniques (Gray and Ritchie, 1986). The current was described by replacing chloride with impermeant anions such as gluconate and measuring the decrease in the magnitude of voltage-activated currents. This current was activated by depolarizations of glial cells past –40 mV. It has been suggested that this chloride channel may play a role during intense potassium depolarizations of glial cells. In this situation, a chloride current would be activated concurrently with an influx of potassium caused by raised extracellular potassium levels. The chloride current would help contain the potassium that flows into the glial cell during the raised potassium levels in the immediate area of the neurons that have lost potassium. This would then enable the glial cells to later return the potassium to the neurons.

### 3.4.2. Inhibitory Synaptic Currents

The GABA-activated inhibitory system has been well studied in the CNS (e.g., Eccles, 1964; Krnjevic, 1980). GABA opens an ionophore that is permeable to chloride and causes an influx of chloride at the soma, which results in a hyperpolarization. GABA applied to the dendrites of neurons such as hippocampal pyramidal cells, however, causes a depolarization (Alger and Nicoll, 1982; Djorup et al., 1981). GABA-mediated changes in chloride permeability cause a decrease in extracellular chloride concentration during intense synaptic activation (Dietzel et al., 1982). Chloride permeability is tested by altering the extracellular chloride concentration by replacing chloride with an impermeant anion, such as gluconate. The chloride-activated channel is also blocked by picrotoxin. The binding of GABA to the GABA-A receptor (which is linked to the chloride channel) can be blocked by bicuculline. There is also a GABA-B receptor for which baclofen is an agonist and that acts through mechanisms independent of chloride permeability changes. Voltage clamp analysis of chloride-activated channels and noise analysis of chloride-activated currents have indicated that the channel has a conductance of 20 pS (Gray and Johnston, 1985b; Segal and Barker, 1984).

### 3.4.3. Voltage-Activated Chloride Currents

Voltage-activated chloride currents were described in the 1960s in a few primitive cell types. Only recently have voltage-activated chloride currents been described in cultured CNS neurons (Owen et al., 1986). Chloride currents are activated by hyperpolarization of neurons and may contribute to the rectification properties of CNS neurons (e.g., Chesnoy-Marchais, 1982, 1983). These currents have been described using techniques to abolish all other currents; e.g., lowering extracellular sodium, replacing potassium with an impermeant ion such as cesium, and pharmacologically reducing calcium currents with divalent cations. There are no agents to inhibit the chloride current, so studies have examined currents with and without chloride present.

## 4. Conclusion

The advent of brain slice techniques and the evolution of pharmacological techniques to isolate ionic currents have immeasurably enhanced our knowledge of the electrophysiological properties of CNS neurons. It has been possible to describe the various ionic currents that are present in neurons. The challenge of the future is to integrate this recently acquired knowledge into a comprehensive picture of how individual neurons behave and how ensembles interact. In some studies, such as those in the inferior olive (Llinas and Yarom, 1981a,b) and the thalamus (Jahnsen and Llinas, 1984a, 1984b), it is possible to correlate the properties of calcium current with the generation of unique neuronal activity. In many cases, however, particularly with respect to potassium currents, the functional role of such a diverse multitude of conductances is a mystery. Computer modeling of electrophysiological properties of individual neurons (Traub, 1982; Traub and Llinas, 1979) and of large groups (Traub and Wong, 1985) will provide a conceptual framework for the future.

## Acknowledgments

We thank Rae Barolet for typing the manuscript. M.O'B. has an Alberta Heritage Foundation for Medical Research (AHFMR) student fellowship and B.A.M. is an AHFMR Scholar and a Sloan Fellow.

## References

Adams D. J., Smith S. J., and Thompson S. H. (1980) Ionic currents in molluscan soma. *Ann. Rev. Neurosci.* **3**, 141–167.

Adams P. R., Brown D. A., and Constanti A. (1982a) M-currents and other potassium currents in bullfrog sympathetic neurons. *J. Physiol.* **330**, 537–572.

Adams P. R., Constanti A., Brown D. A., and Clark R. B. (1982b) Intracellular $Ca^{2+}$ activates a fast voltage-sensitive $K^+$ current in vertebrate sympathetic neurons. *Nature* **296**, 746–749.

Alger B. E. and Nicoll R. A. (1980) Epileptiform burst after-hyperpolarization: Calcium-dependent potassium potential in hippocampal CA1 pyramidal cells. *Science* **210**, 1122–1124.

Alger B. E. and Nicoll R. A. (1982) Pharmacological evidence for two kinds of GABA receptor on rat hippocampal pyramidal cells studied *in vitro*. *J. Physiol.* **328**, 125–141.

Andrew R. D. and Dudek F. E. (1983) Burst discharge in mammalian neuroendocrine cells involves an intrinsic regenerative mechanism. *Science* **221**, 1050–1052.

Armstrong W. E. and Sladek C. D. (1982) Spontaneous "phasic-firing" in supraoptic neurons recorded from hypothalamo-neurohypophysiol explants *in vitro*. *Neuroendocrinology* **34**, 405–409.

Barish M. E. (1983) A transient calcium-dependent chloride current in the immature *Xenopus* oocyte. *J. Physiol.* **342**, 309–325.

Bartlett W. P. and Banker G. A. (1984) An electron microscopic study of the development of axons and dendrites by hippocampal neurons in culture. I. Cells which develop without intercellular contacts. *J. Neurosci.* **4**, 1944–1953.

Bird M. M. (1985) Establishment of synaptic connections between explants of embryonic neural tissue in culture: Experimental ultrastructural studies. *Exp. Brain Res.* **57**, 337–347.

Bourque C. W. and Renaud L. P. (1983) A perfused *in vitro* preparation of hypothalamus for electrophysiological studies on neurosecretory neurons. *J. Neurosci. Meth.* **7**, 203–214.

Brown D. A. and Griffith W. H. (1983) Calcium-activated outward current in voltage-clamped hippocampal neurones of the guinea-pig. *J. Physiol.* **372,** 221–244.

Chesnoy-Marchais D. (1982) A Cl⁻ conductance activated by hyperpolarization in *Aplysia* neurons. *Nature* **299,** 359–361.

Chesnoy-Marchais D. (1983) Characterization of a chloride conductance activated by hyperpolarization in *Aplysia* neurones. *J. Physiol.* **342,** 277–308.

Chun L. L. Y., Barres B. A., and Corey D. P. (1986) Induction of a calcium channel in astrocytes by cAMP. *Soc. Neurosci. Abst.* **12,** 1346.

Colquhoun D., Neher E., Reuter H., and Stevens C. F. (1981) Inward current channels activated by intracellular Ca in cultured cardiac cells. *Nature* **294,** 752–754.

Connors B. W. and Prince D. A. (1982) Effects of local anesthetic QX-314 on the membrane properties of hippocampal pyramidal neurons. *J. Pharmacol. Exp. Ther.* **200,** 476–481.

Connors B. W., Gutnick M. J., and Prince D. A. (1982) Electrophysiological properties of neocortical neurons *in vitro*. *J. Neurophysiol.* **48,** 1302–1320.

Connor T. A. and Stevens C. F. (1971) Voltage clamp studies of a transient outward membrane current in gastropod neural somata. *J. Physiol.* **213,** 21–30.

Constanti A. and Galvan M. (1983) M-current in voltage-clamped olfactory neurones. *Neurosci. Lett.* **39,** 65–70.

Cull-Candy S. G. and Ogden D. C. (1985) Ion channels activated by L-glutamate and GABA in cultured cerebellar neurons of the rat. *Proc. Roy. Soc. Lond. B* **224,** 367–373.

Dietzel I., Heinemann U., Hofmeier G., and Lux H. D. (1982) Stimulus-induced changes in extracellular Na⁺ and Cl⁻ concentration in relation to changes in the size of the extracellular space. *Exp. Brain Res.* **46,** 73–84.

Dingledine R., ed. (1984) *Brain Slices* Plenum, New York.

Dingledine R., Dodd J., and Kelly J. S. (1980) The *in vitro* brain slice as a useful neurophysiological preparation for intracellular recording. *J. Neurosci. Meth.* **2,** 323–362.

Djorup A., Jahnsen H., and Mosfeldt-Laursen A. (1981) The dendritic response to GABA in CA1 of the hippocampal slice. *Brain Res.* **219,** 196–201.

Eccles J. C. (1964) *The Physiology of Synapses* Springer-Verlag, New York, pp. 167–171.

Gahwiler B. H. (1981) Organotypic monolayer cultures of nervous tissue. *J. Neurosci. Meth.* **4,** 329–342.

Gahwiler B. H. (1984a) Development of the hippocampus *in vitro:* Cell types, synapses and receptors. *Neuroscience* **11,** 751–760.

Gahwiler B. H. (1984b) Slice cultures of cerebellar, hippocampal and hypothalamic tissue. *Experientia* **40**, 235–243.

Gahwiler B. H. and Brown D. A. (1985) Functional innervation of cultured hippocampal neurones by cholinergic afferents from co-cultured septal explants. *Nature* **313**, 577–579.

Gray R. and Johnston D. (1985a) Macroscopic calcium currents in acutely-exposed granule cells from adult hippocampus. *Soc. Neurosci. Abst.* **11**, 792.

Gray R. and Johnston D. (1985b) Rectification of single GABA-gated chloride channels in adult hippocampal neurons. *J. Neurophysiol.* **54**, 134–142.

Gray R. and Johnston D. (1986) Multiple types of calcium channels in acutely-exposed neurons from adult hippocampus. *Biophys. J.* **49**, 432a.

Gray P. T. A. and Ritchie J. M. (1986) A voltage-gated chloride conductance in rat cultured astrocytes. *Proc. Roy. Soc. Lond. B.* **228**, 267–288.

Griffith W. H., Brown T. H., and Johnston D. (1986) Voltage-clamp analysis of synaptic inhibition during long-term potentiation in hippocampus. *J. Neurophysiol.* **55**, 767–775.

Gustafsson B., Galvan M., Grafe P., and Wigstrom H. (1982) A transient outward current in a mammalian central neurone blocked by 4-aminopyridine. *Nature* **299**, 252–254.

Haas H. L., Schaerer B., and Vosmansky M. (1979) A simple perfusion chamber for the study of nervous tissue slices *in vitro*. *J. Neurosci. Meth.* **1**, 323–325.

Halliwell J. V. (1983) Caesium-loading reveals two distinct Ca-currents in voltage-clamped guinea-pig hippocampal neurones *in vitro*. *J. Physiol.* **341**, 10P.

Halliwell J. V. (1986) M-current in human neocortical neurones. *Neurosci. Lett.* **67**, 1–6.

Halliwell J. V. and Adams P. R. (1982) Voltage-clamp analysis of muscarinic excitation in hippocampal neurones. *Brain Res.* **250**, 71–92.

Hatton G. I., Doran A. D., Salm A. K., and Tweedle C. D. (1980) Brain slice preparation. Hypothalamus. *Brain Res. Bull.* **5**, 405–414.

Hermann A. and Gorman A. L. F. (1981a) Effects of 4-aminopyridine on potassium currents in a molluscan neuron. *J. Gen. Physiol.* **78**, 63–86.

Hermann A. and Gorman A. L. F. (1981b) Effects of tetraethylammonium on potassium currents in a molluscan neuron. *J. Gen. Physiol.* **78**, 87–110.

Herron C. E., Lester R. A. J., Coan E. J., and Collingridge G. L. (1986)

Frequency-dependent involvement of NMDA receptors in the hippocampus: A novel synaptic mechanism. *Nature* **322,** 265–268.

Higgins D. and Burton H. (1982) Electrotonic synapses are formed by fetal rat sympathetic neurons maintained in a chemically defined medium. *Neuroscience* **7,** 2241–2253.

Hodgkin A. L. and Huxley A. F. (1952a) The components of membrane conductance in the giant axon of Loligo. *J. Physiol.* **116,** 473–476.

Hodgkin A. L. and Huxley A. F. (1952b) A quantitative description of membrane current and its application to conduction and excitation in nerve. *J. Physiol.* **117,** 500–544.

Hotson J. R. and Prince D. A. (1980) A calcium-activated hyperpolarization follows repetitive firing in hippocampal neurons. *J. Neurophysiol.* **43,** 409–419.

Hotson J. R., Prince D. A., and Schwartzkroin P. A. (1979) Anomalous rectification in hippocampal neurons. *J. Neurophysiol.* **42,** 889–895.

Hounsgaard J. and Nicholson C. (1983) Potassium accumulation around individual Purkinje cells in cerebellar slices. *J. Physiol.* **340,** 359–388.

Jahnsen H. (1986) Extracellular activation and membrane conductance of neurones in the guinea-pig deep cerebellar nuclei *in vitro. J. Physiol.* **372,** 149–168.

Jahnsen H. and Laursen A. M. (1983) Brain Slices, in *Current Methods in Cellular Neurobiology* vol. 3 *Electrophysiological Techniques* (Barker J. L. and McKelvy J. F., eds.) John Wiley, New York.

Jahnsen H. and Llinas R. (1984) Electrophysiological properties of guinea-pig thalamic neurones: An *in vitro* study. *J. Physiol.* **344,** 205–226.

Jefferys J. G. R. and Haas H. L. (1982) Synchronized bursting of CA1 hippocampal pyramidal cells in the absence of synaptic transmission. *Nature* **300,** 448–450.

Kerkut G. A. and Wheal H. V., eds. (1981) *Electrophysiology of Isolated Mammalian CNS Preparation* Academic, New York.

Kessler J. A., Spray D. C., Saez J. C., and Bennett M. V. L. (1984) Determination of synaptic phenotype: Insulin and cAMP independently initiate development of electrotonic coupling between cultured sympathetic neurons. *Proc. Natl. Acad. Sci. USA* **81,** 6235–6239.

Kimura J., Noma A., and Irisawa H. (1986) Na-Ca exchange current in mammalian heart cells. *Nature* **319,** 596–587.

Konnerth A. and Heinemann U. (1983) Effects of GABA on presumed presynaptic $Ca^{2+}$ entry in hippocampal slices. *Brain Res.* **270,** 185–189.

Konnerth A., Heinemann U., and Yaari Y. (1984) Slow transmission of neural activity in hippocampal area CA1 in absence of active chemical synapses. *Nature* **307,** 69–71.

Kramer R. H. and Zucker R. S. (1985) Calcium-dependent inward current in *Aplysia* bursting pacemaker neurones. *J. Physiol.* **362**, 107–130.

Krnjevic K. (1980) Principles of Synaptic Transmission, in *Antiepileptic Drugs: Mechanisms of Action* (Glaser G. H., Penry J. K., Woodbury D. M., eds.) Raven, New York.

Lancaster B. and Adams P. R. (1986) Calcium-dependent current generating the afterhyperpolarization of hippocampal neurons. *J. Neurophysiol.* **55**, 1268–1282.

Lanthorn T., Storm J., and Andersen P. (1984) Current-to-frequency transduction in CA1 hippocampal pyramidal cells: Slow potentials dominate the primary range firing. *Exp. Brain Res.* **53**, 431–443.

Lemos J. R., Nordmann J. J., Cooke I. M., and Stuenkel E. L. (1986) Single channels and ionic currents in peptidergic nerve terminals. *Nature* **319**, 410–412.

Llinas R. and Hess R. (1976) Tetrodotoxin-resistant dendritic spikes in avian Purkinje cells. *Proc. Natl. Acad. Sci. USA,* **73**, 2520–2523.

Llinas R. and Sugimori M. (1978) Dendritic calcium spiking in mammalian Purkinje cells: *In vitro* study of its function and development. *Soc. Neurosci. Abst.* **4**, 66.

Llinas R. and Sugimori M. (1980a) Electrophysiological properties of in vitro Purkinje cell somata in mammalian cerebellar slices. *J. Physiol.* **305**, 171–195.

Llinas R. and Sugimori M. (1980b) Electrophysiological properties of *in vitro* Purkinje cell dentrites in mammalian cerebellar slices. *J. Physiol.* **305**, 197–213.

Llinas R. and Yarom Y. (1981a) Electrophysiology of mammalian inferior olivary neurones *in vitro*. Different types of voltage dependent ionic conductances. *J. Physiol.* **315**, 549–567.

Llinas R. and Yarom Y. (1981b) Properties and distribution of ionic conductances generating electroresponsiveness of mammalian inferior olivary neurones *in vitro*. *J. Physiol.* **315**, 569–584.

Llinas R., Sugimori M., and Walton K. (1977) Calcium dendritic spikes in the mammalian Purkinje cells. *Soc. Neurosci. Abst.* **3**, 58.

Llinas R., Yarom Y., and Sugimori M. (1981a) Isolated mammalian brain *in vitro:* New technique for analysis of electrical activity of neuronal circuitry function. *Fed. Proc.* **40**, 2240–2245.

Llinas R., Steinberg T. Z., and Walton K. (1981b) Presynaptic calcium currents in squid giant synapse. *Biophys. J.* **33**, 289–322.

MacDermott A. B. and Weight F. F. (1982) Action potential repolarization may involve a transient, $Ca^{2+}$-sensitive outward current in a vertebrate neuron. *Nature* **300**, 185–188.

MacDermott A. B., Mayer M. L., Westbrook G. L., Smith S. J., and Barker J. L. (1986) NMDA-receptor activation increases cytoplasmic calcium concentration in cultured spinal cord neurones. *Nature* **321**, 519–522.

MacVicar B. A. (1984a) Infrared video microscopy to visualize neurons in the in vitro brain slice preparation. *J. Neurosci. Meth.* **12**, 133–139.

MacVicar B. A. (1984b) Voltage dependent $Ca^{2+}$ channels in glial cells. *Science* **226**, 1345–1347.

MacVicar B. A. (1985) Depolarizing prepotentials are $Na^+$-dependent in CA1 pyramidal neurons. *Brain Res.* **333**, 378–381.

MacVicar B. A. (1986a) Modulation of calcium channels in astrocytes by neurotransmitters and cAMP. *Nature,* submitted.

MacVicar B. A. (1986b) Neurotransmitters modulate calcium channels in astrocytes. *Proc. IUPS* **16**, P520.17.

MacVicar B. A. and Dudek F. E. (1982) Electronic coupling between granule cells of the rat dentate gyrus: Physiological and anatomical evidence. *J. Neurophysiol.* **47**, 579–592.

MacVicar B. A. and Dudek F. E. (1980a) Local synaptic circuits in rat hippocampus. Interactions between pyramidal cells. *Brain Res.* **184**, 220–223.

MacVicar B. A. and Dudek F. E. (1980b) Dye-coupling between CA3 pyramidal cells of the rat hippocampus. *Brain Res.* **196**, 494–499.

MacVicar B. A. and Dudek F. E. (1981) Electrotonic coupling between pyramidal cells: A direct demonstration in rat hippocampal slices. *Science* **213**, 782–783.

MacVicar B. A., Baker K., Burnard D., and O'Beirne M. (1986) Intracellular calcium release regulates sodium-dependent slow depolarizations in rat hippocampal neurons. *J. Neurophysiol.,* submitted.

Madison D. V. and Nicoll R. A. (1986) Actions of noradrenaline recorded intracellularly in rat hippocampal CA1 pyramidal neurones, *in vitro. J. Physiol.* **337**, 221–244.

Maruyama Y. and Peterson O. H. (1982) Single channel currents in isolated patches of plasma membrane from basal surface of pancreatic acina. *Nature* **299**, 159–161.

Masukawa L. M. and Prince D. A. (1984) Synaptic control of excitability in isolated dendrites of hippocampal neurons. *J. Neurosci.* **4**, 217–227.

Mayer M. L. (1985) A calcium-activated chloride current generates the after-depolarization of rat sensory neurones in culture. *J. Physiol.* **364**, 217–239.

Mayer M. L., Westbrook G. L., and Guthrie P. B. (1984) 'Voltage-dependent' block by $Mg^{++}$ of NMDA responses in spinal cord neurones. *Nature* **309**, 261–263.

McCormick D. A. and Prince D. A. (1986) Acetylcholine induces burst firing in thalamic reticular neurons by activating a potassium conductance. *Nature* **319**, 402–405.

McIlwain H. (1961) Techniques in tissue metabolism. 5. Chopping and slicing tissue samples. *Biochem. J.* **27**, 213–218.

Mechmann S. and Pott L. (1986) Identification of Na-Ca exchange current in single cardiac myocytes. *Nature* **319**, 597–599.

Miledi R. (1982) A calcium-dependent transient outward current in *Xenopus laevis* oocytes. *Proc. Roy. Soc. B.* **215**, 491–497.

Monaghan D. T., Holets V. R., Toy D. W., and Cotman C. W. (1983) Anatomical distributions of four pharmacologically distinct $^{3}$H-L-glutamate binding sites. *Nature* **306**, 176–179.

Muhlethaler M. and Llinas R. (1984) Studies on the optimal condition for long-term survival of mammalian brain *in vitro. Soc. Neurosci. Abstr.* **10**, 31.

Narahashi T. (1974) Chemicals as tools in the study of excitable membrane. *Physiol. Rev.* **54**, 813–889.

Newberry N. R. and Nicoll R. A. (1984) Direct hyperpolarizing action of baclofen on hippocampal pyramidal cells. *Nature* **308**, 450–452.

Newman E. A. (1985) Voltage-dependent calcium and potassium channels in retinal glial cells. *Nature* **317**, 809–811.

Nicholl R. A. and Alger B. E. (1981) A simple chamber for recording from submerged brain slices. *J. Neurosci. Meth.* **4**, 153–156.

Nicholson C. (1980) *Dynamics of the Brain Cell Microenvironment* NRP Bulletin Vol. 18, MIT Press, Boston, Massachusetts, pp. 177–322.

Nicholson C. and Kraig R. P. (1981) The Behaviour of Extracellular Ions During Depression, in *The Application of Ion-Selective Microelectrodes* (Zeuther T., ed.) Elsevier, Amsterdam.

Nowak L., Bregestovski P., Ascher P., Herbert A., and Prochiantz A. (1984) Magnesium gates glutamate-activated channels in mouse central neurones. *Nature* **307**, 462–465.

Nowycky M. C., Fox A. P., and Tsien R. W. (1985) Three types of neuronal calcium channels with different calcium agonist sensitivity. *Nature* **316**, 440–443.

Owen D. G., Segal M., and Barker J. L. (1984) A Ca-dependent Cl⁻ conductance in cultured mouse spinal neurones. *Nature* **311**, 567–570.

Owen D. G., Segal M., and Barker J. L. (1986) Voltage-clamp analysis of a $Ca^{2+}$- and voltage-dependent chloride conductance in cultured mouse spinal neurons. *J. Neurophysiol.* **55**, 1115–1135.

Pumain R. and Heinemann U. (1985) Stimulus- and amino acid-induced calcium and potassium changes in rat neocortex. *J. Neurophysiol.* **53**, 1–16.

Quandt F. N. and MacVicar B. A. (1986) Calcium activated potassium channels in cultured astrocytes. *Neuroscience* **19**, 29–41.

Quandt F. N. and Narahashi T. (1984) Isolation and kinetic analysis of inward currents in neuroblastoma cells. *Neuroscience* **13**, 244–262.

Requena T. (1983) Calcium transport and regulation in nerve fibers. *Ann. Rev. Biophys. Bioeng.* **12,** 237–257.

Schubert P., Heinermann U., and Kolb R. (1986) Differential effect of adenosine on pre- and postsynaptic calcium fluxes. *Brain Res.* **376,** 382–386.

Schwartzkroin P. A. (1975) Characteristics of CA1 neurons recorded intracellularly in the hippocampal *in vitro* slice preparation. *Brain Res.* **85,** 423–436.

Schwartzkroin P. A. and Slawsky M. (1977) Probable calcium spikes in hippocampal neurons. *Brain Res.* **135,** 157–161.

Segal M. and Barker J. L. (1984) Rat hippocampal neurons in culture: Properties of GABA-activated Cl⁻ conductance. *J. Neurophysiol.* **51,** 500–515.

Stafstrom C. F., Schwindt P. C., Chubb M. C., and Crill W. E. (1985) Properties of persistent sodium conductance and calcium conductance of layer V neurons from cat sensorimotor cortex *in vitro. J. Neurophysiol.* **53,** 153–170.

Stafstrom C. F., Schwindt P. C., and Crill W. E. (1982) Negative slope conductance due to a persistent subthreshold sodium current in cat neocortical neurons *in vitro. Brain Res.* **236,** 221–226.

Sugimori M. and Llinas R. (1983) Voltage clamping of Purkinje cells in vitro: A study in guinea pig cerebellar slices. *Soc. Neurosci. Abst.* **9,** 681.

Swandulla A. D. and Lux H. D. (1985) Activation of a nonspecific cation conductance by intracellular $Ca^{2+}$ elevation in bursting pacemaker neurons of *Helix pomatia. J. Neurophysiol.* **54,** 1430–1443.

Taylor C. P. and Dudek F. E. (1982) Synchronous neural afterdischarges in rat hippocampal slices during blockade of chemical synapses. *Science* **218,** 810–812.

Thompson S. H. (1977) Three pharmacologically distinct, potassium channels in molluscan neurones. *J. Physiol.* **265,** 465–488.

Thompson S. M. and Prince D. A. (1986) Activation of electrogenic sodium pump in hippocampal CA1 neurons following glutamate-induced depolarization. *J. Neurophysiol.* **56,** 507–522.

Traub R. D. and Llinas R. (1979) Hippocampal pyramidal cells: Significance of dendritic ionic conductances for neuronal function and epileptogenesis. *J. Neurophysiol.* **42,** 476–496.

Traub R. D. and Wong R. K. S. (1985) Synchronized burst discharge in disinhibited hippocampal slice. II. Model of cellular mechanism. *J. Neurophysiol.* **49,** 459–471.

Van Harreveld A. (1978) Two mechanisms for spreading depression in the chicken retina. *J. Neurobiol.* **9,** 419–431.

Walz W. and Hertz L. (1983) Functional interactions between neurons and astrocytes. II. Potassium homeostasis at the cellular level. *Prog. Neurobiol.* **20**, 133–183.

White W. F., Nadler J. V., and Cotman C. W. (1978) A perfusion chamber for the study of CNS physiology and pharmacology *in vitro*. *Brain Res.* **152**, 591–596.

Wolinsky E. J., Patterson P. H., and Willard A. L. (1985) Insulin promotes electrical coupling between cultured sympathetic neurons. *J. Neurosci.* **5**, 1675–1679.

Wong R. K. S., Prince D. A., and Basbaum A. I. (1979) Intradendritic recordings from hippocampal neurons. *Proc. Natl. Acad. Sci. USA* **76**, 986–990.

Yamamoto C. and Chujo T. (1978) Visualization of central neurons and recording of action potentials. *Exp. Brain Res.* **31**, 299–301.

Yamamoto C. and McIlwain H. (1966) Electrical activities in thin sections from the mammalian brain maintained in chemically defined media *in vitro*. *J. Neurochem.* **13**, 1333–1343.

Yarom Y., Sugimori M., and Llinas R. (1986) Ionic currents and firing patterns of mammalian vagal motoneurons in vitro. *Neuroscience* **16**, 719–737.

Yau K.-W. and Nakatani K. (1984) Electrogenic Na-Ca exchange in retinal rod outer segment. *Nature* **311**, 661–663.

Yellen G. (1982) Single $Ca^{2+}$-activated non-selective cation channels in neuroblastoma. *Nature* **307**, 462–465.

Zbicz K. L. and Weight F. F. (1985) Transient voltage and calcium-dependent outward currents in hippocampal CA3 pyramidal neurons. *J. Neurophysiol.* **53**, 1038–1058.

# Calcium Ions

## R. Pumain

## 1. Introduction

The involvement of calcium (Ca) in living organisms has long been recognized, particularly its presence in invertebrate shells and vertebrate bones. It was not, however, before the studies of S. Ringer (1882), who showed that its presence was necessary for normal heart beating, that its role in physiological processes was assessed. Since that time, it has become more and more apparent that the number of cellular functions that are triggered, regulated, or otherwise influenced by calcium ions is quite large, ranging from activation of a whole palette of enzymes to playing a pivotal role as an intracellular messenger in the physiology of excitable cells (Carafoli and Crompton, 1978; Campbell, 1983).

Calcium is not evenly distributed within nervous tissue. In the mammalian central nervous system, the free $Ca^{2+}$ concentration is, in the extracellular space (which represents about 20% of the total volume; Nicholson et al., 1979), between 1.2 and 1.3 m$M$ (Ames et al., 1964; Heinemann et al., 1977; Nicholson et al., 1978; Pumain and Heinemann, 1985), whereas in the cytoplasm of quiescent cells, it is around $10^{-7}$ $M$. Yet, the total quantity of cellular calcium may be as high as 1 m$M$ (Kretsinger, 1979); therefore, calcium is distributed unevenly in the various cellular components. Indeed, calcium is present in different states within living tissues, namely, stored calcium, bound calcium, and free ionized calcium.

Calcium ions, like other divalent cations, have a screening effect on the negative surface charges of cell membranes, and decreases in extracellular $Ca^{2+}$ enhance the excitability of nerve cell membranes. Such decreases may also influence the efficiency of chemical synaptic transmission, which is reduced by about 80% when extracellular $Ca^{2+}$ falls to a level of 0.6 m$M$ and is totally abolished at levels below 0.2 m$M$ (Dingledine and Somjen, 1981). Indeed, it has been shown that very large extracellular free calcium decreases occur during synchronized neuronal activities such as epilepsy (Heinemann et al., 1977; Pumain et al., 1985) or spreading depression (Kraig and Nicholson, 1978), as well as during anoxia

(Hansen, 1985) or applications of excitatory amino acid transmitters (Heinemann and Pumain, 1980; Pumain and Heinemann, 1985).

The extremely low value for free $Ca^{2+}$ in the intracellular compartment results partly from $Ca^{2+}$ binding to various ligands, namely macromolecules attached to membranes (sialic acid residues, acidic phospholipids), cytosolic proteins, and various small organic molecules (citrate, glutamate, and so on). The affinity constants of $Ca^{2+}$ for such ligands are, however, not such that they could account for the low levels of free $Ca^{2+}$ observed in cells (the pKd values of these ligands is usually larger than 7.0). Some other mechanisms must therefore be involved in the regulation of intracellular free $Ca^{2+}$. It has been shown that sequestration of $Ca^{2+}$ occurs in intracellular organelles and, for instance, endoplasmic reticulum and mitochondria do possess active uptake systems for $Ca^{2+}$ (Carafoli, 1974). Further mechanisms of regulation involve transport systems through the plasma membrane, such as the $Na^+/Ca^{2+}$ exchange system or extrusion of $Ca^{2+}$ by a $Ca^{2+}$-dependent ATPase.

The very large difference in concentration of free $Ca^{2+}$ between extra- and intracellular compartments (nearly four orders of magnitude) results in a very large driving force for $Ca^{2+}$ across cellular membranes, of about $+120$ mV. Any increase in the permeability for calcium of cytoplasmic membranes will therefore give rise to a large intracellular influx of calcium ions.

Indeed, it has been shown that central neurons possess voltage-dependent $Ca^{2+}$ channels through which a depolarizing inward $Ca^{2+}$ current may flow. At least two different currents have been described. The first is a high-threshold current that is activated when the membrane is depolarized at around $-30$ mV and is maximum at about $+20$ mV (Bossu et al., 1985; Brown et al., 1984); this current is slowly inactivated and blocked by divalent cations such as $Ni^{2+}$, $Co^{2+}$, or $Mn^{2+}$, and by organic antagonists of the verapamil series (Boll and Lux, 1985). The second one is a transient current that is largely inactivated at rest and cannot be activated through depolarization (Carbone and Lux, 1984), but can be relieved from inactivation by hyperpolarization and is not blocked by verapamil. Such currents can induce endogenous rhythmic activity in neurons (Jahnsen and Llinas, 1984).

Calcium ions may likewise enter neurons through transmitter-activated channels. It has been shown that the excitatory amino acid transmitter promotes a depolarizing inward current, which is partially carried by $Ca^{2+}$ ions (Pumain and Heinemann, 1985;

MacDermott et al., 1986), through channels activated specifically by the glutamate agonist N-methyl-D-aspartate (Pumain et al., in press). Such receptors have been considered to play a key role in chronic events such as long-term potentiation and epilepsy (Harris et al., 1984; Meldrum et al., 1983; Pumain et al., 1985).

The intracellular $Ca^{2+}$ accumulation resulting from transmembrane $Ca^{2+}$ currents or release from intracellular stores can produce cascades of metabolic effects (Rasmussen and Barrett, 1984) and also activate $Ca^{2+}$-dependent membrane currents, such as the repolarizing $Ca^{2+}$-dependent $K^+$ current (Krnjevic and Lisiewicz, 1972; Meech, 1978). Because of the high buffering capacity of the cytoplasm, however, only a small fraction of the $Ca^{2+}$ ions that penetrates the cells accumulates as free $Ca^{2+}$. For example, it was estimated that only 2% of the $Ca^{2+}$ ions entering Aplysia neurons during the initial part of a long voltage-clamp depolarizing command contributed to the resulting increase in cytoplasmic free $Ca^{2+}$ concentration (Gorman and Thomas, 1980).

It is clear, from this far-from-complete list of events in which $Ca^{2+}$ is implicated, that a thorough understanding of these issues depends critically on the possibility of measuring accurately extra- as well as intracellular free $Ca^{2+}$ concentrations and to monitor possible alterations in $Ca^{2+}$. For many years, however, estimations of baseline $Ca^{2+}$ and of possible activity-induced extra- and intracellular changes in $Ca^{2+}$ concentration were based on indirect methods, such as relating cell metabolism to total $Ca^{2+}$ content, tracer fluxes, membrane currents, alterations of extracellular $Ca^{2+}$ concentrations, permeabilization of membranes by ionophores, and the use of pharmacological agents (*see* Thomas, 1982). About 20 years ago there appeared the first techniques allowing for direct measurements of free $Ca^{2+}$ concentrations with metallochromic indicators (Jobsis and O'Connor, 1966) and photoproteins (Ridgway and Ashley, 1967). The growing interest in the role of $Ca^{2+}$ in cellular functioning promoted a wealth of research aimed at improving already existing techniques and developing new methods. It ensues that $Ca^{2+}$ ions, despite the drawbacks linked with the fact that their intracellular free concentration is very low, are presently the ions for which techniques of measurement are the most elaborate.

Total $Ca^{2+}$ quantity can be measured using many techniques, including flame photometry, atomic absorption spectrometry, or X-ray microanalysis, but such techniques will not be considered here. This chapter is devoted to the various methods currently used for measuring concentration of free $Ca^{2+}$ in biological preparations; it is divided into five parts, concerning, respectively, the photoluminescent indicators, the metallochromic indicators, the fluorescent indicators, the $Ca^{2+}$ selective microelectrodes, and a few other methods that are less popular.

## 1.1. Free Concentration or Activity?

The parameter that is measured by most of the methods presented in this review is really the activity of $Ca^{2+}$ ions instead of their free concentration. Indeed, in aqueous solutions, the electrostatic interactions between the various ions lower the free energy of each, and, at high ionic strengths, the activity of the ions, which is a function of the ionic strength, may be appreciably different from the corresponding free concentrations. It is therefore tempting to calculate the activity coefficients by means of the semi-empirical equations derived from the original Debye-Huckel formula. There is no general agreement, however, on how to compute single ion activity coefficients, and uncertainties exist concerning the determination of the activity coefficients in poly-ionic solutions, and even more in biological fluids. Therefore, since the values of ion concentrations are obtained by comparing the signals measured during the experiments to those obtained from calibration solutions in which ion concentrations instead of activities are known, and further since the ionic strengths of biological fluids are usually approximately constant, it appears advisable to refer to free concentrations instead of activities. Further, the comparison of results from various laboratories is then facilitated (*see* reviews by Mohan and Bates, 1975; Blinks et al., 1982; Meier, 1982; Bates et al., 1983; Tsien, 1983; Ammann, 1986).

## 2. Photoproteins

The bioluminescence of living luminous organisms has long attracted attention, but it was not before the studies of Dubois on the firefly and on the bivalve Pholas at the end of the nineteenth century that the chemical steps involved in the reaction were disclosed. He showed that the reaction involved a heat-stable factor that he named luciferine, which, when combined with a heat-labile factor, the luciferase (now known as an enzyme) in the presence of oxygen, gave rise to light emission (Dubois, 1887). In contrast, the luminescence produced by coelenterates was observed in the absence of oxygen, and the underlying mechanisms were a mystery to biologists for more than 70 years after the work of Dubois. This mystery was resolved by the studies of Shimomura et al. (1962, 1963). They show that a protein extracted from the hydrozoan jelly fish *Aequora forskalea* could be stimulated to emit light by the addition of $Ca^{2+}$, without requiring oxygen or ATP. Similar substances were subsequently extracted from various coelenterates, for instance the hydroid Obelia, and it was proposed that such precharged luminescent proteins should be called photoproteins (Shimomura and Johnson, 1966). Although the purified protein emits a blue light, it is associated, in many parent organisms, with a green fluorescent protein that converts the emitted light into a green light.

## 2.1. Principles

### 2.1.1. Origin of Luminescence

The binding of $Ca^{2+}$ does not appear to directly provide the energy required for the light emission, but rather activates conformational changes of the binding sites of the prosthetic group. This allows oxidation of the chromophore by bound oxygen and subsequent emission of photons while $CO_2$ molecules are produced (Fig. 1). Native aequorin exhibits a slight fluorescence around 340 nm and emits an intense blue fluorescence (at a wavelength of 469 nm) on the continued binding of $Ca^{2+}$ ions. When the $Ca^{2+}$ is removed, the oxidized chromophore dissociates from the protein. In spite of impressive work (Campbell et al., 1979; Shimomura and Johnson, 1979; Prendergast, 1982), not all the steps involved in the production of the $Ca^{2+}$-triggered luminescence are yet fully understood. One molecule of aequorin needs to react with two $Ca^{2+}$ ions to produce luminescence, and the quantum yield of the reaction, that is, the number of photons emitted per number of reacting molecules, is 0.15 at 25°C (Shimomura and Johnson, 1970). Once having reacted, the molecules of photoproteins are inactivated and cannot usually be regenerated. However, regeneration of the protein can be achieved by incubation of the apoprotein in the presence of coelenterazine and molecular oxygen (Fig. 1) (Shimomura and Johnson, 1975).

### 2.1.2. Relationship Between Concentration of Free $Ca^{2+}$ and Luminescence

The intensity of the emitted light increases by more than one million-fold when free $Ca^{2+}$ is augmented from $10^{-7}M$ to $10^{-4}M$. Therefore the relationship between luminescence and free $Ca^{2+}$ is

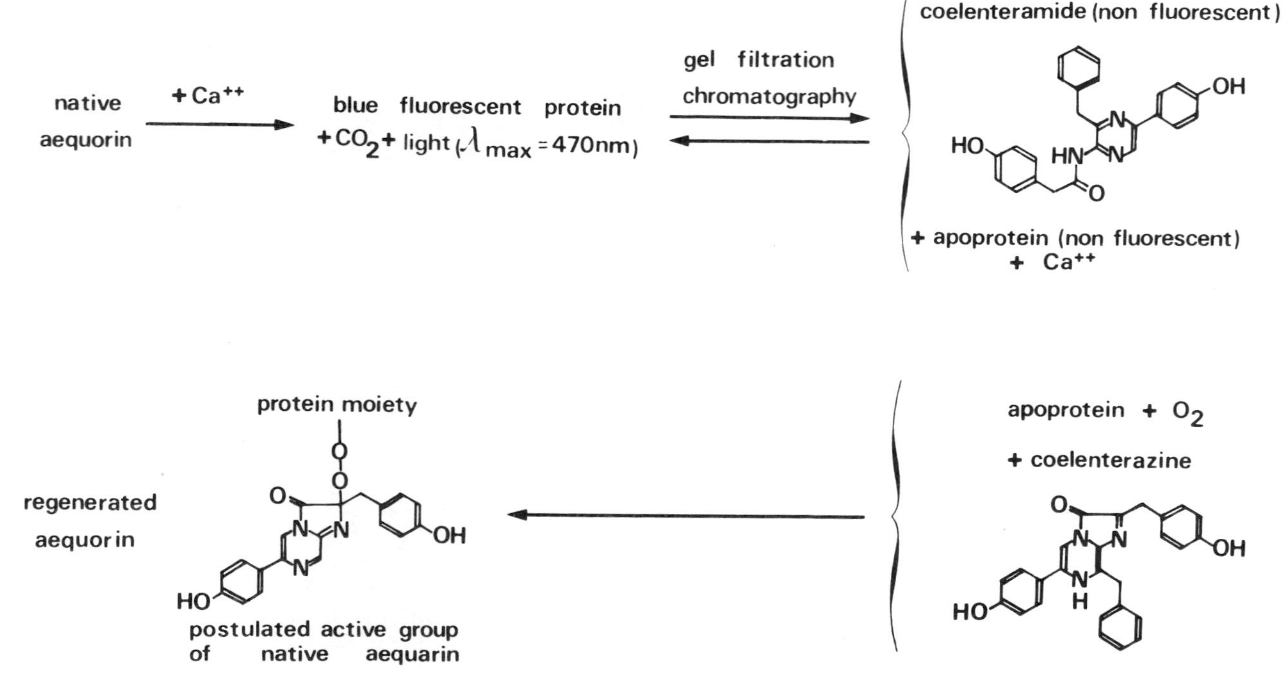




Fig. 1. Schematic drawing of the main steps of the light-emitting reaction between aequorin and $Ca^{2+}$.

often given as the logarithm of the ratio of the peak light intensity $(L)$ to the peak light intensity obtained with a saturating free $Ca^{2+}$ concentration $(L_{max})$ versus the logarithm of the free $Ca^{2+}$ concentration (Fig. 2). The detection limit is between $10^{-7}$ and $10^{-8}M$. It is dependent upon the level of a residual $Ca^{2+}$-independent luminescence, which can be influenced by temperature (it is about 70 times larger at 40°C than at 0°C) (Fig. 2A) and by the presence of cations such as $K^+$ and $Mg^{2+}$ (Fig. 2B and D) (Allen et al., 1977). Above the $Ca^{2+}$ concentration corresponding to the detection limit, light intensity increases until reaching saturation for concentrations between $10^{-4}$ and $10^{-3}M$, and the resulting graph in double logarithmic coordinates is a sigmoid curve (Fig. 2). The maximum slope of the linear part of the curve is between 2.5 and 3 (Allen et al., 1977; Cobbold, 1980). Such a steep power relationship makes it sometimes difficult to precisely determine the absolute values of free $Ca^{2+}$ concentrations, but ensures a high sensitivity for values between $0.5 \times 10^{-6}M$ and $10^{-5}M$. Another consequence of such a power law is that the measurements of $Ca^{2+}$ will be biased in favor of the highest $Ca^{2+}$ concentrations.

The sensitivity of the binding reaction may be influenced by cations that do not by themselves initiate light emission. Increasing $K^+$ produces a reduction of the sensitivity of the reaction, although large concentrations are required to induce a noticeable effect (Fig. 2B). Although pH changes in the physiological range have only a slight effect (Fig. 2C), the influence of $Mg^{2+}$ ions cannot be neglected because they compete with $Ca^{2+}$ for binding to the protein, and therefore millimolar concentrations of $Mg^{2+}$ (i.e., in the physiological range) can decrease the intensity of the $Ca^{2+}$-induced aequorin response (Fig. 2D) (Allen and Blinks, 1979).

## 2.2. Properties of Photoproteins

### 2.2.1. Structural Characteristics

It is generally agreed that the molecular weight of aequorin is about 20,000 (Shimomura and Johnson, 1979). The protein is negatively charged at physiological pH, and about 20% of its amino acids are either glutamate or aspartate (Shimonura and Johnson, 1969; Prendergast and Mann, 1978). The native protein contains two components, an apoprotein and a prosthetic group that is a low molecular weight chromophore. This compound is oxidized into coelenteramide in the course of the luminescence process, when the protein is mixed with $Ca^{2+}$ (see above).

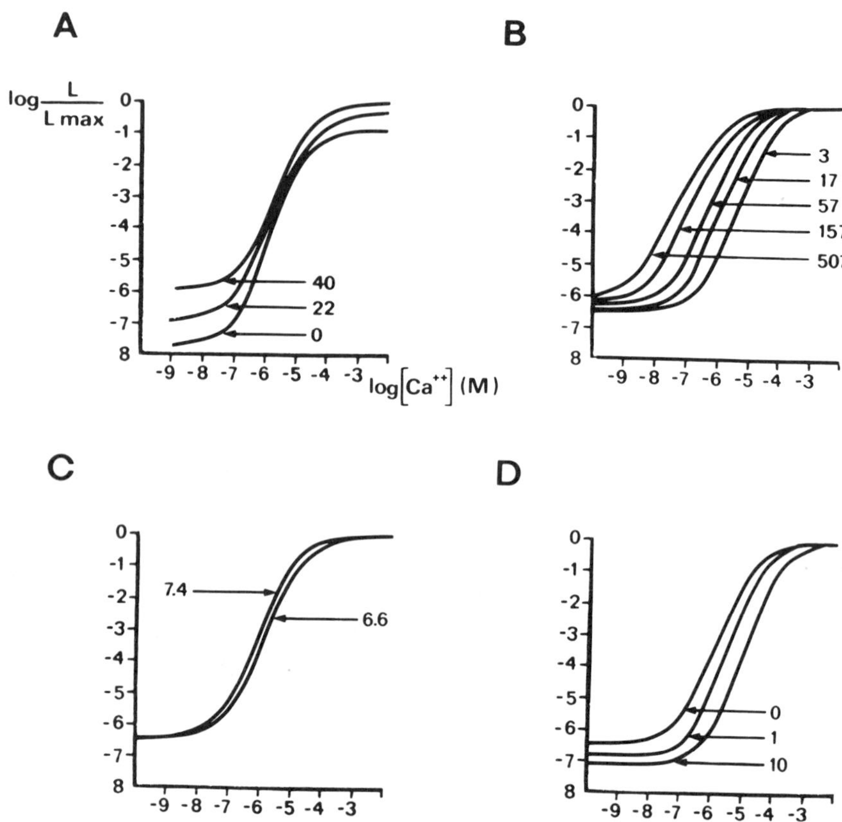



Fig. 2. Influence of physical conditions on the aequorin $Ca^{2+}$-induced light emission. (A) Effect of temperature: curves determined at 0, 22, and 40°C in solutions containing 150 m$M$ KCl, 5 m$M$ PIPES (pH 7.0). The $Ca^{2+}$ buffer was 1 m$M$ EGTA. (B) Effect of increasing concentrations of $K^+$ (at 3, 17, 57, 157, and 507 m$M$). All the solutions contained 5 m$M$ PIPES (pH 7.0), at 21°C. (At 3 m$M$ KCl, the PIPES concentration was 1 m$M$.) (C) Effect of pH: curves determined in solutions containing 150 m$M$ KCl, 5 m$M$ PIPES (pH 6.6 and 7.4), at 21°C. The $Ca^{2+}$ buffers were both EGTA (1 m$M$) and CDTA (1 m$M$). (D) Effect of increasing concentrations of $Mg^{2+}$ (without $Mg^{2+}$, 1 and 10 m$M$). The solutions contained 150 m$M$ KCl and 5 m$M$ PIPES (pH 7.0), at 21°C. The $Ca^{2+}$ buffer was 1 m$M$ EGTA. Modified and redrawn from Allen and Blinks (1979).

### 2.2.2. Selectivity

Calcium ions are not the only cations that are able to induce light emission: lanthanides are more effective than $Ca^{2+}$, whereas $Sr^{2+}$ and $Ba^{2+}$ are less potent. These cations, however, are not likely to interfere with measurements in biological samples. As mentioned above, $Mg^{2+}$ ions do not initiate any light emission.

### 2.2.3. Detection and Measurement of Luminescence

The $Ca^{2+}$-induced luminescence is initially directly proportional to the number of chromophore molecules that have reacted, to the quantum yield, and to the rate constant of the reaction (about 1.5/s for aequorin). It subsequently decays following an exponential function. The signal emitted from the preparation is measured by a photomultiplier located such that the photocathode is just above the fluid surface or connected to the preparation via a fiber optic system whose tip may be immersed in the perfusion fluid, and the measurements are performed against a dark background. For maximum sensitivity, high-gain, low dark-current photomultiplier tubes are preferred. The anode current is measured via a current-to-voltage converter and can be displayed directly on an oscilloscope or on a chart recorder. It is often convenient to record the total luminescence counts during a defined time interval and meanwhile store the transient signals on tapes for subsequent analysis (*see* Campbell et al., 1979). The signals recorded from photoproteins have an excellent long-term stability and, further, appear to be practically insensitive to movement artifacts, which makes them particularly suitable for studying muscle fibers (Hainaut et al., 1975).

Image intensifier techniques, which can provide unique information concerning the spatial distribution of $Ca^{2+}$, have been applied to photoprotein signals (Reynolds, 1979; Reynolds, 1980; Thomas, 1982).

### 2.2.4. Response Time

The kinetics of the $Ca^{2+}$-dependent light emission have been examined in detail for aequorin (Hasting et al., 1969; Löschen and Chance, 1971). The main observation was that upon a rapid step change in $Ca^{2+}$, the corresponding light intensity signal had a rise time of 6–10 ms and a fall time of about 10 ms at 20°C. Besides, it was not clear how the kinetics were dependent on the initial $Ca^{2+}$ concentration.

During actual experiments, however, aequorin did not appear to respond accurately to step $Ca^{2+}$ changes with rise times shorter than 15–20 ms (Stinnakre, 1981). The decay phases of light signals were approximately exponential, with half-decay time between 0.04 and 0.14 s. When $Ca^{2+}$ is injected by ionophoresis in a cell loaded with aequorin, the relationship between the ionophoretic current and the peak amplitude of light responses is linear (Kusano et al., 1975; Eusebi et al., 1985), in spite of the steep power law already mentioned. A likely explanation is that the free $Ca^{2+}$ concentration just around the pipet tip may have been high enough to saturate all the immediately available aequorin (Eusebi et al., 1985). Besides, several other factors such as increasing $Mg^{2+}$ concentration (Blinks et al., 1978) or ionic strength and decreasing temperature will delay the rate of photon emission.

## 2.3. Use of Photoproteins in Biological Experiments

### 2.3.1. Availability

The commercially available products have been, for a long time, of too low purity to be useful for intracellular measurements, but more highly purified products are now available. Most users of aequorin were initially given samples prepared by Drs. O. Shimomura and J. F. Johnson. A few researchers have prepared the proteins from the source animals, however, which are either small (Aequora) or very small (Obelia). The protein purification process is usually based on that initially designed by Shimomura and colleagues (1962). Several groups have since developed multistage purification schemes. The description of such methods is beyond the scope of this review, and the interested reader is referred to the following articles for further details: Hainaut et al. (1975); Blinks et al. (1978); Johnson and Shimomura (1978); Campbell et al. (1979).

### 2.3.2. Solutions

For use, the photoproteins are dissolved in a pH-buffered solution containing variable amounts of potassium, magnesium chloride, and calcium buffers (for instance, 5–50 $\mu M$ EDTA for a volume of 100 $\mu L$). The solution is then usually "desalted" using a Sephadex G-25 column.

### 2.3.3. Introduction into Cells

Photoproteins are too large to cross plasma membranes and have usually been incorporated into cells by microinjections. The protein is usually introduced into the tip of a clean glass micropipet and pressure-injected, since ionophoresis is not possible. Gas-filled ejection systems are advisable to avoid contamination of the photoprotein by $Ca^{2+}$. For small cells, it is necessary to use a fine tip micropipet (to cause minimal damage to the cell) connected to a precisely controlled pressure generator to avoid excess injection (Stinnakre, 1979) (between 1 pL and 1 nL, according to the size of the cell under study). Alternative methods for loading cells with photoproteins have been described, including liposome fusion to cytoplasmic membranes, use of aequorin-loaded red cell ghosts, or hypoosmotic shocks. Such methods have been thus far of limited use, however, for the study of nerve cells (Ashley and Campbell, 1979; Prendergast, 1982; Tsien, 1983).

### 2.3.4. Intracellular Diffusion

After injection, adequate time must be allowed for intracellular photoprotein diffusion to obtain a uniform distribution in the cytoplasm, a feature required to obtain quantitative information. In free solution, the reported diffusion coefficient for aequorin is about $7 \times 10^{-7}$ cm$^2$/s at 20°C (Campbell et al., 1979). The speed of diffusion in the cytoplasm is substantially lower, however; in muscle fibers, aequorin diffuses about eight times more slowly than in free solution, and the apparent diffusion coefficient is between $10^{-7}$ and $5 \times 10^{-8}$/s (Blinks et al., 1982). The photoproteins do not appear to enter mitochondria or sarcoplasmic reticulum, but penetration into nuclei of cells from insect salivary glands has been reported (*see* Blinks et al., 1982).

### 2.3.5. Intracellular Inactivation of Photoproteins

Photoproteins are inactivated in the course of the light-emission process, and the binding to $Ca^{2+}$ inside the cell could therefore significantly decrease the intracellular concentration of native protein during an experiment. It can be calculated, however, that for measurements of resting $Ca^{2+}$ levels, the quantity of photoprotein inactivated during the experiments is negligible in comparison to its total quantity within the cell. During stimulation, however, the intracellular free $Ca^{2+}$ increases to levels at which a significant proportion of the intracellular photoprotein might be inactivated (Blinks et al., 1978).

### 2.3.6. Calibration of Fluorescence Signals

Several calibration procedures have been used, each assuming a uniform distribution of free $Ca^{2+}$ within the cell. For resting level measurements, the method has been usually a "null point method": calcium buffers, at various known free $Ca^{2+}$ concentrations, are injected into aequorin-loaded cells to determine the buffer that just does not alter the resting luminescence (Baker et al., 1971; DiPolo et al., 1976). Such a procedure helps to eliminate the need to know the concentration of aequorin inside the cell, and, further, the calibration is performed *in situ*, in the actual ionic environment. The resting free $Ca^{2+}$ concentration must be high enough to give detectable signals, however. Besides, the various $Ca^{2+}$ buffers must mimic as closely as possible the ionic composition of the cytoplasm, since pH and $K^+$ and $Mg^{2+}$ ions can influence the binding of $Ca^{2+}$ to EGTA.

A different approach depends on comparing the light intensity at rest or during activity, with the signal measured subsequently when exposing the intracellular aequorin to a saturating concentration of $Ca^{2+}$, the result obtained by making the plasma membrane permeable to $Ca^{2+}$. An in vitro calibration is first performed with calibrating solutions as close as possible to the ionic content of the cytoplasm, in order to determine the corresponding double-logarithmic plot of the relationship between luminescence and free $Ca^{2+}$ (Fig. 2). The saturation peak light intensity ($L_{max}$) is measured at the end of the experiment by adding a detergent such as Triton X-100 to the perfusion medium, a procedure that rapidly destroys cell membranes and exposes the intracellular aequorin to a saturating quantity of $Ca^{2+}$. The time course of such a light signal is slower than those of signals observed during the in vitro calibration procedures, but the total amount of light collected by the photomultiplier can be measured, and this value allows one to determine the saturation peak light intensity (Allen and Blinks, 1979).

For in vitro calibrations, a light-tight chamber in which fast and reproducible mixing of products can be performed is necessary. Although adequate mixing times around 0.1 s can be achieved using simple injectors, more sophisticated devices have been designed for performing detailed kinetic studies (mixing times as low as 1 ms) (Hastings et al., 1969).


### 2.3.7. Possible Side Effects

The intracellular concentrations of photoproteins obtained after injection appear not to be such as to markedly disturb the cytosolic $Ca^{2+}$ balance, even in the smallest cells, and the physiological cellular responses are not affected by the protein (Ashley and Ridgway, 1970; Baker et al., 1971; Ashley and Campbell, 1978).

# 3. Metallochromic Indicators

Metallochromic indicators are substances that undergo changes in color and absorbance when binding to metal ions. For several such compounds, the difference of absorption spectra (i.e., the variations of absorbance as a function of the wavelengths of the incident light) between the free form and the form bound to $Ca^{2+}$ is large and can therefore be used for quantitative measurements of the concentration of free $Ca^{2+}$. The first report of such measurement in biological experiments was published by Ohnishi and Ebashi in 1963, and since then a number of different compounds have been proposed for the same purpose. The most widely used have been the bi-azo dyes murexide, arsenazo III, and antipyrylazo III (Fig. 3).

## 3.1. Principles

### 3.1.1. Absorbance

The absorbance is defined as the negative logarithm of the ratio of the intensity of the transmitted light to that of the incident light. Therefore, when measuring absorbance changes of bi-azo dyes in a biological sample, the total absorbance will be the sum of the natural absorbance of the sample, that of the free dye, and that of the bound dye. Since the two last terms of the sum are small in comparison to the first, the significant signal will be only a very small fraction of the total signal, often close to the detection limits of the recording system.

It ensues further that the natural absorbance of the biological sample has to be determined prior to the application of the dye, and its spectral dependence (see below) determined. Most cellular pigments absorb at wavelengths shorter than 400 nm, however, and measurements performed at longer wavelengths (600–800 nm) are therefore preferable (Scarpa, 1979).

### 3.1.2. Differential Measurements

According to the Lambert-Beer law, the free $Ca^{2+}$ concentration can be determined from measurements of absolute absorption values if the dissociation constant of the dye, the light path length, the molar extinction coefficients (i.e., the absorbance of a solution of $1M$ of the dye, with a light path length of 1 cm), and the dye concentration are known. The intracellular concentration of the indicator is often difficult to assess, however, and an error on this

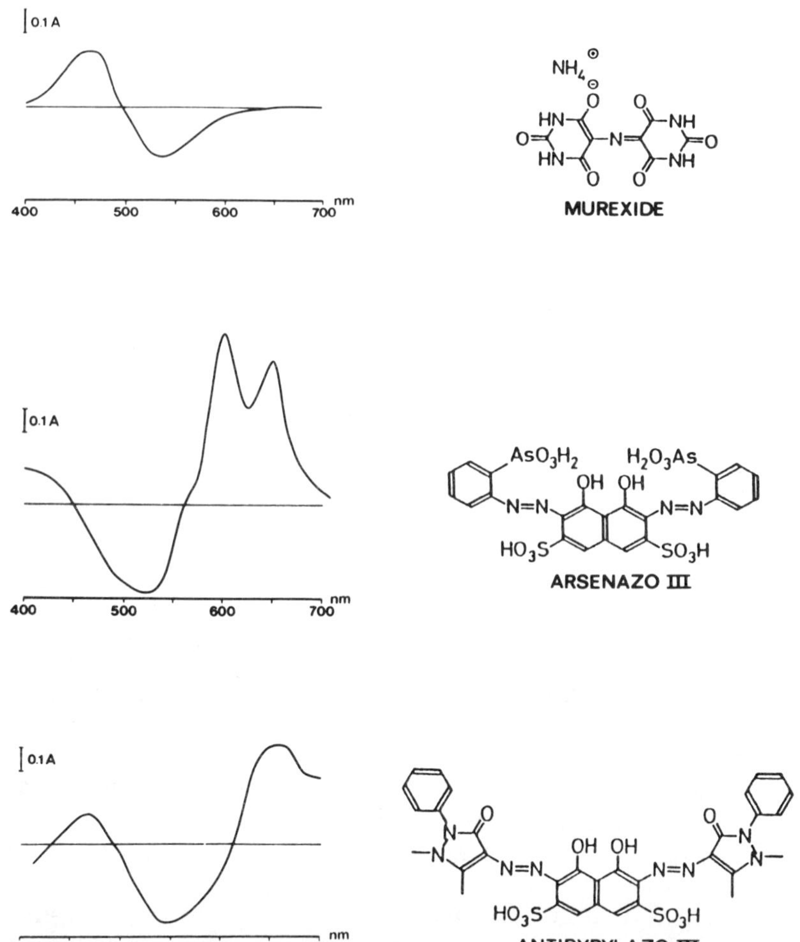


Fig. 3. (Left) Chemical structures of various bi-azo indicators. (Right) Difference absorption spectra between the forms bound to $Ca^{2+}$ and the free forms, corresponding to the various indicators. For murexide, the solution contained 100 m$M$ KCl, 20 m$M$ MOPS, and 50 μ$M$ murexide; 500 μ$M$ $Ca^{2+}$ was subsequently added. For arsenazo III, the solution contained 500 m$M$ KCl, 10 n$M$ MOPS (pH 7.4), and 100 μ$M$ arsenazo. 200 μ$M$ $Ca^{2+}$ was subsequently added. For antipyrylazo III, the solution contained 500 m$M$ KCl, 20 m$M$ HEPES (pH 7.0), and 50 μ$M$ antipyrylazo. 500 μ$M$ $Ca^{2+}$ was subsequently added (modified and redrawn from Scarpa, 1979).

parameter can be misleading, especially if there are uncertainties about the stoichiometry of the reaction (see below).

The absorbance of the free and the $Ca^{2+}$-bound indicator varies with the wavelength (Fig. 3) and, in principle, the concentration changes should be measured using the wavelength for which the difference in absorbance is maximum. It is advantageous to perform the measurements at two different wavelengths, however, assuming that the free and bound form are in equilibrium, in order to minimize possible nonspecific absorbance changes. Such nonspecific changes may arise from light scattering (corresponding to the light detected by the recording system, which have not actually passed through the sample), instability of light sources, change in volume or in refractivity indices occurring simultaneously with the specific absorbance changes, or change in absorbance of pigments present in the biological material. Besides, by choosing adequate wavelengths, differential measurements can help to minimize possible contributions of interfering ions to the observed absorbance changes (see below). Such measurements can be achieved by subtracting the absorbance of the indicator at the wavelength for which the change upon binding to $Ca^{2+}$ is maximum, from the signal corresponding to an isosbestic wavelength, i.e., a wavelength for which the absorbance for the free and bound forms are identical. Alternatively, two neighboring wavelength pairs corresponding, respectively, to a high and low absorbance state may be chosen.

### 3.1.3. Linearity of Response

Absorbance will be linearly related to the free $Ca^{2+}$ concentration when the ratio of the concentration of bound indicator to that of free $Ca^{2+}$ remains constant. The relationship between the concentration of the bound dye to that of free $Ca^{2+}$ is approximately linear over a certain range of free $Ca^{2+}$ concentrations, and errors may therefore arise from measurements performed outside this range. Besides, the position of the linear part of the curve is independent of the dye concentration when the stoichiometry of the binding reaction is 1 to 1, but not when it is 1 to 2. Possible intracellular free $Ca^{2+}$ gradients and nonhomogeneous distribution of the intracellular dye will therefore make the interpretation of results more complicated.

## 3.2. Properties of Bi-azo Indicators

The main characteristics of bi-azo indicators are indicated in Table 1.

Table 1<br>
Properties of Bi-azo Indicators[a]

|  | Murexide | Arsenazo III | Antipyrylazo III |
|---|---|---|---|
| Molecular weight | 284 | 776 | 746 |
| Water solubility, mM | 20 | 20 | 20 |
| Affinity constants for calcium ions | ca. 2 mM | ca. 30 $\mu M$ | ca. 150 $\mu M$ |
| Wavelength of maximal differencial absorbance, nm | 470 | 595 and 658 | 630 |
| Pairs of wavelength optimal for measurements | 507–540 | 675–685; 650–685 | 670–690; 720–790 |
| Maximal differences in molar extinction coefficient (free dye vs. dye in saturated $Ca^{2+}$ solution) | 6000 | 25000 | 7000 |
| Optimal working range, $M$ | $10^{-3}$ to $10^{-4}$ | $10^{-6}$ to $10^{-7}$ | $10^{-5}$ to $10^{-6}$ |
| Relaxation times (temperature jump experiments) | ca. 2 $\mu s$ | 2.8 ms | 180 ms |
| Interference from $Mg^{2+}$ | – | – | +++ |
| Interference from $H^+$ | ++ | ++ | ++ |

[a]Most values are taken from Scarpa (1979) and Thomas (1982).

### 3.2.1. Solubility

Bi-azo indicators are readily soluble in water, but poorly soluble in apolar solvents, which implies that they cross biological membranes very poorly.

### 3.2.2. Affinity for Calcium

The affinity of the indicators for free $Ca^{2+}$ should be relatively low, such that they will be predominantly in the free form during measurements because (1) it results in linearity of the calibration curve and (2) the indicator will not markedly alter the baseline free $Ca^{2+}$ concentration. Therefore, the range of the $Ca^{2+}$ concentrations to be measured should be below the dissociation constant $K_d$ that characterizes the affinity of the indicator for $Ca^{2+}$. It follows that the various indicators should be in a limited concentration range, and further, that several have to be used when the concentration of $Ca^{2+}$ is expected to vary widely in the course of an experiment. Thus, murexide ($K_d$ approximately 2 m$M$) is suited for measurements of changes in $Ca^{2+}$ at high concentrations ($10^{-4}$ to $10^{-3}$ $M$) and arsenazo III ($K_d$ approximately 30 $\mu M$ in the presence of $Mg^{2+}$) is suited for lower concentrations ($10^{-6}$ to $10^{-7}$ $M$); antipyralo III ($K_d$ approximately 150 $\mu M$) should be used for concentration ranges in between those two (Table 1). Measurements of low resting concentrations are therefore difficult to perform with such indicators and can usually be determined only using a null point method and internal dialysis in giant cells (DiPolo et al., 1976; Mullins and Requena, 1979). It follows that measurements of $Ca^{2+}$ concentrations below 100 n$M$ require indicators with higher affinity.

### 3.2.3. Stoichiometry

The stoichiometry of the dye–$Ca^{2+}$ binding reaction has been studied in detail for arsenazo III. Early studies indicated that the binding of $Ca^{2+}$ involved one molecule of dye for one calcium ion (Michaylova and Ilkova, 1971; Kendrick et al., 1977), implying that the concentration of the dye–$Ca^{2+}$ complex was linearly related to the concentration of both dye and free $Ca^{2+}$. It was shown in later studies, however, that, under conditions comparable to those used in actual experimental situations, i.e., when the dye concentration far exceeded that of free $Ca^{2+}$, two dye molecules complexed one $Ca^{2+}$ ion (Gorman and Thomas, 1978; Thomas, 1979), and the concentration of such complexes would be expected to vary with the square of free $Ca^{2+}$. Thus, simultaneous equilibria between 1:1, 2:2, and 1 $Ca^{2+}$: 2 $Ca^{2+}$–dye complexes have to be taken into account to fit the data. Such a behavior requires that the dye should be calibrated at a concentration very close to that achieved during physiological experiments, or else significant errors will result.

### 3.2.4. Spectral Characteristics

Examples of difference absorption spectra for the various indicators, i.e., the difference, at different wavelengths, between the values of absorbtion of the dye–$Ca^{2+}$ complex and those of the free form, are shown in Fig. 3. For arsenazo III, such spectrum display two peaks, around 600 and 660 nm, respectively, and an isosbestic point near 580 nm. Therefore, to perform differential recordings (see above), the wavelength pair 580–660 nm should give the largest signals. It appears that the pair 660–690 nm gives greater stability, however, and is preferred for measurement of small absorbance changes (Gorman and Thomas, 1978). Isosbestic points for murexide and antipyrylazo were determined to be at 479 and 613 nm, respectively. Several suitable pairs of wavelengths have been proposed for the various dyes (Scarpa, 1982): 507–540 nm for murexide; 675–685 or 660–690 nm for arsenazo III; 670–690 or 720–790 nm for antipyrylazo III (the last pair giving a better selectivity over $Mg^{2+}$; see below).

### 3.2.5. Molar Absorption Coefficients

The maximal values of the differences between the molar extinction coefficients of the free and bound forms of the dyes have been reported (Thomas, 1982): about 6000/$M$/cm for murexide, 7000/$M$/cm for antipyrylazo III and 25000/$M$/cm for arsenazo III.

### 3.2.6. Selectivity

Besides $Ca^{2+}$, other cations may bind to the various indicators, thereby interfering with $Ca^{2+}$ measurements. Interference may result from absorption spectra properties of the interfering ion similar to that of $Ca^{2+}$, from competition with $Ca^{2+}$ for binding to the indicator or from a stoichiometry ratio with the indicator different than that of $Ca^{2+}$ (Blinks et al., 1982). The wavelength dependence of the absorbance changes of these complexes differs for the different cations, however. The apparent selectivity can therefore be increased by an appropriate choice of the wavelength pair used for measurements. In practice, the only interfering ions are $Mg^{2+}$ and $H^+$. A measure of the selectivity of an indicator for $Ca^{2+}$ over another cation can be taken as the ratio of the sensitivity for $Ca^{2+}$ at a given wavelength (i.e., the slope, at that wavelength, of the curve relating the signal to the free $Ca^{2+}$ concentration) to the sensitivity for the cation at the same wavelength.

Concerning murexide, the selectivity for $Ca^{2+}$ over $Mg^{2+}$ has thus been estimated to be 1845 using a wavelength pair of 500–544 nm (Ohnishi, 1978). Besides, the addition of $Mg^{2+}$ at a concentration as high as 10 m$M$ does not alter significantly the absorption spectrum for $Ca^{2+}$ (Ogawa et al., 1980). Interference from $Mg^{2+}$ does not therefore appear to be a major problem. In addition, the interference from protons is small.

Concerning arsenazo III, there is a significant binding with

$Mg^{2+}$, and the resulting difference absorption spectrum shows a single broad peak, centered around 608 nm, whereas the signal is much attenuated at wavelengths of above 675 nm. It is therefore possible, by choosing the wavelengths pair of 675–685 nm, to minimize the interference from $Mg^{2+}$, although the signals may be smaller than those obtained using shorter wavelengths. The selectivity for $Ca^{2+}$ over $Mg^{2+}$ has been determined, and several values have been reported: 39, 87, 2000, 4000 (Scarpa, 1979; Blinks et al., 1982). Besides, although the $Mg^{2+}$ concentration *per se* will not influence the differential measurements, changes in free $Mg^{2+}$ will alter the equilibrium between the free and $Ca^{2+}$-bound indicator. The spectrum corresponding to pH changes is superficially similar to the spectrum for $Mg^{2+}$ at long wavelength, except that the peak signal is at 605 nm. Shifts in pH can be minimized, however, by including some pH buffer in the dye. Besides, the time course of pH changes inside cells appears to be much slower than that of pH changes and therefore would affect only the falling phase of possible $Ca^{2+}$ transients (Parker, 1979).

Concerning antipyrylazo III, $Mg^{2+}$ produces significant shifts in the $Ca^{2+}$ spectrum at concentrations in the m$M$ range (Scarpa et al., 1978). Such interferences are negligible for wavelengths above 660 nm, however. Selecting a wavelengths pair of 720–790 nm therefore makes possible the use of this indicator for $Ca^{2+}$ measurements. The interference caused by protons is similar to that observed for arsenazo III.

### 3.2.7. Response Time

The kinetics of the $Ca^{2+}$ binding reaction have usually been determined using the temperature jump technique: the indicator is in equilibrium with a known concentration of $Ca^{2+}$ at a constant temperature. The equilibrium is subsequently displaced by a rapid increase in temperature, and the corresponding absorbance change is recorded. The relaxation times thus measured have been 2 μs for murexide, approximately 2.8 ms for arsenazo III, and 180 μs for antipyrylazo III (Scarpa, 1979). A major advantage of such indicators is therefore that they make possible the recording of very short $Ca^{2+}$ transients.

It has been shown, however, that the decay time of signals induced, in large fibers, by short ionophoretic $Ca^{2+}$ injections is about five times slower with arsenazo than with aequorin (Eusebi et al., 1985). Therefore, during experiments the overall response might be slowed down by the intrinsic buffering capacity of arsenazo III or by unspecific effects on the various intracellular buffering systems.

## 3.3. Use of Bi-azo Dyes for Measurements

### 3.3.1. Availability

The three above-mentioned dyes can be obtained from several drug companies. Batches of dyes can be contaminated by $Ca^{2+}$ or organic material, however, and they often have to be purified before use. Cation exchange chromatography is an efficient way to remove $Ca^{2+}$, and it has been proposed that double recrystallization be a routine procedure (Scarpa, 1979). Batches of arsenazo III with a high degree of purity are now commercially available, however. Purification of murexide and antipyrylazo III is usually considered unnecessary.

### 3.3.2. Preparation of Solutions

Use of dilute solutions (20–110 $\mu M$) is recommended to avoid precipitation or polymer formation, and they should be pH-buffered to minimize possible interference from protons.

### 3.3.3. Introduction into Cells

Most often, metallochromic indicators have been introduced into pipets and injected into cells by pressure or by ionophoresis (Parker, 1979; Ahmed and Connor, 1979). In the latter case, arsenazo may be prepared in a solution containing 10 m$M$ KHCO$_3$ (which increases the solubility of the dye), at pH 7, and negative currents have to be used.

### 3.3.4. Intracellular Diffusion

The time required for arsenazo III or antipyrylazo III to reach diffusion equilibrium is that expected from consideration of the molecular weight.

### 3.3.5. Binding to Cellular Constituents

The various dyes appear not to bind to intracellular organelles, in particular not to mitochondria (Scarpa, 1979). It has been shown, however, that arsenazo may bind to cellular components inside muscle fibers (Beeler et al., 1980).

### 3.3.6. Apparatus

Under conditions of biological experiments, the expected absorbance changes are very small, approximately 0.001. A very sensitive spectrophotometer is therefore required that, as mentioned above, must provide the possibility of performing multiwavelength measurements. Such apparatus is commercially available. A schematic drawing of a chopped-beam, multiwavelength, differential spectrophotometer is shown in Fig. 4. The light from a tungsten iodine lamp is passed through a rapidly spinning disk (several hundreds revolutions per s) that contains several narrow-band interference filters. The resulting beam, which consists of a series of pulses of light of various wavelengths, is focussed and directed to the preparation or, when necessary, to the calibration device. The transmitted light is then converted to voltage pulses using a photomultiplier or photodiodes with high quantum efficiency at long wavelengths and a current-to-voltage converter. Each of the pulses corresponding to the different filters is directed by a sampling device to separate amplifiers. Synchrony between each amplifier and its appropriate filter is maintained by a reference pulse corresponding to each revolution of the disk and detected by a phototransistor (Fig. 4). Absorbance changes are measured by subtracting the output of one amplifier from a fraction of the output of a second amplifier, such that the average difference signal is zero. The resulting signal can be displayed on an oscilloscope or stored for subsequent analysis.

### 3.3.7. Artifacts Caused by Movement

Changes in cell volume or cellular movements induce variations in absorption that are not easy to eliminate, and further the amplitude of the artifact may be dependent on wavelength (Kowacs et al., 1979). There is apparently no simple method to compensate for such artifacts.

### 3.3.8. Intracellular Buffering Capacity for $Ca^{2+}$

It can be calculated that, for the usual concentrations of arsenazo III injected into cells, the ratio of free to dye-bound $Ca^{2+}$ is about 0.1 (Brindley, 1979). Therefore, if intracellular free $Ca^{2+}$ concentration rises in the cell, a significant amount of $Ca^{2+}$ will be bound to the dye, which is thus acting as a buffer for $Ca^{2+}$. Calculations of the total intracellular $Ca^{2+}$ buffer capacity in muscle fibers indicate, however, that for an intracellular dye concentration below 1 m$M$, the amount of $Ca^{2+}$ bound to the dye is very much smaller than that bound to intracellular $Ca^{2+}$ buffers (Baylor et al., 1982).

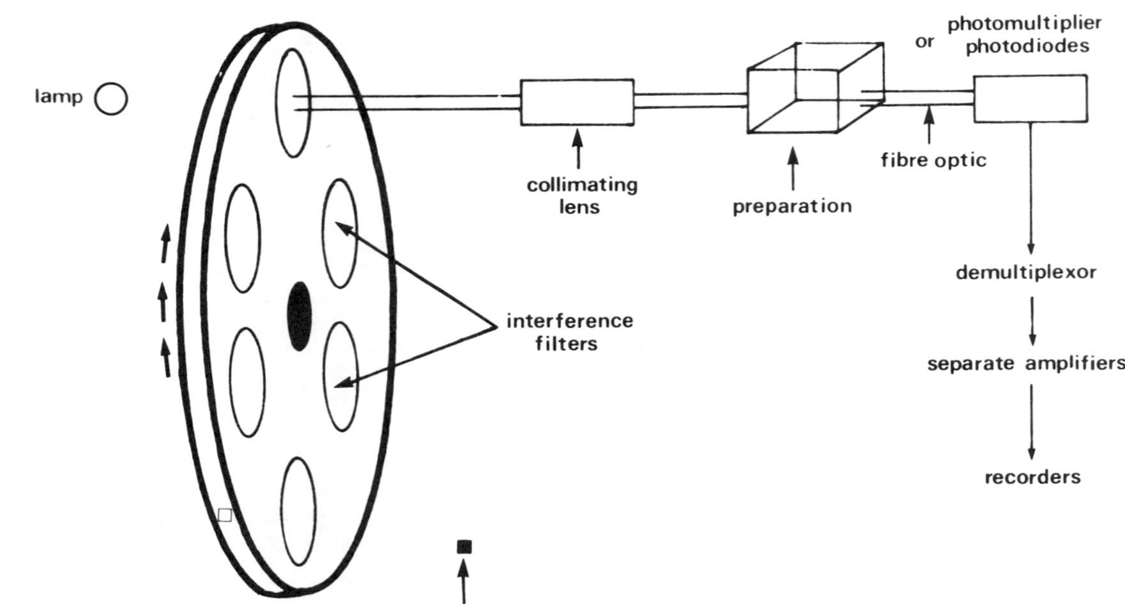


Fig. 4. Schematic drawing of a chopped beam, multi-wavelength spectrophotometer for measuring intracellular $Ca^{2+}$ concentrations using bi-azo indicators (see text) (modified from Gorman and Thomas, 1978).

### 3.3.9. Toxicity

Possible toxicity on cellular processes has been tested using relatively high concentrations of the three indicators. They have been found to be devoid of any significant effect on oxidative phosphorylation of isolated mitochondria, $Ca^{2+}$ transport, ATPase activity, glycolysis rates, or voltage-dependent conductance changes in squid axons (Scarpa et al., 1978; Brown and Pinto, 1979).

### 3.3.10. Calibration of the $Ca^{2+}$ Signal

As mentioned above, the difficulties in determining precisely the intracellular concentration of dye (although it can be estimated by measuring during injection the absorbance increase at an isosbestic wavelength), the values of the extinction coefficients, and those of dissociation constants in the experimental conditions usually prevents making absolute absorbance measurements. Absorbance changes are therefore usually calibrated in terms of free $Ca^{2+}$ changes. In practice, calibration solutions should mimic as closely as possible the ionic environment in which the measurements will be performed, and the concentrations of dye used for calibration should be close to those injected during experiments. It has been shown, however, that in muscle fibers the wavelength dependence of the intracellular arsenzo III signal did not match the one obtained in a cuvet using comparable concentrations of dye and free $Ca^{2+}$. Such a discrepancy was revealed by performing several measurements, each at different wavelengths (Baylor et al., 1982). Therefore, when possible, it appears advisable to perform calibration in vivo by adding to the sample known concentrations of $Ca^{2+}$ or EGTA and to measure the resulting absorbance changes.

## 4. Fluorescent Indicators

Fluorescent indicators have long been used for measurements of $Ca^{2+}$. The fluorescent indicators that were initially used, however, such as chlorotetracycline (Caswell and Hutchinson, 1971) or calcein, will not be considered in detail because they do not appear to discriminate appropriately between $Ca^{2+}$ and $Mg^{2+}$ and therefore lend unreliable quantitative data (see details in the review of Blinks et al., 1982). This section will be devoted to tetracarboxylate indicators, which have been more recently developed.

Tetracarboxylate indicators evolved from a new family of calcium ligands (Tsien, 1983) related to the well-known $Ca^{2+}$ chelator EGTA (Fig. 5). Such substances undergo large changes in fluorescence when binding to $Ca^{2+}$ ions. Their main advantage is that there is an easy method to incorporate controlled quantities of such compounds into the cytoplasm of cells. Indeed, these indicators do not appear to cross or bind to cell membranes significantly; however, the carboxylate groups that make the compounds membrane-impermeant can be esterified with special chemical groups (usually acetoxymethyl groups), such that the resulting derivatives are uncharged and lipid-soluble, and can therefore readily permeate the membrane of intact cells. Once inside cells, the cytoplasmic esterases cut back the esterifying groups and restore the molecules to their original, membrane-impermeant state. The most extensively used of these dyes has been quin-2 (Fig. 5) (Tsien et al., 1982), but new derivatives with improved properties, of which fura-2 is the prototype, have recently appeared (Fig. 5) (Grynkiewicz et al., 1985). Most measurements performed with such indicators have been obtained in cell suspension experiments (Tsien and Rink, 1982; Rink and Pozzan, 1985), but recently, they have been applied to measurements of intracellular $Ca^{2+}$ from isolated muscle or nerve cells (Williams et al., 1985; Kudo and Ogura, 1986) or from slices of nervous tissue (Kudo et al., 1987).

## *4.1. Principles*

Fluorescent indicators emit light of defined spectral properties (emission spectrum) when illuminated with radiations corresponding to a narrow range of wavelengths (excitation spectrum). The peak wavelength of the excitation spectrum need not be identical to that of the emission spectrum and, therefore, by using adequate filters, the latter can be separated from the former. Interaction with $Ca^{2+}$ ions can result in a change of fluorescence intensity or a shift in the emission or excitation spectrum, both of which can be used to measure changes in $Ca^{2+}$ concentration. The usefulness of such indicators depends, as is the case for the metallochromic dyes, on the $K_d$, the fractional change of the signal upon binding, and on the selectivity and sensitivity to possible interferences.

A particular drawback, however, arises from the fact that most cells contain substances that absorb in the UV range and emit in the blue or green range. It is therefore advisable to use indicators that

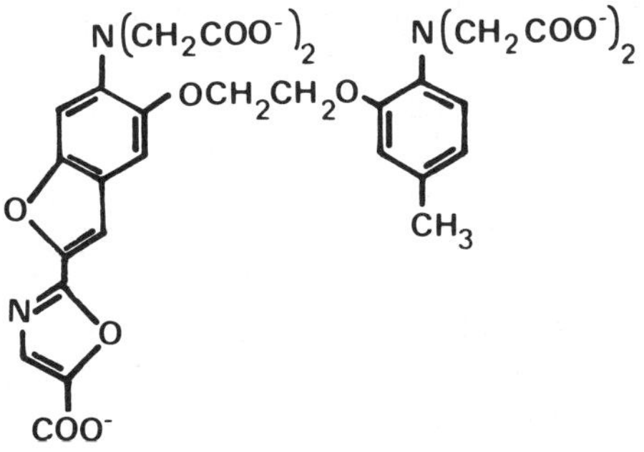

Fig. 5. Chemical formula of the $Ca^{2+}$ chelator EGTA and $Ca^{2+}$ indicator derivatives Bapta, quin-2, and fura-2.

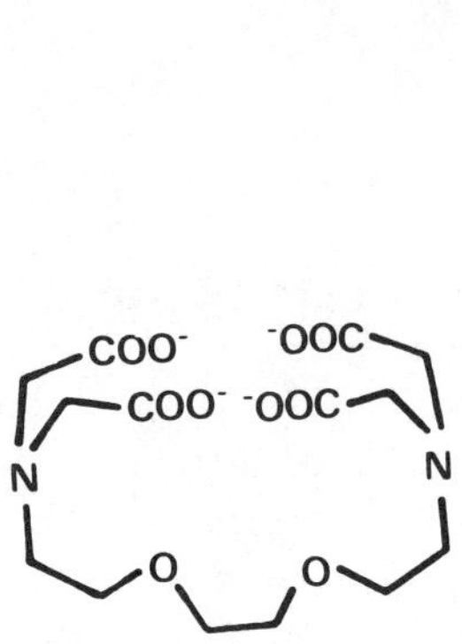

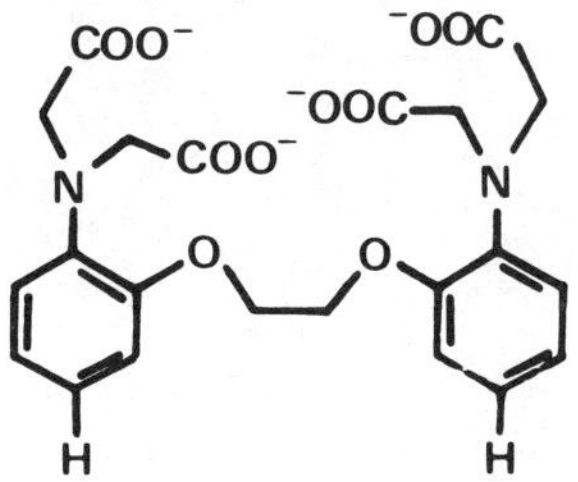

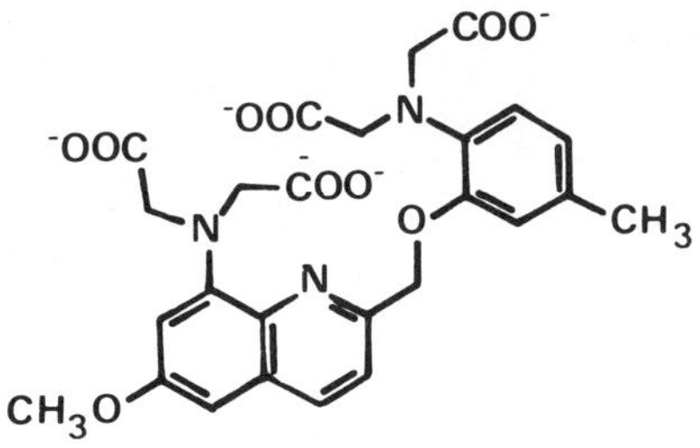

have their emission spectra at wavelengths different than those of the intrinsic fluorescence spectra, if possible with excitation spectra in the shorter wavelengths of the visible spectrum and emission spectra toward the red end of the spectrum.

Another problem might arise from what is called the inner filter effect, which occurs when the absorption spectrum of the substance overlaps partially with its fluorescence emission spec-

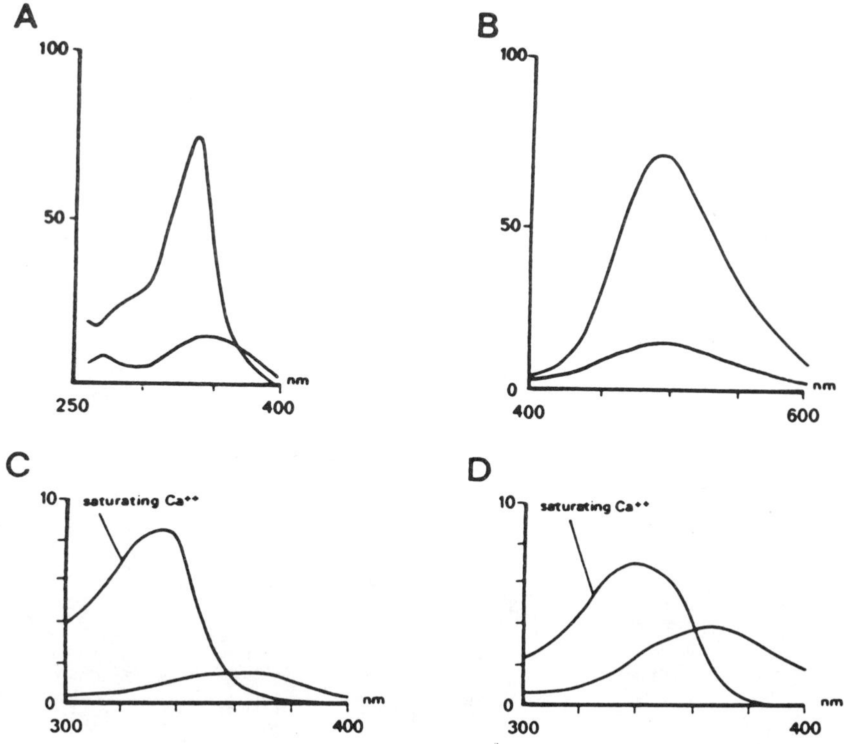

Fig. 6. Spectral characteristics of quin-2 and fura-2. (A, B) Respectively, excitation and emission spectra of 20 μM quin-2 at zero $Ca^{2+}$ and at a concentration of 200 μM $Ca^{2+}$, in 120–135 mM $K^+$, 20 mM $Na^+$, 1 mM free $Mg^{2+}$, pH 7.05, 37°C. Ordinate, normalized fluorescence intensity; abscissa, wavelength in nm (redrawn from Tsien et al., 1982). (C, D) Respectively, excitation spectra of 30 μM quin 2 and of 1 μM fura-2 determined under the same experimental conditions: 130 mM KCl, 10 mM MOPS, 1 mM EDTA, pH 7.20, 20°C at zero $Ca^{2+}$ and after addition of a saturating concentration (2 mM) of $Ca^{2+}$. Ordinate, fluorescence intensity; abscissa, wavelength in nm. The emission signal was measured at a wavelength of 500 nm (redrawn from Grynkiewicz et al., 1985).

trum such that part of the emitted light may be absorbed by the substance. Therefore, a marked overlap between absorption and emission spectra, which usually results in the use of large quantities of indicator, gives rise to non-linear responses.

## 4.2. Properties of Quin-2 and Fura-2

### 4.2.1. Properties of the Fluorescence Signal

Upon binding to $Ca^{2+}$, the intensity of quin-2 fluorescence intensity increases by a factor of 6.2, with little spectral shift in the emission spectrum. Its quantum efficiencies with and without $Ca^{2+}$ are 0.14 and 0.029, respectively (Tsien, 1980), and its excitation spectrum has an isosbestic point (independent of $Ca^{2+}$ binding) at 360 nm and a maximum at 335 nm (Fig. 6).

For fura-2, the intensity increase is much larger than for quin-2, by about 30 times (Grynkiewicz et al., 1985), since it has about a sixfold higher extinction coefficient and about a fivefold greater quantum efficiency than quin-2 (Table 2); besides, there is a shift of the peak value of the excitation spectrum on binding to $Ca^{2+}$ from 360 to 340 nm, the isosbestic point being at a wavelength slightly inferior to 360 nm (Fig. 6).

Such characteristics of the excitation spectra were taken advantage of to perform measurements of free $Ca^{2+}$ independent of dye concentration (see below).

### 4.2.2. Dissociation Constants

Dissociation constants ($K_d$) and maxima of excitation and emission spectra are indicated in Table 2. It should be noted that the $K_d$ of quin-2 and of fura-2 are significantly shifted to higher values in the presence of 1 m$M$ $Mg^{2+}$ (*see* Fig. 7).

### 4.2.3. Stoichiometry

The complexes that the dyes form with $Ca^{2+}$ have a stoichiometry of one over one.

### 4.2.4. Measurements of $Ca^{2+}$ Concentrations

For quin-2, the parameter that is measured is the increase in the fluorescence intensity on binding to $Ca^{2+}$. For a defined quantity of quin-2, the $Ca^{2+}$ concentration can be given by the following formula (Tsien et al., 1982):

$$[Ca^{2+}] = K_d \frac{(F - F_{\min})}{F_{\max} - F}$$

Table 2
Properties of Fluorescent Ca$^{2+}$ Indicators[a]

| | Apparent $K_d$ for Ca | Absorption maxima | | Emission maxima | | Fluorescence quantum efficiency | |
|---|---|---|---|---|---|---|---|
| | | Free anion | Ca complex | Free anion | Ca complex | Free anion | Ca complex |
| BAPTA | 110 | 254(16) | 274(4,2) | 363 | 363 | | |
| Quin-2 | 80 | 261(37) | 240(36) | 492 | 498 | 0.14 | 0.029 |
| | 115 | 354(5) | 332(5) | | | | |
| Fura 2 | 135[b] 224[c] | 362(27) | 335(83) | 512 | 505 | 0.23 | 0.49 |

[a]Absorption maxima measured at 22°C in 100 m$M$ KCl. The first number indicates the wavelength in nm, and the number in parentheses refers to the corresponding extinction coefficient $\times 10^{-3}$, per $M$ per cm. Emission maxima and quantum efficiencies measured in 100 m$M$ KCl at 22°C.
[b]In 100 m$M$ KCl, 20°C, pH 7.1–7.2.
[c]In 115 m$M$ KCl, 20 m$M$ NaCl, 10 m$M$ K-MOPS, pH 7.05, 1 m$M$ free Mg$^{2+}$, 37°C.



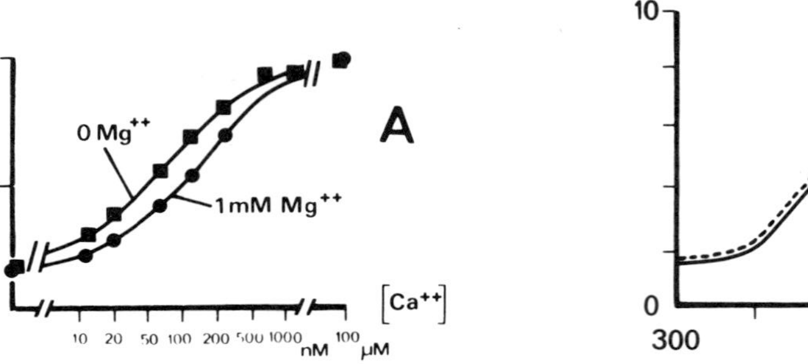


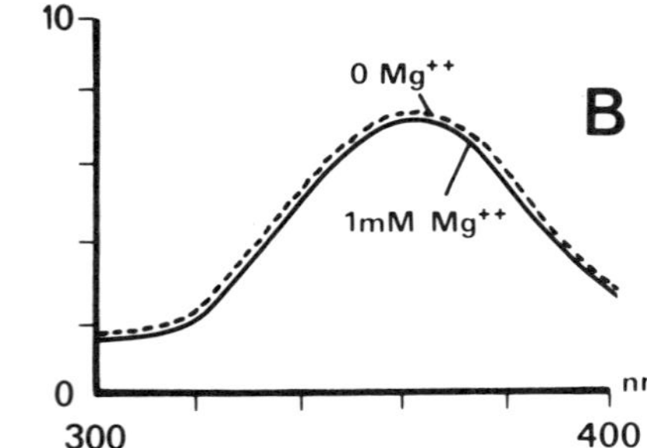

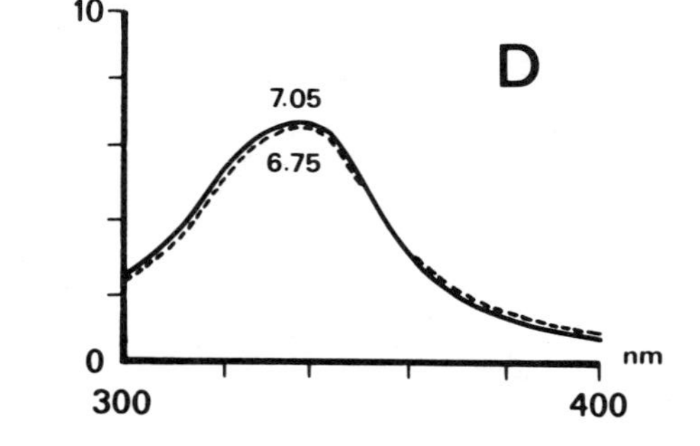
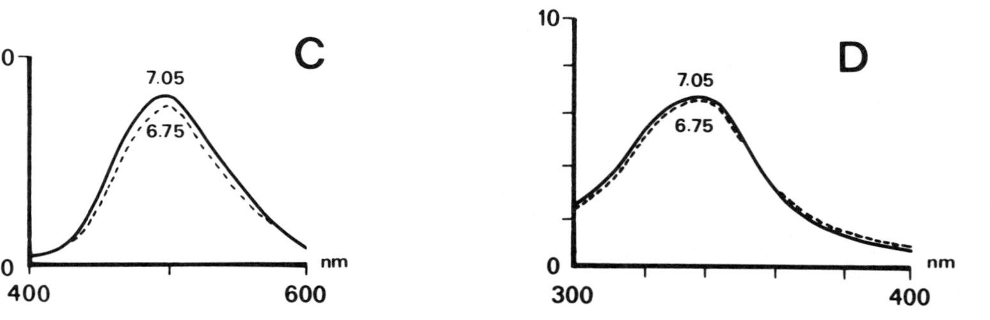


Fig. 7. Selectivity of quin-2 and fura-2 for $Ca^{2+}$ *vis-a-vis* $Mg^{2+}$ and pH. (A) Dose–response curve of 20 $\mu M$ quin-2 with increasing concentrations of $Ca^{2+}$ at zero $Mg^{2+}$ and at 1 m$M$ free $Mg^{2+}$, in 125 m$M$ KCl, 20 m$M$ Na$^+$, pH 7.05, 37°C. Ordinate, normalized fluorescence intensity; abscissa, free $Ca^{2+}$ in n$M$, except for the highest value, which is expressed in $\mu M$. In $Mg^{2+}$-free medium, the apparent dissociation coefficient is 80 n$M$, whereas in 1 m$M$ free $Mg^{2+}$, its value is 115 n$M$. (B) Excitation spectrum of 2 $\mu M$ fura-2 in zero $Mg^{2+}$ and in 1 m$M$ free $Mg^{2+}$, in 132 m$M$ KCl, 10 m$M$ MOPS, 1 m$M$ EGTA, pH 7.18, and 20°C. Emission was recorded at 510 nm. Ordinate, fluorescence intensity; abscissa, wavelengths in nm. (C) Emission spectra from 20 $\mu M$ quin-2 in 100 n$M$ $Ca^{2+}$ at two different pH values, 7.05 and 6.75, in 125 m$M$ KCl, and 20 n$M$ Na$^+$. Ordinate, normalized fluorescence intensity; abscissa, wavelengths in nm. (D) Excitation spectra from 0.5 $\mu M$ fura-2 in 200 n$M$ $Ca^{2+}$ at two different pH values, 7.05 and 6.75, in 135 m$M$ KCl, 5 m$M$ MOPS, 2 m$M$ EGTA, 37°C. Emission recorded at 505 nm. Ordinate, fluorescence intensity; abscissa, wavelengths in nm. (A) and (C) redrawn from Tsien et al. (1982); (B) and (D) redrawn from Grynkiewicz et al. (1985).

where $F_{min}$ and $F_{max}$ are the fluorescence of quin-2 in a $Ca^{2+}$-free medium and in a medium containing a saturating concentration of $Ca^{2+}$ respectively.

The intracellular concentration can be estimated by recording the peak response of the emission spectrum, which shifts from 430 to about 490 nm as quin-2/AM is converted into quin-2.

The same method could be used with fura-2. Since it is sometimes difficult to determine precisely the concentration of the dye inside cells, however, it appears advisable to use a dual wavelength technique: if the indicator exists only in the free form and the form bound to $Ca^{2+}$, and if the transition occurs with a shift in the excitation or emission spectra, it is sufficient to perform measurements at two wavelengths to determine the ratio of bound to free form and, therefore, the free $Ca^{2+}$ concentration. For quin-2 and fura-2, there is a shift in the excitation spectrum on binding to $Ca^{2+}$, from a wavelength of about 340 to about 360 nm. Thus, the ratio of the values of fluorescence intensity at 500 nm obtained with excitation wavelengths of 340 and 360 nm or 380 nm (depending on the recording system) should yield the value of the free $Ca^{2+}$ concentration (Kruskal et al., 1984; Grynkiewicz et al., 1985) independently of intracellular dye concentration, path length, or absolute sensitivity of the recording system, provided that calibration curves had been previously determined. The ratio is given by the following formula (Kruskal et al., 1984):

$$\frac{I_{340}}{I_{360}} = \frac{R_0 + R_{sat}([Ca^{2+}]/K_d)}{1 + ([Ca^{2+}]/K_d)}$$

where $R_0$ and $R_{sat}$ represent the values of this ratio at 0 $[Ca^{2+}]$ and at a saturating concentration of $Ca^{2+}$, respectively, and $K_d$ representing the dissociation constant.

The main advantage of this method is that it is not necessary to know the intracellular concentration of the dye, making unnecessary the final lysis step normally required for calibration. This method has seldom been used with quin-2, however, because the intensity of quin-2 fluorescence at long wavelengths is low and the measurements can therefore be biased by the background fluorescence, and, further, it is too sensitive to $Mg^{2+}$.

### 4.2.5. Selectivity

Quin-2 has a selectivity for $Ca^{2+}$ over $Mg^{2+}$ of about 10,000:1, and its affinity for $Ca^{2+}$ is not significantly affected by pH in the physiological range (Fig. 7). Concerning fura-2, titrations with $Mg^{2+}$ yielded apparent dissociation constants of about 5.6 m$M$ at

37°C and 9.8 m$M$ at 20°C at 0.1–0.15 m$M$ ionic strength (Gryn-kiewicz et al., 1985). Besides, the effect of $Mg^{2+}$ on the excitation spectrum of fura-2 is weak compared to that of $Ca^{2+}$ (Fig. 7).

Fura-2 binds zinc with a dissociation constant of around 1.6–2 m$M$, whereas the effect of pH within the physiological range seems to be very restricted (Fig. 7) (Grynkiewicz et al., 1985).

### 4.2.6. Response Time

Quin-2 binds $Ca^{2+}$ more quickly than EGTA. Recent experiments using a stop-flow fluorescence apparatus (Quast et al., 1984) have shown that the half-time for $Ca^{2+}$ binding was around 1 ms at 37°C with 3 m$M$ $Mg^{2+}$ in the milieu and the half-time for relaxation about 10 ms.

## 4.3. Use in Biological Preparations

### 4.3.1. Availability

These compounds and their esterified derivatives (quin-2/AM and fura-2/AM) are commercially available from several companies, one of which is Molecular Probes, Inc., Rapid Junction, Ore, USA.

### 4.3.2. Cell Loading

Quin-2 or fura-2 could be introduced into cells using micropipets and pressure injections, but by far the most popular procedure has been to incubate the preparation with the esterified derivatives quin-2/AM and fura-2/AM, as already described. Concerning quin-2, because of the low extinction coefficient, it is necessary to obtain intracellular concentrations in the millimolar range, and, for instance, incubation may be performed in solutions containing 50 μ$M$ of quin-2/AM for 20–40 min. In practice, about 15–30% of quin-2 introduced into the incubating medium accumulates into the cells, and the final intracellular concentration should be around 0.3–1 m$M$. Concerning fura-2, intracellular concentrations of some tens of micromoles are sufficient for measurements. Thus, platelets incubated at 37°C for 1–2 with concentrations of 7 and 0.6 μ$M$ of fura-2/AM were estimated to reach a final intracellular concentration of fura-2 of 150 and 10 μ$M$, respectively (Tsien et al., 1985). In hippocampus slices, cell loading was done by incubating the slice in a solution of 10 μ$M$ fura-2 for 30 min at 37°C (Kudo et al., 1987). A useful procedure to demonstrate a satisfactory loading is to wash the cell suspension after incubation, to lyse the cells and to demonstrate that the indicator thus released has the same properties as the nonesterified dye. The available evidence, although obtained from few cell types, indicates that nearly all intracellular quin-2 or fura-2 is in the cytoplasm and nucleus (Williams et al., 1985), and they do not appear to bind to biological membranes to a great extent.

A factor that is to be taken into account, however, is that in cell suspensions, or even more in slices, all cells may not be equally loaded.

### 4.3.3. Possible Dye Leakage

In cell suspension experiments, some dye may leak out of the cells, a factor that has to be controlled for. One possibility is to centrifuge the suspension gently; another is to measure the immediate drop in the intensity of fluorescence that follows addition of EGTA to the preparation. For fura-2, dye leakage appears to be small and slow.

### 4.3.4. Fluorescence Quenching

The fluorescence of these indicators may be quenched inside the cells by transition metals such as $Zn^{2+}$, and such an effect may be particularly prominent when low quantities of indicators are used. In such a case, it is possible to introduce lipid-soluble heavy metal chelators such as TPEN ($N,N,N'N'$-tetrakis(2-pyridylmethyl)ethylenediamine), which permeates artificial and biological membranes. Thus, in Ehrlich and Yoshida ascite tumor cells, addition of TPEN at a concentration of 20–100 $\mu M$ to quin-2-loaded cells markedly enhanced the baseline fluorescence, indicating that the resting $Ca^{2+}$ value had been previously underestimated (Arslan et al., 1985).

### 4.3.5. Dye Bleaching

To reduce bleaching of the dye that occurs during measurements (thus often inducing a slight constant baseline shift), it is advisable to expose the sample for short periods of time by interposing a shutter in the path of the excitation beam.

### 4.3.6. Apparatus

Measurements of fluorescence from cell suspensions can be carried out with commercial fluorometer with good sensitivity and stability. One needs a light source, a set of monochromators or filters for selecting the excitation wavelengths, a thermostatted chamber, a means for selecting emission wavelengths, and a photomultiplier. Such apparatus will be inappropriate for a whole range of applications, however. For example, for single-cell recording or for spacial measurements from slices of nervous tissue, it will be necessary to record signals using a microscope. For thick preparations, epi-illumination techniques and dual wavelengths measurements will be useful (Tsien et al., 1985). Image intensification techniques may also be worthwhile to consider. Recently, sophisticated techniques combining microfluorometry and image processing techniques have been developed for single-cell measurements, and the reader is referred to the original publications for further information (Williams et al., 1985; Kudo et al., 1986).

### 4.3.7. Side Effects

The hydrolysis of the esterified derivatives by cytoplasmic esterases produces byproducts such as protons, acetate, or formaldehyde. Although the process does not appear to be very toxic, several cellular functions are significantly affected. For example, the ATP concentration is lowered (Tsien et al., 1982), resulting in inhibition of the $Ca^{2+}$ pump (Tiffert et al., 1984), the phosphorylases are activated in hepatocytes (Charest et al., 1983), or stimulus-induced platelet aggregation is inhibited (Rao et al., 1985). Therefore, for each new preparation, possible metabolic actions should not be overlooked. Possible side effects should be less problematic for fura-2 than for quin-2, however, since smaller quantities of the former are required.

### 4.3.8. Stimulus-Induced Changes

When looking at possible changes in intracellular $Ca^{2+}$ during stimulation, two more factors at least have to be considered. The first is that the indicator may be in itself a significant buffer of intracellular $Ca^{2+}$. The problem is more serious for quin-2 than for fura-2, since the $K_d$ of the former is smaller and larger quantities are required for measurements. Although this effect appears not to affect markedly steady-state measurements (Tsien et al., 1984), it should result in reduced amplitudes and slowing down of the stimulus-induced signals (Rink and Pozzan, 1985). Besides, if intracellular $Ca^{2+}$ rises to levels in the micromolar range, the dyes may approach saturation and the responses are no longer linear. The second factor is that autofluorescence of cells may change during stimulation, and it is therefore advisable to check possible

influence of the experimental procedures on the autofluorescence
of unloaded cells and to correct, if necessary, the fluorescence
signals obtained from loaded cells.

### 4.3.9. Calibration

Concerning quin-2, it is necessary to determine the intracellular concentration of the dye (see above) and the value of
$F_{min}$ and $F_{max}$. For cell suspensions, the cell membranes are disrupted at the end of the experiment by adding digitoxin or Triton
X-100 to the preparation or through sonicating, thus releasing
trapped quin-2. Such procedures often alter the endogenous autofluorescence, however, and it may be preferable to use a non-
fluorescent $Ca^{2+}$ ionophore such as ionomycin. $F_{min}$ and $F_{max}$ are
subsequently determined through adding excess EGTA, in alkaline pH (8.4), or some quenching cations such as $Mn^{2+}$ (Roger et
al., 1983), and through adding excess $Ca^{2+}$, respectively. It is
sometimes difficult to obtain consistent values for $F_{max}$, however,
because of the presence of heavy metals that can quench quin-2
fluorescence. In such case, the use of a heavy metal chelator (di-
ethylenetriaminepentaacetic acid or DTPA) may be useful. For
fura-2, the ratio of the signals obtained for the two wavelengths
used is computed, and calibrations can be determined by comparing such ratios to ratios obtained from reference samples of fura-2
in presence of known concentrations of $Ca^{2+}$ in solutions stabilized
with gelatin. More details are given in the original publications
(Rink and Pozzan, 1985; Williams et al., 1985; Tsien et al., 1985).


# 5. $Ca^{2+}$-Selective Microelectrodes

## 5.1. Principles

### 5.1.1. Potentials Generated Across Ion-selective Membranes

The potential difference that is generated across an ion-
selective membrane is the sum of a diffusion potential within the
membrane and of two interfacial potential differences at the phase
boundaries. Assuming that the diffusion potential is constant, the
activity of the particular ion species for which the membrane is
selectively permeable can be deduced from the value of the sum of
such interfacial potentials. This value can be described by the
Nicolsky-Eisenmann equation, which, for $Ca^{2+}$ ions, if the concentration of these ions is maintained constant on one side of the
membrane, can be written:

$$E = E_0 + \frac{RT}{zF} \cdot \ln\left[ a_{ca} + \sum K_{CaM}^{Pot} \cdot (a_M)^{2/zM} \right]$$

where $E_0$ is a constant reference potential, $a_{ca}$ and $a_M$ are, respectively, the activities of calcium and of the interfering ions in the sample solution, $R$ is the gas constant (8.314 J/K/mol), $T$ the absolute temperature, $F$ is the Faraday equivalent ($9.6487 \times 10^4$ C/mol), $z_M$ is the charge number of the interfering ion, and $K_{CaM}^{Pot}$ is the potentiometric selectivity coefficient for $Ca^{2+}$ with respect to ion M. For an ideal $Ca^{2+}$ sensor, the selectivity coefficient is zero (no interference) and the equation reduces to the well-known Nerst equation:

$$E = E_o + s \log a_{ca}$$

where $s$ is the slope factor, which indicates the potential change to a tenfold increase in $a_{ca}$ (29.58 mV at 25°C).

### 5.1.2. Development of $Ca^{2+}$-Selective Membranes

$Ca^{2+}$-selective electrodes became available with the development of two classes of $Ca^{2+}$-selective sensors that were included in liquid solvents that could be introduced into microelectrodes to form a liquid membrane at the tip. The first class was based on an electrically charged lipophilic ester of phosphoric acid (Christoffersen and Johansen, 1976; Brown et al., 1976). Such membranes are composed of a water-immiscible solvent containing lipophilic anions trapped in the membrane phase, such that the transmembrane ion movements are restricted to cations. The selectivity of the membrane depends on the mobilities of the cations in the solvent and on their partition coefficients between the aqueous and organic phases and, therefore, entirely on the characteristics of the organic solvent *vis-a-vis* the various cations present in the sample solution. The second class was based on a synthetic neutral carrier (ETH 1001) (Fig. 8A), which is a lipophilic, multidentate molecule that forms a stable conformation that provides a cavity for the selective uptake of a cation (Ammann et al, 1976). The coordination number and cavity size determine the preference for a defined cation, and the selectivity depends on the characteristics of the carrier. The cation–carrier complex in the membrane is positively charged, and transport experiments suggest that maintenance of electroneutrality within the membrane phase requires the presence of $H_2O$ clusters containing hydroxylic anionic groups (Thoma et al., 1977). Besides, the selectivity for divalent over monovalent

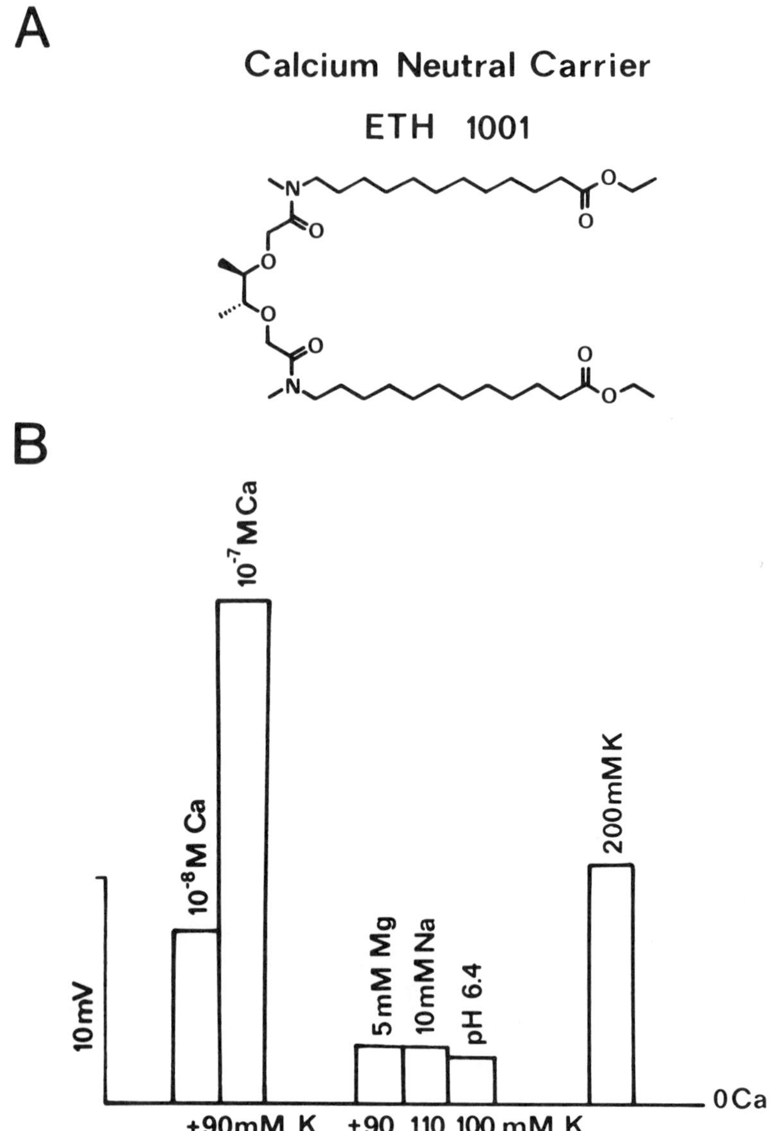



Fig. 8. (A) Chemical formula of the neutral carrier ETH 1001. (B) Values of responses of a PVC-neutral carrier microelectrode in a solution containing virtually no $Ca^{2+}$ (pH 7.8), upon addition of various cations. The solutions contained the amount of KCl specified under each block and 10 m$M$ HEPES (except for the solution corresponding to $10^{-7}$ m$M$ $Ca^{2+}$, which contained 10 m$M$ MOPS, pH 7.29, and the solution at pH 6.4, which contained 10 m$M$ PIPES) and various quantities of EGTA. Temperature, 22–24°C (redrawn from Tsien and Rink, 1981).

cations is further influenced by the dielectric constant of the solvent, which should be as high as possible (Fiedler, 1977).

### 5.1.3. Composition of the Liquid Membranes

Concerning the resins based on the neutral carrier, extensive studies have been carried out to determine the composition of the liquid membrane that would supply the best results. The membrane is composed of a solvent, the carrier, and additives that improve the response. So far, the solvent that has been found to be adequate is *o*-nitrophenyl-*n*-octyl ether (*o*-NPOE). The most commonly used additives are alkali salts with lipophilic anions, such as sodium tetraphenylborate. The incorporation of such mobile cation-exchange sites improved the performance of the liquid membranes in several respects: increased divalent over monovalent cation selectivity (Simon et al., 1978), lowered electrical membrane resistance (Steiner et al., 1979), and decreased electrode response time (Lindner et al., 1978). The commercially available resin (Fluka AG, CH-9470 Buchs, Switzerland) has the following composition by weight: 10% neutral carrier ETH 1001, 89% *o*-NPOE, and 1% sodium tetraphenylborate. For electrodes with a tip diameter of less than 1 μm, it is advisable to mix the resin with polyvinyl chloride in the proportion of 86–14%, respectively, and to dissolve the mixture in three times its weight of tetrahydrofuran to fill the electrodes and subsequently let the tetrahydrofuran evaporate. Such procedures, as discussed below, increase the performance of the electrodes. Resins of various compositions have recently been developed. For further information, the reader is referred to the extensive review of Ammann (1986).

### 5.1.4. Differential Recordings

The ion-selective microelectrode detects a potential difference between the site of recording and a distant, stable reference electrode. Such a potential is the sum of a dc potential existing between the tip of the microelectrode and the distant reference electrode and a potential resulting from the ion-selective properties of the liquid membrane. It is therefore necessary to subtract the dc potential from the total potential in order the measure the pure ionic signal. The value of the dc potential to be subtracted is determined by measuring the local dc potential with respect to the same distant reference through a second, nonselective microelectrode whose tip is positioned as close as possible to the tip of the ion-selective electrode. For extracellular recordings, the closeness of the tips is most often achieved by pulling double-barreled microelectrodes, whereas for intracellular measurements, both two separate and double-barreled microelectrodes configurations have been employed.

In order to have a reliable recording, it is necessary that the distant reference electrode be as stable as possible and not influenced by any activity-induced or ionic variations and that the different junction potentials be small and stable.

## 5.2. Properties of Ca²⁺-Selective Microelectrodes

Ion-selective microelectrodes based on both the charged sensor and the neutral carrier have been successfully used for intracellular measurements, but it appears that the latter ones are superior, mainly because of their better selectivity for $Ca^{2+}$ over $Mg^{2+}$ and a lesser sensitivity to pH (Table 3).

### 5.2.1. Relationship Between the Potential Difference and Free Ca²⁺ Concentration

Several published calibration curves from microelectrodes based on the neutral carrier are shown in Fig. 9. At $Ca^{2+}$ concentrations between $10^{-2}$ and $10^{-6}M$, the potential is a rectilinear function of the logarithm of the free $Ca^{2+}$ concentration, with a slope of 25–30 mV for a ten-fold change in concentration. At lower $Ca^{2+}$ values, the response becomes smaller and smaller (24 mV between $10^{-6}$ and $10^{-7}M$, and 10 mV between $10^{-7}$ and $10^{-8}M$; Marban et al., 1980), until the potential signal is flat (Fig. 9). The detection limit is the concentration corresponding to the point of intersection of the two linear parts of the curve. The initial studies showed that the responses of microelectrodes with tip diameter below 1 μm often departed from the expected behavior in that they showed super-nerstian slopes between $Ca^{2+}$ concentrations of $10^{-4}$ and $10^{-6}M$ (Fig. 9), a reduction in the selectivity and detection limit, and hysteresis (i.e., the signals recorded at the same concentration during the descending and ascending sequences of the calibration procedure were systematically slightly different). Tsien and Rink (1981) showed that incorporation of polyvinyl chloride (PVC) to the liquid resin (see below) suppressed the super-nerstian behavior and restored the detection limit for $Ca^{2+}$ to levels below $10^{-7}M$ (Fig. 9). After dissolution in tetrahydrofuran and polymerization, PVC forms an inert matrix in which the resin is trapped. Tsien and Rink suggested that the improvement was mainly caused by coating the

Table 3

Selectivity Factors of $Ca^{2+}$-Selective Microelectrodes Filled with Resins Containing a Charged (DTMBPP–) and a Neutral (ETH 1001) Sensor *vis-a-vis* the Most Significant Physiological Ions[a]

| Membrane composition | log $K_{CaM}^{Pot}$ | | | | Reference |
|---|---|---|---|---|---|
| | $Na^+$ | $K^+$ | $Mg^{2+}$ | $H^+$ | |
| *DTMBPP–* | | | | | |
| DOPP | –5.6 | –5.6 | –3 | 4 | Brown et al. |
| PVC | FIM | FIM | SSM | FPM | (1976), corrected by Simon et al. (1978) |
| *ETH 1001* | | | | | |
| *o*-NPOE | | | | | |
| NaTPB | –5.5 | –5.4 | –4.9 | –0.05[b] | Ammann (1986) |
| PVC | FIM | FIM | FIM[c] | FIM[c] | |

[a]DTMBPP–: di-(4-(1,1,3,3-tetramethylbutyl)-phenyl)-phosphoric acid.
DOPP: di-(*n*-octyl)-phenylphosphonate.
PVC: poly(vinyl chloride).

ETH 1001: (-)-(R,R)-*N*,*N*'-bis-(11-(ethoxycarbonyl)-undecyl)-*N*,*N*',4,5-tetra-methyl-dioxaoctane diamide.
*o*-NPOE: *o*-nitrophenyl-n-octyl ether.
NaTPB: sodium tetraphenylborate.

FIM: Fixed interference method.
SSM: Separate solution method (0.1$M$ solutions).
FPM: Fixed primary ion method.
[b]Value obtained from a macroelectrode (Simon et al., 1978).
[c]Unbuffered $Ca^{2+}$ solutions were used.

outside of the tip with the PVC-resin mixture, thereby eliminating the electrical shunts through the extremely thin hydrated glass walls of the very tip of the micropipet. As the $Ca^{2+}$ concentration is decreased in the sample, the resistance at the interface sample increases, therefore enhancing the shunt through the glass walls. The value of the electrode potential becomes more and more influenced by that of the potential generated through the glass. The super-nerstian slope indicates that the potential generated across the glass has more negative value than that generated across the ion-selective membrane. To benefit from the PVC treatment, it is therefore necessary to fill the electrode through the tip.

The influence of the PVC treatment is apparent in the calibration curves of Fig. 9. Not all investigators found improvement of the electrode responses using the PVC-resin mixture, however (Morris et al., 1985). Besides, the super-nerstian slopes are most apparent for freshly prepared electrodes, and after an overnight storage, such responses are usually markedly reduced (Dagostino and Lee, 1982; Fozzard et al., 1985).

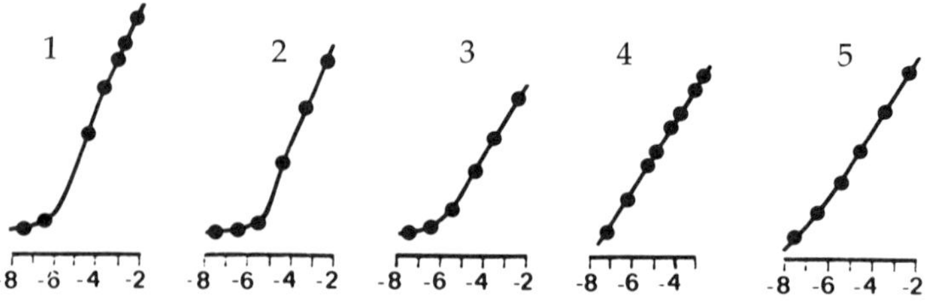

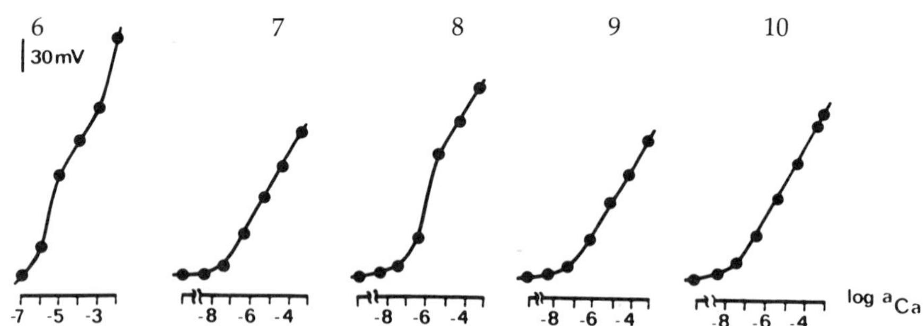

Fig. 9. Published calibration curves of $Ca^{2+}$-selective microelectrodes based on the neutral carrier ETH 1001. The tip diameter of all electrodes was smaller than 1 μm, and the silanization had been performed using vapors, except for curve 7, for which the tip diameter was over 1 μm and the silanization was performed using liquid silane. The responses of electrodes 8 and 9, which contained the original liquid resin, can be compared to the responses of electrodes for which the resin had been mixed with PVC (modified from Ammann, 1986). (1) Coray et al., 1980. (2) Tsien and Rink, 1981. (3) Tsien and Rink, 1980. (4) O'Doherty et al., 1980. (5) Lee et al., 1980b. (6) Morris et al., 1985. (7) Oehme et al., 1976. (8) Lanter et al., 1982. (9) Lanter et al., 1982. (10) Tsien and Rink, 1981.

### 5.2.2. Selectivity

The selectivity of a microelectrode is expressed by its potentiometric selectivity coefficients. Three methods are currently used to determine these coefficients: the separate solution method, the fixed interference method, and the fixed primary ion method. These methods are considered in another chapter and will not be described in detail here. In brief, in the separate solution method, the selectivity coefficient is determined from potential measurements in two solutions, one containing only $Ca^{2+}$ ions at a defined concentration and the second containing only the interfering ion at a known concentration. In the fixed interference method, the potential responses are measured in solutions containing a fixed concentration of the interfering ion with varying concentrations of calcium ions. In the fixed primary ion method, the responses are determined from solutions with a fixed concentration of $Ca^{2+}$ and varying concentrations of the interfering ion. For the three methods, the selectivity coefficients are subsequently determined using the Nicolsky equation. Therefore, the selectivity factors of electrodes whose behavior cannot be described by this equation have a meaning only for the set of experimental parameters used to determine the selectivity factors, and these parameters should then be explicitly defined. For electrodes used for intracellular $Ca^{2+}$ measurements, at very low $Ca^{2+}$ concentrations, the method of choice to determine the selectivity coefficients is the fixed interference method, using $Ca^{2+}$-buffered solutions of ionic strength similar to that of the intracellular medium.

The selectivity coefficients corresponding to microelectrodes filled with different $Ca^{2+}$-selective resins are listed in Table 3. Resins based on the charged and neutral sensors have a high preference for $Ca^{2+}$ over $Na^+$ and $K^+$. The neutral carrier shows a distinctive higher selectivity for $Ca^{2+}$ over $Mg^{2+}$ and $H^+$ when compared to the electrically charged sensor, however. Indeed, for the latter, the detection limit during intracellular recording is limited by the interference from $Mg^{2+}$, and a direct experimental comparison of the performance of the two liquid membranes has shown a distinct advantage of the neutral carrier sensor (Tsien and Rink, 1981). Figure 8B shows the voltage shifts obtained when adding to solutions containing "zero" $Ca^{2+}$, at about constant background levels of KCl, various amounts of interfering ions such as $Na^+$, $Mg^{2+}$, $H^+$, and $K^+$, using a microelectrode based on the neutral carrier (Tsien and Rink, 1981). It has further been reported that at a concentration of $10^{-7}M$ and with a background of 127 m$M$

of $K^+$, increasing $Na^+$ ions by 10 m$M$ produces a voltage response of 4 mV in a $Ca^{2+}$-free medium (Marban et al., 1980).

As discussed above, precise values of selectivity coefficients are difficult to determine. Ammann (1986), however, calculated the influence of $K^+$ and $Mg^{2+}$ ions on the response of an electrode based on the neutral carrier ETH 1001, in a solution mimicking the intracellular medium (free $Ca^{2+}$, 400 n$M$; $K^+$, 120 m$M$; $Mg^{2+}$, 2 m$M$), using the selectivity factors listed in Table 3. The value of the slope is then 22.2 mV, and a change of 1 mV of the potential is approximately equivalent to a 11% change in the $Ca^{2+}$ activity.

### 5.2.3. Detection Limit

The detection limit is usually between $10^{-7}$ and $10^{-8}$ $M$. For instance, a detection limit of around 40 n$M$, measured in a solution containing 127 m$M$ $K^+$, has been reported (Marban et al., 1980). Although the detection limits may be higher during actual experiments than those measured in calibration solutions, substantial decreases in resting intracellular $Ca^{2+}$ (from 210 n$M$ to about 140 n$M$) have been observed following intracellular injection of EGTA (Gorman et al., 1984; *see also* Alvarez-Leefmans et al., 1981).

### 5.2.4. Electrical Resistance and Response Time

Because of the high specific resistivity of the resins, the electrode resistance is very high, usually between 1 and 10 G$\Omega$. The response times are therefore of rather long duration, and the values are between 40 ms and several seconds, most reported values being, however, between 100 and 200 ms. Attempts have been made to reduce the resistance, either by decreasing the specific resistivity of the liquid resins or shortening the length of the column of resin in the electrode tip. The latter approach has been used by Ujec et al. (1980) by inserting inside a conventional ion-selective microelectrode a thinner coaxial pipet such that the tip of the inner pipet is very close to the tip of the outer pipet (10 μm or less) (Fig. 10). The recordings were then performed between the inner pipet filled with a solution containing the ion to be measured, therefore reducing the effective thickness of the membrane to less than 10 μm, and the local reference electrode. The longitudinal resistance is then much reduced and response times as low as 7 (Ujec et al., 1980) or 4 ms (Pumain et al., 1983) have been obtained (Fig. 10).

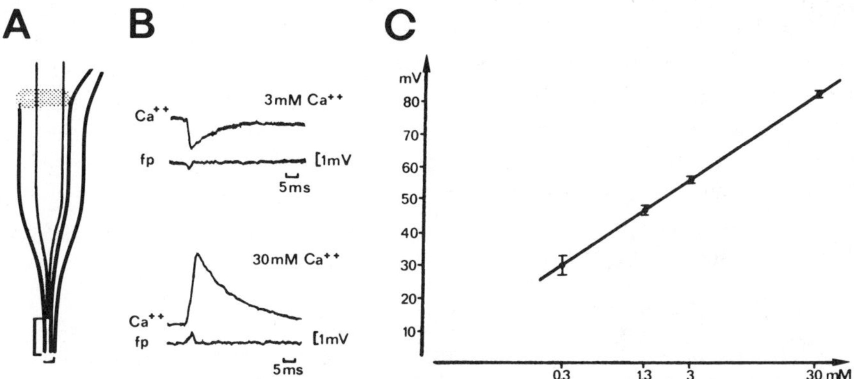

Fig. 10. (A) Schematic drawing of a double-barreled, concentric, fast-responding, $Ca^{2+}$-selective electrode. The distance between the tip of the inner pipet and that of the external pipet must be less than 10 μm to obtain satisfactory responses. The overall diameter of the electrode tip is about 2 μm. (B) Examples of the speediness of responses obtained with such electrodes. In the upper part is seen the electrode response to a fast decrease in $Ca^{2+}$ concentration, and in the lower part, the response to an increase in $Ca^{2+}$. Note the value of the time base. fp, corresponding dc field potential recorded through the reference barrel. (C) Calibration curve of several fast electrodes corresponding to variations in $Ca^{2+}$ concentrations between 0.3 and 30 m*M*. Vertical bars represent SD.

### 5.2.5. Stability and Lifetime

Oehme et al. (1976) have reported stabilities of 0.3 mV over periods of 12 h in calibration solutions. The electrodes can be used for several days, usually with an increase in stability and a slight decrease in sensitivity. The resins can be used for several months when kept refrigerated in the dark. Higher sensitivities are obtained with freshly made resins, however.

## 5.3. Preparation of $Ca^{2+}$-Selective Microelectrodes

Many details concerning the preparation of selective microelectrodes are available in the book of R. C. Thomas (1978).

### 5.3.1. Glass

Micropipets can be made from various batches of glass, such as soda lime glass, aluminosilicates, and borosilicates. The borosilicate Pyrex type seems to be suitable for the preparation of fine-tip pipets, however, because of its high reactivity with silanizing agents. Tubing of 1–3 mm outer diameter is suitable, and it is advisable to use tubing with internal glass filaments to help subsequent filling of the electrode. The glass is usually cleaned by immersion in strong acids, for instance 60% $HNO_3$ or fuming sulfuric acid, to which an oxidizing agent can be added (for example, concentrated hydrogen peroxide), followed by a careful rinsing in water and acetone. Treatment of glass with acid is useful not only for cleaning, but also because it increases its reactivity with the silanizing agents (Munoz et al., 1983).

### 5.3.2. Pulling of Electrodes

The glass capillaries are pulled in a conventional puller to obtain the desired tip shape and diameter. In many experiments, however, double-barreled microelectrodes are prepared (see below). It is then possible to use $\theta$ glass, which is a capillary with a separation in the middle, or to glue together two single capillaries. $\theta$ tubing is easier to pull, but sometimes the glass wall separating the two barrels is too thin and may leak. Two capillaries can be glued using epoxy resin. The electrode assembly must subsequently be pulled in a vertical puller, with a twist of the glass during pulling to stick the two channels together, as is done for multibarreled ionophoretic electrodes. The tips of double-channel electrodes prepared for extracellular recordings are generally broken back under microscopic control to obtain an overall diameter of about 2 $\mu$m for the two channels. The electrodes for intracellular recordings may be bevelled to obtain a more favorable outer diameter/inner diameter ratio at the tip.

### 5.3.3. Silanization and Filling Procedures

Glass surfaces are quite hydrophilic and repel organic liquids. The hydrophobic phase of the liquid sensor is therefore easily displaced by the aqueous solutions present in and around the micropipet. To avoid such drawbacks, it was proposed (Walker, 1971) to screen the hydroxyl groups of the glass surface by silicon compounds to produce a lipophilic surface. The original method has been modified and improved and can be achieved using either liquids or vapors containing the silanizing agent.

Silanizing can be achieved by introducing into the microelectrode tip, from the back or the tip, a liquid containing, for instance, 5% trimethylchlorosilane (or dimethyldichlorolsilane, aminosilane, or siloxane) dissolved in carbon tetrachloride or trichlorotrifluoorethane, and then baked in an oven at about 200°C for half an hour.

For electrodes with tip diameters of above 1 μm, the ion-selective channel can first be filled from the back with a solution of $Ca^{2+}$ ions, for instance 100 m$M$ $CaCl_2$, and the silanizing agent can be then introduced through the tip by suction to produce a column of about 100–300 μm length in the tip, and afterward be expelled by applying pressure to the channel, under visual control. Such a procedure has to be repeated several times (Lux and Neher, 1973; Heinemann et al., 1977). It is very useful to have the pressure and suction leads controlled through electromagnetic valves. For double-barreled microelectrodes, it is advisable to fill the reference channel (either with 150 m$M$ NaCl for extracellular recordings or with, for example, 0.5$M$ KCl or a mixture of 0.1$M$ KCl/0.5$M$ $K_2SO_4$ for intracellular use) prior to silanization in order to avoid contamination of the reference channel by the silanizing agent.

For electrodes with very fine tips, however, it is sometimes difficult to fill the tip using such methods, and silanization with vapors is advisable. A simple procedure is to place the pipets covered with a beaker turned upside-down in an oven for half hour and to afterward introduce a small droplet of the silanizing agent in the beaker (Tsien and Rink, 1980). Such a procedure is not suitable for double-barrelled micropipets, however: the ion-selective side has to be silanized separately (Deyhimi and Coles, 1982), the reference channel being protected meanwhile from contamination by blowing pure nitrogen through it continuously during the process. In a detailed study, Coles and coworkers (Deyhimi and Coles, 1982; Munoz et al., 1983) recommended, for fine tip microelectrodes, the use of borosilicate or aluminosilicate glass and (dimethylamino)-trimethyl-silane instead of chlorosilane as silanizing agent. Care should be taken to eliminate any trace of water from the tip because the silanizing agent can react with water and form polymeric silicon compounds, which can then plug the tip.

The sensor can be introduced into the electrode from the tip or through the back. When the microelectrode has been silanized through the tip, the resin may spontaneously fill the tip to the level the silane has previously reached. It is often necessary to apply a reduced suction to get the resin into the tip, however. When the electrode is filled from the back, the height of the column of resin in the tip can be several millimeters. The electrode is then filled with the $CaCl_2$ solution, when it has not been done previously, and care must be taken to avoid the formation of air bubbles.

Microelectrodes with three or four channels can be manufactured to measure more than one ion concentration at the same time. Such electrodes are very difficult to prepare because there is often cross-contamination of the different ion-selective resins in the various channels. It is easier to make the various ion-selective electrodes separately and to glue them together afterward.

### 5.3.4. Connections

The ion-sensitive channel is subsequently connected to the recording device by means of a carefully chlorided silver wire, and it is advisable to seal the channel with, for instance, dental wax, to reduce the hydrostatic pressure and avoid evaporation of the filling solution.

## 5.4. Use in Biological Experiments

### 5.4.1. Storage

The electrodes may be stored in the refrigerator, their tip dipping in a solution mimicking the extracellular milieu to reduce equilibration times.

### 5.4.2. Shielding

In spite of their high resistance, the ion-sensitive $Ca^{2+}$ microelectrodes are not usually very noisy, but detect very easily any nearby object in motion. A local shielding of the headstage of the amplifier and the electrode is usually sufficient to get rid of most noise problems. Instability of the electrode assembly is often caused by the local reference electrode or by the distant reference electrode.

### 5.4.3. Quality of Recordings

For intracellular experiments, it is essential that the membrane potential be correctly subtracted from the signal of the ion-selective channel; if not, the measurements are largely erroneous. Besides, since the equilibrium potential for $Ca^{2+}$ ions is very much positive (around $+120$ mV), any leakage of the membrane results in massive entry of $Ca^{2+}$ into the cell. The quality of the impalement must therefore be carefully checked.

### 5.4.4. Amplifiers

Ion-selective microelectrodes record dc potentials, and stable dc amplifiers with precise dc compensations are therefore necessary. These electrodes also have very high resistances values, from $10^{-9}$ to $10^{-10}$ $\Omega$, and to avoid attenuation of the signals, amplifiers with input impedances of $10^{13}$–$10^{14}$ $\Omega$ are required. It is also necessary to subtract the signal recorded through the reference channel from that of the ion-selective side to obtain the pure ionic signal (see above). Therefore, it is necessary that the amplifier record both signals against a distant reference electrode and subtract them electronically. The time constant of response of the ion-selective side is usually much larger than that of the reference side, and it is useful to have capacity compensation features on both amplifier inputs to minimize artifactual subtractions of signals during transients. Other features, like an active bridge on the reference channel, are useful for intracellular recordings. Amplifiers suitable for ion-selective measurements can be purchased from WPI Instruments, New Haven, USA or from Meyer and Renz, Muenchen, West Germany.

Because of their high input impedance, the headstage of the amplifiers may be easily damaged by high electrostatic voltages. It is therefore advisable to ground oneself to earth before connecting the electrodes to the headstage.

### 5.4.5. Calibration

In most cases, the microelectrodes are calibrated in solutions containing known concentrations of $Ca^{2+}$, at an ionic strength and with levels of interfering ions as close as possible to that of the sample medium. The slope(s) of the electrode is then evaluated (the potential difference between the local reference and the bath reference electrode should stay about constant). The signal recorded subsequently in the sample medium can be converted to concentration of free $Ca^{2+}$ using the following formula:

$$C_{sample} = C_{cal} \cdot 10^{\,(E_{sample} - E_{cal})/s}$$

where $E_{cal}$ and $E_{sample}$ are the differences of potential between the ion-selective and the local reference channels corresponding, respectively, to the calibration solution (of known $Ca^{2+}$ concentration, $C_{cal}$) and to the sample solution, both potentials being recorded against the same distant reference electrode.

The factor $(E_{sample} - E_{cal})$ also includes possible changes in the liquid junction potential, which are usually negligible.

Electrodes for intracellular recordings have to be calibrated in a range of concentrations of $10^{-2}$ to $10^{-8}$ $M$, since the extracellular and intracellular concentrations are around $10^{-3}$ and $10^{-7}$ $M$, respectively, in vertebrates. It is almost impossible to obtain $Ca^{2+}$ concentrations of less than $10^{-5}$ $M$ just through diluting standard solutions, because of contamination from reagents, water, and vials. It is therefore advisable to use mixtures of $Ca^{2+}$ and calcium buffers: The corresponding equilibrium constants are defined in terms of concentration, which allows one to prepare solutions of precise $Ca^{2+}$ concentrations. At pH 7.0 and 36°C, the association constant for CaEGTA is $3.47 \times 10^{-6}M$, and will buffer $Ca^{2+}$ over the pCa range of 5.9–6.9. Outside that range, EDTA and CDTA, which have affinities for $Ca^{2+}$ higher than that of EGTA, and HEDTA or NTA, which have lower affinities than that of EGTA, should be used. However, when determining the apparent association constants of the $Ca^{2+}$-buffers in mixed solutions from the absolute binding constants (Schwartzenbach et al., 1957), the influences of pH (as referring to the activity of the $H^+$ ions) (see Tsien and Rink, 1980), temperature, and ionic strength have to be taken into account. For further details, the reader is referred to the article of Tsien (1980) and to the review of Blinks et al. (1982). A good example of a series of calcium-buffered solutions is given in two papers of Tsien and Rink (1980, 1981). The stability constants for the various calcium buffers were listed from the values reported by Martell and Smith (1974). Each solution, which is pH-buffered, contains a background level of 125 m$M$ KCl, to account for intereference by $K^+$ and for ionic strength. More realistic calibration solutions should also include $Na^+$ and $Mg^{2+}$ in adequate concentrations, however. Further suitable calibration solutions have been described (Bers, 1982; Otto and Thomas, 1984).

In practice, there is some variability in the performance of different electrodes, and each electrode must therefore be tested individually. The most reliable results are obtained when the calibration curves thus obtained are identical before and after the experiment.

## 6. Other Methods

### 6.1. Null Point for the Permeabilized Plasma Membrane Method

The permeabilized plasma membrane method is based on the fact that the resting permeability of plasma membrane for $Ca^{2+}$ is very low, but can be markedly increased by various procedures. Dissociated cells are incubated in solutions of various free $Ca^{2+}$ concentrations. The null point corresponds to the concentration of free $Ca^{2+}$ for which the cells in suspension rapidly rendered permeable to $Ca^{2+}$ do not alter the concentration of this ion in the incubation medium. Such a method therefore represents an average measurement for the cell population. Plasma membranes can be made permeable using digitoxin (Murphy et al., 1980) or saponin (Wakasugi et al., 1982). Digitoxin has the advantage of having few effects on mitochondria. The possible changes in extracellular free $Ca^{2+}$ following permeabilization can be detected using arsenazo III (Murphy et al., 1980) or ion-selective microelectrodes. However, such a method has special drawbacks, however. Since intracellular free $Ca^{2+}$ is very low, the cells are exposed during the procedure to $Ca^{2+}$ concentrations that are very far from physiological levels, and the resting intracellular $Ca^{2+}$ levels to be measured might be altered. Further, to obtain very low free $Ca^{2+}$ concentrations, the incubation solutions must be buffered with EGTA such that any subsequent change in $Ca^{2+}$ is difficult to detect. Therefore, the values obtained with this method should be considered as upper limits of intracellular free $Ca^{2+}$.

## 6.2. $^{19}F$ Nuclear Magnetic Resonance ($^{19}F$ NMR)

The recent development of fluorine-labeled derivatives of fluorescent indicators has permitted use of nuclear magnetic resonance (NMR) techniques for measurements of intracellular free $Ca^{2+}$. The prototype of such derivatives is the 5F-Bapta (see above), which is a tetracarboxylate $Ca^{2+}$ ligand. When $Ca^{2+}$ is added to 5F-Bapta, two distinct resonance peaks are observed in the $^{19}F$ NMR spectrum, at chemical shifts corresponding to free *(F)* and bound *(B)* 5F-Bapta. The surfaces of these two peaks are proportional to the concentration of the free and bound forms, respectively, and therefore the free $Ca^{2+}$ concentration can be determined by the following formula (Metcalfe et al., 1985):

$$[Ca^{2+}]_{free} = (B/F) \cdot K_d$$

$K_d$ being the dissociation constant of the complex $Ca^{2+}$ 5F-Bapta (around 708 n*M*).

Such a method therefore provides a direct measurement of free $Ca^{2+}$ (Metcalfe et al., 1985). The measurements do not appear to be markedly affected by $Mg^{2+}$ at physiological concentrations nor by minor intracellular cations, and the kinetics of the reaction are conveniently rapid. As indicated above, however, tetracarboxylate indicators may affect cell metabolism, and possible side effects have to be investigated in the various preparations under investigation. Such methods have been used primarily with cell suspensions, but also, on a few occasions, with organs. Cell loading can be achieved as indicated above by introducing into the perfusion medium the esterified derivative, which is able to cross the cell plasma membranes and is subsequently trapped in the cytoplasm after the ester groups have been cut off by intracellular esterases. To date, signals with a time course of few minutes can be followed by this technique.

NMR techniques appear to have a large development potential, but are far from being used for single cells. In addition, the instrumentation is extremely expensive.

## 7. Comparison of the Various Available Methods

Data collected using the various methods described in this chapter have shown that, in the mammalian central nervous system, the extracellular concentration of free $Ca^{2+}$ is around 1.25 m$M$ and can vary to a large degree under defined circumstances. The intracellular concentrations, measured in a variety of cells, were in most cases between 100 and 200 n$M$ (Table 4), but could rise to high values (up to several μ$M$) during intense activity. Each method has advantages and disadvantages that have to be taken into account when designing experiments in which free $Ca^{2+}$ measurements are considered.

### 7.1. Availability and Costs

Most of the necessary materials may be purchased, including aequorin, which is nevertheless very expensive. The costs of apparatus are of the same order, except for NMR, for which they are much higher.

### 7.2. Sensitivity

The detection limit is of the same order for the different methods, about 50 n$M$, except for the tetracarboxylate indicator quin-2, for which the detection limit may be lower. It has been reported, however, than aequorin detects lower free $Ca^{2+}$ values than does arsenazo (Eusebi et al., 1985) or $Ca^{2+}$-selective electrodes (Requena et al., 1984), even though values as low as 30 n$M$ have been measured with selective electrodes (Lopez et al., 1983). A distinct advantage of $Ca^{2+}$-selective electrodes is that they have no upper limit to their range of sensitivity, which makes them the only method currently used for monitoring extracellular $Ca^{2+}$ changes.

## 7.3. Selectivity

All the optical indicators may be subject to interference by $Mg^{2+}$ and pH, although the new fluorescent indicator fura-2 is probably less than the others. $Ca^{2+}$-selective microelectrodes are not very sensitive to $Mg^{2+}$ and pH, whereas for intracellular recordings, $K^+$ is the most strongly interfering ion.

## 7.4. Response Time

The response time of optical indicators is such that they can all detect fast transients. $Ca^{2+}$-selective electrodes have a slower response time, in the order of seconds for intracellular recordings. Coaxial microelectrodes with response times of less than 10 ms have been obtained, however. For NMR, the time of response is in the order of minutes.

## 7.5. Introduction into Cells

For most methods, the measurement process includes penetration of cell membranes with a micropipet, which may result in entry of $Ca^{2+}$ into cells. A distinct advantage of fluorescent indicators is therefore that the esterified form can readily permeate cell membranes, whereas subsequently the de-esterified compound was trapped in cells. This method is particularly suitable for the study of cellular suspensions, but its application to isolated neurons or slices of nerve tissue is at present limited.

## 7.6. Quantitative Measurements of Signals

Photoproteins and metallochromic indicators do not easily yield absolute measurements, except for very large cells. For fluorescent indicators, calibration involves permeabilization of plasma membrane to $Ca^{2+}$, a method that is not easily applicable to single cells under microscopic control or to slices of nervous tissue. $Ca^{2+}$-selective microelectrodes readily give absolute values and are easy to calibrate.

Although optical indicators usually give average values (except when using techniques such as light intensification or digital imaging), $Ca^{2+}$-selective electrodes perform measurements at a single point. Moreover, measurements using $Ca^{2+}$ microelectrodes also yield other useful physiological parameters such as membrane potential or resistance.

Table 4
A few examples of values of intracellular free $Ca^{2+}$ concentration
measured in a variety of cells using different methods

| Fluorescent indicators | nM | Reference |
|---|---|---|
| Intact lymphocytes (quin-2) | 120 | Tsien et al., 1982 |
| Hepatocytes (quin-2) | 120 | Charest et al., 1983 |
| Neuroblastoma × glioma hybrid cell line (quin-2) | 80 | Perney et al., 1984 |
| Toad smooth muscle cells | 137 | Williams et al., 1985 |
| (fura-2) | 245 | Williams et al., 1985 |
| (cytoplasm) | | |
| (nucleus) | | |
| *Aequorin* | | |
| Heart | 250 | Cobbolt and Bourne, 1984 |
| Squid | 20 | DiPolo et al., 1976 |
| *$^{19}F$ NMR* | | |
| Heart (rat) | 580 | Metcalfe et al., 1985 |
| *Ion-selective electrodes* | | |
| Helix neurons | 170 | Alvarez-Leefmans et al., 1981 |
| Frog skeletal muscle | 120 | Lopez et al., 1983 |
| Sheep cardiac muscle | 272 | Sheu and Fozzard, 1982 |
| Necturus kidney cells | 341 | Lee, 1980a |
| Heart fibers | | |
| (Purkinje fibers) | 240 | Weingart and Hess, 1984 |
| (Ventricular fibers) | 270 | Weingart and Hess, 1984 |
| Frog skeletal muscle | 52 | Weingart and Hess, 1984 |
| Squid | 106 | DiPolo et al., 1983 |
| Rabbit cardiac fibers | 119 | Lee, 1981 |
| Ferret heart | 119 | Marban et al., 1980 |

## 7.7. Side Effects

Most optical indicators have few side effects. Quin-2, however-
er, which has to be introduced at a high concentration inside cells,
has been reported to influence a number of cellular functions and
to act as a significant intracellular $Ca^{2+}$ buffer. Such effects appear
to be reduced with fura-2. $Ca^{2+}$ electrodes do not appear to have
toxic effects.

## 8. Conclusions

In spite of the ingenuity of many researchers, there is still not a unique ideal method for measuring the concentration of free $Ca^{2+}$ in living organisms. Although $Ca^{2+}$-selective microelectrodes are the method of choice for extracellular measurements, each method has its advantages and its drawbacks if intracellular studies are required. Whereas photoproteins and metallochromic indicators are suitable for detecting fast transients, $Ca^{2+}$-selective electrodes measurements readily give absolute values. Development of tetracarboxylic indicators was an important breakthrough that allowed for measurements in small cells that otherwise could not be investigated, and further applications to nerve cells is foreseen. NMR appears to have a large potential for further development, but has to date been used little.

## References


Ahmed Z. and Connor J. A. (1979) Measurements of calcium influx under voltage clamp in molluscan neurones using the metallochromic dye Arsenazo III. *J. Physiol.* **286,** 61–82.

Allen D. G. and Blinks J. R. (1979) The Interpretation of Light Signals from Aequorin-Injected Skeletal and Cardiac Muscle Cells: A New Method of Calibration, in *Detection and Measurement of Free $Ca^{2+}$ in Cells* (Ashley C. C. and Campbell A. K., eds.) Elsevier/North Holland, Amsterdam.

Allen D. G., Blinks J. R., and Prendergast F. G. (1977) Aequorin luminescence: Relation of light emission to calcium concentration-a calcium-independent component. *Science* **196,** 996–998.

Alvarez-Leefmans F. J., Rink T. J., and Tsien R. Y. (1981) Free calcium ions in neurones of *Helix aspersa* measured with ion-selective electrodes. *J. Physiol.* **315,** 531–548.

Ames III A., Sakanoue M., and Endo S. (1964) $Na^+$, $K^+$, $Mg,^{2+}$ and $Cl^-$ concentrations in choroid plexus fluid and cisternal fluid compared with plasma ultrafiltrate. *J. Neurophysiol.* **27,** 672–681.

Ammann D. (1986) *Ion-Selective Microelectrodes. Principles, Design and Application* Springer-Verlag, Berlin, Heidelberg, New York, Toronto.

Ammann D., Bissig R., Cimerman Z., Fiedler V., Guggi M., Morf W. E., Oehme M., Osswald H., Pretsch E., and Simon W. (1976) Synthetic Neutral Carrier for Cations, in *Ion and Enzyme Electrodes in Biology and Medicine* (Kessler M., Clark L. C., Lubbers D. W., Silver I. A., and Simon W., eds.) University Park Press, London.

Ashley C. C. and Campbell A. K. (1978) Calcium transients in barnacle muscle induced by the putative excitatory transmitter, L-glutamate. *Biochim. Biophys. Acta* **512**, 429–435.

Ashley C. C. and Campbell A. K. (1979) *Detection and Measurement of Free $Ca^{2+}$ in Cells* Elsevier/North Holland, Amsterdam.

Ashley C. C. and Ridgway E. B. (1970) Aequorin-Calcium Luminescence and Its Application to Muscle Physiology, In *Calcium and Cellular Function* (Cuthbert A. W., ed.) MacMillan, London.

Arslan P., Di Virgilio F., Beltrame M., Tsien R. Y., and Pozzan T. (1985) Cytosolic $Ca^{2+}$ homeostasis in Ehrlich and Yoshida carcinomas. *J. Biol. Chem.* **260**, 2719–2727.

Baker P. F., Hodgkin A. L., and Ridgway E. B. (1971) Depolarization and calcium entry in squid giant axon. *J. Physiol.* **218**, 709–755.

Bates R. G., Dickson A. G., Gratzl M., Hrabeczy-Pall A., Lindner E., and Pungor E. (1983) Determination of mean activity coefficient with ion-selective electrodes. *Anal. Chem.* **55**, 1275–1280.

Baylor S. M., Chandler W. K., and Marshall M. W. (1982) Use of metallochromic dyes to measure changes in myoplasmic calcium activity in frog skeletal muscle fibers. *J. Physiol.* **331**, 139–178.

Beeler T. J., Schibesi A., and Martonosi A. (1980) The binding of arsenazo III to cell components. *Biochim. Biophys. Acta* **629**, 317–327.

Bers D. M. (1982) A simple method for the accurate determination of free [Ca] in Ca-EGTA solutions. *Am. J. Physiol.* **242**, C404–C408.

Blinks J. R., Mattingly P. H., Jewell B. R., van Leewen M., Harre G. C., and Allen D. G. (1978) Practical aspects of the use of aequorin as a calcium indicator: Assay, preparation, microinjection, and interpretation of signals. *Meth. Enzymol.* **57**, 292–328.

Blinks J. R., Wier W. G., Hess P., and Prendergast F. G. (1982) Measurements of $Ca^{2+}$ concentrations in living cells. *Prog. Biophys. Molec. Biol.* **40**, 1–114.

Boll W. and Lux H. D. (1985) Action of organic antagonists on neuronal calcium currents. *Neurosci. Lett.* **56**, 335–339.

Bossu J. L., Feltz A., and Thomann J. M. (1985) Depolarization elicits two distinct calcium currents in vertebrate sensory neurones. *Pfluegers Arch.* **396**, 154–162.

Brindley (1979) Techniques for Measuring Free Calcium In Situ in Single Isolated Cells Using Aequorin and Metallochromic Indicators, in *Detection and Measurement of Free $Ca^{2+}$ in Cells* (Ashley C. C. and Campbell A. K., eds.) Elsevier/North Holland, Amsterdam.

Brown J. E. and Pinto L. H. (1979) The Measurement of Intracellular Free Calcium Concentrations in Squid Axons and Limulus Ventral Receptor Using Arsenazo III, in *Detection and Measurement of Free $Ca^{2+}$ in Cells* (Ashley C. C. and Campbell A. K., eds.) Elsevier/North Holland, Amsterdam.

Brown H. M., Pemberton J. P., and Owen J. D. (1976) A calcium sensitive microelectrode suitable for intracellular measurement of calcium (II) activity. *Anal. Chim. Acta* **85**, 261–276.

Brown A. M., Wilson D. L., and Lux H. D. (1984) Activation of calcium channels. *Biophys. J.* **45**, 125–127.

Campbell A. K. (1983) *Intracellular Calcium: Its Universal Role as Regulator* John Wiley, Chichester, Brisbane, Toronto, Singapore.

Campbell A. K., Lea T. J., and Ashley C. C. (1979) Coelenterate Photoproteins, in *Detection and Measurement of Free $Ca^{2+}$ in Cells* (Ashley C. C. and Campbell A. K., eds.) Elsevier/North Holland, Amsterdam.

Carafoli, E. (1974) Mitochondrial uptake of calcium ions and the regulation of cell function. *Biochem. Soc. Symp.* **39**, 89–109.

Carafoli E. and Crompton M. (1978) The regulation of intracellular calcium by mitochondria. *Ann. NY Acad. Sci.* **307**, 269–284.

Carbone E. and Lux H. D. (1984) A low voltage-activated calcium conductance in embryonic chick sensory neurons. *Biophys. J.* **46**, 413–418.

Caswell A. H. and Hutchinson J. D. (1971) Selectivity of cation chelation to tetracycline: Evidence for special conformation of calcium chelate. *Biochem Biophys. Res. Commun.* **43**, 625–630.

Charest R., Blackmore P. F., Berthon B., and Exton J. H. (1983) Changes in free cytosolic $Ca^{2+}$ in hepatocytes following α-adrenergic stimulation. *J. Biol. Chem.* **258**, 8769–8773.

Christoffersen G. R. J. and Johansen E. S. (1976) Microdesign for a calcium-sensitive electrode. *Anal. Chim. Acta* **8**, 191–195.

Cobbold P. H. (1980) Cytoplasmic free calcium and amoeboid movement. *Nature* **285**, 441–446.

Cobbold P. H. and Bourne P. K. (1984) Aequorin measurements of free calcium in single heart cells. *Nature* **312**, 446–448.

Coray A., Fry C. H., Hess P., Mcguigan J. A. S., and Weingart R. (1980) Resting calcium in sheep cardiac tissue and in frog skeletal muscle measured with ion-selective microelectrodes. *J. Physiol.* **305**, 60P–61P.

Dagostino M. and Lee C. O. (1982) Neutral carrier $Na^+$ and $Ca^{2+}$ selective microelectrodes for intracellular application. *Biophys. J.* **40**, 199–208.

Deyhimi F. and Coles J. A. (1982) Rapid silylation of a glass surface: Choice of reagent and effect of experimental parameters on hydrophobicity. *Helv. Chim. Acta* **65**, 1752–1759.

Dingledine R. and Somjen G. (1981) Calcium dependence of synaptic transmission in the hippocampal slice. *Brain Res.* **207**, 218–222.

DiPolo P. L., Requena F., Brindley F. J., Mullins L. J., Scarpa A., and Tiffert T. (1976) Ionized calcium concentrations in squid axon. *J. Gen. Physiol.* **67**, 433–467.

Dipolo R., Rojas H., Vergara J., Lopez R., and Caputo C. (1983). Measurements of intracellular ionized calcium in squid giant axons using calcium-selective electrodes, *Biochim. Biophys. Acta* **728**, 311–318.

Dubois R. (1887) Fonction photogénique chez le Pholas dactylus. *C. R. Soc. Biol.* **39**, 564–565.

Eusebi F., Miledi R., and Stinnakre J. (1985) Post-synaptic calcium influx at the giant synapse of the squid during activation by glutamate. *J. Physiol.* **369**, 183–197.

Fiedler V. (1977) Influence of the dielectric constant of the medium on the selectivities of neutral carrier ligands in electrode membranes. *Anal. Chim. Acta* **89**, 11–18.

Fozzard H. A., Chapman R. A., and Friedlander I. R. (1985) Measurements of intracellular calcium ion activity with neutral exchanger ion sensitive microelectrodes. *Cell Calcium* **6**, 57–68.

Gorman A. L. F. and Thomas M. V. (1978) Changes in the intracellular concentration of free calcium ions in a pacemaker neurone, measured with the metallochromic indicator dye arsenazo III. *J. Physiol.* **275**, 357–376.

Gorman A. L. F. and Thomas M. V. (1980) Intracellular calcium accumulation during depolarization in a molluscan neurone. *J. Physiol.* **308**, 259–285.

Gorman A. L. F., Levy S., Nasi E., and Tillotson D. (1984) Intracellular calcium measured with calcium-sensitive microelectrodes and arsenazo III in voltage-clamped "Aplysia" neurones. *J. Physiol.* **353**, 127–142.

Grynkiewicz G., Poenie M., and Tsien R. Y. (1985) A new generation of $Ca^{2+}$ indicators with greatly improved fluorescence properties. *J. Biol. Chem.* **260**, 3440–3450.

Hainaut K., Brunko E., Kats R., Bourguet M., and Desmedt J. E. (1975) Aequorin and intracellular calcium movements in skeletal muscle fibres. *Arch. Int. Physiol. Biochem.* **83**, 15–18.

Hansen A. K. (1985) Effect of anoxia on ion distribution in the brain. *Physiol. Rev.* **65**, 101–148.

Harris E. W., Ganong A. H., and Cotman C. W. (1984) Long-term potemtiation in the hippocampus involves activation of $N$-methyl-D-aspartate receptors. *Brain Res.* **323**, 132–137.

Hastings J. W., Mitchell G., Mattingly P., Blinks J. R., and Van Leewen M. (1969) Response of aequorin bioluminescence to rapid changes in calcium concentration. *Nature* **222**, 1047–1050.

Heinemann U. and Pumain R. (1980) Extracellular calcium activity changes in cat sensory cortex induced by iontophoretic applications of excitatory amino acids. *Exp. Brain Res.* **40**, 247–250.

Heinemann U., Lux H. D., and Gutnick M. J. (1977) Extracellular free calcium and potassium during paroxysmal activity in the cerebral cortex of the cat. *Exp. Brain Res.* **27**, 237–243.

Jahnsen H. and Llinas R. (1984) Ionic basis for the electroresponsiveness and oscillatory properties of guinea-pig thalamic neurones in vitro. *J. Physiol.* **349,** 227–247.

Jobsis F. F. and O'Connor M. J. (1966) Calcium release and reabsorption in the sartorius muscle of the toad. *Biochem. Biophys. Res. Commun.* **25,** 246–252.

Johnson F. H. and Shimomura O. (1978) Introduction to the bioluminescence of medusae, with special reference to the photoprotein aequorin. *Meth. Enzymol.* **57,** 271–291.

Kendrick N. C., Ratzlaff R., and Blaustein M. P. (1977) Arsenazo III as an indicator for ionized calcium in physiological salt solutions: Its use for the determination of the CaATP dissociation constant. *Anal. Biochem.* **83,** 433–450.

Kowacs L., Rios E., and Schneider M. F. (1979) Calcium transients and intramembrane charge movements in skeletal muscle fibers. *Nature* **279,** 391–396.

Kraig R. P. and Nicholson C. (1978) Extracellular ionic variations during spreading depression. *Neuroscience* **3,** 1045–1059.

Kretsinger R. H. (1979) The informational role of calcium in the cytosol. *Adv. Cyclic Nucleotide Res.* **11,** 1–26.

Krnjevic K. and Lisiewicz A. (1972) Injection of calcium ions into spinal motoneurones. *J. Physiol.* **225,** 363–390.

Kruskal B. A., Keith C. H., and Maxfield F. R. (1984) Thyrotropin-releasing hormon-induced changes in intracellular $[Ca^{2+}]$ measured by microspectrofluorometry on individual quin2-loaded cells. *J. Cell. Biol.* **99,** 1167–1172.

Kudo Y. and Ogura A. (1986) Glutamate-induced increase in intracellular $Ca^{2+}$ concentration in isolated hippocampal neurones. *Br. J. Pharmacol.* **89,** 191–198.

Kudo Y., Ito K., Miyakawa H., Izumi Y., Ogura A., and Kato H. (1987) Cytoplasmic calcium elevation in hippocampal granule cell induced by perforant path stimulation and L-glutamate application. *Brain Res.* **407,** 168–172.

Kudo Y., Ozaki K., Miyakawa A., Amato T., and Ogura A. (1986) Monitoring of intracellular $Ca^{2+}$ elevation in a single neuron using a microscope/video-camera system. *Jpn. J. Pharmacol.* **41,** 345–351.

Kusano K., Miledi R., and Stinnakre J. (1975) Post-synaptic entry of calcium induced by transmitter action. *Proc. Roy. Soc. Lond. B.* **189,** 49–56.

Lanter F., Steiner R. A., Ammann D., and Simon W. (1982) Critical evaluation of the applicability of neutral-based calcium selective microelectrodes. *Anal. Chim. Acta* **135,** 51–59.

Lee C. O. (1981) Ionic activities in cardiac muscle cells and application of ion-selective microelectrodes. *Am. J. Physiol.* **241,** 459–478.

Lee C. O., Taylor A., and Whinhager E. E. (1980a) Cytosolic calcium ion activity in epithelial cells of necturus kidney. *Nature* **287,** 859–861.

Lee C. O., Uhm D. Y., and Dresdner K. (1980b) Sodium-calcium exchange in rabbit heart muscle cells: Direct measurement of sarcoplasmic Ca-activity. *Science* **209,** 699–701.

Lindner E., Toth E., Morf W. E., and Simon W. (1978) Response time studies on neutral carrier ion-selective membrane electrodes. *Anal. Chem.* **50,** 1627–1631.

Lopez J. R., Alamo L., Caputo C., DiPolo R., and Vergara J. (1983) Determination of ionic calcium in frog skeletal muscle fibers. *Biophys. J.* **43,** 1–4.

Löschen G. and Chance B. (1971) Rapid kinetic studies of the light emitting protein aequorin. *Nature New Biol.* **233,** 273–274.

Lux H. D. and Neher E. (1973) The equilibration time course of $[K^+]_0$ in cat cortex. *Exp. Brain Res.* **17,** 190–205.

MacDermott A. B., Mayer M. L., Westbrook G. L., Smith S. J., and Barker J. L. (1986) NMDA-receptor activation increases cytoplasmic calcium concentration in cultured spinal cord neurones. *Nature* **321,** 519–522.

Marban E., Rink T., Tsien R. W., and Tsien R. Y. (1980) Free calcium in heart muscle at rest and during contraction measured with $Ca^{2+}$-sensitive microelectrodes. *Nature* **286,** 845–850.

Martell A. E. and Smith R. M. (1974) *Critical Stability Constants* Plenum, New York.

Meech R. W. (1978) Calcium-dependent potassium activation in nervous tissues. *Ann. Rev. Biophys. Bioeng.* **7,** 1–18.

Meier P. C. (1982) Two parameter Debye-Huckel approximation for the evaluation of mean activity coefficient of 109 electrolytes. *Anal. Chim. Acta* **136,** 363–368.

Meldrum B. S., Croucher M. J., Badman G., and Collins J. F. (1983) Antiepileptic action of excitatory amino acid antagonists in the photosensitive baboon, *Papio papio. Neurosci. Lett.* **39,** 101–104.

Metcalfe J. C., Hesketh T. R., and Smith G. A. (1985) Free cytosolic $Ca^{2+}$ measurements with fluorine labelled indicators using $^{19}FNMR$. *Cell Calcium* **6,** 183–195.

Michaylova V. and Ilkova P. (1971) Photometric determination of micro-amounts of calcium with arsenazo III. *Anal. Chim. Acta* **53,** 194–198.

Mohan M. S. and Bates R. G. (1975) Calibration of ion-selective electrodes for use in biological fluids. *Clin. Chem.* **21,** 864–872.

Morris M. E., Krnjevic K., and MacDonald J. F. (1985) Changes in intracellular free $Ca^{2+}$ ion concentration evoked by electrical activity in cat spinal cord neurons in situ. *Neuroscience* **14,** 563–580.

Mullins L. J. and Requena J. (1979) Calcium measurements in the periphery of an axon. *J. Gen. Physiol.* **74,** 393–413.

Munoz J.-L., Deyhimi F., and Coles J. A. (1983) Silanization of glass in the making of ion-sensitive microelectrodes. *J. Neurosci. Meth.* **8,** 231–247.

Murphy E., Coll K., Rich T. L., and Williamson J. R. (1980) Hormonal effects on calcium homeostasis in isolated hepathocytes. *J. Biol. Chem.* **255,** 6600–6608.

Nicholson C., Phillips J. M., and Gardner-Medwin A. R. (1979) Diffusion from an iontophoretic point source in the brain: Role of tortuosity and volume fraction. *Brain Res.* **169,** 580–584.

Nicholson C., Ten Bruggencate G., Stockle H., and Steinberg R. (1978) Calcium and potassium changes in the extracellular microenvironment of cat cerebellar cortex. *J. Neurophysiol.* **41,** 1026–1039.

O'Doherty J., Youmans S. J., Armstrong W. M., and Stark R. J. (1980) Calcium regulation during stimulus secretion coupling: Continuous measurements of intracellular calcium activities. *Science* **209,** 510–513.

Oehme M., Kessler M., and Simon W. (1976) Neutral carrier $Ca^{2+}$-microelectrode. *Chimia* **30,** 204–206.

Ogawa Y., Harafugi H., and Kurebayashi N. (1980) Comparison of the characteristics of four metallochromic dyes as potential calcium indicators for biological experiments. *Anal. Biochem.* **87,** 1293–1303.

Ohnishi S. T. (1978) Characterization of the murexide method: Dual-wavelength spectrophotometry of cations under physiological conditions. *Anal. Biochem.* **85,** 165–179.

Ohnishi T. and Ebashi S. (1963) Spectrophotometrical measurements of instantaneous calcium binding of the relaxing factor of muscle. *J. Biochem.* **54,** 506–511.

Otto M. and Thomas J. D. R. (1984) Models for specification of alkali and alkaline earth metal ions in body and intracellular fluids. *Anal. Proc.* **21,** 369–371.

Parker I. (1979) Use of Arsenazo III for Recording Calcium Transients in Frog Skeletal Muscle Fibers, in *Detection and Measurement of Free $Ca^{2+}$ in Cells* (Ashley C. C. and Campbell A. K., eds.) Elsevier/North Holland, Amsterdam.

Perney T. M., Dinerstein R. J., and Miller R. J. (1984) Depolarization-induced increases in intracellular free calcium detected in single cultured neuronal cells. *Neurosci. Lett.* **51,** 165–170.

Prendergast F. G. (1982) The use of photoprotein in the detection and quantification of $Ca^{2+}$ in biological systems. *Trends Anal. Chem.* **1,** 378–383.

Prendergast F. G. and Mann K. G. (1978) Chemical and physical properties of aequorin and the green luminescent protein isolated from *Aequora forskalea. Biochemistry* **17,** 3448–3454.

Pumain R. and Heinemann U. (1985) Stimulus- and amino acid-induced calcium and potassium changes in rat neocortex. *J. Neurophysiol.* **53,** 1–16.

Pumain R., Kurcewicz I., and Louvel J. (1983) Fast extracellular calcium transients: Involvement in epileptic processes. *Science* **222**, 177–179.

Pumain R., Kurcewicz I., and Louvel J. (1987) Ionic changes induced by excitatory amino acids in the rat cerebral cortex. *Can. J. Physiol. Pharmacol.* **65**, in press.

Pumain R., Menini C., Heinemann, U., Louvel J., and Silva-Barra C. (1985) Chemical transmission is not necessary for epileptic seizures to persist in the baboon *Papio papio. Exp. Neurol.* **89**, 250–258.

Quast U., Labhardt A. M., and Doyle V. M. (1984) Stopped-flow kinetics of the interaction of the fluorescent calcium indicator quin-2 with calcium ions. *Biochem. Biophys. Res. Commun.* **123**, 604–611.

Rao G. H. R., Peller J. D., and White J. G. (1985) Measurements of ionized calcium in blood platelets with new generation calcium indicators. *Biochem. Biophys. Res. Commun.* **132**, 652–657.

Rasmussen H. and Barrett P. Q. (1984) Calcium messenger system: An integrated view. *Physiol. Rev.* **64**, 939–984.

Requena J., Hittenbury J., Tiffert T., Eisner D. A., and Mullins L. J. (1984) A comparison of measurements of intracellular Ca by Ca electrode and optical indicators. *Biochem. Biophys. Acta* **805**, 393–404.

Reynolds G. T. (1979) Localization of Free Ionized Calcium in Cells by Means of Image Intensification, In *Detection and Measurement of Free* $Ca^{2+}$ *in Cells* (Ashley C. C. and Campbell A. K., eds.) Elsevier/North Holland, Amsterdam.

Reynolds G. T. (1980) Application of image intensification to low level fluorescence studies of living cells. *Microsc. Acta* **83**, 55–62.

Ridgway E. B. and Ashley C. C. (1967) Calcium transients in single muscle fibers. *Biochem. Biophys. Res. Commun.* **29**, 229–234.

Ringer S. (1882) Concerning the influence exerted by each of the constituents of the blood on the contraction of the ventricle. *J. Physiol.* **3**, 380–393.

Rink T. J. and Pozzan T. (1985) Using quin2 in cell suspension. *Cell Calcium* **6**, 133–144.

Roger J., Hesketh T. R., Smith G. A., Beaven M. A., Metcalfe J. C., Johnson P., and Garland P. (1983) Intracellular pH and free calcium changes in single cells using quene2 and quin2 probes and fluorescence microscopy. *FEBS Lett.* **161**, 21–27.

Scarpa A. (1979) Measurement of Calcium Ion Concentrations with Metallochromic Indicators, In *Detection and Measurement of Free* $Ca^{2+}$ *in Cells* (Ashley C. C. and Campbell A. K., eds.) Elsevier/North Holland, Amsterdam.

Scarpa A. (1982) Cell Ion Measurement with Metallochromic Indicators, in *Techniques in Cellular Physiology* (Baker, P. F., ed.) Elsevier/North Holland, Amsterdam.

Scarpa A., Brindley F. J., and Dubyak G. (1978) Antipyrylazo III, a middle range metallochromic indicator. *Biochemistry* **17**, 1378–1386.

Schwartzenbach G., Senn H., and Anderegg G. (1957) Komplexone. XXIX. Ein grosser chelateffekt besonderer art. *Helv. Chim. Acta* **40**, 1886–1900.

Sheu S. S. and Fozzard H. A. (1982) Transmembrane $Na^+$ and $Ca^{2+}$ electrochemical gradients in cardiac muscle and their relationship to force development. *J. Gen. Physiol.* **80**, 325–351.

Shimomura O. and Johnson F. H. (1966) Partial Purification and Properties of the Chaetopterus Luminescence System, in *Bioluminescence in Progress* (Johnson F. H. and Haneda, Y., eds). Princeton University Press, Princeton.

Shimomura O. and Johnson F. H. (1969) Properties of the bioluminescent protein aequorin. *Biochemistry* **8**, 3991–3997.

Shimomura O. and Johnson F. H. (1970) Calcium binding, quantum yield, and emitting molecule in aequorin bioluminescence. *Nature* **227**, 1356–1357.

Shimomura O. and Johnson F. H. (1975) Regeneration of the photoprotein aequorin. *Nature* **256**, 236–238.

Shimomura O. and Johnson F. H. (1979) Chemistry of the Calcium-Sensitive Photoprotein Aequorin, in *Detection and Measurement of Free $Ca^{2+}$ in Cells* (Ashley C. C. and Campbell A. K., eds.) Elsevier/North Holland, Amsterdam.

Shimomura O., Johnson F. H., and Saiga Y. (1962) Extraction, purification and properties of aequorin, a bioluminescent protein from the luminous hydromedusan, Aequora. *J. Cell Comp. Physiol.* **59**, 223–239.

Shimomura O., Johnson F. H., and Saiga Y. (1963) Further data on the bioluminescent protein, aequorin. *J. Cell Comp. Physiol.* **62**, 1–8.

Simon W., Ammann D., Oehme M., and Morf W. E. (1978) Calcium-selective electrodes. *Ann. NY Acad. Sci.* **307**, 52–69.

Steiner R. A., Oehme M., Ammann D., and Simon W. (1979) Neutral carrier sodium ion-selective microelectrodes for intracellular studies. *Anal. Chem.* **51**, 351–353.

Stinnakre J. (1979) Pressure Injection of Aequorin into Molluscan Neurones, in *Detection and Measurement of Free $Ca^{2+}$ in Cells* (Ashley C. C. and Campbell A. K., eds.) Elsevier/North Holland, Amsterdam.

Stinnakre J. (1981) Detection and measurement of intracellular calcium. A comparison of techniques. *Trends Neurosci.* **4**, 46–50.

Thoma A. P., Viviani-Nauer A., Arvanitis S., Morf W. E., and Simon W. (1977) Mechanism of neutral carrier mediated ion transport through ion selective bulk membranes. *Anal. Chem.* **49**, 1567–1572.

Thomas M. V. (1979) Arsenazo III forms 2:1 complexes with Ca and 1:1 complexes with Mg under physiological conditions. *Biophys. J.* **25,** 541–548.

Thomas M. V. (1982) *Techniques in Calcium Research* Academic, London.

Thomas R. C. (1978) *Ion-sensitive Intracellular Microelectrodes. How To Make and Use Them.* Academic, New York.

Tiffert T., Garcia-Sancho J., and Lew V. L. (1984) Irreversible ATP depletion caused by low concentrations of formaldehyde and of calcium chelator esters in intact red cells. *Biochem. Biophys. Acta* **773,** 143–156.

Tsien R. Y. (1980) New calcium indicators and buffers with high selectivity against magnesium and protein: Design, synthesis and properties of prototype structures. *Biochemistry* **19,** 2396–2404.

Tsien R. Y. (1983) Intracellular measurements of ion activities. *Ann. Rev. Biophys. Bioeng.* **12,** 91–116.

Tsien R. Y. and Rink T. J. (1980) Neutral carrier ion-selective microelectrodes for measurement of intracellular free calcium. *Biochim. Biophys. Acta* **599,** 623–638.

Tsien R. Y. and Rink T. J. (1981) $Ca^{2+}$-selective electrodes: A novel PVC-gelled neutral carrier mixture compared with other currently available sensors. *J. Neurosci. Meth.* **4,** 73–86.

Tsien R. Y., Pozzan T., and Rink T. J. (1982) Calcium homeostasis in intact lymphocytes: Cytoplasmic free calcium monitored with a new, intracellularly trapped fluorescent indicator. *J. Cell. Biol.* **94,** 325–334.

Tsien R. Y., Pozzan T., and Rink T. J. (1984) Measuring and manipulating cytosolic $Ca^{2+}$ with trapped indicators. *Trends Biochem. Sci.* **9,** 263–266.

Tsien R. Y., Rink T. J., and Poenie M. (1985) Measurements of cytosolic free $Ca^{2+}$ in individual small cells using fluorescence microscopy with dual excitation wavelengths. *Cell Calcium* **6,** 145–157.

Ujec E., Keller E. E. O., Kriz N., Pavlik V., and Machek J. (1980) Low-impedance, coaxial, ion-selective, double-barrel microelectrodes and their use in biological measurements. *Bioelectrochem. Bioenerget.* **7,** 363–369.

Wakasugi H., Kimura T., Haase W., Kribben A., Kaufmann R., and Schulz I. (1982) Calcium uptake into acini from rat pancreas: Evidence for intracellular ATP-dependent calcium sequestration. *J. Memb. Biol.* **65,** 205–220.

Walker J. L. (1971) Ion specific liquid ion exchanger microelectrodes. *Anal. Chem.* **43,** 89A–92A.

Weingart R. and Hess P. (1984) Free calcium in sheep cardiac tissue and frog skeletal muscle measured with $Ca^{2+}$ selective micraelectrodes. *Pfluegers Arch.* **402,** 1–9.

Williams D. A., Forgarty K. E., Tsien, R. Y., and Fay F. S. (1985) Calcium gradients in single smooth muscle cells revealed by the digital imaging microscope using fura-2. *Nature* **318,** 558–561.

# Acid–Base Balance

Bo K. Siesjö, Fredrik Boris-Möller,
and Eduardo Martins

## 1. Introduction

Changes of cerebral acid–balance, notably of extracellular pH
($pH_e$), have long attracted the attention of physiologists. This is
mainly because both cerebral blood flow (CBF) and pulmonary
ventilation are exquisitely sensitive to alterations in $pH_e$, whether
these are caused by changes in $pCO_2$ or by addition of strong acid
or base to cerebral extracellular fluids (*see* reviews by Kuskchinsky
and Wahl, 1978; Siesjö and Ingvar, 1983).

A corresponding interest in changes of intracellular pH ($pH_i$)
has been slower to develop. This is understandable since $pH_i$ has
been much more difficult to measure, and the intracellular fluids
more difficult to manipulate. For many years, though, it has been
recognized that cerebral hypoxia/ischemia must lead to dramatic
changes in $pH_i$, related to the production/accumulation of metabolic acid, chiefly lactic acid (for literature, *see* Siesjö 1982, 1985; Kraig
et al., 1985, 1986; von Hanwehr et al., 1986). Two recent developments have caused a surge of interest in intracellular acid–
base events. First, with the advent of micro-pH electrodes, which
allow for measurements in isolated neurons and muscle cells, it
became possible to characterize molecular mechanisms achieving
regulation of $pH_i$ by transmembrane transport of $H^+/HCO_3^-$ (*see*
Roos and Boron, 1981; Boron, 1983; Thomas, 1984). Second, it has
become evident that conditions that cause excessive accumulation
of lactic acid in the hypoxic/ischemic brain lead to enhanced brain
damage (for reviews, *see* Myers, 1979; Siesjö, 1981, 1984; Plum,
1983; Kraig et al., 1985).

In this chapter, we will describe methods suited for studies of
cerebral acid–base balance in vivo. It seems justified to begin with a
clarification of terms, i.e., to define independent and dependent
variables in acid–base alterations of extra- and intracellular fluids.
We will then proceed by discussing both time-honored methods
for estimating $pH_i$, by the so-called weak acid methods ($CO_2$,
DMO), and those that represent unconventional variations of the

basic theme. Since another chapter in this volume is devoted to microelectrode measurements of ionic activities in cerebral interstitial fluid (*see* chapter by Nicholson and Rice, in this volume), our own discussion of measurements of $pH_e$ will be brief. The final sections are devoted to noninvasive measurements of acid–base events in experimental animals and humans with positron emission tomography (PET) and nuclear magnetic resonance (NMR) techniques.

## 2. Acid–Base Theory

In discussing methods for assessing acid–base balance in extra- and intracellular fluids, we must define those variables that can be measured, or derived by calculation. Since these calculations require the use of a series of constants, it seems necessary to briefly summarize the equations that relate independent and dependent acid–base variables. For more detailed discussions, the reader is referred to previous reviews (Siesjö and Messeter, 1971; Siesjö, 1978, 1985; Roos and Boron, 1981; Stewart, 1981).

### 2.1. Terminology

Of key interest in acid–base physiology is the hydrogen ion activity ($a_{H^+}$) or, to a less precise approach, the hydrogen ion concentration ($H^+$). It is customary to take the negative logarithm of this term and use the resulting term—pH—as the chief dependent variable in acid–base balance. The current definition of pH is operational and rests on the measurements of differences in electrochemical potentials between the unknown solutions and certain standard buffer solutions (Waddell and Bates, 1969; Roos and Boron, 1981). With a stringent approach, therefore, it is not possible to define quantitatively the $H^+$ activity (or concentration) of a solution, nor can its negative logarithm be precisely equated with pH of that solution. In this chapter, though, and for the purpose of discussing alterations in $H^+$ activity, we will assume that $a_{H^+}$ and pH can be interconverted by the equation $pH = -\log a_{H^+}$.

If $pH_e$ is the dependent variable, which are the independent ones? Clinical acid–base chemists recognize two: $pCO_2$ and the buffer base (BB) concentration or, as some have preferred to call it, the base excess/base deficit (*see* Siggaard-Andersen, 1966, 1974;

Siesjö, 1978, 1985). The $pCO_2$ of a fluid or a tissue is by definition that of a gas in diffusion equilibrium with the fluid or tissue. It is easy to define *changes* in BB since addition of strong acid (e.g., HCl) or strong base (e.g., NaOH) will cause a mole-to-mole decrease and increase in BB, respectively. We define a strong acid as one that ionizes completely, or nearly completely (e.g., $pK_{Ha} < 4$) at physiological pH values. In order to define the absolute BB concentration, assume an isotonic $Na^+/Cl^-$ solution that is equilibrated with a $pCO_2$ of 40 mm Hg. This would lead to an acid solution with a pH of 4.5. Maintaining the $pCO_2$, we can now titrate the solution with isotonic NaOH to a pH of 7.3. In titrating the solution we raise the $HCO_3^-$ concentration from close to 0 to about 20 m$M$. From an acid–base point of view, the solution now mimicks CSF or ECF.

The equation for electrical neutrality can be written as

$$Na^+ + H^+ = Cl^- + OH^- + HCO_3^- + 2\,CO_3^{2-} \tag{1}$$

*Before* titration, the term $[Na^+] - [Cl^-]$ is zero, and the equation can be simplified into

$$H^+ - OH^- = HCO_3^- + 2\,CO_3^{2-} \tag{2}$$

Clearly, the term $H^+ - OH^-$ is small (about 0.03 m$M$) and so is, therefore, $HCO_3^- + 2\,CO_3^{2-}$.

In the final solution, i.e., *after* titration, the addition of NaOH has raised the $Na^+$ concentration. It is convenient to rewrite the equation of neutrality as

$$Na^+ - Cl^- = OH^- - H^+ + HCO_3^- + 2\,CO_3^{2-} \tag{1}$$

At $pH = 7.3$ the term $OH^- - H^+$ is much smaller than $HCO_3^- + 2\,CO_3^{2-}$ and can be neglected. We can then write

$$Na^+ - Cl^- = HCO_3^- + 2\,CO_3^{2-} \tag{3}$$

The sum of the $HCO_3^-$ and $CO_3^{2-}$ concentrations is now equal to the BB concentration. Thus, $HCO_3^-$ and $CO_3^{2-}$ are the only anions present that can combine with $H^+$ at physiological pH values. We observed that the difference for $Na^+ - Cl^-$, i.e., the amount of strong base added, is equal to the BB concentration.

Suppose now that we add a nonbicarbonate buffer acid (Ha, e.g., with $pK_{Ha}$ close to 7) to the original solution and then titrate it with NaOH until pH is close to 7.0. The solution would mimick an intracellular fluid. For this solution, Eq. (3) must now be written as

$$Na^+ - Cl^- = HCO_3^- + 2\,CO_3^{2-} + a^- \tag{4}$$

In other words, BB is composed of bicarbonate, carbonate, and the anion of the buffer acid. It follows from this that a change in BB, the independent variable, will no longer be quantitatively reflected in changes of $HCO_3^-$.

Stewart (1981) has recently published a detailed account of acid–base equilibria in systems mimicking both extra- and intracellular fluids. He lists as independent variables the strong ion difference (SID), the $pCO_2$, and $C$, the latter expressing the total concentration of a nonbicarbonate buffer system ($C = Ha + a^-$). SID is the difference between the sum of all "strong" cations ($Na^+$, $K^+$, and so on) and all "strong" anions ($Cl^-$, lactate$^-$, and so on). Thus, SID is, for all practical purposes, the same as BB. As dependent variables, Stewart lists pH (or $a_{h+}$), $HCO_3^-$, $CO_3^{2-}$, and $a^-$. Stewart's approach leads to the same equations as those proposed earlier (e.g., Siesjö and Messeter, 1971), with the term SID replacing BB. However, his description is useful in its strict adherence to basic physiochemical events and its focus on changes in strong ion concentrations as causes of acid–base alterations (for a discussion, *see* Siesjö, 1985).

## 2.2. Quantitative Acid–Base Relationships

Quantitative acid–base relationships have been worked out for extracellular fluids containing, as the only buffer system, the $CO_2/HCO_3^-$ system, and for intracellular fluids containing, in addition, one or several nonbicarbonate systems (*see* Siesjö and Messeter, 1971; Siesjö, 1972, 1978, 1985; Stewart, 1981). We recall that BB is equal to SID. For an extracellular fluid the acid–base relations are quantitatively described by four equations. Within physiological ranges, the relationship between BB, $pCO_2$, and $H^+$ is given by the equation

$$BB = \frac{pCO_2 SK_1'}{H^+}\left(1 + \frac{2K_2'}{H^+}\right) \tag{5}$$

where $S$ is the solubility cofficient for $CO_2$ in the fluid (at 37°C, $S$ is $3.1 \times 10^{-3}$ m$M$/L/mm Hg); $K_1'$ and $K_2'$ are the apparent ionization constants of $H_2CO_3$ and $HCO_3^-$, respectively (at 37°C and pH = 7.3, $pK_1' = 6.13$ and $pK_2' = 9.88$). If the $CO_3^{2-}$ concentration is neglected, the term within the parenthesis disappears.

For an intracellular fluid the corresponding relationship is

$$BB = \frac{pCO_2 SK_1'}{H^+}\left(1 + \frac{2K_2'}{H^+}\right) + \frac{CK_{Ha}}{H^+ + K_{Ha}} \tag{6}$$

where $C$ is the total concentration of nonbicarbonate buffer ($a^- +$ Ha), and $pK_{Ha}$ is the ionization constant of the buffer acid. The equation can easily accommodate several buffer acids, with varying $pK_{Ha}$ and $C$ values (Siesjö and Messeter, 1971: Siesjö, 1978, 1985). Thus,

$$BB = \frac{pCO_2 SK_1}{H^+}\left(1 + \frac{2K_2}{H^+}\right) + \sum_i \frac{C_i K_{Ha_i}}{H^+ + K_{Ha_i}} \tag{7}$$

Equation (7) is presented here to show that there exists a single relationship between BB, $pCO_2$, and $H^+$, provided, however, that the physical properties of the fluid and the nonbicarbonate buffers are known. Conversely it is possible to evaluate the physical properties provided we know BB, $pCO_2$, and $H^+$.

## 2.3. Determinants of pH In Vivo

Accepting that $pCO_2$ and BB are the main determinants of pH in vivo, we probe into the causes of a change in BB, i.e., in SID. Many years ago, three mechanisms were defined that accomplish regulation of $pH_i$ in hyper- and hypocapnia (Siesjö and Messeter, 1971);

> Physiochemical buffering
> Consumption or production of metabolic acids
> Transmembrane transport of $H^+$ or $HCO_3^-$

The schematic diagram in Fig. 1 helps to illustrate the last two of these mechanisms. Thus, a change in BB either means that acid has been produced or consumed within the intracellular fluids, or that $H^+$ or $HCO_3^-$ has been translocated across the membrane. We recognize that $H^+$ crosses the membrane in exchange for $Na^+$, and $HCO_3^-$ in exchange for $Cl^-$. The electrochemical force of the distribution of $Na^+$ across the membrane is so much stronger than the $H^+$ force that it is $Na^+$ that determines how the $Na^+/H^+$ antiporter will behave when activated. The sodium ion will make the $Na^+/H^+$ antiporter function as a $H^+$ pump with the flux directed out of the cell.

In a similar manner, the coupling of $Cl^-$ and $HCO_3^-$ in the antiporter will result in a flux of $HCO_3^-$ out of the cell. This is equivalent to a $H^+$ leak into the cell. Thus, if the cell is acidified by an increase in $pCO_2$, pH regulation can occur by consumption of metabolic acids, such as lactic acid, and/or by increased $Na^+/H^+$ exchange across the membrane. If, instead, pH rises because of

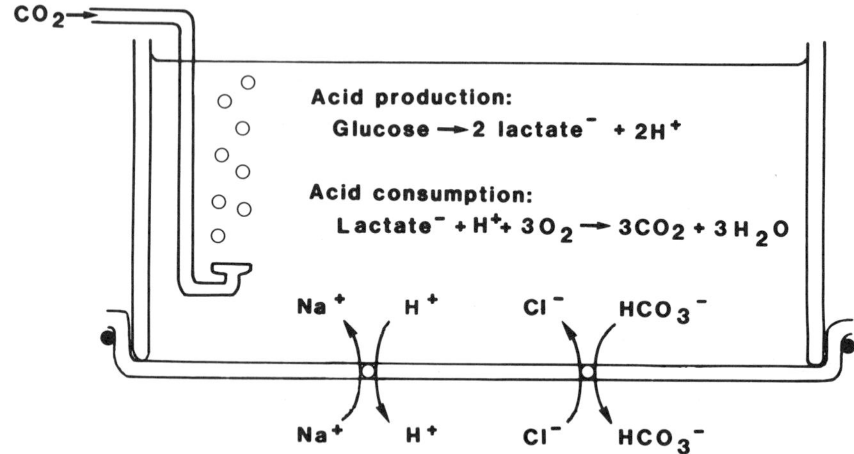


Fig. 1. The changes in $H^+$ concentration in intracellular fluid can be regulated by acid production/consumption within the cell or transmembrane ion fluxes via the $H^+/Na^+$ and $HCO_3^-/Cl^-$ antiporters. The arrows denote the directions of ion fluxes under normal physiological conditions when the antiporters are activated (reproduced, with permission, from Siesjö, 1985).

hypocapnia, regulation can occur by production of metabolic acids, and/or by increased $HCO_3^-/Cl^-$ exchange.

It has been customary to regard acidosis during anaerobiosis as arising from lactic acid production, with 2 lactate⁻ and 2 $H^+$ being formed from one glycosyl unit in glucose or glycogen. Actually, this is not so, since the $H^+$ released during glycolysis stems from other reactions, mainly that causing hydrolysis of ATP (*see* Alberti and Cuthbert, 1982; Hochachka and Mommsen, 1983). However, at constant ATP concentration the net result of glycolysis is, under all circumstances, the production of 2 lactate⁻ and 2 $H^+$ per glycosyl unit metabolized.

The diagram of Fig. 1 helps us to define $\Delta$BB and to understand why it is identical to $\Delta$SID. For example, if a strong acid, such as lactic acid ($pK < 4$), is released within the cell, so that its concentration is increased by 5 m$M$, BB is reduced by 5 m$M$. Since the concentration of lactate increases by 5 m$M$, SID decreases by the same amount. Furthermore, if $Na^+/H^+$ or $Cl^-/HCO_3^-$ exchange proceeds so that the internal $Na^+$ increases by 5 m$M$, or $Cl^-$ decreases by 5 m$M$, BB and SID have both increased by 5 m$M$. Similarly, if lactic acid (Hla) diffuses from the cell so that the external lactate⁻ increases by 5 m$M$, $\Delta$BB$_e$ is 5 m$M$. In this case, $\Delta$BB

= $\Delta HCO_3^-$ since extracellular fluids are virtually devoid of anions other than $HCO_3^-$.

It follows from these considerations that an adequate analysis of acid–base events should encompass the major independent variables (BB and $pCO_2$), and the most important dependent variable (pH). It is also clear that changes in BB are associated with (or caused by) changes in concentrations of strong cations and anions. We also recognize that in cerebral extracellular fluids, $\Delta HCO_3^- = \Delta BB$, hence measurements of $pCO_2$ and $HCO_3^-$ concentration directly characterize the independent variables. For this reason, and since it influences membrane transport via the $Cl^-/HCO_3^-$ antiporter, it seems justified that the $HCO_3^-$ concentration is measured or calculated.

## 3. Estimation of pH$_i$ by Weak/Acid Weak Base Methods

### 3.1. Theory

The principle is based on the following considerations. If a weak acid or a weak base has a pK' close to the physiological pH values, if only the unionized species of the acid or base can penetrate cell or intracellular membranes, and if the species are neither bound to tissue constituents nor metabolized, then the ionized species will distribute across cellular and intracellular membranes according to the pH gradient (*see* Waddell and Bates, 1969; Roos and Boron, 1981). Although a number of weak base methods have been published, weak acids have dominated. Assume an acid (Ha), whose $pK'_{Ha}$ is known and equal in all extra- and intracellular fluids, and whose total concentration is $C = a^- + Ha$. For any compartment we can write

$$pH = pK' + \log \frac{a^-}{Ha} = pK' + \log \frac{C - Ha}{Ha} \qquad (8)$$

Suppose that this compartment is plasma, in which pH ($pH_{pl}$) and C ($C_{pl}$) are measured. Clearly, Eq. (8) allows us to calculate the H$a$ concentration that, following equilibration, is the same in plasma, ECF, and ICF water. We will now measure C also in tissue ($C_t$) and calculate its concentration per unit weight of tissue water. Since we know H$a$ and $C_t$, tissue pH can be calculated according to Eq. (8). However, a simple calculation of this type on a two-compartment system (blood and tissue) will give some kind of "aggregate" tissue pH ($pH_t$).

If we wish to calculate $pH_i$ in a tissue, we must consider it composed of three components, namely blood and extracellular and intracellular parts. The total is the sum of the components:

$$C_t V_t = C_{pl} V_{pl} + C_e V_e + C_i V_i \tag{9}$$

$$V_t = V_{pl} + V_e + V_i \tag{10}$$

For many tissues, such as muscle, one can assume that plasma and ECF have similar composition and, accordingly, measure $pH_e$ and $C_e$ in plasma. For brain tissues, one must sample CSF, or measure $pH_e$ in interstitial fluid by microelectrodes. We recall that since Ha should be the same in plasma and ECF, one can use measured $pH_e$ and Eq. (8) to derive $a_e^-$. Whether CSF is sampled or $pH_e$ measured, though, the volume of ECF must be estimated (see below).

Since weak acids (or bases) equilibrate across all biological membranes, the $pH_i$ values derived will be some sort of average for different cells and different cellular compartments. Thus, even if the different cell types of a tissue had the same cytosolic pH values, one must assume that cell compartments differ; for example, mitochondria are supposed to be more alkaline and lysosomes more acid than the cytoplasm (*see* Roos and Boron, 1981). It is no longer recommended that one attempt to interpret the distribution of weak acids or weak bases in terms of the mean $OH^-$ or mean $H^+$ concentrations (Waddell and Bates, 1969; Roos and Boron, 1981). What can be stated is that when such inhomogeneities exist, weak acids give a higher pH value than a weak base, and these two values will bracket the volume-weighted average $pH_i$, defined as

$$pH_i = \sum_{j=1}^{n} f_j pH_j \tag{11}$$

where $n$ is the number of compartments and $f_j$ is the fractional volume of the jth compartment (equal to $V_j/V_t$).

Although weak bases have been applied to nervous tissues, we will center the discussion on two weak acids, $CO_2$ and DMO (dimethyloxazolidinedione). This is because these techniques have been used extensively, and because they have been adopted for measurements of $pH_i$ in humans with PET techniques. Measurements of $pH_i$ with another weak acid (umbelliferone) will then be considered. For a detailed review of weak base methods, the reader is referred to a recent review article (Roos and Boron, 1981).

## *3.2. Extracellular Fluid Volume ($V_e$)*

$V_e$ varies substantially in a number of situations associated with major ionic shifts across membranes, particularly in ischemia and spreading depression (van Harreveld and Ochs, 1956; Hansen and Olsen, 1980; Hossmann, 1982), in severe hypoglycemia (Pelligrino et al., 1981a), and during epileptic seizures (Elazar et al., 1966; Dietzel et al., 1980, 1982). It is necessary, therefore, to attempt to estimate these variations when $pH_i$ is derived. Many methods have been described for measuring $V_e$. Three of these will be discussed, the first two of which may be used for routine measurements. Since measurements of $V_e$ are reviewed in another chapter of this handbook, we will restrict this discussion to a few brief comments.

### *3.2.1. Tissue Impedance*

Measurements of tissue impedance give information on changes in $V_e$. The specific impedance of tissues, measured at low frequencies, is mainly determined by the amount of electrolytes in the extracellular fluids (*see* van Harreveld et al., 1963; Ranck, 1963; Tarby et al., 1968). Thus, it has been estimated that only about 10% of an applied current passes through vascular elements; the part passing through glial and neuronal elements is probably also small because of the high resistivity of the plasma membranes. Some uncertainty will remain regarding the contributions played by these current pathways, especially in situations in which flow is altered and/or the neuronal membranes are depolarized (*see* Pelligrino et al., 1981a). There is no evidence to suggest, though, that the values obtained are grossly in error. In fact, values for ECF measured by impedance techniques are very similar to those derived from tracer data and, in one instance, the methods were directly compared and shown to give similar results (Fenstermacher et al., 1970).

In theory, impedance data can be used to calculate absolute values for ECF volume. Two equations have been described for this purpose (Cole et al., 1969). One, the Maxwell equation, considers the passage of a current through a uniform suspension of spheres:

$$\frac{\rho}{2} = \frac{1 - r_1/r_2}{2 - r_1/r_2} \tag{12}$$

The other, the Rayleigh equation, is based on current flow through a uniform suspension of cylinders normal to the axis of the cylinders:

$$\rho = \frac{1 - r_1/r_2}{1 + r_1/r_2} \tag{13}$$

In these equations $\rho$ = fractional volume of the cells in the suspension, $r_1$ = specific impedance of the fluid surrounding the cells, and $r_2$ = specific impedance of the tissue as a whole. The ECS volume is then calculated as $1 - \rho$. The ECF has a value of 55–56 $\Omega$ cm at 37°C (Fenstermacher et al., 1970).

The absolute values obtained by such calculations are somewhat uncertain, mostly because the geometrical arrangements of cellular processes are so vastly different from those assumed when the equations are used. It appears more appropriate, therefore, that one assumes a normal value for ECF volume (e.g., 15% of tissue weight) and then employs the impedance measurements to estimate *changes* in $V_e$, using the Maxwell or Rayleigh equations (Pelligrino et al., 1981a).

### 3.2.2. Distribution of Tracers

One widely used technique for assessing the size of ECF is to administer systematically a tracer that penetrates cell membranes poorly, and therefore remains localized to ECF, wait until equilibrium has been reached, and then calculate $V_e$ from tracer concentrations in plasma and tissue. For brain, such techniques have failed for reasons that are to be found in the poor penetration of such markers (inulin, mannitol, $SO_4^{2-}$, and so on) across the blood–brain barrier.

An alternative technique is to administer the tracer via the CSF pathway. Even then, special problems arise because of the length of perfusion required for diffusion into the tissue to occur. Woodward et al. (1967) perfused the CSF system in rats with an artificial CSF containing $^{14}$C-labeled inulin, using a perfusion system with an inflow needle in one lateral ventricle and an outflow needle in the cisterna magna. Although this system does not necessarily perfuse the entire CSF system of the rat, it delivers the isotope to the ECF side of the blood–brain barrier. Equilibrium between brain and perfusate in terms of a constant $^{14}$C-inulin ratio was found after 6 h of perfusion, yielding ECF volumes of 13.5–14%.

Other workers proved it possible to reduce the time of perfusion considerably, using a graphical procedure to allow for estimation of tissue concentration close to the source, i.e., CSF. Rall et al. (1962) maintained a steady concentration of tracer compounds ($^{14}$C-inulin, $^{35}$S-sulfate, or $^{14}$C-dextran) in the CSF by ventriculocisternal perfusion, and measured the relative concentration of tracer as a function of both the distance from the brain-CSF boundary and time. In order to allow measurements on neocortex, later perfusion was performed of the cortical subarachnoid space via inflow cannulae inserted through the superior sagittal sinus, and outflow proceeding through a cisternal needle (Fenstermacher et al., 1970; Levin et al., 1970).

In practice, following perfusion the tissue must be sliced in such a way that tracer concentration as a function of distance to the surface facing CSF can be determined. The curve for diffusion of a substance remaining in the ECF is such that roughly the first half of the curve shows a linear fall in the tracer concentration. This makes it permissable to extrapolate the tissue tracer concentration at the tissue-fluid boundary. Here we know that the tracer concentration is equal in ECF and CSF. Therefore the relative extracellular volume is simply given by the ratio cpm (tissue)/cpm (fluid). Probably, the first tissue slice should be corrected for an adsorbed, 15 $\mu$m thick, layer of CSF containing a high concentration of tracer (Levin et al., 1970).

Using these procedures, Levin et al. (1970) measured a cortical sucrose space of 19.5–19.9% in four different species, and an inulin space of 17.6–18.8%. Correction of the values of the first samples for an absorbed layer of CSF would lower these values by 2–3% (Fenstermacher et al., 1970). It would seem justified to conclude that, in a variety of species, cortical ECF volumes are normally in the range 15–20%. Similar values are obtained for caudate nucleus (Rall et al., 1962), whereas, as will be discussed below, white matter has a lower space.

CSF perfusion, with subsequent slicing of a core of tissue perpendicular to an even surface, is too complex a procedure to allow for routine measurements of ECF volume. In large animals, such as dogs, simplified procedures have been used that nonetheless seem to approximate ideal conditions for distribution of ECF tracers. Thus, Arieff et al. (1976) administered $^{35}$S-sulfate (and $^{14}$C-DMO) to functionally nephrectomized dogs, both by iv and intracisternal injection, employing radioactive doses that would give approximately equal concentrations in plasma and CSF. Following an equilibration period of 3 h, *cortical* CSF was then sampled together with a 0.4-mm thick superficial slice of cortical tissue, and ECF volume was calculated from the $^{35}$S-sulfate concentrations of these samples. So calculated, $V_e$ was 22% in cortex and 7.2% in white matter.

A similar procedure was used by Pelligrino et al. (1981b), also working in dogs. The authors administered $^{35}$S-sulfate via both the blood and CSF routes; however, in order to maintain a CSF source, they perfused the ventriculocisternal system for 2.5 h. Following sampling of cortical CSF, and of tissue, the latter was sliced for counting and extrapolation to zero thickness, thus allowing for calculation of ECF volume, as described above. The authors obtained values for $V_e$ of 22% in cortex, and about 5% in white matter. The calculated ECF volume for caudate nucleus was $16.7 \pm 1.2\%$ (mean $\pm$ SEM), a value significantly lower than for cortex.

In general, tracer methods to assess $V_e$ have a sound theoretical basis. However, they are based on assumptions regarding the ideal behavior of tracers that are not necessarily valid in a given experimental situation. The tracers should move through ECF by free diffusion, neither binding to surface structures nor being translocated into intracellular fluids. The major problems in applying tracer methods are probably the difficulty of allowing sufficient equilibration between CSF and tissue and that associated with graphical extrapolation of tracer concentrations of tissue slices to zero thickness. The methods are unsuitable for assessment of rapid changes in $V_e$.

### 3.2.3. Ion-Selective Electrodes

The third method for determining $V_e$ is one in which the concentration of ions for which cell membranes are largely impermeable is measured. The method is based on the principle that the concentration of a substance is inversely proportional to the volume within which it can distribute. As the concentration of such ions, e.g., tetramethylammonium and choline ions, is inversely proportional to the volume fraction occupied by the ECF, one can calculate a relative volume change from the concentration changes measured (Nicholson et al., 1979; Hansen and Olsen, 1980; Dietzel et al., 1980). One method of introducing such ions into the extracellular fluid is by repeated iontophoretic injections of constant amounts of these ions. For details of calculation of changes in $V_e$, the reader is referred to the original publications.

### 3.3. The $H_2CO_3/HCO_3^-$ Method

The $CO_2$ method was the first weak acid method to be used on solid tissues, notably muscle (Fenn and Maurer, 1935; Wallace and Hastings, 1942). This method is now used much less frequently than the DMO method. This is because $CO_2$, being a gas and thereby labile, is more exacting to measure in tissues. It has the advantage, though, that $CO_2$ equilibrates more rapidly between blood and tissue than does DMO. Furthermore, it is possible to measure independently the concentration of the acid, i.e., $pCO_2$. Thus, the $CO_2$ method can be used also when arrest of the circulation prevents equilibration between plasma and tissue (Kraig et al., 1985, 1986; Smith et al., 1985; von Hanwehr et al., 1986).

The acid–base pair of the $CO_2$ system is $H_2CO_3/HCO_3^-$. However, since the $H_2CO_3$ concentration is proportional to the amount of dissolved $CO_2$, and since the latter is proportional to $pCO_2$, one can use the expression $pCO_2 S$ to denote the concentration of acid, where $S$ is a solubility coefficient (mmol/L/mm Hg). The $HCO_3^-$ concentration cannot be measured directly, but the total $CO_2$ content ($tCO_2$) can. $tCO_2$ is composed of dissolved $CO_2$, bicarbonate, and carbonate. However, since the carbonate concentration is so low that it can be neglected, we obtain $HCO_3^- = tCO_2 - pCO_2 S$. In order to derive $pH_i$, we can then use the following equations (e.g., Siesjö et al., 1972):

$$[HCO_3^-]_t = tCO_2 - p_t CO_2 \cdot S_1 \tag{14}$$

$$[HCO_3^-]_i = \frac{[HCO_3^-]_t - [HCO_3^-]_e V_e - [HCO_3^-]_{bl} V_{bl}}{V_i} \tag{15}$$

$$pH_i = pK_1' + \log \frac{[HCO_3^-]_i}{p_t CO_2 S_2} \tag{16}$$

Here, $PtCO_2$ is the mean tissue $CO_2$ tension, a variable that will be discussed below. $S_1$, which is the solubility of $CO_2$ in whole tissue, has the value of 0.0292 mmol/kg/mm Hg at 37.5°C (Siesjö, 1962a). A slightly different value is used for $S_2$, the solubility of $CO_2$ in intracellular water (0.0314 mmol/kg/mm Hg). $V_e$, $V_i$, and $V_{bl}$ are the fractional volumes of ECF, ICF, and blood, respectively. Finally $pK_1$, the first apparent ionization constant of carbonic acid, has a value of 6.12 at 37°C (Siesjö, 1962b). Its variation with temperature and pH has been assessed in the CSF (Mitchell et al., 1965).

The blood content of the tissue is low (about 3%) (*see* Everett et al., 1956), and the correction for blood $HCO_3^-$ content is hardly critical. Important variables are $tCO_2$, $p_tCO_2$, $[HCO_3^-]_e$, and $V_e$. These are discussed in turn.

### 3.3.1. Total CO₂ Content

This is usually measured after acidification that converts $HCO_3^-$ (and $CO_3^{2-}$) to the diffusible $CO_2$. Convenient methods for measuring $CO_2$ involve vacuum extraction and measurements of gas pressures, or microdiffusion, trapping in base, and titration of the excess base with acid. One method, based on the microdiffusion techniques of Conway (1950), has been used extensively in our laboratory. The method was originally used to measure $tCO_2$ in CSF (Siesjö, 1962a) and subsequently adopted for tissue (Ponten and Siesjö, 1964). The following is a brief description of the latter technique.

For measuring tissue $CO_2$ content it is advisable to freeze the tissue *in situ*, e.g., with liquid nitrogen (Ponten et al., 1973). Freezing arrests $CO_2$ production; in addition, it allows the tissue to be broken up by crushing. There is a price to be paid, though. Thus, liquid nitrogen attracts $CO_2$ from ambient air. It is necessary, therefore, to crush the tissue and transfer the aliquots to the diffusion units in a glove box that has been previously perfused with $CO_2$-free air and/or provided with efficient $CO_2$ absorbers. It is also advisable to titrate excess base in a stream of $CO_2$-free air.

Figure 2 shows a convenient diffusion unit. The main flask contains 5% trichloracetic acid and the side arm contains 0.3 mL of a $0.1M$ $Ba(OH)_2$ solution. The flasks must be flushed with $CO_2$-free air before the $Ba(OH)_2$ is pippeted into the side arm. The stoppered tube on the side arm allows the unit to be perfused with $CO_2$-free air during titration.

Following freezing of the tissue *in situ* and storage of the brains at –90°C, the samples are brought into the glove box and kept in a Dewar flask with liquid nitrogen. It is advisable to filter the liquid $N_2$ through glass wool in the box to remove flakes of ice and $CO_2$. Samples (100–200 mg) of cortical tissue, or other parts of the brain, are then chipped off and crushed in a mortar in liquid $N_2$. Aliquots (50–100 mg) are then transferred with a spoon to the diffusion units just as the liquid $N_2$ has evaporated, i.e., when the samples are in powder form. The diffusion units are then immediately restoppered. When all samples have been transferred, the units are reweighed to determine the weight of the sample. The

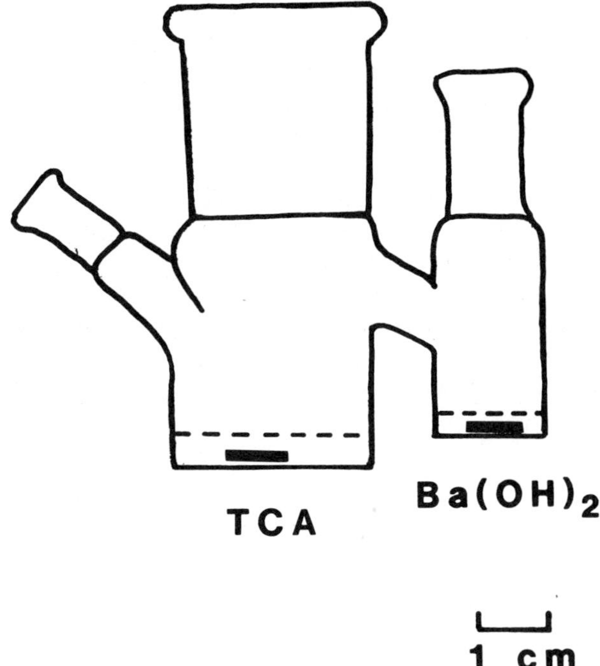



Fig. 2. Diffusion chamber used for the determination of the total carbon dioxide content of tissue samples. See text for explanation of the procedure of measuring.

acid is stirred for 2–3 h to allow diffusion of $CO_2$, and the excess $Ba(OH)_2$ is titrated with a 0.02$M$ HCl solution. The amount of $CO_2$ in the sample can then be calculated as

$$CO_2\ (\mu mol/g) = 0.02(T_b - T_s)/(2W) \qquad (17)$$

where $T_b$ and $T_s$ are $\mu L$ of acid titrated for blank and sample, respectively, and $W$ is the weight of the sample in g. The blanks are treated in the same way as the samples, except that no tissue is added to the flasks after opening the stoppers. The accuracy of the technique is conveniently checked by treating frozen samples of a sodium carbonate solution (10 mmol/L) as tissue.

Alternative methods for measuring $tCO_2$ exist. At the time when the classical method of van Slyke and Neil (1924) was used to measure $CO_2$ in blood, attempts were made to liquefy solid tissue samples. Anrep et al. (1936) devised one such method in which tissue samples were added to a 0.125$N$ NaOH solution, maintained at 60–65°C under a layer of paraffin. Alkaline hydrolysis for 30 min was found to liquefy most samples tested. An aliquot of the hydrolyzate was then used for manometric analysis, the $CO_2$ being blown off with $0.4N$ sulfuric acid containing 1% of copper sulfate, the latter added to precipitate volatile sulfides liberated during acid hydrolysis.

The method of Anrep et al. (1936) was used by Weyne et al. (1968, 1970). These authors modified the technique somewhat since they froze the brain *in situ*. The tissue was then removed under $CO_2$-free air and submerged in the NaOH solution under paraffin at 0°C. Nowadays, few laboratories use the manometric technique of van Slyke. However, if properly controlled, the method of alkali hydrolysis could be used as an introductory step for liquefaction of tissue, followed by determination of $CO_2$ with nonmanometric techniques, e.g., gas chromatography.

A simple but attractive technique was recently described for $tCO_2$ determination with a micro-$pCO_2$ electrode (Kraig et al., 1986). The authors stored pieces of tissue, frozen *in situ*, at $-25°C$ for 2–5 h, a procedure that allowed any environmental $CO_2$ deposited on the samples to disappear. The weighed samples were then added to a 10-mL test tube containing 300 $\mu$L of 150 m$M$ sodium fluoride (pH 7), followed by 2 mL of light mineral oil. Then 100 $\mu$L of $1M$ HCl was gently added to the aqueous layer. Finally, the active surface of the $pCO_2$ electrode was lowered into that layer, and the rise in $pCO_2$ recorded. $pCO_2$ is calculated from the rise in $pCO_2$ and the solubility coefficient at the temperature of the solution, the electrode being calibrated in streaming gas before and after the measurement.

### 3.3.2. Tissue $CO_2$ Tension

The mean tissue $CO_2$ tension can be defined from diffusion theory (Gleichmann et al., 1962; Ponten and Siesjö, 1966). As Fig. 3 shows, one can define geometrical conditions for $CO_2$ diffusion from the tissue to the blood. Let us consider a simple Krogh model, i.e., in which the geometry of the capillary vessels is such that they are parallel and at constant distances from each other. Furthermore, the blood flow is unidirectional. Then it is possible to define the mean capillary $CO_2$ tension as the arithmetical mean of the arterial and venous $CO_2$ tensions, halfway the distance from the arterial to the venous end, and the mean tissue $CO_2$ tension can then be obtained by calculation. At known values for $CO_2$ diffusion, rate of $CO_2$ production, and intercapillary distances, mean

$P_tCO_2$ should be only 0.5–1.0 mm Hg higher than the mean capillary $CO_2$ tension, i.e., lower than the venous $CO_2$ tension.

The predictions made from diffusion theory can be tested in two ways. First, one can assume that the relatively stagnant CSF pools come into diffusion equilibrium with the tissue, i.e., that they can be used as tonometers in vivo. Second, it is possible to use a $CO_2$ electrode to measure $p_tCO_2$ directly.

The principle of a $pCO_2$ electrode is that pH is measured in a dilute bicarbonate solution, separated from the medium by a membrane that is permeable to $CO_2$, but not to ions (Gertz and Loeschke, 1956; Severinghaus and Bradley, 1958). Measurements

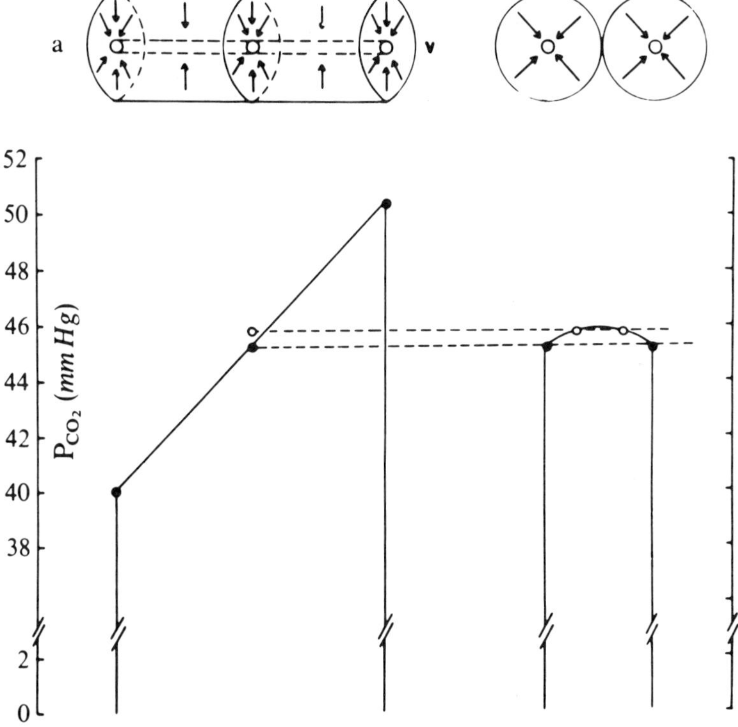

Fig. 3. $CO_2$ diffusion in tissue. The left part of the diagram shows the linear increase in $pCO_2$ in the capillary as the blood passes from the arterial to the venous side. The right part shows a cross-section of two adjacent capillaries midway between the arterial and venous side. The arrows denote the direction of diffusion of $CO_2$ from the tissue into the capillary. The mean value of the tissue $pCO_2$ is 0.5–1.0 mm Hg higher than the capillary $pCO_2$ at this point (reproduced, with permission, from Siesjö, 1972).

on the surface of tissues are made possible by the use of a flat,
Mac-Innes and Dole type of pH electrode (Siesjö, 1961). The glass
and calomel reference electrodes were subsequently built into a
thermostat-controlled glass housing, and were successfully used
in experiments on cats (Ponten and Siesjö, 1966). This electrode is
commercially available (Eschweiler and Co., Kiel, West Germany).
A smaller electrode (outer diameter of $pCO_2$ electrode 3 mm, of pH
electrode 2 mm) can be purchased from Microelectrodes Inc., Lon-
donderry, New Hampshire. Its characteristics have been described
by Kraig et al. (1986).

For measuring $p_tCO_2$ on the surface of the cerebral cortex, one
can take up a circular craniotomy and fit the electrode into the
craniotomy in such a way that the electrode replaces the bony
covering. In that case, the shoulder of the end of the electrode can
make a tight fit with the bone via a rubber washer (Ponten and
Siesjö, 1966). A small electrode is conveniently fitted into a craniotomy with a larger diameter. It is then advisable that a well be
constructed at the bony edges and that mineral oil, equilibrated
with $CO_2$, be placed in the well to minimize diffusional losses
of $CO_2$ and drying out of the surface (Kraig et al., 1986). The
electrode should make good contact with the tissue, but should
neither dimple nor be separated from the surface by fluid. One way
to ensure adequate pressure is to use a lever and a counter-bal-
ance; another is to hook the electrode onto a spring. The electrode is conveniently calibrated in streaming gas, warmed to the
temperature of the tissue. Note that the relationship between log
$pCO_2$ and pH is essentially linear over the physiological $pCO_2$
range.

Extensive measurements of $p_tCO_2$ in dogs (Gleichmann et al.,
1962) and cats (Ponten and Siesjö, 1966) showed that both the
$pCO_2$ measured by a surface electrode, and that measured in
cisternal CSF, agree with that calculated from diffusion theory.
According to these results, the relationship between tissue, arterial
(a) and venous (v) $CO_2$ tensions can be approximated by the
equation

$$PtCO_2 = \frac{p_aCO_2 + p_vCO_2}{2} + 1 \qquad (18)$$

Under physiological conditions, the arteriovenous $pCO_2$ difference across the brain is close to 10 mm Hg (e.g., Siesjö, 1972). This
means that the expected $p_tCO_2$ is about 6 mm Hg higher than

$p_aCO_2$. The expected $\Delta pCO_2$ ($=p_tCO_2 - p_aCO_2$) depends on blood flow; accordingly, $\Delta pCO_2$ is higher in hypocapnia and lower in hypercapnia.

Some results do not fit this relationship. In one study, CSF $pCO_2$ in dogs was found to vary as predicted from the presence of a Wien effect (Davies and Gurtner 1973). This refers to the increased dissociation of weak acids under the influence of electrical fields. It is postulated that capillary surface charges cause dissociation of $H^+$ from blood proteins, and that the $H^+$ released reacts with bicarbonate, the $CO_2$ formed acting as a diffusion barrier for $CO_2$ produced in the tissue. The effect was reported to be larger at low plasma pH and increased CBF (Davies and Gurtner, 1973). However, experiments in rats designed to study this effect failed to confirm its presence: rather, the experiments supported the results predicted by diffusion theory (Caronna et al., 1974; Brzezinski et al., 1967).

An elegant mass spectrometric technique has given much higher $\Delta pCO_2$ values than those predicted by Eq. (18), with $p_tCO_2$ exceeding $p_aCO_2$ by more than 20 mm Hg (Seylaz et al., 1983). These results were obtained in the caudate nucleus of rabbits. One must conclude that either local $p_tCO_2$ is increased in the vicinity of the 0.2 mm probe or that the relationships are different in the caudate nucleus. Thus, species differences seem very unlikely, especially since $p_tCO_2$ measured with a micro-$pCO_2$ electrode in the neocortex of rabbits was as predicted by diffusion theory (Hogg et al., 1984).

The balance of evidence thus favors the $pCO_2$ relationships expressed by Eq. (18). This is of practical importance since it means that, for most purposes, $p_tCO_2$ can be derived without the use of a $pCO_2$ electrode. We recall that the electrode is exacting to apply and that its use carries the risk of some tissue trauma (Ponten and Siesjö, 1966; Kraig et al., 1986). However, as already stated, the electrode is indispensable at grossly reduced tissue perfusion.

### 3.3.3. Extracellular $HCO_3^-$ Concentration

What one needs to know is the $HCO_3^-$ concentration of extracellular fluid proper, i.e., of interstitial fluid (ISF). Elegant experiments in gout support the contention that, at steady state, the acid–base composition of bulk CSF is similar to that of ISF. If this is so, analyses of $tCO_2$ in cisternal CSF would give a good estimate of ISF $HCO_3^-$ content. However, in non-steady-state conditions, bulk CSF $HCO_3^-$ concentration must deviate from that of ISF. It is desirable, therefore, to obtain more direct estimates of ISF $HCO_3^-$

concentration. Measurements of ISF pH ($pH_e$) provide such estimates. Thus, if $p_tCO_2$ is known, $[HCO_3^-]_e$ is derived by calculation from the Hendersen-Hasselbalch equation, conveniently expressed in the following form

$$[HCO_3^-]_e = p_tCO_2S_210^{\,pH_e - pK_1'} \qquad (19)$$

Many authors have reported measurements of pH on the cortical surface with membrane electrodes. Although these have given estimates of changes in $pH_e$ similar to those obtained with spear-shaped electrodes, the surface electrodes measure pH in a fluid whose composition may differ from that of interstitial fluid proper. Direct measurements of ISF pH ($pH_e$) were first performed with semimicro-electrodes (Wang and Sonnenschein, 1955; Astrup et al., 1978; Heuser 1978). It soon became possible to construct glass electrodes with a tip diameter of a few micrometers (Urbanics et al., 1978; Schneider et al., 1977; Nemoto and Frinak, 1981; Siemkowicz and Hansen, 1981), and measurements of $pH_e$ were greatly facilitated by the advent of liquid ion exchanger electrodes (Amman et al., 1981; Kraig et al., 1983; Mutch and Hansen, 1984).

In cats, having a $p_aCO_2$ of about 32 mm Hg, $pH_e$ measured with micro-glass electrodes is close to 7.31 (Schneider et al., 1977). Assuming a $p_tCO_2$ of 38 mm Hg, one can calculate a $[HCO_3^-]_e$ of about 19 mM/L. In rats, with a $p_aCO_2$ of 38 mm Hg (assumed tissue $pCO_2$ of 44 mm Hg), the corresponding values were $pH_e = 7.33$, and $[HCO_3^-]_e = 23$ mM/L (Mutch and Hansen, 1984). In both species, the calculated $[HCO_3^-]_e$ is lower than that derived from $tCO_2$ measurements on bulk CSF, suggesting that ISF $HCO_3^-$ is indeed somewhat lower than that of CSF. Whatever the reason for this difference, measurements of $pH_e$ and $p_tCO_2$ should provide valid estimates of changes in $[HCO_3^-]_e$ for calculation of $pH_i$.

### 3.4. DMO Method

The introduction of the DMO method (Waddell and Butler, 1959) was a significant methodological advance. DMO (dimethyloxazolidinedione), a metabolite of the antiepileptic drug trimethadione, is an acid with a $pK_{Ha}$ of 6.1. The compound can be readily labeled with $^{14}C$ to yield a radioisotope method for estimation of $pH_i$, or of the internal pH of organelles such as mitochondria and lysosomes (*see* Waddell and Bates, 1969; Roos and Boron, 1981). The compound has the required features: the unionized form readily penetrates membranes, whereas the ionized does not, and the two forms are neither bound to macromolecules nor metabolized. However, equilibration between plasma and tissue is relatively slow.

### 3.4.1. Theory and Procedures

Waddell and Butler (1959) mathematically expressed $pH_i$ in terms of measurable quantities, or values easily derived from such quantities, as

$$pH_i = pK' + \log\left\{\left[\frac{C_t}{C_e}\left(1 + \frac{V_e}{V_i}\right) - \frac{V_e}{V_i}\right]\right.$$

$$\left. \times \; [10^{(pH_e - pK')} + 1] - 1\right\} \tag{20}$$

In this equation, $C_t$ and $C_e$ are the total tissue and the extracellular concentration of DMO, respectively, each expressed per unit weight of sample water. $pK'$, the ionization constant of HDMO, was estimated to be 6.13 at 37°C (*see* also Boron and Roos, 1976). $V_e$ and $V_i$ are the extra- and intracellular volumes. In subsequent studies, the DMO was labeled with $^{14}C$, and the method was subjected to a detailed and careful analysis (Roos, 1965, 1971; Arieff et al., 1976).

The principle of the DMO method is very similar to that of the $CO_2$ method. However, since the compound is nonvolatile, the handling of the tissue is greatly simplified. As with the $CO_2$ method, a major problem is to assess the volume of the extracellular fluid and the DMO concentration of that fluid. Determination of the former must be performed, as discussed previously. If CSF can be sampled, and if this CSF comes into diffusion equilibrium with plasma, it is possible to measure both pH and total DMO activity in the sample, and to calculate $DMO^-$ and HDMO activities (Roos, 1965; Arieff et al., 1976; Pelligrino et al., 1981b). It is then easy to derive corresponding intracellular values, since HDMO concentration should be identical in ICF and CSF, and $[DMO^-]_i$ is obtained from the difference between the total tissue and the extracellular $DMO^-$ contents.

If CSF sampling is not feasible, or if CSF lags behind in equilibrating with plasma, one must use $[HDMO]_e$ from measured plasma values, and derive $[DMO^-]_e$ from the equation

$$[DMO^-]_e = [HDMO]_e \cdot 10^{pK' - pH_e} \tag{21}$$

As the equation shows, however, it is then necessary to know $pH_e$. It is hardly satisfactory to assume that plasma pH values can substitute for $pH_e$. Thus, unless extracellular acid–base changes can be adequately assessed on CSF samples, $pH_e$ must be measured by microelectrodes. Situations exist in which CSF cannot be sampled for adequate analysis or $pH_e$ measurements cannot be performed. In these cases, the DMO method only gives an aggregate tissue pH, i.e., a volume-weighted average of intra- and extracellular pH.

When $pH_i$ is derived from DMO data, the following practical points should be taken into consideration (*see* Roos, 1965, 1971; Arieff et al., 1976; Pelligrino et al., 1981b). First, sufficient time must be allowed for steady-state distribution of DMO between blood and tissue. Since the rate of this process must depend on several factors, including cerebral blood flow, it seems advisable to collect tissue first when plasma activity has been reasonably steady for *at least* 30 min, preferably 60 min (*see* Roos, 1965; Arieff et al., 1976; Kobatake et al., 1984). Second, when DMO has been measured in tissue and plasma, its activity should be expressed in terms of unit weight of water. For CSF, water content is 99% of the CSF weight. For plasma, one can usually assume a water content of 93%. The corresponding figure for cortical tissue is normally around 78%, but water content is lower in tissues containing white matter (*see* Pelligrino et al., 1981b). If water content of tissue (or plasma) is determined, it should be recalled that heat drying of samples can lead to appreciable loss of $^{14}C$-DMO (Roos, 1965). For that reason, it may be advisable to determine water and DMO contents on separate samples.

### 3.4.2. Autoradiography

In routine applications of the DMO method, tissue DMO concentration is determined by blunt dissection of regions of interest, followed by liquid scintillation counting. Clearly, tissue concentrations could instead be determined by quantitative autoradiography, which gives greatly improved spatial resolution. Procedures for this have been described (Kobatake et al., 1984; Arnold et al., 1985). As applied to rats weighing 200–250 g, the experiments involved iv injection of $^{14}C$-DMO (60–70 $\mu$Ci), with sampling of blood for plasma $^{14}$-DMO determination after 60 and 120 min, followed by decapitation for $^{14}C$ autoradiography. The problem with this variant of the $^{14}C$-DMO technique is to accurately assess tissue DMO concentration. Usually, quantification is obtained by the use of a calibrated set of $^{14}$C-methyl methacrylate standards, applied to each X-ray film (*see* Sokoloff et al., 1977). However, Kobatake et al. (1984) found it necessary to use internal standards from the $^{14}$C-DMO-injected rats in the autoradiographic procedure. This was because $^{14}$C-DMO was somewhat unstable, tending to be lost from the sections even when the films were exposed to room temperature and normal pressures. Specifically, spinal cords from the animals were homogenized, and the homogenates were used for either liquid scintillation counting or autoradiography, following freezing and sectioning.

As mentioned, DMO data obtained on a two-compartment system (blood and brain) give an aggregate pH ($pH_t$). Kobatake et al. (1984) derived gray matter $pH_t$ values of 7.03–7.08 and, assuming 15% ECF volume and $pH_e$ = pH plasma, $pH_i$ values of 6.87–6.96. Although somewhat lower, these values are close to those estimated from conventional $^{14}$C-DMO measurements (e.g., Roos, 1971). The authors also derived regional $pH_t$ values for rats that had one middle cerebral artery occluded 2 h prior to decapitation. The results highlight the advantages of autoradiographic evaluation of $^{14}$C-DMO concentrations, but also the difficulty of estimating $pH_i$ values. Thus, in pathological tissue it is difficult to make valid estimations of $V_e$ and $pH_e$.

### 3.5. Fluorescence Methods

Fluorescence is a property of certain classes of compounds, in particular organic molecules with delocalized electrons formally present in conjugate double bonds. Upon excitation to an electronically excited state, the molecule returns to the ground state with the emission of photons. Characteristic of fluorescence is the fact that there is a shift to longer wavelengths of the emission relative to absorption (Stoke's shift) and that certain fluorophors display pH-dependent emission spectra.

Among these compounds, umbelliferone (7-hydroxycoumarin) has been shown to exhibit a relatively good sensitivity for changes in pH at physiological ranges (6.8–7.8), and both the neutral and ionic forms of the molecule are fluorophors with a good quantum yield upon excitation (Fink and Koehler, 1970). The favorable $pK'$, apparently independent of binding, the free diffusibility through blood–brain barrier and cell membranes, and the absence of toxic effects led Sundt et al. (1978), and Sundt and Anderson (1980a,b) to employ umbelliferone for microfluorometric determination of pH in vivo. The principle of the measurement is that a nomogram is created relating pH to the ratio of the 450 nm fluorescent curves of the indicators from the 340 and 370 nm excitation. For measuring brain pH, the fluorophore is injected into the carotid circulation and its clearance from tissue is monitored with a microspectrofluorometer viewing an avascular field about 100 μm in diameter. In brief, after operation and removal of the bone, the dura is replaced with a thin sheet of plastic with known transmission characteristics, and the microscope objective with a beam splitter mirror is then set in place. The compound is rapidly injected as a bolus and the ratio of the emission at 450 nm from the 370 and 340 nm excitation is measured. Endogenous fluorophors such as pyridine nucleotides and hemoglobin contribute to the inaccuracy of the method. NADH and NADPH fluorescence can be corrected for with appropriate calibration. Hemoglobin displays variable absorption of both the excitation energy and emission signal, related both to its concentrations and to its redox potential. Its effects can be minimized by focusing on an avascular field of the cortex.

The microspectrofluorometric umbelliferone technique is exacting and its application requires both access to the relatively complicated equipment and its proper use and handling. Its use has given pH values surprisingly close to those derived by completely different techniques. The absolute values suggest that they reflect the intracellular compartment, possibly because umbelliferone is highly soluble in fat and poorly soluble in salt solutions. It will probably remain a valuable tool for those who have access to and master the use of double-beam microspectrofluorometers. It has obvious advantages, especially since the experiments can be carried out in conjunction with CBF measurements. Its drawbacks are the complicated and expensive equipment required and the difficulty of obtaining regional data. Clearly, the method does not lend itself to measurements of pH in complete ischemia.

A different technique based on umbelliferone was developed to yield a topographic picture of pH distribution within the tissue (Csiba et al., 1983). In this application the authors froze the tissue *in situ* and applied 20 μm sections, cut in a cryostat at –20°C, to umbelliferone-impregnated filter paper. Impregnation was done by immersing the filter paper (Electrophoresis strips, Sartorium-Membrane filter GMbH, Gottingen, West Germany) for 1 min in 0.125% wt/vol umbelliferone (Serra, Heidelberg, West Germany). The sections with the umbelliferone-soaked paper were then quickly transferred to a nonreflecting black box in an ice bath under a photographic camera. They were then illuminated with UV light, first at 370 nm, then at 340 nm. After each illumination fluorescence was recorded at 450 nm with a 35mm photographic camera. pH standards (0.1$M$ phosphate buffer, 1.5% bovine albumin, and 0.01% umbelliferone, pH 6–7.7) were recorded together with the brain samples. Films were then developed and scanned in a rotating microdensitometer. The digital maps of optical densities of film negatives for 370 and 340 nm excitation were subtracted, and the results displayed with a digital computer and image processing system.

Since this is a technique in which umbelliferone is not brought into contact with the tissue until sectioning of the frozen sample, interference from other, endogenous fluorophors is reduced. Quantitative resolution seems good. Since relatively inexpensive equipment is required (except for the image analyzer used by the authors), the method may find relatively wide application. A further advantage is that it can be used on ischemic tissue. One drawback is that extra- and intracellular pH cannot be differentiated; in addition, the pH measured is related to $pH_i$ and $pH_e$ in an unknown way.

## 4. Noninvasive Techniques

So far, the discussion has centered on invasive techniques for estimating $pH_t$ and $pH_i$, unsuited for application to humans or for repeated measurements on the same brain. During recent years two noninvasive technologies have been developed for use in humans. One of them, PET, allows for continuous measurements of tissue concentrations of $^{11}C$-labeled compounds. Accordingly, methods have been described that utilize either $^{11}C$-labeled $CO_2$ or DMO. The second technology, nuclear magnetic resonance (NMR) spectroscopy, yields truly unique estimates of $pH_i$. We describe, in turn, PET and NMR techniques. Since these techniques are rather specialized, and since the necessary equipment is available to relatively few groups, we will confine the description to general principles.

### 4.1. Positron Emission Tomography

A positron is the "antiparticle" of an electron. When a positron and an electron meet, they are both annihilated and the rest product is two equally energetic γ-rays. These photons travel in almost exactly the opposite directions of each other.

These are the principles of PET: The tracer compound emits positrons that almost immediately combine with its "antiparticle": the electron. Both the electron and the position are annihilated and the two resulting γ-rays, of opposite direction, are recorded in a circular set of detectors. These detectors are arranged in pairs opposite to each other and only react to signals from γ-rays if these arrive simultaneously. A computer reconstructs the image of the tracer concentration and presents it in the form of an axial cross section.

The advantage of PET is that one can choose the tracer's chemical characteristics according to the intended studies. However, the technique has its drawbacks. To provide the positron emitting tracer requires an expensive cyclotron. The tracer is so short-lived (halftime of minutes) that the time for its incorporation into a chemical compound may pose great problems. Finally, the spatial resolution still needs improvement.

Since PET allows for noninvasive measurements of regional tracer concentrations, which can be related to corresponding concentrations in plasma, the technique is obviously suited for estimation of regional pH. Two techniques have been described, adapted to $^{11}C$-$CO_2$ and $^{11}C$-DMO, respectively.

### 4.1.1. $^{11}C$-$CO_2$

It was suggested by Raichle et al. (1979) that brain pH could be measured by the use of $CO_2$ labeled with $^{11}C$, and PET. The feasability of such measurements was examined by Buxton et al. (1984) in a novel approach. These authors based their procedure on the following arguments. $CO_2$, which is produced by the cells, has a steady-state distribution between tissue and blood, implying, among other things, a difference between tissue and plasma $CO_2$ tensions. In contrast, if $^{11}C$ is maintained at constant activity in plasma, it will reach equilibrium distribution, with equal partial pressures of $^{11}C$ in blood and tissue, provided that the following conditions are fulfilled: $^{11}C$-bicarbonate flux across the blood–brain barrier can be neglected, and the rate of $^{11}CO_2$ fixation is low (see below). At this idealized equilibrium, the partition of total $CO_2$ is given by the equation

$$L \equiv \frac{[HCO_3^-]_t + [CO_2]_t}{[HCO_3^-]_a + [CO_2]_a} \text{ eq} = \frac{S_t(1 + 10^{pH_t - pK_t'})}{S_a(1 + 10^{pH_a - pK_a'})} \tag{22}$$

where $L$ is the equilibrium partition coefficient of total $CO_2$ between tissue and arterial blood. $S_t$ and $S_a$ are the solubilities of $CO_2$ in tissue and blood, respectively. Clearly, if $S_t$ is equal to $S_a$, and $pK_t'$ to $pK_a'$, the value of $L$ will only depend on the pH difference between plasma and tissue. When $L$ is known, $pH_t$ can be calculated from the equation.

$$pH_t = 6.12 + \log[L(1 + 10^{pH-6.12}) - 1] \tag{23}$$

In reality, $CO_2$ fixation from $^{11}CO_2$ (or $^{14}CO_2$) occurs at a significant rate, with incorporation of label into acid-soluble compounds, lipids, proteins, and nucleic acids (Siesjö and Thompson, 1965). In order to correct for this, Buxton et al. (1984) developed a kinetic model for the distribution of $^{11}C$ between blood and tissue, encompassing two tissue compartments.

These authors give guidelines for experimental design and describe optimal conditions for administering the $^{11}C$ activity. It is required that plasma is analyzed after sampling of blood and anaerobic centrifugation, and that acid-labile $^{11}CO_2$ is determined, i.e., that nonbicarbonate $^{11}C$ compounds are eliminated. The $^{11}C$ activity should be continuously infused to minimize the fraction of $^{11}C$ that is fixed in brain nonbicarbonate compounds, preferably in such a way that the arterial concentration is increasing. The authors emphasize that bolus injection or a single breath inhalation, which will lead to a rapidly decreasing arterial activity, is a poor choice since such administration increases the fraction of $^{11}C$ that is fixed in nonbicarbonate compounds. The authors suggest that, under ideal conditions, it should be possible to estimate $pH_t$ with an error of about 0.04 pH units.

Measurements of regional brain pH in human subjects were reported by Brooks et al. (1984). Using the model of Buxton et al. (1984), these authors derived $pH_t$ values in the following manner. A constant $^{11}CO_2$ activity was administered by face mask. Arterial samples were obtained at 30-s intervals, the samples were centrifuged, and plasma was injected into $0.1N$ NaOH for counting. PET scans were performed during 15 min of $^{11}CO_2$ inhalation. The plasma and brain $^{11}C$ uptake curves were then analyzed with a program designed to give calculated brain uptake curves from measured plasma curves. Values were assigned for the rate constants of the three-compartment model, which gave an optimum fit between calculated and observed brain uptake curves.

In this way, a mean whole brain pH of 6.99 was obtained (Table 1). This was lowered somewhat (to 6.94) if tissue values were corrected for a 4% blood volume. As Table 1 shows, linear increases in the amount of $^{11}C$ fixed by plasma by 0.5 and 1% per min would increase calculated pH by 0.05 and 0.10 pH units, respectively.

We recognize that the $pH_t$ calculated from PET data, a composite value reflecting both extra- and intracellular fluids, is somewhat lower than that derived from $CO_2$ measurements or DMO data in experimental animals (*see* Roos, 1965, 1971; Messeter and Siesjö, 1971; Siesjö et al., 1972; Arieff et al., 1976; Pelligrino et al., 1981b). Thus, if animal data are calculated for zero extracellular space volume, $pH_t$ is around 7.10, or slightly higher. It remains to be shown whether the difference between animals and human, in itself trivial, is caused by a significant rate of $CO_2$ fixation in plasma.

The advantages of the $^{11}CO_2$-PET method are obvious. Thus, the tracer is easy to synthesize and to administer and, since $CO_2$ diffuses rapidly between compartments, measurements can be performed at short intervals. This is important since it minimizes the potential error of $CO_2$ fixation in the tissue. One disadvantage with the technique is that it cannot yield information on $pH_i$, but only on $pH_t$. Furthermore, it shares with many other techniques the disadvantage that it cannot be applied with grossly reduced tissue perfusion, e.g., in ischemia.

### 4.1.2. $^{11}C$-DMO

A DMO method was developed and used in nine patients with stroke by Syrota et al. (1983). A similar method was reported by Rottenberg et al. (1985), who studied patients with primary and metastatic tumors.

Table 1
Calculated Effects of $^{11}C$ Fixation in Plasma and Cerebral Blood Volume (CBV) on Estimated Cerebral pH

|  | Mean pH, four normal subjects |
| --- | --- |
| Standard fit | 6.99 |
| 4% CBV | 6.94 |
| 0.5% $^{11}C$ Fixation/min in plasma | 7.04 |
| 1% $^{11}C$ Fixation/min in plasma | 7.09 |
| 4% CBV + 0.5% $^{11}C$ Fixation/min | 7.00 |
| 4% CBV + 1% $^{11}C$ Fixation/min | 7.04 |

Synthesizing $^{11}$C-DMO, Syrota et al. (1983) administered a dose of 15–25 mCi with a specific activity of 180 Ci/mmol to patients who had suffered a stroke 10–34 d earlier. Images were recorded for 70 min in a system yielding a spatial resolution of 17 mm. A concentration ratio ($C_1/C_2$) was obtained between infarcted tissue ($C_1$) and the contralateral mirror locus ($C_2$), and mathematically related to $pH_e$, $pH_i$, the amount of water in the tissue region of interest *(f)*, and the fraction of total water present in the extracellular space *(x)*. Venous blood samples were obtained to yield plasma $^{11}$C concentration.

Equilibrium was achieved in the brain about 40 min following iv injection. The authors found that none of the patients had a reduced $C_1/C_2$ ratio, and in four of the nine patients, the ratio was significantly increased (>20%). Acidosis in the infarcted tissue, which would have caused a reduced $C_1/C_2$ ratio, could thus be excluded. A raised $C_1/C_2$ ratio must have been caused either by intracellular alkalosis or by increased values for $f_1$, or $x_2$, i.e., for total or extracellular water. Clearly, since infarcted tissue at that stage of the disease must show neuronal necrosis, proliferation of astrocytes, and invasion by macrophages, this issue is not easily settled.

Rottenberg et al. (1985) followed a similar procedure, administering 13–66 mCi to their patients. Measured values for $^{11}$C concentration in blood were corrected for the pH-dependent distribution of DMO between erythrocytes and plasma, to yield concentrations per unit weight of plasma water. The authors chose to consider a two-compartment system comprising blood and tissue, and thus derived an aggregate tissue pH ($pH_t$). Their values were corrected for intravascular $^{11}$C-DMO.

Rottenberg et al. (1985) reported grey matter $pH_t$ values of 6.74–7.09 and tumor pH of 6.88–7.26. Tumor tissue was consistently more alkaline than normal tissue. Such results raise the question of whether intracellular alkalosis prevails or if the raised $pH_t$ values reflect an increase in extracellular fluid water.

There is no question that $^{11}$C-DMO PET studies give valid estimates of regional $pH_t$ for situations in which equilibration between plasma and tissue is achieved. There are some disadvantages. The method for synthesizing $^{11}$C-DMO is relatively complicated and equilibration is relatively slow. In common with the DMO and $CO_2$ techniques is the difficulty of interpreting altered $^{11}$C-DMO concentrations in terms of intra- and extracellular acid–base changes.

## 4.2. Nuclear Magnetic Resonance Spectroscopy

It was first suggested in 1973 that [31]P NMR spectroscopy can be used to determine intracellular pH (Moon and Richards, 1973). Since then, the technique has been extensively applied, yielding information on pH in subcellular compartments, isolated cells, intact but isolated tissues, and tissues *in situ* (for literature, *see* Gadian et al., 1979; Meyer et al., 1982; Petroff et al., 1985). The technique has some unique features that will likely increase its usefulness in future studies.

Measurements of pH with [31]P NMR spectroscopy are based on the fact that phosphorous nuclei are magnetic dipoles that, if placed in a magnetic field, will occupy one of two energy levels with respect to the direction of the field (*see*, e.g., Meyer et al., 1982). Radiant energy of the appropriate resonant frequency (in the radio frequency range) can induce transitions of nuclei between the two states. What is important is that the electronic structure of a molecule containing a particular nucleus will determine by what amount the field at the location of the nucleus $(H)$ differs from the applied field $(H°)$. The following relationship holds

$$\nu = \frac{\gamma H°}{2\pi}\,(1 - \sigma) \tag{24}$$

where $\nu$ is the resonant frequency, and $\gamma$ is a constant for all [31]P nuclei (the so-called gyromagnetic ratio). We see that at a given applied field there is a tight relationship between the resonant frequency and $\sigma$, the chemical shift caused by the shielding of the nuclei by their molecular environment.

Determination of pH involves pulse NMR experiments (*see* Meyer et al., 1982). These are based on the fact that application of a magnetic field gives rise to a net magnetization vector $(M)$, parallel to $H°$, which behaves as a gyroscope. This means that if $M$ is tipped relative to $H°$, it will rotate around $H°$ with a frequency of rotation that is given by $\nu$, the resonant frequency.

Application of a pulse of radio frequency energy will induce such rotation. If the pulse is properly timed it can rotate $M$ by 90°, and $M$ will spin at the resonant frequency. The rotation of $M$ in this plane induces a voltage signal in the detector coil of the spectrometer. The latter will then transform the signal to a spectrum, i.e., a plot of the magnitude of $M$ as a function of frequency. If several nuclei are present, each will give a signal whose amplitude is proportional to the number of nuclei with that chemical shift.

$^{31}$P signals display chemical shifts that are sensitive to pH in the regions in which the phosphorous compounds have their p$K$ values. In biological systems, the chemical shift of inorganic phosphate (P$_i$) is usually employed and measured relative to a phosphorous compound that does not alter its ionization with pH, such as phosphocreatine (Fig. 4). The chemical shift difference between these two $^{31}$P resonances can be converted to a pH value by the use of the Henderson-Hasselbalch equation and a suitable calibration curve (*see* Petroff et al., 1985).

Thus, pH can be calculated from the equation

$$pH = 6.77 + \log \frac{P_i \text{ shift} - 3.29}{5.68 - P_i \text{ shift}} \tag{25}$$

In this equation 6.77 is the p$K'$ of the second ionization of phosphoric acid, 5.29 and 5.68 are constants of the Henderson-

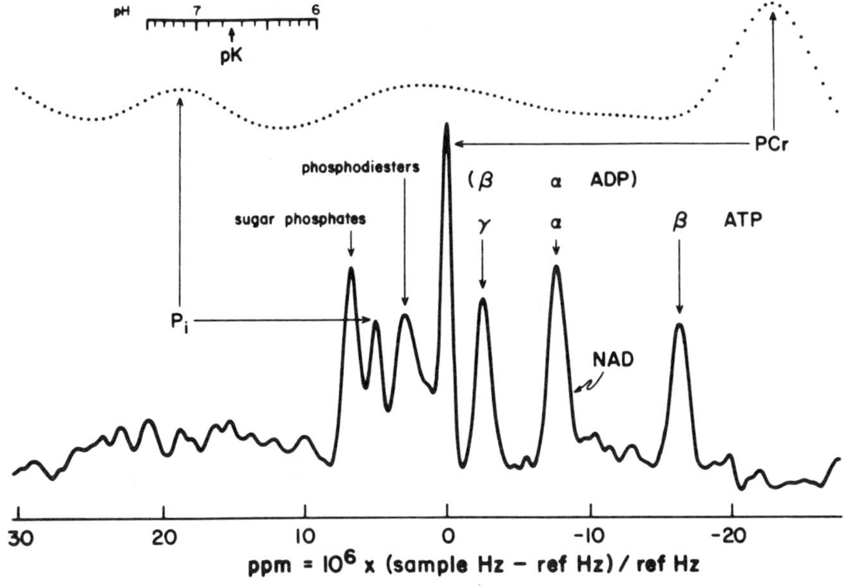


Fig. 4. $^{31}$P-Phosphorous spectrum of rabbit brain. The dotted curve is an expanded portion of the lower curve between the resonances of P$_i$ and PCr. PCr resonance is independent of pH, but P$_i$ resonance varies with pH. A scale showing how the pH is related to the chemical shift difference between P$_i$ and PCr is placed in the upper left part of the figure (reproduced, with permission, from Petroff et al., 1985).

Hasselbalch equation (Petroff et al., 1985), and $P_i$ shift is the chemical shift ($\sigma$) between the PCr and $P_i$ resonances.

Since $P_i$ molecules are found in extracellular fluid, cytosol and organelles such as mitochondria, the question arises as to what pH is measured. At first sight, it would appear that the measured pH is a weighted average of various cellular and subcellular compartments. If the compartments had moderately different pH values, one would expect to see a broadening of the $P_i$ peak and, if the pH values were more different (>0.20 pH units), a splitting of the peak. In general, though, only mobile ions give sharp lines, and if the water phase is structured and viscous the $^{31}P$ nuclei present could give such broad lines that they merge with the background noise. It is possible, therefore, that the influence of pH signals from such organelles as mitochondria, lysosomes, and synaptic vesicles is small. At least for muscle tissue, evidence exists that it is primarily the cytoplasmic pH that is measured (Bailey et al., 1981).

Brain pH values obtained by $^{31}P$-NMR are, with few exceptions, in the range 7.10–7.20 (Petroff et al., 1985). These values are similar to those derived by weak acid methods ($CO_2$ or DMO) by lumping intra- and extracellular fluids (see Roos and Boron, 1981). It is then tempting to assume that $^{31}P$-NMR gives a pH value that is a weighted average of extra- and intracellular fluids. Thus, $P_i$ concentrations in extra- and intracellular fluids are relatively similar. If $pH_e$ does not contribute much, two alternative explanations can be suggested. First, $^{31}P$-NMR measures pH in both the cytosol and an alkaline cell compartment. Second, the pH obtained with weak acid methods is weighted in favor of a more acid cell compartment.

Differences in pH obtained with $^{31}P$-NMR and other techniques must be considered trivial. What is more important is that NMR techniques hold great promise for the future. This is because the techniques are nondestructive and noninvasive, thus allowing for repeated measurements within a relatively short period. One unique feature is that since the pH signal emanates from an endogenous marker, pH can be measured even if blood flow is reduced to zero during ischemia (*see* Garlick et al., 1979; Bailey et al., 1981; Thulborn et al., 1982).

## 5. Conclusion

For many years, tissue acid–base balance was a relatively exclusive field. This is partly because acid–base theory has been burdened with intricacies and conceptual difficulties, and partly because existing methodologies were both exacting and indirect. The situation has now changed dramatically. The main reason is the rapid advent of techniques for precise and continuous measurement of intracellular pH in cells studied in suspension or culture. This has led to an almost explosive increase in the knowledge of intracellular pH regulation. The results obtained have made it clear that $pH_i$ enters as an important determinant of many physiological and pathophysiological events. This development, in turn, has undoubtedly contributed to refinements of older techniques to assess acid–base changes in composite tissue, allowed for precise measurements of $pH_e$, and stimulated the development of new techniques, including PET and NMR. These latter techniques carry the important advantages that they can be applied to humans. Of these, NMR offers a truly unique method of measuring $pH_i$ continuously in experimental animals and in humans.

## Acknowledgments

Work from the authors' own laboratory was supported by continuing grants from the Swedish Medical Research Council and from US Public Health Service via the NIH. The authors wish to thank Erna Björkengren for skilled secretarial work.

## References

Alberti K. G. M. M. and Cuthbert C. (1982) The Hydrogen Ion in Normal Metabolism: A Review, in *Metabolic Acidosis* (Ciba Foundation Symposium 87) Pitman, London.

Amman D., Lanter F., Steiner R. A., Schultero P., Shijo Y., and Simon W. (1981) Neutral carrier based hydrogen ion selective microelectrode for extra- and intracellular studies. *Anal. Chem.* **53,** 2267–2269.

Anrep G. V., Ayadi M. S., and Talaat M. (1936) A method for determination of carbon dioxide applicable to blood and tissues. *J. Physiol.* **86,** 153–161.

Arieff A. I., Kerian A., Massry S. G., and DeLima J. (1976) Intracellular pH of brain: Alterations in acute respiratory acidosis and alkalosis. *Am. J. Physiol.* **230,** 804–812.

Arnold J. B., Junck L., and Rottenberg D. A. (1985) In vivo measurement of regional brain and tumor pH using $^{14}C$ dimethyloxazolidinedione and quantitative autoradiography. *J. Cereb. Blood Flow Metab.* **5,** 369–375.

Astrup J., Heuser D., Lassen N. A., Nilsson B., Norberg K., and Siesjö B. K. (1978) Evidence Against $H^+$ and $K^+$ as Main Factors for the Control of Cerebral Blood Flow: A Microelectrode Study, in *Cerebral Vascular Smooth Muscle and its Control* Ciba Foundation Symposium 56 (new series), Elsevier/Excerpta Medica/North-Holland, Amsterdam.

Bailey I. A., Williams S. R., Radda G. K., and Gadian D. G. (1981) Activity of phosphorylase in local ischemia in the rat heart. *Biochem. J.* **196,** 171–178.

Boron W. F. (1983) Topical review: Transport of $H^+$ and of ionic weak acids and bases. *J. Membrane Biol.* **72,** 1–16.

Boron W. F. and Roos A. (1976) Comparison of microelectrode, DMO, and methylamine methods for measuring intracellular pH. *Am. J. Physiol.* **231,** 799–809.

Brooks D. J., Lammertsma A. A., Beaney R. P., Leenders K. L., Buckingham P. D., Marshall J., and Jones T. (1984) Measurement of regional cerebral pH in human subject using continuous inhalation of $^{11}CO_2$ and positron emission tomography. *J. Cereb. Blood Flow Metab.* **4,** 458–465.

Brzezinski J., Kjällquist Ä., and Siesjö B. K. (1967) Mean carbon dioxide tension in the brain after carbonic anhydrase inhibition. *J. Physiol.* **188,** 13–23.

Buxton R. B., Wechsler L. R., Alpert N. M., Ackerman R. H., Elmaleh D. R., and Corrcia J. A. (1984) Measurement of brain pH using $^{11}CO_2$ and positron emission tomography. *J. Cereb. Blood Flow Metab.* **4,** 8–16.

Caronna J. J., Plum F., and Siesjö B. K. (1974) $PCO_2$ gradients between blood and CSF in rat during alterations of acid–base balance. *Am. J. Physiol.* **227,** 1173–1177.

Cole K. S., Li C.-L., and Bak A. F. (1969) Electrical analogues for tissues. *Exp. Neurol.* **24,** 459–473.

Conway E. J. (1950) *Microdiffusion Analysis and Volumetric Error* C. Lockwood, London.

Csiba L., Paschen W., and Hossmann K.-A. (1983) A topographic quantitative method for measuring brain tissue pH under physiological and pathophysiological conditions. *Brain Res.* **289,** 334–337.

Davies D. G. and Gurtner G. H. (1973) CSF acid–base balance and the Wien-effect. *J. Appl. Physiol.* **34,** 249–254.

Dietzel I., Heinemann U., Hofmeier G., and Lux H. D. (1980) Transient changes in the size of the extracellular space in the sensorimotor cortex of cats in relation to stimulus-induced changes in potassium concentration. *Exp. Brain Res.* **40,** 432–439.

Dietzel I., Heinemann U., Hofmeier G., and Lux H. D. (1982) Stimulus-induced changes in extracellular $Na^+$ and $Cl^-$ concentration in relation to changes in the size of the extracellular space. *Exp. Brain Res.* **46,** 73–84.

Elazar Z., Kado R. T., and Adey W. R. (1966) Impedance Changes During Epileptic Seizures, in *Epilepsia* vol 7, Elsevier Amsterdam.

Everett N. B., Simmons B., and Lasher E. P. (1956) Distribution of blood ($Fe^{59}$) and plasma ($I^{131}$) volumes of rats determined by liquid nitrogen freezing. *Circ. Res.* **4,** 419–424.

Fenn W. O. and Maurer F. W. (1935) The pH of muscle. *Protoplasma* **24,** 337–345.

Fenstermacher J. D., Li C.-L., and Levin V. A. (1970) Extracellular space of the cerebral cortex of normothermic and hypothermic cats. *Exp. Neurol.* **27,** 101–114.

Fink D. W. and Koehler W. R. (1970) pH effects on fluorescence of umbelliferone. *Anal. Chem.* **42,** 990–993.

Gadian D. G., Radda G. K., Richards R. E., and Seeley P. J. (1979) $^{31}P$ NMR in Living Tissue: The Road from a Promising to an Important Tool in Biology, in *Biochemical Applications of Magnetic Resonance* (Shulman R. G., ed.), Academic, New York.

Garlick P. B., Radda G. K., and Seeley P. J. (1979) Studies of acidosis in the ischaemic heart by phosphorus nuclear magnetic resonance. *Biochem. J.* **184,** 547–554.

Gertz K. H. and Loeschke H. H. (1956) Electrode zur Bestimmung des $CO_2$-Drucks. *Naturwissenschaften* **45,** 160–161.

Gleichmann U., Ingvar D. H., Lubbers D. W., Siesjö B. K., and Thews, G. (1962) Tissue $pO_2$ and $pCO_2$ of the cerebral cortex, related to blood gas tensions. *Acta Physiol. Scand.* **55,** 127–138.

Hansen A. J. and Olsen C. E. (1980) Brain extracellular space during spreading depression and ischemia. *Acta Physiol. Scand.* **108,** 355–365.

Heuser D. (1978) The Significance of Cortical Extracellular $H^+$, $K^+$ and $Ca^{2+}$ Activities for Regulation of Local Cerebral Blood Flow Under Conditions of Enhanced Neuronal Activity, in *Cerebral Vascular Smooth Muscle and its Control* Ciba Foundation Symposium 56 (new series) Elsevier/Excerpta Medica/North-Holland, Amsterdam.

Hochachka P. W. and Mommsen T. P. (1983) Protons and anaerobiosis. *Science* **219,** 1391–1397.

Hogg R. J., Pucacco L. R., Carter N. W., Loptook A. R., and Kokko J. P. (1984) In situ $PCO_2$ in the renal cortex, lever, muscle, and brain of the New Zealand white Rabbit. *Am. J. Physiol.* **247,** F491–F493.

Hossmann K.-A. (1982) Treatment of experimental cerebral ischemia. *J. Cereb. Blood Flow Metab.* **2,** 275–297.

Kobatake K., Sako K., Masahiro I., Yamamoto Y. L., and Hakim A. M. (1984) Autoradiographic determination of brain pH following middle cerebral artery occlusion in the rat. *Stroke* **15,** 540–547.

Kraig R. P., Ferreira-Filho C. R., and Nicholsson C. (1983) Alkaline and acid transients in the cerebellar microenvironment. *J. Neurophysiol.* **49,** 831–849.

Kraig R. P., Pulsinelli W. A., and Plum F. (1985) Heterogenous Distribution of Hydrogen and Bicarbonate Ions During Complete Brain Ischemia, in *Progress in Brain Research* vol. 63 (Kogure K., Hossmann K.-A., Siesjö B. K., and Welsh F. A., eds.) Elsevier, Amsterdam.

Kraig R. P., Pulsinelli W. A., and Plum F. (1986) Carbonic acid buffer changes during complete brain ischemia. *Am. J. Physiol.*, 250, R348–R357.

Kuschinsky W. and Wahl M. (1978) Local chemical and neurogenic regulation of cerebral vascular resistance. *Physiol. Rev.* **58,** 656–689.

Levin V. A., Fenstermacher J. D., and Patlak C. S. (1970) Sucrose and inulin space measurements of cerebral cortex in four mammalian species. *Am. J. Physiol.* **219,** 1528–1533.

Messeter K. and Siesjö B. K. (1971) The intracellular pH in the brain in acute and sustained hypercapnia. *Acta Physiol. Scand.* **83,** 210–219.

Meyer R. A., Kushmerick M. J., and Brown T. R. (1982) Application of $^{31}$P-NMR spectroscopy to the study of striated muscle metabolism. *Am. J. Physiol.* **242,** C1–C11.

Mitchell R. A., Herbert D. A., and Carman C. T. (1965) Acid–base constants and temperature coefficients for cerebrospinal fluid. *J. Appl. Physiol.* **20,** 27–30.

Moon R. B. and Richards J. H. (1973) Determination of intracellular pH by $^{31}$P nuclear magnetic resonance. *J. Biol. Chem.* **248,** 7276–7278.

Mutch W. A. C. and Hansen A. J. (1984) Extracellular pH changes during spreading depression and cerebral ischemia: Mechanisms of brain pH regulation. *J. Cereb. Blood Flow Metab.* **4,** 17–27.

Myers R. (1979) Lactic Acid Accumulation as Cause of Brain Edema and Cerebral Necrosis Resulting from Oxygen Deprivation, in *Advances in Perinal Neurology* (Korobkin R. and Guilleminault G., eds.) Spectrum, New York.

Nemoto E. M. and Frinak S. (1981) Brain tissue pH after global brain ischemia and barbiturate loading in rats. *Stroke* **12,** 77–82.

Nicholson C., Phillips J. M., and Gondner-Medwin A. R. (1979) Diffusion from an iontophoretic point-source in the brain. *Brain Res.* **169,** 580–584.

Pelligrino D., Almquist L.-O., and Siesjö B. K. (1981a) Effects of insulin-induced hypoglycemia on intracellular pH and impedance in the cerebral cortex of the rat. *Brain Res.* **221,** 129–147.

Pelligrino D. A., Musch T. I., and Dempsey J. A. (1981b) Interregional differences in brain intracellular pH and water compartmentation during acute normoxic and hypoxic hypocapnia in the anesthetized dog. *Brain Res.* **214,** 387–404.

Petroff O. A. C., Prichard J. W., Behar K. L., Alger J. R., den Hollander J. A., and Shulman R. G. (1985) Cerebral intracellular pH by $^{31}$P nuclear magnetic resonance spectroscopy. *Neurology* **35**, 781–788.

Plum F. (1983) What causes infarction in ischemic brain? The Robert Wartenberg lecture. *Neurology* **33**, 222–233.

Ponten U. and Siesjö B. K. (1964) A method for the determination of the total carbon dioxide content of frozen tissues. *Acta Physiol. Scand.* **60**, 297–308.

Ponten U. and Siesjö B. K. (1966) Gradients of $CO_2$ tension in the brain. *Acta Physiol. Scand.* **67**, 129–140.

Ponten U., Ratcheson R. A., Salford L. G., and Siesjö B. K. (1973) Optimal freezing conditions for cerebral metabolites in rats. *J. Neurochem.* **21**, 1127–1138.

Raichle M. E., Grubb R. L. Jr., and Higgins C. S. (1979) Measurements of brain tissue carbon dioxide content in vivo of emission tomography. *Brain Res.* **166**, 413–417.

Rall D. P., Oppelt W. W., and Patlak C. S. (1962) Extracellular space of brain as determined by diffusion of inulin from the ventricular system. *Life Sci.* **2**, 43–48.

Ranck Jr. J. B. (1963) Analysis of specific impedance of rabbit cerebral cortex. *Exp. Neurol.* **7**, 153–174.

Roos A. (1965) Intracellular pH and intracellular buffering power of the cat brain. *Am. J. Physiol.* **209**, 1233–1246.

Roos A. (1971) Intracellular pH and buffering power of rat brain. *Am. J. Physiol.* **221**, 176–181.

Roos A. and Boron W. F. (1981) Intracellular pH. *Physiol Rev.* **61**, 296–434.

Rottenberg D. A., Ginos J. Z., Kearfott K. J., Junck L., Dhawan V., and Jarden J. O. (1985) In vivo measurement of brain tumor pH using $^{11}$C DMO and positron emission tomography. *Ann. Neurol.* **17**, 70–79.

Schneider W., Wahl M., Kuschinsky W., and Thurau K. (1977) Instruments and techniques: The use of microelectrodes for measurement of local $H^+$ activity in the cortical subarachnoidal space of cats. *Pflugers Arch.* **372**, 103–107.

Severinghaus J. W. and Bradley A. E. (1958) Electrodes for blood $PO_2$ and $PCO_2$ determination. *J. Appl. Physiol.* **13**, 515–520.

Seylaz J., Pinard E., Meric P., and Correze J.-L. (1983) Local cerebral $PO_2$, $PCO_2$, and blood flow measurements by mass spectroscopy. *Am. J. Physiol.* **245**, H513–518.

Siemkowicz E. and Hansen A. J. (1981) Brain extracellular ion composition and EEG activity following 10 minutes ischemia in normo- and hyperglycemic rats. *Stroke* **12**, 236–240.

Siesjö B. K. (1961) A method for continuous measurement of the carbon dioxide tension on the cerebral cortex. *Acta Physiol. Scand.* **51**, 297–313.

Siesjö B. K. (1962a) The solubility of carbon dioxide in cerebral cortical tissue of cats. With a note on the solubility of carbon dioxide in water, 0.16 *M* NaCl and cerebrospinal fluid. *Acta Physiol. Scand.* **55**, 325–341.

Siesjö B. K. (1962b) The bicarbonate/carbonic acid buffer system of the cerebral cortex of cats, as studied in tissue homogenates. II. The $pK_1'$ of carbonic acid at 37.5°C, and the relation between carbon dioxide tension and pH. *Acta Neurol. Scand.* **38**, 121–141.

Siesjö B. K. (1972) The regulation of cerebrospinal fluid pH. *Kidney Intl.* **1**, 360–374.

Siesjö B. K. (1978) *Brain Energy Metabolism* Wiley, Chichester, New York.

Siesjö B. K. (1981) Cell damage in the brain: A speculative synthesis. *J. Cereb. Blood Flow Metab.* **1**, 155–185.

Siesjö B. K. (1982) Lactic Acidosis in the Brain: Occurrence, Triggering Mechanisms and Pathophysiological Importance, in *Metabolic acidosis*. Ciba Foundation Symposium 87, Pitman, London.

Siesjö B. K. (1984) Cerebral circulation and metabolism. *J. Neurosurg.* **60**, 883–908.

Siesjö B. K. (1985) Acid–Base Homeostasis in the Brain: Physiology, Chemistry, and Neurochemical Pathology, in *Progress in Brain Research* vol. 63 (Kogure K., Hossmann K.-A., Siesjö B. K., and Welsh F. A., eds.) Elsevier, Amsterdam.

Siesjö B. K. and Ingvar M. (1983) Blood Flow, in *Handbook of Neurochemistry* vol. 3, 2nd Ed. (Lajtha A., ed.) Plenum, New York.

Siesjö B. K. and Messeter K. (1971) Factors Determining Intracellular pH, in *Ion Homeostasis of the Brain* (Siesjö B. K. and Sørenson S. C., eds.) Munksgaard, Copenhagen.

Siesjö B. K. and Thompson W. O. B. (1965) The uptake of inspired $^{14}CO_2$ into the acid-labile, the acid-soluble, the lipid, the protein and the nucleic acid fractions of rat brain tissue. *Acta Physiol. Scand.* **64**, 182–192.

Siesjö B. K., Folbergrova J., and MacMillan V. (1972) The effect of hypercapnia upon intracellular pH in the brain, evaluated by the bicarbonate-carbonic acid method and from the creatine phosphokinase equilibrium. *J. Neurochem* **19**, 2483–2495.

Siggaard-Andersen O. (1966) Titrable acid or base of body fluids. *Ann. NY Acad. Sci.* **133**, 41–58.

Siggaard-Andersen O. (1974) *The Acid-Base Status of the Blood*. 4th Ed. Munksgaard, Copenhagen.

Smith M.-L., von Hanwehr R., and Siesjö B. K. (1985) Cerebral acid–base changes in ischemia: Influence of preischemic plasma glucose concentration. *J. Cereb. Blood Flow Metab.* **5** (suppl. 1), S239–240.

Sokoloff L., Reivich M., Kennedy C., Les Rosiers M. H., Pathley C. S., Pettigrew K. D., Sakurada O., and Shinahara M. (1977) The $^{14}$C-deoxyglycose method for the measurement of local cerebral glucose utilization: Theory, procedure, and normal values in the conscious and anaesthetized albino rat. *J. Neurochem.* **28,** 897–916.

Stewart P. A. (1981) *How To Understand Acid–Base. A Quantitative Acid–Base Primer for Biology and Medicine* Edward Arnold, London.

Sundt T. M. and Anderson R. E. (1980a) Intracellular brain pH and the pathway of a fat soluble pH indicator across the blood–brain barrier. *Brain Res.* **186,** 355–364.

Sundt T. M. and Anderson R. E. (1980b) Umbelliferone as an intracellular pH-sensitive fluorescent indicator and blood–brain barrier probe: Instrumentation, calibration, and analysis. *J. Cereb. Blood Flow Metab.* **44,** 60–75.

Sundt T. M., Anderson R. E., and van Dyke R. A. (1978) Brain pH measurements using a diffusible, lipid soluble pH sensitive fluorescent indicator. *J. Cereb. Blood Flow Metab.* **31,** 627–635.

Syrota A., Castaing M., Rougemont D., Berridge M., Baron J. C., Bousser M. G., and Pocidalo J. J. (1983) Tissue acid–base balance and oxygen metabolism in human cerebral infarction studied with positron emission tomography. *Ann. Neurol.* **14,** 419–428.

Tarby T. J., Costin A., and Ross Adey W. (1968) Effects of tetrodotoxin on impedance in normal and asphyxiated cerebral tissue. *Exp. Neurol.* **22,** 517–531.

Thomas R. C. (1984) Experimental displacement of intracellular pH and the mechanism of its subsequent recovery. *J. Physiol.* **354,** 3P–22P.

Thulborn K. R., du Boulay G. H., Duchen L. W., and Radda G. (1982) A $^{31}$P nuclear magnetic resonance in vivo study of cerebral ischaemia in the gerbil. *J. Cereb. Blood Flow Metab.* **2,** 299–306.

Urbanics R., Leniger-Follert E., and Lubbers D. W. (1978) Time course of changes of extracellular $H^+$ and $K^+$ activities during and after direct electrical stimulation of the brain cortex. *Pflugers Arch.* **378,** 47–53.

van Harreveld A. and Ochs S. (1956) Cerebral impedance changes after circulatory arrest. *Am. J. Physiol.* **187,** 180–192.

van Harreveld A., Murphy T., and Nobel K. W. (1963) Specific impedance of rabbit's cortical tissue. *Am. J. Physiol.* **205,** 203–207.

van Slyke D. D. and Neill J. M. (1924) The determination of gases in blood and other solutions by vacuum extraction and manometric measurement. *J. Biol. Chem.* **61,** 523–575.

von Hanwehr R., Smith M.-L., and Siesjö B. K. (1986) Extra- and intracellular pH during near-complete forebrain ischemia in the rat. *J. Neurochem.* **46,** 331.

Waddell W. J. and Bates R. G. (1969) Intracellular pH. *Physiol. Rev.* **49,** 285–329.

Waddell W. J. and Butler T. C. (1959) Calculation of intracellular pH from the distribution of 5,5-dimethyl-2,4-oxazolidinedione (DMO). Application to skeletal muscle of the dog. *J. Clin. Invest.* **38,** 720–729.

Wallace W. M. and Hastings A. B. (1942) The distribution of the bicarbonate in mammalian muscle. *J. Biol. Chem.* **144,** 637–649.

Wang R. I. H. and Sonnenschein R. R. (1955) pH of cerebral cortex during induced convulsions. *J. Neurophysiol.* **18,** 130–137.

Weyne J., Demeester G., and Leusen I. (1968) Bicarbonate and chloride shifts in rat brain during acute and prolonged respiratory acid–base changes. *Arch. Int. Physiol. Biochim.* **76,** 415–433.

Weyne J., Pannier J. L., Demeester G., and Leusen I. (1970) Bicarbonate and chloride of rat brain during infusion-induced changes in bicarbonate concentration of blood. *Pflügers Arch.* **320,** 45–63.

Woodward D. L., Reed D. J., and Woodbury D. M. (1967) Extracellular space of rat cerebral cortex. *Am. J. Physiol.* **212,** 367–370.

# Magnesium Ions

**Jerry G. Chutkow**

## 1. Introduction

Magnesium (Mg), the second most abundant soft tissue intracellular cation in vertebrates, is an essential nutrient for all organisms. Because of its small ionic radius and relatively large charge, $Mg^{2+}$ functions within living cells primarily as a reversible chelator, forming relatively stable complexes, particularly with phosphoryl and carbonyl groups. In so doing, it activates or inhibits enzyme substrates involved in the metabolism of carbohydrates and lipids; is an essential cofactor in the synthesis of proteins and nucleic acids; facilitates the transfer of high-energy phosphate bonds; combines directly with certain enzymes and membrane-bound transport systems; forms stabilizing complexes with ribosomes, phospholipids, and components of the cytoskeleton; and promotes cellular cohesion. (*see* Aikawa, 1971; Guenther, 1981; Wacker, 1980). Although most of its biochemical functions have been worked out in vitro or in vivo in isolated enzyme and other biologic systems and in cells from nonnervous tissues, there is every reason to believe $Mg^{2+}$ functions similarly in the cells in the nervous system. For example, CNS hexokinase, a critical enzyme in aerobic metabolism of carbohydrates, probably requires $Mg^{2+}$ for allosteric activation when the concentration of ATP falls below that of free $Mg^{2+}$ in the cytosol (Lustyik and Nagy, 1985).

In addition to its other functions, $Mg^{2+}$ contributes to cellular excitability. Extracellularly, it participates in peripheral somatic motor and autonomic chemical neurotransmission, and there is increasing evidence that it has a similar role in the central nervous system (CNS) (Chutkow, 1972, 1974, 1980, 1981). In fact, all the clinical consequences of an increase in extracellular concentrations of $Mg^{2+}$ ($[Mg^{2+}]$) to toxic levels reflect progressive peripheral neuromuscular blockade and autonomic hypofunction primarily caused by impaired influx of calcium into presynaptic nerve terminals. Peripheral and, more dramatically, central neurologic hyperirritability are, furthermore, the commonest and most serious clinical manifestations of Mg deficiency in animals and humans. Centrally originating dysfunction correlates well experimentally with a combined decrease in the $[Mg^{2+}]$ in the cerebrospinal and CNS cellular fluid compartments (Chutkow, 1974, 1981; Chutkow and Grabow, 1972; Chutkow and Myers, 1968).

Accurate quantitation of the total content, reliable mapping of the distribution, and direct measurement of the chemical state of interstitial and intracellular $Mg^{2+}$ are prerequisite for characterizing the dynamics and regulation of Mg's cellular metabolism and for defining Mg's aforementioned biological importance in vivo. However, much work remains to be done, particularly in the nervous system, in large part because the tools needed to perform the experiments properly have not been available. Fortunately, rapidly expanding analytic technology promises to remedy the situation.

In this chapter, I shall detail the standard methods for quantitating $Mg^{2+}$ and for investigating its metabolism in the CNS. What works for the CNS usually works for peripheral nerve and muscle, with a few technical modifications. I shall also discuss in more general terms those more-recently developed techniques whose inorganic neurochemical usefulness appears to have potential, but which have not yet been applied to any extent on nervous tissues or for which there are as yet no widely accepted analytic protocols for $Mg^{2+}$. In either case, I have included references documenting applications to other organ systems. Methods can be extracted and modified from these as needed for neurochemical purposes. The descriptions of each neuromethod will include a summary of the principles involved, established and potential applications, and critical comments.

Finally, as a caveat to the reader, I relied on publications of work and review articles when I did not have personal experience with a technique, realizing full well that these descriptions almost never include all the procedural pitfalls (and proven remedies) that invariably appear in practice.

## 2. Quantitation and Distribution of Magnesium Ions in Biological Materials

### 2.1. Atomic Absorption Spectrophotometry (AAS)

#### 2.1.1. Principles and Instrumentation

The solution containing the analyte is heated to a temperature sufficient to evaporate the solvent, to dissociate the chemical bonds of the $Mg^{2+}$-containing compounds, and to reduce $Mg^{2+}$ to its free ground (unexcited) state. Electrons of the resulting atoms absorb photons of light at a resonant spectral line characteristic for Mg. All other factors being equal, the amount of incident light absorbed is proportional to the number of Mg atoms in the atomic vapor.

Most often, a combustion flame burner with an oxidizing flame (air, oxygen, or nitrous oxide combined with acetylene) provides the thermal energy ("flame AAS"). There are two basic types of burners: (1) total consumption, in which the sample is mixed with the gases as they enter the flame and (2) a premix burner, in which the sample is volatilized with the gases in a mixing-expansion chamber before being burned. In the former, relatively large droplets are produced in the flame. These increase signal noise by light scattering. Although it requires a larger volume of aspirate because of the additional waste involved, a premix burner fitted with a triple slotted (Boling) head reduces signal noise, has greater sensitivity, and is easier to keep clean. Flame AAS is adequate for most purposes.

One can increase the sensitivity and considerably decrease the amount of the biological fluid sample or the tissue needed for analysis by replacing the burner with an electrothermal atomizing chamber ("flameless AAS"). This uses some type of graphite, tantalum, or tungston boat or filament to vaporize the solvent from microliter volumes of analyte, to dry-ash the resulting deposit, and to atomize efficiently the inorganic residue in a controlled atmosphere of inert gas (Donega and Burgess, 1970; Fassel, 1978; Hwang et al., 1972).

A lamp with a hollow cathode containing metallic Mg generates the stable, high-intensity, monochromatic light to be absorbed by the Mg atoms in the flame. AAS measures the peak absorption of this sharp spectral line. Anything that broadens the line width (e.g., the quenching effect of foreign molecules or self-absorption) will affect sensitivity.

The spectrophotometer basically consists of mirrors to split the incident light into two beams, one of which passes through the flame, the other of which is transmitted unchanged; a prism or grating monochromator to isolate the wavelength of interest; and photomultipliers to amplify and convert both beams into electrical current. The difference in intensity of the two beams measures losses caused by atomic absorption.

AAS is specific, accurate, and sensitive and has a high degree of precision (Chutkow and Grabow, 1972; Fassel, 1978; Hwang et al., 1972). The detection limit (EPA, 1979) for Mg using flame AAS is approximately 0.1 ppb (w/v); using flameless AAS, $\sim 1 \times 10^{-13}$ g (absolute detection limit) (Fassel and Kniseley, 1974; Hwang et al., 1972). In flame AAS, the linear dynamic range for $[Mg^{2+}]$ in the analyte usually extends to $\sim 0.6$ ppm at a sensitivity of 8 ppb (concentration of Mg producing 1% absorption signal).

Interelemental and matrix interferences (EPA, 1979; Fassel, 1978) plague both flame and flameless AAS. Chemical or condensed-phase interference appears when $Mg^{2+}$ combines with an anion, usually phosphate, to form a molecule that does not dissociate. This affects not only the sensitivity of the procedure, but the accuracy as well, since the interfering substance is not present in the same concentration in the standards. Because its solubility product with phosphate is lower than that of $Mg^{2+}$, adding lanthanum $(La^{3+})$ in sufficient quantity to the aspirate eliminates this problem, at least in flame AAS. Alternatively, a high-temperature flame (nitrous oxide-acetylene) also reduces this type of interference by increasing the dissociation of $(Mg)_3(PO_4)_2$.

Matrix interference, a mismatch between the physical characteristics of the sample and the $Mg^{2+}$ standards caused by a difference in salt or acid contents, is a minor problem with analytes derived from soft tissues. Ionization interference arises when the vaporizing-atomizing temperatures are high enough to excite the Mg atoms. The resulting ions do not absorb light photons at the same resonance line as their parent atoms. This complication is overcome by adding an excess of a more easily ionized element, such as sodium, to the standards and the sample. Background absorption—a combination of flame absorption, molecular absorption, and light scattering—is rarely a problem if the flame has a sufficiently high temperature and if the instrument manufacturer's recommendations on the flow of oxidant and fuel are followed.

### 2.1.2. Methods

2.1.2.1. GLASSWARE AND CONTAINERS.        Glassware and polyethylene or polypropylene containers are cleaned with dichromate-sulfuric acid solution, rinsed thoroughly with distilled-deionized water, soaked in a bath of reagent grade dilute hydrochloric or nitric acid for several hours, then rerinsed with deionized-distilled water, air-dried, and sealed until used.

All solutions and solubilized specimens are stored, preferably acidified, in polyethylene or polypropylene containers. This minimizes the risk of possible changes in Mg content by leaching from, or adsorptive loss to, the container wall.

2.1.2.2. REAGENTS AND STOCK SOLUTIONS.    Acids are ultrahigh purity grade or its equivalent; all other chemicals are reagent grade; and water is deionized and distilled in quartz distillers. A stock $Mg^{2+}$ reference standard (1000 ppm) of suitable purity and accuracy for use in AAS can be purchased commercially from several chemical suppliers or prepared by dissolving either Mg metal or magnesium oxide in dilute HCl. To make a stock solution of lanthanum, 50,000 ppm, dissolve purified lanthanum oxide in a 1:4 (v/v) dilution of concentrated HCl ($12M$).

2.1.2.3. COLLECTION AND PREPARATION OF BIOLOGIC SAMPLES. Although the following discussion applies primarily to flame AAS (Chutkow and Grabow, 1977; Chutkow and Meyers, 1968), the recommendations are applicable with modification to flameless AAS.

A. *Fluids.* Despite the precision of flame AAS, it is wise to run duplicate determinations whenever possible. Five to ten milliliters of aspirate are usually required, depending upon the burner and the stability of the spectrophotometer. Therefore, one needs a sample of sufficient volume to allow for duplicate aliquots, each of which can provide a $[Mg^{2+}]$ in the analyte that will fall within the span of the analytic curve.

The $[Mg^{2+}]$ in mammalian CSF is greater than that of ultrafilterable plasma and, when normal, usually averages from 18 to 30 ppm, depending on the species (Chutkow, 1971). Store the CSF in polyethylene test tubes of a size that will minimize evaporation of water from smaller quantities. A sample of less than 0.2 mL can be collected directly into a dry, tared, 5-mL volumetric flask, sealed, and immediately reweighed. The fluids are stored frozen.

B. *Tissues.* Tissues must be in solution for flame AAS. This is not necessarily the case for flameless AAS. In most mammals so studied, the $[Mg^{2+}]$ in the brain is about 145–175 ppm wet wt (Chutkow, 1971), of which about 98% is cellular-intracellular. The $[Mg^{2+}]$ in gray matter is about 16 ppm greater than that of areas containing mostly white matter (Chutkow, 1972). The spinal cord falls into this latter category.

C. *Ashing and Solubilizing Tissues.* The methods detailed below are applicable to both whole tissue samples and adequate quantities of cellular or subcellular fractions.

a. *Wet Ashing.* Nitric-sulfuric-perchloric acid (Chutkow, 1972; Chutkow and Grabow, 1972) is particularly useful for lipid-rich samples. Wash the specimen with small volumes of 0.9% saline at the time it is removed from the animal. Place it immediately in a tared, dry weighing bottle, weigh, dry at 110°C for 18 h, cool to ambient temperature in a dessicator, and reweigh. If desired, lipids may be removed at this point: finely grind the dried specimen, if necessary; extract the lipids in a Soxhlet apparatus for 24 h with a 2:1 (v/v) mixture of chloroform-methanol or similar solvent system; and redry the residue in a tared weighing bottle. Then transfer a measured quantity of the dried, lipid-rich or extracted tissue to a 100-mL Kjeldahl flask. Add 7 mL of concentrated nitric acid (16*M*), 1 mL of concentrated perchloric acid (12*M*), 1 mL of concentrated sulfuric (12*M*), 1 mL of concentrated sulfuric acid (18*M*), and a few glass bumping beads. Gently heat in a micro-Kjeldahl apparatus, shaking the flask frequently, until the perchloric acid boils off and sulfuric acid fumes appear.

Quantitatively transfer the resulting clear, colorless, sulfate-containing digest to a 25-mL volumetric flask, using Pasteur pipets and rinsing with multiple small volumes of water, dilute to mark, and store frozen in polyethylene tubes until analyzed. Carry two reagent blanks throughout the entire procedure.

*b. Nitric Acid* (Clegg et al., 1981; Hershey et al., 1983, 1984). In my experience, one oxidizing acid does not digest lipid-rich substances completely; however, some workers feel it is superior to the three-acid system (Clegg et al., 1981). It is most appropriate for use with small samples and cellular-subcellular material. The tissue can be incubated with concentrated nitric acid, either in an Erlenmeyer flask or a Teflon-lined acid-bomb. Again, carry two reagent blanks through the entire procedure.

*c. Dry Ashing.* Transfer a weighed, dry sample to a quartz or, preferably, platinum crucible and heat it for 24 h at 500°C in a muffle furnace. Dissolve the resulting ash in dilute HCl and transfer the digest quantitatively to a suitable volumetric flask, dilute to mark with deionized-distilled water, and store frozen in a polyethylene container. Carry water, similar in volume to the sample, through the ashing procedure.

2.1.2.4. ANALYTIC PROCEDURES.     To iterate, duplicate determinations are advisable.

*A. Fluids (CSF).* If an adequate amount of CSF is available to provide a concentration in the final analyte of approximately 0.1–0.5 ppm, dilute an appropriate volume of the sample aliquot and 1 mL of stock $La^{3+}$ to a final volume of 10 mL with a solution of $Na^+$ (120 ppm, w/v) as NaCl. The $[Mg^{2+}]$ in the analyte must be spanned by the standards and, ideally, should fall within the linear dynamic range of the analytic curve. Specimens stored in volumetric flasks can be diluted directly to mark. The reagent blank consists of the $Na^+$ diluent with $La^{3+}$ 5000 ppm.

*B. Digests.* Prepare 10 mL of analyte containing an aliquot of the solubilized tissue sample and $La^{3+}$ as described for CSF, using distilled deionized water as the diluent. The reagent blank contains the same amount of digest and $La^{3+}$ as the tissue analyte.

*C. Standards.* Prepare three to five working standards (10 mL) from the stock $Mg^{2+}$ standard, spanning the range of Mg concentrations anticipated in the final analytes of the specimen samples and containing $La^{3+}$ (5000 ppm). For CSF samples, use the $Na^+$ diluent. For solubilized tissue, add concentrated sulfuric acid equal in volume to that of an aliquot of digest. Water is used as the blank to set the null-point ($[Mg^{2+}] = 0$). The manufacturer of the spectrophotometer provides specific directions for operating the burner and spectrophotometer.

If necessary, the results can be corrected for residual $Mg^{2+}$ in the vascular system and interstitial fluid of the CNS (from the CSF) in order to get a good approximation of the cellular-intracellular Mg pool (Chutkow and Grabow, 1972).

### 2.1.3. Applications

One can use flame AAS to measure the mean total $[Mg^{2+}]$ in CSF; whole and regional samples of brain, spinal cord, peripheral nerve, and muscle; and cellular and subcellular components of the nervous system obtained by differential centrifugation (Ashley and Ellory, 1972; Bogucka and Wojtczak, 1971, 1976; Chutkow, 1972, 1978; Donega and Burgess, 1970; Flatman and Lew, 1977; George and Heaton, 1975; Heaton and George, 1979; Rugolo and Zoccarato, 1984; Saetersdal et al., 1980; Veloso et al., 1973). Flame AAS, alone or in combination with radioisotopes ([28]Mg, [C[14]]-sucrose, and tritiated water), has sufficient sensitivity and precision to be used in studies on $Mg^{2+}$ flux across cellular and subcellular membranes (Diwan et al., 1979; Flatman and Lew, 1977; Rugolo and Zoccarato, 1984). To date, flameless AAS has been used primarily in trace metal research. However, this technique permits the measurement of $[Mg^{2+}]$ in microliter volumes of fluid and has considerable potential for studying minute variations in the regional distribution of Mg—in the ventriculo-subarachnoid system and the cellular and subcellular components of the parenchymal CNS (perhaps without prior ultracentrifugation)—as well as the regulation of Mg exchange across cellular membranes, including the epithelial cells of the choroid plexi.

### 2.1.4. Comments

Flame AAS is the quantitative analytic technique against which all other methods for measuring the $Mg^{2+}$ content of biologic materials are now compared. It has achieved this status because it is accurate, sensitive, precise, and easy to use; has a large number of applications, and, in terms of equipment costs, is relatively inexpensive. However, there are limitations:

1. It takes considerable time to solubilize tissue—a particularly onerous problem with a large number of samples.
2. It measures the total mean $[Mg^{2+}]$. In order to map quantitative variations in the regional, cellular, and subcellular distribution of the cation, at least with flame AAS, one must resort to differential ultracentrifugation. This increases the risk of contamination when the sample is homogenized either from the substance used as the gradient or from the repeated washings or both. Of equal importance is the fact that the cells are disrupted, introducing the potential for translocational loss of Mg.
3. One cannot determine the chemical state of intracellular Mg directly.
4. Only one element can be analyzed at a time. Transmembranous shifts in $Mg^{2+}$ in response to perturbations in its own metabolism or in the metabolism of other inorganic and organic compounds frequently lead to changes in the extracellular or intracellular concentrations of potassium, calcium, and phosphate.

## 2.2. Inductively Coupled Argon Plasma—Atomic Emission Spectrometry (ICAP-AES)

### 2.2.1. Principles and Instrumentation

A plasma is a gas, a large fraction of whose atoms or molecules are ionized. In ICAP-AES (Craelius et al., 1980; Donega and Burgess, 1970; EPA, 1979; Fassel, 1978; Fassel and Kniseley, 1974), argon is seeded with electrons by a Tesla discharge as it flows through a quartz tube. Thereby, it is converted from a neutral, nonconducting gas to a plasma. The plasma is then inductively coupled to oscillating electromagnetic fields produced by a radio frequency generator. These fields accelerate the electrons and induce them to flow in axially-oriented loops or eddy currents. The electrons meet with resistance to flow as they pass through the argon stream, causing Ohmic heating and additional ionization. As a consequence, the plasma becomes self-sustaining and forms a torch ranging from $10,000^0$ K at the level of the induction coils to $6500^0$ K approximately 15 mm above the coils—temperatures considerably higher than found in combustion flames.

A pneumatic crossed-flow needle or ultrasonic nebulizer converts the analyte into an aerosol. Alternatively, microliter aliquots of the unknown can be vaporized and atomized electrothermally (Nixon et al., 1974). The aerosol is heated, and its constituent solutes are eventually ionized in the plasma. The ions then emit element-specific spectra as they return to ground state. The spectral lines are dispersed by a diffraction grating, isolated by exit slits, and detected by a rapid sequential scanning monochromator or a multichannel polychromator programmed to analyze for a large number of elements. A number of systems are available on the commercial market.

The detection limit for Mg is 0.03–0.07 ppb. High-quality instruments reportedly provide accuracy, sensitivity, and precision comparable to those of AAS.

### 2.2.2. Methods

The methods (Craelus et al., 1980; Hershey et al., 1983, 1984) are essentially the same as those detailed for AAS with two exceptions: water is the diluent in the preparation of all analytes, and there is no need to add $La^{3+}$. The linear dynamic range is broad for $Mg^{2+}$ and many other elements (Fassel, 1978).

### 2.2.3. Applications

There are very few publications using this technique to analyze biological materials. However, ICAP-AES should have the same general applications as flame and flameless AAS (Craelius et al., 1980; Hershey et al., 1983, 1984).

### 2.2.4. Comments

ICAP-AES allows 20 to 40 elements to be measured simultaneously over a large range of concentrations in 2–3 min. Included are a number of elements, such as phosphorus, which are otherwise refractory to analysis by standard absorption and emission techniques. Because the aerosol is heated to an ionic state at relatively high temperatures without interacting to any extent with the plasma, most of the interference problems encountered with AAS are minimal or absent. The reported detection limits for Mg and most other elements are better than for flame AAS, sometimes by several orders of magnitude.

There are four major limitations to this method at the present time: (1) most biological tissues must be solubilized, (2) differential centrifugation must be utilized to map the approximate cellular-subcellular distribution of Mg, (3) the chemical state of intracellular Mg cannot be determined directly, and (4) equipment is very expensive.

## 2.3. Electron Probe Microspectrometry

### 2.3.1. Principles and Instrumentation

Electron probe (X-ray) microspectrometry (Forslind et al., 1985; Lustyik and Nagy, 1985; Moreton, 1981; Panessa-Warren, 1983; Saubermann and Scheid, 1985; Somlyo and Walz, 1985; Somlyo et al., 1977; Wroblewski and Wroblewski, 1984; Wroblewski et al., 1983) measures elements in average or high concentrations located in very small cellular compartments. For this reason, it is well suited for use on $Mg^{2+}$. It is a histophysical instrument that combines electron microscopy (EM) with quantitative spectrometry. An electron beam is focused on an area of interest in an histologic section, identified by EM, where it penetrates a short distance into the sample. The incident beam collides with and displaces one or more orbital electrons from a target atom, creating a void(s) that an electron(s) from a higher energy shell drops down to fill. In so doing, the electron emits an X-ray photon at a wavelength and with an energy that is characteristic for the target element. In addition, incident electrons are slowed or change direction as they pass through the target leading to emission of other X-ray photons of variable energy and wavelength. This background continuum (Bremsstrahlung) is a function, among other factors, of the atomic number of the elements in the target and accelerating energy of the incident beam.

An X-ray spectrometer, affixed to the EM, detects the various element-specific X-rays. There are two general types of spectrometers. (1) The first is an energy-dispersive unit in which the X-ray photons are detected by a large silicon crystal "doped" with lithium, and their energy level is converted into current measured by a pulse-height analyzer. Each element-specific X-ray appears as a Gaussian peak superimposed on a sloping gradient of background continuum. Because of its large collecting power, an energy dispersive spectrometer is able to collect the entire emitted X-ray spectrum at one time, thereby allowing for multiple simultaneous analyses. However, its resolution is low ($\sim$150 eV). (2) The second type of spectrometer, a wavelength-dispersing (diffracting) unit, uses a special crystal as a diffraction grating, provides excellent resolution (1–10 eV), and eliminates spectral line interference. However, its low collecting power leads to long count time, and it can analyze for only one element at a time.

The electron microscope—either scanning (SEM) or transmission (TEM)—also affects the analysis. The TEM provides superior spatial resolution, but is restricted in use to tissue sections that are 200 nm or less in thickness. Although the SEM has a lower resolution than TEM, it has greater versatility, can be used with thick histologic sections, has a large specimen chamber, and can be fitted with a cold-storage unit that allows one to work with frozen-hydrated material at temperatures as low as –150°C.

The absolute detection limit ranges from $10^{-16}$ to $10^{-18}$ g for atomic numbers greater than 8 (Mg = 12). Identification of elements is easy and very specific, because X-ray spectral lines progress in an orderly manner that depends on the atomic number of their emitting atom. Analytic sensitivity is high.

### 2.3.2. Methods

#### 2.3.2.1. PREPARATION OF SAMPLES.

*A. The Sample.* This is a critical step. The sections must be suitable for analysis under conditions of high vacuum and electron bombardment and, yet, remain similar to the live tissue in order to preserve *in situ* structural relationships and the distribution of soluble substances. Second, preparation must be kept to a minimum to avoid contamination.

There is no uniform approach at present for the analysis of Mg or other cations in nervous tissues. Based on work with other organ systems, freeze-dried or, better, freeze-hydrated, tissues are desirable (Wroblewski and Wroblewski, 1984; Wroblewski et al., 1983). Rapidly freezing small pieces of hydrated tissues produces a thin layer (10–20 $\mu M$) in which cytosol ice is either vitreous or in a very fine microcrystalline state. Image quality is sufficiently good to recognize structures, spaces, and larger organelles. Most importantly, ions are frozen *in situ* rather than trapped in a meshwork of dehydrated cytosol. Thin or ultrathin cryosections are cut dry in a special cryostat chamber at temperatures of −80 to −130°C, using a cryoultramicrotome with a glass or steel knife.

*B. Standards.* Many workers recommend that standards be prepared in solutions that simulate the physical-electrolytic state of the analyte. Therefore, the solutions should contain appropriate amounts of saline, potassium, and phosphate in a protein (albumin, gelatin) or carbohydrate matrix, to which is added known amounts of $Mg^{2+}$. A drop of the standard is frozen and treated exactly the same as the tissue. The standards must be assayed at the same time and under the same conditions as the unknown. In order to achieve uniformity, some workers suggest immersing the specimen in a standard solution for a few minutes before it is frozen. This deposits a layer of the solution on the periphery of the block, thereby providing each cryosection with its own "internal standard."

The reported absolute detection limit for Mg falls in the range of $10^{-16}$ to $10^{-18}$ g. The degree of accuracy depends in large part on the quality of the standards. Precision should be comparable to other quantitative methods described above, when repetitive analyses are performed on the same location in the section.

### 2.3.3. Applications

These techniques are designed to map and quantitate the cellular-subcellular distribution of Mg and many other elements in the nervous system. They have been used in a unique manner to analyze for diffusible $Mg^{2+}$ in the cytosol (see below) (Maughan, 1983).

### 2.3.4. Comments

Electron probe microanalysis was introduced over 30 yr ago to increase the usefulness of SEM and TEM. The number of laboratories acquiring scopes with both capabilities is increasing. When spectroscopic and cold-storage capabilities are added, one has the instruments needed to localize and to assay for Mg *in situ* regardless of its chemical state, using either wavelength or energy-dispersive spectrometry. Within limits, the analysis is nondestructive and, therefore, can be repeated. The method is quantitative and allows one to determine the concentration of an element in thick or thin sections directly, if the volume of the analyte is measured accurately. On paper, it is an excellent tool for measuring the qualitative and quantitative effects of a number of inorganic and organic metabolic changes on the subcellular distribution of Mg. There are other variations on the theme of particulate probe microanalysis, including the proton microprobe (Forslind et al., 1985) and secondary ion mass microanalysis (Galle et al., 1983), which show promise for the microquantitation of $Mg^{2+}$.

On the other hand, one cannot determine the chemical or isotopic state of Mg in the samples with these methods. There is a reported nonlinearity of the standard curves at high concentrations—a common problem with absorption and emission spectrometry as well. Used to study Mg in other organs, the rapidly expanding technology of electron probe microanalysis has yet to become a popular inorganic neurochemical tool. Therefore, procedural details have yet to be worked out.

## 2.4. Laser Microprobe Mass Spectrometry

### 2.4.1. Principles and Instrumentation

A UV wavelength laser light (265 nm) of extremely high intensity is focused through the objective of a high-quality light microscope on a very small spot (about 1.0 $\mu m^2$) of a microscopic section of the sample (Gottesberge-Orsulakova and Kaufmann, 1985; Heinen and Holm, 1984; Heinen et al., 1980). This causes electrically neutral thermal vaporization, ashing, and eventual ionization of elements in the analyte to produces anions and cations with equal efficiency. The resulting ions are analyzed with a time-in-flight mass spectrometer.

### 2.4.2. Methods

A thin or ultrathin histologic section, weighing approximately $10^{-12}$ g, taken from a sample processed for EM by rapid freezing or chemical fixation is prepared on a standard, 3-mm EM grid. The grid is fixed in a sample holder behind a quartz slip in the vacuum chamber of the mass spectrometer. The quartz slip serves as a window for microscopic visualization of the section, as well as for focusing and passage of the laser beam.

The reported absolute detection limit for $Mg^{2+} = 4 \times 10^{-20}$ g in a volume of $3 \times 10^{-13}$ $cm^3$, and a relative limit of 0.4 ppm (Heinen et al., 1980). The precision is $\pm 5\%$.

### 2.4.3. Applications

The applications are similar to those listed for electron probe microspectrometry.

### 2.4.4. Comments

This method analyzes simultaneously for all elements and measures isotopes and molecules—opening a number of research vistas into Mg metabolism. The analytic procedure is very rapid, once the microscopic section is in place.

The analyte encompassed by the laser spot is destroyed, preventing repeated studies of the same area. The method provides no direct information about Mg's chemical state in the cells and is semiquantitative at the present time. For precise quantitation, one must be able to measure the exact volume of the analyte, a problem yet to be solved. Furthermore, standards must be prepared in solutions with the same physical and chemical properties and, at the time of analysis, the same geometry as the unknown specimen—another unresolved technical issue.

# 3. Chemical State of Intracellular Magnesium

## 3.1. Introduction

The concentration of free cytosolic $Mg^{2+}$ ($[\text{free } Mg^{2+}]i$) may be about 0.4 $\mu M/mL$ (Guenther, 1981). It presumably regulates the level of activity of Mg-dependent enzymes in vivo as it does in vitro; determines the amount of $Mg^{2+}$ bound to soluble and fixed anionic sites attached to the cellular cytoskeleton; controls the net flux of $Mg^{2+}$ across the plasmalemma and between the cytosol and subcellular organelles; and contributes to transmembranous electrochemical gradients (Aikawa, 1971; Guenther, 1981).

Until recently, the methods used to quantitate the concentrations of free and bound forms of cellular $Mg^{2+}$ were largely indirect, usually relied on assumptions that were not necessarily valid, were technically unsuitable for use in cells from most areas of the CNS, or all three. These include:

(1) Calculating $[\text{free-}Mg^{2+}]i$ from the total cellular $[Mg^{2+}]$ and either theoretical or experimentally determined binding constants of a few major intracellular components forming complexes with $Mg^{2+}$ (Nanninga, 1961; Veloso et al., 1973). To be accurate, this approach depends on reliable data about all the

ionic species that complex with $Mg^{2+}$ in the cell, their dissociations constants, and cellular pH—information that is not available.

(2) Determining the $[Mg^{2+}]$ needed to produce the same equilibrium concentrations of the substrate and product of a $Mg^{2+}$-dependent enzymatic reaction in vitro as that found experimentally for the reaction in vivo (Lowry et al., 1965; Veech et al., 1979; Veloso et al., 1973). It is important to establish that the enzymatic reaction in question is at equilibrium in the cell. Because the intracellular concentrations of the reactants are often very low, accurate measurement may be difficult. Furthermore, one does not know what, if any, portion of the reactant pool is kinetically inactive.

(3) Estimating the diffusible $Mg^{2+}$ in the cytosol of a cell after replacement microinjection (Ashley and Ellory, 1972), an approach with a very limited number of applications.

(4) Using the plasmalemma as a semipermeable membrane for some variant of equilibrium dialysis. On a positive note in this regard, micropipets containing a few microliters of distilled water have been applied to the surface of a cell, and changes in the $[Mg^{2+}]$ of the water measured over time by electron probe microanalysis (Maughan, 1983).

There are two, nondestructive, methods available for measuring $[Mg^{2+}]$ directly *in situ* in living cells. Although some workers reported data in muscle fibers using privately fabricated ion-selective glass microelectrodes (Hess and Weingart, 1981), no reliable electrodes are manufactured commercially at the present time (M. E. Anbar, personal communication; *see* chapters by Nicholson and Rice; Schule and Dietmar, this volume, for general methodologic considerations).

## 3.2. Eriochrome Blue SE (EBSE)

### 3.2.1. Principles, Instrumentation, and Methods

EBSE (Brinley and Scarpa, 1975; Brinley et al., 1977; Casillas et al., 1981; Scarpa, 1974; Scarpa and Brinley, 1981), a mono-arylazo derivative of chromotropic acid (mw = 518), is a metallochromic indicator sensitive to changes in [free $Mg^{2+}$]. It is water-soluble, as well as physiologically and chemically inert. It does not bind to any extent with cells or cell fractions, nor is it transported by them; it forms and dissociates from its chelate with $Mg^{2+}$ rapidly; and it has a low affinity for $Mg^{2+}$. When microinjected into the cytosol of a cell, it measures [free $Mg^{2+}$]$i$. The basic measuring device is a double beam microspectrophotometer fitted with a continuous readout to record the effects of changes in [$Mg^{2+}$] on dual wavelength absorbance of EBSE (554–592 nm). Measuring this particular wavelength pair eliminates interference by calcium and other divalent cations.

Scarpa and Brinley (1981) detailed the instrumentation for a pulsed multiwavelength microspectrophotometer to record moment-to-moment fluctuations in [free $Mg^{2+}$]$i$ and a null-point titration of EBSE to eliminate effects of pH and medium composition on the EBSE–$Mg^{2+}$ complex. Null-point titration identifies that extracellular [$Mg^{2+}$] that produces no change in the observed differential absorbance of intracellular EBSE, using the plasmalemma as a dialyzer between the intra- and extracellular fluid compartments—a value presumably equal to that of [free $Mg^{2+}$]$i$. EBSE is introduced by displacement microinjection.

### 3.2.2. Applications

EBSE has been used successfully to measure [free $Mg^{2+}$]$i$ in large axons, muscle fibers, and ova.

### 3.2.3. Comments

The EBSE technique is minimally invasive and nondestructive. It is suitable for recording changes in [free $Mg^{2+}$]$i$ in large, living cells. It is affected by small pH changes. Finally, one may need special optical equipment for some types of microspectrophotometric studies.

## 3.3. [31]Phosphorus Nuclear Magnetic Resonance Spectroscopy (NMRS)

### 3.3.1. Principles and Instrumentation

The general analytic principles and instrumentation are discussed in detail elsewhere in this volume (chapter by Go), and in several excellent references (Ackerman et al., 1980; Bottomley et al., 1984; Burt et al., 1976; Cady et al., 1983; Cohen and Burt, 1977; Garfinkel and Garfinkel, 1984; Gupta and Moore, 1980; Gupta and Yushok, 1980; Gupta et al., 1978, 1983).

Those compounds containing resonating natural abundance phosphorous ($^{31}$P) are dissolved unbound in the cytosol and can be identified by their position in a nuclear magnetic resonance spectrum. The three phosphoryl groups of ATP give rise to characteristic major resonances ($\alpha$P, $\beta$P, and $\gamma$P). About 95% of cytosolic ATP is complexed with $Mg^{2+}$ (Brinley and Scarpa, 1975; Burt et al., 1976; Cohen and Burt, 1977; Ozawa et al., 1966).

$Mg^{2+}$ causes the $^{31}$P ATP resonance peaks to shift chemically. The magnitude of the shifts and of changes in ATP's $^{31}$P spin–spin coupling constants at various [free $Mg^{2+}$]$i$ are proportional to the fraction of ATP chelated with $Mg^{2+}$. The intracellular dissociation constant for MgATP has been determined reliably (Garfinkel and Garfinkel, 1984; Gupta and Moore, 1980; Gupta et al., 1978, 1983). Gupta and associates developed mathematical equations that allow one to use the $^{31}$P NMR spectrum to measure the fraction of ATP chelated with Mg and, from this, to calculate [free $Mg^{2+}$]$i$ noninvasively in otherwise unperturbed cells in vivo (Gupta and Moore, 1980; Gupta and Yushok, 1980; Gupta et al., 1978, 1983). The data are reportedly accurate and reproducible.

### 3.3.2. Applications

This appears to be the procedure of choice with which to measure [free $Mg^{2+}$]$i$ and its complexes with soluble phosphoryl-containing compounds in the cytosol. It should also be useful in experiments on the dynamics of $Mg^{2+}$ exchange across the plasmalemma and within the cell.

### 3.3.3. Comments

$^{31}$P NMR chemical shift spectroscopy is noninvasive and does not harm living cells. Once the details are worked out, strong field NMR systems, combined with high-resolution planar imaging, can provide easily defined $^{31}$P ATP resonance peaks with which to study anatomically well-localized metabolism of $Mg^{2+}$ in the superficial layers of the cerebral cortex *in situ* in unoperated animals and humans (Bottomley et al., 1984; Cady et al., 1983; Shoubridge et al., 1982).

There are limitations. The $^{31}$P NMR ATP spectrum consists of a single set of resonances, indicating one large MgATP "compartment," whereas data obtained by other means are consistent with a multicompartmental distribution of Mg and ATP within the cell. Although the reasons are not clear, none of the alternative explanations for this disparity invalidates this as a method for direct measurement of cytosolic [$Mg^{2+}$] (Gupta and Moore, 1980).

## 4. Kinetics of Magnesium Exchange in Tissues, Cells, and Cellular Organelles

### *4.1. Introduction*

The $[Mg^{2+}]$ in the CSF is maintained at a level greater than $[Mg^{2+}]$ in ultrafilterable plasma by an energy-dependent transport system in the endothelial cells of the choroid plexi that moves the cation into the CSF against an electrochemical gradient (Chutkow, 1965, 1971, 1978; Ginsburg et al., 1966; Woodward and Reed, 1969). No such transport system has been demonstrated in the "blood–brain barrier" created by the endothelial cells of the cerebrospinal microvasculature (Chutkow, 1971, 1974, 1978; Woodward and Reed, 1969).

Mg movement across cellular, mitochondrial, and microsomal membranes is accelerated. The roles of facilitated versus active mechanisms and their relationship to influx and efflux of Mg are still under investigation. Most of the current data were obtained in nonneurologic tissues. Therefore, very little is known about cellular–subcellular transmembranous kinetics and the physiologic compartmentalization of Mg within the CNS, particularly in mammals.

### *4.2. Quantitative Analytic Methods*

$^{31}P$ NMR described above is probably the best method currently available for the investigation of $Mg^{2+}$ exchange at a cellular level in the CNS. I am not aware of any published reports employing the technique for this specific purpose.

### *4.3. Radioactive Magnesium ($^{28}Mg$)*

#### *4.3.1. Principles, Methods, and Applications*

Many of the theoretical and technical aspects of exchange kinetics using radioactive tracers are covered elsewhere in this volume (chapters by Walz and Kettenmann). $^{28}Mg$ ($t_{1/2} = 21.3$ h) emits one beta particle and four gamma rays of 1.35 Mev (70%), 0.95 Mev (30%), 0.40 Mev (30%), and 0.032 Mev (100%) as it decays to $^{28}Al$. The latter ($t_{1/2} = 2.24$ min) decays to stable $^{28}Si$ by emitting one beta particle and one gamma ray of 1.78 Mev (Hine and Brownell, 1956). Therefore, the activity of $^{28}Mg$ can be assayed by standard gamma or, less practically, by beta counting techniques. Currently, high specific activity $^{28}$Mg ($\sim$30 $\mu$Ci/$\mu$g of Mg$^{2+}$ at time of shipping) is available commercially only from Brookhaven National Laboratories (NY, USA).

For the reasons noted in the introduction to this section (section 4.1.), the exchange of Mg$^{2+}$ between the blood and the parenchymal fluid compartments of the CNS is very slow (Chutkow, 1965, 1978; MacIntyre, 1959; Woodward and Reed, 1969); therefore, it is difficult as well as expensive to use $^{28}$Mg to study Mg$^{2+}$ flux and compartmentalization in the CNS in the intact animal (*see* the following references, in which $^{28}$Mg was used to study selected aspects of Mg$^{2+}$ compartmentation and exchange in the CNS (Aikawa, 1965; Chutkow, 1965, 1978; Ginsburg et al., 1966; Lazzara et al., 1963; MacIntyre, 1959; Oppelt et al., 1963; Rogers and Mahan, 1959; Rogers et al., 1964; Woodward and Reed, 1969).

Cellular–intracellular exchange of Mg$^{2+}$ has not been investigated radioisotopically in appropriate preparations from the CNS or cultures of neurons and glia, despite availability of the isotope and standardized procedures reported in published work on other tissues, including muscle and peripheral nerve. The latter methods and those used to characterize the exchange kinetics for such cations as calcium in the CNS should be suitable for Mg$^{2+}$. Because of its relatively short half-life, $^{28}$Mg can be used with such fluid compartment "markers" as [$^{14}$C]-sucrose or inulin and $^3$H$_2$O. It must be combined with one of the aforementioned quantitative techniques for measuring nonradioactive Mg.

### 4.3.2. Comments

The recent improvement in the specific activity of $^{28}$Mg is an asset. Its relatively short half-life, limited source of distribution, and cost are detriments. On balance, it is the method for choice for characterizing cellular-level Mg exchange kinetics and physiologic compartmentalization.

## 5. Summary

In this chapter I reviewed the more important current or yet-to-be-applied methods for quantitating the normal and pathologic distribution, intracellular chemical state, and exchange kinetics of Mg$^{2+}$ in the CNS. Atomic absorption spectrophotometry is the standard technique for determining mean total concentrations of Mg in biological specimens. However, inductively coupled argon plasma emission spectroscopy not only equals AAS in this regard, but is also a much more powerful tool for investigating Mg's interrelationships with other elements, particularly calcium, potassium, and phosphate. I believe electron probe microanalysis is the preferred method for mapping quantitatively the cellular–subcellular microdistribution of Mg. Similarly, $^{31}$P NMR resonance spectroscopy coupled with AAS and reliable dissociation constants for the soluble and cytoskeletal compounds whose phosphoryl groups chelate with $Mg^{2+}$ is the best approach with which to determine the mean chemical state of intracellular Mg *in situ* and in vivo. Finally, tracer studies with $^{28}$Mg remain the best way to characterize Mg exchange kinetics and compartmentalization at a cellular level.

## Acknowledgment

The author thanks Nora Post, PhD, for her editorial review and advice.

## References

Ackerman J. J. H., Grove T. H., Wong G. G., Gadian D. G., and Radda G. K. (1980) Mapping of metabolites in whole animals by $^{31}$P NMR using surface coils. *Nature* **283,** 167–170.

Aikawa J. (1965) $Mg^{28}$ Studies of Magnesium Metabolism, in *Radioisotopes in Animal Nutrition and Physiology* International Atomic Energy Agency, Vienna.

Aikawa J. K. (1971) The Biochemical and Cellular Functions of Magnesium, in *1st International Symposium on Magnesium Deficit in Human Pathology* (Durlach J., ed.) SGEMV, Vittel, France (Imprimerie Amelot, Brionne).

Ashley C. C. and Ellory J. C. (1972) The efflux of magnesium from single crustacean muscle fibers. *J. Physiol.* **226,** 545–565.

Bogucka K. and Wojtczak L. (1971) Intra-mitochondrial distribution of magnesium. *Biochem. Biophys. Res. Commun.* **44,** 1330–1337.

Bogucka K. and Wojtczak L. (1976) Binding of magnesium by proteins of the mitochondrial intermembrane compartment. *Biochem. Biophys. Res. Commun.* **71,** 161–176.

Bottomley P. A., Hart H. R., Edelstein W. A., Schenck J. F., Smith L. S., Leue W. M., Mueller O. M., and Redington R. W. (1984) Anatomy and metabolism of the normal human brain studied by magnetic resonance at 1.5 Tesla. *Radiology* **150,** 441–446.

Brinley F. J. and Scarpa A. (1975) Ionized magnesium concentration in axoplasm of dialyzed squid axons. *FEBS Lett.* **50,** 82–85.

Brinley F. J., Scarpa A., and Tiffert T. (1977) The concentration of ionized magnesium in barnacle fibers. *J. Physiol.* **266,** 545–565.

Burt C. T., Glonek T., and Barany M. (1976) Analysis of phosphate metabolites, the intracellular pH, and state of adenosine triphosphate in intact muscle by phosphorus nuclear magnetic resonance. *J. Biol. Chem.* **251,** 2584–2591.

Cady E. B., Costello A. M., Dawson M. J., Delpy D. T., Hope P. L., Reynolds E. O., Tofts P. S., and Wilkie D. R. (1983) Noninvasive investigation of cerebral metabolism in newborn infants by phosphorus nuclear magnetic resonance spectroscopy. *Lancet* **ii,** 1059–1062.

Casillas E., Harry D. J., and Kenny M. (1981) Measurement of ionized magnesium in biological fluids. *Anal. Biochem.* **116,** 319–324.

Chutkow J. G. (1965) Studies on the metabolism of magnesium in the magnesium-deficient rat. *J. Lab. Clin. Med.* **65,** 912–926.

Chutkow J. G. (1971) Role of Magnesium in Neuromuscular Physiology, in *1st International Symposium on Magnesium Deficit in Human Pathology* (Durlach J., ed.) SGEMV, Vittel, France (Imprimerie Amelot, Brionne).

Chutkow J. G. (1972) Distribution of magnesium and calcium in brains of normal and magnesium-deficient rats. *Mayo Clin. Proc.* **47,** 647–653.

Chutkow J. G. (1974) Metabolism of magnesium in central nervous system: Relationship between concentration of magnesium in the cerebrospinal fluid and brain in magnesium deficiency. *Neurology* **24,** 780–787.

Chutkow J. G. (1978) Uptake of magnesium into the brain of the rat. *Exp. Neurol.* **60,** 592–602.

Chutkow J. G. (1980) The Neurophysiologic Functions of Magnesium: Effects of Magnesium Excess and Deficits, in *Magnesium in Health and Disease* (Cantin M. and Seelig M., eds.) Spectrum, New York.

Chutkow J. G. (1981) The neurophysiologic function of magnesium: An update. *Magnesium Bull.* **3,** 115–120.

Chutkow J. G. and Grabow J. D. (1972) Clinical and chemical correlations in magnesium-deprivation encephalopathy of young rats. *Am. J. Physiol.* **223,** 1407–1414.

Chutkow J. G. and Meyers S. (1968) Chemical changes in the cerebrospinal fluid and brain in magnesium deficiency. *Neurology* **18,** 963–974.

Clegg M. S., Keen C. L., Lonnerdal B., and Hurley L. S. (1981) Influence of ashing techniques on the analysis of trace elements in animal tissue. I. Wet ashing. *Biol. Trace Element Res.* **3**, 107–115.

Cohen S. M. and Burt C. T. (1977) [31]P nuclear magnetic relaxation studies of phosphocreatine in intact muscle: Determination of intracellular free magnesium. *Proc. Natl. Acad. Sci. USA* **74**: 4271–4275.

Craelius W., Jacobs R. M., and Jones A. O. L. (1980) Mineral composition of brains of normal and multiple sclerosis victims. *Proc. Soc. Exp. Biol. Med.* **165**, 327–329.

Diwan J. J., Daze M., Richardson R., and Aronson, D. (1979) Kinetics of $Mg^{2+}$ flux into rat liver mitochondria. *Biochemistry* **18**, 2590–2595.

Donega H. M. and Burgess T. E. (1970) Atomic absorption analysis by flameless atomization in a controlled atmosphere. *Anal. Chem.* **42**, 1521–1524.

Environmental Protection Agency (1979) Guidelines establishing test procedures for the analysis of pollutants: Proposed regulations. Appendix IV. Inductively coupled plasma optical emission spectrometric method for trace element analysis of water. *Fed. Reg.* **44** (233), 69559–69564.

Fassel V. A. (1978) Quantitative elemental analyses by plasma emission spectroscopy. *Science* **202**, 183–191.

Fassel V. A. and Kniseley R. N. (1974) Inductively coupled plasma—optical emission spectroscopy. *Anal. Chem.* **46**, 1110A–1120A.

Flatman P. and Lew V. L. (1977) Use of ionophore A23187 to measure and to control free and bound cytoplasmic Mg in intact red cells. *Nature* **267**, 360–362.

Forslind B., Kunst L., Malmqvist K. G., Carlsson L. E., and Roomans G. M. (1985) Quantitative correlative proton and electron microprobe analysis of biological specimens. *Histochemistry* **82**: 425–427.

Galle P., Berry J. P., and Escaig F. (1983) Secondary ion mass microanalysis: Applications in biology. *Scan. Electron Microsc.* **II**, 827–839.

Garfinkel L. and Garfinel D. (1984) Calculation of free-$Mg^{2+}$ concentration in adenosine 5'-triphosphate containing solutions in vitro and in vivo. *Biochemistry* **23**, 3547–3552.

George G. A. and Heaton W. F. (1975) Changes in cellular composition during magnesium deficiency. *Biochem. J.* **152**, 609–615.

Ginsburg S., Graziani L., Escriva A., and Katzman R. (1966) Exchange of magnesium between blood, brain and CSF. *Physiologist* **9**, 186.

Gottesberge-Orsulakova M. and Kaufmann R. (1985) Recent advances in laser microprobe mass analysis (LAMMA) of inner ear tissue. *Scan. Electron Microsc.* **I**, 393–405.

Guenther Th. (1981) Biochemistry and pathobiochemistry of magnesium. *Magnesium Bull.* **3**, 91–101.

Gupta R. K. and Moore R. D. (1980) [31]P NMR studies of intracellular free $Mg^{2+}$ in intact frog skeletal muscle. *J. Biol. Chem.* **255**, 3987–3993.

Gupta R. K. and Yushok W. D. (1980) $^{31}$P NMR probes of free $Mg^{2+}$, MgATP, and MgADP in intact Ehrlich ascites tumor cells. *Proc. Natl. Acad. Sci. USA* **77**. 2487–2491.

Gupta R. K., Benovic J. L., and Rose Z. B. (1978) The determination of the free magnesium level in human red blood cells by $^{31}$P NMR. *J. Biol. Chem.* **253**, 6172–6176.

Gupta R. K., Gupta P., Yushok W. D., and Rose Z. B. (1983) Measurement of the dissociation constant of MgATP at physiologic nucleotide levels by a combination of $^{31}$P NMR and optical absorbance spectroscopy. *Biochem. Biophys. Res. Commun.* **117**, 210–216.

Heaton F. W. and George G. A. (1979) Submitochondrial distribution of magnesium and calcium: Changes during magnesium deficiency. *Int. J. Biochem.* **10**, 275–277.

Heinen H. J. and Holm R. (1984) Recent development with laser microprobe mass analyzer. *Scan. Electron Microsc.* **III**, 1129–1138.

Heinen H. J., Hellenkamp F., Kaufmann R., Schroder W., and Wechsung R. (1980) LAMMA: A New Laser Microprobe Mass Analyzer for Biomedicine and Biological Materials Analysis, in *Recent Developments in Mass Spectrometry in Biochemistry and Medicine,* 6. (Figerio A. and McCamish M., eds.) Elsevier, Amsterdam.

Hershey C. O., Varnes A. W., Hershey L. A., and Strain W. H. (1983) Multiple element analysis of human cerebrospinal fluid and other tissues by inductively coupled argon plasma emission spectrometry. *Neurotoxicology* **4**, 157–160.

Hershey L. A., Hershey C. O., Varnes A. W., Wongmonkolrit T., and Strain W. H. (1984) Zinc Content in CSF, Brain, and Other Tissues in Alzheimer Disease and Aging, in *The Neurobiology of Zinc. B. Deficiency, Toxicity, and Pathology* Alan R. Liss, New York.

Hess P. and Weingart R. (1981) Free magnesium in cardiac and skeletal muscle with ion-selective micro-electrodes. *J. Physiol.* **318**, 14P–15P.

Hine G. J. and Brownell G. L. (1956) *Radiation Dosimetry* Academic, New York.

Hwang J. Y., Mokeler C. J., and Ullucci P. A. (1972) Maximization of sensitivities in tantalum ribbon flameless atomic absorption spectrometry. *Anal. Chem.* **44**, 2018–2021.

Lazzara R., Hyatt K., Love W., Cronvich J., and Burch G. E. (1963) Tissue distribution, kinetics and biological half-life of Mg28 in the dog. *Am. J. Physiol.* **204**, 1086–1094.

Lowry O. H., Passonneau J. V., Hasselberger F. X., and Schulz D. W. (1964) Effect of ischemia on known substrates and cofactors of the glycolytic pathway in brain. *J. Biol. Chem.* **239**, 18–30.

Lustyik Cy and Zs. Nagy I. (1985) Alteration of intracellular water and ion concentration in brain and liver cells during aging revealed by energy dispersive X-ray microanalysis of bulk specimen. *Scan. Electron Microsc.* **I**, 323–337.

MacIntyre I. (1959) Some aspects of magnesium metabolism and magnesium deficiency. *Proc. Roy. Soc. Med.* **52**, 212–214.

Maughan D. (1983) Diffusible magnesium in frog skeletal muscle. Electron probe microanalysis of 0.2 nl of liquid samples of sarcoplasm. *Biophys. J.* **43**, 75–80.

Moreton R. B. (1981) Electron-probe X-ray microanalysis: Techniques and recent applications in biology. *Biol. Rev.* **56**, 409–461.

Nanninga L. B. (1961) Calculation of free magnesium, calcium, and potassium in muscle. *Biochim. Biophys. Acta* **54**, 338–344.

Nixon D. E., Fassel V. A., and Kniseley R. N. (1974) Inductively coupled plasma—optical emission analytical spectroscope: Tantalum filament vaporization of microliter samples. *Anal. Chem* **46**, 210–213.

Oppelt W., MacIntyre I., and Rall D. (1963) Magnesium exchange between blood and cerebrospinal fluid. *Am. J. Physiol.* **205**, 959–962.

Ozawa K., Seta K., and Honda H. (1966) The effect of magnesium on brain mitochondrial metabolism. *J. Biol. Chem.* **60**, 268–273.

Panessa-Warren B. J. (1983) Basic biological X-ray microanalysis. *Scan. Electron Microsc.* **II**, 713–723.

Rogers T. and Mahan P. (1959) Exchange of radioactive magnesium in the rat. *Proc. Soc. Exp. Biol. Med.* **100**, 235–239.

Rogers T., Simesen M., Lunaas T., and Luick J. (1964) The exchange of radioactive magnesium in the tissues of the cow, calf, and fetus. *Acta Vet. Scan.* **5**: 209–216.

Rugolo M. and Zoccarato F. (1984) Magnesium transport by brain mitochondria: Energy requirement and dependence on $Ca^{2+}$ fluxes. *J. Neurochem.* **42**, 1127–1130.

Saetersdal T., Engedal H., Roli J., and Myklebust R. (1980) Calcium and magnesium levels in isolated mitochondria from human cardiac biopsies. *Histochemistry* **68**, 1–8.

Saubermann A. J. and Scheid V. L. (1985) Elemental composition and water content of neuron and glial cells in the central nervous system of the North American medicinal leech (*Macrobdella decora*). *J. Neurochem.* **44**, 825–834.

Scarpa A. (1974) Indicators of free magnesium in biological systems. *Biochemistry* **13**, 2789–2794.

Scarpa A. and Brinley F. J. (1981). *In situ* measurements of free cytosolic magnesium ions. *Fed. Proc.* **40**, 2646–2652.

Shoubridge E. A., Briggs R. W., and Radda G. K. (1982) [31]P NMR saturation transfer measurements of steady state rates of creatine kinase and ATP synthetase in rat brain. *FEBS Lett.* **140**, 288–292.

Somlyo A. P. and Walz B. (1985) Elemental distribution in Rosia pipiens retinal rods: Quantitative electron probe analysis. *J. Physiol.* **358,** 183–195.

Somlyo A. V., Shuman H., and Somlyo A. P. (1977) Elemental distribution in striated muscle and the effects of hypertonicity. *J. Cell. Biol.* **74,** 828–855.

Veech R. L., Lawson J. W. R., Cornell N. W., and Krebs H. A. (1979) Cytosolic phosphorylation potential. *J. Biol. Chem.* **254,** 6538–6547.

Veloso D., Guynn R. W., Oskarsson M., and Veech R. L. (1973) The concentrations of free and bound magnesium in rat tissues. *J. Biol. Chem.* **248,** 4811–4819.

Wacker W. E. C. (1980) *Magnesium and Man.* Harvard University Press, Cambridge, Massachusetts.

Woodward D. and Reed D. (1969) Uptake of $^{28}$Mg and $^{45}$Ca by tissues of magnesium-deficient rabbits. *Am. J. Physiol.* **217:** 1483–1486.

Wroblewski R. and Wroblewski J. (1984) Freeze drying and freeze substitution combined with low temperature embedding: Preparation techniques for microprobe analysis of soft tissues. *Histochemistry* **81,** 469–475.

Wroblewski J., Muller R. M., Wroblewski R., and Roomans G. M. (1983) Quantitative X-ray microanalysis of semi-thin cryosections. *Histochemistry* **77,** 447–463.

# Index